E-commerce 2016

Amir Manzoor

Bahria University, Pakistan

E-commerce 2016

First Edition 2015
ISBN-13: 978-0-692-35927-3

Printed in the United States of America

Trademark Acknowledgements

Full acknowledgement is given of all proprietary trademarks and registered trademarks that are mentioned in this book. In addition, terms suspected of being trademarks or series marks have been appropriately capitalized, author cannot attest to the accuracy of this information. Use of a term in this book should not be regarded as affecting the validity of any trademark or service mark.

Amir Manzoor

DEDICATION

To my loving brother Dr. Tahir Manzoor M.D.

ABOUT THE AUTHOR

Amir Manzoor holds a bachelor's degree in engineering from NED University, Karachi, an MBA from Lahore University of Management Sciences (LUMS), and Chartered Banker MBA from Bangor University, UK. He also holds more than 12 international IT industry certifications. He has served at senior managerial and consulting positions in USA and Pakistan for many years. He is currently serving as Faculty member in Management Sciences department of Bahria University, Karachi. He has published five book and many research articles in international referreed journals of repute. His research interests include electronic business and business use of information technology for competitive advantage.

PREFACE

The growth of the Internet continues to influence our lives and businesses. Irrespective of their type and size, all firms and organizations are rethinking their strategies and operations. An increasing number of businesses are using e-commerce to gain competitive advantage. However, doing e-commerce is complex and requires understanding a vast array of topics. Rapid advancement of technologies has made things possible we never thought of. Simulation of physical touch, augmentd reality, Internet-of-Things (IoT), and semantic web are just some examples. This book provides coverage of essential topics in E-commerce i.e. technology infrastructure, building blocks of e-commerce, online marketing, online payment systems, online business models, online business environment issues, website usability, e-commerce strategy, mobile commerce, e-CRM/e-SCM/e-SRM, and e-commerce site development. Compared with available texts on e-commerce, the perspective of this book is global e-commerce. The book is written in simple language, provides up-to-date coverage of material and associated research, and filled with examples to support material presented. This book is useful for undergrad and graduate students, professionals, and anyone looking to gain a solid foundation to continue their learning of dynamic e-commerce environment.

Amir Manzoor

ACKNOWLEDGEMENTS

The final shape of this book would not have been possible without the contribution of so many people that invested their time and energy to provide valuable ideas and suggestions. Collective wisdom, work, and experience of many professors, researchers, students, and practitioners have enriched the text. All the references to the published work have been provided in the bibliography section of each chapter. I am highly indebted and thankful to all those people who helped shape the development of this text. My special thanks and appreciation for all those people helped shape the development of this text by providing valuable material and suggestions for this book. I would like to thank my colleagues and friends at many universities in Pakistan, India, United States, Canada, Germany, France, Japan, UK, China, Netherlands, Turkey, Russia, Belgium, Australia, Austria, Denmark, Sweden, Spain, and New Zealand.

I also want to thank you, the reader, for investing the time and effort to read and study this text. I have put significant energy and effort to make sure this text provides you the knowledge required to underatand, formulate, implement, and evaluate e-commerce solutions for any organization with which you become associated. Finally, I want to welcome and invite your suggestions, ideas, thoughts, comments, and questions regarding any part of this text or the supplementary materials. Please contact me at amir@amirmanzoor.com or write me at the Management Sciences Department, Bahria University, 13, National Stadium Road, Karachi, Pakistan 75260. I sincerely appreciate and need your input to continually improve this text in future editions. I will especially appreciate if you are willing to draw my attention to specific errors or deficiencies in coverage or exposition.

Thank you for using this text.

Amir Manzoor

Amir Manzoor

List of Figures

List of Tables

CONTENTS

1

INTRODUCTION TO

E-COMMERCE

Learning Objectives

After reading this chapter, reader should be able to:

- Define e-commerce and describe how it differs from e-business and Internet economy.

- Understand dimensions of E-commerce.

- Understand the difference between Internet and Non-Internet E-commerce.

- Describe various categories of E-commerce.

- Describe Electronic Data Interchange.

- Describe typical forms of E-commerce organizations.

- Understand the difference among Intermediation, Disintermediation, Reintermediation, Hypermediation, and Syndication.

- Discuss the development and growth of E-commerce including its generations.

- Identify the critical success factors of E-commerce.

- Identify E-commerce technology trends.

- Identify the various international cooperation and collaboration efforts to promote E-commerce.

- Identify the suitability of products and services for E-commerce.

- Describe advantages and disadvantages of E-commerce.

- Understand how E-commerce can help businesses respond to various pressures created by digital economy.

- Understand E-commerce outlook.

E-commerce: The Definition

E-commerce refers to the use of electronic means and technologies to conduct commerce (sale, purchase, transfer, or exchange of products, services, and/or information), including within business, business-to-business, and business-to-consumer interactions. Delivery of a product or service may occur over or outside of the Internet (Whinston, Choi, & Stahl, 1997).

E-commerce and E-business

Sometimes the terms e-commerce and e-business are used interchangeably but they are distinct concepts. E-commerce is the term used to describe the process of transacting business over the Internet. E-business, on the other hand, involves the fundamental reengineering of the business model into an Internet based networked enterprise. The difference in the two terms is the degree to which an organization transforms its business operations and practices thorough the use of the Internet (Hackbarth & William, 2000; Mehrtensb, Cragg, & Mills, 2001; Poon, 2000; Poon 2000; Poon & Swatman, 1997). E-business can include any process that a business organization conducts using Internet including internal processes such as employee services and training.

Compared with E-business, E-commerce is relatively easy to implement because it involves only three types of integration: vertical integration of front-end Web site applications to existing transaction systems; cross-business integration of a company with Web sites of customers, suppliers or intermediaries such as Web-based marketplaces; and integration of technology processes for order handling, purchasing or customer service (Bartels, 2000).

E-business is more difficult to implement because it involves four types of integration: vertical integration between Web front- and back-end systems; lateral integration among a company and its customers, business partners, suppliers or intermediaries; horizontal integration among E-commerce and Enterprise Resource Planning systems,; and downward integration through the enterprise, for integration of new technologies with redesigned business processes for E-business.

However, e-commerce and e-business share some common characteristics. Both E-commerce and E-business:

- Provides s higher payoff in the form of more efficient processes
- Lowers costs and provide potentially greater profits
- Utilize a technology infrastructure of databases, application servers, security tools, systems management and legacy systems
- Involve the creation of new value chains between a company and its customers and suppliers, as well as within the company itself
- Involve major and potentially disruptive organizational change

The risks of failure and the consequences from limited success are higher in an e-business than in e-commerce. A wise choice for a company may be to first consolidate gains and establish its e-commerce, ERP and other systems before becoming and e-business. Move made in a hurry can be as disastrous as a late move.

Selling through e-commerce can include many things (see Table 1-1).

TABLE 1-1: E-COMMERCE BUSINESS TYPES

Retailing	Myntra (myntra.com)
Marketplace	Snapdeal (snapdeal.com)
Sharing Economy	AirBnB (airbnb.com)
Aggregators	Taxi for Sure (taxiforsure.com)
Group buying	Groupon (groupon.com)
Digital goods / downloads	Apple iTunes (apple.com/itunes)
Virtual goods	Zynga (zynga.com)
E-learning and Training	Coursera (coursera.org), SimpliLearn (simplilearn.com)
Pay what you want	Instamojo (instamojo.com)
Auction commerce	eBay (ebay.com)
Crowd sourced Services	Elance (elance.com), oDesk (odesk.com)

E-commerce History

E-commerce's history is brief but fascinating. Over the course of a few decades, networking and computing technology has improved at exponential rates. Powerful personal computers linked with global information networks have powered a completely new world of intellectual, social and financial interactions. In addition, this is only the beginning. The Table 1-2 provides a brief history of e-commerce covering important events.

TABLE 1-2: E-COMMERCE HISTORY

Year	Event
1960s	This is the initiation of the history of ecommerce, when what is called Electronic Data Interchange permits companies to carry out electronic transactions. Electronic data interchange allows companies to carry out electronic transactions - a precursor to online shopping
1979	English inventor Michael Aldrich connected a TV set to a computer with a phone line and created 'teleshopping' - meaning shopping at a distance
1981	- Thomson Holidays submits the first ever B2B electronic transaction using online technology - The first business-to-business transaction from Thompson Holidays
1982	France Telecom invents Minitel that is considered the most successful pre-World Wide Web online services
1984	- Format ASC X12 provides a dependable means to conduct electronic business - Gateshead SIS/Tesco is the first B2C online shopping - The 'Electronic mall' is launched by Compuserve - users could purchase items directly from 110 online merchants.
1985	Nissan UK sells cars and finance with credit checking to customers online from dealers' lots

1987	Swreg creates the first electronic Merchant account to let software developers sell online.
1989	Peapod brings the grocery store to the home PC
1990	Time Berners-Lee creates the first World Wide Web server and browser, using a NeXT computer
1991	The National Science Foundation (NSF) lifts restrictions on the commercial use of the Internet, clearing the way for ecommerce.
1992	J.H. Snider and Terra Ziporyn publish 'Future Shop: How new technologies will change the way we shop and what we buy'.
1993	Marc Andreesen at the National Center for Supercomputing Applications (NCSA) introduced the first widely distributed Web browser called Mosaic.
1994	- Netscape unveils SSL encryption. - Jerry and David's Guide to the www is renamed Yahoo - Third-party payment services for processing online credit card sales begin to appear - The first online bank opens. - Online retailer NetMarket makes the 'first secure retail transaction on the web' - a copy of Sting's album Ten Summoner's Tales. - Joe McCambley ran the first ever online banner ad. It went live on HotWired.com and promoted 7 art museums.
1995	- Amazon sold its first item - a science textbook - eBay sold its first item - a broken laser pointer - Amazon.com starts selling each and everything online - AuctionWeb launches a site soon to be rechristened eBay - A company called Verisign begins developing digital IDs
1998	- PayPal launches pay service - Google debuts on eCommerce - Yahoo launches Yahoo Stores
1999	- Zappos launches web-only shoe store - Internet Retailer debuts - Global Sports launches out-sourced ecommerce platform -Victoria's Secret debuts site - The first online-only shop, Zappo's, opens
2000	- The ongoing dot-com investment bust - OsCommerce is started in Germany as The Exchange Project
2001	Amazon.com blazes a trail launching a mobile commerce site
2002	- PayPal is acquired by eBay - CSN Stores and NetShops begin selling products through several targeted domains.
2003	- Apple launches iTunes store - Congress passes the Can SPAM Act - ZenCart branches from OsCommence
2004	Credit card companies create PCI data security standards
2005	- First internet Retailer Conference and Exhibition - Launches YouTube

	- Web 2.0 takes hold making sites more interactive - The final release of VirtueMart - Social commerce (people using social media in their buying decisions) is born thanks to networks like Facebook
2006	Google debuts Google Checkout
2007	- Apple launches the IPhone with full web browsing and downloadable apps, advancing mCommerce - Free open-source Prestashop software is founded
2008	- Groupon is launched - Amazon Introduces TextBuyIt - Magento eCommerce solution is launched by Varien
2009	- Arnaz.com and Overstock.com lose New York Online Sales tax battle. -Total ecommerce sales amount to $143.4 billion
2010	- Magento mobile is released allowing store owners to create native mobile storefront apps - ecommerce gets serious about social media and more personal conversations taking place between businesses and consumers
2014	With mobile commerce gaining speed, mobile payments are expected to quadruple by 2014, reaching $630 billion in value

Commercialization of the Internet

The internet has revolutionized the computer and communication world like nothing before. The internet provides worldwide broadcasting capabilities a mechanism of information decentralization, and a medium for collaboration and interaction between individuals and their computers without any regard of geographical location. The internet represents one of the most successful examples of the benefits of sustained investment and commitment to research and development of information infrastructure. Commercialization of the Internet has provided an extremely effective transition of research results into a broadly deployed and available information infrastructure.

Internet access is malleable as technology and as an economic unit and because of that privatization fostered attempts to adapt the technology in new uses, new locations, different market settings, new applications and in conjunctions with other lines of business. The NSF who actually commercialized the Internet was lucky in one specific sense that the time they commercialized Internet was a propitious moment, at the same time as the growth of an enormous new technological opportunity, the World Wide Web. One drawback of commercialization was that the commercializing Internet access did not give rise to many of the anticipated technical and operational challenges.

Role of Privatization in Commercialization of Internet

Privatizing Internet access was tantamount to giving a boom in Internet access technology to a wide variety of location, circumstances and users. Privatization transferred the operation of the technology to a new set of decision makers who had different ideas about what could be done with it. Businesses adopted the technology in innovative uses, new locations, different market settings, new applications. While many of these attempts failed, a large number of them also succeeded. While some, if not many, people imagine that the Internet is and should still be the product of pure, free market competition, this view ignores reality. The U.S. government funded much of the initial investment in the Internet and continued to be a major player in its evolution. Government and industry have all shifted the size and direction of their investments in the Internet over time as the potential applications of the rapidly evolving computing and networking technologies became apparent. Public investment in the early days (1960-1985) was based primarily on three objectives i.e. 1) establishing a secure, reliable communications and control system for national defense purposes, 2) Facilitating cooperative research among government agencies and among academic institutions, and 3) to advance the computing and networking technologies themselves. Nevertheless, the emergence of unanticipated, yet extremely attractive, applications forced a shift from public funding and public management of its research results towards commercialization and privatization. The Internet will have different technological changes if it has more widely commercialization network in order to have demanded of new technologies of the Internet. However, the most pressing question for the future of the internet is not how the technology will change, but how the process of change of evolution itself will be run. The architecture of the Internet has traditionally been driven by a core group of designers, but the form of that group has changed as the number of interested parties has grown. The success of the Internet has brought a propagation of stakeholders with both economic as well as intellectual investment in the network.

Driving Forces of E-commerce

There are at least three major forces fueling e-commerce: economic forces, marketing and customer interaction forces, and technology forces (Mohapatra, 2012).

ECONOMIC FORCES: One of the most evident benefits of e-commerce is economic efficiency. This efficiency results from the reduction in communications costs, low-cost technological infrastructure, speedier and more economical electronic transactions with suppliers, lower global information sharing and advertising costs, and cheaper customer service alternatives.

MARKET FORCES: Corporations are encouraged to use e-commerce in marketing and promotion to capture international markets, both big and small. The Internet is likewise used as a medium for enhanced customer service and support. It is easier for companies to provide their target consumers with more detailed product and service information using the Internet.

TECHNOLOGY FORCES: The development of Information and Communication Technologies (ICT) is a key factor in the growth of e-commerce. For instance, technological advances in digitizing content, compression and the promotion of open systems technology have paved the way for the convergence of communication services into one single platform. This in turn has, made communication more efficient, faster, easier, and more economical as the need to set up separate networks for telephone services, television broadcast, cable television, and Internet access is eliminated. From the standpoint of firms/businesses and consumers, having only one information provider means lower communications costs. Digital technology has enabled it to convert characters, sounds, pictures and motion videos into bit streams that can be combined, stored, manipulated and transmitted quickly and efficiently and in large volume without loss of quality. As a result, E-commerce and the multimedia revolution are driving the previously disparate industry such as communication, entertainment, publishing, and computing world into ever-closer contact forcing industries with traditionally different histories and cultures to compete and cooperate.

Functions of E-commerce

There are four important functions of E-commerce.

COMMUNICATION: It is the first essential function. It is aimed at the delivery of information and/or documents to facilitate business transactions. For example, use of e-mail can facilitate business information exchange.

PROCESS MANAGEMENT: This function covers the automation and improvements of business processes. One example is computer networking that can enhance the process of interaction among different departments.

SERVICE MANAGEMENT: This function aims at the application of technology to improve the quality of service. For instance, a logistics services company can establish a website to provide shipment-tracking services to its customers.

TRANSACTION CAPABILITIES: This function aims to provide ability to buy or sell on the Internet or some other online services.

Digital Economy, E-commerce, and E-business

Digital Economy (also called the Internet economy, the new economy, or the Web economy) can be defined as an economy that is based on digital technologies, including digital communication networks, software, and other related ICT. The concept of the Internet economy is broad in nature and includes both e-commerce and e-business. The CREC (Center for Research and Electronic Commerce) at the University of Texas developed a conceptual framework for how the Internet economy works (Barua, Pinnell, Shutter, & Whinston, 2000). According to this framework, Internet economy comprises the value-added chain of firms/institutions that wholly or partially transact business via the electronic networks of the Internet and the array of institutions that provide the necessary infrastructure and peripheral services to make these transactions happen. According to the framework, Internet economy composed of companies operating at four layers. The figure 1-1 shows the layers of the digital economy.

FIGURE 1-1: THE DIGITAL ECONOMY

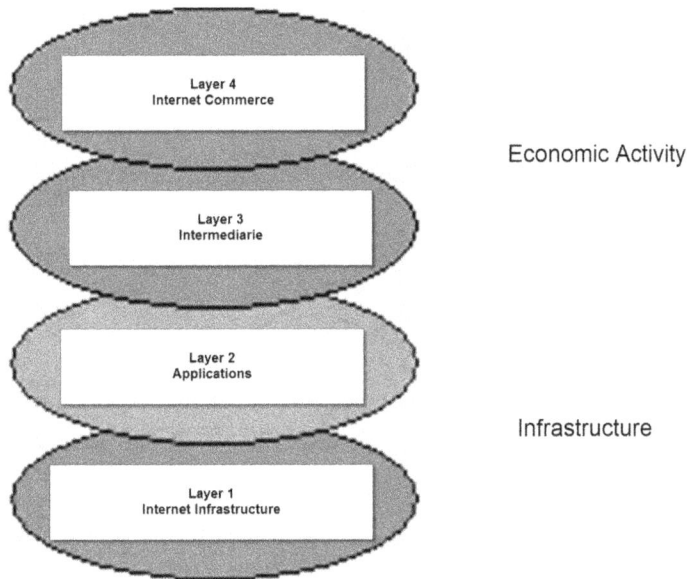

Layer 4
Internet Commerce

Economic Activity

Layer 3
Intermediarie

Layer 2
Applications

Infrastructure

Layer 1
Internet Infrastructure

Internet infrastructure layer consists of companies that provide enabling hardware and software for Internet and World Wide Web. Such hardware and software may include networking hardware/software, PCs and servers, and network media. Examples of companies proving such hardware and software are Cisco, IBM, and AT&T.

Internet Applications Infrastructure layer consists of companies that provide software to help facilitate transactions on the World Wide Web. Examples of this software are web development software, web-enabled databases, and search engines. Examples of companies offering this type of software are Adobe, Microsoft, and Oracle.

Internet intermediaries' layer consists of companies that link e-commerce buyers and sellers, provide web content, and provide e-marketplaces. Examples of such companies are Travelocity, E-trade, and Yahoo.

Internet commerce layer comprises of companies that sell products of services directly to the consumers or business. Examples of such companies are Dell and Amazon.

Technology and the layer of associated services are the foundation of the digital economy. The technology infrastructure needs to be spread adequately and at high quality standards at all levels for the digital economy to flourish. This is described in figure 1-2.

FIGURE 1-2: THE FORMULA OF DIGITAL ECONOMY

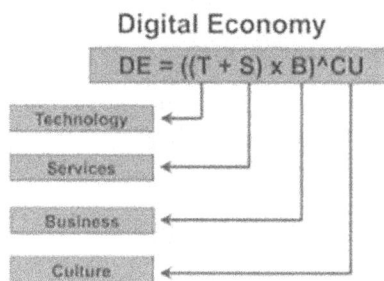

Digital Economy

$$DE = ((T + S) \times B)^{CU}$$

- Technology
- Services
- Business
- Culture

Courtesy (Salvatella, 2013)

The digital economy is developing rapidly worldwide. It is the single most important driver of innovation, competitiveness and growth, and it holds huge potential for entrepreneurs and small and medium-sized enterprises (SMEs). How businesses adopt digital technologies will be a key factor of their future growth. New digital trends such as cloud computing, mobile Web services, smart grids, and social media, are radically changing the business landscape, reshaping the nature of work, the boundaries of enterprises and the responsibilities of business leaders. These trends enable more than just technological innovation. They spur innovation in business models, business networking and the transfer of knowledge and access to international markets.

CLOUD TECHNOLOGY creates value for consumers and businesses by making the digital world simpler, faster, more powerful, and more efficient. By delivering Internet-based services and applications, it provides a more productive and flexible way for companies to handle their information technology. This has the potential to disrupt entire business models, giving rise to new approaches that are asset-light, highly mobile, and flexible. Furthermore, cloud technology is an enabler of other highly impacting emerging technologies, such as Big Data or the Internet of Things. **BIG DATA** has been at the core of ICT-led innovation based on measurement, experimentation, sharing and scaling up. Over 9 billion devices are currently connected to the Internet, and this number is expected to increase dramatically within the next decade to an estimated 50 billion to 1 trillion devices. This is the expanding **INTERNET OF THINGS**, where nearly every aspect of human life and economic activity is provided with networked sensors and actuators that monitor the surrounding environment, report their status, receive instructions, and even take action based on received information. *3D printing* has the potential for disruptive influence on how products are designed, built, distributed, and sold. 3D printers are commonplace for designers, engineers, and architects, who use these printers to create product designs and prototypes; they are becoming popular for personal use and gaining traction for direct production of tools, moulds, and even final products. Such uses could enable unprecedented levels of mass customization, smaller and cheaper supply chains, and even the "democratization" of manufacturing by allowing consumers or entrepreneurs to print their own products.

Two billion people are now linked to the internet (World Internet Stats, 2015). By 2016, this number is expected to exceed 3 billion (i.e. almost half of the world's population). Businesses that fail to be digitally connected will become excluded from the global market. The huge potential of the digital economy is under exploited in many parts of the world. Modern digital opportunities create new business opportunities. With rising youth unemployment rates in various regions, the growth prospects offered by the digital economy are promising.

Over the last five years, the development of mobile applications alone has created nearly 500 000 new jobs in the US, implying strong employment growth prospects. According to an estimate, 1.5 million additional jobs could be created in the EU digital economy if it mirrors the performance of the US or Sweden. Just engaging with customers online seems to make growth. Statistics show small and medium enterprises (SMEs) from various countries that have actively engaged with customers on the internet have experienced sales growth rates that are up to 22 percentage points higher over three years than those companies in countries with low or no internet presence. By not taking full advantage of digital technologies businesses miss out on the chance to expand and create jobs.

New technologies also hold potential for the manufacturing industries. Digitization of manufacturing can transform the entire industry, offering prospects for the re-location of industries. The digital economy will reach EUR 3.2 trillion in the G-20 economies and already contributes up to eight per cent of GDP, powering growth and creating jobs. In addition, over 75% of the value added created by the Internet is in traditional industries, due to higher productivity gains (European Commission, 2015).

Dimensions of E-commerce

Electronic commerce can take several forms depending on the degree of digitization of the product or service, the process and the delivery agent (or intermediary). Whinston, Choi, and Stahl (1997) created a framework, shown in figure 1-3 that explains the possible scenarios on these three dimensions. A product, process, and delivery agent can be physical or digital. These possibilities create eight scenarios, each with three dimensions. In traditional commerce, all three dimensions are physical and in pure e-commerce all dimensions are digital. All other scenarios include a mix of digital and physical dimensions.

A situation is considered part e-commerce if there is at-least one digital dimension. Purchase of a book from Amazon is an example of partial e-commerce because the transaction is processed electronically but the book is delivered physically. Purchase of an e-book or a piece of software from Amazon is an example of pure e-commerce, because the product itself, delivery mechanism, payment, and transfer agent are all digital.

FIGURE 1-3: DIMENSIONS OF E-COMMERCE

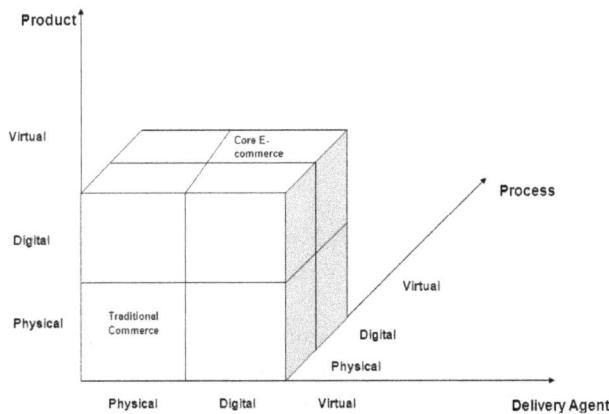

Internet vs. Non-Internet E-commerce

Not all e-commerce is conducted over the Internet. E-commerce can be conducted on private networks (e.g. local area networks). Buying food from vending machines is another example of non-Internet E-commerce (Chung, Lee, & Turban, 2006).

Categories of E-commerce

E-commerce is commonly classified based on nature of transaction or the relation among participating entities. Following classifications are commonly used.

Business-to-Business (B2B)

B2B E-commerce includes companies dealing with each other. One example is a lumber supplier selling wood to the furniture manufacturer. In B2B, prices can often be negotiated because they are a function of the quantity of the order. B2B e-commerce is the most important category of E-commerce by far and expected to grow faster than B2C (Schneider, 2009). In the United States, size of B2B E-commerce market was US$ 3.1 trillion in 2009 (US Cenus Bureau, 2009). However, B2C and C2C have been growing rapidly in recent years, mainly because of rapid expansion and diffusion of the Internet and enhanced broadband access. There are two primary components of B2B market: E-infrastructure and e-markets. E-marketplace is an online electronic market where buyers and sellers meet in order to exchange goods, services, money, or information. E-infrastructure is the architecture of B2B, which primarily consists of logistics, application service providers, outsourcing of functions in the process of e-commerce etc.

CHARACTERISTICS OF B2B E-COMMERCE

Following are some typical characteristics of B2B E-commerce (Cornish, 2014).

MULTIPLE DECISION MAKERS: In B2B, there is often four or more decision makers involved in the purchase process. In practice, this may require multiple user roles in the checkout /cart process with multiple stages taking many days (or weeks).

LONGER DECISION CYCLE: The B2B buying cycle is much longer than for B2C. As such, the lead-time between initial contact and receiving any payment is longer.

CUSTOMER SPECIFIC DISCOUNTS: In B2B, the variations in price lists, discounts and even available products is generally more complex than for B2C.

CONFLICT WITH DIRECT SALES CHANNELS: Many B2B businesses have an established sales team who will be unhappy with online competition that can be seen to decrease their performance bonus.

INTERNATIONAL MARKETS: B2B ecommerce is often used as a way of reaching international markets, maybe in small numbers. Regulations, legal and cultural issues can cause more of an impact than for B2B ecommerce and this influence is exaggerated if products are insignificant in number and high in value.

B2B MARKET SIZE

With B2C ecommerce still at the center stage of media attention, B2B ecommerce is gaining ground and generating bigger revenues both globally and in the United States. The entry of ecommerce giants such as Alibaba and Amazon into B2B has accelerated the trend of B2B websites becoming more like B2C. Online B2B sellers now recognize that the customer experience in a B2B environment is just as important as the customer experience for B2C. Consequently, expectations have grown and more B2B buyers require a simple ecommerce experience that mimics the consumer purchase model. Detailed specifications and product descriptions are crucial. Amazon Supply is the Amazon's B2B portal. Amazon Supply offers free two-day shipping on orders above $50, detailed specifications, and a customer review section.

The current size of the global B2C e-commerce market is $3.2 trillion. The global B2B E-commerce market is expected to become twice as large as the B2C market. According to Forrester Research (Forrester Research, 2015), China is expected to emerge as the largest online B2B market with $2.1 trillion in sales by 2020. In USA, B2B market is already twice as large as the B2C E-commerce market. The US B2B E-commerce market is expected to reach $780 billion and represent 9.3% of all B2B sales by the end of 2015. It is expected that US B2B ecommerce will exceed $1.1 trillion and comprise 12% of all B2B sales in the United States by 2020. Forrester Research forecasts a compound annual growth rate of 7.7 percent in B2B ecommerce over the next five years. This growth would result from the need for B2B companies to reduce their costs.

Buyers also benefit from the self-service automated approach to purchasing. According to Forrester Research, average conversion rate on B2B ecommerce sites is 7.3% while the average conversion rate on consumer retail sites is 3%. Within B2B E-commerce, petroleum and petroleum products, as well as pharmaceuticals and druggist sundries, represent the largest categories today. The same industries are expected to claim the largest share of all B2B ecommerce through 2020. The pharmaceutical industry will see 20% of total sales coming from online sources by 2020, the highest penetration of all industries. In B2B E-commerce, durable goods categories, motor vehicles, motor vehicle parts and supplies, electrical and electronic goods, and machinery, equipment, and supplies are expected to be the fastest growing segments between 2015 and 2020 (Kaplan, 2015).

B2B ECOMMERCE BUSINESS MODELS

Different B2B business models exist and the best choice is dependent on the size and complexity of a B2B company and its available expertise. Some generic examples of B2B models are IBM, Hewlett Packard (HP), Cisco, Marks and Spencer, and Dell. Cisco, for instance, receives over 90% of its product orders over the Internet (Andam, 2003) and 70 -80 percent of their customer service requests are also dealt with online. Orders are routed to contract electronics manufacturers who build the products to Cisco's specifications. Most B2B business models deal with supplier management, inventory management, distribution management, channel management, and payment management. Following is two major classes of B2B E-commerce business models.

ONE-TO-MANY

In this model, companies have their own B2B online store, where customers can purchase goods. Industrial supplier Grainger, one of the leading catalog companies to move online, uses this straightforward model. In a private consortium model, companies set up their own network that includes suppliers, distributors, retailers, and shippers. Only companies with substantial purchasing clout and a sophisticated supply chain infrastructure can make this work. Walmart Retail Link is an example of a private consortium. Through its Retail Link, Wal-Mart has given all its store suppliers access to an exclusive view of its inventory status in stores/warehouses, etc. Using predictive analytics, the suppliers are now able to plan their logistics better and ensure supplies free of hiccups.

MANY-TO-MANY

The many-to-many model involves companies joining an extensive online B2B marketplace. This can be a private marketplace in which several companies choose to form a closed network, or a public marketplace model, which is open to all suppliers (some criteria for joining may exist) and is usually administered by a third-party with a recognized name and marketing and logistics expertise. Amazon Supply and Alibaba are an example of public marketplaces.

FIGURE 1-4: B2B BUSINESS MODELS

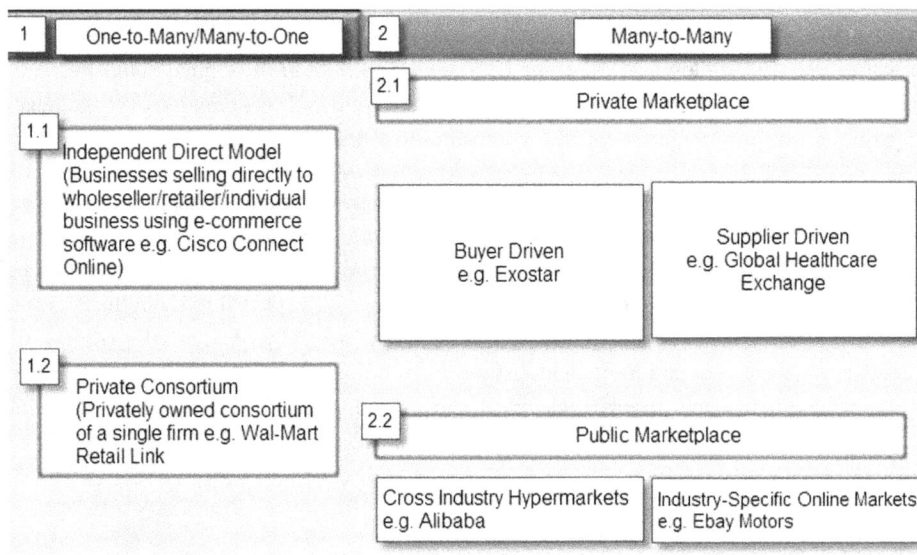

Source: Frost & Sullivan (www.frost.com)

B2B E-COMMERCE TRENDS

B2B online buying is expected to grow as much enterprise and procurement software vendors are making mobile versions of their products available and employees of the purchasing companies will use smart phones and tablets to buy goods anywhere and anytime. According to Forrester Research (Forrester Research, 2015), B2B companies can substantially decrease sales costs (up to 90%) by guiding customers to an online self-service ecommerce environment. In a 2013 Forrester survey (Briggs, 2014), 52% of B2B E-commerce executives surveyed reported that migrating formerly offline-only customers to online purchasing reduced their customer support costs. Grainger, a pioneer in B2B ecommerce, stated that 94% of its 2014 revenue growth came from online sales (eoscloudstore.com, 2015). Differences still exist between B2B and B2C ecommerce. Prices fluctuate often with customer-specific pricing and volume-based discounts, and impulse buying is not available. Today, B2B websites share many characteristics of B2C websites such as personalization and customization; interactive catalogs; customer reviews; and real-time inventory availability. Many B2B buyers now prefer to both research products and buy online. Strong growth is expected to continue during 2015-2020, especially purchases via mobile devices (Kaplan, 2015).

Basware, the leading provider of e-invoicing and purchase-to-pay solutions, predicts four key trends will transform B2B commerce over the next twelve months (Tihila, 2014). New real-time, online payment and financing options will drive a new era in consumerisation of B2B. These online and real-time trade-financing solutions, along with new, innovative financing models, are the disruptive technologies for 2015, changing the old way of doing business. They are removing suppliers' dependency on buyers for cash flow, and radically transforming the one-sided relationship that has favored the buyer, to a mutually beneficial B2B commerce experience. Social, mobile, and the cloud are the new norms. Accounts Payable (AP) departments will rely more heavily on social capabilities within business commercial networks to interact with trading partners and address issues in real time. Finance departments are more readily relying on cloud technology to cost-effectively manage their applications and resources, and access them seamlessly anytime and from any device. There is a continued push from governments around the world to adopt electronic invoicing to reduce costs, improve tax compliance and efficiency and reduce fraud. Governments are receiving up to 30 per cent of their Gross Domestic Product from VAT and sales taxes. Therefore, ensuring tax compliance is also a fundamental issue that is spurring e-invoicing adoption. Organizations will continue to rely on analytics more heavily than ever before to drive business results. Analytics solutions are identifying procurement, spend and payment trends; optimizing processes and technology; and uncovering opportunities for accelerated payments and other strategies. By leveraging analytics, organizations are realizing improved cash-to-cash conversion cycles, days sales outstanding (DSOs) and days payment outstanding (DPOs). Additionally, the insight into spending, cash flow and working capital enabled by automated analytics will propel a closer collaboration among internal departments, such as AP, Finance, Procurement and Treasury and extend to external trading partners.

Business-to-Consumer (B2C)

B2C includes businesses selling to the public typically through catalogs utilizing shopping cart software (e.g. Amazon.com). According to eMarketer's definition (eMarketer.com, 2014), B2C ecommerce sales includes all products and services ordered or booked via the internet on any device, including leisure and unmanaged business travel. B2C was the second largest and the earliest form of e-commerce in 2004 (Andam, 2003). Despite the dot-com failures of the 2000s, B2C e-commerce has continued to expand steadily (Slyke, Belanger, & Comunale, 2004). This growth has occurred both in the developed and developing world resulting in an increasingly global community of online shoppers (Cyr, Bonanni, & ilsever, 2004). In 2006, size of European B2C E-commerce market was 106 billion euros ($133 billion) and expected to grow at an annual growth rate of 25% over the next 5 years (Marketingcharts.com, 2007). In the United States, size of B2C E-commerce market was US$ 133 billion in 2008 (Census Bureau (US), 2009).

Prevalent B2C business models are the online retailing companies such as Drugstore.com, Beyond.com, Barnes and Noble.com and ToysRus.com. B2C e-commerce reduces transaction costs (particularly search costs) and market barriers to entry. Transaction costs are reduced because customers can access information and find the most competitive price for a product or service. Market entry barriers are reduced because cost of putting up and maintaining a web site is much cheaper than establishing a traditional business structure for a firm. B2C e-commerce even saves firms additional costs of establishing a physical distribution network.

According to eMarketer.com (2014) B2C ecommerce sales worldwide will reach $1.471 trillion in 2014, increasing nearly 20% over 2013. On a regional basis, North America (which includes only the US and Canada) will remain the leading region in B2C ecommerce sales share in 2014, accounting for 33% of the global B2C E-commerce market. Asia-Pacific is expected to become the leading region for e-commerce sales in 2015, representing 33.4% of the total, compared with 31.7% in North America and 24.6% Western Europe. These three regions combined will continue to share around 90% of the global ecommerce market (eMarketer.com, 2014). It is expected that by 2015, nearly 70% of internet users in both Western Europe and North America will purchase items on digital devices compared with just over 50% in Asia-Pacific. In North America and Western Europe, ecommerce continues to expand at double-digit rates and will do so for several more years. In markets as large as these, this point to the fact that individual buyers are making high value purchases more frequently and consumer behaviors are relatively consistent across countries in both regions. Consumer behaviors are disparate across Asia-Pacific countries. China alone will account for more than half of the entire region's ecommerce sales in 2015, and by 2018, its share will top 70%. In less mature markets of the Asia-pacific region, like India and Indonesia, there are large absolute numbers of digital buyers, but many are new to the market. These new digital buyers prefer to buy less costly purchases, due to product availability or simply to income constraints.

CHARACTERISTICS OF B2C E-COMMERCE

The main characteristics of B2B e-commerce are the fact that it usually involves building and maintaining relationships over time and the B2C market is a relatively small and easily identified target market. There exist multiple buying influences, and a larger number of people making buying decisions. B2C E-commerce involves a longer buying process and decision-making is based on service and after-care more than price. B2C buyers are generally more knowledgeable.

Business-to-Business-to-Consumer (B2B2C)

B2B2C uses B2B to support B2C companies. In B2B2C, a business provides some product or service to a client business (Chung, Lee, & Turban, 2006). The client business takes responsibility to maintain its own customers. One example is that of a company that pays AOL to provide its employees with Internet access (rather than having each employee pay an access fee directly to AOL). Another example is Maytag.com, a home and commercial appliance company. Maytag allows consumers to select the features they want in durable goods. Maytag retailers act as service centers. Maytag ships the goods ordered by the customers to the retailers and customers can then either collect them from the retailer personally or have them shipped to their home.

Business-to-Government (B2G)

B2G E-commerce includes transactions between businesses and government. It can include the use of the Internet for public procurement, licensing procedures, and other government-related operations. An example of

B2G is Companies Office Website (www.companies.govt.nz) in New Zealand. Companies can be register themselves online at this website. The facility makes the process much faster and greatly reduces the costs. One popular form of B2G is e-procurement service. By choosing this service, businesses can learn about the purchasing needs of various government agencies and agencies can request proposals from businesses. The eProcurement.gov.in is the e-procurement web site of government of Andhra Pradesh, India that provides real-time bidding for buyers and sellers.

Consumer-to-Business (C2B)

C2B E-commerce involves commerce between consumers and businesses in which consumers decide what they want to pay, and the vendors decide whether to accept. C2B business model is built on three players: a consumer acting as seller, a business acting as buyer and an intermediary dealing with the connection between sellers and buyers. One example is Fotolia (fotolia.com). Fotolia is an international stock photography agency where people can buy and sell images to illustrate professional and commercial brochures, advertisements, magazines, websites, etc. Another example is the best travel and ticket offer from airlines. Airlines extend this offer to travelers in response to travelers' post that they want to fly from e.g. San Francisco to Chicago. Another example is consumer posting his project with a set budget online. Companies respond to this post and bid on the project. Upon review, the customer can select a company to complete the project. Elance.com empowers consumers around the world by providing the meeting ground and platform for such transactions.

Consumer-to-Consumer (C2C)

C2C E-commerce involves commerce between consumers in which consumers interact with other consumers online. C2C business model is based on three players: a consumer acting as seller, a consumer acting as buyer and a platform provider (or intermediary) that connects buyers and sellers to facilitate transactions. Consumers are increasingly using the Internet as a means to sell goods and services through their personal websites, e-mail, auction sites, and sites providing classified advertising services. Over the years a number of C2C online platforms have developed, including classified ads (e.g. Craigslist.com and classifieds.excite.com), online marketplaces (e.g. eBay), peer-to-peer systems (such as the Napster, Kazaa and eDonkey), and sites offering both auction and classified ads (e.g. synthzone.com).

Active interactions and relationships among members are crucial to the success of C2C E-commerce. According to Xu, Zhang, and Chen (2009) there exist two types of trust in C2C E-commerce: Mutual trust among members and members' trust in the platform provider. Both the emotional and informational interaction among customers boosts mutual trust among members. This trust among members, in turn, boosts their trust in and loyalty to the platform provider.

Business-to-Employee (B2E)

B2E e-commerce uses an intra-business network, which allows companies to offer products and/or services to their employees. In a broad sense, B2E encompasses everything that businesses do to attract and retain well-qualified staff in a competitive market (Darrell-hill.com, 2009).Typical use of B2E E-commerce includes automation of employee-related corporate processes such as online insurance policy management, corporate announcement dissemination, special employee offers, employee benefits reporting, and 401(k) Management. For example, Merrill Lynch provides a program called Benefits OnLine. Employers can use this program to provide an employer-sponsored 401K plan for employees. Employees can access their plan information 24 hours a day through Benefits OnLine Website (www.benefits.ml.com).

Government-to-Government (G2G)

G2G is the online non-commercial interaction between Government (local and central) organizations, departments, and authorities and other government organizations, departments, and authorities. One example of G2G E-commerce is the European Union (EU) State Aid system to provide support to SMEs & business start-ups. Another example is the Schengen Information System (SIS) which is a governmental database used by several European countries to maintain and distribute information on individuals and property of interest for the sake of national security, border control and law enforcement.

Government-to-Employee (G2E)

G2E E-commerce includes services provided by the government to its employees such as the provision of human resource training and development that improve the bureaucracy's day-to-day functions and dealings with citizens. For example, employees of the US department of labor can interact with their department to obtain and submit information using labor department's intranet and automated employee service systems.

Government-to-Business (G2B)

G2B e-commerce includes transactions between government and the businesses. In these transactions, businesses can be suppliers, partners, or customers of the government. In addition, businesses must comply with government regulations while they maintain these roles (e.g., OSHA rules). G2B transactions may include various services exchanged between government and the business community, including dissemination of policies/regulations, obtaining current business information, downloading application forms, renewing licenses, registering businesses, obtaining permits, and payment of taxes. Examples of how businesses interact with the government include obtaining and submitting information using various government agencies home pages and e-Procurement activities.

Government-to-Citizen (G2C)

G2C includes information dissemination to the public, basic citizen services (such as license renewals, ordering of birth/death/marriage certificates and filing of income taxes), and citizen assistance for such basic services as education, health care, hospital information, libraries, and the like.

Other Forms of E-commerce

INTRA-BUSINESS E-COMMERCE: Intra-business E-Commerce includes all internal organizational activities that involve the exchange of services, or information among various units and individuals in that organization (Pušara & Katić, 2004). Such activities can include selling corporate products to one's employees, online training, and collaborative design efforts.

NON-BUSINESS E-COMMERCE: Non-business organizations, such as academic institutions, not-for-profit organizations, religious organizations, social organizations, and government agencies also use e-commerce to reduce their expenses or to improve their general operations and customer service (Pušara & Katić, 2004).

Electronic Data Interchange (EDI)

By definition, EDI (Electronic Data Interchange) is the transfer of data from one computer system to another by standardized message formatting, without the need for human intervention. EDI systems help businesses manage their supply chains. Traditional EDI systems use a combination of computers and communications equipment. Using traditional EDI, businesses can conduct secure, reliable transactions electronically. EDI is primarily used by large companies to have a uniform processing system, enabling efficiency. Cost, speed, accuracy and efficiency are the major benefits of EDI. The system is expensive to implement and usually requires help from a consultant that specializes in the field. Today EDI has become a vital part of businesses and many well-known organizations, such as Home Depot., Toys R Us and Wal-Mart, are using it. EDI has become a vital component of their business. Electronic data interchange technology is particularly important for international commerce where paperwork required for international trade creates costs that can be more than the value of the items being traded. With electronic data interchange technology, on the other hand, shippers, carriers, custom agents, and customers all can send and receive documents electronically, thereby saving both time and money for international transactions.

EDI is the computer-to-computer exchange of routine business data between trading partners in standard data formats. This definition includes three key concepts about EDI.

1) COMPUTER-TO-COMPUTER: EDI is most efficient when data flows directly from sender's computer system into receiver's computer system without any human intervention.

2) ROUTINE BUSINESS DATA: EDI applies to documents such as purchase orders, invoices, shipping notices and commission sales reports, as well as other important or classified information. For example, an insurance company can verify that an applicant has a driver's license through an EDI exchange. EDI is not used for non-routine business documents like complicated contracts or information meant for humans to read and analyze.

3) STANDARD DATA FORMATS: A standard definition of the location and structure of the data is provided.

The conventional paper process can be considerably slower than the EDI process. Additionally, a high-level of human intervention in conventional paper process is required to move business information from one company to another. The conventional process requires someone to manage a printed computer generated form and mail it. Then, the recipient re-keys the data back into another computer for their internal processing. In EDI process, a computer transmits the information directly to another computer, eliminating the paperwork and human intervention.

Before EDI, communication between trading partners occurred on phone and fax and the transaction process took several hours. On the buyer's side, when a request for purchase is sent then a purchase order was placed by the purchasing department, which is then sent to the finance department as well, for payment upon receipt of the product. The sales department received this purchase order, at the seller side. The sales department then sends it to the manufacturing or the warehouse. The product was subsequently sent to the shipping, which delivers it to the buyers receiving department. Figure 1-5 shows flow of information before EDI. Figure 1-6 shows flow of information after EDI.

FIGURE 1-5: FLOW OF INFORMATION BEFORE EDI

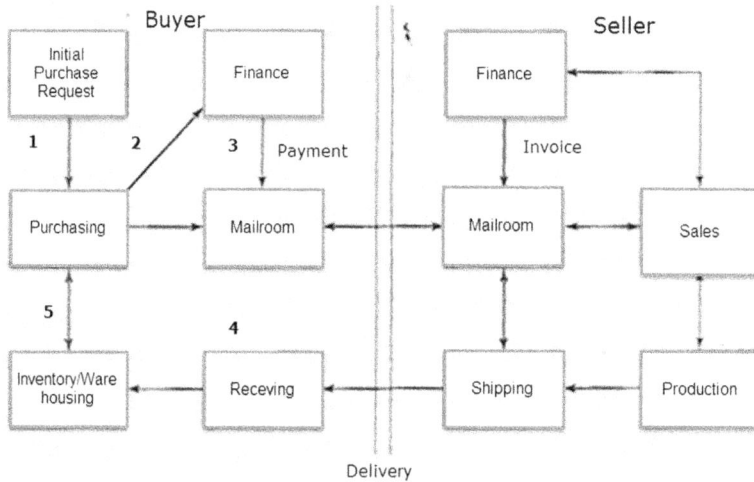

FIGURE 1-6: FLOW OF INFORMATION AFTER EDI

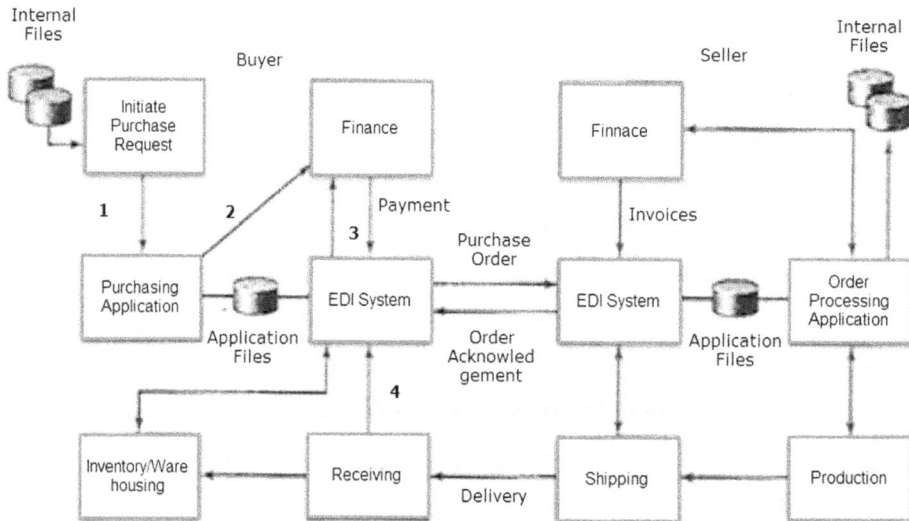

EDI transforms this labor demanding and time consuming process. When the buyer takes a decision to order the product with EDI, then the buyer's EDI computer generates the purchase order in its purchase application. This purchase order transmission is directly transmitted to the sellers EDI system in machine-readable EDI standard generated by the buyers EDI system. An acknowledgement notice is sent to the buyer's EDI system, after it passes the order information to the receiving order entry application for processing. This processing is handled like an incoming purchase order. The seller's EDI system communicates with the company's network to send shipping and billing information. A shipping confirmation is also generated and a copy is sent to the buyer. In all this process, the transmission takes place through the dedicated lines or VANs are used.

Components of EDI

There are four basic components involved in establishing and running a successful EDI (JANAKIRAMAN & SARUKESI, 2008).

1) Necessary hardware and communication devices that link the computes

2) EDI standards

3) EDI software that enables smooth conduct of the EDI process

4) Communication service providers

HARDWARE AND COMMUNICATION DEVICES

For successful implementation of ED, it is essential that all data processing activities in the company be completely computerized. This is the basis for the creation of templates. As far as hardware and communication devices for EDI are concerned, the existing hardware devices with minor changes in the hardware configuration can be effectively used in the process. All one needs additionally is a high-speed network connection that connects both the computers.

EDI STANDARDS

When sending an EDI document, both parties must stick to the same set of rules. These standards define where and how the information from the document will be found. EDI cannot succeed unless and until the parties involved agree to transfer data in electronic form. Moreover, trading partners do use heterogeneous computer hardware and software necessitating the need for standards. EDI standards stipulate the methods and format by which data can be shared between the trading partners. In respect of EDI standards, the transportation industry in the USA initially took the lead in forming standards. Later, the grocery industry segment devised an EDI standard namely Uniform Communication Standard (UCS). Every other branch of trade devised its own EDI standards. For example, warehousing industry devised Trade Data Interchange (TDI), shipping industry devised DISH (Data Interchange for Shipping), and European automobile industry developed ODEITE (Organisation for Data Exchange by Tele Transmission in Europe).

With EDI gaining popularity, efforts were made to bring in some standards, which are acceptable to all industry segments. In USA, the American National Standards Institute (ANSI) developed a standard called ANSI X12 for use by all US businesses. These standards are needed because different industries, and even different companies within the same industry, have adopted their own formats for documents. The ANSI X12 provides a standard layout and sequence of information for each type of document, as well as agreement on a standard. Automotive, retail, and transportation industries are the major users of EDI. Therefore, ANSI X12 provides specific standards for them. For example, AIAG (for automotive industry), VICS (for the retail industry), and TDCC (for transportation industry).

In respect of International Standards, UN/EDIFACT (EDI for Administration, Commerce and Transport) standards are gaining popularity. This standard is the result of the consistent efforts of the United Nations Economic Commission and International Standards Organization (ISO). For evolving standards, there exist six EDIFACT boards, viz. Western Europe, Pan America, Australia/New Zealand, Asia, and Africa. India is an active member of the Asia EDIFACT board whose other member countries are China, Hong Kong. Japan, Korea, Malaysia, Singapore, and Taiwan.

Different EDI standards meet the needs of specific industries or regions or other specifications. ANSI ASC X12 (X12) is routinely used in the U.S., and EDIFACT (Electronic Data Interchange for Administration, Commerce and Transport) is used outside of the U.S. In addition, numerous other standards relate to specific industries, such as VDA for the German automotive industry. The Data Interchange Standards Association (DISA), a company that develops and proliferates the standards for interchanging electronic business across multiple industries, supports ASC X12. X12 members work together across various industry verticals such as finance and healthcare to develop new versions of the standard.

EDI MESSAGES

The figure 1-7 presents the basic structure of an EDIFACIT. An EDI is possible only when there is a connection between two computers. When a connection is set up, there is an interchange of information. An EDI message contains a string of data elements, each of which is a singular fact, such as a price, product model number, and so forth, separated by delimiter. The entire string is referred to as a data segment. One or more data segments framed by a header and trailer form a transaction set, which is the EDI unit of transmission (equivalent to a message). A transaction set often consists of what would be contained in a typical business document or form.

FIGURE 1-7: STRUCTURE OF EDIFACT

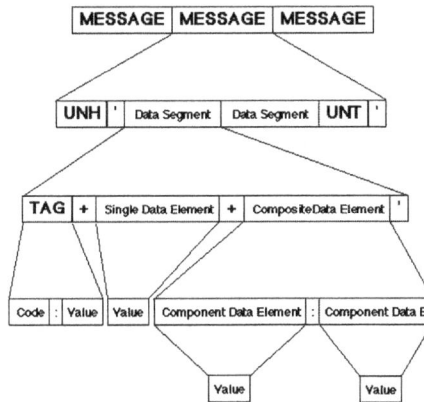

EDI SOFTWARE

EDI software refers to that software which helps in translating the requirements of a customer into an electronic form. It also assists in translating data from any format available to EDI format. The three main tasks of an EDI software are conversion of data into EDI formats, formatting the required data into structured means/templates, and establishing communication between computers and ensuring that the data is transferred properly. EDI gateway software is used to convert application system data into a standard format and it is used to send and receive message. EDI gateway consists of hardware platform and EDI translation software. The mapping function converts data from a standard file format into EDI standard and an EDI standard format into a standard computer system file so that it can be processed through business application software. The translation function adds standard enveloping and delimiter to mapped data to permit an EDI message to be directed appropriately to the designated partner. Translation software processes the information differently for sent and received messages and performs a complete audit of each step to ensure information is sent or received in EDI format. When the translator on the receiving computer reads a document, it knows where to find the buyer's company name, order number, purchase items and price, for example. This information is subsequently sent to the receiver's order entry system without necessitating manual order entry.

COMMUNICATION SERVICE PROVIDER

These third parties are the backbone for the success of EDI and are responsible for providing electronic communication services. To transfer EDI documents electronically a communication network is used by the trading partners. EDI messages can be transferred through direct connect or value added network (VAN). Different types of EDI are implemented to suit the business's needs, capabilities and budget. Methods include direct EDI (point-to-point); EDI via VAN or EDI network services provider; EDI via ES2; EDI via FTP/VPN, SFTP or FTPS; and Managed Service. With Direct EDI, one connection is created between two business partners. Larger businesses might choose this method if they have numerous transactions with the same partner per day. EDI via VAN protects businesses from the complications that come from supporting the many communication protocols that are required when dealing with multiple business partners. It is a secure network where documents are transmitted between business partners. EDI via Action Script Version 2 (AS2) is a communications protocol used to securely exchanges data over the Internet. Two computers (a client and a server) connect in a point-to-point mode through the Internet. AS2 is transmitted securely in an envelope that uses digital certificates and encryption. File Transfer Protocol (FTP) over Virtual Private Network (VPN), Secure File Transfer Protocol (SFTP) or File Transfer Protocol Secure (FTPS) exchange documents through the Internet, connecting business partners directly. These protocols encode data during transmission from one business to the other, to safeguard sensitive information. Data is decrypted upon arrival. Managed services systems outsource the EDI document control to a third party provider. The figure 1-8 shows how data coding and encoding work in EDI. Figure 1-9 explains how EDI translator system works.

FIGURE 1-8: DATA ENCODING AND DECODING IN EDI

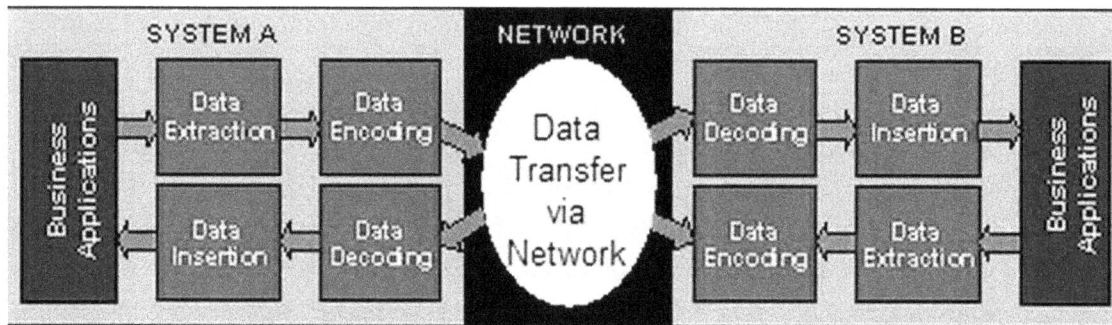

FIGURE 1-9: TRANSLATION SYSTEM IN EDI

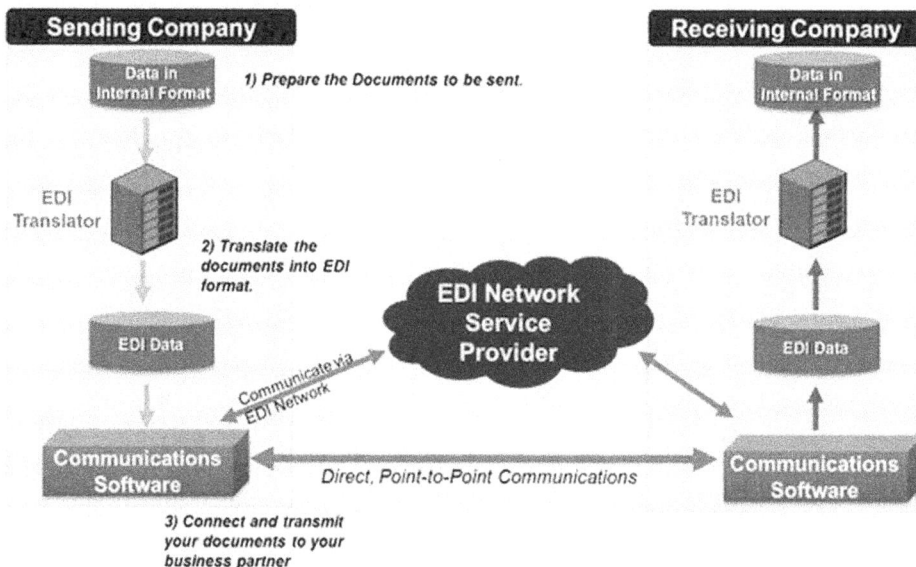

Advantages and Disadvantages of EDI

EDI, when implemented, would ensure quick response to queries and demands of customers. In the majority of modern day businesses where EDI is likely to be implemented, profit margins are thin and companies survive by responding to demands immediately.

When the query of a customer is answered immediately, it goes a long way in improving the customer relationship. By adopting EDI, companies do reduce the transaction costs (which include stationery cost, expenses towards telephone and fax messages, mailing costs and printing costs), labor costs, errors due to incorrect keying-in, and lead times in a business transaction. Some additional benefits of EDI include reduced cycle time, better inventory management, increased productivity, improved accuracy, improved business relationships, enhanced customer service, increased sales, and increased cash flow.

In spite of all these advantages, businesses do shy away from EDI for some reasons. Since standards have not evolved fully, businesses are reluctant to adopt EDI. With global economic recession, small businesses are not prepared for investment in operational areas they consider not vital. Costs involved in EDI are high. Businesses argue that transformation to EDI would necessitate changing their systems to suit EDI that involve heavy expenditure and the returns are not commensurate with the investments. EDI is beneficial for businesses, which have a large volume of transactions. For businesses having small volume, EDI is more expensive than conventional means of business transactions. Fear over data security is another matter of concern. Firms still have a fear that by adopting EDI, their secure data might become accessible to all and it takes time and convincing talk to remedy this phobia.

Security in EDI

Security is of prime concern in EDI. The key issues of EDI security are confidentiality, Integrity, authenticity and non-repudiation. Confidentiality means the information contained in the EDI message must be guarded and protected from unauthorized users. Integrity means that the message sent through EDI must appear the same at the destination. Authenticity refers to provisions that must be made to prevent any person from masquerading as an authentic user. Non-repudiation means the system must provide facilities such that the sender of an EDI message must not be, at a later point of time, deny having sent the same. Security in EDI can be ensured by applying various encryption algorithms. We shall discuss encryption in detail in later chapters.

Selecting an EDI System

Business needs and processes determine which EDI system to implement. Before moving to EDI in your organization, establish a managerial structure. Members of the team should have EDI and executive experience. Next, set priorities for adoption and then analyze which areas of the business will benefit most from EDI. Select an EDI network provider that focuses on your business necessities. Then, assimilate EDI and data into the business. To outline how the data in the EDI transaction correlate to the data in the internal system, create a map that will determine where each incoming field goes and if the data needs to be reformatted. Test the system before implementing it in live environment.

Future of EDI

The UN/EDIFACT is still widely accepted in Europe and EDIFACT message volumes hold considerably high percentage of all EDI message exchange. Outsourcing service offerings are covering more and more business processes EDI is still playing an important role regardless of the technology being utilized. From the technology point of view, the most convenient way of exchanging EDI messages would be using one message standard specification, which also contains the message transport. EDI is evolving into more and more industries. The EDI software of today is quite powerful providing customized solutions based on what a customer needs. Taking into account today's 24/7 business climate, EDI will continue to thrive because companies are relying more and more on EDI systems to help run their businesses profitably. At one point, it was forecast that XML would replace EDI. However, many businesses that have invested in EDI have found that it is efficient and works well, so they do not see any need to spend the money to reinvent this particular wheel. Thus, EDI remains a mainstay for business and not much change in this situation is expected in the near future.

Value Added Networks (VAN)

Traditional EDI uses a closed network called VAN (Value-added Network). A VAN connects all members of a production process (every supplier, manufacturer and distributor) to EDI. For example, an automobile manufacturer may have an EDI system in place to manage its supply and distribution relationships. In a given day, the auto manufacturer might receive thousands of yards of sheet metal, countless shipments of electronic equipment and dozens of parts from various suppliers worldwide. Each of these shipments must pass through complex distribution channels. Since the shipments are essential to the timely completion of an automobile, the manufacturer must make all efforts to ensure that the products will be delivered on time. Operations personnel at the manufacturing plant use the EDI system to purchase supplies, track shipments and keep an accurate inventory count. This process is achieved through a standardized transfer of electronic documentation that verifies each party in a transaction, records the terms/conditions of the transaction, and processes the order. Figure 1-10 shows working of a value-added-network.

FIGURE 1-10: A VALUE-ADDED NETWORK

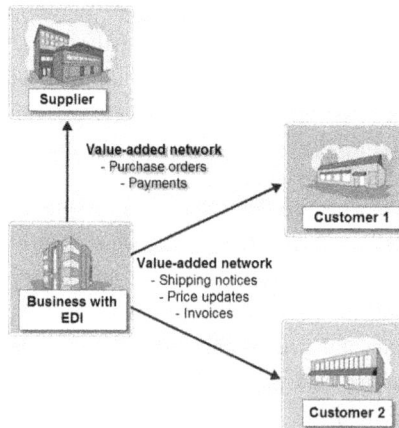

26

Value Added Networks (VANs) are no longer just facilitating transmission of EDI transactions. In a recent trend, large companies are acquiring these VANs and adding value by layering their own technology and services on top of these VANs. Business process management, advanced information management, and Big Data Analytics are some of the notable technologies being applied to VANs. One example of VAN acquisition is OpenText's acquisitions of EasyLink and GXS. OpenText Corporation is headquartered in Canada and is Canada's largest software company. It develops and sells Enterprise Information Management (EIM) software solutions for large corporations across all industries. EasyLink is a leading global provider of EDI and B2B solutions, and cloud fax that enables you to receive and send faxes via email, and notifications services. GXS is the Global leader in B2B Integration Services. On the GXS acquisition, OpenText acquired 600 thousand connections between which 16 billion B2B transactions and $6 trillion worth of B2B commerce occurred annually. As part of OpenText, GXS acquired 2 billion fax pages per year from EasyLink that could be incorporated into their network. Looking at the data in the transactions (EDI transactions, faxes, email, SMS, and voice data) OpenText is building business processing functions on top of the network, to handle specific document types including orders, invoices, shipping notices, claims, contracts, statements, and reports for added transaction context and traceability for regulatory protection. In another example, IBM purchased the Sterling network, services and software products. Sterling Network is a provider of flexible cloud B2B integration solution. Liaison purchased NuBridges along with a few other value-adding service providers. Liasion is a software developer of Electronic Document Delivery and workflow automation systems for mid-market accounting systems. Liaison coveted nuBridges for its cloud-based offerings, including nuBridges' token-as-a-service (TaaS) offering and its managed file transfer (MFT) software, which Liaison will use to bolster its position in the emerging market for cloud B2B software and services. SAP acquired the Ariba network. Ariba is the creator and provider of the Ariba Network, a cloud-based B2B marketplace where buyers and suppliers can find each other and do business within a single, networked platform. This is evidence that there is value in B2B integration. It is evident that there is a need for more value, on top of value added networks (Bond, 2014).

Advantages and Disadvantages of VAN

Some advantages of VAN are safety, reliability, and modern features.

SAFETY AND RELIABILITY

As a system specifically designed for secure data transfer, information architecture behind most VANs, including everything from encryption methods to safe data transfer protocols, tends to be far more sophisticated than an average firewall or network security set-up. Naturally, such a level of security is probably unnecessary for most data, but may be a critical advantage for an e-commerce business that handles confidential data like credit card information, bank account details or proprietary design information for manufacturers. VANs also tend to be more reliable than secure e-mail servers or other systems that experience periodic outages.

MODERN FEATURES

In today's Internet age, VANs have struggled to keep up with online security solutions like XML coding. VAN providers have introduced a number of new services that can be very useful for e-commerce businesses. Transaction Delivery Networks (TDNs), for example, are a new variety of Internet-based VAN that guarantees secure data transactions from one point to another, with added features like enhanced encryption, guaranteed server availability and delivery success notifications. Many modern VANs can also be programmed for automatic generation of certain types of data transfers, like factory orders or customer notifications. These can save an e-commerce business money and improve relationships with suppliers and clients.

Some disadvantages of VAN are costs, installation, and communication that is more complex.

COSTS AND INSTALLATION

The added features available on a VAN are not free. In fact, many of the most sophisticated VANs can be quite expensive, charging subscription costs or data-transfer rates. Setting up a VAN in an e-commerce business can also be rather complex and costly, often requiring new equipment or employee training as data management processes change. These added costs can be worthwhile for some businesses that are particularly concerned with data security.

COMPLEXITY OF COMMUNICATION

Given the added cost of contracting the service, VAN systems are most often found in larger corporations and e-commerce sites. A small business with a VAN may be able to streamline communication and transactions with the bigger players in the field. This is a considerable advantage in some sectors, such as e-commerce resellers. Having a VAN, however, can also make communication more complicated with small players that rely on simpler data-transfer methods. Small businesses are often forced to keep their old systems running after contracting a VAN in order to communicate with some of their smaller partners and affiliates.

Examples of Value-added Networks

Beyond the basic levels of service, EDI-specific VANs provide a number of additional services, which enable businesses to maintain their EDI programs, such as translation software, network interconnection, trading partner profiling, supplemental communications support and network monitoring, and education. Some of the larger VANs are provided by GE Information Services, IBM, AT&T, MCI, and Sterling Commerce. Other, more market-focused VANs serve specific industry groups and excel in their respective niches. These include Kleinschmidt, RailLinc, SNS, and others.

The basic concept of a VAN is a hub that links many pathways. It is a gateway to other networks. For example, IVANS links to worldwide Brussels-based Reinsurance and Insurance Network (RINET), the London Insurance Market's LIMNET, and others. When IVANS implements a new EDI transaction set, such as the sets for First Report of Injury and Subsequent Reports for Workers Compensation Claims, participants do not have to make any change in their own systems. SITA is a leading ICT provider of integrated IT business solutions and communication services for the air transport industry. SITA provides communication services across the world's largest global network in over 200 countries, supporting more than 3,200 customers, 24 hours a day, seven days a week and 95% of all international destinations are covered by SITA's extensive network. The truly global SITA air transport industry (ATI) cloud is currently pre-connected to 380 airports, 17,000 air transport sites and 15,000 aircraft.

The services provided by value added networks (VANs) ease the adoption of EDI by smaller organizations with lower levels of technical expertise. Large organizations with several trade partners may also find VANs quite attractive as VANs, in essence, provide a common trading ground for many traders. The selection of VAN by a business may depend upon the services offered, experience, reliability, and availability of other related trading partners. In case of smaller organizations and ancillary units, the decision to join a VAN is often governed by their dominant partners. There are many third party VAN providers the marketplace. Some of them are included here (Bhasker, 2008).

CABLE & WIRELESS WORLDWIDE: Highly reliable, with a subscriber base of over 2000 top companies of the world, cable & wireless held nearly 8 % market share of the global VAN market.

GELS: GEIS is operated by General Electric of USA. GEIS has presence in over 50 countries. GE as the major trader (buyer as well as supplier) of goods from top corporations of the world has brought major trade partners on a VAN.

GNS: It is one of the largest value added network, and has presence in around 36 countries.

INFONET: It is a VAN service jointly owned and operated by WorldComm, Singapore Telecom, and Transpac. The owning organizations themselves offer VAN services in the local domains and cover rest of the world through the Infonet.

TRANSPAC: A France based EDI VAN provider, Transpac owns the largest domestic VAN market share and has a strong presence in Europe. It uses the Infonet for offering VAN services outside the domestic domain.

Internet-based (or web-based) EDI

Although EDI systems improve efficiency and promote better accounting practices, they can be costly to operate. Another issue is integration with business partners. Many suppliers and distributors are small businesses, which do not possess the technology to link themselves into a traditional EDI system. If a supplier standardizes its information systems with a single manufacturer, it may become more difficult to deal with other manufacturers that have incompatible systems.

The basic operational aspect of Internet based EDI is simply to use already existing communication protocols available on the Internet to exchange EDI data. EDI is directed via the Internet browser, duplicating paper-based documents into Web forms that contain fields where users enter information. It is then automatically converted into an EDI message and sent via secure Internet protocols. Internet based EDI uses the SMTP (email), HTTPS and Secure FTP communication protocols to exchange data. These protocols are often referred to through a set of acronyms that were established in the mid-90s. Accordingly, when using SMTP for Internet based EDI the common terminology is AS1, while HTTPS is AS2 and Internet based EDI through secure FTP is known as AS3.

The Internet can support EDI in a variety of ways. First, Internet e-mail can be used to transport EDI messages in place of a VAN. To this end, standards for encapsulating messages within Secure/ Multipurpose Internet Mail

Extensions (S/MIME) have been established. Second, a company can create an extranet that enables its trading partners to enter information into a web form, the fields of which correspond to the fields in an EDI message or document. Third, companies can use a Web-based EDI hosting service in much the same way that companies rely on third parties to host their e-commerce sites. Sun Java System Web Server is an example of the type of Web-based EDI software that enables a company to provide its own EDI services over the Internet. Harbinger Express is an example of a company that provides third-party hosting services. Web-based EDI is frequently XML based to ease integration among different business partners.

Internet EDI in this context is really software as a service where the EDI software is delivered in a hosted multitenant or ASP (Application Service Provider) environment over the Internet. This implies that in contrast to most EDI software installations, Internet EDI is rented or leased and delivered over a browser. Internet EDI is in contrast to the types of software that is installed on a computer behind a firewall. There are some notable difficulties with Internet EDI, but most of them are centered on the difficulty of doing high volume EDI transactions. In some cases, Internet EDI can be integrated with backend systems. Even when Internet EDI is integrated with accounting or ERP systems, high volume transactions are hard to attain. Internet EDI has another major issue that EDI data is not really controlled by the company. This is a significant issue in disputes where the Internet EDI vendor can withhold your data from you until the dispute is resolved, effectively crippling your business. Though there are few examples of Internet EDI vendors having this as a practice, it may become more prevalent with time.

Selecting a suitable Internet EDI solution can be quite a challenge for any business. It is essential for a business to ensure that the selected Internet EDI provider is a company that is reputable, stable and reliable. A small business should be aware that just because the software it selects is capable of processing Internet EDI, this does not necessarily mean that its trading partner will be satisfied with that particular Internet EDI solution. Internet EDI solutions are available either through the open source community or via private software. The best Internet EDI solutions are required to go through a certification process to ensure that the Internet EDI communications are functioning according to accepted standards. This method of certification of Internet EDI software is often referred to as Drummond Certification. Drummond Group Inc. provides interoperability software testing. AS2 (Applicability Statement 2) is one of the most widely recognized messaging standards for B2B e-commerce with thousands of implementations around the globe. It enables users to connect, deliver and reply to all forms of data (including EDI and XML) securely and reliably, thereby delivering cost savings and providing flexibility and control on how the data are utilized. AS2 is broadly adopted among the world's major retailers and manufacturers (e2open.com, 2014). Internet EDI software, however, does not necessarily have to be certified. Many retailers will allow Internet EDI solutions that use unsecured means – such as FTP connections. Internet EDI solutions that use insecure protocols are typically much cheaper and much more cost effective. Some examples of web-based EDI include DiWeb (http://www.dicentral.com/web_based_edi/), EDIPX (http://www.1edisource.com/edi-solutions/web-based), CovalentWorks EDI (http://www.covalentworks.com/web-based-edi.asp), and HighJump EDI (http://www.highjump.com/solutions/truecommerce/web-based-edi).

The future of Internet EDI holds a lot of promise. Internet EDI is being adopted more frequently by more and more retailers. Many research firms estimate that Internet EDI will become the dominant EDI exchange method in the next few years. In the early days, software that allowed for AS1, AS2 or AS3 communications were extremely expensive. Today, there are solutions in the market place that make using Internet-based EDI significantly more cost effective making it a viable technology not only for large organizations but also for smaller businesses. Because of these developments, many small businesses are beginning to convert to this technology and are seeking software solutions that either includes this technology or that have it available as an add-on at reasonable prices.

Working of Internet-based EDI

It works in the same way as the traditional EDI works with few differences. EDI is directed via Internet browser, duplicating paper-based documents into Web forms that contain fields where users enter information. It is then automatically converted into an EDI message and sent via secure Internet protocols. EDIUSA.net is a simple web-based EDI application. Figure 1-11 shows how EDIUSA.net works.

FIGURE 1-11: WORKING OF WEB-BASED EDI

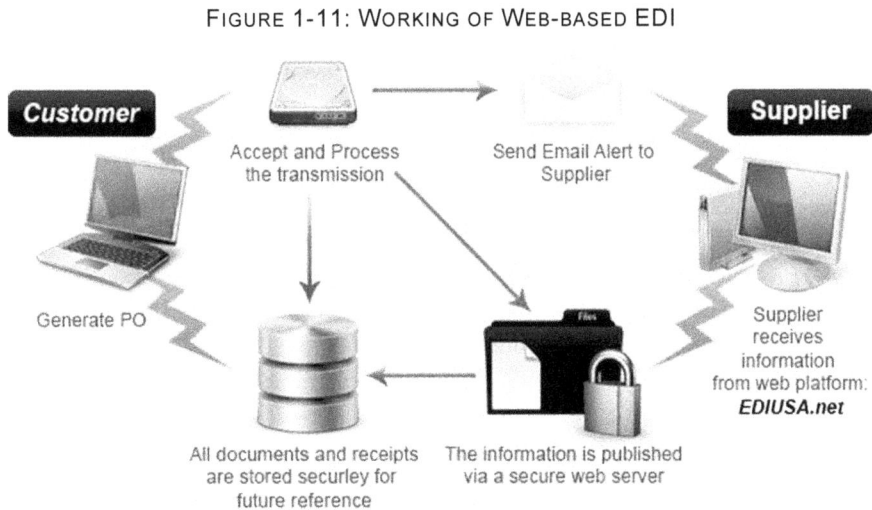

Source: www.ediusa.net

31

FIGURE 1-12: WEB-BASED EDI TRANSACTION PROCESSING

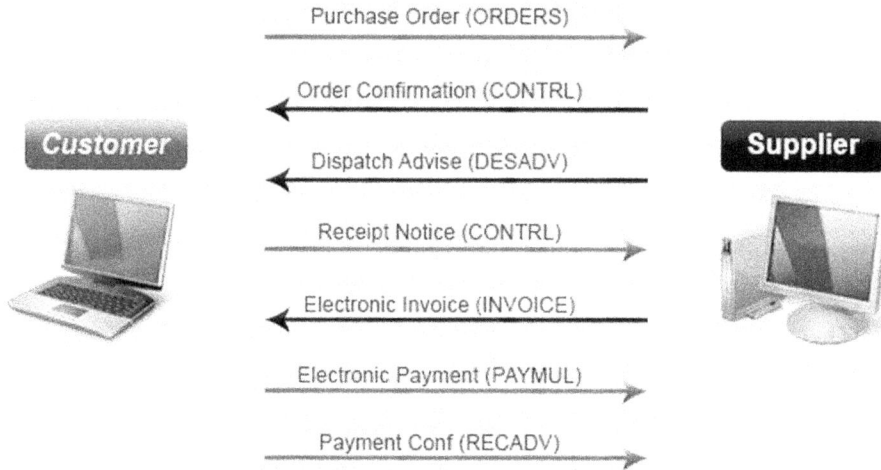

Source: www.ediusa.net

Benefits of Internet-based EDI

Internet based EDI provides a number of benefits over VAN based EDI. The charge of using VAN based EDI is based on the amount of data that is used and transmitted over EDI. By employing Internet based EDI, a direct link can be created with trading partners bypassing third parties and sending and receiving data directly. It was this benefit, coupled with the low cost of using the public Internet, which convinced Wal-Mart to switch their entire supplier network to Internet-based EDI. For businesses, utilizing Internet-based EDI can save a great deal of money and provide significant return on investment.

Applications of Internet-based EDI

The combination of the Web and XML makes EDI worthwhile even for small, infrequent transactions. The following examples demonstrate the use of Web-based EDI. After Compucom Systems started using web-based EDI, their average transaction increased from 5,000 to 35,000 transactions. Tradelink of Hong Kong provided export/import transactions with government agencies. Utilizing web-based EDI, Tradelink of Hong Kong increased their registered member companies from several hundred to many thousands. Atkins Carlyle Corp in Australia is a wholesaler of industrial, electrical, and automotive parts. By moving to a web-based EDI system, the company significantly increased its number of collaborative partners and its transaction cost per message reduced by approximately $2. By moving to a web-based EDI, Proctor & Gamble was able to get a many fold increase in their number of suppliers (Turban et al., 2010).

XML and EDI

Any methodology that enables the computer-to-computer exchange of business documents falls under the heading of Electronic Data Interchange (EDI). The Internet is improving EDI standards by making it more accessible to a broader group of manufacturers, distributors and retailers. Since the transfer of data is conducted through a common system (the web), compatibility is less of an issue. XML (eXtensible Markup Language) can now be employed to improve compatibility between disparate systems, creating new market opportunities.

There are two basic approaches to EDI. The first approach is to create business documents that adhere to one of the EDI standards such as ANSI X12, or EDIFACT. These standards define exactly where each piece of data is to be located in the electronic business document (Gxsblogs.com, 2014). The second approach enables the creation of electronic business documents in a more flexible way, one that is not bound by the strict rules of data location. The XML language is designed to provide such flexibility. XML is not a standard at all. It is a powerful language that gives a company a great deal of flexibility in defining and constructing documents, such as the types of business documents defined by ANSI and EDIFACT (e.g., purchase orders, invoices, remittance advices). For example, RosettaNet is a standard that uses the XML language. It was developed by a consortium of major computer, consumer electronics, semi-conductor manufacturers, and telecommunication and logistics companies. It facilitates some industry-wide global supply chain processes. In addition, within an enterprise, XML is heavily used for sharing data among various system components. Moreover, internal system integration software is often designed based on XML.

Document based on EDI standards and the one that is based on XML have major structural differences. The ANSI or EDIFACT document is based upon strict rules governing the position of data within a file, whereas the data in an XML file is not bound to a specific location but is instead identified by tags, such as "<quantity>300 </quantity>" to indicate a quantity value of 300. These tags result in XML files being much larger than their comparable ANSI or EDIFACT files. However, these tags also tend to make a XML-based business document more readable because the tag identifies the type of information that follows. This readability feature comes in handy for troubleshooting when human intervention is necessary.

Forms of E- Commerce Organizations

Businesses involved in E-commerce can take various forms. The following table describes some of the common forms.

Brick and Mortar Business	An old-economy business, which performs most of its business off-line. They sell physical products using physical delivery agents.
Click-and-Mortar Business	Organizations that conduct some of their business activities online but do their primary business in the physical world.
Virtual Business (Pure-Play Business)	Organizations that conduct their business activities solely online.
Electronic market (E-marketplace)	An online electronic market where buyers and sellers meet to exchange goods, services, money, or information.

BRICK-AND-MORTAR BUSINESS

It is a traditional business that deals with its customers face to face in an office or store that the business owns or rents. Physical stores offer a trustworthy staff and opportunities for customers to examine items before purchasing. Products are sold in physical stores and most products are tangible. These businesses use paper catalogs to promote their products and make use of physical marketplaces. These businesses use VANs and traditional EDI and customers are billed using paper bills. Production is mass production, which starts with a demand forecast. Marketing of product is physical which is commission-based and utilize word-of-mouth publicity with slow and limited advertisement. A brick-and-mortar business requires large amounts of capital for startup and fixed costs of operations are significant.

CLICK-AND-MORTAR BUSINESS

It is a non-traditional business that exists both online and in the physical world. Retailers get the most benefit when they leverage the benefits of their physical presence and the benefits of their online presence. Barnes & Noble and wallmart.com are examples of click-and-mortar business. Customers have access to the same inventory online as they do in the physical stores. In the event that a customer is dissatisfied with a purchase made online, the product can be returned to a local brick-and-mortar Barnes & Noble store. The major advantages of the click-and-mortar businesses are that they have an already established brand name, an established customer base, an established sales force, and the resources to operate on very thin margins associated with the retail industry. That also makes it less expensive for them to acquire new customers. Web sites frequently offer better prices and selection. However, click-and-mortar businesses face new competition in an extremely competitive environment from new firms who may have more expertise at building credible Web sites, and who can focus exclusively on building rapid response order systems.

VIRTUAL BUSINESS

It is an Internet-only business. By way of computers, the Internet and networking, virtual organizations exist today primarily in the high-tech industry. In this context, a virtual organization is an organization, which is flexible, connected, and focused. This organization focus on best-in-class core capabilities and outsource slices of their activities that others do better, thus allowing the virtual organization to remain focused, and yet, nimble enough to shift, as it needs to move to the next project. Virtual organizations do not have any tie to a physical location and have low barriers to entry into the web market. Amazon.com is an example of a virtual business. A virtual organization aims to offer products or services instantly in response to customer demands. A virtual corporation can arrange more resources than it actually possesses by means of both inner and outer partnerships. Ensemble- a division of Hallmark is an example of the virtual corporation that relies on outsourcing to handle their manufacturing and distribution. One of the main advantages Ensemble gains are that they only pay for the space and services they require. Through this, they control costs, have flexibility, and still produce dynamic sales and marketing programs. Jobs.com is a virtual company that brings together job seekers with employers. Jobs.com is a website that links individuals with job openings. An individual can search for a job and submit a resume on line. Employers can search job seekers to find a suitable candidate. Virtual businesses sell mostly digital products online. These businesses use electronic catalogs to promote their products and make use of electronic marketplaces. Customer billing is electronic and mass customization of products is possible. Small amount of capital is needed for operations and fixed costs of operations are lower. The overhead costs of running a virtual organization are also generally lower than costs of running traditional brick-and-mortar businesses. Virtual organizations offer the convenience of doing business from home and often reduce costs for the consumer. Virtual organizations face several challenges as well. A virtual organization functions round the clock every day of the year. One unique challenge is how to manage such an always-on business. Another challenge is customer relations and satisfaction. Internet-only banks can struggle to build a customer base, because many consumers demand Face-to-Face communication when managing their money. Name recognition is another significant challenge because Internet shoppers must rely on on-screen information of a product when they are making purchasing decisions. Texture, true color and quality are often difficult to determine. General characteristics of virtual organizations are that these organizations are infinite and without boundaries, complement each other's competitive advantages/sources sharing, provide knowledge sharing, geographic distribution, and collaboration in an electronic context, temporary source sharing for common goals and quick and effective response to customers' demands, another common feature of virtual organizations is geographic distribution (Poorkhanjani et al., 2013).

E-MARKETPLACES

E-marketplaces are commonly classified based on their ownership. Some examples of e-marketplaces include Covisint (Automobile industry), Worldwide Retail Exchange (WWRE) (Retail industry), GlobalNet (GNX) (Retail industry), and Transora (Consumer Packaged Goods (CPG) industry) (Bidgoli, 2004). There are three types of E-marketplaces: private, public, Independent, and consortia. The marketplace is physical, face-to-face and requires an infrastructure, such as a store. A Marketspace is an online space that facilitates bi-directional commerce. Here not only sellers can list their goods, but buyers can list their needs. The role of the Marketspace is to match buyers and sellers whose contexts have sufficient similarity - if the product being offered and being requested has a high degree of match and the buyer and seller can come to agreement on price, location and timing, then a successful match can be made that will culminate in a transaction. An example of this is Pikaba.com.

PUBLIC E-MARKETPLACES are B2B markets that are owned by a third party (not a seller or a buyer) and include markets with many sellers and many buyers. These markets are also known as exchanges (e.g., a stock exchange). They are open to the public and regulated by the government or the exchange's owners. An exchange may be owned by one company that may be either the seller or the buyer or in some cases, they may be owned by a consortium. Electronic exchanges are also called B2B hubs. For example, the Automotive Network Exchange (ANX) is expected to link the entire North American auto supply chain. Benefits of public

PRIVATE E-MARKETPLACES, on the other hand, are trading platforms set up by individual companies in order directly reach their key suppliers or customers. Examples include Cisco, Dell, and Walmart's private exchange. Private e-marketplaces are usually set up by companies having a dominant industry position or leading supply chain management capabilities.

INDEPENDENT E-MARKETPLACE is an e-marketplace established by third parties distinct from the e-marketplace's targeted buyers and sellers. Independent e-marketplaces such as Chemdex and Partminer, are owned by industry outsiders and typically commence market development without the buy-in commitment of key market participants. While the e-marketplace owners often hire a management team with industry experience and existing relationships with some of these market participants, these relationships tend to be fragile.

CONSORTIA E-MARKETPLACE is established by a group of key market participants, typically buyers or sellers but not both. These consortia offer major industry players the opportunity to obtain the benefits of embedding an e-marketplace within the industry network but retaining a substantial portion of the value generated through the e-marketplace amongst them rather than allowing it to flow to a third party. The founders of these marketplaces typically represent a substantial portion of the industry's trading volume, and thus possess the potential to marginalize competing e-marketplaces. For example, the twenty-one founding members of Quadrem, the e-marketplace for mining, minerals, and metals companies, represent two-thirds of the industry's total market capitalization and more than 25% of its buying power.

Intermediation, Disintermediation, Reintermediation, Hypermediation, and Syndication in E-commerce

INTERMEDIATION

Direct interaction between producers and consumers in an e-marketplace is possible. However, direct interactions are sometimes undesirable or unfeasible because of certain limitations. In such cases, intermediation is needed. Intermediaries, whether human or electronic, can address the following important limitations of direct interaction.

First, some intermediaries maintain databases of customer preferences, and they can predict demand and reduce search costs by selectively routing information from providers to consumers and by matching customers with products and/or services. Second, intermediaries can relay messages and make pricing and allocation decisions without revealing the identity of buyer and seller. Third, intermediaries can gather more complete product information by using many sources (including independent evaluators and other customers) other than the product provider. Many third-party Web sites provide such information (e.g., bizrate.com, mysimon.com, and consumerguide.com). Fourth, intermediaries can reduce contract risk (i.e. the risk that a party in the contract does not fulfills its obligations) because intermediary can act as a police officer at its own by either becoming the guarantor or publicize information about the bad behavior of providers and consumers. Fifth, intermediaries can use pricing mechanisms that induce just the appropriate trades, thereby reducing price inefficiency.

DISINTERMEDIATION

When intermediaries are removed in a supply chain the phenomenon is called disintermediation. When disintermediation occurs, companies may now deal with every customer directly, for example via the Internet. In the same way, customers can deal with companies directly. Disintermediation may result in significant cost savings for both the customers and the companies. One example of E-commerce-supported disintermediation is healthcare industry. Before the World Wide Web, local health-care providers often served as a primary resource of health-related information. More than often, a doctor's opinion was accepted with little or no question. The explosion in online information had a significant impact on information-hungry consumers, encouraging them to seek answers about healthcare from various Web sites instead of from local professionals. Other examples of disintermediation include online bookstores and Dell computers.

REINTERMEDIATION

Reintermediation refers to the reintroduction of an intermediary in a supply chain. Reintermediation can occur due to many reasons e.g. high cost of shipping small orders, customer service issues, requirement of large pool of resources, and dealing with disgruntled conventional retailers and supply chain partners. One example of Reintermediation is photograph brokers such as CORBIS and Getty images.

HYPERMEDIATION

Hypermediation refers to adding more intermediaries in the supply chain. It can also refer to extensive use of both human and electronic intermediation in all phases of E-commerce.

SYNDICATION

Syndication involves the sale of the same good to many customers, who then integrate it with other offerings and resell it or give it away free (Werbach, 2000). Software and other digitizable items, logistics, security, and system integration tools are among the popular choices for syndication. For example, companies syndicate E-Commerce services such as payments and shopping-cart ordering systems used by e-tailers.

CYBERMEDIATION

It refers to the creation of new kinds of intermediaries that simply could not have existed prior to the advent of E-business and the Internet, in categories including Searching, Price Discovery, Logistics, Settlement and Trust. Obvious examples include comparison-shopping sites such as Kelkoo and bank account aggregation services like Citibank (Giaglis, Klein, O'Keefe, & O'keefe, 1999).

Development and Growth of E-commerce

The term Electronic Commerce (E-Commerce) was first introduced in the early 1990s. Over the last three decades, E-commerce has transformed. E-Commerce applications were first developed in the early 1970s e.g. Electronic Funds Transfer (EFT) in which funds could be routed electronically from one organization to another. However, the extent of the applications was limited to large corporations, financial institutions, and a few other businesses. Introduction of Electronic Data Interchange (EDI) extended electronic transfers from financial transactions to other types of transaction processing (such as ordering) and expanded the pool of participating companies from financial institutions to manufacturers, retailers, services, and many other types of businesses. Rapid development of new networks, network protocols, and E-commerce software was one major reason this rapid expansion of E-commerce.

Starting in mid-90's, many innovative applications developed ranging from online direct sales to e-learning experiences. Most medium- and large-sized organization now have a Web site, and most large U.S. corporations have comprehensive portals through which employees, business partners, and the public can access corporate information. Many of these sites contain tens of thousands of pages and links.

Over the years, we have witnessed the emergence of a large number of dedicated E-commerce firms. There were both failures and success stories. eToys, Drkoop.com, Webvan.com, and Xpeditor are examples of some well-known B2C failures. Similarly, Verticalnet.com and Chemdex.com are examples of some well-known B2B failures. Cisco, IBM, Intel, and General Electric are examples of some big E-commerce success stories. Alloy.com, Drugstore.com, and FTD.com are examples of success stories of start-up companies.

Generations of E-commerce

E-commerce development can broadly be divided into two generations. The first generation of E-commerce (from 1995 to 2000) was dominated by companies operating is USA. Web pages were primarily in English, particularly on commerce sites. The second generation of E-commerce (from 2000 onwards) is international in its scope, with global enterprises doing business in many countries and in many languages. Second generation E-commerce websites are available in multiple languages. Problems of language translation and handling currency conversion were the main issues that hindered the efficient conduct of business.

First generation of e-commerce businesses enjoyed easy access to start-up capital. This resulted in an overemphasis on creating new large enterprises to exploit e-commerce opportunities. Excited about e-commerce and willing to participate, investors were not concerned how much it cost or how bad the underlying ideas were. Second generation e-commerce businesses are established companies that used their own internal funds to finance gradual expansion of e-commerce opportunities. Their measured and carefully considered investments helping electronic commerce grow more steadily, though more slowly. The second generation of E-commerce is anticipated to include a larger proportion of smaller businesses. Providing services that help smaller companies will also be a substantial area of online business.

First generation of E-commerce used slow and inexpensive Internet technologies. These technologies were integrated into B2B transactions and internal business processes. The technologies utilized included bar codes and scanners to track parts, assemblies, inventories, and production status. Most first generation e-commerce consumers connected to the Internet using dial-up modems. In the second generation of E-commerce, there was a significant increase in broadband connections in homes. Although expensive, broadband connections are more than 10 times faster than dial-up. This increased speed not only makes Internet use more efficient, it can alter the way people use the web. The technologies used in second-generation e-commerce are increasingly integrated with each other and with communication systems that allow companies to communicate with each other and share transaction, inventory level, and customer demand information effectively.

In the first generation e-commerce, e-mail was used as a tool for relatively unstructured communication. In second-generation e-commerce, it is being used as an integral part of marketing and customer contact strategies of the companies.

First generation of E-commerce is characterized by over-reliance on simple online advertising as the main revenue source. Second generation e-commerce is characterized by use of multiple sophisticated advertising approaches and making the Internet as an effective advertising medium. As a result we see many new forms of Internet advertising emerging and growing rapidly e.g. Ezine Advertising (advertising in online newsletters).

First generation e-commerce businesses did not have ways to distribute digital products on the web. As a result, they faced difficulties in selling digital products and digital piracy flourished. Second generation of E-commerce is characterized by various new ways for legal distribution of digital products (music, video, and other digital products) on the Web. Apple Computer's iTunes site was one of the first second-generation e-commerce business to attempt digital product distribution.

First generation e-commerce businesses tried to take first-mover advantage because many companies and investors believed that being the first web site to offer a particular type of product or service would give them a competitive advantage in every type market or industry. Studies on e-commerce firms revealed that such moves led to success only for some companies in certain specific markets and industries. Successful first movers tended to be large companies that had an established reputation (or brand) and that had marketing, distribution, and production expertise. First mover companies that were either smaller or lacked the expertise in these areas appeared to be unsuccessful. In addition, first-mover companies that entered highly volatile markets or in industries with high rates of technological change often did not do well. Fewer second-generation e-commerce companies rely on a first-mover advantage when they take their businesses online.

E-Commerce: Critical Success Factors (CSF)

Profit making is one major goal of every business. Critical success factor is defined as an essential factor that helps the business to achieve its desired goals. Consequently, critical success factor (CSF) is one of the issues, which needs to be identified for the success in market. Some factors are internal and in control of the companies while other factors are external and beyond the control of companies [See Jones & Field, 2000; Atchariyachanvanich & Okada, 2001; Cullen & Taylor, 2009]. Following are some of the critical success factors for E-commerce businesses.

Internal Factors

QUALITY ASSURANCE

Since buyers cannot see and feel a product before they purchase it, quality becomes a key issue in E-commerce. The issue of quality is linked to the issues of trust and consumer protection. Quality assurance can be provided through a number of ways e.g. a trusted third-party intermediary (e.g. TRUSTe and the BBBOnLine), by providing free samples, by offering returns if the buyer is not satisfied, and by offering insurance, escrow, and other services. However, not all the options are feasible in all situations and company must analyze options carefully before deciding [For details see Whinston, Choi, & Stahl, 1997; Choi & Whinston, 2000].

BUYING AIDS

Increasing number of products and services available online is to be accompanied by improved trading mechanisms, search engines, online shopping aids, intermediary services, and presentations in multiple languages so that more and more sellers and buyers become willing to give E-Commerce a try.

PURCHASING INCENTIVES

With advancement in e-commerce, it is expected that many innovative options will be available, and electronic shopping may even become a social trend. Companies need to come up with incentives for consumers to convince them buy their products.

INCREASED SECURITY AND TRUST

Customer perception of inadequate security and privacy, coupled with a lack of customer trust, are the major inhibitors of growth of B2C and B2B E-commerce. Customers' resistance to the change from a real to a virtual store, and lack of customer trust in paperless/faceless transactions is another issue.

EFFICIENT CUSTOMER INFORMATION HANDLING

With increasing amount of customer information collected on E-commerce websites, companies are looking for ways to handle this information efficiently. Companies are using data warehouses, data mining, and intelligent agents in order constantly learn about their customers, steering marketing and service activities accordingly.

INNOVATIVE ORGANIZATIONS

The number of business models in E-commerce is limited only by the human imagination. New business models are being invented daily and organizations are using IT heavily to restructure and reengineer themselves (Turban, McLean, & Wetherbe, 2004; El Sawy, 2000; Hammer, 1995). Innovative organizations with a strong appetite for process reengineering and organizational creativity will probably be more inclined to use E-Commerce.

OUTSOURCING

Outsourcing of primary activities (e.g., distribution, manufacturing, and human resources) can bring numerous benefits e.g. reduced costs and become more responsive to the ever-changing marketplace. Most significantly, outsourcing enable an e-business create the virtual enterprise.

External Factors

INTERNET USAGE

A major factor fueling the e-commerce growth worldwide is the increasing number of people connected to the Internet. The 22% of the world's population used the Internet in 2007. In 2014, 40% of the world population uses the Internet. Two-third of this Internet population belonged to the developing world. The uptake has been significantly strong in the developed world where 62% of the population used the Internet in 2007 while in developing world 17% of the population used the Internet. The number of Internet users is rising sharply. With the integration of computers and television, Internet access via mobile devices, increased availability of access kiosks, increased publicity about the Internet, and availability of inexpensive computers (such as the simputer, short for simple computer), the number of Internet surfers will continue to increase.

BROADBAND AND MOBILE DEVICES

A more recent broadband penetration and the growing use of mobile devices have further boosted e-commerce. There were 430.7 million fixed broadband subscribers worldwide in 2009, a 13% increase over 2008 (Buckley, 2010). By end 2014, fixed-broadband penetration reached almost 10 per cent of world population. Consumers with broadband Internet access have been more active in e-commerce compared with those who do not have broadband Internet access. In OECD (Organization for Economic Co-operation and Development) countries, the number of cellular mobile subscribers grew at an average compound rate of 30% per year between 1993 and 2007. In the United States, in 2007, there were 263 million US mobile subscribers representing an 87.1% penetration rate of the US population. In Japan, 3G CDMA subscribers exceeded 100 million in April 2009. The combined spending on consumer and business mobile applications was expected to surpass US$ 13 billion worldwide by 2012 (OECD, 2009).

E-RETAILERS

The presence of e-retailers is another important factor influencing E-commerce. Success of many large Internet-only retailers (such as Amazon.com) and presence of most major brick-and-mortar retailers online is fueling E-commerce in USA (OECD, OECD Communications Outlook 2009, 2009). Countries with a high proportion of retailers selling online had more consumers purchasing products and services online (European, 2009).

CRITICAL MASS OF BUYERS AND SELLERS

This critical mass is one of the major success factors for a start-up B2B vendor (Ramsdell, 2000). Critical mass of buyers and sellers is necessary for profitability, for markets to be efficient, and to develop strong and fair competition.

COMPETITION

In E-commerce, reduced search costs and increased bargaining power of consumers may reduce long-term profit margins for sellers because of the increased number of transactions. The industries will have to become highly competitive with low barriers to entry so that sellers remain willing to participate in E-commerce.

CONSUMER CHARACTERISTICS

In e-markets, consumers are required to make a certain degree of effort (e.g. searching, analyzing, and making a buy decision). Industries where a sizable percentage of the purchases are made by impulse buyers may not be suitable for E-commerce.

MOBILE COMMERCE

With 4 billion predicted mobile cellular subscribers by late 2008 (ITU, 2008) the ease with which one can connect from them to the Internet, and the introduction of 3G capabilities, it is clear that m-commerce will play a major role in E-commerce. In 2009, 56% of adult Americans accessed the internet by wireless means, such as using a laptop, mobile device, game console, or MP3 player. The fact that one does not require a computer to go online will bring more and more people to the Web.

VIRTUAL COMMUNITIES

Virtual communities can enhance commercial activities online and communities organized around professional areas of interest can facilitate B2B and B2C E-commerce.

PAYMENT SYSTEMS

Use of e-cash and micro payments online are increasing and B2B payment systems have matured. As international standards become the norm and these systems are implemented on a large scale, it is expected electronic payments will extend globally, facilitating global e-commerce.

TECHNOLOGY INTEGRATION

B2B e-commerce is expected to continue dominating the E-Commerce field for the intermediate future. More sellers, more buyers, and more services are expected to appear and the rapid growth is likely. The development of information and communication technology (ICT) is a critical factor in the growth of E-commerce. However, the success of B2B to a large scale will be dependent on the success of integrating technology with business processes and with conventional information systems.

E-COMMERCE LEGISLATION

The legislative process is slow, especially when multiple countries are involved. The development of more non-regulatory legislation will be a key factor in growth of global e-commerce.

E-Commerce: Technology Trends

Current trends in E-commerce technologies generally focus on developing technologies that recues costs considerably and provide improvements in capabilities, ease of use, increased availability of software, ease of site development, and improved security and accessibility. Specific technology trends include the following.

CLIENTS: PCs of all types are becoming inexpensive, smaller, and more capable. Use of network computer (a computer whose processing and storage is performed by servers on the network) could bring the price of a PC to that of a television. Tablet PCs and Pocket PCs are two examples of mobile PCs. These PCs are smaller with many capabilities of a modern desktop PC.

EMBEDDED CLIENTS: An embedded client, a typical device in pervasive computing, is a client (e.g. car or a microwave) with an embedded microchip. In many cases, an expert system is embedded with rules that client can use to become more responsive to changes in the environment.

PERVASIVE COMPUTING: According to IBM, pervasive computing refers to convenient access, through a new class of appliances, to relevant information with the ability to easily take action on it when and where you need it. Pervasive computing is affecting e-commerce positively and will be facilitated by improvements in wireless communication and wearable devices.

WIRELESS COMMUNICATION AND M-COMMERCE: For countries without fiber-optic cables, wireless communication can save considerable installation time and money. Price of wireless devices and wireless Internet access is dropping. Wireless communications are changing the nature of E-commerce from content to context, reaching customers whenever and wherever they are ready to buy.

WEARABLE DEVICES: With advances in technology, the number of wearable computing devices is increasing. Wearable devices provide a new way of human-computer interaction. Wrist worn computer (e.g. Zypad WL 1000) is a powerful computer that can be worn on the user's wrist. Google Glass is a wearable, voice-controlled Android device that resembles a pair of eyeglasses and displays information directly in the user's field of vision. A wearable device is always on and always ready and accessible. Wearable devices will enhance collaborative commerce, B2E, and Intrabusiness E-commerce.

SERVERS AND OPERATING SYSTEMS: Linux and Windows are the prevalent platforms for the majority of companies and these two operating systems will take more market share from UNIX in the future (Flood, 2008). Server Clustering is gaining in popularity. By clustering servers, a business can gain increased processing power economically. Server cloud is another important technology. A cloud server is a logical server that is built, hosted and delivered through a cloud-computing platform over the Internet.

NETWORKS: The use of e-commerce frequently involves rich multimedia (such as color catalogs or samples of movies or music) which requires large bandwidth for delivery. Several broadband technologies have been developed to provide increased bandwidth.

E-COMMERCE SOFTWARE AND SERVICES: Growing availability of all types of e-commerce software is making it possible to set up online stores and conduct all types of trades. Hundreds of sites are offering inexpensive web pages for a variety of business activities. Other support services, such as escrow companies are developing rapidly.

SEARCH ENGINES: Search engine technology is getting smarter and better. Use of the search engine is becoming a norm for B2C and B2B consumers to find and compare products and services in an easier and faster way.

PEER-TO-PEER TECHNOLOGY: Peer-to-peer technology is developing rapidly. Peer-to-peer technology is making knowledge sharing, communication, and collaboration activities better, faster, less expensive, and more convenient.

INTEGRATION: The computer-telephony and computer-TV integration is increasing Internet accessibility. Microsoft Windows XP Media Center Edition software with bundled DVR and numerous media center PCs (e.g. HP Pavilion Media Center PC) is available.

WEB SERVICES: Web Services (see Chapter on Elements of E-commerce) are being developed rapidly. Web Services resolve major issues in e-commerce systems development and integration and are especially helpful in complex B2B systems and exchanges. Web Services are enabling companies to build more efficient and cheap e-commerce applications more quickly.

SOFTWARE AGENTS: Software agents are being developed that consumers can use to search, match, negotiate, and conduct many other tasks that will facilitate e-commerce activities.

INTERACTIVE TV: Interactive TV has shown few signs of success in the past and it is expected to impact e-commerce significantly in the future. In 2008, Microsoft acquired Navic Networks, an interactive television ad developer. Navic uses audience measuring data to determine which broadcast commercials might be of interest to households. Many regard it as a beginning of a trend that may eventually reach much further. (Raphael, 2008)

INTERNET2: Many research institutions around the world are working on tomorrow's Internet. Although projects such as Internet2 are slow to progress, eventually these efforts will greatly advance e-commerce applications.

RE-PLATFORMING: Many e-commerce organizations choose their first-generation e-commerce platform to get online quickly. However, as time goes by, they realize how critical it is to get a broader scope of capabilities than what their original platform offers. While re-platforming is a huge undertaking that involves a lot of downstream systems integration, the new capabilities it affords (like support for mobile and omnichannel commerce) will position the company for future growth and competitive advantage.

RESPONSIVE WEB DESIGN: Today's consumers demand a consistent online experience, regardless of what device they use. With the mobile channel arguably becoming the channel of choice, e-commerce sites that are not optimized for mobile via a responsive web design risk losing customers from the outset. For some sites, an adaptive design or native mobile site might make more sense, but regardless, a "mobile-centric" site will be mandatory.

OMNICHANNEL INTEGRATION: The connected or mobile consumer is driving a change in the retail landscape that is based on flexibility. Consumers want to access a variety of online resources using a range of devices to research and make purchases at their convenience whether in store or online. Omnichannel integration enables this and more: It also provides opportunities for retailers to engage with consumers at any point in their shopping journey, which in turn creates greater revenue and brand loyalty.

REAL KPIS FOR REAL PERFORMANCE: Many organizations are still using old-world metrics to measure the effectiveness of their e-commerce site. However, this approach does not make much sense in an omnichannel retail environment. For example, if a consumer researches a product online but goes to the store to purchase it, it might be attributed as physical store revenue, despite that it was triggered by an online experience. Figuring out the right metrics to attribute value to different customer experiences is challenging but necessary to guide investment decisions. Retailers will need to keep this in mind and put more thought into the right measurements for omnichannel commerce.

RUM TOOLS: Synthetic measurement uses a simulation of users to gain theoretical insight into web and mobile performance. Synthetic measurement is still used by most e-commerce organizations. However, there is a shift from using synthetic measurement to real-user measurement (RUM). RUM tools provide more accurate performance measurements, as they leverage data from actual customers for analysis. RUM is a recent capability available to e-commerce organizations. RUM tools are expected to take off in the near future as companies demand accurate, real-world data.

SECURITY: The list of companies victimized by credit card breaches certainly grew longer in 2014, with Target, Home Depot and Staples among those companies that suffered extensive brand damage and expense. In years to come, e-commerce companies will tighten their security by deploying technologies that defend attacks in the cloud, not just at their network perimeter.

MOBILE PERFORMANCE AND MOBILE EXPERIENCE: While getting a mobile site up and running was a good first step to capture that growing mobile traffic, the challenge now is how to get this mobile traffic to convert. Adopters of responsive web design sites have the advantage addressing this, as they can dive right into solving performance issues with a software-defined application delivery solution that ensures ultra-fast, visually immersive user experiences.

These trends provide a broad overview of what e-commerce organizations should prepare for in the years to come. Mobile has for years been an important channel to watch, and e-commerce organizations that fail to make mobile a priority will suffer lost market share and diminished customer loyalty. Beyond 2015, we expect to watch many trends that are more interesting. One interesting trend is the role of virtual reality in e-commerce. Although virtual reality has been hyped for decades, its only practical application to date has been in gaming. However, as technologies continue to advance, virtual reality in e-commerce may be closer to fruition than we think. No one knows for sure what the future will bring. To elevate its brand for continued success, an E-commerce organization will need to keep ahead of trends or set the trends by investing in new technologies will surely elevate an e-commerce organization's brand and put it squarely on the path to continued success (Mital, 2015).

E-Commerce: Suitability of Products and Services

Not all businesses are well suited for E-commerce. Businesses must consider the ability of their organizations to function online. Restaurants, for example, cannot exist solely on the Web, but they can offer a web site for providing directions and making reservations. Menus, entertainment, and special events can be displayed on the site. The same might apply to many other businesses in service industries. Auto and home repair, beauty and medical assistance are among the industries that cannot provide their services online.

Just as not all markets and industries are suitable for E-commerce, not all products are good candidates for E-commerce. With all else being equal in online environment, goods with following characteristics are more suitable for E-commerce.

GOODS WITH HIGH BRAND RECOGNITION (E.G. SONY, DELL, AND NIKE): Due to strong brand identity, these goods are easier to sell online than an unbranded item. Brand's reputation reduces the buyer's concerns about quality when buying that item which customer has not seen.

GOODS WITH GUARANTEE FROM HIGHLY RELIABLE VENDORS: For example Dell computers, L.L Beans products, Microsoft Software are products from highly reliable vendors. Due to high level of trust customers have on these vendors, the products from these vendors can be sold online.

RELATIVELY INEXPENSIVE GOODS: For example, office supplies.

FREQUENTLY PURCHASED AND COMMODITY ITEMS (E.G. GROCERIES, BOOKS, CDS, AND PRESCRIPTION DRUGS, GASOLINE): A commodity item is a product or service that is hard to distinguish from the same products or services provided by other sellers. The features of these goods are standardized and well known. Customers do not need to experience the physical characteristics of the particular item before they buy it. For example, each copy of the book is identical to others.

PRODUCTS WITH SUITABLE SHIPPING PROFILE: Product shipping profile can make a product well suited for E-commerce. A product's shipping profile is the collection of attributes that affect how easily that product can be packaged and delivered. With a high value-to-weight ratio, overall shipping cost can become a small fraction of selling price of the product. An airline ticket is an excellent example of an item that has a high value-to-weight ratio. Products with consistent size, shape, and weight can make warehousing and shipping much simpler and less costly. However, not all high value-to-weight ratio products are suited for e-commerce. For example, expensive jewelry has a high value-to-weight ratio, but many people are reluctant to buy it without examining it in person unless the jewelry is sold under a well-known brand name and with a generous return policy.

Some other items suitable for selling online include Computer hardware and software, consumer electronics, sporting goods, office supplies, music, toys, health and beauty products, entertainment products, apparel products, cars, services (e.g. travel, stock trading, electronic banking, real estate, and insurance), and others (e.g. flowers, food, pet supplies etc.). Other items that are well suited to electronic commerce are those that appeal to small, but geographically dispersed, groups of customers. Collectible comic books are an example of this type of product.

Some products may never be good candidates for e-commerce. Following products are not suitable for E-commerce.

PERISHABLE FOODS AND HIGH-COST, UNIQUE ITEMS (E.G. CUSTOM-DESIGNED JEWELRY): This is because it might be impossible to inspect adequately from a remote location, regardless of any technologies that might be devised in the future.

PRODUCTS THAT REQUIRE PERSONAL SELLING: When personal selling skills are a factor in making the sale happen e.g. sale of commercial real estate.

PRODUCTS THAT REQUIRE PHYSICAL INSPECTION: When the condition of the products is difficult to determine without making a personal inspection (e.g. high-fashion clothing, antiques, or perishable food products) pure E-commerce might not be a suitable option.

A combination of electronic and traditional commerce strategies works best when the business process includes both commodity and personal inspection elements. One possible combination is using traditional commerce for product/service inspection and using E-commerce to sell the items or services. For example, many people are finding information on the Web about new and used automobiles.

Suitability of products/services for E-commerce is dependent on the current state of available technologies, and this will change as new tools emerge for implementing electronic commerce. For example, low-denomination transactions are not well suited to e-commerce because no standard method for transferring small amounts of money on the web has become generally accepted. If a company or group of companies could create a standard that gains general acceptance among buyers and sellers, low-denomination transactions could become e-commerce transactions.

Advantages and Disadvantages of E-commerce

E-commerce provides significant benefits to the economy and to consumers. Consumers benefit significantly through enhanced capacity to research and compare products, expanded choice in products that may be purchased at any time and from anywhere, and more possibilities to customize products to better meet personal preferences. Increased transparency in e-commerce intensifies competition and oftentimes results in lower prices for consumers.

Advantages of E-commerce

EQUALIZER AND ENABLER

E-commerce serves as an equalizer by enabling start-up and small- and medium-sized enterprises to reach the global market. E-commerce serves as an enabler because it makes mass customization possible by including an easy-to-use ordering systems that allow customers to choose and order products according to their personal and unique specifications e.g. Dell.com that allows customers to order a PC with customer-chosen specifications.

NETWORK PRODUCTION

Network production refers to distributing production process to contractors who are geographically dispersed but who are connected to each other via computer networks. Network production can provide reduced costs, increased strategic marketing, and can facilitate sale of add-on products. In network production, a company can assign tasks within its non-core competencies to factories all over the world that specialize in such tasks (e.g., certain assembly operations).

E-COMMERCE AND NETWORK EFFECTS

Network effect is defined as a change in the benefit, or surplus, that someone derives from a good when the number of other people consuming the same kind of good changes. Network effects can occur when widespread use of the product makes it more valuable for the consumers. For example, a telephone becomes more valuable if every other consumer has a phone that can be accessed by that telephone. Another example is e-mail account. An e-mail account provides access to a network of people with e-mail accounts. In smaller network, this e-mail is generally less valuable. However, if you have an Internet e-mail account, it is far more valuable than single-organization e-mail account due to the network effect. In the beginning, network effect is often the result of word-of-mouth testimonials but in later stages analysis of a network's size and projected growth plays an important role. For example, you may adopt a service initially because someone you know uses it; later, you may adopt a service because everyone uses it.

BROADENED CONSUMER CHOICES AND ENHANCED BUSINESS RELATIONSHIPS

E-commerce allows consumers to shop or perform other transactions from anywhere in the world round the clock 365 days a year. Consumers have more choices. They have a wide choice of vendors and products and can conduct quick comparisons. They can also have more influence over what and how products are manufactured and how services are delivered. E-commerce allows for a faster and more open process with greater customer control. E-commerce makes information on products and the market readily accessible and available and increases price transparency so that customers to make more appropriate purchasing decisions. In the cases of digitized products, e-commerce allows for quick delivery. E-commerce transforms old business relationships from vertical (or linear) relationships to integrated or extended relationships. Companies can interact more closely with customers even if through intermediaries.

NEW BUSINESS MODELS AND EFFICIENT PROCESSES

E-Commerce allows for many innovative business models and reduces the time between the inception of an idea and it commercialization. E-commerce enables efficient processes by reducing administrative overhead, process simplification, increased process flexibility, decreased cycle time, and by providing new business partners.

ECONOMIC ADVANTAGES

E-commerce has economic advantages over traditional commerce. Consider the following examples. The variable cost per unit of digital products is very low (in most cases) and almost fixed, regardless of the quantity. Therefore, total cost per unit will decline as quantity increases, as the fixed costs are spread (pro-rated) over more units. This, in turn, will result in increased returns and increased sales. In terms of production, introduction of E-commerce lowers the amount of labor and/or capital needed to produce the same level of production. In terms of transaction costs (the costs associated with conducting a sale), E-commerce makes it possible to have low transaction costs with a smaller firm size or to enjoy much lower transaction costs when firm size increases. In terms of agency costs (expenses or other costs associated with an agency relationship), with e-commerce companies can significantly expand their business without too much increase in administrative costs (Evans & Wurster, 2000).

REACH AND RICHNESS

Another economic impact of E-Commerce is the trade-off between the numbers of customers a company can reach (called "reach") and the amount of interactions and information services they can provide to customers (called "richness") (Chung, Lee, & Turban, 2006). When E-commerce is introduced, a business can increase its reach and richness at the same time. One good example is Charles Schwab brokerage house. Initially, Schwab attempted to increase its reach. To do so, the company reduced its richness. However, with additional value-added services on its Website (such as Mutual Fund Screener), Schwab drastically increased its reach and at the same time provide richness in terms of customer service and financial information to customers. (Slywotzky & Morrison, 2001)

BENEFITS FOR SOCIETY

E-commerce also provides advantages to the society. E-commerce provides more individuals work at home. This reduces traveling for work, which in turn results in less traffic on the roads and reduced air population. Using E-commerce, some merchandise can be sold at lower prices. That can allow less wealthy people to buy more and increase their standard of living. People in developing countries and rural areas can be in a position to enjoy products and services that were unavailable in the past. Public services, such as health care, education, and distribution of government social services, can be done at a reduced cost and/or improved quality.

Disadvantages of E-commerce

Most of the disadvantages of E-commerce today are because technology is new and rapidly changing. It is expected that with increased acceptance of E-commerce by businesses and individuals many of these disadvantages will be resolved or become less problematic.

A DIFFERENT BUSINESS LANGUAGE

Online business is just a different language to most people with traditional business experience. You do not see the customer, money is transacted through unsecured processes at times, and refunds can be complicated.

LEGAL AND TAXATION ENVIRONMENT

The legal and taxation environment of E-commerce (both national and international) has many unclear and conflicting laws.

RETURN-ON-INVESTMENT (ROI) CALCULATIONS

Calculating return-on-investment for E-commerce ventures has been historically difficult because the costs and benefits have been hard to quantify and there are no mature measurement methodologies. Costs, which are a function of technology, can change dramatically even for short-term E-commerce projects because the underlying technologies are changing so rapidly.

HUMAN RESOURCES

Many firms have had trouble recruiting and retaining employees with the technological, design, and business process skills needed to generate and operate an effective E-commerce business.

SOFTWARE INTEGRATION

One other issue with E-commerce is the integration of existing databases and transaction-processing software designed for traditional commerce into the E-commerce software. A number of companies offer software integration services and promise seamless integration of existing information systems and E-commerce software. However, these services can be expensive and businesses need to do their homework before selecting a service provider.

UNIVERSALLY ACCEPTED STANDARDS

There is lack of universally accepted standard for quality, security, and reliability of E-commerce operations.

OTHER ISSUES

Insufficient bandwidth (especially for M-commerce), still evolving software development tools, Culture, B2B interfaces, difficult-to-get venture capital, and online fraud are some other disadvantages.

Impacts of Digital Economy and Organizational Responses

Digital economy provides a vast array of digital products, digital transactions, digital currencies, and physical goods with processing and networking capabilities. The digital economy is creating a digital economic revolution and web-based E-commerce systems are accelerating the digital revolution by providing competitive advantage to organizations. Enhancing competitiveness or creating strategic advantage is the number-one benefit of web-based systems (Hemmatfar & Salehi, 2010).

The New Business Environment

Digital economy is influencing the business environment in many significant ways and creating various pressures on organizations. (Hsu, Kraemer, & Dunkle, 2006), studying US firms, reports that the pressure of business partners and the government was major factor behind businesses' drive to deepen and diversify their use of e-business. To remain competitive, organizations need to respond to these pressures. E-commerce has a potential role in supporting these responses. Various environmental factors, such as economic, legal, societal, and technological factors have created a highly competitive business environment in which customers are becoming more powerful. Technology advancements are creating specific knowledge that feeds on itself resulting in more and more technology. Rapid growth in technology is resulting in a large variety of more and more complex systems. These environmental factors can change quickly sometimes in an un-predictable manner. As a result, we see the emergence of a more turbulent business environment with more problems and opportunities, and stronger competition. The organizations need to make decisions more frequently and quickly because more factors (market, competitive, political, and global) need to be considered and more information and/or knowledge needed for making.

Today's corporate giants are finding themselves under attack from modern-day small start-up companies, which overwhelm their established rivals with new technologies. The old powers of market incumbents - massive scale, control over distribution, brand power, millions of customer relationships - are no longer seen as the obstacles they once were to agile rivals with innovative business models. The top executives of many a corporate giant must feel like the fictional character Gulliver, waking up to find themselves under attack from modern-day Lilliputians, small start-up companies which overwhelm their established rivals with new technologies. The old powers of market incumbents - massive scale, control over distribution, brand power, millions of customer relationships - are no longer seen as the obstacles they once were to agile rivals with innovative business models. A new survey, conducted by a research center at top-ranked Swiss business school IMD, the International Institute for Management Development, with backing from Internet equipment maker Cisco (IMD & Cisco, 2015), found that business leaders believe four out of 10 top-ranked companies in their industries worldwide won't survive the next five years (Auchard, 2015). According to these business leaders, accelerating change in technology, shifting business models and a need to merge to cut costs were essential for their company survival. Not just lone companies, but entire industries are being sideswiped by these effects.

Digital disruption now has the potential to overturn incumbents and reshape markets faster than perhaps any force in history. The industry that will experience the highest level of digital disruption between now and 2020 is technology products and services. Industries with the highest number of top-rated companies at risk were hospitality/travel, media and entertainment, retail, financial services and consumer goods/manufacturing, in that order, the survey showed. Industries with the least number of top-rated companies at risk were hospitality/travel, media and entertainment, retail, financial services and consumer goods/manufacturing. Pharmaceuticals, utilities and the oil and gas sectors are likely to experience the least amount of digital disruption. However, all industries will see competitive upheavals as innovations become increasingly exponential. New threats are coming from start-ups analyzing big data to offer a personalized approach to medicine, for example. Aggregators in travel e-commerce have taken millions of customers from direct bookings with hotels and airlines already struggling with a decade of decline in business travel amid the economic and structural challenges. Disruptive players are coming from out of nowhere. For example, home rental service Airbnb is used by individuals to offer their homes and vehicles for rent. There also exist office-sharing firm LiquidSpace and similar sharing economy start-ups. The airline industry missed the rise of mobile travel apps, the top dozen of which now collectively have a value around 88 billion euros ($99 billion). Airlines are seeking ways to woo back customers who expect them to deliver more than boarding passes to their mobile phones. Airlines are planning free on-board Wi-Fi available on every plane and new apps to help frequent travelers make new travel bookings in mid-air. In the banking sector, we see the emergence of a large number of FinTech startups from mortgage lending to wealth management to small business loans. FinTech (financial technology) is a line of business based on using software to provide financial services. FinTech companies are generally startups founded with the purpose of disrupting incumbent financial systems and corporations that rely less on software. These startups are using cloud computing and smart phones to deliver services to the customers. With technology support, these startups do not need the conventional branch network to tie customers to their companies. Lending, to a large part, is still an archaic process for banks that are based on paper forms designed to give customers a poor experience. Research shows that four out of five banking customers will happily leave their banks for a better customer experience (IMD & Cisco, 2015).

For different industries, the big digital threats could be different. For automakers and transport companies, Tesla, the luxury electric car company, or Uber, the online taxi service, are the big digital threats. Uber uses your phone's GPS to detect your location and connects you with the nearest available driver. Tesla recently announced "energy storage" business, which aims to produce batteries capable of solving the elusive problem of storing electrical energy produced at optimum times for use at other times. In industry circles, it is believed that Tesla (that initially started as a car company) is now going to become an alternative energy company. For hotels and airlines, it is Airbnb or Trivago. Trivago is a travel metasearch engine focusing on hotels. The site compares prices for over 730,700 hotels from more than 200 booking sites, such as Expedia, Booking.com, Hotels.com, Jovago and Priceline.com. Airbnb is a website for people to rent out lodging. It has over 1,000,000 listings in 34,000 cities and 190 countries. The major share of Airbnb is now owned by Expedia.

There is also mounting wave of digitally inspired, cross-border mergers, stepped-up corporate venture funding. According to Mini, a British carmaker owned by BMW, future customers would be able to offer their private vehicles for car sharing. This prediction came from a trend amongst younger drivers not to have their own cars. New digital startups offer massive improvements in how customers use the products or services of established businesses and find innovative ways to slash costs and enter markets without investing heavily to own physical assets or distribution infrastructure. Uber, the online taxi-hailing service, started a new service by signing up drivers to deliver everything from groceries to heavy equipment. This new service is a great challenge for logistics giants like FedEx and UPS. After the advent of the Web in the mid-1990s, approximately 25% of the Fortune 100 top U.S. companies were still in existence 15 years later. Can we draw a parallel here? Only time will tell.

Business Pressures

The Digital environment, created by digital economy, is producing various pressures on businesses. These pressures can be divided into following categories: market and economic pressures, societal pressures, and technological pressures.

MARKET AND ECONOMIC PRESSURES

Competition among companies is getting stronger and government de-regulation along with shrinking government subsidies is making the competition fiercer. Managers need to think and devise their strategies in a global economic perspective. The regional trade agreements (e.g., NAFTA and SAFTA) possess another challenge because these agreements make competition fiercer for out-of-region companies. Labor costs in some countries are extremely low and markets are undergoing frequent and significant changes. Consumers have more power because they can choose from a wide variety of products and vendors.

SOCIETAL PRESSURES

Nature of workforce is changing and the concept of virtual teams is gaining popularity. There is an increased emphasis on ethical and legal issues and social responsibility of organizations. Rapid political changes in many countries are another concern because of the risk they pose to business operations.

TECHNOLOGICAL PRESSURES

Technology is changing rapidly and innovations are coming quickly. The risk of technological obsolescence and increases in information overload are forcing companies continuously evaluate their current capabilities and future technology needs.

Organizational Responses

To succeed in such a dramatically changing business environment and pressures, organizational responses must not only be traditional e.g. lowering cost and closing unprofitable facilities, but also innovative responses such as customizing or creating new products or providing superb customer service (Carr, 2001). Organizational response activities can take place in some or all organizational processes, from daily processing of payroll to strategic activities such as the acquisition of a company. Organizational response activities can also take place in the supply chain. Organizational response activity can be a reaction to a specific pressure already in existence, it can be an initiative that will defend an organization against future pressures, or it can be an activity that exploits an opportunity created by changing conditions.

Numerous organizational response activities can be greatly facilitated by E-commerce. In some cases, e-commerce is the only solution to these business pressures [Tapscott, Lowy, & Klym, 1998; Callon, 1996; Turban, McLean, & Wetherbe, 2004]. Following are some examples of E-commerce supported organizational response initiatives.

STRATEGIC SYSTEMS

There is a variety of E-commerce-supported strategic systems. One example is FedEx's tracking systems, which allows FedEx to identify the status of every individual package, everywhere in the system. Most of the FedEx's competitors have already copied the FedEx systems. Another example is Merrill Lynch's Cash Management Account (CMA), a financial product. This product acted as both a bank account and brokerage account combined. The customer could write checks against the account and even receive a bank charge card. Merrill Lynch gained a significant competitive advantage with its cash management account system by attracting a large number of small investors. The technology made it possible for Merrill Lynch to offer a new service that helped expand the firm's market share and increased the size of its liquid assets fund.

CONTINUOUS IMPROVEMENT EFFORTS

Continuous improvement programs are being used by the companies to improve their productivity, quality, and customer service. An example of E-commerce supported continuous improvement initiative is DELL computers. Dell has integrated its enterprise resources planning (ERP) software and just-in-time assembly operations. The orders are taken electronically and then moved immediately via ERP software to the assembly line.

BUSINESS PROCESS REENGINEERING (BPR)

BPR initiatives are needed when business pressures are strong and radical structural changes are needed. E-Commerce is frequently interrelated with BPR, because a BPR may be needed for implementation of e-commerce initiatives such as electronic procurement. An example of E-commerce supported BPR initiatives is Ducati Motorcycles (Italy), which started to sell exclusive motorcycles through the Internet in 1998, but later streamlined this process drastically to attract new customers (Jelassi & Leenen, 2003).

CUSTOMER RELATIONSHIP MANAGEMENT (CRM)

Digital revolution has brought stronger than ever bargaining power of customers. This bargaining power is growing continuously. Availability of information and comparisons online are fueling this trend. Organizations must keep their customers happy in order to keep them. This may be achieved by customer relationship management (CRM). E-commerce supports CRM through multiple technologies, ranging from computerized call centers to various types of intelligent agents.

BUSINESS INTELLIGENCE (BI)

E-commerce heavily supports various types of alliances e.g. by providing electronic transmission of information to real-time collaboration between trading partners. One example of alliance supported by E-commerce is B2B marketplaces e.g. Covisint (created by General Motors, Ford, and others in the automotive industry). In 2001, General Motors., Suzuki Motors, Fuji Heavy Industries and Isuzu Motors formed a joint venture company named Japan AutoWeb Services Inc to develop e-business synergies and leverage Internet-based technologies. In March 2007 Ingram Micro, Cisco, Level Platforms, Intel and Microsoft jointly formed an industry alliance known as MSP Partners. MSP Partners was set up to educate channel partners and resellers about managed services.

ELECTRONIC MARKETS

Electronic markets, private or public, can optimize trading efficiency; enable their members to compete globally. Most electronic markets require the collaboration of diverse companies that may even include competitors. This collaboration can also be supported by E-commerce. In 2009, Walmart established its Walmart Marketplace in which it collaborated with several retailers. The marketplace enabled merchants to sell their products at Walmart.com. CSN Stores, Pro Team, and eBags were among the first group of retailers that started selling via Walmart.com. In the same year, Sears launched its third-party marketplace. The marketplace was integrated with the Sears website (Sears.com) and allowed merchants to list their products on the Sears site.

EMPOWERMENT OF EMPLOYEES

E-commerce supported CRM enables employee empowerment. Empowered employees (e.g. salespeople and customer service employees) can be given the authority to make customers happy and do it quickly, helping to increase customer loyalty. E-commerce allows the decentralization of decision-making and authority via empowerment and distributed systems, but simultaneously supports a centralized control.

SUPPLY CHAIN IMPROVEMENTS

E-commerce can help reduce supply chain delays, reduce inventories, and eliminate other inefficiencies.

MASS CUSTOMIZATION

In mass customization, items are produced in a large quantity but are customized to fit the desires of each customer. Mass customization is not easy to achieve but E-commerce can help (for details see Zipkin, 2001). E-commerce can provide mass customization by enabling interactive communication between buyers and designers so customers can do quick and correct configuration of the products they want. Automated ordering can reach the production facility in minutes. Dell Computers are a good example of a company that adopted E-commerce supported mass customization strategy.

INTRABUSINESS APPLICATIONS

Intrabusiness applications can include support provided to employees in the field, warehouse employees, and other employees. E-commerce supports Intrabusiness applications by providing support that improves the productivity of the workers.

KNOWLEDGE MANAGEMENT

Knowledge management is the process of creating or capturing knowledge, storing and protecting it, updating and maintaining it constantly and using it whenever necessary. Using e-commerce, corporate portals can be used to capture knowledge, assist users or to teach employees.

Promoting Global E-commerce: International Cooperation, Collaboration, and Regulations

International cooperation and collaboration is essential for the growth and development of global e-commerce and development of international guidelines for E-commerce. In this regard, many important initiatives have been taken by governments through bilateral or multilateral arrangements, and in other international forums and organizations. The Guidelines for Consumer Protection by OECD (2003), Electronic Commerce Directive by European Union (2003), Directive on Consumer Rights by European Union (2008), Interpretive Guidelines on Electronic Commerce and Information Property Trading by Japan's Ministry of Economy, Trade and Industry (2007), E-commerce directives by Office of Fair Trading (OFT) of UK, US FTC guidelines on the application of existing law to online advertising and marketing, US Congress CAN-SPAM Act of 2003, and US Safe Web Act of 2006 are examples of such initiatives. However, developing international guidelines for E-commerce face a few issues. While these guidelines may be appropriate for e-commerce environment in developed countries, they may not be fully relevant to developing economies where e-commerce laws are not fully developed, or have not developed in the same manner as in developed countries. Another issue is that the guidelines can be too general in nature and therefore require continuous and regular updating to keep up with the evolution of new technologies. The guidelines may require the participation of a range of stakeholders, which will complicate the process. Enforcement of these guidelines may also require resources and training needed to enforce various e-commerce cases. Limitations on information sharing and the lack of harmonization of applicable laws and jurisdiction are some other issues.

The regulatory frameworks for E-commerce vary among countries and countries have different approaches toward regulation. This can have implications for both businesses and consumers especially those involved in cross-border trade. Consumer rights and obligations vary considerably from one jurisdiction to another. Treatment of the role and responsibilities of intermediaries (e.g. Internet Service Providers) to consumers (e.g. information disclosures and fraudulent activities such as cyber fraud and counterfeiting) also varies among countries. OECD countries apply limited liability rules to intermediaries. In USA, under certain circumstances, intermediaries are not liable for the actions or omissions of sellers that harm buyers or vice versa or for content that other users post on their websites. In France and Germany, online communication services that act as intermediaries are responsible for making sure that their platform is not used for illegal purposes, such as selling counterfeit or pirated products. To address e-commerce issues, some countries use generic regulation developed in other consumer protection contexts, while the other countries use regulations that are more specific. Some countries use a mix of both approaches. Examples of specific legislation include the Loi Chatel, which France introduced in 2008, and a standard contract developed by the Consumer Ombudsman of Norway in 2009.

To many European policy makers, the significant reasons why region's homegrown Internet companies have not earned their due position in the market include a patchwork of tax, copyright and e-commerce rules that have stunted growth of Internet companies. Another factor, pointed out by European policy makers, is allegedly unfair business practices by U.S.-based competitors. "The Digital Single Market Strategy" of the European Union is meant to tackle both issues. "The Digital Single Market Strategy" is a plan to unify Europe's fragmented digital market and crack down on potential abuses of market power by U.S. Web giants (European Commission, 2015). The strategy includes a set of targeted actions to be delivered by the end of 2016. It is built on three pillars: (1) better access for consumers and businesses to digital goods and services across Europe; (2) creating the right conditions and a level playing field for digital networks and innovative services to flourish; (3) maximizing the growth potential of the digital economy (European Commission, 2015). The plan called for an overhaul of EU telecommunications rules, reconciling tax and copyright rules within the European Union, and simplifying regulations for companies that sell goods electronically or send data across European borders. One key goal is to help European consumers shop online in other EU countries as easily as in their own, and to get the best products at the best prices (European Commission, 2015). At the heart of the project is Europe's battle against the dominance of U.S.-based Web companies. The plan calls for several major inquiries into possible abuses by U.S. companies, including a comprehensive analysis of the role of online platforms such as search engines and price-comparison websites, as well as a previously signaled investigation by antitrust regulators into whether e-commerce companies such as Amazon.com Inc. are restricting cross-border trade.

Companies on both sides of the Atlantic generally favored the EU move toward a single online market but feared that new barriers would be erected. For the success of this plan, however, European Commission will need concrete legislative proposals that will be debated and modified by national governments and the European Parliament, a process that usually takes years. The legal landscape is already shifting in response to European fears that big U.S. tech firms have become too dominant. National regulators in the Netherlands, Spain and other countries are investigating Facebook's privacy practices (Schechner, 2015). The EU's antitrust regulator, meanwhile, has filed formal charges against Google alleging it has abused its dominance as a search engine to promote its own businesses. Google denied the allegations by saying it faces fierce competition in a fast changing marketplace (Brussels et al., 2015). Facebook said that it follows data protection rules of European Union and performs repeated audits by the data-protection authority in Ireland, home of its European headquarters (Fairless & Schechner, 2015). Among the most controversial of these EU strategy proposals was one to better fight "illegal content on the Internet." Under the EU's current rules, tech companies must comply with valid requests from governments and copyright holders to take down copyrighted material, but the companies generally are not liable for content to which they have not been alerted. In the changes being considered as part of the new plan, the EU could require "Internet intermediary services" to monitor their systems for illicit content. However, executives of technology firms think it could have serious consequences for the right of free speech (Dow Jones Business News, 2015). Technology firms believed these new rules could constrain Europe's ability to attract technology companies and to develop its own. Some of these proposals take on real problems, but others also take on perceived problems where there is very little evidence base. Europe's telecom operators welcomed the plan but called for a more ambitious timetable, and urged regulators to allow more mergers between telecom operators (Fairless & Schechner, 2015).

E-commerce Outlook

According to Wikipedia, Book Stacks Unlimited of Cleveland opened their first online shop in 1992, followed by Amazon.com in 1995. This is how far the E-commerce has come. Despite many disadvantages, E-commerce is expanding rapidly. Internet economy is growing dramatically and opening up new commercial opportunities for business and consumers. For many industries, Internet has become a significant mean to conduct their business. Copyright-based industries (e.g. film, music, games, and news) are developing new types of products and content and playing a leading role in promoting E-commerce. In United States, sales of selected service industries amounted to 1.8 % of total revenue in 2007.

Acceptance of the Internet as a trading platform is growing and more consumers are making purchases online. However, the acceptance varies among sectors. The most popular and purchased goods over the Internet worldwide in 2007 included books, clothing, videos, DVDs, games, airline tickets and electronic equipment. In the United States, consumers spent US$ 83.9 billion in 2008 on online leisure travel while spending US$ 141.3 billion on other online retail items (Forrester, 2010). Music and videos sales online was 74 % of the total online sales of all categories (US Census, 2007). According to a survey, 66% of American online users made online purchase of a product in 2007 (Pew Internet, 2008). In 2008 in the European Union (EU), 36% of Internet users compared goods or services through price comparison websites. Out of these 36% users, 17% purchased a product on-line (European, 2009). According to a study by Nielson, consumers in the Asia-Pacific region were the world's most prolific online shoppers. Many Asian-pacific consumers relied on Internet reviews when making purchases and South Koreans were the heaviest online buyers in Asia. These consumers gave more importance to pinions posted online when buying products such as electronics, cosmetics, cars, software and food. Compared with consumers elsewhere, these consumers were most likely to share dissatisfaction at a product on the Internet (adoimagazine, 2010). Social networking sites such as Facebook and Twitter or blogging sites were the powerful tools influencing what people buy and urged businesses to embrace the trend (Yahoo, 2010).

The global e-commerce industry saw impressive growth in 2014 with goods and services worth $1.5 trillion bought by shoppers via desktops, tablets and smart phones. Advertisers are expanding their Internet marketing budgets and their ad spending is expected to surpass $160 billion in 2015. Mobile shopping is maturing further and consumer mindshare continues to be split across multiple devices. Billions of online transactions are providing marketers with unique insights into consumer online shopping behavior. Mobile share of online sales grew steadily in 2014 and in 2015 it is expected that mobile devices would account for 40% of e-commerce transactions globally (criteo.com, 2015).

Mobile purchases are becoming easier with the growing deployment of HTML and mobile optimized websites. In addition, new mobile payment services, particularly Apple Pay, will accelerate consumer willingness to make purchases via their mobile phones in 2015. In future, mobile purchases are expected to attract a greater share of digital ad spending. Since consumers are using a multi-device path to purchase, marketers still confused to tell if the person who saw an ad on a smart phone and later made the purchase on a tablet was the same user or a new customer. In 2015, the availability of more precise exact-match methods will make cross-device advertising easier and subsequently result in increased sales for retailers. Retailers will be able to differentiate between existing users and new prospects through a single view of consumers' shopping behavior across desktops, tablets and smart phones. Marketers will be able to deliver relevant, personalized ads to users across devices with accuracy and scale (Aziz, 2015). This would further boost advertiser confidence in performance marketing. Performance Marketing is a comprehensive term that refers to online marketing and advertising programs in which advertisers (also known as "retailers" or "merchants") and marketing companies (also known as "affiliates" or "publishers") are paid when a specific action is completed; such as a sale, lead or click. It will also help advertisers meet the growing consumer expectation that ads are relevant to them. In 2015, native ads will become every bit as scalable and measurable as IAB standard ad units. Native advertising is a form of online advertising that matches the form and function of the platform on which it appears. For example, an article written by an advertiser to promote their product but that uses the same form as an article written by the editorial staff. Therefore native ad inventory will grow significantly. An increasing number of technology providers are now offering services to publishers to help them integrate native ads that look consistent with their website or mobile app, in the exact same way Facebook sells its "News Feed" ads. This phenomenon is expected to scale in 2015 and driven by demand generated through programmatic buying. Programmatic buying describes online display advertising that is aggregated, booked, analyzed and optimized via demand side software interfaces and algorithms. With programmatic buying, native ads implementation will become a lot easier than it used to be. In addition, publishers will be able to charge higher CPMs because native ads perform better as compared with IAB standard ad units, especially the mobile ads (Aziz, 2015).

For brick-and-mortar retailers, 2015 will be a 'do or die" year. This year, brick-and-mortar retailers would experience the increasing impact of shopper webrooming (the process of researching products online and then visiting a store to make a purchase) and showrooming (the practice of visiting a shop or shops in order to examine a product before buying it online at a lower price) behavior. Webrooming is emerging as a stronger trend than showrooming, and many brick-and-mortar retailers will start deploying beacons and tablets and offer free Wi-Fi in their stores to accelerate this trend in 2015. This will also help retailers create more touch points with consumers, and develop insights on how they can engage with store visitors before they enter the store, in the store and after they leave the store. While app installs will continue to be important in 2015, mobile app marketers will start to focus more on re-engaging with users who have previously installed the app but are not using it (Digital News Asia, 2015). There was significant increase in the cost of driving an app install during 2014 as demand increased on Facebook and RTB platforms. Further, more advertisers have entered the mobile domain, pushing up CPMs, CPCs and CPIs. A greater share of mobile app marketing budgets in 2015 will be spent on promoting the app those who installed the app but not using it. Retailers, who until now focused heavily on increasing their app-installed bases, will pay more attention to improving app usage and re-visits from existing users (Warrington, 2015). In the past, advertisers had to choose between scalability and access to inventory. With preference for scalability, ad formats that are not IAB standards are not used. Such ads account for a significant share in markets like China, which has roughly 170,000 different ad formats. In addition, a big reason for the ad format changes is driven by users shifting their media consumption to mobile, especially the mobile in-app environment (criteo.com, 2015). The network effects of first movers with solid business models will help them gain commercial and technological strength that in turn will enable them to use acquisitions to increase their talent, innovative technology and coverage. This consolidation will make it simpler for e-commerce marketers to identify marketing solutions that meet their objectives. Marketers will be able to access and use actionable data in huge volumes to optimize spend across global channels, through fewer partners with more comprehensive solutions (bizshifts-trends.com, 2014; criteo.com, 2015).

Today, many consumers opt to buy online for convenience, price, and broad product choices available. Still many consumers refrain from online purchases due to security and privacy concerns or the inability to touch and feel products. Online businesses are continuing to add new content and other features to mitigate these concerns. In 2015, there will be risks and opportunities for marketers as online shopping becomes more complex. Success and failure of marketers will depend on their ability to use key technologies and partners to reach always-on consumers with personalized messages, across devices and at scale. Advertisers that focus on conversion and sales will get better results from their online campaigns than those using broad audience reach metrics. Retailers are investing in mobile and cross-device targeting solutions will achieve a significant boost in sales at the same cost. They will also benefit at the expense of competitors who are slow to embrace mobile and performance marketing (Pardeep, 2015; Goward, 2015; criteo.com, 2015). With growing opportunities to sell internationally, e-commerce companies should collaborate with technology companies that can enable them to run online ad campaigns with rich targeting and optimization capabilities at a truly global scale. Cross-device advertising will allow retailers to identify a consumer reaching them via multiple devices as a unique user, and to differentiate between a customer and a prospect. This insight will enable retailers to deliver relevant messages and optimize ad spending for engagement with existing customers, and for acquiring new customers.

The ecommerce experience is due to become decidedly more physical, sensory and immersive. Virtual reality technologies akin to Oculus Rift (recently acquired by Facebook) and Sony's Morpheus – both virtual reality displays, worn like goggles, are indicative of how consumers, may soon quite literally step into more life-like shopping and service experiences online. By simply donning a relatively inexpensive virtual reality headset, shoppers could soon be able to transport themselves to any store in the world, browsing and buying in a very natural, intuitive and highly experiential way. One of the inherent drawbacks of online shopping is the inability to touch and feel items that you are considering buying. That is likely to change. Touch is something we may soon widely replicate with technology. Researchers are have been successfully mapping patterns of vibration in order to accurately simulate the feel of various materials, allowing users to sense, through a glide pad and stylus, the unique tactility of different surfaces and textures. Using haptography, vibrations are transmitted into a stylus that accurately simulates the feeling of denim. Soon, feeling something online before you buy it will likely be commonplace (Stephens, 2014). Wearable, smart technologies like Google Glass in combination with location-based augmented reality applications (like the one from Layar) will enable consumers to see and interact with digital information, making entire cities and the things in them clickable. By simply saying a sentence, the location-based augmented reality application would immediately detect any digital data present in the field of vision. Once detected, it initiates that content and allows the user to interact with it. For example, by instructing Google Glass to scan a street lined with shops, hotels and restaurants, the user may then be able to interact with data attached to those various physical places. It may be possible to browse hotel reviews, make reservations at a restaurant or download coupons or offers from stores. Layar for Google Glass enables a user to see which apartments on the street are available for rent by enabling augmented reality content. Accessing this new phy-gital data layer will allow us to move seamlessly between real world and online commerce, when and where it makes the most sense to do so. Online retailers therefore, will have to begin thinking in more contextual terms, placing digital opportunities to buy products not simply in online catalogues but also throughout the physical world, where it makes the most geographic and contextual sense for consumers. For example, an electronics retailer could easily give conference attendees the ability to order from an assortment of the most commonly needed electronic accessories directly from an augmented reality store located right at the conference center. Overall, it is not hard to see how technologies like these will bring e-commerce infinitely closer to feeling alive and experiential.

With many retailers in the big box channel struggling in a post-Internet world, it's reasonable to assume that even Amazon, eBay and other digital behemoths could find themselves being challenged by a new breed of competition – competition born out of a new generation of technologies. Industry experts, in general, agree that the future of E-commerce is bright. It is expected that E-commerce will become an increasingly important method of reaching customers, providing services, and improving operations of organizations. However, there is no consensus about the growth rate of E-commerce, the time frame for E-commerce to become a substantial portion of the economy, and the industry segments that will grow the fastest. Overall, the growth of E-commerce will continue into the near future. Despite the failure of individual companies and initiatives, the total volume of E-commerce is expected to increase.

Review Questions

1. Differentiate between E-commerce and E-business.

2. How Internet economy is related to E-commerce and E-business?

3. Distinguish between pure and partial E-commerce.

4. Differentiate between a) B2C and C2B b) B2G and B2B c) C2C and B2E d) C2C and B2E e) G2G and G2E f) G2B and G2C.

5. Differentiate between Internet and Non-Internet E-commerce.

6. Discuss the role of EDI in E-commerce. Do you think Internet-based EDI will eventually replace traditional EDI completely? Why? Why Not?

7. Differentiate between intra-business and non-business E-commerce.

8. Differentiate among brick-and-mortar, click-and-mortar, and virtual business.

9. Define three types of E-marketplaces.

10. How intermediaries support E-commerce?

11. List some E-commerce successes and failures.

12. Describe some distinguishing characteristics of first and second generation of E-commerce.

13. List some critical success factors for E-commerce.

14. List some major technology trends in E-commerce.

15. List some international cooperation efforts to promote global E-commerce.

16. List some products, which are suitable for E-commerce. Also, list some products which are not suitable for E-commerce.

17. List some advantages and disadvantages of E-commerce.

18. List the major business pressures faced by organizations today due to digital economy.

19. List the major organizational responses to business pressures.

20. Describe how E-commerce supports organization responses to business pressures.

21. List some important E-commerce trends.

Bibliography

adoimagazine. (2010). *Asia Pacific Consumers Are the World's Most Prolific Online Shoppers: Nielsen.* Retrieved October 12, 2010, from adoimagazine.com: http://www.adoimagazine.com/newhome/index.php?option=com_content&view=article&id=5980:-asia-pacific-consumers-are-the-worlds-most-prolific-online-shoppers-nielsen&catid=1:breaking-news&Itemid=5

Andam, Z. R. (2003, May). *e-Commerce and e-Business.* Retrieved October 09, 2010, from e-ASEAN Task Force, UNDP-ADIP: www.apdip.net/publications/iespprimers/eprimer-ecom.pdf

Atchariyachanvanich, K., & Okada, H. (2001). *Critical Success Factors of E-Commerce: External factors beyond the context of corporate.* Retrieved October 10, 2010, from National Institute of Informatics, Tokyo, Japan: www.stkc.go.th/stportalDocument/stportal_1118720587.doc

Auchard, E. (2015, June 25). Many big companies live in fear for their future in digital age. Reuters. Retrieved from http://www.reuters.com/article/2015/06/25/management-digital-threats-idUSL8N0ZA2Q020150625

Aziz, K. (2015, February 10). Mobile Commerce Trends in 2015. Retrieved July 10, 2015, from http://www.sitepoint.com/mobile-commerce-trends-2015/

Bartels, A. (2000, October 30). *The difference between e-business and e-commerce.* Retrieved October 10, 2010, from computerworld.com: http://www.computerworld.com/s/article/53015/The_difference_between_e_business_and_e_commerce

Barua, A., Pinnell, J., Shutter, J., & Whinston, A. B. (2000). *Measuring the Internet Economy.* Retrieved October 10, 2010, from Center for Research in Electronic Commerce, The University of Texas at Austin: http://ai.kaist.ac.kr/~jkim/cs492a/internet_economy-UT.pdf

Bhasker. (2008). Electronic Commerce Framework Technologi. New Delhi: Mcgraw Hill Education.

Bidgoli, H. (2004). The Internet Encyclopedia. John Wiley & Sons.

bizshifts-trends.com. (2014, September). First Mover or Fast Follower: What is Right Business Strategy? Think Hard About; Pros-Cons of Being-- First, Fast, Late... Retrieved July 10, 2015, from http://bizshifts-trends.com/2014/09/24/first-mover-fast-follower-right-business-strategy-think-hard-pros-cons-first-fast-late/

Bond, S. (2014, June 18). Value Added Networks are finally adding Value. Retrieved June 30, 2015, from https://www.linkedin.com/pulse/20140618224222-16074186-value-added-networks-are-finally-adding-value

Briggs, B. (2014, November 10). B2B Market Trends - B2B buyers want B2C features, experts say at hybris Game Plan event - Internet Retailer. Retrieved July 9, 2015, from https://www.internetretailer.com/2014/11/10/b2b-buyers-want-b2c-features

Brussels, T. F. in, Winkler, R., & Francisco, A. B. in S. (2015, April 16). EU Files Formal Antitrust Charges Against Google. Wall Street Journal. Retrieved from http://www.wsj.com/articles/eu-files-formal-charges-against-google-1429092584

Buckley, S. (2010, April). *ABI research: Fixed broadband access is still growing*. Retrieved October 10, 2010, from fiercetelecom.com: http://www.fiercetelecom.com/story/abi-research-fixed-broadband-access-still-growing/2010-04-29

Callon, J. D. (1996). *Competitive Advantage Through Information Technology*. McGraw-Hill Higher Education.

Carr, N. G. (2001). *The Digital Enterprise*. Harvard Business Press.

Choi, S. Y., & Whinston, A. (2000). *The Internet Economy, Technology and Practice*. SmartEcon Publishing.

Chung, M., Lee, J. K., & Turban, E. (2006). *Electronic Commerce: A Managerial Perspective*. Prentice Hall.

Cornish, R. (2014, June 24). Characteristics of B2B e-commerce. Extra Digital. Retrieved June 29, 2015, from http://www.extradigital.co.uk/articles/internet-marketing/b2b-ecommerce.html

criteo.com. (2015, June). Criteo eCommerce Industry Outlook 2015. Retrieved July 1, 2015, from http://www.criteo.com/resources/criteo-ecommerce-industry-outlook-2015/

Cullen, A. J., & Taylor, M. (2009). Critical success factors for B2B e-commerce use within the UK NHS pharmaceutical supply chain. *International Journal of Operations & Production Management , 29* (11), 1156 – 1185.

Cyr, D., Bonanni, C., & ilsever, J. (2004). Design and e-loyalty across cultures in electronic commerce. *Proceedings of the 6th International Conference on Electronic Commerce* (pp. 351-360). Delft: ACM.

D'Onfro, J. J. (2014, June 1). Here's What It's Really Like Cooking With Blue Apron — The NYC Food Startup That's Worth Half-A-Billion Dollars. Retrieved July 11, 2015, from http://www.businessinsider.com/blue-apron-review-cooking-startup-2014-6

Darrell-hill.com. (2009). *Business to Employee*. Retrieved October 10, 2010, from http://www.darrell-hill.com/e-business/business-to-employee-b2e

Deviceatlas. (2015, March 27). 10 great examples of adaptive web design. Retrieved July 17, 2015, from https://deviceatlas.com/blog/adaptive-web-design-examples

Digital News Asia. (2015, February 20). E-commerce will be "do or die" for brick-and-mortar retailers: Criteo | Digital News Asia. Retrieved July 10, 2015, from https://www.digitalnewsasia.com/digital-economy/ecommerce-will-be-do-or-die-for-brick-and-mortar-retailers-criteo

Dow Jones Business News. (2015, May 6). U.S. Tech Giants Set to Fight EU Proposal for New Tech Regulation - NASDAQ.com. Retrieved July 11, 2015, from http://www.nasdaq.com/article/us-tech-giants-set-to-fight-eu-proposal-for-new-tech-regulation-20150506-00669

Dressler, A. (2014, November 20). 5 Surprising Sites Using Adaptive Web Design - 'Net Features - Website Magazine. Retrieved July 17, 2015, from http://www.websitemagazine.com/content/blogs/posts/archive/2014/11/20/5-surprising-sites-using-adaptive-web-design.aspx

E. C. (2009). *The Consumer Markets Scoreboard, Second Edition*. Retrieved October 10, 2010, from ec.europa.eu: http://ec.europa.eu/consumers/strategy/docs/2nd_edition_scoreboard_en.pdf

e2open.com. (2014, November 20). E2open Earns Drummond Certification in Fall 2014 AS2 Interoperability Test Event | Press Releases | News & Events. Retrieved July 9, 2015, from http://www.e2open.com/news/article/e2open-earns-drummond-certification-in-fall-2014-as2-interoperability-test-event

El Sawy, O. A. (2000). *Redesigning Enterprise Processes for E-Business*. McGraw-Hill.

eMarketer.com. (2014a, July 23). Worldwide Ecommerce Sales to Increase Nearly 20% in 2014 - eMarketer. Retrieved June 27, 2015, from http://www.emarketer.com/Article/Worldwide-Ecommerce-Sales-Increase-Nearly-20-2014/1011039

eMarketer.com. (2014b, December 23). Retail Sales Worldwide Will Top $22 Trillion This Year - eMarketer. Retrieved July 1, 2015, from http://www.emarketer.com/Article/Retail-Sales-Worldwide-Will-Top-22-Trillion-This-Year/1011765

eoscloudstore.com. (2015). Decreasing Your Sales Costs By Up To 90%. Retrieved July 9, 2015, from http://info.eoscloudstore.com/blog/decreasing-your-sales-costs-by-up-to-90

European Commission. (2015a). The importance of the digital economy - European Commission. Retrieved July 19, 2015, from http://ec.europa.eu/growth/sectors/digital-economy/importance/index_en.htm

European Commission. (2015b, May 6). European Commission - PRESS RELEASES - Press release - A Digital Single Market for Europe: Commission sets out 16 initiatives to make it happen. Retrieved July 1, 2015, from http://europa.eu/rapid/press-release_IP-15-4919_en.htm

Evans, P., & Wurster, T. S. (2000). *Blown to Bits: How the New Economics of Information Transforms Strategy*. Harvard Business Press.

Fairless, T., & Schechner, S. (2015, May 7). EU Makes Play for Leverage Over E-Commerce. Wall Street Journal. Retrieved from http://www.wsj.com/articles/eu-announces-sweeping-plans-to-create-a-digital-single-market-1430906432

Flood, G. (2008). *The server OS: Present and future trends*. Retrieved October 10, 2010, from zdnet.co.uk: http://www.zdnet.co.uk/news/it-strategy/2008/05/30/the-server-os-present-and-future-trends-39424186/

Forrester Research. (2015, April 2). US B2B eCommerce Forecast: 2015 To 2020. Retrieved July 9, 2015, from https://www.forrester.com/US+B2B+eCommerce+Forecast+2015+To+2020/fulltext/-/E-RES115957

Forrester, R. (2010). *US Online Retail Forecast, 2008 To 2013*. Retrieved October 12, 2010, from 199it.com: http://www.199it.com/wp-content/uploads/2010/06/002284.forrester.usonlineretailforecast.pdf

Giaglis, G. M., Klein, S., O'Keefe, R. M., & O'keefe, R. M. (1999). Disintermediation, reintermediation, or cybermediation? The future of intermediaries in electronic marketplaces. In Global Networked Organizations, Proceedings 12 th Electronic Commerce Conference, Moderna organizacija. Citeseer.

Google.co.uk. (2013, August). Examples of Great Adaptive Sites – Think Insights – Google. Retrieved July 17, 2015, from http://www.google.co.uk/think/articles/examples-of-great-mobile-experience.html

Goward, C. (2015, January 16). WiderFunnel Marketing Conversion Optimization – The Top 7 Conversion Optimization Trends for 2015. Retrieved July 10, 2015, from http://www.widerfunnel.com/conversion-rate-optimization/conversion-optimization-trends-2015

Gxsblogs.com. (2014, April 21). What is the relationship between EDI and XML? Retrieved July 9, 2015, from http://www.gxsblogs.com/rochellecohen/2014/04/edi-and-xml-are-they-related.html

Hackbarth, G., & William, J. K. (2000). Building an E-Business Strategy. *Information Systems Management* , 78-94.

Hamburger, E. (2014, April 22). "Sleep startup" Casper dreams of overturning the mattress racket. Retrieved July 11, 2015, from http://www.theverge.com/2014/4/22/5638400/casper-dreams-of-overturning-the-mattress-racket

Hammer, M. (1995). *The Reengineering Revolution*. Harper Paperbacks.

Hemmatfar, M., & Salehi, M. (2010). Competitive Advantages and Strategic Information Systems. *International Journal of Business and Management* , 5 (7), 158-169.

Hooven, C. (2015). Re-platform Your Ecommerce Website. Retrieved July 2, 2015, from /re-platform-ecommerce-website-article

Hsu, P. F., Kraemer, K. L., & Dunkle, D. (2006). Determinants of E-Business Use in U.S. Firms. *International Journal of Electronic Commerce*, *10* (4), 9.

IMD, & Cisco. (2015, June). Digital Vortex: How Digital Disruption Is Redefining Industries. Retrieved from http://www.imd.org/uupload/IMD.WebSite/DBT/Digital_Vortex_06182015.pdf

Internet, P. (2008). *Pew Internet & American Life Project, 2008*. Retrieved October 12, 2010, from pewinternet.org: http://www.pewinternet.org/Reports/2008/Online-Shopping/03-Trends-in-Online-Shopping/03-The-number-of-online-users-buying-or-researching-products-online-since-2000-has-roughly-doubled.aspx

ITU. (2008, September). *ITU Press Release*. Retrieved October 10, 2010, from itu.int: http://www.itu.int/newsroom/press_releases/2008/29.html

JANAKIRAMAN, V. S., & SARUKESI, K. (2008). DECISION SUPPORT SYSTEMS. PHI Learning Pvt. Ltd.

Jelassi, T., & Leenen, S. (2003). An E-Commerce Sales Model for Manufacturing Companies. *European Management Journal*, *21* (1), 38-47.

Jones, N., & Field, R. (2000). *Ecommerce: Critical Success Factors That Will Make or Break Your Online Business*. CDG Books Canada, Inc.

Kaplan, M. (2015, April 9). B2B Ecommerce Growing; Becoming More Like B2C. Retrieved from http://www.practicalecommerce.com/articles/85970-B2B-Ecommerce-Growing-Becoming-More-Like-B2C

Kettinger, W. J., & Hackbartha, G. (2000). Building an e-business strategy. *Information Systems Management*, 78-94.

Levy, K. (2014, November 3). I Tried Plated, The DIY Food-Delivery Site Started By A Couple Of Wall Street Guys Who Didn't Want To Get Fat. Retrieved July 11, 2015, from http://www.businessinsider.com/how-to-cook-using-plated-2014-11

Marketingcharts.com. (2007, September). *European E-Commerce to Reach 323 Billion Euros in 2011*. Retrieved October 10, 2010, from marketingcharts.com: http://www.marketingcharts.com/direct/european-e-commerce-to-reach-323-billion-euros-in-2011-1239/

Mehrtensb, J., Cragg, P. B., & Mills, A. M. (2001). A miodel of Internet adoption by SMEs. *Information & Management*, 165-176.

Mercer, E. (n.d.). Advantages & Disadvantages of Using a Value-Added Network for Electronic Commerce Communication. Retrieved June 29, 2015, from //www.ehow.com/info_12226113_advantages-disadvantages-using-valueadded-network-electronic-commerce-communication.html

Mital, M. (2015, February 26). E-Commerce 2015: Seven Technology Trends That Will Define the Year Ahead - Innovation Insights. Retrieved July 10, 2015, from http://insights.wired.com/profiles/blogs/e-commerce-2015-seven-technology-trends-that-will-define-the

Mohapatra, S. (2012). E-Commerce Strategy: Text and Cases. Springer Science & Business Media.

OECD. (2009). *2009 Conference on Empowering E-Consumers: Strengthening Consumer Protection in the Internet Economy, Washington, 8-10 Dec. 2009, Background report.* Retrieved October 10, 2010, from OECD.org: http://www.oecd.org/dataoecd/44/13/44047583.pdf

OECD. (2009). *OECD Communications Outlook 2009.* Retrieved October 10, 2010, from http://browse.oecdbookshop.org/oecd/pdfs/browseit/9309031E.PDF

Onfro, J. J. D'. (2014, June 1). Here's What It's Really Like Cooking With Blue Apron — The NYC Food Startup That's Worth Half-A-Billion Dollars. Retrieved July 11, 2015, from http://www.businessinsider.com/blue-apron-review-cooking-startup-2014-6

Pardeep. (2015, January 8). 10 Digital Marketing Trends that will Impact Your Business in 2015 - Grepsr. Retrieved July 10, 2015, from http://www.grepsr.com/blog/10-digital-marketing-trends-will-impact-business-2015/

Poon, S. (2000). Business environment and internet commerce benefit – a small business perspective. *European Journal of Information Systems* , 72-82.

Poon, S. (2000). Small business use of the Internet: Findings from Australian case studies. *International Marketing Review* , 72-82.

Poon, S., & Swatman, P. M. (1997). Small business use of the Internet: Findings from Australian case studies. *International Marketing Review* , 385-402.

Poorkhanjani, A. M., Ziaidoostan, H., Ghaneh, H., Janatpoor, A., Gholipoor, M., & Talebi, S. (2013). Study of Virtual Organization and Information System.

Pušara, K., & Katić, M. (2004). Adoption Of eCommerce Terminology. *17th Bled eCommerce Conference* (p. Paper 4). Bled: Bledconference.

Ramsdell, G. (2000). The real business of B2B. *The McKinsey Quarterly* (3), pp. 175-6.

Raphael, J. (2008). *Microsoft Steers Toward Interactive TV With Navic Buy*. Retrieved October 10, 2010, from ecommercetimes.com: http://www.ecommercetimes.com/story/Microsoft-Steers-Toward-Interactive-TV-With-Navic-Buy-63462.html?wlc=1286951230

rjmetrics.com. (2015, April 2). The Five Indicators of Breakout Ecommerce Growth. Retrieved July 1, 2015, from https://blog.rjmetrics.com/2015/02/04/the-five-indicators-of-breakout-ecommerce-growth/

Salvatella, J. (2013, April 1). The Digital Economy Formula. Retrieved July 19, 2015, from http://www.rocasalvatella.com/en/digital-economy-formula

Schechner, S. (2015, May 6). Silicon Valley Slams EU Web Regulation Plan. Wall Street Journal. Retrieved from http://www.wsj.com/articles/u-s-tech-giants-set-to-fight-eu-proposal-for-new-tech-regulation-1430916181

Schneider, S. (2009). *Electronic Commerce*. Course Technology.

Slyke, C. V., Belanger, F., & Comunale, C. L. (2004). Factors influencing the adoption of web-based shopping: the impact of trust. *DATA BASE for Advances in Information Systems , 35* (2), 32-49.

Slywotzky, A. J., & Morrison, D. J. (2000). *How Digital Is Your Business?* Crown Publishing Group.

Stephens, D. (2014, April 15). The Near Future Of Ecommerce -. Retrieved from http://www.retailprophet.com/blog/the-near-future-of-ecommerce/

Tapscott, D., Lowy, A., & Klym, N. (1998). *Blueprint to the Digital Economy: Creating Wealth in the Era of E-Business*. McGraw-Hill Professional.

Tihila, E. (2014, December 16). Basware Predicts Advances in Online, Real-Time Payment Solutions Will Drive Global B2B Commerce in 2015 | Basware. Retrieved July 17, 2015, from http://www.basware.com/about-us/news/basware-predicts-advances-in-online-real-time-payment-solutions-will-drive-global-b2b-commerce-in

Turban, E., King, D., & Lang, J. (2010, October). Introduction to Electronic Commerce (3rd Edition) (Pearson Custom Business Resources): Efraim Turban, David King, Judy Lang: 2900136109234: Amazon.com: Books. Retrieved July 11, 2015, from http://www.amazon.com/Introduction-Electronic-Commerce-Business-Resources/dp/0136109233/ref=sr_1_2?s=books&ie=UTF8&qid=1436597615&sr=1-2&keywords=E-commerce+by+Turban

Turban, E., McLean, E., & Wetherbe, J. (2004). *Information Technology for Management: Transforming Organizations in the Digital Economy*. Wiley.

US Census, B. (2007). *E-commerce 2007 Sector Highlights*. Retrieved October 12, 2010, from census.gov: www.census.gov/econ/estats/2007/2007reportfinal.pdf

US, C. B. (2009, August). *Quarterly Retail E-Commerce Sales, 2nd Quarter 2009*. Retrieved October 9, 2010, from www.census.gov/retail/mrts/www/data/pdf/09Q2.pdf

UT Dalls. (2015). Multi-modal 3D Tele-Immersion Research Project holds its Annual Meeting. Retrieved July 1, 2015, from http://cs.utdallas.edu/multmodal3dteleimmersionresearchprabha/

Warrington, G. (2015, February 5). Top 7 predictions of digital ad industry for 2015. Retrieved July 10, 2015, from https://www.linkedin.com/pulse/top-7-predictions-digital-ad-industry-2015-glenn-warrington

Werbach, K. (2000, May-June). Syndication: The Emerging Model for Business in the Internet Era. pp. 85-93.

Whinston, A. B., Choi, S. Y., & Stahl, D. O. (1997). *The Economics of Electronic Commerce*. Macmillan Technical Pub.

Xu, Y., Zhang, C., & Chen, J. (2009). The Role of Mutual Trust in Building Members' Loyalty to a C2C Platform Provider. *International Journal of Electronic Commerce , 14* (1), 147-171.

Yahoo. (2010). *Life Style of Internet Shoppers*. Retrieved July 2010, 2010, from news.yahoo.com: http://news.yahoo.com/s/afp/20100713/tc_afp/lifestyleasiatechnologyinternetretail

Yell.com. (2012, March 16). Does E-Commerce Vary From B2B to B2C? Retrieved from https://business.yell.com/knowledge/does-e-commerce-vary-from-b2b-to-b2c/

Zipkin, P. (2001). The Limits of Mass Customization. *MIT Sloan Management Review , 42* (3), p. 81.

2

E-COMMERCE INFRASTRUCTURE

Principal Components of E-commerce Infrastructure

E-commerce infrastructure needs to be able to handle a large volume of data, a large volume of customer interactions, and to provide very quick response time. Lacking any of these capabilities can render the infrastructure less effective. The principal components of E-commerce infrastructure include:

WEB BROWSER: A web browser is a software program that users can use to access World Wide Web. Internet Explorer and Mozilla Firefox, and Netscape Navigator are examples of popular web browsers.

WEB SERVERS: These special computers host the E-commerce websites. Users access these E-commerce websites using their web browser. A Web server can host one or more Web sites.

PAYMENT SYSTEM: This is the mechanism to process online transactions. Primary methods of electronic money exchange include credit cards, electronic checks, smart cards, digital cash and electronic funds transfer.

E-COMMERCE SOFTWARE: The basic functions of E-commerce software include customer registration and authentication, provide online catalog viewing, order receiving and processing, provide customer service, and reporting and analysis tools. Optional functions include order fulfillment and payment processing etc.

E-MAIL: E-mail supports E-commerce website by providing a way of communication with the customers.

MAILING LIST SERVER: This is an optional component, which is used to automate the management of mailing lists for groups of users.

DATABASE SERVER: This special computer provides storage and retrieval of all the data used by E-commerce site. The data stored by database server may include customer financial and personal information as well as information related to transaction.

TECHNOLOGIES FOR CUSTOMER AND ORGANIZATIONAL SUPPORT: These technologies can include search engine, intelligent agents, knowledge management technologies, electronic customer relationship management, electronic supply chain management etc.

Broadband Internet Access: Internet is the medium through which customers interact with the E-commerce sites.

Figure 2-1 shows an example of a typical E-commerce infrastructure.

FIGURE 2-1: A TYPICAL E-COMMERCE INFRASTRUCTURE

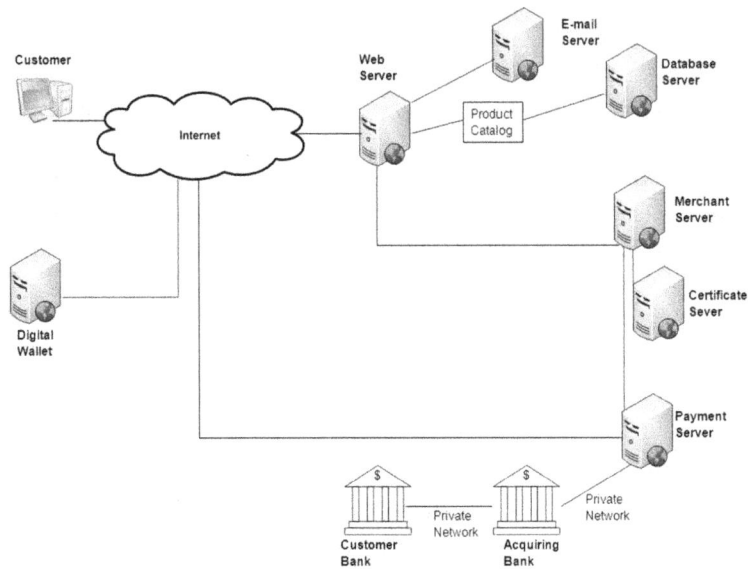

Key Ecommerce-Infrastructure Decisions

Every business requires an infrastructure (including facilities, equipment, and processes to support all the functional areas of your business) to support its customers and operations. An infrastructure that matches your business strategies provides efficient operations and vice versa. Take an example. For a business whose value proposition is to provide the highest level of customer service, the infrastructure would include processes to deliver quick and responsive service (such as live chat and self-service tools). For a business whose value proposition is become the cost leader, an infrastructure would be required that provide lowest possible infrastructure costs. Typically, ecommerce businesses try to maintain a high degree of flexibility in their infrastructure to keep fixed costs low and to be able to react quickly to market changes or competitive pressures. A key infrastructure decision is whether to outsource or manage operations in house. When deciding on your business infrastructure and operations, be sure to evaluate what your core strengths are. You should outsource those activities that are not your core strength or ones that require high levels of skill or specialization.

Following are some key infrastructure decisions that an ecommerce businesses faces (Traxler, 2012).

MARKETING

Marketing is probably the most important infrastructure element of your ecommerce business. To succeed, visitors must be able to come to your website and then you need to keep them there and compel them to buy from you. Marketing effort may involve many aspects such as website design, social media, search marketing, merchandising, email, or other forms of advertising. Effective management of marketing activities in-house is very challenging. Most small ecommerce businesses outsource some element of marketing.

FACILITIES

One big advantage of an online business is that they do not require investment in their physical offices and warehouses. In many cases, a small space may be sufficient to run your ecommerce business provided you drop ship or outsource fulfillment. Even when your organization grows, you will not need a fancy office in the right location. E-commerce business should try to host their businesses at a place that has a wide variety of spaces in different sizes. This would allow them to start small and move up to a larger spaces, as your needs change.

CUSTOMER SERVICE

To deliver high-quality customer service you can either manage those activities in-house or outsource to a third party. To provide basic customer service for sales and post-sales activities, you can use email and a toll-free number for more extensive phone support. Larger ecommerce businesses can use a customer-management system to make those activities easier.

INFORMATION TECHNOLOGY

For an ecommerce business, selection of the right ecommerce platform is one of the most important decisions. You can build and host your own system, outsource the development and then manage the system going forward, or use a hosted, software-as-a-service (SaaS) platform that is more turnkey and externally managed. To build and host your own system, you may need more financial and in-house human resources upfront. By using a SaaS platform, you will save effort to host or manage the system in-house, but you may still need in-house web developers. If you outsource the development and hosting, your staffing costs will reduce but you will incur higher costs for any future enhancements or changes to your websites. Ecommerce businesses need careful analysis of each option and gauge possible impacts of each on both their staffing, cash flow, and bottom line before they move forward.

FULFILLMENT

E-commerce businesses also need to decide whether they will manage their own inventory or outsource those activities to a fulfillment house or through drop shipping arrangements with their suppliers. Businesses have a high degree of control if they manage their inventory themselves but it requires significant investment in inventory, warehouse space, and own fulfillment staff. For some businesses, such as the ones dealing in precious stones, managing your own inventory would be the most logical choice because most items are purchased in bulk and are very small. These businesses cannot trust preparation and fulfillment to an outside service. Before moving forward, E-commerce businesses must understand the costs involved and analyze the other options.

FINANCE AND ADMINISTRATION

Similar to other infrastructure elements, you will need to decide if you want to manage your finance and administration activities in-house, outsource, or a hybrid of the two. With a tightly integrated ecommerce platform with the accounting system, little or no in-house bookkeeping staff would be needed. If you wish to focus only on sales, marketing, and customer service, you should consider outsourcing payments, payroll, and other basic accounting activities.

HUMAN RESOURCES (HR)

For many ecommerce businesses, recruitment, compensation, compliance and other HR activities are specialized and time consuming. Therefore, it is a better option to outsource those activities. Businesses should keep their target market and value proposition in mind while making decision about any infrastructure element of their ecommerce business. All elements of their ecommerce infrastructure should support their value proposition. Businesses should be cautious and must optimize their financial and human resources across their ecommerce infrastructure.

E-commerce and Business Processes

Using their computers, clients, partners, and suppliers can connect to the E-commerce business using a variety of Internet connection methods ranging from slow dial-up connections to faster broadband connections. The Internet connections can be established through an Internet Service Provider (ISP) or directly using a dedicated line. Customers first identify their needs and then try to find product/service information that can fulfill these needs. Corporate website is the primary source, used by an E-commerce firm, to provide information about their product/service offerings. There are also a variety of sources that can be used e.g. Newsgroup, Net communities, chat rooms etc. Customers evaluate the offerings and then make a purchase decision. Finally, they make a purchase using a variety of payment methods. Figure 2-2 describes the business processes involved in E-commerce.

FIGURE 2-2: E-COMMERCE AND BUSINESS PROCESSES

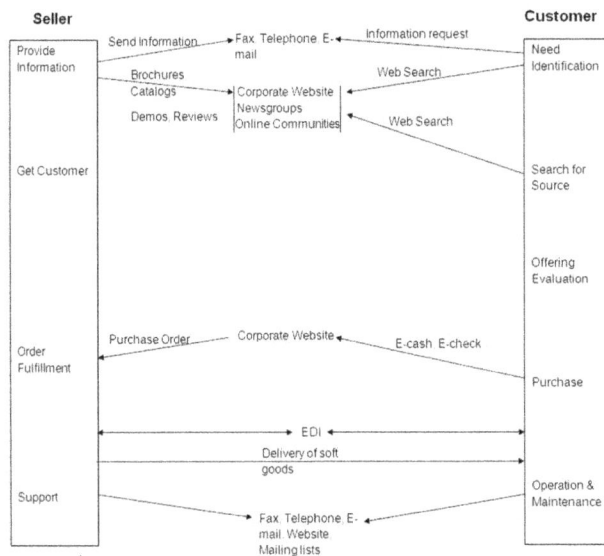

Working of a Typical E-commerce System

Following is a simplified description of how a sophisticated e-commerce system might work. Not all e-commerce systems work in exactly this way.

FIGURE 2-3: A TYPICAL E-COMMERCE SYSTEM

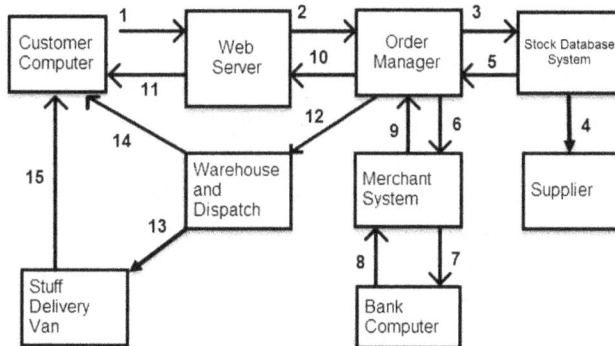

1. Sitting at a **computer**, a customer tries to place an online order to purchase a book. The client's Web browser communicates back-and-forth over the Internet with a Web server that manages the store's website.
2. The **Web server** sends the client's order to the order manager. This is a central computer responsible to manage all orders through every stage of processing from submission to dispatch.
3. The **order manager** queries a product database to find out whether the product (the book) the customer wants is actually in stock.
4. If the item is not in stock, the **Stock Database system** can order new supplies from the suppliers. This might involve real-time communication with supplier order systems to find out estimated supply times while the customer is still sitting at his computer.
5. The stock database confirms whether the item is in stock or suggests an estimated delivery date when product will be received from the supplier.
6. Assuming the item is in stock, the order manager continues to process it. Next, it communicates with a **merchant system** to take payment using the customer's credit or debit card number. This merchant system is run by a credit-card processing firm or linked to a bank.
7. The merchant system might make extra checks with the customer's own bank computer.
8. The **bank computer** confirms whether the customer has enough funds.
9. The merchant system authorizes the transaction to go ahead, though funds will not be completely transferred until several days later.
10. The order manager confirms the successful transaction processing and notifies the Web server.
11. The Web server shows the customer a Web page confirming that his order has been processed and the transaction is complete.
12. The order manager sends a request to the warehouse to dispatch the goods to the customer.
13. The delivery truck collects the goods from the warehouse and delivers them to the customer.

14. Once the goods have been dispatched, the warehouse computer e-mails the customer to confirm that her goods are on their way.
15. The goods are delivered to the customer

This whole process is invisible to the customer. All the customer sees is the message that transaction has been successfully completed and the delivery truck that arrives at his door with goods that he ordered.

Major Players in E-commerce

A conducive E-commerce environment requires the E-commerce infrastructure with many players and prerequisites. A weak or less developed player is a barrier to the increased uptake of E-commerce as a whole. For instance, a country with an excellent Internet infrastructure will not achieve high E-commerce trade volumes if banks do not offer support and fulfillment services to e-commerce transactions. A typical e-commerce transaction involves the following major players.

SELLER

A seller requires:

- A corporate Web site with e-commerce capabilities (e.g., a secure transaction server)

- A corporate intranet so that orders are processed in an efficient manner

- IT-literate employees to manage the information flows and the E-commerce system.

TRANSACTION PARTNERS

Transaction partners can be banking institutions that offer transaction-clearing services. Another partner is national and international freight companies that enable the movement of physical goods within, around and out of the country. For B2C transactions, the system must offer a means for cost-efficient transport of small packages (e.g. books). Transaction partners also include authentication authority that serves as a trusted third party to ensure the integrity and security of transactions.

CONSUMERS

Consumers in E-commerce refer to a critical mass of the population with access to the Internet and disposable income enabling widespread use of credit cards. They possess a mindset for purchasing goods over the Internet rather than by physically inspecting items.

FIRMS AND BUSINESSES

Firms and businesses together form a critical mass of companies with Internet access and the capability to place and take orders over the Internet.

GOVERNMENT

Governments establish legal frameworks governing e-commerce transactions (including electronic documents, signatures etc.) and provide legal institutions that would enforce the legal framework and provide businesses and general consumer protection from fraud.

Networks and Internet

The successful use of Internet depends on a robust and reliable Internet infrastructure and a pricing structure that does not penalize consumers for spending time on and buying goods over the Internet. Businesses also need robust intranets and extranets to facilitate collaboration among employees and other firms.

The Internet

Internet is a global collection of networks, both big and small, which can connect together in many different ways. Nobody owns the Internet however; it is monitored and maintained in different ways. The Internet Society (www.isoc.org) is a not-for-profit organization that administers the formation of the policies and protocols that relates to use and interaction with Internet. The internet has the potential to transform an ordinary brick and mortar business model or strategy into a competitive and eventually a successful online business. Some of the reasons for this are:

LOW COST: Online businesses can be setup and operated at comparatively low costs. The marginal cost of conducting an online transaction is almost zero.

TIME: Online businesses can provide readily accessible information about the products and services offered by them. This can save a lot of time for customers and the business.

INTERACTIVITY: An online business can provide their customers with increased interactivity. The users can get information from the website, post reviews, customize their requirements, and use features such as order tracking to see where their order is at a particular stage.

UNIVERSAL REACH: Internet removes geographic constraints. A customer anywhere in the world can order a product or service from an online business.

THE STRUCTURE OF THE INTERNET

The Internet follows a hierarchal structure that allows any Internet connected device talk to another Internet connected device irrespective of the geographic locations where the devices are located. The way that the information is transmitted varies greatly. Open Systems Interconnection (OSI) is an effort to standardize computer networking and communication. The OSI model defines a networking framework to implement protocols in seven layers. Control is passed from one layer to the next, starting at the application layer in one station, and proceeding to the bottom layer, over the channel to the next station and back up the hierarchy. The main concept of OSI is that the process of communication between two endpoints in a telecommunication network can be divided into seven distinct groups of related functions, or layers. Each communicating user or program is at a computer that can provide those seven layers of function. So in a given message between users, there will be a flow of data down through the layers in the source computer, across the network and then up through the layers in the receiving computer. The seven layers of function are provided by a combination of applications, operating systems, and networking hardware that enable a system to put a signal on a network cable or out over Wi-Fi or other wireless protocol).

FIGURE 2-4: OPEN SYSTEMS INTERCONNECTION (OSI) MODEL

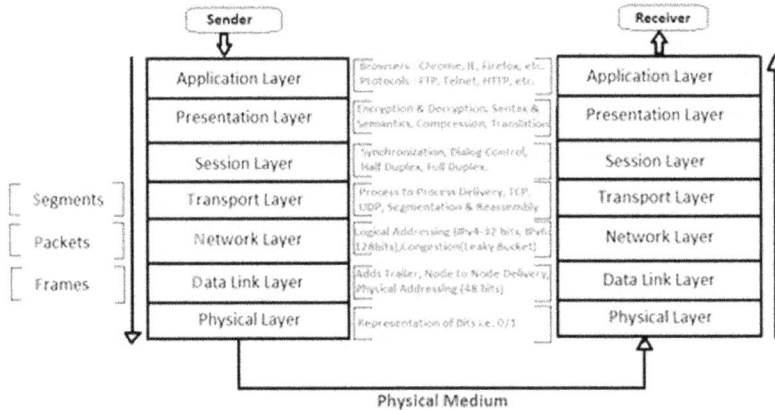

Internet Tools and Services

USENET

Usenet is a worldwide discussion system distributed mainly (but not exclusively) over the Internet (its formation actually predates the Internet). It consists of thousands of publicly accessible so-called newsgroups each of which deals with a specific topic. Everyone can post messages (or news articles, as they are called) to any of these groups and read what others have posted. Everyone even can - observing certain rules - create a new newsgroup dealing with a new topic. Usenet is a service on the Internet, as are Email, FTP or the WWW. However, it has a different communication structure. Usenet is public and egalitarian. There is no asymmetry between vendors and consumers. Every participant in Usenet is both reader and contributor. In this respect, Usenet, more than the other services, is what the Internet is really about (Ritual.org, 2015).

FTP

FTP is an acronym for File Transfer Protocol. FTP is used to transfer files between computers on a network. You can use FTP to exchange files between computer accounts, transfer files between an account and a desktop computer, or access online software archives. Graphical FTP clients simplify file transfers by allowing you to drag and drop file icons between windows. When you open the program, enter the name of the FTP host (e.g., ftp.microsoft.com) and your username and password. If you are logging into an anonymous FTP server, you may not have to enter anything. Two common FTP programs are Cyberduck (for Mac) and WinSCP (for Windows). You can use a web browser to connect to FTP addresses exactly as you would to connect to HTTP addresses. Using a web browser for FTP transfers makes it easy for you to browse large directories and read and retrieve files (Indiana University, 2015).

MAILING LIST

A mailing list is a collection of names and addresses used by an individual or an organization to send material to multiple recipients. The term is often extended to include the people subscribed to such a list, so the group of subscribers is referred to as "the mailing list", or simply "the list" (Vishanth, 2012).

GOPHER

Gopher is a system that pre-dates the World Wide Web for organizing and displaying files on Internet servers. A Gopher server presents its contents as a hierarchically structured list of files. With the ascendance of the Web, many gopher databases were converted to Web sites, which can be more easily accessed via Web search engines (Chick, 2011).

EMAIL

Electronic mail, most commonly referred to as email or e-mail, is a method of exchanging digital messages from an author to one or more recipients. Modern email operates across the Internet or other computer networks. Today's email systems are based on a store-and-forward model. Email servers accept, forward, deliver, and store messages. Neither the users nor their computers are required to be online simultaneously. An email message consists of three components, the message envelope, the message header, and the message body. The message header contains control information, including, minimally, an originator's email address and one or more recipient addresses. Usually descriptive information is also added, such as a subject header field and a message submission date/time stamp (Geocaching, 2015).

ARCHIE

Archie was an early search program that indexed files on anonymous File Transfer Protocol (FTP) servers and allowed users to search for specific files. Archie was created by a team of students at McGill University in 1990. Archie helped users find files in the directories of public hosts that they might never have otherwise discovered. Archie ultimately became less important with the growth of the World Wide Web. It was perhaps of most use for serious researchers who already know that the topic of their search is likely to be found on FTP servers (Kent & Williams, 2002).

VERONICA

Veronica is a program similar to Archie that indexed and searched the files on Gopher servers. Veronica is a program that allowed you to search the files of the Internet's Gopher servers for a particular search string. Veronica is an indexing spider that visits the Gopher sites, reads the entire directory and files names, and then indexes them in one large index. However, with the almost complete demise of Gopher servers (most content has probably been put on the Web), Veronica has become a relic of the early 1990s (S.K.Bansal, 2009).

WAIS

WAIS (Wide Area Information Servers) is an Internet system in which specialized subject databases are created at multiple server locations, kept track of by a directory of servers at one location, and made accessible for searching by users with WAIS client programs. The use of WAIS requires a list of distributed databases. The user enters a search argument for a selected database and the client then accesses all the servers on which the database is distributed. The results provide a description of each text that meets the search requirements. The user can then retrieve the full text (JOSEPH, 2011).

FINGER

Finger is a program that tells you the name associated with an e-mail address. It may also tell you whether they are currently logged in at their system or their most recent logon session and possibly other information, depending on the data that is maintained about users on that computer. To finger another Internet user, you need to have the finger program on your computer or you can go to a finger gateway on the Web and enter the e-mail address. The server at the other end must be set up to handle finger requests (Rouse, 2015).

TELNET

Telnet is a user command and an underlying TCP/IP protocol for accessing remote computers. Through Telnet, an administrator or another user can access someone else's computer remotely. On the Web, HTTP and FTP protocols allow you to request specific files from remote computers, but not to actually be logged on as a user of that computer. With Telnet, you log on as a regular user with whatever privileges you may have been granted to the specific application and data on that computer. Telnet is most likely to be used by program developers and anyone who has a need to use specific applications or data located at a particular host computer (BHATNAGAR, 2008).

WWW

The World Wide Web is a system of Internet servers that support specially formatted documents. The documents are formatted in a markup language called HTML (Hypertext Markup Language) that supports links to other documents, as well as graphics, audio, and video files. This means you can jump from one document to another simply by clicking on hot spots. Not all Internet servers are part of the World Wide Web (Neogy, 2009).

SEARCH ENGINE

Search engines are programs that search documents for specified keywords and return a list of the documents where the keywords were found. A search engine is really a general class of programs; however, the term is often used for specific description of systems like Google, Bing and Yahoo! Search that enable users to search for documents on the World Wide Web. Typically, Web search engines work by sending out a spider to fetch as many documents as possible. Another program, called an indexer, then reads these documents and creates an index based on the words contained in each document. Each search engine uses a proprietary algorithm to create its indices such that, ideally, only meaningful results are returned for each query (Kent & Williams, 2002).

Switching

A switch on a telecommunication network refers to a device that receives data from an input port and transfers it to an output port that would take the data to its intended destination. A switch determines the destination port of the data by either looking at physical device address (Media Access Control or MAC address) in each incoming message frame or at the IP address in each packet. The trip from one switch to another in the network is called a hop. The time a switch takes to figure out where to forward a data packet is called its latency. Switches can be either desktop switches or core switches. Core switches connects networks with each other while desktop switches are used to connect workstations, which are close to each other.

CIRCUIT SWITCHING VS. PACKET SWITCHING

In circuit switching, a network's paths can be dedicated for certain duration for its exclusive use by two or more parties. The particular route a data unit travels does not change. Circuit switching is used by ordinary voice phone call systems. In packet switching, a network's paths are shared by many parties at the same time and the particular route a data unit travels can be varied as conditions change. The travelling message is divided into packets and packets in the same message may travel different routes and may not arrive in the same order that they were sent. When these packets reach at their destination, they are reassembled to create the original message.

Routing

In telecommunication network, routing refers to process of moving information across an inter-network from a source address to a destination address. The routing algorithm is the software part of the router responsible for choosing a path over which to send the packets. Routing algorithm uses various metrics to evaluate what path will be the best for a packet to travel. A metric is a standard of measurement; such as path bandwidth, reliability, delay, current load on that path etc. Routing algorithm initializes and maintains a routing table containing route information to aid the process of path determination. Routing algorithm decides the route taken by a particular packet during its travel.

TCP/IP

TCP/IP (Transmission Control Protocol/Internet Protocol) is the basic protocol of the Internet that can also be used as a communications protocol in a private network (either an intranet or an extranet). TCP/IP uses the client/server model of communication in which a client requests a service (such as sending a Web page) by a server in the network.

Working of Internet

The Internet is a combination various network. Every computer connected to Internet is part of a network. Users can use a variety of connection methods to connect to the Internet e.g. dial-up modem or through a LAN via broadband DSL. In either case, users connect to an Internet Service Provider (ISP) to become part of ISP network. The ISP network then connects to a larger network. Most large ISPs have their own dedicated lines connecting various regions. In each region, the ISP has a Point of Presence (POP), which is the place for local users to access the company's network. Users often access POP through a local phone number or dedicated line. The high-level networks connect to each other through Network Access Points or NAPs. A NAP is one of several major Internet interconnection points in a country that connect all the ISPs together. Very large amounts of data flow between the individual networks at NAPs. The Internet backbone refers to the principal data routes between large interconnected networks. Internet backbones are typically high capacity trunk lines that consist of multiple fiber optic lines. Fiber optic cables are designated OC for optical carrier e.g. and OC-3 line can transmit 155 Mbps while an OC-48 can transmit 2,488 Mbps (2.488 Gbps). Modern optical carriers such as OC-192, OC-256, and OC-7868 are capable of data transmission speed of 10 Gbps, 13.271 Gbps, and 40 Gbps respectively. Internet backbones are hosted by commercial firms, governments, academic institutions and other high-capacity network centers. Internet backbones interchange Internet traffic between the countries and continents. ISPs participating in the Internet backbone, exchange traffic by privately negotiated interconnection agreements. Figure 2-3 shows how POP, NAP, and Internet connects together.

FIGURE 2-5: POP, NAP, AND INTERNET BACKBONE

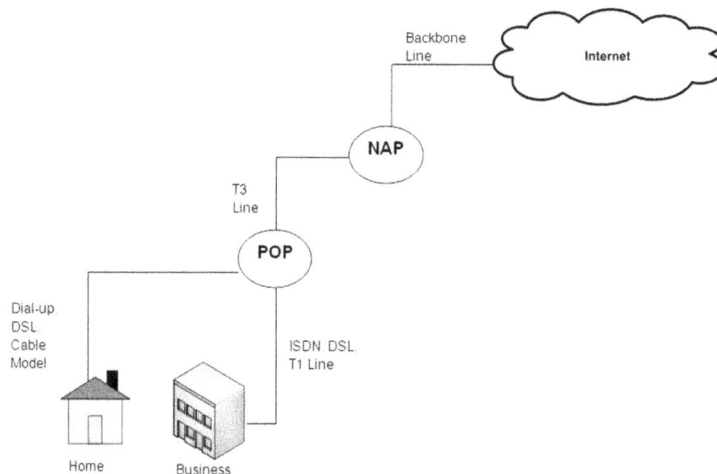

FUNCTION OF ROUTER

A router is a specialized computer that joins two networks, passing information from one to the other, and preventing the traffic on one from unnecessarily spilling over to the other. The basic operation and function of the router remains the same regardless of how many networks are attached.

All the networks on Internet rely on NAPs, backbones and routers to talk to each other. On the Internet, routers determine where to send information from one computer to another. A message can leave one computer, travel to the other part of the world through several different networks, and arrive at another computer in a fraction in no time. Given the huge size of the Internet use of routers is an absolute necessity to make sure that information does make it to the intended destination.

PROTOCOL

In telecommunication networks, a protocol refers to a special set of rules that is used by the end-points when they communicate. These rules define formatting, ordering, and error checking of the data that is sent during the communication. For example, protocols determine how the sending device indicates that it has finished sending a message, and how the receiving device indicates that it has received (or not received) the message. A protocol also includes rules about what is allowed in a transmission and how it is formatting. Computers that communicate with each other must use the same protocol for data transmission.

IP ADDRESSES

IP address is a unique identification number assigned to every computer on the Internet. The IP (Internet Protocol) is a very popular network protocol used by computers to communicate over the Internet. A typical IP address is expressed in decimal format as a dotted decimal number like this:

154.31.16.13

This format is easier to understand by humans but computers communicate in binary form. If we convert the above IP address in binary form it will look like this:

10011010.00011111.00010000.00001101

Each number in an IP address is called an octet (each has eight positions in binary form). IP addresses are called 32-bit because the sum of all positions is 32. Since each of these eight positions can take one of two possible states (zero or 1), the total number of possible combinations per octet is 2^8 or 256. By combining the four octets, we can get a possible 4,294,967,296 unique values. Out of these values, certain values are restricted from use as typical IP addresses. Take for example, the IP address of 0.0.0.0 (which is reserved for the default network) and the address 255.255.255.255 (which is used for broadcasts).

The octets are also used to create classes of IP addresses that can be assigned to a particular business, government or other entity based on size and need. The octets are split into two sections: Network and Host. The Network section always contains the first octet and identifies the network to which the computer belongs. Host identifies the actual computer on the network. Last octet always contains the host section. There exist five classes of IP addresses.

Each computer connected to the network needs to have a unique IP address. When two computers have the same IP address, both will not be able to connect to the network and access network resources or perform other network operations. To avoid IP address conflict, the public registration of IP addresses is administered by Network Information Centre (NIC). NIC also manages the registration of domain names within the top-level domains for which it is responsible.

IP addresses can also be divided into two general categories: Public and Private. A public network is a network, which is configured with public IP addresses, and the devices on the network are visible to devices outside the network (from the Internet or another network). In a private network, computers are configured with private IP address and the devices outside the network cannot see or communicate directly to them. A private network can be one or a combination of two networks joined using a dedicated connection. Private networks are commonly used at home, office, and enterprise local area networks (LANs) where public IP addresses are either not needed or not available. NIC has reserved the private IP addresses in the following ranges:

From 10.0.0.0 to 10.255.255.255

From 172.16.0.0 to 172.31.255.255

From 192.168.0.0 to 192.168.255.255

IPV6 is version 6 of the Internet Protocol. Based on RFC 2373, version 6 of the Internet Protocol has a vastly increased address space over the IPV4 version. IPV6 allows for 128 bit addressing, or 2 to the power of 128, or 3.40282e+38 devices on the network. IPV6 notation is slightly different from IPV4 in that hex numbers (base 16) are used rather than decimal (base 10) numbers. An example of IPV6 is shown in Figure 2-6.

FIGURE 2-6: IPV6 ADDRESS

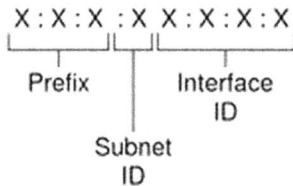

X : X : X : X X : X : X : X

Prefix

Interface ID

Subnet ID

Example:

2001:0db8:3c4d:0015:0000:0000:1a2f:1a2b

Site Prefix

Subnet ID

Interface ID

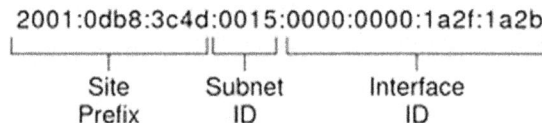

DOMAIN NAME SYSTEMS (DNS)

A computer on Internet can connect to another computer by providing the IP address of the computer. This arrangement works fine if there were only a few computers but it will not be feasible if there are millions of computers online. The first solution to the problem was a simple text file maintained by the Network Information Center (NIC) that mapped computer names to IP addresses. Soon this text file became so large and too cumbersome to manage. Domain Name System (DNS) was created by the University of Wisconsin and it maps text names to IP addresses. This way you only need to remember the domain name (e.g. www.cnn.com) instead of its IP address.

DNS servers are computers that convert domain names into IP addresses. When a request to convert domain name into IP address comes in, the DNS server can do one of four things with it:

1. It can answer the request with the IP address of requested domain.

2. It can contact other DNS server(s) and try to find the IP address for the domain name requested.

3. It can provide the IP address of another DNS server whom it thinks might have the IP address for the domain name requested.

4. It can return an error message if the requested domain name did not exist or was invalid.

A DNS EXAMPLE

Assume that you type www.cnn.com into your web browser. The browser will contact a DNS server to get the IP address of cnn.com. A DNS server would start its search by contacting one of the root DNS servers and asking for IP address of www.cnn.com. The root servers keep the IP addresses for all of the DNS servers that handle the top-level domains (.COM, .NET, .ORG, etc.). The root server would respond by providing the IP address of another DNS server e.g. .COM DNS server. Your DNS server will then send a query to the .COM DNS server to ask for the IP address for www.cnn.com. The .COM DNS server knows the IP addresses for the name server handling the www.cnn.com , so it will return the IP address of this server let's call it CNN DNS. Your DNS server will then contact the CNN DNS and ask if it knows the IP address for www.cnn.com. It actually does, so it will return the IP address to your DNS server, which returns it to the browser, which can then contact the server for www.cnn.com to retrieve the Web page.

One key aspect of this process is redundancy. Multiple DNS servers are provided at each level so that if one server fails there is always another server that could handle the requests. The other key aspect is caching. Once a DNS server resolves a request, it caches the IP address it receives so it does not have to bug the root DNS servers again for that information. Even though operation of DNS servers is invisible, DNS servers handle billions of requests every day and they are essential to the smooth functioning of Internet.

URL: UNIFORM RESOURCE LOCATOR

A Uniform Resource Locator (URL) defines the unique address of a file or other resource on the network and the mechanism to access it. A typical URL consists of some of the following: the protocol name followed by a colon, then, depending on protocol, a hostname or IP address, a port number, the path of the resource to be fetched or the program to be run, then, for programs such as Common Gateway Interface (CGI) scripts, a query string, and with HTML documents, an anchor (optional) for where the page should start to be displayed.

The combined syntax is: scheme://username:password@domain:port/path?query_string#anchor

For example, if we type the following URL at the address bar of Internet browser, the browser will perform an HTTP request to the host example.org, at the port number 80.

http://example.org:80

On Internet, the Top-Level-Domain (TLD) is a system where .COM, .NET, .ORG or other suffix at the end of a domain determines what type of entity is using that name for its website. In the beginning of public use of the internet, A TLD classification was created to organize information and users into categories appropriate for their distinct needs.

A country TLD or country code top-level domain (ccTLD) is a TLD generally used or reserved for a country (a sovereign state or a dependent territory). All ccTLD identifiers are two letters long and are created and delegated by Internet Assigned Numbers Authority (IANA). Table 2-1 shows the list of various TLDs.

TABLE 2-1: GENERAL AND COUNTRY TLDS

General TLDs		Country TLDs	
TLD	Used By/For	TLD	Use
.com	Commercial Firms	.au	Australia
.org	Non-profit Organizations	.in	India
.net	General Purpose	.pk	Pakistan
.edu	Educational Institutes	.ca	Canada
.gov	US Government	.de	Germany
.mil	US Military	.fr	France
.name	Individual persons	.jp	Japan
.biz	Businesses	.se	Sweden
.info	General purpose	.nz	New Zeeland
.pro	Professionals	.fi	Finland
.int	Treaty Organizations		

A list of all valid top-level domains is maintained by the IANA and is updated from time to time. This list is available at http://data.iana.org/TLD/tlds-alpha-by-domain.txt.

Internet Client-Server Model

The client–server model is a distributed application structure that partitions tasks or workloads between the providers of a resource or service, called servers, and service requesters, called clients. All computers on Internet are either servers or clients. The computers that provide services to other computers are servers and computers that utilize these services are clients. There can be many types of servers e.g. Web server, E-mail server, FTP server etc. Often clients and servers communicate over a computer network on separate hardware, but both client and server may reside in the same system. A server host runs one or more server programs, which share their resources with clients. A client does not share any of its resources, but requests a server's content or service function. Clients therefore initiate communication sessions with servers, which await incoming requests. Suppose you connect to www.cnn.com using your computer. Your computer is a client accessing the cnn.com web server. The server finds the page you requested and sends it to you. Clients direct their requests to specific software running on the server computer. For example, if you are running a Web browser on your computer, it will want to talk to the Web server on the server machine, not the e-mail server. A server has a static IP address that does not change very often while user's computer at home typically has an IP address assigned by the ISP that changes every time user connects to the ISP.

Examples of computer applications that use the client–server model are Email, network printing, and the World Wide Web. Servers are classified by the services they provide. For instance, a web server serves web pages and a file server serves computer files. The sharing of resources of a server constitutes a *service*. Whether a computer is a client, a server, or both, is determined by the nature of the application that requires the service functions. For example, a single computer can run web server and file server software at the same time to serve different data to clients making different kinds of requests. Client software can also communicate with server software within the same computer. Clients and servers exchange messages in a request-response messaging pattern: The client sends a request, and the server returns a response. The language and rules of communication are defined in a communications protocol. A server may receive requests from many different clients in a very short period. To prevent abuse and maximize uptime, the server's software limits how a client can use the server's resources.

Ports and HTTP

Any server computer makes its services available using numbered ports. One port number is reserved for each service that is available on the server. For example, web server is typically available at port 80 and FTP servers at port 21. When clients connect to a service provided by a server they connect to a specific IP address on a specific port. Once connected, the clients can then access the service using a specific protocol.

E-mail Protocols

Most organizations use a client/server structure to handle e-mail. They have a computer called an e-mail server that is devoted to handling e-mail. That computer has software to store and forward e-mail messages. People in the organization might use a variety of programs, called e-mail client software, to read and send e-mail. These programs include Microsoft Outlook, Mozilla Thunderbird, Netscape Messenger, Pegasus Mail, Qualcomm Eudora, and many others. The e-mail client software communicates with the e-mail server software on the e-mail server computer to send and receive e-mail messages. An increasing number of people use web-based e-mail services e.g. Yahoo! Mail or Hotmail. In web-based e-mail service, the e-mail servers and the e-mail clients are operated by the owners of the Web sites and individual users only see the e-mail client software (and not the e-mail server software) in their Web browsers when they log on to the Web mail service. Standardization and rules become very important when there are so many different e-mail client and server software choices available. If e-mail messages did not follow standard rules, an e-mail message created by a person using one e-mail client program could not be read by a person using a different e-mail client program.

Email is based around the use of electronic mailboxes. When an email is sent, the message is routed from server to server, all the way to the recipient's email server. More precisely, the message is sent to the mail server tasked with transporting emails (called the MTA, for Mail Transport Agent) to the recipient's MTA. On the Internet, MTAs communicate with one another using the protocol SMTP, and so are logically called SMTP servers (or sometimes outgoing mail servers).The recipient's MTA then delivers the email to the incoming mail server (called the MDA, for Mail Delivery Agent), which stores the email as it waits for the user to accept it. An email message consists of two or three parts. All email messages have headers and a body, and some messages may have attachments. The headers specify certain standard information about the message, such as the Subject (what the message is about), the Date that the message was composed, and the address of message receiver. Some of the headers contain technical information, such as what character set the message is in (Latin, Greek, Hebrew, and Unicode), how the attachments are encoded, or how and when the mail servers passed it along to its destination. The body is the text of the message itself. The attachments, if there are any, are additional computer files that are being sent along with the message. These can be anything – pictures, word processing documents, computer software, or anything else that can be stored in a computer file. However, the recipient can only usefully open the attachments if he has software installed for opening that kind of file. Some file types, such as PNG images, are standard enough that almost everyone can open them.

SMTP and POP

Simple Mail Transfer Protocol (SMTP) and Post Office Protocol (POP) are two common protocols used for sending and retrieving e-mail. SMTP specifies the format of a mail message and describes how mail is to be administered on the e-mail server and transmitted on the Internet. The SMTP protocol is used by the Mail Transfer Agent (MTA) to deliver your email to the recipient's mail server. The SMTP protocol can only be used to send emails, not to receive them. Depending on your network / ISP settings, you may only be able to use the SMTP protocol under certain conditions. Outgoing mail for both POP and IMAP clients uses the Simple Mail Transfer Protocol (SMTP). An e-mail client program running on a user's computer can request mail from the organization's e-mail server using the POP. A POP message can tell the e-mail server to send mail to the user's computer and delete it from the e-mail server; send mail to the user's computer and not delete it; or simply ask whether new mail has arrived.

Interactive Mail Access Protocol (IMAP)

Interactive Mail Access Protocol (IMAP) performs the same basic functions as POP but can also perform additional features. For example, IMAP can instruct the e-mail server to send only selected e-mail messages to the client instead of all messages. IMAP can also allow the user to view only the header and the e-mail sender's name before deciding to download the entire message. IMAP can let users create and manipulate mail folders, delete messages, and search parts of a message. IMAP lets users manipulate and store their e-mail on the e-mail server and access it from any number of computers. The main drawback to IMAP is that users' e-mail messages are stored on the e-mail server so with increasing number of users the size of the e-mail server's disk drives must also increase. The IMAP has also been criticized for being insufficiently strict and allowing behaviors that effectively negate its usefulness. For instance, the specification states that each message stored on the server has a "unique id" to allow the clients to identify the messages they have already seen between sessions. However, the specification also allows these UIDs to be invalidated with no restrictions, practically defeating their purpose.

When using POP, clients typically connect to the e-mail server briefly, only as long as it takes to download new messages. When using IMAP4, clients often stay connected as long as the user interface is active and download message content on demand. For users with many or large messages, this IMAP4 usage pattern can result in faster response times.

Multiple clients simultaneously connected to the same mailbox. The POP protocol requires the currently connected client to be the only client connected to the mailbox. In contrast, the IMAP protocol specifically allows simultaneous access by multiple clients and provides mechanisms for clients to detect changes made to the mailbox by other, concurrently connected, clients.

Messaging Application Program Interface (MAPI)

MAPI is a Microsoft Windows program interface that allows you to send e-mail from within a Windows application and attach the document you are working on to the e-mail note. Applications that make use of MAPI include word processors, spreadsheets, and graphics applications. MAPI-compatible applications typically include a Send Mail or Send in the File pull-down menu of the application. Choosing one of these sends a request to a MAPI server.

Internet Service Provider (ISP)

An ISP provides service of Internet. They can provide this service free or may charge a fee. Each ISP has a service area, offers certain types of connections, and provides customers with other benefits (such as email, security against viruses and other harmful programs, webpage space etc.). Finding the ISP that fits your needs and is available in your area can be a daunting task. A Virtual ISP (VISP) is an operation, which purchases services from another ISP, which allows the VISP's customers to access the Internet using services and infrastructure owned and operated by the wholesale ISP. Free ISPs provide service free of charge. Many free ISPs display advertisements while the user is connected; like commercial television, in a sense they are selling the users' attention to the advertiser. Other free ISPs, often called free nets, are run on a nonprofit basis, usually with volunteer staff.

Selecting an ISP

Ideal ISP offers a number of different connections, including dial-up, DSL, cable and satellite, and they offer these services to a large number of people. The ISP should provide ample protection to customers using firewalls, parental controls and virus and spyware protection. It is necessary that ISP provide convenient and helpful customer service.

Below are some criteria that can be used to evaluate ISPs.

FEATURES OF ISP: ISP should include certain elements as a basic part of their service, including virus and spyware protection, firewalls, parental controls, email accounts, webpage space and more.

SERVICE AREA: ISP should offer service to a large region since larger ISPs could usually offer more for less. The ISP website should supply tools that show if their service is available in your area.

CONNECTION SPEEDS: ISP that offers a choice of bandwidths allows you more flexibility when choosing an internet service plan.

CUSTOMER SERVICE/SUPPORT: The ISP should have excellent customer service, including a FAQs page, phone support, a support email address, and live chat. The ISP should respond to customer questions promptly.

World Wide Web (WWW)

World Wide Web, as defined by the World Wide Web consortium, is a universe of network-accessible information. It is the collection of users and resources on the Internet that use Hypertext Transfer Protocol (HTTP). To access the Web you need a web browser, such as Microsoft Internet Explorer, Mozilla Firefox, or Google Chrome.

Internet and the World Wide Web is not the same thing, although both terms are often used synonymously. Internet is an infrastructure providing interconnectivity between network computers while the web is one of the services of the Internet. Web is a collection of documents that can be shared across Internet using hyperlinks. Hyperlinks provide an efficient cross-referencing system and create a non-linear form of text. Web servers form the backbone of the World Wide Web and Hyper Text Transfer Protocol (HTTP) is used to access the web. Web browser can distinguish between web pages and other types of data on the Internet because web pages are written in a computer language called Hypertext Markup Language (HTML). When web browser makes a request for a particular web page to the web server, web server responds with the requested web page and its contents. Web browser then displays the web page as rendered by HTML or other web languages used by the page. Each resource on the web has a unique address and can be accessed with the help of a web browser.

Hyperlinks, global access to the content and universal readership are some of the prominent features of the World Wide Web. The information on the web is available all the time across the globe. While some websites require user login to access, most websites are open to all users. Using HTML as a common format for rendering web content and using HTTP as a common access method, the web has achieved universal readership.

Sometimes in a URL, the "WWW" is followed by a number, such as "WWW1" or "WWW2." The number that follows the "WWW" indicates that the data being retrieved by the Web browser is gathering the information from a different Web server than the one that serves the typical "WWW" address. Web sites, especially dynamic Web sites that handle large amounts of traffic, often need more than one server to accommodate the many requests they receive as one server often cannot handle the multitude of requests. The numbers that follow the "WWW" refer to different Web servers, often as elements of a server farm, that all contain the exact same information. The servers are used in coordination with each other for load balancing. An example of this system is www.google.com, which uses multiple servers to handle all its traffic. Sometimes the user's physical location determines which server receives the routed requests, and sometimes the different servers are used when one or more of the servers are taken offline for information update.

Markup Languages and the Web

Web pages can include many elements (e.g. graphics, photographs, sound clips, and small programs that run in the Web browser). Each element is stored on the Web server as a separate file. However, the most important parts of a Web page are page structure and the text that makes up the main part of the page. The page structure and text are stored in a text file that is formatted, or marked up, using a text markup language. Markup languages are designed for the processing, definition and presentation of text using language-specific codes. The code used to specify the formatting is called tags. To understand how a markup language works, consider the following example.

Let's say you have a friend and you want him to give you his family details. You asked him to write Name, Age, Date of birth and Qualification of his father, mother and sister. You would be unsure whether he write down all the details in a paragraph and send it to us or not. In other words, what is the probability that he will write the

> Name of my father is Mr. A, he is 55 years old. He was born on 24th Feb 1955 and he told me that he is a music graduate from XYZ University. My mother's name is B and she is ……….. My sister's name is C and she is 25 years old. She was born on ……….

information in this form:

We could safely assume that most people would provide information in the format shown in Figure 2-7.

One reason most people would provide information in this format is that it is easy to understand and is more structured and well formed. The same concept is the base for the development and use of markup languages. Markup languages help us present and store information in a structured and well-formed manner. So that we can process the information in a markup language based on some specific rules.

If we write this information using a markup language, it can be written in the following manner:

<Father>

<Name>Mr. A</Name>

<Age>55 years</Age>

<DOB>24 Feb 1955 </DOB>

<Qualification>Graduate in music</Qualification>

</Father>

<Mother>

<Name>Mrs. B</Name>

<Age>50 years</Age>

<DOB>2 Feb 1960 </DOB>

Father:

Name: Mr. A

Age: 55 years

DOB: 24 Feb 1955

Qualification: Graduate in music.

Mother:

Name: Mrs. B

Age: 50 years

DOB: 2 Feb 1960

Qualification: Graduate in Economics

Sister:

Name: C

Age: 25 years

DOB: 4 Feb 1985

Qualification: Medical student

FIGURE 2-7: FORMATTED TEXT

<Qualification>Graduate in Economics</Qualification>

</Mother>

<Sister>

<Name>C</Name>

<Age>25 years</Age>

<DOB>4 Feb 1985 </DOB>

<Qualification>Medical student</Qualification>

</Sister>

This way of providing information is known as a markup. If we understand how to read information present in the father block, we can easily read the whole document. This is because the same set of rules will apply for reading information in the mother block also and in the sister block. In technical terms <Father></Father> is a node. <Father> is the starting of this node and </Father> is the ending of this node (It has a /, forward slash, in front of the word Father). Anything that lies inside the <Father> and </Father> node will be considered information for this node. Then this information is further written using nodes. The name of the node tells us the kind of information provided The <Name></Name> node tells us that the information written inside this node is the name of that person. Understanding that we have nodes and these nodes can have child nodes is enough to understand and write a markup language. The style of writing information above is known as a markup language.

HTML is an example of a widely known and used markup language. HTML is a subset of a much older and far more complex text- markup language called Standard Generalized Markup Language (SGML). Figure 2-8 shows the classification of markup languages.

FIGURE 2-8: CLASSIFICATION OF MARKUP LANGUAGES

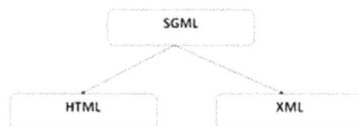

Standard Generalized Markup Language (SGML)

SGML is a meta language (i.e. a language that defines other languages). SGML is nonproprietary and platform independent and offers user-defined tags. SGML is extremely flexible and can separate content from appearance. However, SGML is costly to set up and maintain, requires the use of expensive software tools, and is hard to learn. SGML is an international standard for defining methods of encoding electronic texts to describe layout, structure, syntax, etc., which can then be used for analysis or to display the text in any desired format. SGML is based on the idea that documents have structural and other semantic elements that can be described without reference to how such elements should be displayed. The actual display of such a document may vary, depending on the output medium and style preferences. SGML is based somewhat on earlier generalized markup language.

SGML divides any document into structure, content and style. For accurate definition of the structure of any document using SGML, a file known as DTD (Document Type Definition), which creates and maintains the logical pattern of a document is used. SGML documents have tags around it. These tags show the structure. These tags are already made when using SGML software. This is to save time and cost. The standard based time sheets used by SGML are the OS (Output Specification) and the DSSSL (Document Style Semantics Understanding SGML should be a major challenge to anyone who does not know its principles by now.

Hypertext Markup Language (HTML4 and HTML5)

HTML is a markup language for describing web pages. HTML is easier to learn and use than SGML. HTML is the prevalent markup language used to create documents on the Web today. In order to perform scripts on a web page, one would need to know HTML. For example, in order to work with PHP more in depth, one would need HTML knowledge to have a foundation to start with. That would be like going to work on C++ without the basic knowledge of programming. In the grand scheme of things, HTML is important. Without HTML, there would be no way of really understanding how to go about creating in other web languages like CSS or working with scripting languages. Today there exist two versions of HTML: HTML 4 and HTML 5. HTML4 has been a standard web development for more than 10 years. HTML4 is approved and ratified as a standard language for browsers by the World Wide Web Consortium (W3C), an organization that specifies and approves standards for web technologies. However, with the advent of smart-phones era, things have totally changed and that change has effected or rather revolutionized the way websites are developed. Now apart from targeting desktop PCs and laptops, web-developers have to keep in mind that their website would be accessed by mobile-user; and with the growing number of mobile users, this concern is growing proportionally. However, HTML4 is still a W3C standard for browser applications; it does not fully cater to the changing trends of the computing industry. Therefore, HTML5 has been developed with intent to cop-up with these new challenges in web industry. HTML5 is more flexible, robust and advanced as compare to its older counterpart. The syntax in HTML5 is extremely clear and simple as compared to HTML4. HTML5 contains built in support, in the form of video and audio tags, for integrated multimedia files into web page. Previously, in HTML4, the multimedia content was integrated in web pages via third party plugins such as Silverlight and flash. In HTML4, it was an extremely cumbersome task to get the geographical locations of the visitors visiting the site. It was even difficult when the website was accessed through mobile devices. On the other hand, in HTML5 is extremely easy to get the user location. In HTML4, in order to store important data on client side, browser's cache was used. However, that cache is limited and does not support relational storage mechanism. In HTML5, this issue has been addressed via Web SQL database and application cache. In HTML4, the communication between the client and server was done through streaming and long polling, since there are no web sockets available in HTML4. On the contrary, HTML5 contains web sockets that allow full duplex communication between clients and servers. In HTML4, JavaScript and the browser interface with which user interacts, run in the same thread, which affects performance. HTML5 allows JavaScript and Browser interface to run in separate threads. HTML4 is compatible with almost all web-browsers. On the other hand, HTML5 is still in the process of evolution. As such, HTML5 lags behind HTML4 in terms of compatibility with the browsers (Shabbir, 2014).

Scripting Languages and Style Sheets

Scripting language is a programming language in which programs, consisting of a series of commands, are interpreted and then executed one by one. The most common scripting languages used on Web pages are JavaScript, JScript, Perl, and VBScript. Scripts written in these languages and embedded on Web pages can execute programs on computers that display those pages.

VBScript (short for Visual Basic Scripting Edition) is a scripting language developed by Microsoft. VBScript is a limited variation of Microsoft's Visual Basic programming language. VBScript is installed by default in every desktop version of Windows operating system and Windows server operating system. VBScript needs a host environment to execute. Many host environments are available with Microsoft Windows including Windows Script Host (WSH), Internet Explorer (IE), and Internet Information Services (IIS).

JavaScript is a client-side scripting language for web pages. JavaScript uses many Java names and naming conventions but it is different from Java. JavaScript is primarily used to write functions that are embedded in or included from HTML pages (e.g. opening or popping up a new window, changing images when mouse cursor moves over them etc.). JavaScript code can only run locally in a user's Internet browser. JavaScript code can detect user actions (such as such as individual keystrokes) which HTML alone cannot. JavaScript has become the de-facto language of the web and is evolving quickly with ECMAScript 6 and beyond.

Jscript is the Microsoft's standardized version of JavaScript. The most current version is called Jscript.NET. Both Jscript and JavaScript are just different names for the same language, and the reason the names are different was to get around trademark issues.

Perl (Practical Extraction and Report Language) is a high-level programming language, which is cross platform and interpreted. Perl was originally developed in 1987 as a general-purpose UNIX scripting language to make report processing easier. The Perl languages borrow features from other programming languages including C, shell scripting (sh), AWK, and sed. There exist two languages in Perl family: Perl 5 and Perl 6. Both are developed independently. In addition to CGI, Perl 5 is used for graphics programming, system administration, network programming, finance, bioinformatics, and other applications.

Cascading Style Sheets (CSS) is style sheet language for adding styles (e.g. fonts, colors, and spacing) to Web documents written in a markup language. Styles are set of instructions. The most common application of CSS is to style web pages written in HTML. Using CSS, programmers can define formatting styles that can be applied to multiple Web pages. CSS styles can be stored in separate files and referenced using the HTML style tag or they can be included as part of a Web page's HTML file. The term cascading means that programmers can apply many style sheets to the same Web page, one on top of the other. For example, a three-stage cascade might include one style sheet with formatting instructions for text within heading 1 tags, a second style sheet with formatting instructions for text within heading 2 tags, and a third style sheet with formatting instructions for text within paragraph tags.

Extensible Markup Language (XML)

Another markup language for use on the Web is Extensible Markup Language (XML).XML and HTML were designed to serve different purposes and one does not offer the replacement of the other. HTML is about displaying data, XML is about describing data, and to focus on what data is. XML does not necessarily restrict the author to certain tags as HTML, XHTML does, and an author could define his own set of tags and document structure. XML tags do not specify how text appears on a Web page; the tags convey the meaning of the information included within them. XML is also a meta language because users can create their own markup elements that extend the usefulness of XML.

XML and HTML Editors

Web designers can create HTML documents in any general-purpose text editor or word processor. However, there are special-purpose HTML editors available that can help Web designers create Web pages much more easily. There are many freeware, shareware, and commercial HTML editors available for download on the Internet, including CoffeeCup, HomeSite, and CuteHTML. HTML editors are also included as part of more sophisticated Web site design and creation programs that are sometimes called Web page builder software e.g. Dreamweaver and Microsoft FrontPage. Using these programs, web designers can create and manage complete Web sites, including features for database access, graphics, and fill-in forms. These programs display the Web page as it will appear in a Web browser in one window and display the HTML-tagged text in another window. The designer can edit in either window and changes are reflected in the other window. For example, the designer can drag and drop objects such as graphics onto the Web browser view page and the program automatically generates the HTML tags to position the graphics. The web site design and creation software also includes maintenance tools that allow the web designer to create a Web site on a PC and then upload the entire site (HTML documents, graphics files, and so on) to a Web server computer. When the site needs to be edited later, the designer can edit the copy of the site on the PC and instruct the program to synchronize those changes on the copy of the site that resides on the Web server. The web page builder software is capable of editing XML files. However, programs designed to make the task of designing and managing XML files easier are also available. These programs include Epic Editor, TurboXML, XMetal, and XML Spy.

XHTML (Extended Hypertext Markup Language)

XHTML stands for Extensible Hypertext Markup Language. XHTML is a family of XML markup languages that mirrors or extends versions of HTML. XHTML is the next generation of HTML. XHTML is in many ways similar to HTML, but is designed to work with XML. XHTML is designed so that different documents, in different languages, can be easily mixed together. While HTML tags can be written in UPPER case, mixed case, or lower case, to be correct, XHTML tags must be all lower case. All XHTML tags must have an end tag. Tags with only one tag, such as <hr> and <img> need a closing slash (/) at the end of the tag. All attributes must be quoted in XHTML. Some people remove the quotes around attributes to save space, but they are required for correct XHTML. XHTML requires that tags be nested correctly. If you open a bold () tag and then an italics (<i>) tag, you must close the italics tag (</i>) before you close the bold (). XHTML attributes must have a name and a value. Attributes that are stand-alone in HTML must be declared with values as well.

Intranets and Extranets

Not all networks connect to the Internet and many organizational networks do not extend beyond their organizational boundaries. An intranet is an interconnected network that does not extend beyond the organization that created it. An extranet is an intranet that has been extended to include specific entities outside the boundaries of the organization, such as business partners, customers, or suppliers.

Intranets

Intranets are an excellent low-cost and efficient solution for distributing internal corporate information. Intranets can save companies large amounts of paper and effort required for processing paper-based information. Intranets are compatible with the Internet Protocol (IP) and often use the same software used on Internet (e.g. web browsers and protocols e.g. FTP, Telnet, HTML, and HTTP). Information from intranet can be shared among departments that use different technologies as well as among external consumers. Companies can use intranets reduce software maintenance and update costs for their employees' computers by automating the software update process.

A primary benefit of an intranet is that enterprises can now easily build custom applications that can be immediately accessed by users anywhere on the intranet, on any platform. Because both development and deployment are much faster, the cost savings over the old desktop-centric models are significant. Further, users do not need extensive training. The applications run in the familiar browser interface. Each company typically has hundreds of other application needs, large and small, many of which have never been addressable because previously there was no universal way for individual employees to instantly access any information resource or application from the desktop, and there was no way to develop and deploy new applications rapidly and universally. Companies running intranets are finding that commercial applications built on an intranet environment address these needs head on. These applications address pressing business needs, such as the following examples:

- The sales force needs up-to-date information about your products and services, and it needs a system that allows orders to be taken and tracked from the road.
- Members of a project team may be distributed across territories and time zones. There needs to be a universal way for each project team to provide information and documentation on its efforts and results so they can work together more efficiently.
- The technical support staff needs a way to track customers and their problems, ensuring successful resolution of each reported problem and providing a way for customers to "help themselves" resolve basic problems.
- A company needs one email system that spans every desktop and every person. Many large companies are running five or more, proprietary legacy email systems, and have huge compatibility and interoperability problems. For example, each email system has its own proprietary directory, and attachments do not transmit seamlessly across email systems, creating islands of users instead of seamless connectivity.
- New employees need a way to find information on company procedures, organization, and benefits as soon as they come onboard.

- The marketing staff needs a way to access, through a consistent interface from any desktop, all the customer and market research databases the company maintains.

Extranets

Extranets are Intranets with extended connectivity to external entities (e.g. suppliers and business partners). Each participant in the extranet can have access based on the access level permissions granted. An extranet can be set up through the Internet, or it can use a separate network.

One famous example of an extranet is Federal Express Tracking System (www.fedex.com). Customers can locate and track any shipment made through FedEx using this system. Using their package shipping service available on FedEx site, customers can enter all the information needed to prepare a shipper form, obtain a tracking number, print the form, and schedule a pick up. Other uses of extranets may include shared product catalogs accessible only to wholesalers or those in the trade and project management and control for companies that are part of a common work project.

An extranet has many applications to your business from lowering costs, to producing faster results and improving the quality of service to customers. Applications can depend to a large degree on your reasons for introducing the extranet in the first place. However, the types of applications that organizations using extranets typically experience include:

- Integrated supply chains through the use of online ordering, order tracking and inventory management
- Reduced costs by making manuals and technical documentation available online to trading partners and customers
- More effective collaboration between business partners - perhaps members of a project team - by enabling them to work online on common documentation
- Improved business relationships with key trading partners because of the close collaborative working that extranets support
- Improved customer service by giving customers direct access to information and enabling them to resolve their own queries
- A single user interface between you and your business partners
- Improving the security of communications between you and your business partners, since exchanges can take place under a controlled and secure environment
- Shared news of product development exclusively with partner companies
- Flexible working for your own staff, as an extranet allows remote and mobile staff to access core business information 24 hours a day, irrespective of location

Extranets are invaluable in partnerships that revolve around just-in-time inventory and manufacturing philosophies. Extranet applications allow real-time, economical, and efficient electronic commerce and EDI (Electronic Data Interchange) with the trading partners of a company. Extranets save considerable money over the operating costs of EDI networks and VANs (Value-Added Networks) in reaching suppliers and customers. Extranets transfer information to trading partners much faster than VANs.

Virtual Private Network (VPN)

A VPN is a private computer network that uses a public telecommunication network (usually the Internet) to connect remote sites or users together. A VPN establish virtual connections which are routed through Internet from company's Intranet to remote employees. VPNs usually work as part of a firewall. A well-designed VPN provides multiple benefits to a company. For example, it can extend geographic connectivity, reduce operational costs versus traditional WANs, reduce transit times and traveling costs for remote users, improve productivity, provide global networking opportunities, provide telecommuter support, and provide faster Return on Investment (ROI) than traditional WAN. Figure 2-9 shows structure of a typical VPN.

FIGURE 2-9: VPN STRUCTURE

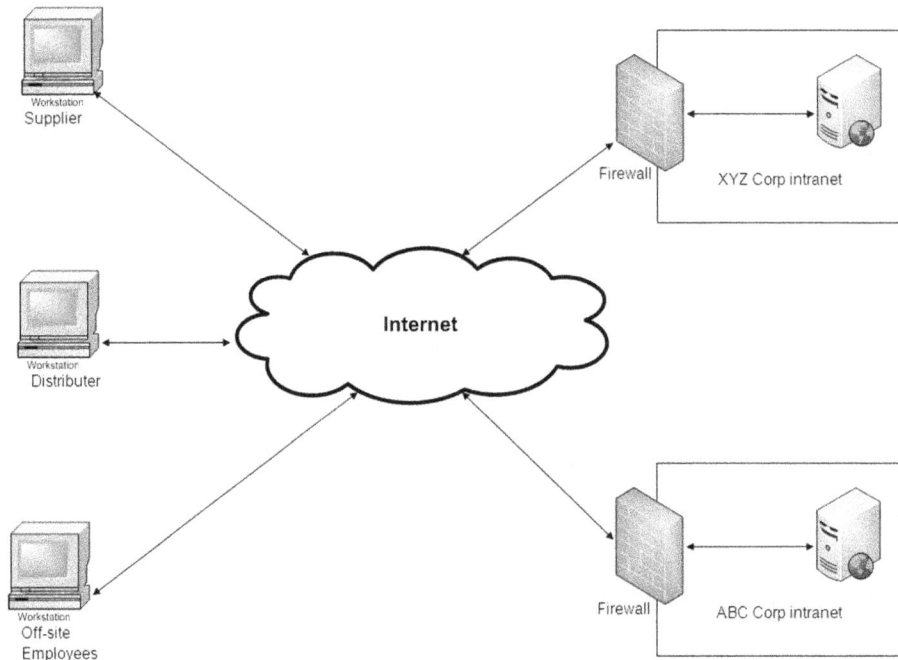

Three technologies can be used to create a VPN. First, many of the firewall products (hardware and software) support VPN functionality. Second, routers can also function as VPN servers. Finally, software solutions are available to handle VPN connections. Many telecommunications carriers (e.g. AT&T VPN Services) and larger ISPs offer VPN services for Internet- based dial-up and site-to-site communications.

Two common types of VPN are remote-access VPN and site-to-site VPN. A remote-access VPN is used by company employees at various remote locations to connect to company's private network. Employees first access the Internet and then use their VPN client software to access the corporate network. Large firms with hundreds of sales people in the field may set-up a remote access VPN to provide its sales force secure and encrypted connection to the company's private network. Users typically just need to know their VPN server/concentrator IP address or host name (a name that resolves on the Internet), their network login name and password, and possibly a secure token (which allows for a 2-factor login and authentication). Once they have these essentials, users can setup their VPN client. Starting with Microsoft Windows 2000 the native VPN client is based on the PPTP protocol. A site-to-site VPN uses dedicated equipment and large-scale encryption. Using a site-to-site VPN, a company can connect multiple fixed sites over a public network such as the Internet. Since each participating sites needs only to connect to the Internet, cost of private leased-lines can be saved. With VPN, everything we send or receive on our computer is encrypted, such as file transfer, email, IRC, ICQ, P2P, as well as surfing.

VPNs are a simple tool. It can be used for a variety of application including the following:

REMOTE CONNECTIVITY: VPNs are frequently used by business travelers to access their business' network, including all its local network resources, while on the road. The local resources do not have to be exposed directly to the Internet, which increases security.

BYPASS NETWORK CENSORSHIP: VPNs can be used to get around the firewall and gain access to the sites that are now allowed for general access by the employees

Storage Area Networks (SANs)

Large online businesses produce huge amounts of data. This data is needed for many purposes including analysis tools, enterprise resource planning (ERP) systems, multimedia Web sites and e-commerce systems. With the increase in amount of data the complexities of maintaining these data in a reliable and fault-tolerant way becomes increasingly difficult.

A Storage Area Network (SAN) is a high-speed sub-network of shared storage devices that provides high capacity, reliable data storage and delivery on a network (LAN or WAN). A SAN storage device is a machine that contains disk(s) for storing large amounts of data. Any additional storage devices added to a SAN also become accessible on network. By offloading data from the network servers, full processing power of the servers is utilized for processing network user requests. A storage-area network is typically assembled using three principle components: cabling, host bus adapters (HBAs) and switches. Each switch and storage system on the SAN must be interconnected and the physical interconnections must support bandwidth levels that can adequately handle peak data activities. A virtual storage-area network (VSAN) is a software-defined storage offering that is implemented on top of a hypervisor (a program that allows multiple operating systems to share a single hardware processor) such as VMware ESXi or Microsoft Hyper-V. Virtual SANs yield a number of benefits such as ease of management and scalability.

The main benefit to using a SAN is that raw storage is treated as a pool of resources that can be centrally managed and allocated on an as-needed basis. SANs are also highly scalable because additional capacity can be added as required. SAN devices may also provide backup and recovery services. For example, using mirroring technology a SAN device can store redundant copies of data, so that if one copy is lost or damaged, a mirrored copy can be used to recover the lost data. SAN uses fiber-channel technology to connect computers on the network to SAN devices. Fibre Channel, or FC, is a high-speed network technology (commonly running at 2-, 4-, 8- and 16-gigabit per second rates) primarily used to connect computer data storage. The main disadvantages to SANs are cost and complexity. SAN hardware tends to be expensive and building and managing a SAN requires a specialized skill set.

Internet Telephony or Voice over IP (VoIP)

Internet telephony refers to the science or technology of integrating telephone services into computer networks. In essence, Internet telephony converts analog voice signals into digital signals, transmits them, and then converts them back again. Voice over IP (VoIP) is a common Internet telephony service. For users who have free or fixed-price Internet access, Internet telephony software essentially provides free telephone calls anywhere in the world. However, Internet telephony does not offer the same quality of telephone service as direct telephone connections. There are many Internet telephony applications available. Some like Cool Talk, NetMeeting, are stand-alone products. Internet telephony products are sometimes called IP telephony, Voice over the Internet (VOI) or Voice over IP (VOIP) products.

Advantages of Internet Telephony

VoIP provides many benefits. The very nature of VoIP technology means that everyone can make significant cost savings for their business, especially if you have multiple branches nationwide or overseas. Cheap calls and free calls may be the attraction for VoIP right now, but the future will be about value-added VoIP services, and cost will take the backseat. VoIP is not distance or location dependent. With VoIP, you could be calling your supplier 1,000 miles away from you or calling your business partner on the other end of town, and it does not make any difference at all, in terms of connectivity and cost. A VoIP phone number, unlike your regular phone number, is completely portable. Most commonly referred to as a virtual number, you can take it with you anywhere you go.

Even if you change your office address to another state, you phone number can go with you. VoIP provides integrated communication. That means you can have many excellent communications tools in one package such as making cheap local and international phone calls, audio conferencing & video conferencing, voice messages sent to your email, call forwarding, call waiting, fax through e-mail, send and receive multimedia files, sharing photos while talking etc.

Methods of Accessing VoIP

There are currently three ways to access VOIP:

ATA (ANALOG TELEPHONE ADAPTOR)

An ATA is a hardware device that converts analog signals to digital to send over the Internet. Using the device to connect your traditional landline handset to your PC or router, you can then make and receive calls via the Internet if you have an account with a telecommunications provider that supports VOIP. When you make a call to someone, the ATA converts the analog signal to digital. The telecommunications company then determines whether the person you are trying to call has VOIP support or has a standard landline and routes the call accordingly, converting it back to analog if necessary.

IP PHONES

An IP phone looks just like a normal telephone but is "out of the box" digital signal compatible. Just plug it in to your computer via a USB port or into your router, open an account with a VOIP provider and you are set to go. Depending on the type of account you open, you can also have a standard telephone number that people can call you on.

COMPUTER-TO-COMPUTER

This is the simplest/cheapest implementation and most standard PC's are ready to use this option. It only requires a headset to use it effectively. You can make free "calls" to any computer also using the same service via software offered by a slew of companies. Depending on the software, you can also make computer to landline/cell phone calls, usually at greatly reduced rates. Using this option, people can call you via a messenger type application, but not via a landline phone. The computer-to-computer option is the best way to go while you are exploring the benefits of VoIP.

VoIP Applications in E-commerce

One of the more costly aspects of running an ecommerce site can be the phone bill. There is not just communications with clients, but also with potential partners and merchants who may be in other countries. Networking via voice, even in this age of email, is still a very important marketing/strategic partner negotiations tool. One good starting point of using VoIP for your E-commerce business is trialing a computer-to-computer service such as Skype. The software is free, you can make voice calls to other Skype users free as well; and their rates for calling landline phones are very reasonable. Setting up the software is easy. Skype software also includes a full featured IM application. The voice quality is generally good BUT it depends which country you are calling. You can keep track of your balance. Skype also keeps a log of all calls. Skype can also be used for conference calls - up to 4 parties simultaneously. Skype also offer a service called SkypeIn. It gives you a personal number so that your customers and business partners can call you on Skype from their landline or cell phones (or computer), wherever you connect to the web around the world. You can also have a number in another country that allows you to have a virtual office in other parts of the world. Skype Voicemail lets you pick up messages when you are offline.

Internet Connectivity Options

To connect to Internet, a business or individual needs a telephone connection or a connection to a LAN or intranet. To connect to the Internet, several connection options are available. These options are provided by Internet Service Providers (ISP). The most common connection options are dial-up, ISDN, various types of broadband connections, leased line, and wireless.

One chief difference among ISPs is their bandwidth management. Bandwidth refers to the amount of data that can travel through a communication line per unit of time. The higher the bandwidth, the faster data files travel and web pages are displayed on computer screen. Each connection option offers different bandwidths, and each ISP offers varying bandwidths for each connection option. Many ISPs purchase bandwidth from other vendors.

ISPs typically pay for purchased bandwidth in dollars and charge their customers for the same bandwidth in local currency.

Net bandwidth is the actual speed with which information travels. Net bandwidth is greatly affected by Traffic on the Internet and at your local Internet service provider. When few people are competing for service from an ISP, net bandwidth approaches the carrier's upper limit and users experience slowdowns during high-traffic periods. In general, with increasing number of customer, ISPs increase their bandwidth proportionately.

Bandwidth can differ for data traveling to or from the ISP (i.e. uplink and down link) depending on the user's connection type. Symmetric connections provide the same uplink and down link bandwidth while asymmetric connections provide different uplink and down link bandwidths. Upstream bandwidth, also called upload bandwidth, is a measure of the amount of information that can travel from the user to the Internet in a given amount of time. Downstream bandwidth, also called download or downlink bandwidth is a measure of the amount of information that can travel from the Internet to a user in a given amount of time (for example, when a user receives a Web page from a Web server).

Dial-up

Dial-up is the most common way to connect to an ISP. A user connects to ISP through a modem connected to local telephone service provider. The bandwidth provided ranges from 28- 56 Kbps. Figure 2-10 shows how a dial-up Internet connection works.

FIGURE 2-10: DIAL-UP INTERNET CONNECTION

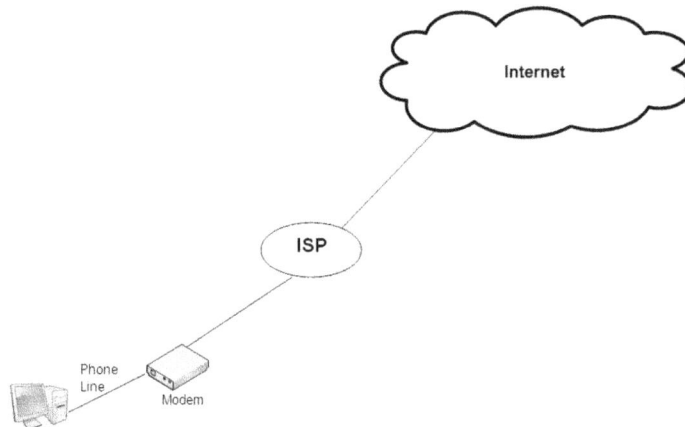

Integrated Services Digital Network (ISDN)

Integrated Services Digital Network (ISDN) is a technology that uses existing telephone lines for combined digital telephone, data, video plus other services. ISDN provides two types of services: Basic Rate Interface (BRI) and Primary Rate Interface (PRI). BRI is most commonly used at homes and most private companies while PRI is used by larger businesses. ISDN is more expensive than regular telephone service and offers bandwidths between 128 Kbps and 256 Kbps. Charges for ISDN can be on pay per call and/or per minute, depending on distance. Technically speaking, ISDN consists of two separate 64-Kbps channels. Load balancing of the two channels into a 128-K single channel is possible if you have compatible hardware on each end of a connection. Since ISDN is a digital service it is not subject to the line noise, that slows most analog connections, and thus offers net bandwidth much closer to its promised bandwidth. ISDN connection can be made either using an ISDN router or ISDN modem (ISDN terminal adapter) that can be attached to the serial port of the router. Figure 2-11 shows how an ISDN Internet connection works.

FIGURE 2-11: ISDN INTERNET CONNECTION

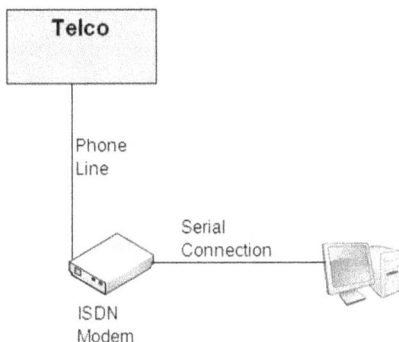

Broadband Connections

Connections that operate at speeds of greater than about 200 Kbps are called broadband connections.

CABLE MODEMS

A cable modem uses the cable television (CATV) infrastructure to provide broadband Internet access. Cable television network provides high bandwidth. Typically, bandwidth provided by cable modem is between 300 Kbps and 1 Mbps from the client to the server. The downlink bandwidth can be as high as 10 Mbps. Cable modems are commonly deployed in Australia, Europe, North America and South America. Cable modem provides a shared media. Therefore, when there are a number of subscribers accessing cable Internet simultaneously the bandwidth of cable modem connection may vary.

DIGITAL SUBSCRIBER LINE (DSL)

The xDSL is the collective name of all DSL technologies. The two main categories of DSL are ADSL and SDSL. xDSL, similar to ISDN, operate over existing copper telephone lines and requires shorter distances to the central telephone office. A DSL connection works better when you are closer to the provider's central office because signals become stronger. The xDSL provides much higher bandwidth (up to 32 Mbps uplink and over 1 Mbps downlink). Researchers at Bell Labs have reached speeds of 10 Gigabit/s, while delivering 1 Gigabit/s symmetrical broadband access services using traditional copper telephone lines. These higher speeds are lab results, however (Owano, 2014). Modulation techniques are used to pack and transfer data on a DSL connection. DSL uses two pieces of equipment, one on the customer end and one at the Internet service provider, Telephone Company or other provider of DSL services. DSL transceiver is the point where data from the user's computer or network is connected to the DSL line. The DSL service provider's DSL Access Multiplexer (DSLAM) receives customer connections from many customers and aggregates them onto a single, high-capacity connection to the Internet. DSLAMs are generally flexible, can support multiple types of DSL, and different protocol and modulation schemes. Figure 2-12 shows how a DSL connection works.

FIGURE 2-12: DSL INTERNET CONNECTION

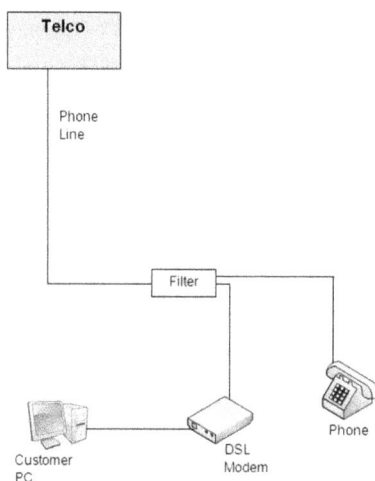

ADSL (asymmetric digital subscriber line) is the most commonly deployed types of DSL in North America. It supports data rates of from 8 to 24 Mbps downstream and from 1 to 3.3 Mbps upstream. ADSL connection requires a special ADSL modem to setup.

SDSL (symmetric digital subscriber line) provides similar uplink and downlink bandwidth. It is a technology that allows more data to be sent over existing copper telephone lines (POTS). SDSL supports bandwidth up to 3 Mbps and requires a special SDSL modem.

HDSL (High-data-rate DSL) and VDSL (Very high DSL) are two other types of DSL technologies. VDSL can provide higher bandwidths at shorter distances. HDSL gave way to new symmetric DSL technologies, HDSL2 and HDSL4, the proprietary SDSL, and G.SHDSL. HDSL2 offers the same data rate over a single pair of copper; it also offers longer reach, and can work over copper of lower gauge or quality. SDSL is a multi-rate technology, offering speeds ranging from 192 Kbps to 2.3 Mbps.

Both DSL and Cable Internet can be shared on a LAN by using either some software (e.g. Microsoft Internet Connection Sharing) or a device (e.g. a router). DSL and cable internet prices are not standard and depend on the level of competition among services providers and the area where service is to be provided.

LEASED-LINE CONNECTIONS

Leased line connections are high bandwidth symmetric telecommunication connections offered by telecommunications companies. Leased lines connect two locations and Telecommunication Company may need to install new cabling. Unlike traditional phone lines it does not have a telephone number and each side of the line is permanently active and connected to the other whether there is network traffic between the two locations or not. Leased line connections require a device called CSU/DSU (channel service unit /digital service unit) which is connected between the router and the leased line. CSU/DSU inputs data in the form understandable to the router and converts it into a form understandable by the leased line. Leased lines are typically used by companies to connect their office at larger geographical distances but they can also be used for telephone or Internet services. In general, a fixed monthly rent is charged for using the leased line. This rate can be affected by distance between end-points and the speed of the circuit. Because the leased line is a dedicated connection, the carrier can assure a given level of quality. All leased line connections are much more expensive than dial-up, ISDN, or DSL connections. Figure 2-13 shows how a leased-line connection works.

FIGURE 2-13: LASED-LINE CONNECTION

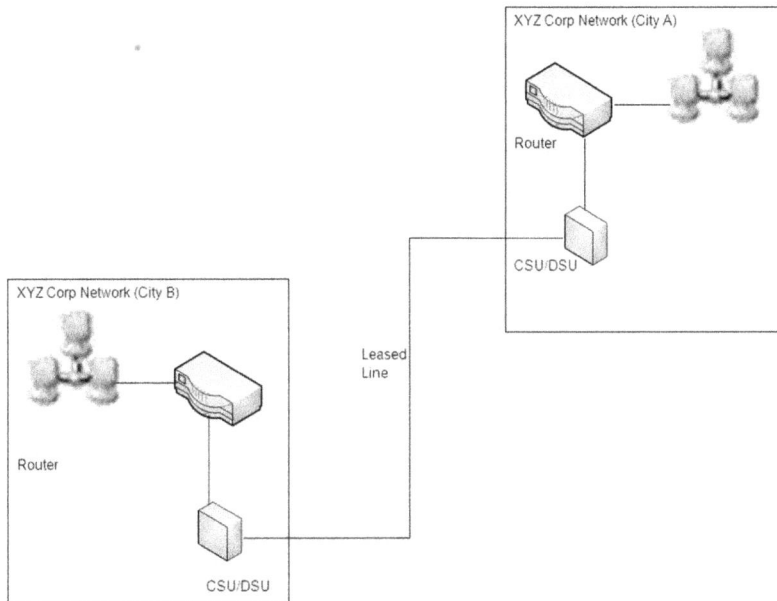

T1 AND T3 LINES

A T1 line is a technology that can be fiber optic or copper line based. T1 line is widely used in North America and Japan to transmit voice and data between devices. Outside North America and Japan, E1 line is used in place of T1 line. A T1 line can carry 24 digitized voice channels, or it can carry data at a rate of 1.544 Mbps. A T1 line is more reliable and can carry roughly 60 times more data than a normal residential modem. A T1 line might cost between $1,000 to 1,500 per month. This charge depends on service provider and the point where the line terminates. With Fractional T1 (FT1), customer can lease any channel of a T1 line and pay only for what they need. Fractional T1 lines look the same as full T1 lines in the sense of termination and overall bandwidth. However, customers can only use bandwidth allocated to them. A T3 line offers bandwidth of 44.736 Mbps (the equivalent of 30 T1 lines). A fractional T3 is functionally the same as a fractional T1, but the bandwidth is in multiples of T1 lines. In the global Internet infrastructure, Network Access Points (NAPs) use T1 and T3 lines.

Wireless Connections

SATELLITE INTERNET

Satellite Internet access is ideal for Internet users who want broadband access but do not have telephone lines or cable systems. Satellite Internet uses a satellite dish for two-way (upload and download) data communication. Upload speed is about 1/10 of the 500 kbps download speed. Satellite systems are about 10 times faster than a normal modem and recent improvement in the accuracy of the Antennas has made the initial installation of the microwave transmitter a lot easier. One disadvantage of satellite Internet is the line-of-sight requirement. Such requirements can make Satellite Internet services unusable for many people living in places where line-of-sight requirements cannot be met. Figure 2-14 shows how satellite Internet connection works.

FIGURE 2-14: SATELLITE INTERNET CONNECTION

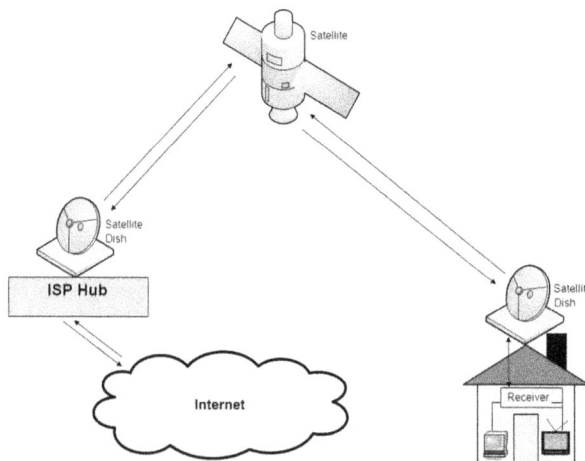

BLUETOOTH

Bluetooth is one of the first wireless protocols designed for personal use over short distances. Bluetooth operates reliably over distances of up to 35 feet and can be a part of up to 10 networks of eight devices each. It is a low-bandwidth technology, with speeds of up to 722 Kbps. Bluetooth is useful for tasks such as wireless synchronization of laptop computers with desktop computers and wireless printing from laptops, PDAs, or mobile phones. Such small Bluetooth networks are called personal area networks (PANs).

Bluetooth technology consumes very little power, which is an important advantage for mobile devices. Bluetooth devices can discover one another and exchange information automatically without logging in to the network or installing software on either device. The devices electronically recognize each other as Bluetooth devices and immediately can begin exchanging information.

The IEEE standardized Bluetooth as IEEE 802.15.1, but no longer maintains the standard. The Bluetooth SIG oversees development of the specification, manages the qualification program, and protects the trademarks. A manufacturer must make a device meet Bluetooth SIG standards to market it as a Bluetooth device.

ULTRA WIDEBAND (UWB)

UWB is a wireless communications technology capable of transmitting data at speeds between 40 to 60 megabits per second and eventually up to 1 gigabit per second over short distances (30 to 100 feet). The UWB is very well suited to short-distance applications e.g. PC Peripherals such as Wireless USB. UWB require low power to operate, difficult to detect, and can be utilized indoor and underground. Slow progress in UWB standards development, the cost of initial implementation, and performance significantly lower than initially expected are several reasons for the limited use of UWB in consumer products.

WIRELESS ETHERNET (WI-FI)

Also called wireless Ethernet, Wi-Fi is the most common wireless connection technology for use on LANs. Wi-Fi works uses radio frequency (RF) technology and requires no physical wired connection between sender and receiver. Computers need a Wi-Fi network card in order to connect to a wireless LAN. A Wi-Fi equipped computer can communicate to other network computers through a wireless access point connected to a LAN. A wireless access point (WAP) is a device that transmits network packets between Wi-Fi-equipped computers and other devices that are within its range. The user must have authorization to connect to the LAN and might be required to login before computer can access the LAN through the access point. Some organizations operate WAPs that are open to the public- called hot spots. Some organizations allow free access to their hot spots while others charge an access fee.

Many devices and application support Wi-Fi technology e.g. PDAs, mobile phones, many operating systems, and other types of consumer electronics products. Products from different manufacturers that are tested and approved as "Wi-Fi Certified" are interoperable with each other. Wi-Fi certified products carry an identification seal on their packaging, which also indicates the radio frequency band used (2.5GHz for 802.11b, 802.11g, for 802.11n, and 5GHz for 802.11a). Many newer consumer devices support the latest 802.11ac standard, which uses the 5 GHz band exclusively and is capable of multi-station WLAN throughput of at least 1 gigabit per second.

According to a study, devices with the 802.11ac specification are expected to be common by 2015 with an estimated one billion spread around the world (Murph, 2011).

LI-FI

The term Li-Fi was coined by pureLiFi's CSO, Professor Harald Haas, and refers to visible light communications (VLC) technology that delivers a high-speed, bidirectional networked, mobile communications in a similar manner as Wi-Fi. Like Wi-Fi, LiFi is wireless and uses similar 802.11 protocols; but it uses visible light communication (instead of radio frequency waves), which has much wider bandwidth. Although Li-Fi can be used to off-load data from existing Wi-Fi networks, implementations may be used to provide capacity for the greater downlink demand such that existing wireless or wired network infrastructure may be used in a complementary fashion. VLC is the use of the visible light portion of the electromagnetic spectrum to transmit information. This is in contrast to established forms of wireless communication such as Wi-Fi, which use traditional radio frequency (RF) signals to transmit data. With VLC, data is transmitted by modulating the intensity of the light, which is then received by a photosensitive detector, and the light signal is demodulated into electronic form. This modulation is performed in such a way that it is not perceptible to the human eye. VLC is a category of Optical Wireless Communications (OWC). OWC includes infrared and ultra-violet communications as well as visible light. However, VLC is unique in that the same visible light energy used for illumination may also be used for communication. In October 2011, companies and industry groups formed the Li-Fi Consortium, to promote high-speed optical wireless systems and to overcome the limited amount of radio-based wireless spectrum available by exploiting a completely different part of the electromagnetic spectrum. In April 2014, the Russian company Stins Coman announced the development of a Li-Fi wireless local network called BeamCaster. Their current module transfers data at 1.25 gigabytes per second but foresee boosting speeds up to 5 GB/second in the near future. One of the advantages of Li-Fi is its ability to function in electromagnetic-sensitive areas such as in aircraft or nuclear power plants, without causing interference. Additionally, Li-Fi signals cannot travel through walls, a welcome feature for many users who wish to increase their digital privacy, although this comes with its obvious drawbacks, such as difficulty in creating "blanket" wireless coverage (michaelaprile.wordpress.com, 2015).

FIXED-POINT WIRELESS

Fixed-Point wireless devices or systems are situated in fixed locations, e.g. an office or home. Power to fixed-point wireless devices is normally provided through utility mains (portable wireless devices normally derive their power from batteries). These devices work with point-to-point signal transmissions that occur through the air over a terrestrial microwave platform and thus fixed-wireless does not require satellite feeds or local phone service. These devices can connect users in remote areas without lying new cabling and provide large bandwidth that is not impeded by fiber or cable capacities.

CELLULAR TELEPHONE NETWORKS

Mobile phones, or cellular phones, broadcast voice signals to and receive signals from antennas. Mobile phones can transmit data as well with data transmission speeds ranging from 10 Kbps to 2 Mbps. Many mobile phones have a small screen and can be used to send and receive short text messages. Some cell phones, including PDAs, include tiny Web browsers that users can use to access web and email. Mobile telephone service providers also sells Internet access through their cellular networks. These companies offer a variety of pricing plans and service levels.

Evolution of Internet and World Wide Web

Internet2

Internet2 (www.internet2.edu) is a not-for-profit advanced networking consortium composed of a large number of universities, corporations, government agencies, laboratories, and other institutions of higher learning. Internet2 operates the Internet2 Network which is a next-generation IP and optical network that delivers production network services and increased bandwidth provided through high bandwidth backbone expected to provide built by Internet2 organization. This high-bandwidth backbone is expected to provide bandwidth of 2.4 gigabits of data per second- a rate 10 to 15 times faster than that provided by the current technology.

Internet2 comprises servers, routers, switches, and computers that are all connected together. Routers decide which way to send information, and servers handle Web site requests and store information for retrieval. Internet2 is limited to a select few, and its backbone is made up entirely of large-capacity fiber-optic cables. Internet2 moves data at 10 gigabits per second and more. i.e. Internet2 moves data 100 to 1,000 times faster than the old-fashioned Internet. More than 200 universities, 70 private companies, 45 government agencies, and 45 international organizations log on to Internet2 every day. The foundations of Internet 2 are comprised of two main technological notions. The first is the *Gigapop*, which is a regional network's interconnection point to the new Internet 2 innovative services. Some types of Gigapops are being constructed so that Internet 2 members can connect solely to Internet 2 services, while other Gigapops are being constructed to connect non Internet 2 members to various other services, such as the old Internet (or Internet 1). The second technological notion is QoS, which stands for Quality of Services and is a new method of sending information around more efficiently. The basis of QoS is to create priorities for the information sent. In this way, crucial medical information will have priority when compared to chess game simulations. The implementation of these two notions, together with other innovative technologies, requires a vast amount of funding, which is partly private but mainly governmental. The methods for budgeting and funding of institutions for the Internet2 project are interesting and are a main force in the shaping of Internet 2. Moreover, they will also influence important infrastructure and technical decisions yet to be made, such as routing methods, protocols and speeds that will shape and mold Internet2.

The Internet2 consortium is working on several advanced applications that will enable interactive access over the Internet. Teleimmersion refers to the sense of shared presence with distant individuals and their environments that feels substantially as if they were in one's own local space. A two-way Internet technology that enables remote users to interact in a simulated environment that gives the feel of being face-to-face. Teleimmersion is an application that manufacturers and designers can use to meet and share 3D images of models from locations around the world. In education, teleimmersion can be used to bring together students at remote sites in a single environment relationship among educational institutions. Already, the academic world is sharing information on research and development to better the results. Teleimmersion will only promote this collaboration. This will be distinct advantage in surgical training. While it will not replace the hands on training, this technology will give surgeons chance to learn complex situations before they treat their patients. With teleimmersion in schools, students could have access to data or control a telescope from a remote location, or meet the students from other countries by projecting themselves into a foreign space.Internet2 will provide access to digital libraries and labs, opening up the lines of communication for students. Teleimmersion will bring to them place, equipment and situations earlier not available, helping them experience what they could have only watched, read or heard about earlier. Multi-modal 3D teleimmersion is a collaborative project that involves the University of Texas at Dallas, the University of Illinois at Urbana-Champaign, the University of California at Berkeley and the Dallas Veterans Affairs Hospital. UT Dallas is the lead institution. The project, funded by the National Science Foundation for $2.4 Million, is tasked to design and develop a collaborative, multi-modal immersive virtual reality environment for medical health care personnel and off-location patients. This multimedia system uses several 3D cameras to create avatars of humans in two different locations and then puts them together in the same virtual space where they can interact. The system has to process large amounts of data due to movement tracking and representation of multiple views, which could cause a serious bottleneck in terms of lag time or transmission delay (UT Dalls, 2015). Educators believe that teleimmersion will eventually do away with textbooks and allow kids to step inside a problem and see it from the inside out, making learning more intuitive. Teleimmersive technology could significantly improve the access to knowledgeable coaches thereby improving the ability to acquire the knowledge and skills necessary for engaging in physical activity without injuries. Virtual laboratory is an application that can enable researchers worldwide to collaborate on projects and share massive computing power, databases, simulations and software. Digital libraries are another advanced application for Internet2 that will be able to store audio and video files. This will make it possible, for example, to search for a movie by a line from its soundtrack.

Web 2.0

Web 2.0 describes second generation of the World Wide Web that is focused on providing people the ability to collaborate and share information online in new ways (e.g. social networking sites, web services, wikis etc.). Web 2.0 is a transition from static web to a more organized dynamic web. Web 2.0 promotes open communication by development of web-based communities of users, and more open sharing of information. Web 2.0 and semantic web are two terms previously used synonymously however they are not the same.

Semantic Web

Semantic Web is a web of data that can be processed directly or indirectly by machines. Most of the current web content is described and structured for human use and are hardly understandable by computers. The goal of the Semantic is to develop expressive languages to describe information in forms understandable by machines. However, the growth of semantic web is still slow mainly because of its steep learning curve. The Semantic Web is a web that is able to describe things in a way that computers can understand. Semantic web enables machines to understand the semantics or meaning of information on the World Wide Web. One can think of it as being an efficient way of representing data on the World Wide Web, or as a globally linked database. Web 2.0 has brought an onslaught of user-generated content in the form of blogs, podcasts, appending comments at the bottom of articles, posting reviews of restaurants, movies, stores, and hotels. Media has become truly interactive, as opposed to the one-way world we were used to. Many more voices are being raised, and heard. The semantic web (or Web 3.0) will organize itself around two different elements: context and the user. By 'context,' we mean the intent that brings someone to the Web, the reason for surfing. Looking for a job is 'context,' as is planning a trip or shopping for clothes. Fundamental to context is the user. And when you fuse a specific user with genuine context, you wind up with truly personalized service."

SEMANTIC WEB TECHNOLOGIES

The design underlying the Semantic Web is very simple. Based on OWL (Web Ontology Language) files provided by the research platform myOntology, it is possible to create RDF (Resource Description Framework) files, which contain information about a service. OWL is a family of knowledge-representing languages for ontologies. RDF was originally designed as a metadata model. It developed into a general method for conceptual description or modelling of information, which is implemented in Web resources. Both RDF and OWL are families of World Wide Web Consortium (W3C) specifications. W3C is the main international standards organization for the World Wide Web.

RDF is a key component of the Semantic Web Activity. It was developed by the W3C. The RDF language is a part of the W3C's Semantic Web Activity. Ontologies sit on top of the RDF framework and are a critical part of making the Semantic Web "intelligent." Ontologies allow computers to communicate with each other by providing a common set of terms (vocabularies) and rules that govern how those terms work together and what they mean. Ontologies define terms and then lay out the relationships among those terms.

In the Semantic Web, computers understand the meaning of Web data by following links from Web pages to topic-specific ontologies. The meaning of vocabulary terms or XML tags used on a particular Web document would be defined by hypertext links from that page to a topic-specific ontology. For example, ontologies offer cross references so a computer understands that "blouse" and "dress shirt" are the same concept. The infrastructure and semantics provided by ontologies make it easier for databases to talk to each other. While the W3C has sponsored the development of XML and RDF technologies, building ontologies to cover every topic addressed on the Web is an enormous challenge. The Semantic Web calls for ontologies that cover everything from factory automation to post-structural philosophy. The Dublin Core Metadata Initiative (dublincore.org) has been working since 1995 to build vocabularies that could overcome that potential bottleneck.

SEMANTIC WEB AND E-COMMERCE

Millions of us rely upon online information to inform purchasing decisions, but the ad hoc fashion in which free-text descriptions of products and services are interpreted and offered up by mainstream search engines makes this a far less accurate process than we might wish. Within specific organizations and supply chains, processes and procedures are in place to provide accurate and unambiguous description of the products. The picture becomes somewhat less clear as data moves out of the enterprise and onto the web. Even on the product website itself, much of that internal structure and richness is inadequately conveyed. For the consumer (or aggregator) wishing to compare and contrast MP3 players from a number of competing providers, it can frequently be difficult to accurately ensure that they really are comparing apples with apples. Imagine that your computer could understand the kinds of data it handles and know how to interlink them in an intelligent way. This is the premise of Semantic Web. Currently, while conventional search engines are able to search for keywords on Web sites, only human users can read and interpret product and service information on the Web. The innovative technology used by the Semantic Web enables intelligent applications and search engines to grasp the meaning of the information, process this data and display it so that it is comprehensible to humans.

To make the Semantic Web work for e-commerce, few things are needed. First, there must be standardized structures, so-called ontologies, which should be easy to implement in various applications. Next, free and easy-to-use tools must be made available to everyone. An example of such a tool is the event platform OpenEvents, which publishes events on the Semantic Web. The final step is to interlink all the data, such as event information, catering offers and accommodation. Because of this intelligent cross-linking between data, many diverse use cases occur. For example, a hotel Web site may contain information about relevant music events, or can link to a map containing restaurant locations. These events or restaurant offers can be customized so that, for example, only premium events are shown on a five-star hotel homepage or only cyclist-friendly restaurants are recommended along bicycle routes. In general, automatic and intelligent integration, combination and connection of offers are possible. US-based consumer electronics retailer Best Buy already embeds GoodRelations RDFa (or microdata) in product page. UK supermarket giant Tesco has begun to experiment with embedding RDFa in product pages (Zaino, 2014).

E-commerce 2.0

The concept of Ecommerce 2.0 means to make shopping over the Internet interactive, innovative and more user-friendly. The concept of E-Commerce 2.0 is based on few principles. The first principle is product availability. Increased availability of the product increases its visibility and marketing. Customers expect to find a good product, which is easily available, marketed, and found on a number of websites and online market sites like Amazon.com. The second principle is to discover niche markets for products. By doing so, marketers can take advantage of newly discovered sources of income. Third principle is that the customer rules. User-generated content created by buyers, like blog pages, product review sites, and social networking produces similar or greater impact on buyers that any advertising or product promotion campaign can have. Forth principle is to make shopping fun and painless. This could be done e.g. by building quick checkout process on the E-commerce sites. The fifth principle is integration and collaboration that provides uninterrupted user access and interaction with E-commerce sites. The environment of E-Commerce 2.0 consists of many connected systems that require a dynamic exchange of information.

Significance of E-commerce 2.0

There is couple of reasons why businesses should seriously think about E-commerce 2.0. First, E-commerce has become strategic. Businesses need to find differentiation. Second, many of the largest retailers are running systems that are many years old. To meet the continuously changing customer demands businesses would need major re-plat forming of their e-commerce sites (Rackspace, 2015). Third, Consumers are buying more online and they are buying products that are more complex. On the other side, they are becoming more precise in their product selection. Fourth, broadband Internet penetration has gotten to a point where retailers can safely start downplaying dial-up traffic and experimenting with richer content and experiences.

E-commerce 2.0 Trends

There are three major trends behind the evolution of E-commerce 2.0

RICH INTERNET APPS (RIAS): Rich Internet Applications are being used by more and more E-commerce businesses. RIA is a Web application that has many of the characteristics of desktop application software, typically delivered by way of a site-specific browser, a browser plug-in, an independent sandbox, extensive use of JavaScript, or a virtual machine. Take example of Gap's quick Look, Amazons & Angara's Diamond Search, My Rate plan's phone chooser, and Harley Davidson's bike configuration. With E-Commerce 2.0, RIAs will dominate and their main goal will be reducing shopping cart & checkout abandonment. These RIAs will also target search integration and optimizing consumers' browsing & product selection experiences to attract a bigger audience.

DISAGGREGATION: This disaggregation is brought about by the dual forces of focusing on core competencies and leveraging network effects. In E-commerce 2.0 era, this disaggregation is in the form of network-wide services that provide only a portion of the e-commerce experience yet benefit from focused network effects. Ratings and reviews are a good example. Bazaar Voice and Power Reviews are examples of E-commerce 2.0 aggregators, which bring ratings and reviews capabilities to all sites.

EMERGING SOCIAL COMMERCE: Social commerce today comes in two flavors: content-driven and interaction-driven or passive vs. active. Content and social interaction continues to influence what people buy. An engaging experience, easily available information and social circle feedback from others greatly influences purchases.

A COMPOSITE FRONT-END: Sites will be developed that provide a composite front end that integrates disaggregated services into a coherent, fluid user experience. Everything will be highly integrated from a data and user experience standpoint. The front end will initially run in parallel to the existing e-commerce site and eventually pieces of the front end will be embeddable into E-commerce 2.0 sites.

AN ARRAY OF TOOLS: There will be a sophisticated array of tools that would provide brand managers and merchandisers' control of content, promotions, design, layout, interactivity and analytics.

E-commerce 2.0 Features

E-commerce 2.0 has many prominent features (Wong, 2014).

SUPERIOR END-TO-END PRODUCT EXPERIENCE: E-commerce 2.0 companies are also focusing on the entire product experience, which includes marketing, packaging, and the post sales experience. Some good examples include Bavel (https://getbevel.com/) and Chubbies (http://www.chubbiesshorts.com/).

STORY BEHIND THE BRAND: Whether it is music, hobbies, or a sense of authentic goodwill, shared interests bring a new dimension to e-commerce. E-commerce 2.0 companies are using brand stories to reach customers at a personal level. Some good examples include House of Marley (http://www.thehouseofmarley.com/), and Herschel Supply Co. (http://www.herschelsupply.com/).

MOBILE AND SOCIAL FRIENDLY: In addition to having mobile friendly websites, E-commerce 2.0 companies are making social media an integrated part of their marketing initiatives for a better end user experience. Some good examples include Shwood (http://www.shwoodshop.com/) and Rent the Runway (https://www.renttherunway.com/rtr_home).

Software Agents

Software agents are a piece of software, which works for the user. It is not just a program but also a system situated within and a part of an environment that senses that environment and acts on it. Agents are classified into different types based on the characteristics they possess. A collaborative agent is a software program that helps users solve problem, especially in complex or unfamiliar domains by correcting errors, suggesting what to do next, and taking care of low-level details. Interface agents are computer programs that employ machine-learning techniques in order to assist a user dealing with a particular application. Mobile agents are used to solve many problem of network computing with minimum bandwidth and connectivity

Software agents are effective tool to virtual organizations since they provide mechanisms to automate several activities like, gathering data, refining information, negotiate business deals and also intelligent agents work like human beings in supplying and buying goods having the artificial machine knowledge. Software have variety of applications, which includes, B2B E-Commerce, Internet based info systems, robotics, smart systems, DSS, data mining and Knowledge discovery. Agent technology helps in finding intranet or internet, Customer relation management, supply chain management and market pricing.

There are many examples of uses of software agents. A buying agent, also known as shopping bot, helps the users to surf while finding the desired products and services. For example, when a person surfs for an item on eBay, at the bottom of the page there is a list of similar products. These are the products that other customers surfed while searching for the same product. This is because it is assumed the user tastes are relatively similar and they will be interested in the same products. A user agent, also known as personal agent, carries out user tasks automatically. For example, some user agents sort emails according to the user's order of preference, assemble customized news reports, or fill out webpage forms with the user's stored information. Monitoring and Surveillance agents, also known as predictive agents, are used to keep track of company inventory levels, observe competitor's prices and report them back to the company, watch stock manipulation by insider trading and rumors, etc. Data-mining agents can be used to find out the modern fashion in information from many different sources. For example, the agent that detects market conditions/changes and relays them back to a user/company so that the user/company can make decisions accordingly.

Internet of Things (IoT) and E-commerce

The Internet of Things is an expansive ecosystem of items connected to the internet, which can transmit and/or receive data to and from other objects. These "smart objects" are tagged with identifiers, which can be instantly tracked and inventoried by computers (Vaghasia, 2014). IoT-enabled devices can include smartphones, computers, cars, traffic lights, and anything else utilizing sensor applications such as ecommerce platforms, social media applications, traffic control systems, or manufacturing systems. For successful integration into the Internet of Things, users must incorporate all or most of the following elements, including network connectivity, user input and/or sensors to create or capture information, and backend or device computational abilities. Regardless of a company's size or industry, the Internet of Things can play a role in optimizing various facets of their products and operations. Because accurate data is collected and instantly transmitted to operating systems, IoT helps businesses from agricultural companies to healthcare organizations continually measure the efficacy of their devices and make adjustments where needed. All of this occurs without the need for additional personnel.

There are few things businesses should know before start using IoT for their advantage. Use of IoT will require considerable investment in additional resources such as security, infrastructure, applications, and analytics in addition to expanded bandwidth. The initial effort for getting started with IoT is cumbersome. Each item must be inventoried, bar-coded, and crosschecked before it can be brought online. Companies need considerable technical expertise at their disposal to set up the network, which can come from inside the organization or from outside experts. A very large amount of data present in virtual environment would pose security and privacy threats. An examination of the consequences of potential data compromises is a crucial step towards smart engagement with IoT. Businesses utilizing IoT can expect to generate more data than they will ever need or want to use. Businesses must prepare and equip themselves to analyze the most pertinent information in the most effective ways possible.

IoT can help ecommerce retailers deliver relevant information and offers to customers at just the right time and place. First, IoT can provide attendant-less smart vending machines with an internet connection that can sell small, fast-selling, or season-specific items. Second, using an internet connection, retailers can communicate to the vending-machine shopper through the mall's network infrastructure and a cloud-based server. Retailers can provide shoppers with information based on browsing history or sales activity and then inform them about future offerings. Shoppers can also communicate with retailers in order to do things like dynamically adjust pricing to move inventory, take advantage of weather conditions, or mall events. The vending machines can also communicate any functionality problems to service representatives and inventory levels to supply chain constituents to service, or restock the machines as needed. Retailers can use location-based technology to draw in customers. Retailers can enhance their loyalty programs and shoppers by sending them pop-up messages or alerts of in-store trends and new arrivals with time-bound discounts. As a result, shoppers will be enticed and more likely to enter a store, which will more likely turn a visit into a purchase. With location-aware sensors that communicate through a cloud-based server to the retailer's data center, a retailer can analyze regional, seasonal, and other micro-trends in a timely fashion. Sales data across all channels can be analyzed and correlated in real-time to leading indicators of demand (ex. extreme weather conditions, or fashion-trend comments on social media channels) to anticipate proper stock levels by style and location. As a result, shoppers will have several purchase options. This increases revenues, decreases markdowns, and develops a strong and reliable relationship between the customer and the retailer. "Smart mirror" functionality can assist shoppers with buying decisions, by displaying suggested items in the mirror of a changing room. This gives the shopper the impression that they are trying on the suggested items all at once, which saves both the shopper and the retailer time and effort. As a result, smart mirrors can make the shopping experience much easier for the shopper, resulting in more selling opportunities for the retailer. Connecting a customer with a retailer's app can increase in-store sales. For example, if a certain sized item does not fit while the shopper is trying it on, they can scan the item's tag. If the certain sized item were available in the store, the nearest sales associate would receive an alert to locate the item in the correct size and deliver it to the customer. If the correct size is not available at that particular store, the customer is instantly given the option to confirm a "hold" on the desired item at another nearby location, or to purchase the item through the retailer's app and having it scheduled for delivery at home (Vaghasia, 2014).

IoT is here to stay. More and more devices are being connected, enabling automated processes and customized offerings. Communication is becoming standardized, and more devices are coming with built-in connections to central systems. Today's shoppers are becoming used to having immediate access to information wherever they are. Adopting IoT is a natural choice for retailers whose value proposition is to meet the consumer's demand for a seamless shopping experience.

Review Questions

1. List some principal components of E-commerce infrastructure.

2. List some business processes involved in E-commerce.

3. List steps involed in processing a transaction in a typical E-commerce system.

4. List some major player s involved in E-commerce.

5. Differentiate between switching and routing.

6. Illustrate working of Internet using a labeled diagram.

7. Define protocol.

8. What is an IP address? Why do we need public and private IP addresses? What is the difference between IPV4 and IPV6 addressing?

9. What is the purpose of Domain Name System (DNS)?

10. Define Uniform Resource Locator (URL).

11. Differentiate between Internet server and clients.

12. List some e-mail protocols.

13. Define a) World Wide Web b) HTML c) HTTP.

14. List some markup languages used on the web.

15. List some scripting languages.

16. Differentiate between intranet, extranet, and virtual private network.

17. List some purposes of storage area networks.

18. List and discuss some Internet connectivity options.

19. Differentiate among Internet2, Web 2.0, Semantic Web, and E-commerce 2.0.

20. Explain the role of Internet of Things (Iot) in E-commerce.

Bibliography

Andam, Z. (2003, May). *e-Commerce and e-Business*. Retrieved October 15, 2009, from apdip.net: www.apdip.net/publications/iespprimers/eprimer-ecom.pdf

BHATNAGAR, S. (2008). TEXTBOOK OF COMPUTER SCIENCE: FOR CLASS XII. PHI Learning Pvt. Ltd.

Bonnema, D., & Stanford, J. (2002). *Internet 2: A White Paper*. Retrieved October 26, 2010, from homepages.wmich.edu: http://homepages.wmich.edu/~stanford/Internet2.htm

CBTDirect. (2010). *Microsoft VBScript*. Retrieved October 26, 2010, from cbtsys.com: http://www.cbtsys.com/online-training/it-training/Microsoft-VBscript.asp

Chick, douglas. (2011). From Computer Tech to Network Administrator (and everything in between) eBook: Douglas Chick: Books. Retrieved July 20, 2015, from http://www.amazon.com/Computer-Network-Administrator-everything-between-ebook/dp/B005JGZAS2/ref=sr_1_1?s=books&ie=UTF8&qid=1437384581&sr=1-1&keywords=9781937485443

Cisco. (2008, October 13). *How Virtual Private Networks Work*. Retrieved October 26, 2010, from cisco.com: http://www.cisco.com/en/US/tech/tk583/tk372/technologies_tech_note09186a0080094865.shtml

dataversity.net. (2014a, July 29). How GS1 is Shaping eCommerce in the Semantic Web. Retrieved from http://www.dataversity.net/gs1-shaping-semantic-web/

dataversity.net. (2014b, July 29). With Web 3.0, the Bots Can Do the Shopping for You. Retrieved from http://www.dataversity.net/web-3-0-bots-can-shopping/

Florida, U. o. (2010). *Using Synchronous Leased Lines*. Retrieved October 26, 2010, from www.stat.ufl.edu: http://www.stat.ufl.edu/system/man/portmaster/config/leased_line.fm.html

Franklin, C. (2000, August 7). *How DSL Works*. Retrieved October 26, 2010, from howstuffworks.com: http://computer.howstuffworks.com/dsl.htm

Geocaching. (2015). Geocaching - The Official Global GPS Cache Hunt Site. Retrieved July 20, 2015, from http://www.geocaching.com/seek/cache_details.aspx?wp=GC5P164&title=e-mail

Gubar, B. (n.d.). Semantic Web for e-Commerce. Retrieved June 28, 2015, from http://ercim-news.ercim.eu/en77/special/semantic-web-for-e-commerce

Housley, S. (2010). *What is Web 2.0?* Retrieved October 26, 2010, from www.rss-specifications.com: http://www.rss-specifications.com/what-is-web-2.htm

HowstuffWorks.com. (2000, May 03). *How does a T1 line work?* Retrieved October 26, 2010, from howstuffworks.com: http://computer.howstuffworks.com/question372.htm

Howstuffworks.com. (2001, April 03). *How does satellite Internet operate?* Retrieved October 26, 2010, from howstuffworks.com: http://www.howstuffworks.com/question606.htm

Hu, Y.-J. (2004, June 01). *The Semantic Web: Current Status and Future Direction.* Retrieved October 26, 2010, from www.cs.nccu.edu.tw: http://www.cs.nccu.edu.tw/~jong/pub/mis0601talk.pdf

Indiana University. (2015). What is FTP, and how do I use it to transfer files? Retrieved July 20, 2015, from https://kb.iu.edu/d/aerg

Internet2.edu. (2010). *About Us.* Retrieved October 26, 2010, from internet2.edu: http://internet2.edu/about/

Jeff, T. (2001, April 03). *How Internet Infrastructure Works.* Retrieved October 26, 2010, from howstuffworks.com: http://www.howstuffworks.com/internet/basics/internet-infrastructure.htm

Jonathan, S. (2007, December 28). *How Web 2.0 Works.* Retrieved October 26, 2010, from howstuffworks.com: http://www.howstuffworks.com/web-20.htm

JOSEPH, P. T. S. J. (2011). E-COMMERCE. PHI Learning Pvt. Ltd.

Kent, A., & Williams, J. G. (2002). Encyclopedia of Microcomputers: Volume 28 (Supplement 7). CRC Press.

Lassila, O., & Hendler, J. (2004, June 01). *Current Status and Future Promise of the Semantic Web", keynote address at Semantics 2006.* Retrieved October 26, 2010, from www.cs.nccu.edu.tw: http://www.cs.nccu.edu.tw/~jong/pub/mis0601talk.pdf

Layer2communications.com. (2010). *T1 Data Lines.* Retrieved October 26, 2010, from Layer2communications.com: http://www.layer2communications.com/T1_Info.asp

Lu, S., Dong, M., & Fotouhi, F. (2002). *The Semantic Web: opportunities and challenges for next-generation Web applications.* Retrieved October 26, 2010, from informationr.net: http://InformationR.net/ir/7-4/paper134..html

Marshall, B. (2010, April 01). *How Domain Name Servers Work.* Retrieved October 26, 2010, from howstuffworks.com: http://www.howstuffworks.com/dns.htm

Merkow, M. (2010). *Extraordinary Extranets.* Retrieved October 26, 2010, from www.webreference.com: http://www.webreference.com/content/extranet/

michaelaprile.wordpress.com. (2015, June 15). Russian LiFi System Uses Light to Transmitt Wireless Data Faster Than Wifi | THE TRUTHSEEKER'S JOURNAL. Retrieved July 20, 2015, from https://michaelaprile.wordpress.com/2015/06/15/russian-lifi-system-uses-light-to-transmitt-wireless-data-faster-than-wifi/

Miller, P. (2010, February 22). Putting the Semantic Web to work in e-Commerce with GoodRelations. Retrieved June 28, 2015, from http://www.zdnet.com/article/putting-the-semantic-web-to-work-in-e-commerce-with-goodrelations/

Murph, D. (2011, February 8). Study: 802.11ac devices to hit the one billion mark in 2015, get certified in 2048. Retrieved August 15, 2015, from http://www.engadget.com/2011/02/08/study-802-11ac-devices-to-hit-the-one-billion-mark-in-2015-get/

Neogy, J. L. (2009). Set-Rapidex Office Secretary Course. Pustak Mahal.

Oak, M. (2010). *How Does the World Wide Web Work.* Retrieved October 26, 2010, from buzzle.com: http://www.buzzle.com/articles/how-does-the-world-wide-web-work.html

Owano, N. (2014, July 10). Alcatel-Lucent sets broadband speed record using copper. Retrieved July 20, 2015, from http://phys.org/news/2014-07-alcatel-lucent-broadband-copper.html

purelifi.com. (n.d.). What is LiFi? Retrieved from http://purelifi.com/what_is_li-fi/

Rackspace. (2015, July 16). Ecommerce Re-Platforming Basics | Knowledge Center | Rackspace Hosting. Retrieved July 20, 2015, from http://www.rackspace.com/knowledge_center/whitepaper/ecommerce-re-platforming-basics

Ritual.org. (2015). What is Usenet? Retrieved July 20, 2015, from http://www.ritual.org/summer/pinn/usenet.htmld/index.html

Roggio, A. (2014, December 23). 6 Ecommerce Design Trends for 2015. Retrieved from http://www.practicalecommerce.com/articles/77729-6-Ecommerce-Design-Trends-for-2015

Rouse, M. (2015). What is finger? - Definition from WhatIs.com. Retrieved July 20, 2015, from http://searchsoa.techtarget.com/definition/finger

S.k.bansal. (2009). Dictionary of it Terms. APH Publishing.

SANS, I. (2001, December 18). *Security Considerations for Extranets.* Retrieved October 26, 2010, from www.sans.org: http://www.sans.org/reading_room/whitepapers/basics/security-considerations-extranets_527

Schneider, G. (2010). *Electronic Commerce, 9th Ed.* Course Technology.

Shabbir. (2014, June 14). HTML4 Vs HTML5 Comparison - Go4Expert. Retrieved July 20, 2015, from http://www.go4expert.com/articles/html4-vs-html5-comparison-t30141/

Shoppingcart.com. (2010). *Principles Behind ecommerce 2.0*. Retrieved October 26, 2010, from www.shoppingcart.com: http://www.shoppingcart.com/principles-behind-ecommerce-2-0/

Traxler, D. (2012, May 18). 7 Key Ecommerce-Infrastructure Decisions | Practical Ecommerce. Retrieved July 17, 2015, from http://www.practicalecommerce.com/articles/3545-7-Key-Ecommerce-Infrastructure-Decisions

Tufnell, N. (2013, November 5). Chinese professor builds DIY Li-Fi system (Wired UK). Retrieved June 28, 2015, from http://www.wired.co.uk/news/archive/2013-11/05/chi-nan-li-fi

University of Maryland, N. R. (2006, November). *Future Promise of the Semantic Web*. Retrieved October 26, 2010, from www.cs.rpi.edu: http://www.cs.rpi.edu/academics/courses/fall07/semantic/Semantics2006-keynote.ppt

University of Miami, M. S. (2010). *Public vs. private networks*. Retrieved October 26, 2010, from it.med.miami.edu: http://it.med.miami.edu/x198.xml

Vaghasia, R. (2014, July 20). How the Internet of Things is Shaking up Ecommerce - AccuWebHosting Blog. Retrieved July 20, 2015, from https://www.accuwebhosting.com/blog/2014/07/20/internet-things-shaking-ecommerce/

Vishanth, W. (2012). E-Government Services Design, Adoption, and Evaluation. IGI Global.

W3Schools.com. (2010). *HTML Introduction*. Retrieved October 26, 2010, from W3Schools.com: http://www.w3schools.com/html/html_intro.asp

W3Schools.com. (2010). *XML Introduction*. Retrieved October 26, 2010, from W3Schools.com: http://www.w3schools.com/XML/xml_whatis.asp

Webopedia. (2005, June 03). *Cable vs. DSL*. Retrieved October 26, 2010, from webopedia.com: http://www.webopedia.com/DidYouKnow/Internet/2005/cable_vs_dsl.asp

Wikipedia. (2015, June 20). Li-Fi. In Wikipedia, the free encyclopedia. Retrieved from https://en.wikipedia.org/w/index.php?title=Li-Fi&oldid=667730707

Wong, K. (2014, June 8). E-Commerce 2.0: Six Companies You Can Learn From. Retrieved July 20, 2015, from http://www.forbes.com/sites/kylewong/2014/08/06/ecommerce-2-0-six-companies-you-can-learn-from/

Zaino, J. (2014, November 20). Web Components: Even Better With Semantic Markup. Retrieved from http://www.dataversity.net/web-components-even-better-semantic-markup/

3

ELEMENTS OF E-COMMERCE

Learning Objectives

After reading this chapter, reader should be able to:

- Understand how client-server communication works.

- Describe types of web content.

- Understand and describe client-side and server-side scripting.

- Undersytand varisous client-side and server-side scripting languages

- Describe software for web server including operating system software, web server software, and database system software.

- Identify and understand the major considerations involved in choosing Web server software.

- Discuss E-mail spam and its solutions.

- Understand content filtering and its techniques.

- Discuss various website and Internet utility programs that can improve Web site performance.

- Understand the considerations involved in choosing the most appropriate hardware for an E-commerce site.

- Describe the basic functions of E-commerce software.

- Understand different models of web server-hardware architecture.

- Describe the advanced functions of E-commerce software.

- Describe various E-commerce software options for small to large businesses.

- Understand service-oriented architecture (SOA), infrastructure-as-a-Service (IaaS), and Platform-as-a-Service (PaaS) and their relationship to E-commerce application development.

- Discuss various E-commerce application outsourcing options, including application service providers (ASPs), software as a service (SaaS), infrastructure-as-a-Service (IaaS), and Platform-as-a-Service (PaaS) and utility computing.

- Discuss the role of cloud computing in E-commerce

Introduction

E-commerce infrastructure contains many pieces of hardware and software. These include client computers, wireless devices, web server software and hardware, operating system software, utility software for web server that are used to perform utility functions such as site maintenance, diagnostics, and e-mail management, and E-commerce software that accomplishes specific electronic commerce functions, such as order entry and processing, content management, content delivery, security, and payment processing.

Clients and Servers

Most often, clients' communication with E-commerce website is based on a client-server communication model. A server is a host on the Internet that performs some kind of service (e.g. network resource management) and fulfills client requests for such services. The clients take the lead in communication. In general, a limited number of clients can all access the server at the same time. User inputs request to the clients and clients process user input and send the request to the server. When server sends back the results of the request, clients show these results to the user.

There are many types of servers, including web servers, e-mail servers, database servers and file servers. Multiple servers may provide a single service or multiple services may be provided by a single server. For example, one server may act both as a Web server and a file server and multiple servers may act as a web server. Web pages are stored at a web server and web server delivers these pages to clients upon request. A Web browser uses the hypertext transfer protocol (HTTP) to request and transfer pages from a Web server. An FTP server stores files and documents and delivers them to the clients upon request. The model of client-server is shown in Figure 3-1.

FIGURE 3-1: CLIENT-SERVER COMMUNICATION

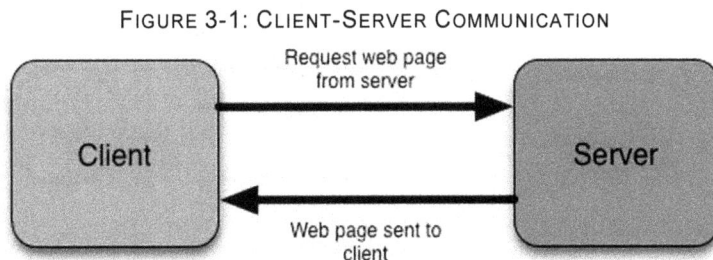

Request web page from server

Client

Server

Web page sent to client

Web Content

Content on a website can be classified into two broad categories: static and dynamic.

Static Content

Static content is stored in a static web page and is always displayed the same way whenever a user requests this page from a browser. An example of a static Web page is a help page that contains contact information (such as return policy or store locations) that does not change frequently. Static web pages are created using HTML. These pages can contain HTML tags and text, images, and animation. However, these pages do not any information contained in a database.

Dynamic Content

Dynamic content is a type of web content that changes frequently. Some examples of dynamic content include animations, video, and audio files. A dynamic page is assembled from back-end databases and internal data on the web site. It is a specific response to a client's request. For example, if a Web client inquires a web server about the status of an existing order by entering a unique order number into a form, the Web server generates a dynamic web page providing status of that particular order. These days most websites on the Internet have dynamic content.

Performance of a web server is affected by the mix and type of Web pages it delivers. The dynamic web pages itself is just a template to display the results of client query and most of the content is provided by website database and no change is made in HTML code of the page. If there is any change in the information, the data is changed in the database.

Contents of static web pages do not change unless the changes are coded into the HTML. Static page delivery requires less computing power than dynamic page delivery. A server delivering mostly static web pages performs significantly better than the same server delivering dynamic web pages does.

The basic idea behind dynamic pages is very simple. Instead of preparing all necessary pages as individual files, you write a program that will create the pages as they are requested by the user. The program itself can be do whatever you want; the only limitation is that upon execution it should produce a HTML page. The Figure 3-2 shows how a dynamic page is processed.

FIGURE 3-2: DYNAMIC PAGE PROCESSING

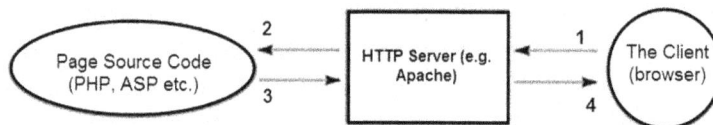

The client asks the server for a specific page. The server finds the file and according to some rules in its configuration determines that it is a dynamic page - that is it should not return the document itself but rather run it to obtain the resulting document. The server executes the source of the page (more on this below) and reads the output. The server then sends the output to the client. For the client it is undistinguishable from a static page.

The most important part is the execution of the script. In the simplest case of CGI, there is a program associated with the page that is responsible for generating the resulting document. In more sophisticated cases, an interpreter of the source code is built directly into the web-server (either hard-coded or as a module) and takes care of the execution inside the server.

DYNAMIC PAGE-GENERATION TECHNOLOGIES

There are two types of technologies used to generate dynamic content: Server side scripting and client side scripting.

SERVER-SIDE SCRIPTING

Server-side scripts are contained in HTML files. Generally, if a client requests from web server an HTML file that contains a server-side script, the script is executed by the server before the file is returned to the browser as plain HTML. A Scripting language is a variety of 'programming' that produces ASCII text-based scripts, which are usually designed for writing small program. Scripting languages support high-level language control features such as selection and iteration (syntax) but they are not considered programming languages as such. Often referred to as 'glue-code' they are instead, seen as being an enhancement of particular software packages or applications. Server-side scripts can perform many functions including dynamically changing or editing web page content, respond to user queries, and providing customized web pages according to client's needs. Server-side scripts provide security because their code cannot be viewed from a browser and user may not even be aware that a script was executed. Client-side scripts may be embedded in documents produced by server-side scripts.

Dynamic Web pages created with server side scripting often use server-side languages such as PHP, Perl, ASP, ASP.NET, JSP, and ColdFusion. Common Gateway Interface (CGI) is typically used by these server-side languages to create dynamic web pages. Server-side scripts are mixed with HTML-tagged text to create the dynamic Web page. Java, a programming language created by Sun, can be used to produce dynamic pages. Such server-side programs are called Java servlets. Server-side scripts require their language's interpreter to be installed on the server in order to produce the same output regardless of the client's browser, operating system, or other system details.

Since the content called up by server-side scripting resides in a database, it can be virtually anything. Content such as product descriptions, price variations, weblog entries, images, and even formatting is stored in the database. It is also possible to nest one server-side script snippet within another, where the first script grabs certain data from the database, but also calls a second script, which accesses secondary data. This is useful, for example, when delivering some custom standard content with additional nested content that may be time sensitive or yet unknown, such as comments on a weblog entry. Web site maintenance, then, involves updating data in the database, which will then affect every page on the website with the included script to call that data.

COMMON GATEWAY INTERFACE (CGI)

In order to deliver dynamic content we need to extend the abilities of the web server so that it can do more than merely send static web pages in response to client requests. The common gateway interface (CGI) provides a mechanism to do this. When serving static web pages the server is normally asked for a file that has a .htm or .html extension e.g. http://mywebsite.com/index.html. If we wish to serve a dynamic page, the extension would be different, for example .cgi or .php. If a request comes in with one of these extensions, the web server knows to pass the request to the CGI, which then interprets it correctly and executes the script in the particular scripting language. Once the script has finished executing the CGI then passes the output back to the web server to be delivered as a response to the client request. The Figure 3-3 illustrates this process.

FIGURE 3-3: PROCESSING OF CGI SCRIPT

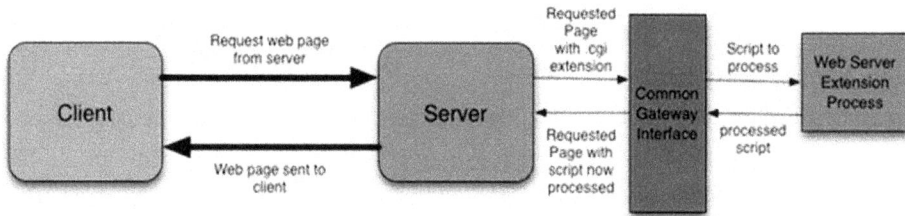

ACTIVE SERVER PAGES (ASP)

ASP stands for Active Server Pages. Active Server Pages are HTML pages with embedded ASP scripts that are processed on the server before the page is sent to the user. ASP allows creating dynamic database driven pages; a user can access data in a database, and interact with page objects such as Active X or Java components. Active Server Pages (ASP), also known as *Classic ASP* or *ASP Classic*, was Microsoft's first server-side script-engine for dynamically generated web pages. ASP.NET has superseded ASP. Web pages with the *.asp* file extension use ASP, although some web sites disguise their choice of scripting language for security purposes (e.g. still using the more common *.htm* or *.html* extension). Pages with the *.aspx* extension use compiled ASP.NET (based on Microsoft's .NET Framework), which makes them faster and more robust than server-side scripting in ASP, ASP.NET pages may still include some ASP scripting. Programmers write most ASP pages using VBScript. JScript (Microsoft's implementation of ECMAScript) is the other language usually available. PerlScript (a derivative of Perl) and others are available as third-party installable Active Scripting engines. When URL is typed in the Address Box or when one clicks on a web page, request is asking the web server to send a file to computer; if the file is standard HTML, then, when web browser receives the web page it will look the same as it did on the web server. However if an ASP file is sent to computer from the web server, firstly, the server will run the HTML code; and then, run the ASP code. Figure 3-4 shows a HTML page is processed. Figure 3-5 shows how an ASP page is displayed.

FIGURE 3-4: PROCESSING OF HTML PAGES

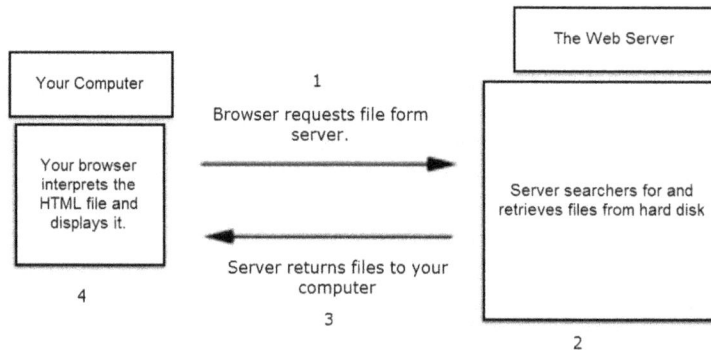

FIGURE 3-5: PROCESSING OF ASP PAGE

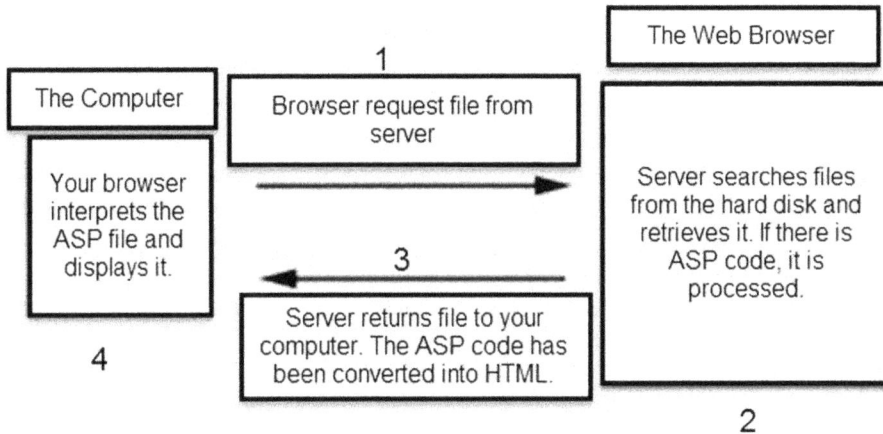

ASP.NET supports three approaches to build web sites: Web pages, Web forms, and MVC (Model View Controller). Web Pages Provides a simple way to seamlessly connect to a database or add in dynamic server code into HTML. Web forms offer a familiar model that allows reuse of controls and incorporate data into Web site, making work reusable and more productive. ASP.NET MVC gives powerful, patterns-based way to build dynamic websites that enables a clean separation of concerns and that gives full control over markup for enjoyable, agile development. ASP.NET MVC includes many features that enable fast, TDD-friendly development for creating sophisticated applications that use the latest web standards.

ASP technology is easy to learn, though programming is required. It is built into Windows 2000 Server family. Professional support for ASP is also available. However, ASP technology is not free (unless you are already using a Windows server). Specialized functionality may require you to purchase commercial components, or develop them yourself in C/C++ or Visual Basic. If you are familiar with C-style syntax (shared by C/C++, Java, Perl, PHP, etc.) you may face some difficulties in familiarizing yourself with VBScript.

ADOBE COLD FUSION

Cold Fusion is tag-based web development platform. Cold Fusion is a commercial server platform. Its use will cost you money to run it whether you set up your own server or rent space on a Web hosting provider. Unlike ASP, however, Cold Fusion is not tied to the Windows platform. Cold Fusion can integrate just as easily with Apache running under Linux as it can with IIS on Windows Server. Cold Fusion is extremely easy to learn, powerful, and very scalable. Cold Fusion also supports both software- and hardware-based server clustering. Professional support is available. Cold Fusion is cross platform. Despite these advantages, Cold Fusion is expensive and programmers may not like the tag-based development methods used in Cold Fusion.

JAVA SERVER PAGES (JSP)

Java Server Pages is a technology for developing web pages that include dynamic content. Unlike a plain HTML page, which contains static content, a JSP page can change its content based on any number of variable items, including the identity of the user, the user's browser type, information provided by the user, and selections made by the user. This functionality is the key to web applications such as online shopping and employee directories, as well as for personalized and internationalized content.

A JSP page contains standard markup language elements, such as HTML tags, just like a regular web page. However, a JSP page also contains special JSP elements that allow the server to insert dynamic content in the page. JSP elements can be used for a variety of purposes, such as retrieving information from a database or registering user preferences. When a user asks for a JSP page, the server executes the JSP elements, merges the results with the static parts of the page, and sends the dynamically composed page back to the browser. JSP pages use XML tags and scriptlets written in the Java programming language to encapsulate the logic that generates the content for the page. It passes any formatting (HTML or XML) tags directly back to the response page. In this way, JSP pages separate the page logic from its design and display. JSP technology is part of the Java technology family. JSP pages are compiled into servlets. These pages may call JavaBeans components or Enterprise JavaBeans components to perform processing on the server. As such, JSP technology is a key component in a highly scalable architecture for web-based applications. JSP pages are not restricted to any specific platform or web server. The JSP specification represents a broad spectrum of industry input.

JSP are extremely powerful and scalable. They are cross-platform. Most Java server plugins are free for personal and development purposes. However, Java takes a lot of work to learn. Most Java server plugins must be paid for if they are to be used to host a commercial Web site. JSP Servlets are Java programs that are already compiled which also creates dynamic web content. Figure 3-6 shows how JSP works.

FIGURE 3-6: WORKING OF JSP PAGE

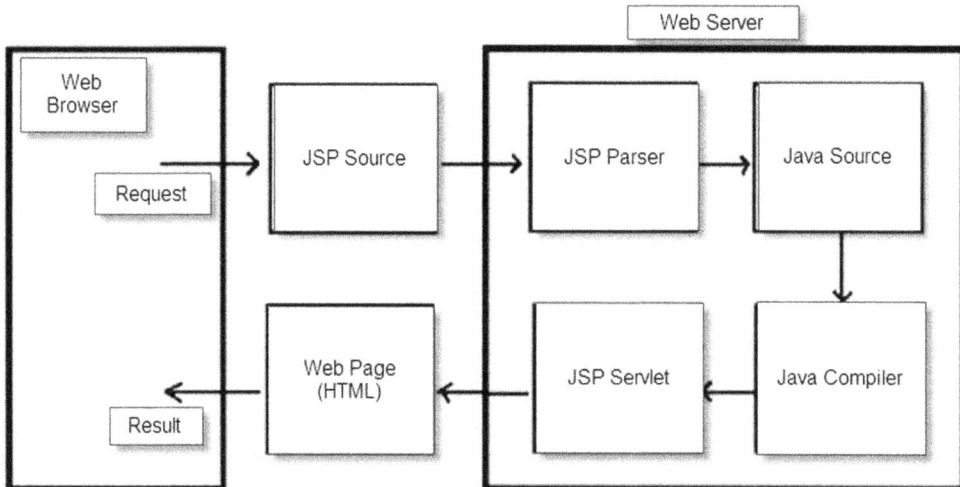

PREHYPERTEXT PROCESSOR (PHP)

PHP is one of the most prevalent Web programming languages for creating dynamic Web content. Its similarity to C's syntax and open-source nature make PHP relatively easy to learn. PHP is a scripting language designed for producing dynamic Web content. PHP originally stood for "Personal Home Page," but as its functionality expanded, it became "PHP: Hypertext Processor" because it takes PHP code as input and produces HTML as output. Although PHP is typically used for Web-based applications, it can also be used for command-line scripting.

PHP is open-source and free. PHP interpreted rather than compiled like C language. It is an embedded scripting language, meaning that it can exist within HTML code. It is a server-side technology; everything happens on the server as opposed to the Web browser's computer, the client. It is cross-platform, meaning it can be used on Linux, Windows, Macintosh, etc., making PHP very portable. PHP does not handle client-side tasks such as creating a new browser window, adding mouse-overs, determining the screen size of the user's machine, etc. Such tasks are handled by JavaScript or Ajax (Asynchronous JavaScript and XML). A large developer community supports PHP and web developers can benefit from a large variety of available PHP libraries.

When a URL is typed in a web browser, the browser sends a request to the web server. The web server than calls the PHP script on that page. The PHP module executes the script, which then sends out the result in the form of HTML back to the browser, which can be seen on the screen. With the static HTML delivery process, the Web server sends the exact content of the requested page to the client without any additional processing. With dynamic Web content creation, PHP acts as a filter by processing PHP code to produce HTML markup that is ultimately sent to the user's browser. Figure 3-7 shows working of PHP script.

FIGURE 3-7: WORKING OF PHP SCRIPT

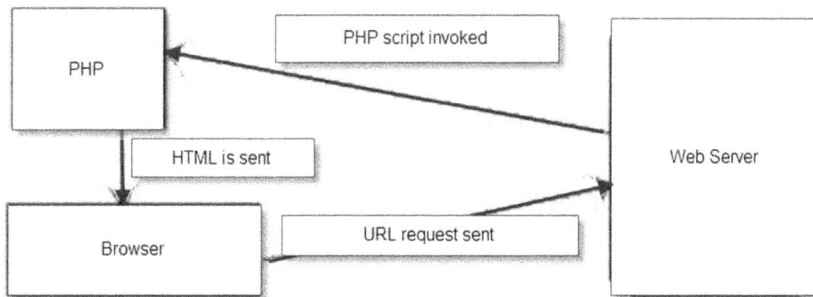

PERL

Perl is a high-level general-purpose programming language used especially for developing Web applications. Perl is an interpreted dynamic programming language. The languages in this family include Perl 5 and Perl 6. Perl was originally developed by Larry Wall in 1987 as a general-purpose UNIX scripting language to make report processing easier. Since then, it has undergone many changes and revisions. The latest major stable revision of Perl 5 is 5.18, released in May 2013. Perl 6, which began as a redesign of Perl 5 in 2000, eventually evolved into a separate language. Both languages are being developed independently by different development teams borrowing ideas from one another. The Perl languages borrow features from other programming languages including C, shell scripting, AWK, and sed. They provide powerful text processing facilities without the arbitrary data-length limits of many contemporary UNIX command line tools, facilitating easy manipulation of text files. Perl 5 gained widespread popularity in the late 1990s as a CGI scripting language, in part due to its parsing abilities. In addition to CGI, Perl 5 is used for graphics programming, system administration, network programming, finance, bioinformatics, and other applications. It is nicknamed "the Swiss Army chainsaw of scripting languages" because of its flexibility and power, and possibly also because of its "ugliness". In 1998, it was also referred to as the "duct tape that holds the Internet together", in reference to both its ubiquitous use as a glue language and its inelegance.

Perl is a mature language with over a decade of history, so the likelihood to run into bugs in Perl is very less. Perl is free and most good Web hosts support Perl. A vast worldwide network of dedicated Perl developers supports Perl. Despite these advantages, Perl does not scale especially well on busy servers. Perl language generally provides multiple ways of doing same thing. That could make learning from the work of others difficult if they did not use the same method you have used. Perl is optimized for the UNIX platform. Perl is not a good choice for Windows servers. Perl is Open Source software, which means there is no formal support for it. Despite the huge support community available to you, ultimately you will have to count on yourself for any unresolved matter.

PYTHON

Python is a widely used general-purpose, high-level programming language. Its design philosophy emphasizes code readability, and its syntax allows programmers to express concepts in fewer lines of code than would be possible in languages such as C. The language provides constructs intended to enable clear programs on both a small and large scale. Python supports multiple programming paradigms, including object-oriented, imperative and functional programming or procedural styles. It features a dynamic type system and automatic memory management and has a large and comprehensive standard library. Like other dynamic languages, Python is often used as a scripting language, but is also used in a wide range of non-scripting contexts. Using third-party tools, Python code can be packaged into standalone executable programs (such as Py2exe, or Pyinstaller). Python interpreters are available for many operating systems.

DJANGO

Django is a Python-based free open framework for web applications. Django's primary goal is to ease the creation of complex, database-driven websites. Django emphasizes reusability and "pluggability" of components, rapid development, and the principle of do not repeat yourself (DRY). In software engineering, do not repeat yourself (DRY) is a principle of software development, aimed at reducing repetition of information of all kinds, especially useful in multi-tier architectures. The DRY principle states, "Every piece of knowledge must have a single, unambiguous, authoritative representation within a system."

RUBY ON RAILS (ROR)

Ruby on Rails, often called Rails, is an open source framework for web application development. RoR runs on the Ruby programming language, which is a dynamic and object-oriented language. It is a full-stack framework: it allows creating pages and applications that gather information from the web server, talk to or query the database, and render templates out of the box. As a result, Rails features a routing system that is independent of the web server. The language (Ruby) on the framework (Rails) means that you can accomplish more with less, better structured code. Since code can be re-used, changes are easy, making iteration and experimentation painless. This means that you can better use your time. For instance, RoR will handle all database communications, provide a template system for handling page sections and layouts, process Ajax updates and a wide set of plugins that make feature implementing easy. In other words, RoR takes care of the boring stuff while you can focus on the cool. Ruby syntax is ideal for those who prefer pattern-matching characters because it uses differing characters as analogues to keywords. It is built for speed and adaptation and provides strong support community both online and offline. Some websites developed using RoR include Hulu, Square, and Airbnb.

Client-Side Scripting

Client-side scripting refers to programming the behavior of a browser. Client-side scripts are attached to HTML documents that run on a user's browser when a user views these HTML documents. Two most commonly used client-side scripting languages are JavaScript and VBScript. There are other scripting languages as well. **Lingo** is the scripting language of Adobe Director. Adobe Director is an authoring system for develop high-performance multimedia content and applications for CDs, DVDs and the Internet. **AppleScript** is a scripting language for the Macintosh that allows the user to send commands to the operating system, for example open applications, and carry out complex data operations. Client-side scripts may also perform actions in response to certain user actions, (e.g., clicking a button). Such actions can be performed without further communication with the server. Client-side scripts provide increased access to information and functionality available on a user web browser. Client-side scripts do not require additional software on the server however; user's web browser must understand the scripting language in which scripts are written. Client side scripts can be used to reduce overhead on server's limited resources if there are a large number of users. In general, security restrictions do not allow client-side scripts to access the user's computer resources other than the web browser software.

Methods Used to Implement Client-side Scripts

Client-side scripts can be embedded within HTML documents in a number of ways. Here we shall discuss some ways by taking JavaScript as example. Client-side JavaScript code can be embedded within HTML documents in a number of ways:

- Between a pair of script tags
- From an external file specified by the src attribute of a script tag
- In an event handler, specified as the value of an HTML attribute such as onclick or onmouseover
- In a URL that uses the special javascript: protocol

Between a Pair of Script Tags

<html>

<head>

```
<title>Today's Date</title>

<script language="JavaScript">

// Define a function for later use

function print_todays_date( ) {

    var d = new Date( );            // Get today's date and time

    document.write(d.toLocaleString( ));  // Insert it into the document

}

</script>

</head>

<body>

The date and time are:<br/>

<script language="JavaScript">

  // Now call the function we defined above

  print_todays_date( );

</script>

</html>
```

Externally Stored Code

The script tag supports a src attribute that specifies the URL of a file containing JavaScript code. It is used like this:

```
<script src="../../scripts/util.js"></script>
```

A JavaScript file typically has a .js extension and contains pure JavaScript, without script tags or any other HTML.

A <script> tag with the src attribute specified behaves exactly as if the contents of the specified JavaScript file appeared directly between the <script> and </script> tags. Any code or markup that appears between these tags is ignored. Note that the closing </script> tag is required even when the src attribute is specified, and there is no JavaScript between the <script> and </script> tags.

There are a number of advantages to using the src attribute. It simplifies your HTML files by allowing you to remove large blocks of JavaScript code from them-that is, it helps keep content and behavior separate. When you have a function or other JavaScript code used by several different HTML files, you can keep it in a single file and read it into each HTML file that needs it. This makes code maintenance much easier. When JavaScript functions are used by more than one page, placing them in a separate JavaScript file allows them to be cached by the browser, making them load more quickly.

Differences between Client-side and Server-side Scripting Environment

Client-side Environment

The client-side environment used to run scripts is usually a browser. The processing takes place on the end users computer. The source code is transferred from the web server to the user's computer over the internet and run directly in the browser. The scripting language needs to be enabled on the client computer. Sometimes if a user is conscious of security risks, they may switch the scripting facility off. When this is the case, a message usually pops up to alert the user when script is attempting to run.

Server-side Environment

The server-side environment that runs a scripting language is a web server. A user's request is fulfilled by running a script directly on the web server to generate dynamic HTML pages. This HTML is then sent to the client browser. It is usually used to provide interactive web sites that interface to databases or other data stores on the server. This is different from client-side scripting where scripts are run by the viewing web browser. The primary advantage to server-side scripting is that you can customize the response based on the user's requirements, access rights, or queries into data stores.

Advantages and Disadvantages of Scripting Languages

Scripting languages are becoming more popular due to the emergence of web-based applications. The market for producing dynamic web content is now expanding rapidly such that new scripting languages have been developed to allow users with little or no programming expertise to develop interactive web pages with minimum effort. Also the increases in computer performance over the past few years has promoted an increase in the power and sophistication of scripting languages that, unlike conventional programming languages, can even have certain security features built-in.

Scripting languages provides many benefits. These languages are easy to learn and use and minimum programming knowledge or experience is required. These languages allow complex tasks to be performed in relatively few steps you can add dynamic and interactive activities to web pages.

The major disadvantage of scripting languages is that executable code can inadvertently be downloaded from a remote server to a web browser's machine, installed and run using the local browser's interpreter. This is easily done by visiting suspicious web sites or downloading programs (with no authenticity) and user is probably unaware of what is happening. This is a major drawback in the formal rules defining scripting languages like JavaScript and VBScript.

Uses of Scripting Languages

Typical uses of scripting languages including JavaScript include:

IMAGE OR TEXT ROLLOVERS: If the user rolls the mouse over a graphic or hypertext then a text or graphic box will appear.

POP-UP WINDOWS: A pop-up window appears in a separate window from the Web page that triggered it. Pop-up displays information and is useful if the user wants to perform a simple calculation or consult a calendar for inputting dates. This is achieved by embedding ActiveX controls or Java applets into the script.

FIELD CONTENT VALIDATION: When filling in forms, each field, especially required fields denoted by an asterisk, are validated for correct input. If the field is left blank or incorrect information entered then a user message will be generated and you may not continue.

Tools for Producing and Running Scripts

You can use Notepad application to produce client-side scripts and server-side scripts. To maximize the full capabilities of what a server-side script can do, it is necessary to use a professional integrated development environment (IDE). There are various IDE's available, depending on the language you will be using and what resources will be accessed. If you plan to use Visual Basic, then the Microsoft tools are ideal such as Microsoft Visual Studio or the VB.Net applications. IDEs help speed development using wizards that auto generate code as you visually link components etc.

On-line Database Accessibility

If you want to link your application to a database then you must make use of server-side scripting. The tools make it easy to link to databases. Web Database follows the client-server database model. A Database Engine sits on some central computer somewhere and serves data to multiple web-based clients. Because we are dealing with web-based clients however, we must also have a Web Server that handles requests from web-browsers and then forwards them to the Database. Likewise, the web server will wait for the database to respond and then pass on that response to the waiting web browsers. Figure 3-8 illustrates this process.

FIGURE 3-8: DATABASE CONNECTIVITY

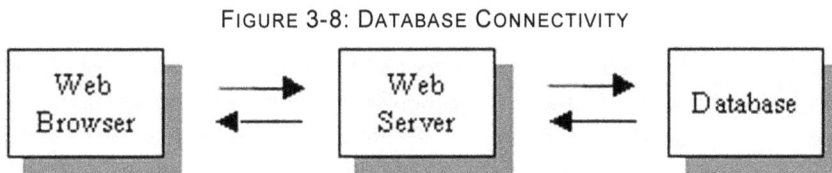

Web Servers are built to talk to Web Browsers and not to talk to Databases. Thus, in order for the Web Server to talk to a Database, it requires a helper (sometimes called Middle Ware). The most basic type of Middle Ware is a CGI script that is written to translate requests from the Web Server to a format that the Database can understand, and to translate Database responses into something the Web Server can send back out to the Web Browser and that the person using the web browser can understand. The CGI Script will be responsible for understanding what the Web Server is saying and will be responsible for knowing how to talk to the Database. CGI script can be written in any computer programming language and can use of a variety of methods to talk to the Database.

FIGURE 3-9: DATABASE CONNECTIVITY USING CGI SCRIPTS

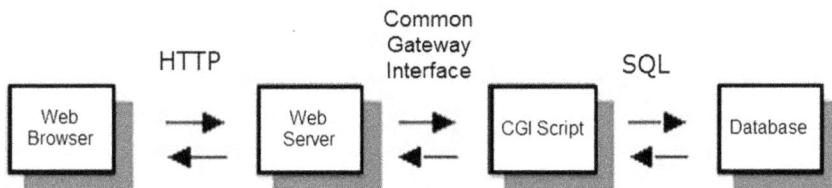

Software for Web Servers

Every piece of software can be broadly classified into two categories: Open-source software and commercial software.

Open Source Software vs. Commercial Software

Open Source is software whose source code is typically made available royalty-free to the users of the software, under terms allowing redistribution, modification and addition, though often with certain restrictions. The support, training, updates and other services for the software may be provided by a range of entities. Open-source software is often developed by a community of programmers. Other programmers then use the software, work with it, and improve it. Those programmers can submit their improved versions of the software back to the community. Open-source software generally costs less than the commercial software.

Commercial Software is developed by a commercial firm and typically licensed for a fee to a customer. The commercial firm often provides support, training, updates and other similar services for the software. Distribution of the source code is available only to certain users through special licensing or other agreements.

Web Server Software

Web server is a computer that delivers content (such as Web pages) using the Hypertext Transfer Protocol (HTTP). Any computer with suitable hardware can be turned into a web server by installing web server software and connecting the machine to the Internet. There exist many web server software such as open-source software from NCSA and Apache, and commercial software from Microsoft, Netscape and others. A client (commonly a Web browser) initiates communication with web server by making a request for a specific resource using HTTP. The server responds with the content of that resource, or returns an error message if unable to do so. The error message may include some information to explain the problem to users. While the primary function of a web server is to serve content, it can also receive content from clients e.g. submitting Web forms and uploading of files. Web servers can also be found in devices such as printers and routers. In these devices, web servers provide administrative user interface in the form of a web page. Web server software requires operating system software to operate. Some Web server software can run on only one operating system, while some can run on several operating systems. Companies often run other programs (e.g. Internet utilities and e-mail software) on web servers as part of their E-commerce operations.

BASIC FUNCTIONS OF WEB SERVER SOFTWARE

Web server software differs in detail but share some basic functionality.

HTTP REQUESTS PROCESSING: Every web server accepts HTTP requests from the clients and provides an HTTP response to the client. The HTTP response may consist of an HTML document, a raw file, an image, or some other type of document.

LOGGING: Web server can record detailed information about client requests and server responses to log files. Website administrators can then run log file analyzer software on these log files to collect useful statistics about site usage.

SECURITY: Web servers can provide security by performing authorization (request of user name and password) before allowing client access to any resource. Web servers can also process digital certificates and provide private/public key information required for credit card processing. Web servers also support HTTPS (by SSL or TLS) to allow secure (encrypted) connections to the server. Web servers also support basic and digest HTTP authentication.

STATIC AND DYNAMIC CONTENT HANDLING: Web server can provide both static and dynamic content. Static content can be saved in server's file system(s) and provided upon client request. Web server provides dynamic content by supporting one or more related technologies such as JSP, PHP, ASP, and ASP .NET etc.). Web servers also perform processing of FTP requests.

CONTENT COMPRESSION: Web servers can provide content compression to reduce the size of the responses. Content compression lowers bandwidth usage because less data is to be sent on network.

VIRTUAL HOSTING: Using virtual hosting, web servers can host multiple websites using a single IP address.

LARGE FILE SUPPORT: Using this functionality, web servers can store files of size greater than 2 GB.

TRAFFIC SHAPING: Also known as "packet shaping", it is a computer network traffic management technique which delays some or all datagrams to bring them into compliance with a desired traffic profile. Traffic shaping is used to optimize or guarantee performance, improve latency, and/or increase usable bandwidth for some kinds of packets by delaying other kinds.

BANDWIDTH THROTTLING: Using this feature, web servers can limit the speed of responses to client queries. This way network is not saturated and server can properly respond to client requests.

SEARCH CAPABILITY: Web servers also provide search capability. Users can use search function to search the site with the keywords.

E-MAIL: Web servers can provide e-mail service. E-mail provides web server users ability to send, receive, and store e-mail.

SITE MANAGEMENT: Site management tools help webmasters manage critical aspects and configuration of several features (e.g. dead links) of websites, and calculating and displaying key site statistics (such as unique visitors, page requests, and the origin of requests).

MODULE SUPPORT: Web server can extend their capabilities by adding or modifying software modules. These modules enhance functionality. These modules can be part of the core server software or added on demand.

REVERSE PROXY FUNCTIONALITY: A reverse proxy is a type of proxy server that retrieves resources on behalf of a client from one or more servers.

CHROOT SUPPORT: A chroot environment can be used to create and host a separate virtualized copy of the software system. This can be useful for testing and development.

URL REWRITTING: Rewritten URLs (sometimes known as short, pretty or fancy URLs, search engine friendly - SEF URLs, or slugs) are used to provide shorter and more relevant-looking links to web pages.

INTERNAL CACHE SUPPORT: To reduce Internet bandwidth consumption, web server stores frequently accessed Web objects (HTML pages, images, sound files, etc.) on the local network, where they can be retrieved by internal users without going out to a server on the Internet every time they want to access them.

IPV6 SUPPORT: Internet Protocol version 6 (IPv6) is the most recent version of the Internet Protocol (IP), the communications protocol that provides an identification and location system for computers on networks and routes traffic across the Internet.

SUPPORT FOR WEBDAV APPLICATIONS: Web Distributed Authoring and Versioning (WebDAV) is an extension of the Hypertext Transfer Protocol (HTTP) that allows clients to perform remote Web content authoring operations. A working group of the Internet Engineering Task Force (IETF) defined WebDAV in RFC 4918.

SUPPORT FOR SERVER NAME INDICATION: Server Name Indication (SNI) is an extension to the TLS computer networking protocol by which a client indicates which hostname it is attempting to connect to at the start of the handshaking process. This allows a server to present multiple certificates on the same IP address and TCP port number and hence allows multiple secure (HTTPS) websites (or any other Service over TLS) to be served off the same IP address without requiring all those sites to use the same certificate. It is the conceptual equivalent to HTTP/1.1 name-based virtual hosting, but for HTTPS.

TLS SUPPORT: Transport Layer Security (TLS) and its predecessor, Secure Sockets Layer (SSL), are cryptographic protocols designed to provide communications security over a computer network.

SUPPORT FOR NON-UNIFORM MEMORY ACCESS (NUMA) HARDWARE: Non-uniform memory access (NUMA) is a computer memory design used in multiprocessing, where the memory access time depends on the memory location relative to the processor. Under NUMA, a processor can access its own local memory faster than non-local memory (memory local to another processor or memory shared between processors).

HTTP/2 SUPPORT: HTTP/2 (originally named HTTP/2.0) is the second major version of the HTTP network protocol used by the World Wide Web.

EVALUATING EFFICIENCY OF YOUR WEBSITE

Efficiency of the website can be measured in the number of hits, the amount of purchases, and the listings by search engines, etc. Most Internet Service Providers (ISPs) offer hit counters. A hit counter typically displays the number of previous visitors on the site or on a particular page and can also report the number of hits per page, the referring search engine, the domain name of the visitors, etc. Some popular hot counter programs include HitBox and WebPosition. Webmasters also need to make sure to continuously check their website for any dead links, and download times. There is service providers available that evaluate the efficiency of your website, some are available free e.g. Netmechanic (www.netmechanic.com)

OPERATING SYSTEMS (OS) FOR WEB SERVERS

Primary task of operating system software include running computer programs and allocating computer resources (such as memory and disk space to programs). An operating system is required to run application programs on a computer. Operating system can be a desktop or network operating system.

Network operations systems, in contrast with desktop operating system, provide extended functionality e.g. managing users logged on the network and providing access to shared resources on the network.

SALIENT FEATURES OF OS FOR WEB SERVER

Some important features of operating systems for web server are:

STABILITY: Scalability of an operating system refers to its ability survive varying load conditions and perform satisfactorily.

SECURITY: It refers to the ability of OS to provide secure access to resources using techniques e.g. authentication.

EASE OF ADMINISTRATION: OS for web servers should provide trouble-free and easy-to-use administration tools.

CORE FEATURES: The required software and features to run a web server should be built-in to core OS software. If not, they should be available as separate module from a third party.

MODULE SUPPORT AND INTEGRATION: OS should have additional software modules available. These additional modules can be provided from either OS developer or third party. Further, integration of these modules should be smooth and trouble-free.

SCALABILITY: The operating system should be scalable to hardware and software up-gradations.

The choice of the operating system for a web server depends on how the website is to be created and what applications will run on the web server. The OS on client desktops should not be a factor affecting the choice.

POPULAR OPERATING SYSTEMS FOR WEB SERVER

Most Web servers run on computers that use one of the following operating systems: Microsoft Windows NT Server, Microsoft Windows 2003 Server, Server 2008, or Server 2012 products, Linux, or one of several UNIX-based operating systems, such as Solaris or FreeBSD.

MICROSOFT WINDOWS

Windows operating systems can be easily integrated into other Microsoft products and their support is widely available. Microsoft server products are generally easier to learn but many concerned about their security risks. Most requirements of web server in Windows OS environment are built-in or come as separate software modules that can be installed quickly and smoothly. Over the years, Microsoft has been able to resolve many problems (such as security and stability) that seriously affected the previous versions of Windows operating systems. All Windows operating systems for web hosting servers support the ASP (Active Server Pages) technology, Cold Fusion and Visual basic scripts. A computer running a Windows operating system can be remotely administered through GUI-based software such as PCAnywhere. These software applications allows you to login to a user's desktop and perform all the functions as you were physically sitting on the computer console. Windows based servers are a good choice for both shared and dedicated web servers. Windows server operating systems come at a price that includes a limited number of user licenses. To connect more computers or user to the server, you would need to purchase additional licenses.

Windows Server 2012, the latest addition to the Microsoft Server operating system line up, was launched recently. Unlike its predecessor, Windows Server 2012 has no support for Itanium-based computers and has four editions. Various features were added or improved over Windows Server 2008 R2 (with many placing an emphasis on cloud computing), such as an updated version of Hyper-V[1], an IP address management role, a new version of Windows Task Manager, and ReFS, a new file system[2].

UNIX AND LINUX

UNIX is a multi-user, multitasking operating system which, by design, is a small, flexible system used exclusively by programmers. UNIX is open-source operating systems software, so anyone who obtained a copy could modify and customize it for his own purposes. Dozens of different versions of UNIX are available today. Due to its portability, flexibility, and power, UNIX has also become a leading operating system for workstations. Historically, UNIX has been less popular in the personal computer market.

Linux is a freely distributable open source operating system that runs on a number of hardware platforms including PCs and Macintoshes. The Linux kernel (the central component of most computer operating systems) is based on UNIX. Linux has become an extremely popular alternative to proprietary operating systems such as Microsoft Windows. There are both open-source and commercial distribution of LINUX. Commercial

---

[1] Hyper-V, codenamed Viridian and formerly known as Windows Server Virtualization, is a native hypervisor; it can create virtual machines on x86-64 systems.
[2] Resilient File System (ReFS) is a Microsoft proprietary file system introduced with Windows Server 2012 with the intent of becoming the "next generation" file system after NTFS.

distributions include useful additional software, such as installation utilities, and a support contract for the operating system. Commercial Linux distributors that sell versions of the operating system with utilities for Web servers include Caldera, Mandrake, Red Hat, and SuSE.

The UNIX and Linux operating systems are specifically designed for web servers and are famous for their stability and reliability even under heavy load conditions. They are less vulnerable to attacks from viruses and hackers and capable of hosting multiple web sites with heavy bandwidth requirements. UNIX and LINUX OS also provide support for various web technologies including Microsoft Technologies. Therefore, if you are developing your web site in Microsoft FrontPage you are not restricted to the Windows Operating system and can run it on Unix/Linux. Servers running Unix/Linux operating systems can be managed easily from any computer connected to Internet. A number of third party software (forums, message boards, chat programs, shopping carts, mailing lists etc.) are available for Unix/Linux based servers that can be easily integrated. Support for UNIX/LINUX based servers is widely available in the form of tons of helpful sites and forums that provide support for LINUX/UNIX.

SELECTING BEST SERVER OPERATING SYSTEM

To know which operating system is best for your server, a comparison is needed based on few factors such as functionality and hardware compatibility, resource requirements, stability and security, as well as cloud-readiness, cost and support (Vugt & Posey, 2014).

OPERATING SYSTEM (OS) FUNCTIONALITY

The most significant characteristics for a server OS include functionality, security and stability, as well as vendor support when things go wrong. Also, consider the workload: Do you need a cloud-ready OS, or does the application have specific hardware requirements? Windows has a reputation as an all-around operating system that is easy-to-manage. Linux is an operating system of choice for some specified services such as data centers. Microsoft has been stressing the concepts of minimization of the server footprint and using dedicated server roles. Licensing costs however are problematic for these dedicated server roles. Today, a Windows Server 2012 or 2012 R2 Datacenter Edition server with Hyper-V virtualization is licensed to run an unlimited number of Windows Server virtual machines. As such, licensing is no longer a consideration for organizations that deploy a large number of virtualized Windows servers.

OS STABILITY

Historically Linux has been regarded as very stable OS as compared to Windows. However, recent Windows Server operating system versions, from 2003 on, are increasingly stable. Microsoft initiated rigorous hardware testing and a logo certification program for compatibility guarantees with various hardware devices. The most probable causes of Windows and Linux failure today is faulty hardware.

CLOUD READINESS

OpenStack is a cloud framework run on Linux. Hundreds of leading companies in the computer industry support OpenStack, including Red Hat. Many vendors have a proprietary cloud offering, such as Windows Azure on the

Windows Server OS. Microsoft markets Windows Server as a cloud operating system, with Microsoft Azure running on top of Windows Server and Hyper-V. Microsoft's software-as-a-service offering of Office 365 also runs on Windows Server and Hyper-V.

COST OF OWNERSHIP IN THE DATA CENTER

While Linux is free, running Linux in the data center is not. For best performance you will need Linux server OS from vendors such as Red Hat and SUSE that charge for support contracts. Linux is less expensive than Windows Server because of the pricing structure. There is no per-user licensing of Linux distributions. As such, potential cost saving of migrating servers from Windows to Linux is significant. On the other hand, Windows OS follows Client Access Licensing. That makes cost of ownership of Windows OS more than Linux.

OS SECURITY

Linux is open source OS but it implements security as part of its kernel and sophisticated mandatory access control systems ,like Security-Enhanced Linux, are built on top of that. Windows OS is proprietary. Microsoft routinely releases security patches for the Windows Server OS as new vulnerabilities are discovered. Microsoft also provides extensive documentation on how to use built-in Windows Server security controls, as well as tips for establishing a network architecture that achieves the best possible security. As such, both Windows and Linux provide comparable levels of security.

HARDWARE NEEDS

Linux requires less hardware than Windows in order run a server in datacenter. For large corporate servers, the operating system overhead of Linux is smaller compared with Windows Server OS. Linux kernel can be tuned to make it even more efficient. The hardware requirements for Windows Server 2012 R2 vary depending on the server's workload, but Windows Server can run on a system that has as little as 512 MB of RAM and a 1.4 GHz (64-bit) CPU.

VENDOR SUPPORT

Vendor support is essential to run the applications that support the business. It is a common belief that Windows offers bigger vendor support than Linux systems. With more and more applications moving from server to the cloud, the general trend is to keep bigger business applications in house such as databases. Microsoft is universally supported by the major server hardware vendors. Nearly all of the hardware manufacturers provide Windows drivers for their products.

CORPORATE AUTHENTICATION

This is one strong area of Windows OS. Active Directory is a full authentication and authorization platform that integrated applications, users, computers and other resources. Linux alternatives to Active Directory do not have the same support of devices and applications. Microsoft Active Directory is the de facto standard for authentication because it is solid, reliable and secure, and in addition is widely supported by an array of third-party products.

Table 3-1 below provides a comparison of Microsoft Windows and Linux OS.

TABLE 3-1: LINUX AND WINDOWS OS COMPARISON

Feature	Linux	Windows
Price	The majority of Linux variants are available for free or at a much lower price than Microsoft Windows.	Microsoft Windows can run between $50.00-$150.00 per each license copy.
Ease	Linux is an OS meant for experienced users.	Microsoft Windows is a much easier to use operating system than Linux.
Reliability	The majority of Linux variants and versions are highly reliable and can often run for long time without needing to be rebooted.	Although Microsoft Windows OS is pretty reliable, it still cannot match the reliability of Linux.
Software	Linux has a large variety of available software programs, utilities, and games.	Because of the large number of Microsoft Windows users, there is a much larger selection of available software programs, utilities, and games for Windows than Linux.
Software Cost	Many of the available software programs, utilities, games, and even complex software (such as OpenOffice) available on Linux are freeware or open source.	Windows does provide free software programs, utilities, and games. However, the majority of the programs are not free.
Hardware	Linux does not enjoy the wide hardware support available to Windows OS.	Windows has a much larger support for hardware devices and almost all hardware manufacturers support their products in Microsoft Windows.
Security	Linux is a very secure operating system.	Microsoft Windows is secure but not as secure as Linux OS.
Open Source	Many of the Linux variants and many Linux programs are open source and enable users to customize or modify the code however, they want to.	Microsoft Windows OS and majority of majority of Windows programs are proprietary.
Support	There are vast amounts of available online documentation and help, available books, and support available for Linux.	Microsoft Windows includes its own help section, has vast amount of available online documentation and help.

WEB SERVER MARKET

Some of the most commonly used Web server programs today are Apache HTTP Server, and Microsoft Internet Information Server (IIS). The other web servers used includes nginx by Igor Sysoev, and Google Web Server. Table 3-2 lists top Web server software vendors published in a Netcraft survey (Netcraft, 2015) in June 2015. Netcraft is a networking consulting company in Bath, England, known throughout the world for its Web server survey. Netcraft continually conducts surveys to tally the number of Web sites in existence and measure the relative popularity of Internet Web server software.

TABLE 3-2: TOP WEB SERVER PRODUCTS

Vendor	Product	Web Sites Hosted	Percent of Total Websites
Apache	Apache	334,731,035	38.78%
Microsoft	IIS	254,408,179	29.48%
Igor Sysoev	nginx	122,965,522	14.25%
Google	GWS	20,130,732	2.33%

APACHE HTTP SERVER

Apache is open source web server developed by a group of volunteer programmers, called the Apache Group. Source code of Apache is freely available. Anyone can adapt the server for specific needs, and there is a large public library of Apache add-ons software. There are now versions of Apache that run under FreeBSD-UNIX, HP-UX, Linux, Microsoft Windows, SCO-UNIX, and Solaris and the hardware that supports them. The name Apache is a tribute to the Native American Apache Indian tribe, a tribe well known for its endurance and skill in warfare. Apache server is known as a powerful and efficient web server. Currently, Apache is used on around 39% of all Web servers in the world, which means it is more widely used than all other Web server software packages combined.

Apache supports a variety of features, many implemented as compiled modules, which extend the core functionality. These can range from server-side programming language support to authentication schemes. Some common language interfaces support Perl, Python, Tcl, and PHP. Other features inOclude Secure Sockets Layer and Transport Layer Security support, a proxy module, a URL rewriter, custom log files, and filtering support. Popular compression methods on Apache include gzip. ModSecurity is an open source intrusion detection and prevention engine for Web applications. Apache logs can be analyzed through a Web browser using free scripts, such as AWStats/W3Perl or Visitors. Virtual hosting allows one Apache installation to serve many different Web sites. For example, one machine with one Apache installation could simultaneously serve www.example.com, www.example.org, test47.test-server.example.edu, etc. Apache features configurable error messages, DBMS-based authentication databases, and content negotiation. It is also supported by several graphical user interfaces (GUIs). It supports password authentication and digital certificate authentication. Because the source code is freely available, anyone can adapt the server for specific needs, and there is a large public library of Apache add-ons. Instead of implementing a single architecture, Apache provides a variety of Multi-Processing Modules (MPMs), which allow Apache to run in a process-based, hybrid (process and thread) or event-hybrid mode. These modules can help better match the demands of each particular infrastructure. This implies that the choice of correct MPM and the correct configuration is important. Where compromises in performance need to be made, the design of Apache is to reduce latency and increase throughput, relative to simply handling more requests, thus ensuring consistent and reliable processing of requests within reasonable periods. According to Netcraft, IIS was running nearly 88 million active sites in January 2015.

Microsoft Internet Information Server (IIS)

Microsoft Internet Information Server (IIS) comes bundled with current versions of Microsoft Windows Server operating systems. IIS is tightly integrated with the Windows operating system and is relatively easy to administer. However, currently IIS is available only for the Windows operating system. IIS is an extensible web server created by Microsoft for use with Windows server operating system family. IIS supports HTTP, HTTPS, FTP, FTPS, SMTP and NNTP.

About 29 % of all web servers run some version of IIS. In recent years, the number of Web sites running IIS has decreased. These losses included many sites running on Microsoft IIS 6.0, which along with Windows Server 2003 will reach the end of its Extended Support period in July 2015. Further abandonment of these platforms is therefore expected in the first half of this year, although Microsoft does offer custom support relationships, which go beyond the Extended Support period. Microsoft claims that the successor of Microsoft IIS 6.0 (i.e. IIS 7.0) provides much stronger security design. An installation of IIS 7.0, with all the default options chosen, provides only the minimal functionality and any additional one, if needed, will have to be explicitly selected and installed by the user. IIS supports the use of ASP, ActiveX Data Objects, and SQL database queries. IIS also includes the Microsoft FrontPage Web site development tool and other reporting tools. Inclusion of ASP in IIS provides the capability to develop dynamic web pages that include ActiveX components, and scripts. All versions of IIS prior to 7.0 running on client operating systems supported only 10 simultaneous connections and a single website. IIS 7.5 was included in Windows Server 2008 R2. IIS 7.5 improved WebDAV and FTP modules as well as command-line administration in PowerShell[3]. It also introduced TLS 1.1 and TLS 1.2 support and the Best Practices Analyzer tool[4]. IIS 8.0 is only available in Windows Server 2012 and Windows 8. IIS 8.0 includes SNI support, Application Initialization, centralized SSL certificate support, and multicore scaling on NUMA hardware, among other new features. IIS 8.5 is included in Windows Server 2012 R2 and Windows 8.1. This version includes idle Dynamic Site Activation, Enhanced Logging, and Automatic Certificate Rebind[5]. IIS 10 is included in Windows Server 2016 and Windows 10. This version includes support for HTTP/2. According to Netcraft, IIS was running nearly 22 million active sites in January 2015.

Google Web Server (GWS)

The Google Web Server is custom-built server software that is used only by Google and runs only on Google's private internet[6] . According to Netcraft, Google Web Server was running nearly 14 million active sites in 2015. This total includes sites solely run by Google and sites that Google operates on behalf of third parties via services like Blogger and Google Docs.

---

[3] Windows PowerShell is a task automation and configuration management framework from Microsoft, consisting of a command-line shell and associated scripting language built on the .NET Framework.
[4] Best Practices Analyzer (BPA) is a server management tool that is available in Windows Server 2008 R2.
[5] This feature, called Certificate Rebind, ensures that a certificate will automatically be rebound to a Web site after the certificate has been renewed.
[6] Google's private internet consists of nearly 40 data centers across the globe and is built atop a number of custom-built and proprietary tools.

NGINX

Nginx ("engine x") is a high-performance and efficient web server. Nginx is a lightweight, stable, and high-performance web server that provides a rich feature set, simple configuration options and consumes low resources. Source code of Nginx is released under a BSD-like license. Nginx runs on UNIX, GNU/Linux, BSD variants, Mac OS X, Solaris, and Microsoft Windows. The system is used by several very large, well-known sites including Github.com, WordPress.com, and hulu.com. According to Netcraft, Nginx was running nearly 24 million active sites in 2015.

LIGHTTPD

Lighttpd is a fast, secure and efficient web server, which is more flexible and configurable. It is open source software licensed under the revised BSD license. The Lighttpd web server is an excellent choice to host small to medium sized web sites.

FINDING WEB SERVER INFORMATION

To know the type of operating system and Web server software that a Web site is running, one can visit the Netcraft Web site (http://news.netcraft.com/). On Netcraft's home page is a link named "What's that site running?"(http://uptime.netcraft.com/up/graph). This page leads to a page with a search function. Visitors can use that search function to find out what operating system and what Web server software a specific site is now running and what the site ran in the past.

Database Management Systems (DBMS) for Web Severs

A database is organized collection of information that a computer program can use for quick selection of desired pieces of data. A database management system (DBMS) involves the data itself and the software that controls the storage and retrieval of data. Relational databases are the most popular database choice today. The relational database model is a logical representation of the data that allows the relationships between the data to be considered independently of the physical implementation of the data structures.

Structured Query Language (SQL) is used with relational database systems to make queries and manipulate data in a database. Some of the more popular commercial enterprise-level relational database systems include Microsoft SQL Server, Oracle, Sybase, IBM DB2, and IBM Informix. These database management systems come with a graphical user interface for their management. These database management applications can be quite expensive and typical installations cost between $5000 and $200,000.

Open-source software is another option available. Increasing number of companies and other organizations are beginning to use open source database packages such as PostgreSQL, and MySQL. These open-source database management systems are free. Large information systems that store the same data in many different physical locations are called distributed information systems, and the databases within those systems are called distributed database systems. These systems are very complex and costly.

DBMS FEATURES

Features commonly offered by database management systems include:

QUERY ABILITY: Querying is the process of requesting attribute information from various perspectives and combinations of factors. A database query language and report writer allow users to interactively interrogate the database, analyze its data and update it according to the users privileges on data.

BACKUP AND REPLICATION: Copies of databases need to be made regularly in case primary disks or other equipment fails. A periodic copy of databases may also be created for a distant organization that cannot readily access the original. DBMS usually provide utilities to facilitate the process of backing up and restoring databases.

RULE ENFORCEMENT: Rules are applied on database attributes so that the attributes are clean and reliable. Ideally, such rules should be able to be added and removed as needed without significant data layout redesign.

SECURITY: Often it is desirable to limit who can see or change which attributes or groups of attributes in a database. This may be managed directly by individual, or by the assignment of individuals and privileges to groups, or through the assignment of individuals and groups to roles, which are then granted entitlements.

COMPUTATION: There are common computations requested on attributes such as counting, summing, averaging, sorting, grouping, cross-referencing, etc.

CHANGE AND ACCESS LOGGING: Often one wants to know who accessed what attributes, what was changed, and when it was changed. Logging services allow this by keeping a record of access occurrences and changes.

AUTOMATED OPTIMIZATION: If there are frequently occurring requests, some DBMS can adjust themselves to improve the speed of those interactions. In some cases, the DBMS provide tools to monitor performance, allowing a human expert to make the necessary adjustments after reviewing the statistics collected.

POPULAR DBMS FOR E-COMMERCE

MICROSOFT SQL SERVER

SQL Server is a relational database management system (RDBM) developed by Microsoft which uses T-SQL and ANSI-SQL as primary languages. SQL Server's ease of use, availability and tight Windows operating system integration makes it an easy choice for firms that choose Microsoft products for their enterprises. Currently, Microsoft promotes SQL Server 2014 as theplatform for both on-premises and cloud databases and business intelligence solutions. Over the years, several performance enhancements have been made in SQL server e.g. client IDE tools, reporting services, data mining services, and messaging services etc. SQL Server aims to make data management self-tuning, self-organizing, and self-maintaining. With the development of SQL Server Always On technologies SQL server aims to provide near-zero downtime.

The current versions of SQL server include SQL server 2012 and SQL Server 2014. SQL Server 2012's new features and enhancements include AlwaysOn SQL Server Failover Cluster Instances and Availability Groups which provides a set of options to improve database availability, Contained Databases which simplify the moving of databases between instances, new and modified Dynamic Management Views and Functions, programmability enhancements including new spatial features, metadata discovery, sequence objects and the THROW statement, performance enhancements such as ColumnStore Indexes as well as improvements to OnLine and partition level operations and security enhancements including provisioning during setup, new permissions, improved role management, and default schema assignment for groups [7].

The latest version of SQL server is SQL Server 2014. SQL Server 2014 has several compelling new features. Without a doubt, the most notable new feature is the new In-Memory OLTP engine. By moving select tables and stored procedures into memory, you can drastically reduce I/O and improve performance of your OLTP applications. SQL Server 2014 provides improved integration with Windows Server 2012 R2 and Windows Server 2012. SQL Server 2014 will have the ability to scale up to 640 logical processors and 4TB of memory in a physical environment. It can scale up to 64 virtual processors and 1TB of memory when running on a virtual machine (VM). SQL Server 2014 also provides a new solid-state disk (SSD) integration capability that enables you to use SSD storage to expand SQL Server 2014's buffer pool. The new buffer pool enhancements can help increase performance in systems that have maxed out their memory capability by using high-speed nonvolatile RAM (NVRAM) in the SSD drives as an extension to SQL Server 2014's standard buffer pool. SQL Server 2014's Resource Governor provides a new capability to manage application storage I/O utilization. SQL Server 2014 also integrates with several new and improved features in Windows Server 2012 R2 and Windows Server 2012. For example, SQL Server 2014 supports the OSs' new Storage Spaces feature. With Storage Spaces, you can create pools of tiered storage to improve application availability and performance. In addition, SQL Server 2014 can take advantage of the OSs' Server Message Block (SMB) 3.0 enhancements to achieve high-performance database storage on Windows Server 2012 R2 and Windows Server 2012 file shares. In SQL Server 2014, Microsoft has enhanced AlwaysOn integration by expanding the maximum number of secondary replicas from four to eight. SQL Server 2014 also provides Windows Azure AlwaysOn integration. This new integration feature enables you to create asynchronous availability group replicas in Windows Azure for disaster recovery. In the event of a local database outage, you can run your SQL Server databases from Windows Azure VMs. Database backups in SQL Server now support built-in database encryption. Previous releases all required a third-party product to encrypt database backups. The backup encryption process uses either a certificate or an asymmetric key to encrypt the data. The supported backup encryption algorithms are Advanced Encryption Standard (AES) 128, AES 192, AES 256, and Triple DES (3DES). SQL Server 2014 also provides new Windows Azure integration to SQL Server's backup capabilities. You can specify a Windows Azure URL as the target for your SQL Server 2014 database backups. This new Windows Azure backup feature is fully integrated into SSMS. Columnstore indexes are another of Microsoft's high performance in-memory technologies. Microsoft introduced the columnstore index in SQL Server 2012 to provide significantly improved performance for data warehousing types of queries. SQL Server Data Tools for BI (SSDT BI) is used to create SQL Server Analysis

---

[7] For more information, see https://msdn.microsoft.com

Services (SSAS) models, SQL Server Reporting Services (SSRS) reports, and SQL Server Integration Services (SSIS) packages. Power BI for Office 365 is Microsoft's cloud-based BI solution that leverages familiar Office 365 and Excel tools. Power BI for Office 365 provides business insights through data visualization and navigation capabilities. Power BI for Office 365 includes Power Pivot, Power View, Power Query, and Power Map (Otey, 2014).

SQL Server 2012 and 2014 also support structured and semi-structured data, and multimedia data. The SQL Server 2012 family comprise of the Enterprise, Business Intelligence, Standard, Developer, and Express editions (For a comparison of features available in different editions of SQL Server 2012, see http://msdn.microsoft.com/en-us/library/cc645993%28v=SQL.110%29.aspx). The SQL Server 2014 family comprise of the Enterprise, Business Intelligence, Standard, Express, and Developer editions. For a list of features comparison of different versions of SQL Server 2014, see https://msdn.microsoft.com/library/cc645993.aspx.

The table 3-3 provides a price comparison of various editions of SQL Server 2012 and 2014.

TABLE 3-3: EDITIONS OF SQL SERVER

SQL Server 2012				
Standard	Enterprise	Business Intelligence	Web and Developer	Express
Per core licenses at $1,793 or purchase a server license at $898 and client access licenses at $209 per client.	$6,874 per core with a maximum of 20 cores.	Server licenses costing $8,592 and client access licenses costing $209 per client.	$50 per license	Free

SQL Server 2014				
Enterprise	Business Intelligence	Standard	Express	Developer
Per core at $14,256	Server + client access license at $8,908	Per core at $3,717 or Server + Client access license at $931	Free	$38

ORACLE

The Oracle Database (commonly referred to as Oracle RDBMS or simply as Oracle) is a RDBMS developed by Oracle Corporation. Oracle provides several native tools to access the database. A variety of third-party tools is also available to access Oracle database. The most commonly used tool for data access is be SQL*Plus (a procedural language extension to SQL).

The latest version of Oracle is Oracle 12c. The current release of Oracle's RDBMS is Oracle 12c. The "c" stands for cloud and is reflective of Oracle's work in extending its enterprise RDBMS to enable firms to consolidate and manage databases as cloud services when needed via Oracle's multitenant architecture and in-memory data processing capabilities. Oracle 12c provides many exciting new features. By consolidating many databases into fewer database servers, both the hardware and operational staff can be more effectively utilized. Oracle's new pluggable database feature reduces the risk of consolidation because the DBA can easily plug or unplug an existing database to or from a container database. There is no need to change any code in the application. When the user connects to a plugged database, the database environment looks exactly as if the user had connected to a traditional database. Further, pluggable databases do lower resource consumption. Memory and processes are owned by the container database and shared by all pluggable databases, improving resource usage overall. It is also easy to unplug a database and convert the pluggable database to a traditional database if required. Version 12c introduces a few useful SQL Optimizer features, and most of these are automatically enabled. Version 12c introduces numerous performance enhancements. Version 12c introduces Global Data Services, which balances the workload not only among instances, but also among the databases. This feature also improves availability because new connections to failed databases can be redirected quickly to a surviving database. Version 12c introduces Partial Indexing, which allows you to create indexes on a partial set of partitions. We can reduce transaction workload by deferring indexes on transaction-intensive partitions and add indexes only when the partitions become less transaction-intensive. In Database 12c, Recovery Manager (RMAN) supports datafile copies over the network with compression. This feature will ease database-cloning efforts tremendously. In addition, the Active Duplicate command supports network compression during the data transfer, enabling faster clones directly from the production database. Many organizations are gearing up to certify IPv6 support as IPv4 address space becomes exhausted. Database 12c supports IPv6 for public network addresses. It does not support IPv6 in private network addresses though. Version 12c uses parallelism to improve the database upgrade to reduce upgrade-related downtime (Shamsudeen, 2013). There are many versions of Oracle Database 12c e.g. Standard edition, Enterprise edition, and Personal edition. The cost per processor of Enterprise edition is US$47,500.

SAP SYBASE ASE

This affordable relational database management system (RDBMS) is designed for high-performance transaction-based applications involving massive volumes of data – and thousands of concurrent users. Sybase is still a major force in the enterprise market after 25 years of success and improvements to its Adaptive Server Enterprise product. Although its market share dwindled for a few years, it has seen a bump in the next-generation transaction processing space following being acquired by Sybase in 2010 and relabeled as SAP Sybase Adaptive Server Enterprise (ASE). Sybase has also thrown a considerable amount of weight behind the mobile enterprise by delivering partnered solutions to the mobile device market.

Sybase is an enterprise software and services company. Companies use Sybase products and services to manage, analyze, and mobilize information. Sybase 365 mCommerce suite is software that provides end-to-end solution for mobile banking, mobile payments, and mobile remittance. Using 365 mCommerce, mobile operators and financial institutions can launch mCommerce services to their customers.

IBM DB2

The IBM DB2 is a RDBMS developed by IBM. It primarily runs on IBM version of UNIX (namely AIX), Linux, System i (also known as AS/400), z/OS, and Windows OS.IBM DB2 for Linux, UNIX, and Windows is available in three separate editions: Express Edition, Workgroup Server Edition, and Enterprise Server Edition. Each edition has different groups of features for different sized workloads. The latest release of DB2, DB2 10.5, runs on Linux, UNIX, Windows, the IBM iSeries and mainframes. According to IBM, its DB2 system is squarely in competition with Oracle, via the International Technology Group, and the results showed significant cost savings for those that migrate to DB2 from Oracle. Companies were able to achieve 34% to 39% savings for comparative installations over a three-year period. IBM DB2 10.5 is also the only database fully optimized for the IBM Power Systems POWER8 processor and the company's Power 8 server systems. DB2 versions include Workgroup, Workgroup Unlimited, and Enterprise Server Edition. The most sophisticated edition for Linux, UNIX and Windows is DB2 Datawarehouse Enterprise Edition (DB2 DWE). This edition is designed for a mixed workload, such as online transaction processing with datawarehousing or business intelligence implementations.

IBM INFORMIX

IBM Informix is a family of relational database management system (RDBMS) developed by IBM. IBM Informix comes in many versions: Informix Developer Edition, Informix Innovator-C Edition, and Informix Ultimate Edition. Each edition has different groups of features for different sized workloads and needs. IBM offers a range of Informix database options, starting with entry-level Workgroup and Express Editions and scaling up to an Enterprise Edition, an Enterprise Hypervisor Edition and finally Advanced Workgroup and Enterprise Editions with the Informix Warehouse Accelerator (IWA). Informix database solutions can be used to manage Internet of Things (IoT) data, and can seamlessly integrate SQL, NoSQL/JSON, timeseries and spatial data. Informix is also regarded as low cost, low maintenance and high reliability database.

POSTGRESQL

PostgreSQL (or simply Postgres) is developed by a global community of developers and companies. PostgreSQL is an open-source object-relational database management system (ORDBMS) that is used in online gaming applications, data center automation suites and domain registries. PostgreSQL's current stable release is 9.4.x. PostgreSQL runs on a wide variety of operating systems, including Linux, Windows, FreeBSD and Solaris. In addition, as of OS X 10.7 Lion, Mac OS X features PostgreSQL as its standard default database in the server edition. PostgreSQL includes enterprise-grade features comparable to Oracle and DB2 such as full ACID compliance for transaction reliability and Multi-Version Concurrency Control for supporting high concurrent loads.

MYSQL

MySQL began as a niche database system for developers but grew into a major contender in the enterprise database market. Sold to Sun Microsystems in 2008, MySQL has since become part of the Oracle empire in 2009 following Sun's acquisition by Oracle. More than just a niche database now, MySQL powers commercial websites by the hundreds of thousands, and it serves as the backend for a huge number of internal enterprise applications.

MySQL is a RDBMS whose source code available under the terms of the GNU General Public License, as well as under a variety of proprietary agreements. MySQL is owned and sponsored by a Swedish company MySQL AB, now owned by Sun Microsystems, a subsidiary of Oracle Corporation. MySQL is being used in various industries and governments around the world e.g. Swedish National Police, Wal-Mart, Apple etc.

Today MySQL remains a very popular option for use in Web applications, and it continues to serve as a central component of the LAMP open-source Web application software stack, along with Linux, Apache and PHP (or Python or Perl). At the same time, MySQL has seen support from users and developers erode over the last few years following the acquisition by Oracle. MySQL's decline has helped fuel the adoption of other open-source database options and forks of MySQL like the fully-open source MariaDB, which doesn't feature closed-source modules like some of those found in newer versions of MySQL Enterprise Edition, as well as Percona and the cloud-optimized Drizzle database system.

MARIADB ENTERPRISE

MariaDB Enterprise is a fully open source database system, with all code released under GPL, LGPL or BSD. MariaDB originated in 2009 and is led by the original developers of MySQL. The current stable series of MariaDB Enterprise is powered by MariaDB 10.x. In 2013 alone, Red Hat Enterprise Linux (RHEL) Fedora opted for MariaDB over MySQL in its Fedora 19 release, and both openSUSE and Slackware Linux made similar switches to MariaDB over MySQL. Wikipedia also adopted MariaDB over MySQL as its backend database in 2013. Another key factor in moving MariaDB ahead of MySQL is its enhanced query optimizer and other performance-related improvements, which give the database system a noticeable edge in overall performance compared to MySQL (Stroud, 2015).

TERADATA

Teradata created the first terabyte database for Wal-Mart in 1992. As a Very Large Database (VLDB) system, Teradata's capabilities have made it a great fit for handling emerging enterprise trends like Big Data analytics, business intelligence (BI) and the Internet of Things (IoT). Teradata released version 15 of its RDBMS in early 2014.

INGRES

Ingres is the parent open source project of PostgreSQL and other database systems, and it is still around to brag about it. Ingres is all about choice, and in this case choosing might mean lowering your total cost of ownership for an enterprise database system. Other than an attractive pricing structure, Ingres prides itself on its ability to ease your transition from costlier database systems. Ingres also incorporates security features required for HIPPA and Sarbanes Oxley compliance.

AMAZON'S SIMPLEDB

Amazon's SimpleDB (Simple Database Service) offers enterprises a simple, flexible and inexpensive alternative to traditional database systems. SimpleDB enables users to store and query data items via web services requests, and it boasts scalability, speed, and minimal maintenance and Amazon services integration. As part of Amazon's EC2 offering, you can get started with SimpleDB free.

MONODB

MongoDB is a cross-platform document-oriented database. Classified as a NoSQL database, MongoDB eschews the traditional table-based relational database structure in favor of JSON-like documents with dynamic schemas (MongoDB calls the format BSON), making the integration of data in certain types of applications easier and faster. Released under a combination of the GNU Affero General Public License and the Apache License, MongoDB is free and open-source software. MongoDB has been adopted as backend software by a number of major websites and services, including Craigslist, eBay, Foursquare, SourceForge, Viacom, and The New York Times among others. As of 2014, MongoDB was the most popular NoSQL database system.

Instead of taking a business subject and breaking it up into multiple relational structures, MongoDB can store the business subject in the minimal number of documents. MongoDB supports search by field, range queries, regular expression searches. Queries can return specific fields of documents and include user-defined JavaScript functions. Any field in a MongoDB document can be indexed (indices in MongoDB are conceptually similar to those in RDBMSes). MongoDB provides high availability with replica sets. A replica set consists of two or more copies of the data. MongoDB can run over multiple servers, balancing the load and/or duplicating data to keep the system up and running in case of hardware failure. MongoDB can be used as a file system, taking advantage of load balancing and data replication features over multiple machines for storing files.

MICROSOFT ACCESS

A popular and low-cost personal relational database is Microsoft Access, which can be used by smaller E-commerce sites. Current version of Microsoft Access is Access 2013, which costs between US$ 110-140. Companies should consider database support as an important factor when evaluating E-commerce software. Usually the database that serves an online store is the same one that is used by the existing corporate clients. If a company has existing inventory and product databases, then it should evaluate only electronic commerce software that supports these systems.

The following table presents a ranking of top database management systems software in 2014. The ranking system (zdnet.com, 2014) calculates the popularity value of a DBMS by standardizing and averaging various individual parameters of DBMS performance.

TABLE 3-4: RANKING OF DBMS

Rank	DBMS	Type	Score
1	Oracle	Relational Database	1470.86
2	MySQL	Relational Database	1281.22
3	SQL Server	Relational Database	1242.50
4	PostgreSQL	Relational Database	249.85
5	MongoDB	Document Store	237.36

SELECTING A DBMS FOR YOUR E-COMMERCE BUSINESS

As companies try to organize and manage an increasing volume of digital information, database systems are becoming a more critical business requirement. However, selecting the right relational database management system (RDBMS) can be tricky, especially for companies that plan to organize or reorganize their business around it. The range of database platforms to choose from is enormous, and there is certainly no one-size-fits-all solution.

If you are new to DBMS world, it is better to seek help from a database consultant to help you with your selection process. New users should generally choose from the market leaders, and that means one of the big three: IBM's DB2, Oracle, or Microsoft SQL Server. There are other options such as an open source DBMS like MySQL or PostgreSQL. For high-end mission-critical applications, you should stick with the big three DBMS. If you are a large organization with a mainframe and want to run your DBMS on that mainframe, you really should go with IBM DB2. Oracle has a mainframe version of their database server, but IBM is by far the market leader. For UNIX and Linux installations, you should pick either Oracle or DB2. Oracle is the market leader on those platforms. IBM is also an important player for these platforms. For Windows development, all big three DBMS are viable options, but Microsoft is the market leader. Sybase, Informix, and Teradata are the next biggest players in the market. Sybase has lost ground in the market, but their DBMS is still solid and they are firmly entrenched in the financial market. Informix was purchased by IBM and it is still being maintained, but DB2 is IBM's primary DBMS. As such, Informix is not a good choice for new companies. Teradata is a high-speed DBMS that is geared for data warehousing and OLAP work and you might want to choose it for those types of projects (Mullins, 2005).

A good way to start choosing the best DBMS software is to consider the size of your organization and the program complexity that your current networks can support. Professionals who work in small or independent operations can benefit from software they can install on their personal computers. These kinds of systems tend to offer basic organizational and retrieval functions and may not be effective for large organizations with great amounts of data to store. Large organizations can choose DBMS software that runs on a mainframe and serves a number of different workstations in a network. An example of this kind of software can be found in airlines. These programs store information about seat reservations, flight schedules, and flight routes. They are often accessed from hundreds, if not thousands, of computers. It also can be helpful to consider the method of database organization you prefer. A relational database, for example, groups data by looking for similar attributes among various pieces of information. A hierarchical database is often helpful for database managers who want to describe relationships between components of a system, such as a computer network. Options for information retrieval are also important for many database managers. Business professionals may prefer to access information that is in the form of a chart or graph. Engineers, on the other hand, might benefit from bills of materials in list format. The quality and cost of DBMS software can vary from product to product. Generally, the more complex software is, the more it costs. Software with the capacity to organize accounts of a multinational corporation is much more costly than an independent accountant's management system. Many experts believe that while it can be tempting to choose an inexpensive system to cut back on short-term costs, this method can backfire in the long term. An expensive software system with a good reputation and a high degree of security can be more cost efficient over a span of years (Wisegeek.com, 2015).

Gartner cites SQL Server strengths as its market vision and capabilities, competitiveness within the DBMS market, its performance and the support offered. The areas, which are cautioned about, are a lack of appliances and pricing. Gartner praises IBM for the broad functionality it provides, its hardware integration and global presence. It cautions against the provider's complexity, pricing, its confusing branding, and poor sales execution with "very aggressive competitor marketing." Gartner praises SAP's vision leadership, strong DBMS offerings and the performance of those offerings. It urges caution regarding the company's marketing communications, the lack of skills available in the market for its DBMS products and its poor provision of support.

There are some other criteria too that should be considered when selecting a DBMS for your business. (IBM, 2001).

MODELS

Probably the most fundamental choice to make in the DBMS hierarchy is the model used to store, manage, and query databases. Besides affecting what software you need to acquire, this affects the very way you will think about the data, and can be a surprisingly hard choice to undo later on.

LANGUAGES

Since most databases are a vital organ for a complete application, the interface between the database and the application development language is quite important. The DBMS of choice should have a natural and efficient API in your programming language of choice. Choose a DBMS that supports the APIs and languages with which you are comfortable.

DEVICES

Your DBMS of choice must work on the platform used by the rest of the application but there might be other platform needs as well. Be sure to consider who might end up using your system, and choose a DBMS that would run on other important platforms in future. Perhaps the users in your near future will want to access your database on the go, from their palm computer or cell phone. In that case, choose DBMS, which are suitable for execution on such devices.

FEATURES

Probably the most important general features to consider in your DBMS hunt are security-related. Consider how thoroughly the DBMS requires authentication from users and keeps an audit trail of the accesses. Be sure your backup supports backup and restore, not just by archiving your raw database files, but also the ability to integrate into incremental backup regimens. Consider whether the DBMS supports access by multiple users at once (multi-user support), which is an important feature in many situations. If so, beware of common problems where database state becomes inconsistent because modifications are interrupted by error conditions, or such modifications affect the state of other programs accessing the database. Solutions to such problems are called transaction and concurrency (locking) control and are important features if your database requires a high volume of access from multiple simultaneous users. Standards support is very important if you want skills and code to be portable to different ventures than the current database development. If using RDBMS, be sure it features broad support for SQL. If you are using an object-oriented DBMS, support for the Object Database Management Group's (ODMG's) standards is must. Regardless of the chosen model, language or platform, investigate what open standards there are for DBMS and look for these in the products under consideration. In addition, of course you should be sure the DBMS is usable. Does it have friendly tools for direct manipulation by the administrator? Does it have good documentation online, with options to get paper documentation if required? If it is a commercial product, is the reputation for technical support a sound one?

A DBMS is not a one-size-fits-all choice by any means. Every nuance of the project you are undertaking can affect DBMS choice. There is less risk in your choice if you can quickly learn from a bite-sized chunk irrespective whether your initial inclinations were right or wrong. You might want to use a free database for such a prototype, or you might want to ask the DBMS vendor of interest for a trial copy. Most vendors make evaluation copies available (although you might just get weekly calls from the regional sales rep after you start your evaluation).

IMPLEMENTING DBMS FOR E-COMMERCE

The following are the steps to be followed for implementing DBMS.

IDENTIFY THE DATABASE ELEMENTS: Identify all the elements that need to be included in the database management system. This will vary depending on the size and type of your business. At the very least, a person will need to be able to access data on raw and finished goods inventories, customer information, shipment tracking, employee data and Accounts Receivable and Payable data.

SET PARAMETERS FOR DATA ACCESS: Not all users need to have access to all data. Establish login credentials for all users and setup a hierarchy of privileges associated with each set of credentials. For larger companies, this usually means establishing specific privileges for a job or position. In smaller companies, the authorizations may focus more on the individual users and less on the position.

TEST THE DATABASE: Before making it operational, run the DBMS for a period through all the projected applications. Enter data, run reports, look up histories, run queries, and conduct any other intended uses of the system. This will provide the chance to correct any small issues with functionality before the system goes live.

EDUCATE THE USERS: In many cases, this is the most time consuming part of the implementation process. Make sure that all users understand how to use the new DBMS. This will minimize downtime and other issues that are commonly associated with a new implementation.

EVALUATE THE SYSTEM: Once the system is up and running for some time, evaluate its performance. This is an often-overlooked aspect of implementation process. Once the system is live, ask users to evaluate the ease of use and to recommend ways that the system could be used to make life easier for everyone in the company.

E-mail

In E-commerce, E-mail is used extensively by buyers and sellers to gather information, execute transactions, confirm the receipt of customer orders and then the shipment of items ordered and perform other tasks related to electronic commerce. E-mail is and easy to use and speedy way of communication. Despite its many benefits, e-mail does have some drawbacks. One disadvantage is the amount of time spent on writing and responding to emails. Second disadvantage is the computer virus that can come as an attachment. Third, more serious disadvantage is SPAM.

SPAM

With respect to E-mail, spam refers to the e-mail message that is both unsolicited and bulk. Unsolicited mail refers to an e-mail for which a verifiable permission to send has not be obtained by the recipient. Some examples of unsolicited e-mails include first contact enquiries and job enquiries. Bulk e-mail means that the message with substantively identical content is sent to a large number of recipients. Some example of bulk e-mail includes subscriber newsletters, customer communications, etc. Both bulk and unsolicited e-mails are considered normal e-mails. An e-mail message is considered Spam only if it is both Unsolicited and Bulk. Spam costs the sender very little to send and most of the costs are paid for by the recipient or ISP.

One particularly disturbing variant of spam is sending spam messages to mailing lists (public or private email discussion forums.). Spammers use E-mail list to send emails. Email lists are inexpensive, easy to purchase, and can contain millions of email addresses. Spammers can also use following methods to create e-mail lists:

- Collect email addresses from newsgroups

- Use software tools to scan web pages for email addresses

- Get e-mail addresses from chat sessions

- Steal e-mail addresses from mail servers

- Randomly generate e-mail addresses

According to spamcop.net (www.spamcop.net), during June 2014-June 2015, the average spam was 7.1 messages per second and total 224574599 spam messages were detected. According to Symantec, there were three top categories of spam: Internet (82%), Leisure (7%), and Fraud (5%). The top three spam content types are HTML (73%), Text (64%), and Multipart (44%) (Symantec, 2015).

SOLUTIONS TO THE SPAM PROBLEM

Spam has grown to become such a serious problem for all users of e-mail that various solutions are being used and proposed to combat spam. Some of the solutions require appropriate legislation, and some require technical changes in the E-mail handling systems of the Internet.

Individual and Corporate Level Solutions

Research establishes that there exists regularities in spamming behavior and spammers behave strategically. The distribution of spam messages reveals a cyclical trend, peaking in mid-week and subsiding on weekends. (Nigel Melville et al., 2006). For organizations the server-based spam control software are more effective and less costly to eliminate spam. A number of companies now offer software that organizations can run on their e-mail server computers to limit the amount of spam. A business interested in a enterprise-wide anti-spam solution can either purchase a corporate anti-spam software that would be installed on the company email server, or employ services of an anti-spam company to filter the spam before coming into their email server. Individual users can install client-based Anti-spam programs on their computers or set filters that might be available within their e-mail client software. The single user can purchase a license of anti-spam software from various sources. Some are free, and others can typically range in price from $20 to $50. In general, the software either deletes the spam immediately or moves the spam into a separate folder. Every time software blocks a piece of spam, it is reported to the software manufacturer company. None of the anti-spam software is 100%, and can generate false alarms i.e. block legitimate emails because they were identified as spam. The way the software handles the spam can vary widely, so it is important to find out the details on it before purchasing. Anti-spam service providers usually filters viruses at the same time, if you do not have a good antivirus solution in place already. Price of the service typically starts at $50 a month, for a small office. This may seem expensive but you need to consider the fact that there is no software to configure and maintain, so your IT labor costs are nil. While many spam filter products are readily available, most free spam filters are much too complex to set up and use or in fact, the better products in this class are not free.

MailWasher Free is the best free option for most users. The program is an email preview utility that allows you to check your email on your mail server before you download it to your PC. The advantage of this approach is that you can kill unwanted messages, including spam, viruses and large attachments before they get anywhere near your computer. MailWasher flags those messages, which it assesses as questionable for you. SPAMfighter Standard is a network-based spam-filtering system that uses the opinions of over two million users' worldwide to help classify spam. Like all network-based spam filters, it requires no training; it is ready to go the minute you install it. Each email is checked after retrieval. If the email has been classified as spam by many other users on the network, then it is deleted from your in-box and placed in a spam mail folder. It works like a charm and is a commanding choice for Outlook users.

Spam control software for businesses is also available. For example, Symantec Mail Security for Microsoft Exchange 7.5 combines Symantec antimalware technology with advanced heuristics to provide real-time email protection against viruses, spyware, phishing, and other malicious attacks while enforcing content filtering policies on Microsoft Exchange Server 2007, 2010 and 2013. In addition, Mail Security leverages Symantec Premium AntiSpam, powered by Brightmail™ technology, to stop 99 percent of incoming spam with less than one in 1 million false positives. It supports Hosted, Microsoft Hyper-V®, or VMware® virtualized Exchange server environments. Symantec Mail Security for Microsoft Exchange complements other layers of protection by preventing the spread of email-borne threats and enforcing data-loss prevention policies. Another product is GFI MailEssentials. MailEssentials uses multiple anti-spam filters that combine SpamRazer technology, greylisting, IP reputation filtering, Bayesian filtering, and other advanced technologies to provide a spam capture rate of more than 99% and minimal false-positives, ensuring the safe delivery of important emails. Granular, user-based email content policy enforcement enables you to control content that enters and leaves your network via email. This function is based on real file type, dictionary keyword checks and regular expression checks, helping to protect your company from accidental or malicious data leaks while assisting with compliance efforts.

Besides using specialized software and technology, spam can also be reduced using few tactics. One technique is to reduce the likelihood that a spammer can automatically generate e-mail addresses. For examples organizations, rather than using combination of first and last name, can use an e-mail address that is more complex, such as qct6hh42e@xyzcompany.com. Spammers use software robots to search the Internet for character strings that include the "@" character (which appears in every e-mail address). Therefore, users can also reduce spam by controlling the exposure of an e-mail address on discussion boards, newsgroups, chat rooms etc. Users also must not respond to spam messages. Doing so simply confirms that the address is live and makes it a more attractive target for spammers. Users should read privacy policies when entering email addresses on web sites. Users should use a separate email address (e.g. a free web-based email account such as Yahoo!) for online purchasing and entering on various sites. Some individuals use multiple e-mail addresses for multiple purposes.

Following are some guidelines for individuals to protect themselves from spam.

- *Never Reply To Or Click On Any Links In A Spam Message.* Do not click the "Unsubscribe" link unless it includes mention of the CAN-SPAM Act. These actions only serve to confirm to spammers that you exist and you are receiving their emails.
- *Read Your Messages as Text and turn off the ability to view pictures, HTML, movies, and formatted text for emails you do not know.* Some email providers like Google and Hotmail automatically block these things from appearing in emails from senders not familiar to you.
- *Preview Your Messages.* Like reading your messages as text, this prevents you from downloading spyware, adware, and viruses without knowing it.
- *View Message Headers and pay special attention to the "From" and "Reply To" addresses:* If they are not the same, this is a warning sign of spam.
- *Use a Complicated Email Address:* This is because spammers' software normally looks for easy and obvious addresses first.
- *Create Alias Email Addresses:* Generate multiple, anonymous email addresses that forward to your real email account.
- *Read Privacy Policies before Disclosing Your Email Address.* Do not register your email address on a website unless you know for sure that you can later opt-out from any emails they send you. Read their privacy policy/
- *Do not Use Your Email Address as Your Screen Name:* If you participate in chat/message boards where you register a username, do not use the section of your email address before the @ sign as your screen name.
- *Disguise Your Email Address:* If you need to publish your email address on a website, disguise it so that spammer's software cannot find it. You can do this by leaving out periods and @ signs/
- *Do not Use a Major Free Email Provider as Your Primary Address.* Spammers will often target common usernames on widely used email domains like Hotmail, Yahoo, AOL, MSN, etc.
- *Use A Spam Filer Or Blocker.*

- *Adjust Your Privacy Settings*: - Make sure the spam filters included in your email service are on their highest setting.

Technical Solutions

When one computer on the Internet sends a message to another computer, it does not send any more messages to the same computer until it receives the acknowledgement. Some vendors, such as IBM, sell software and access to a large database that tracks spammer computers continually. Other vendors sell software that identifies multiple e-mail messages coming from a single source in rapid succession. Companies can also use a technique called teergrubing. A teergrube is a computer server set as a trap for spammers trying to steal email addresses from e-mail servers. In teergrube server, fake email addresses are created in places where address harvester software look for e-mail addresses. These e-mail addresses contain a human-readable warning not to send messages to these addresses. The address harvester software, unable to read the warning, collects the addresses and the spammer duly sends spam. The teergrube server accepts the spammer's messages but very slowly. A teergrube server keeps the session alive so that session is not times out. This extended time can be used to detect the source of message and to launch a return attack sending e-mail messages back to the computer that originated the suspected spam.

Most industry observers agree though that adoption of new e-mail protocols capable of providing absolute verification of e-mail messages is the ultimate solution for spam control. However, this will require all mail servers on the Internet to be upgraded. Proposals for such identification standards have been made by companies such Microsoft, Yahoo!, and others. The Internet Engineering Task Force (IETF) working group that has responsibility for e-mail standards is working on a set of standards that will accomplish sender authentication. See also Anti-Spam Recommendations for SMTP MTAs from IETF (www.ietf.org/rfc/rfc2505.txt). The Messaging Anti-Abuse Working Group (MAAWG) is a global organization focusing on fighting spam. The IETF is working to convert selected MAAWG documents into a set of industry-adopted best common practices.

Legal Solutions

Most ISPs have an Acceptable Use Policy (AUP), which describes unacceptable behaviors while you are using the ISP's services, its reasons, and penalties for violation of AUP.

Sending spam is violation of AUP but ISPs willingness or ability to enforce their AUP varies. Many factors can be attributed to this behavior including unwillingness, lack of personnel or technical skills for enforcement, and reluctance to enforce restrictions against profitable customers. Some industry observers believe that strict legal enforcement of spam control laws is the permanent solution.

Various jurisdictions have implemented spam control legislation. Each law addresses spam in different ways and offer different definitions of what the law covers. For example, the United States CAN-SPAM Act of 2003 requires that each email have a way for the recipient to opt-out of the senders list. Most believe that by doing so you are just confirming to the spammers that they have reached a working email address. If the message contains only commercial content, its primary purpose is commercial and it must comply with the requirements of CAM-SPAM. If it contains only transactional or relationship content, its primary purpose is transactional or relationship. In that case, it may not contain false or misleading routing information, but is otherwise exempt from most provisions of the CAN-SPAM Act (FTC.gov, 2015).

The primary purpose of an email is transactional or relationship if it consists only of content that:

- facilitates or confirms a commercial transaction that the recipient already has agreed to;
- gives warranty, recall, safety, or security information about a product or service;
- gives information about a change in terms or features or account balance information regarding a membership, subscription, account, loan or other ongoing commercial relationship;
- provides information about an employment relationship or employee benefits; or
- delivers goods or services as part of a transaction that the recipient already has agreed to.

It is common for email sent by businesses to mix commercial content and transactional or relationship content. When an email contains both kinds of content, the primary purpose of the message is the deciding factor. Here is how to make that determination: If a recipient reasonably interpreting the subject line would likely conclude that the message contains an advertisement or promotion for a commercial product or service or if the message's transactional or relationship content does not appear mainly at the beginning of the message, the primary purpose of the message is commercial. Therefore, when a message contains both kinds of content – commercial and transactional or relationship – if the subject line would lead the recipient to think it is a commercial message, it's a commercial message for CAN-SPAM purposes. Similarly, if the bulk of the transactional or relationship part of the message does not appear at the beginning, it's a commercial message under the CAN-SPAM Act.

Here is an example:

MESSAGE A:

TO: Jane Smith

FR: XYZ Distributing

RE: Your Account Statement

We shipped your order of 25,000 deluxe widgets to your Springfield warehouse on June 1st. We hope you received them in good working order. Please call our Customer Service Office at (877) 555-7726 if any widgets were damaged in transit. Per our contract, we must receive your payment of $1,000 by June 30th. If not, we will impose a 10% surcharge for late payment. If you have any questions, please contact our Accounts Receivable Department.

Visit our website for our exciting new line of mini-widgets!

MESSAGE A is most likely a transactional or relationship message subject only to CAN-SPAM's requirement of truthful routing information. One important factor is that information about the customer's account is at the beginning of the message and the brief commercial portion of the message is at the end.

MESSAGE B:

TO: Jane Smith

FR: XYZ Distributing

RE: Your Account Statement

We offer a wide variety of widgets in the most popular designer colors and styles – all at low, low discount prices. Visit our website for our exciting new line of mini-widgets!

Sizzling Summer Special: Order by June 30th and all waterproof commercial-grade super-widgets are 20% off. Show us a bid from one of our competitors and we'll match it. XYZ Distributing will not be undersold.

Your order has been filled and will be delivered on Friday, June 1st.

MESSAGE B is most likely a commercial message subject to all CAN-SPAM's requirements. Although the subject line is "Your Account Statement" – generally a sign of a transactional or relationship message – the information at the beginning of the message is commercial in nature and the brief transactional or relationship portion of the message is at the end.

If an email advertises or promotes the goods, services, or websites of more than one marketer, there is a straightforward method for determining who is responsible for the duties the CAN-SPAM Act imposes on "senders" of commercial email. Marketers whose goods, services, or websites are advertised or promoted in a message can designate one of the marketers as the "sender" for purposes of CAN-SPAM compliance.

Each separate email in violation of the CAN-SPAM Act is subject to penalties of up to $16,000, so non-compliance can be costly. However, following the law is not complicated. Here is a summary of CAN-SPAM's main requirements:

- Do not use false or misleading header information. Your "From," "To," "Reply-To," and routing information – including the originating domain name and email address – must be accurate and identify the person or business who initiated the message.
- Do not use deceptive subject lines. The subject line must accurately reflect the content of the message.
- Identify the message as an ad. The law gives you a lot of leeway in how to do this, but you must disclose clearly and conspicuously that your message is an advertisement.
- Tell recipients where you are located. Your message must include your valid physical postal address. This can be your current street address, a post office box you have registered with the U.S. Postal Service, or a private mailbox you have registered with a commercial mail-receiving agency established under Postal Service regulations.
- Tell recipients how to opt out of receiving future email from you. Your message must include a clear and conspicuous explanation of how the recipient can opt out of getting email from you in the future.

Construct the notice in a way that is easy for an ordinary person to recognize, read, and understand. Creative use of type size, color, and location can improve clarity. Give a return email address or another easy Internet-based way to allow people to communicate their choice to you. You may create a menu to allow a recipient to opt out of certain types of messages, but you must include the option to stop all commercial messages from you. Make sure your spam filter does not block these opt-out requests.

- Honor opt-out requests promptly. Any opt-out mechanism you offer must be able to process opt-out requests for at least 30 days after you send your message. You must honor a recipient's opt-out request within 10 business days. You cannot charge a fee, require the recipient to give you any personally identifying information beyond an email address, or make the recipient take any step other than sending a reply email or visiting a single page on an Internet website as a condition for honoring an opt-out request. Once people have told you they do not want to receive more messages from you, you cannot sell or transfer their email addresses, even in the form of a mailing list. The only exception is that you may transfer the addresses to a company you have hired to help you comply with the CAN-SPAM Act.

- Monitor what others are doing on your behalf. The law makes clear that even if you hire another company to handle your email marketing, you cannot contract away your legal responsibility to comply with the law. Both the company whose product is promoted in the message and the company that actually sends the message may be held legally responsible.

Content Filtering

Content filtering is the technique whereby content is blocked or allowed based on analysis of content, rather than source or other criteria. Content filtering is commonly used to filter web and email content on the Internet.

E-mail Content Filtering

E-mail content filters act either on the content (the information contained in the mail body) or on the mail headers (like "Subject:") to either classify, accept or reject access to email message. Email filtering software inputs email. For its output, it might pass the message through unchanged for delivery to the user's mailbox, redirect the message for delivery elsewhere, or even throw the message away. Some mail filters are able to edit messages during processing. Common uses for mail filters include organizing incoming email and removal of spam and computer viruses. A less common use is to inspect outgoing email at some companies to ensure that employees comply with appropriate laws. Users might also employ a mail filter to prioritize messages, and to sort them into folders based on subject matter or other criteria. Mail filters can be installed by the user, either as separate programs (see links below), or as part of their email program (email client). In email programs, users can make personal, "manual" filters that then automatically filter mail according to the chosen criteria. Most email programs now also have an automatic spam filtering function. Internet service providers can also install mail filters in their mail transfer agents as a service to all of their customers. Due to the growing threat of fraudulent websites, Internet service providers filter URLs in email messages to remove the threat before users click. Corporations often use filters to protect their employees and their information technology assets.

Web Content Filtering

Web content filtering is used to filter content viewable to users surfing the web. Users can be disallowed from viewing inappropriate web sites or content. Filtering rules may be implemented via software on individual computers or at a central point on the network such as the proxy server or internet router. Network-based filters block objectionable material before it enters the computers of users. Network-based content control solutions may be more difficult to bypass than individual desktop software solutions, since they are less easily removed or disabled by the local user. At the most basic level, the filter uses a database to identify and block websites that match certain unwanted categories. That does not help when a site has not been categorized, or has added objectionable material since the database was last updated. Better content filters analyze page content to determine if a site is undesirable. Filter software matches up sites with multiple potentially objectionable categories. It is conceivable that an employee might have a legitimate need to access a blocked site. Perhaps a report requires information that can be obtained from LinkedIn whose use is banned on company network. In that case, employee can click a link in the content filter software to request an exception. You will receive an email notification. If the request has merit, you can log in to the online console and create an exception that will take effect right away. Good content filtering software cannt be disabled through use of network commands and can can identify and block unwanted sites even when they use a secure HTTPS connection. Content filtering software can also filter HTTPS traffic. Suppose you block porn but allow access to secure anonymizing proxy sites. If an employee tries to surf to a porn site using a secure anonymizing proxy, content filtering software will still block it.

Bypassing Content Filtering

Content filtering can be bypassed. Some content filtering software may be bypassed successfully by using alternative protocols such as FTP or telnet or HTTPS, conducting searches in a different language, or using a proxy server. Cached web pages by search engines, and alternate paths to the content provided by web syndication services can bypass content filters. A poorly designed content filtering application can be shut down by killing its processes. Many content filter software have an option, which allows authorized people to bypass the content filter.

Content Filtering Software

There exist many good content filtering software that can be used by both individuals and businesses. Net Nanny is one of the most popular content filtering systems. Net nanny is a powerful solution that categorizes in real time (so it does not rely on white/black lists), offers remote management, has a flexible alert/reporting tool, and can handle the usual suspects for parental controls/content filtering (time controls, profanity masking, IM management, and more). Net Nanny is one of the easiest tools in the category to use and although it is primarily targeted to home/parental-control use, it can be deployed in a business environment as well. K9 is another outstanding solution focused primarily on the protection of children. With this tool, you can block entire categories of specific content, block specific websites, take advantage of "Safe Search", and even rate content. K9 also includes a powerful anti-malware tool that will protect your machine from malicious software. Unlike Net Nanny, K9 uses a web-based interface to configure and monitor the system. Safe Squid begins to dive into waters more business-oriented. Safe Squid offers a much more powerful set of filtering tools as well as more detailed logs, user authentication filtering, redundant-level content filtering, re-programmable content filters, programmable templates, caching and pre-fetching, and much more. Safe Squid is an HTTP 1.1 Proxy server and can help you prevent employees from misusing resources. Unlike both Net Nanny and K9, Safe Squid is much more geared toward businesses. Do understand that Safe Squid is not nearly as easy to install, as is Net Nanny or K9. Safe Squid is also available for both Linux and Windows. OpenDNS offers solutions for everyone from households to enterprise businesses. OpenDNS offers industry-leading maleware and botnet protection, web filtering, fast/reliable DNS, a globally distributed cloud, and an incredibly easy web-based administration interface. In addition, for the larger clients, OpenDNS offers Enterprise Insights, which is enterprise-grade security and control delivered through the cloud (Wallen, 2012).

Website and Internet Utility Programs

In addition to Web server software, website may use a number of utility programs, or tools. Some of these programs run on the Web server itself, while others run on the client computers.

Finger and Ping Utilities

Finger is a program that takes an e-mail address as input and returns information about the owner of that e-mail address. Finger can also tell whether the user is currently logged on, user's full name, address, and telephone number etc. (provided the information be in the system). A number of e-mail software has built-in Finger utility so you can send the command while reading your e-mail. Many organizations disable the Finger command on their systems for privacy and security reasons. If you send Finger command to one of such systems, you will receive no response.

Ping is a utility that determines accessibility of a specific IP address. Primarily used to troubleshoot network connections, ping works by sending a packet to the specified IP address and then wait for a reply. On a Windows PC, you can send a Ping command by opening an MS-DOS window and typing "ping" followed by the IP address.

Traceroute and Other Route-Tracing Programs

Traceroute is a utility that can be used to trace route taken by a packet from your computer to an Internet host. Traceroute can show how many hops the packet requires to reach the host and how long each hop takes. For a slow responding website, a user can use traceroute command to find out the bottleneck. Nearly all operating systems have a variant of traceroute command. In Windows, you can run traceroute command by opening an MS-DOS window and typing "tracert" followed by the host name. For example:

tracert www.pcwebopedia.com

Traceroute utilities work by sending packets with low time-to-live (TTL) fields. The TTL value specifies the number of hops (a hop is the trip a packet takes from one point in a network to another) a packet could take. One the limit is reached the host (router or some other device) in the last hop returns the packet and identifies itself. By sending many packets with incrementing TTL values, we can determine the hosts available between two end-points in a network.

GUI-based traceroute programs can also provide a map of packets' route. Network engineers can use this map to determine the location of the greatest delays on the Internet. One such GUI-based route-tracing program is Visual IP Trace by Visualware (visualiptrace.com). You can download the evaluation version to test the program.

Telnet and FTP Utilities

Telnet is a terminal emulation program that allows users to log on to a computer that is connected to the Internet or any network. After logging in, users can then enter commands through the Telnet program. Commands will be executed as if you entered them directly on the system's console. Using telnet, users can control the computer and communicate with other computers on the network. To start a Telnet session, you need a valid username and password. Telnet is commonly used for remotely controlling the web servers.

FTP (File Transfer Protocol) works in the same way as HTTP or SMTP and is used to exchange files over the Internet. FTP uses the TCP/IP protocol to enable data transfer. Most common application of FTP is downloading and uploading files from and onto Internet servers. An FTP connection to a computer on which the user has an account is called full-privilege FTP. FTP connection to a computer as a guest is called Anonymous FTP.

Indexing and Searching Utility Programs

Searching and indexing utilities are important elements of many Web servers. Search tools can search either a specific site or the entire Web for requested information. Indexing program can provide a full-text index of website outlining all documents stored on the server. When a user search a website by providing some keywords the search engine compares the index terms to the user-supplied keywords to see which documents contain matches for the requested keyword(s). More advanced search engine software uses complex algorithms to perform searches. Many Web server software products also contain indexing software. While dedicated indexing programs can be used in indexing web sites, these programs are specifically designed to aid indexers working with web sites and other HTML documents. Two indexing software are HTML Indexer™ (for Windows) (http://www.html-indexer.com) and XRefHT32 (freeware) (http://publish.uwo.ca/~craven/freeware.htm). Zoom (http://www.wrensoft.com/zoom/index.html) is a commercial software package that creates a search engine for your website. It provides fast and powerful full-text searching by indexing your website in advance with a user-friendly desktop application that allows you to configure and index your site, from the convenience of your Windows computer. Zoom provides autocomplete search box, date range searching, URL indexing, search by site, greater customizability, an overhauled interface, SSL and HTTPS spidering indexing support and indexing support for additional file formats. Yioop is a GPLv3, open source, PHP search engine. Yioop comes with a crawler, which can be used to crawl the open web or a selection of URLs of your choice. Once you have created Yioop indexes of your desired data sources, Yioop can serve as a search engine for your data. It supports "crawl mixes" of different data sources. Yioop also provides tools to classify and sculpt your data before being used in search results.

Data Analysis Software

Web servers can capture a lot of information about visitors (e.g. visitor's URL, session duration, the date and time of each visit, and pages viewed). This data is placed into a Web log file. Careful analysis of the log file can reveal useful statistics about site visitors and their preferences. The study of consumer clickstream behavior underlies Web site designs, marketing strategies, on-line advertising prices, and other Web-based initiatives (Kalczynski, Senecal, & Nantel, 2006). Web log-file analysis software can analyze log files and return detailed information (e.g. number of visits, visits duration, days of week and rush hours, domains and countries of host's visitors, most viewed pages, and operating systems used). Popular Web log file analysis programs include WebLog Expert, Webalizer, Analog, and Sawmill, and UrchinWeb Analytics.

Link-Checking Utilities

Link checking programs can automatically check all the links on a page or on a website to confirm that there are no broken links. A broken link does not point to a valid web page. Broken links can arise for a number of reasons e.g. because incorrect definition of the link. Link checking utilities are a practical and efficient solution for large websites containing hundreds of thousands of links. Ensuring that your site is free of dead links is vital because if visitors encounter too many dead links on a site they can switch to another site. Some Web site development and maintenance tools, such as Macromedia's Dreamweaver, include link-checking features. However, most link-checking programs (such as Elsop LinkScan) run as separate programs. The link checker software can either show the results in a Web browser or send via email to a recipient. Besides checking links, Web site validation programs sometimes check spelling and other structural components of Web pages. A reverse link checker checks on sites with which a company has entered a link exchange program and ensures that link exchange partners have included a link back to the company's Web site. LinxCop is one of several reverse link checkers available.

Remote Server Administration Software

With remote server administration, a Web site administrator can control a Web site from any Internet-connected computer. For example, an administrator can install monitoring tool on any Internet-connected Windows computer and monitor/change anything on the Web site from that computer. NetMechanic (www.netmechanic.com/) offers a variety of link checking, HTML troubleshooting, site monitoring, and other programs that can be useful in managing the operation of a Web site. Remote Server Administration Tools for Windows 8.1 enables IT administrators to manage roles and features that are installed on computers that are running Windows Server 2012 or Windows Server 2012 R2 from a remote computer that is running Windows 8.1 Pro or Windows 8.1 Enterprise.

E-commerce Hardware

Companies use a wide variety of computer brands, types, and sizes to host E-commerce operations. Some small companies can run Web sites on desktop PCs. However, most E-commerce Web sites are operated on computers designed for website hosting.

Web Server Hardware: Components and Requirements

E-business requires a web server that is able to run reliably and efficiently 24/7 for months without any service interruption. This requirement needs to be taken into account in web server designing right from the start. You need to pick hardware components that are not only reliable but also fit the purpose of the web server.

The type of server you will need will be partially determined by the type of technologies you plan to deploy on the web server e.g. a web server employing Common Gateway Interface (CGI), Perl, Active Server Pages (ASP), or PHP will need a faster processor. We can use following guidelines while deciding the type of server to be used. These guidelines are flexible and can be changed.

A basic website with few pages requires no special hardware. Websites that include FTP or streaming media required high bandwidth Internet connection. Websites containing CGI scripts and ASP require faster processors.

In general, it is a good idea to have a web server powerful and robust enough to handle resource intensive software technologies. You should consider getting a server with more memory, larger and faster hard disk drives, fast processors, and fault tolerant features like multiple power supplies, redundant disk drives, and network cards. Many Web server computers use multiple processors and are usually much more expensive than workstation PCs. Your Web server should also have an uninterruptible power supply to remain functional in times of power outages. Another technique called clustering in used to increase the availability of web sites. It is a hardware redundancy technique. A cluster is a group of two or more identical computers that work together closely to share the traffic the site receives. Each member of the cluster is equipped with failover capability. If one machine fails for any reason, the web site is instantly served from the remaining machine(s). Failed machine rejoins the cluster, once repaired. Physical facility where your Web server is installed must be secure, have adequate environment, and adequate and reliable power supply. Regular backup of web servers, using tapes or some other backup media, is essential. The backup media should preferably be kept at a separate physical location. Web server bandwidth is another important consideration because the larger the bandwidth available, the more customers can simultaneously access you E-commerce site.

If you decide to host your own website, your server hardware must be able to easily support your highest anticipated traffic to your site and 6 -12 months of anticipated growth. It is a good idea to have a scalable hardware for the E-commerce site. Using hardware scaling, you can meet varying levels of services demand at your E-commerce site by increasing site's size. Hardware scaling can be either vertical, horizontal, or a combination of both. In vertical scaling, servers can be upgraded e.g. from a single processor or hard disk to multiple processors and hard disks. However, this method is expensive and your E-commerce site becomes dependent on one or few powerful servers. In horizontal scaling, multiple small, single processor servers can be added to E-commerce site and perform load balancing. Dedicated servers can also be created and fine-tuned to handle a set of specific tasks. For example one server can handle requests for static content and one sever can handle dynamic content. Horizontal scaling is less expensive because the smaller servers used cost less than the large more powerful servers used in vertical scaling. Many a times you can use older PCs as servers and if one machine fails, another one can step in and perform its role.

Web Server Performance

Performance evaluation of a particular web server requires determination of factors that are relevant to the expected use of the web server. Several factors can affect overall performance of a web server e.g. hardware, operating system software, server software, bandwidth of Internet connection, user capacity, and type of content delivered (static or dynamic), number of users a server can handle, and whether content delivered is cached or not. Two factors relevant to measuring a server's capability to deliver web pages are throughput and response time.

Throughput is the number of HTTP requests that a particular web server can process in a unit of time. Response time is the amount of time a server requires to process one request. These values for an operational web server should be well within the anticipated server loads. One way to choose Web server hardware configurations is to run tests on various hardware combinations. If you do not have the hardware and software set up to perform this testing, third-party services are available to perform this testing. Mindcraft (mindcraft.com) provides testing of software, hardware systems, and network products. Mindcraft site contains reports and statistics comparing combinations of application server platforms, operating systems, and Web server software products.

WEB Server Overloading

With defined load limits, a web server can handle a specified number of concurrent client connections and can serve a certain number of requests per second. Some of the factors responsible for server overloading include:

HIGH VOLUME OF WEBSITE TRAFFIC: A large number of clients connecting to a website in a short period can cause server overload especially if server hardware is inadequate.

DISTRIBUTED DENIAL OF SERVICE (DDOS) ATTACKS: These attacks can cause server to respond slowly or become unresponsive.

VIRUSES AND WORMS: Viruses and worms can corrupt web server software including web server operating systems and other applications.

LOW BANDWIDTH OF INTERNET CONNECTION: A low bandwidth can cause server to respond slowly.

Some symptoms of an overloaded web server include long server response times, errors returned to clients (e.g. HTTP 500, 502, 503, and 504 errors), and refused or reset connections before any content is sent to the client.

TECHNIQUES TO PREVENT OVERLOADING

Webmasters can use a variety of techniques to prevent server overloading.

NETWORK TRAFFIC MANAGEMENT: Webmaster can utilize multiple technologies to manage network traffic e.g., firewalls to block unwanted traffic from suspicious IP addresses and bandwidth and network traffic management applications to drop or redirect bad HTTP requests

WEB CACHE TECHNIQUES: Using web cache techniques, webmasters can cache content to save network bandwidth.

DIFFERENT WEB SERVERS: Web masters can setup different web server with different domain names to serve different (static or dynamic, large or small files) content by separate Web servers. Webmasters can also use separate web server for each network segment.

LOAD BALANCING: In this technique, webmasters can use many web servers grouped together so that they act as one big web server.

SERVER HARDWARE/SOFTWARE TUNING: Webmaster can also add more hardware resources (e.g. memory or processor) and tune operating system parameters to use hardware more efficiently.

Web Server Hardware Architectures

Web server architecture defines the way servers can be connected to each other and to related hardware, such as routers and switches. Configuring this architecture is important because large E-commerce websites processing large number of transactions require many servers.

CENTRALIZED VS. DECENTRALIZED OR DISTRIBUTED ARCHITECTURE

A decentralized or distributed architecture is also called server farm. A server farm also called a server cluster is a group of networked servers that are located in one place. A server farm combines servers and processing power into a single entity. A server farm distributes the workload between member servers of the farm and expedites computing processes. The server farm relies on load balancing software. Load balancing software track demand of processing tasks from different machines, prioritize them, and schedule/reschedule them depending on priority of user's demand. When one server in the farm fails, another server steps in as a backup. Server farms are excellent way of handling the enormous amount of computerization of tasks and services. A Web server farm, or Web farm, can be either a Web site that runs off more than one server or an ISP that provides Web hosting services using multiple servers. The decentralized architecture of server farm uses smaller servers. These smaller servers are less expensive than the large servers used in the centralized architecture. However, the decentralized architecture does require additional networking hardware (hubs or switches) to connect the servers to each other and to the Internet.

BLADE SERVER AND THIN CLIENTS

A blade is literally a self-contained server, which collectively fits into an enclosure with other blades. Sometimes known as a chassis, this enclosure provides the power, cooling, connectivity, and management to each blade server. The blade servers themselves contain only the core processing elements, making them hot- swappable. HP refers to the entire package as a BladeSystem. To get a better idea of what a single blade contains, a blade can hold hot-plug hard-drives, multiple I/O cards, memory, multi-function network interconnects, and Integrated

Lights Out remote management (HP.com, 2015). A blade server can take virtually any workload from client to cloud. They can be used for server virtualization, building virtual desktops, provide cloud infrastructure, support big data applications, and provide IT Infrastructure solutions for collaboration. Blade servers can provide savings of time, money, and energy costs to create a competitive advantage over your competitors to achieve up to 68% reduction in data center costs over traditional rack environment, 90% reduction in downtime and speed delivery of new applications and services (HP.com, 2015). Some famous blade server manufacturers include HP, Dell, and IBM.

FIGURE 3-10: BLADE SERVER

Source: Rackmountmart.com

A thin client is a stateless, fanless desktop terminal that has no hard drive. All features typically found on the desktop PC, including applications, sensitive data, memory, etc., are stored back in the data center when using a thin client. A thin client running Remote Desktop Protocols (RDP), like Citrix ICA and Windows Terminal Services, and/or virtualization software, accesses hard drives in the data center stored on servers, blades, etc. Thin clients, software services, and backend hardware make up thin client computing, a virtual desktop computing model. Thin clients are used as a PC replacement technology to help customers immediately access any virtual desktop or virtualized application. Thin clients provide businesses a cost-effective way to create a virtual desktop infrastructure (VDI). Thin clients are utilized in various industries and enterprises worldwide that all have different requirements but share common goals. The cost, security, manageability, and scalability benefits of thin clients are all reasons that IT personnel are exploring –and switching– to thin clients. Cost wise, the price per seat of a thin client deployment has dropped to the point where it is more cost effective than regular PCs. (Devonit.com, 2015). Some popular manufacturers of thin clients include HP, Dell, and Lenovo.

FIGURE 3-11: THIN CLIENT

Source: Thinclientbrasil.com

E-commerce Software

E-commerce software accomplishes specific E-commerce functions, such as order entry and processing, content management and delivery, user verification and security, and payment processing.

Selecting E-commerce Software

An organization's choice of E-commerce software depends on several factors e.g. the expected size of the business, estimated site traffic, allocated budget, and expected sales revenue. A high-traffic E-commerce site, e.g. Amazon.com, requires different software than a small online business. Overall cost of creating an online store including its infrastructure and startup cost can be much less compared with building a chain of traditional retail stores.

Basic Functions of E-commerce Software

Irrespective of their size, E-commerce software performs some basic functions including:

- Product Catalog

- Shopping Cart

- Transaction Processing

In addition to the basic functionality, larger and more complex E-commerce software provide additional features and capabilities. These additional features can include:

- Middleware (software that integrates the E-commerce system with existing information systems of the company).

- Enterprise Application Integration

- Web Services

- Integration with Enterprise Resource Planning (ERP) software

- Supply Chain Management (SCM) software

- Customer Relationship Management (CRM) software

- Content Management Software

- Knowledge Management Software

Product Catalog

A catalog is a listing of goods and services. Electronic online catalogs is the backbone of most e-commerce sites. Such catalogs consist of a product database, directory and search capabilities, and a presentation function. Online product catalog is hosted on the merchant server in the form of a database. A database is a part of the merchant server designed to store and report on large amounts of information. For example, a database for an online clothing retailer would typically include such product specifications as item description, size, availability, shipping information, stock level and on-order information, and customer information (e.g. name, address, credit card numbers etc.).

Merchants use online catalogs to advertise and promote products and services. Customers use these catalogs as a source of information on products and services. Electronic catalogs have significant advantages such as ease of updating, ability to include a single product in multiple categories, ability to integrate with the purchasing process, coverage of a wide spectrum of products, and quick search with the help of search engines. One disadvantage of online catalogs is that customers need computers and the Internet in order to access online catalogs. As the electronic catalogs are being integrated with shopping carts, order taking, and payment, the tools for building online catalogs are being integrated with merchant sites (e.g., see store.yahoo.com). With computers and Internet access spreading rapidly, we can expect a large portion of paper catalogs to be supplemented by, if not actually replaced by, online catalogs. However, in B2B, paper catalogs may disappear more quickly.

Online catalogs can be divided into two categories: static and dynamic. A static catalog is a simple list written in HTML that appears on a Web page or a series of Web pages. Any modification in the catalog requires editing of HTML code of one or more pages. Larger E-commerce sites are more likely to use a dynamic catalog, which stores the information about items in a database. A dynamic catalog can feature multiple photos of each item, detailed descriptions, and a search tool that allows customers to search for an item and determine its availability. A static catalog is sufficient for a small E-commerce company that that offers only a small number of items and organization of the items is not particularly important. Larger E-commerce companies offer large number of items and organization of the items and their navigation is particularly important. These companies use dynamic catalogs that provide more sophisticated navigation aids and better product organization tools.

Good E-commerce sites also provide buyers alternative ways to find products .Besides offering a well-organized catalog, large E-commerce sites with many products also provides a search engine that allows customers to enter descriptive search terms, such as "men's shirts," so they can quickly find the Web page containing what they want to purchase.

Shopping Cart

In early days of E-commerce, customers used fill-in forms on E-commerce sites to purchase items. These forms used text box and list box form controls to indicate customer's choices. However, this system was not suitable for ordering one or two items at a time. Shoppers had to write down product codes, unit prices, and other information about the product before going to the order form. The forms-based method of shopping was confusing and error prone. Figure3-12 illustrates a fill-in form.

FIGURE 3-12: A FILL-IN ORDER FORM

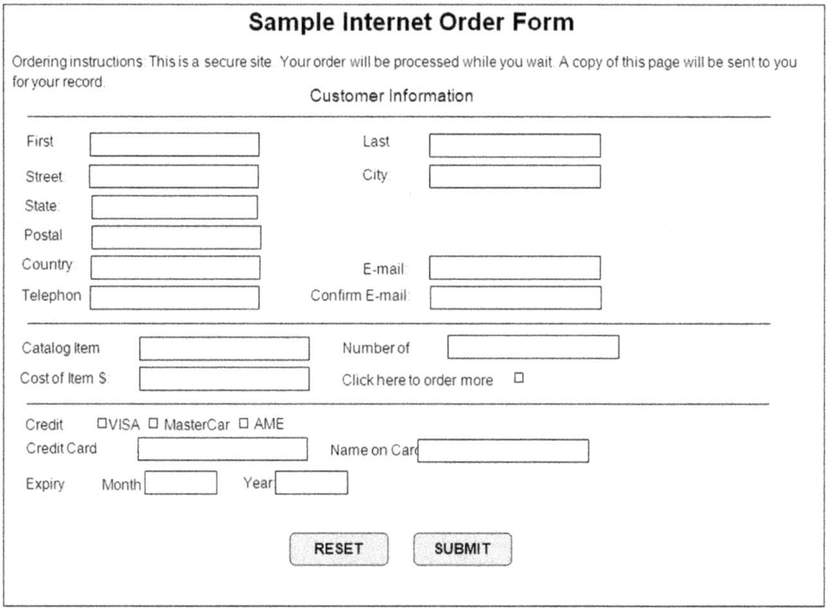

Sample Internet Order Form

Ordering instructions. This is a secure site. Your order will be processed while you wait. A copy of this page will be sent to you for your record.

Customer Information

First		Last	
Street		City	
State			
Postal			
Country		E-mail	
Telephon		Confirm E-mail	

Catalog Item | | Number of | |
Cost of Item $ | | Click here to order more ☐ |

Credit ☐VISA ☐ MasterCar ☐ AME
Credit Card | | Name on Card | |
Expiry Month | | Year | |

RESET   SUBMIT

Because of inconvenience associated with forms-based method of ordering, electronic shopping cart has become very popular. An electronic shopping cart is an order-processing software that allows customers to select items, review what has been selected, make changes, and then finalize the list. When the customer clicks on "buy" button, actual purchase process begins. Electronic shopping cart is supported by a product catalog, which is hosted on the merchant server.

Looking for a particular product on an E-commerce site, you typically enter some keywords in search feature provided on the site to look for the desired product. Once you find the product, you can click on the hyperlink of the product to go the product page. You can use "add to shopping cart" option if you would like to purchase the item. The shopping-cart technology processes the information and displays a list of the products customer has placed in the shopping cart. An electronic shopping cart automatically keeps track of the items the customer has selected and allows customers to view the contents of their carts, add new items, change item quantity, or remove items. To order an item, the customer simply clicks that item. You than have the option to check out or continue shopping. When the customer is ready to conclude the shopping session, the click of a button executes the purchase transaction. Figure 3-13 shows a typical shopping cart page at a site. All of the details about the item, including its price, product name, product picture, and a link to product page are stored automatically in the cart.

FIGURE 3-13: A TYPICAL SHOPPING CART

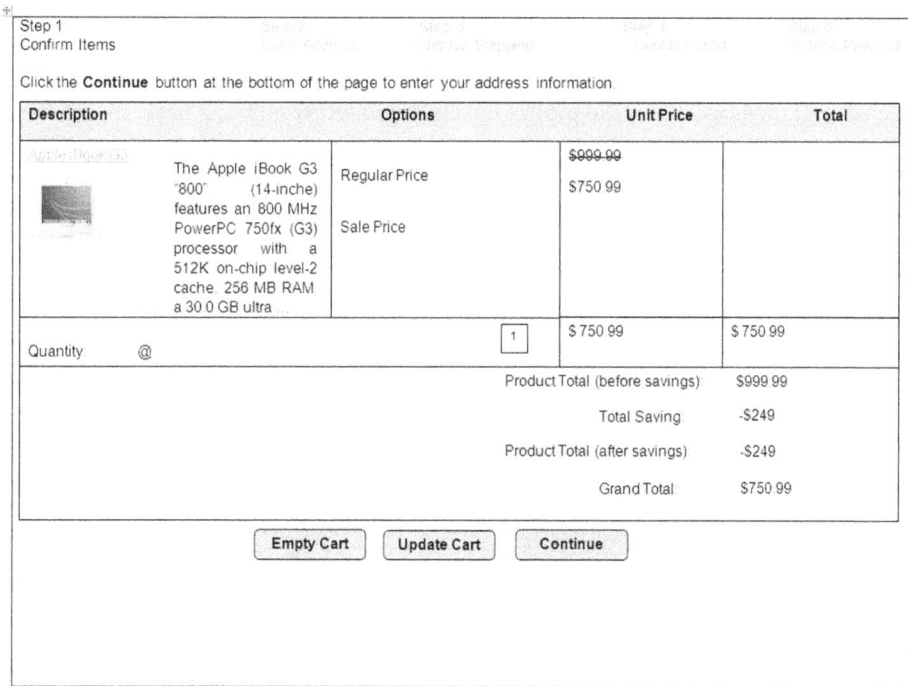

Clicking the Continue button usually displays a screen that asks for billing and shipping information and that confirms the order. Note that the shopping cart software keeps a running total of each type of item. The shopping cart calculates a total as well as sales tax and shipping costs.

As a first-time visitor, you will generally be prompted to fill out a personal-identification form with information including your name, billing address, shipping address, shipping preference and credit card information. You will also be asked to enter a password that you will use to access your account data for all future transactions. You can place your order once you confirm your information. When you have finished placing your order, E-commerce site sends a confirmation to you by e-mail and then send a second email once the product is shipped. Shopping-cart software is sold or provided free as an independent component (e.g.monstercommerce.com,

179

edubiz.bizhosting.com, e-shopping-cart-software.com, and actinic.com). Shopping cart software is also embedded in merchant servers, such as store.yahoo.com.

ADVANTAGES OF SHOPPING CART

Some advantages of shopping carts for E-commerce are as follows (Payloadz.com, 2014).

MULTIPLE PAYMENT MODE COMPATIBILITY: A well-designed online shopping cart allows you to process payments many ways. Besides credit card processing, good shopping carts include all types of online payment methods, such as PayPal and Google Checkout. Multiple payment options make the purchase possible for all types of visitors to the website, which boosts sales.

DIGITAL PRODUCT STORAGE AND DELIVERY SYSTEMS: When people make a payment for a digital product, they expect to receive it immediately. The digital product storage and delivery systems good shopping carts ensure quick and prompt delivery and protect the digital product from being freely shared or stolen.

TRACKING SALES TRENDS AND VISITOR PATTERNS: Shopping carts provide ability to track and assess trends, visitor patterns and buyer behavior. The shopping cart can provide data and analysis regarding which content is performing better and driving sales, the time users take to make their purchase decisions, where the user drops from the page and past sales. Tracking sales trends and visitor patterns can help e-commerce website better meet buyer needs, which can lead to greater profits.

THE USE OF AUTO RESPONDERS: Email marketing is one of the most effective modes of attracting buyers. Email marketing is usually done through auto responders, which are software programs or scripts that can automatically send people emails. Auto responders integrated in a shopping cart can be used for marketing, for delivering digital products and for giving payment confirmations.

IMPROVED PRODUCT MANAGEMENT: As an e-commerce website grows, its offerings also grow. This makes managing them more difficult. Good shopping carts come with product management tools designed to help e-commerce entrepreneurs cope with growth issues.

FEATURES OF SHOPPING CART

Following are some of the desired features of shopping carts that either improve the shopping experience for the consumer or make it easier for the merchant (Roggio, 2011).

Large, Functional Product Images

Product images are among the most effective ways to communicate with customers on an ecommerce site. In online shopping, the product image is the only opportunity you have to see that product. If this image not impressive from the photo, customers probably will not buy the item. Although product images are essentially a standard feature on all shopping carts or ecommerce platforms, it is important to find a solution that is flexible enough to allow you to resize the images, since bigger is usually better. It is also helpful if the cart supports product image zooming. The Victorinox site, the makers of Swiss Army knives, is a good example of how powerful a big, beautiful product image can be.

WELCOME TO VICTORINOX

Another example is clothing retailer Roxy. Roxy uses 417 by 561 pixel product images on its product pages. Shoppers may also see a larger version of the image, which are 683 by 792 pixels.

Uncrate is a digital magazine for men who love to buy stuff such as cars, tools, movies, music, and books. Uncrate also uses large product images. Uncrate often uses product images the full width of the page.

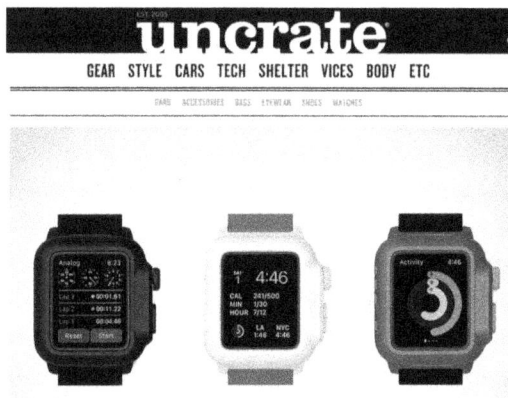

Product Reviews

A study from The Nielsen Company found that some 70 percent of respondents trusted consumer opinions posted online (Nielsen, 2012). While there may not be a perfect one-to-one relationship between "consumer opinions posted online" and product reviews on an ecommerce site, the study nonetheless provides some data.

With this in mind, it is imperative that a shopping cart either includes support for product reviews right out of the box or has an easy way to implement third-party product reviews. There are many examples of successful online stores that make excellent use of product reviews. One example is For example, Newegg (newegg.com), the electronics retailer, frequently has more than 75 reviews for individual products. These reviews can help shoppers make good buying choices.

Layered and Faceted Navigation

Layered and faceted navigation divides products into rational sub-categories, and displays those sub-categories as product filters. Such navigation scheme makes it much easier for shoppers to find just what they are looking for on an ecommerce site. Shoppers may drill down to products based on price, color, features, or attributes. In a good shopping cart, this feature should be data driven and programmatic. Zappos, the footwear retailer, is a good example of an ecommerce business that makes use of layered and faceted navigation.

Single-Page, Fast Checkout

A shopper interested in buying wants to complete the purchase as quickly as possible. Just like an offline supermarket, people are also looking for shortest lines for quick purchase. One of the simplest methods to speeding checkout is to limit the checkout form to as few fields as possible and keep the entire form on a single page to avoid loading a new page at each stage of the checkout process. A single page checkout boosts sales conversions, reduces abandoned shopping carts, enhances user-friendliness, and streamlines customer experience. One good example of single page checkout is MailingBags.ie. On the checkout page, all the details are entered on a single page.

On gift giving sites, a multiple page checkout is better because a buyer may be sending items to different addresses. In that case, a single age checkout could be too cluttered and confusing. For example, if a buyer were sending 10 gift baskets to 10 different addresses, a single page checkout would force buyer to place 10 different orders. This is not efficient.

Search

A good shopping cart should also provide product search facility. Shopping cart should make this function prominent. The Amazon.com site provides a very prominent search feature right at the top of the home page.

Coupons and Discounts

Many sites ,such as Groupon, Living Social, Google Deals, Retail Me Not, Rimbambo, and others offer shoppers daily deals, hourly deals, and coupons. A good shopping cart must be able to process coupon and discount codes, so that deal crazy shoppers can get their savings fix. A good ecommerce platform will have couponing and discounting built in. One good example is the olympusamerica.com, the camera retailer. This site gives buyers an opportunity to enter discount or coupon codes as soon as they start the checkout process.

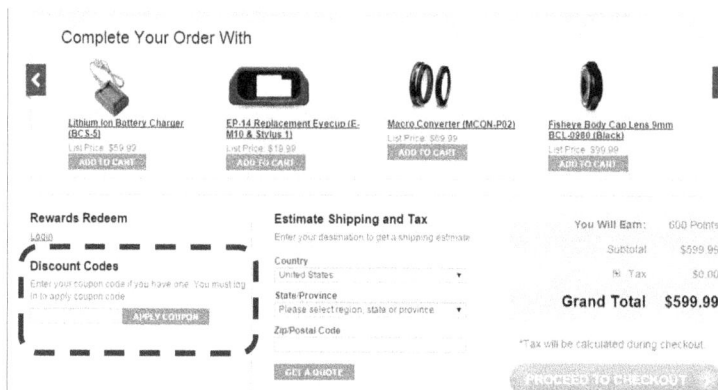

Product Import and Export

A shopping cart should have good product import and export capabilities. Whether one is exporting product data to a shopping comparison site or a Facebook store or synchronizing inventory from an online store to a physical location, the ability to easily transfer price, inventory, or product updates in to or out of a shopping cart is essential.

Easy Integration with Third Party Solutions

A shopping cart can provide many ways to integrate with third-party solutions. A modular shopping cart allows site owners to easily add third-party solutions. For example, if you would like to use QuickBooks for accounting, it can be integrated with your shopping cart with no hassle. Likewise, you can integrate your e-newsletter software (such as MailChimp) with your shopping cart.

Analytics and Sales Reporting

A good shopping cart must have built-in sales reporting and analytics capabilities. The shopping cart should be able to track product sales down to a significant level of detail. Sales reports should also be easy to export. The shopping cart software should offer a number of options to alter the basic appearance. Integration to shopping engines and shopping sites such as eBay, Amazon and Google offers businesses the greatest advantage when using shopping cart software.

SOME POPULAR SHOPPING CART SOFTWARE

Magento (www.magentocommerce.com) is one good shopping cart that provides most of the desired features of shopping. Some other good software include ATG (www.atg.com/), Demandware (www.demandware.com/), LemonStand (lemonstandapp.com/), Shopify (www.shopify.com/), and Volusion (www.volusion.com/ecommerce/online-store-feature-details/).

STORING SHOPPING CART INFORMATION

Many shopping cart software allows the customer to fill a shopping cart with purchases, put the cart in virtual storage, and come back days later to confirm and pay for the purchases. Many companies e.g. BIZNET Internet Services (biznet.net), CartIt! (cartit.com), SalesCart (salescart.com), and WebGenie (webgenie.com) Software sell shopping cart software that sellers can add to their Web sites. These software packages range in price from a few hundred dollars to several thousand dollars, plus an ongoing monthly fee.

It is important for E-commerce websites to store the shopping cart information. It has two purposes. First, the information can be retrieved by the shopper later and second one shopper can be distinguished from another so that the purchases are not mixed up. One way to uniquely identify users and store information about their choices is to create and store cookies. Cookies are bits of information stored on a client computer. When a customer returns to a site that issued a particular cookie, the shopping cart software reads either the cookie from the customer's computer or the database record from the merchant's server.

Another way of preserving shopping cart information is to assign a temporary ID to the shopper. This technique can be used if a shopper's browser does not allow storage of cookies. For example, ShopSite (shopsite.com), an E-commerce software package does this by automatically assigning a shopper a temporary number. The number is added to the end of the shopper's URL and persisted as he or she navigates from one Web site to another. When the customer returns, the URL still contains the bits of information about his or her shopping cart. However if the customer closes the browser this temporary number is discarded and cannot be used even if the customer later reopens the browser and returns to the same Web site.

Shopping carts for B2C are fairly simple (e.g. amazon.com) shopping carts for B2B are more complex. A B2B shopping cart could enable a business customer to shop at several sites while keeping the cart on the buyer's Web site to integrate it with the buyer's e-procurement system. A special B2B cart was proposed for this purpose by (Lim & Lee, 2002) where, in addition to the cart offered at the seller's site, there is a buyers' cart ("b-cart") that resided on the buyers' sites and is sponsored by the participating sellers.

TYPES OF SHOPPING CART SOFTWARE

There essentially are two types of carts available: licensed software and hosted solutions. Knowing the difference often can help an online merchant determine which type is right for their particular needs.

Licensed Shopping Cart

A licensed ecommerce cart refers to software that can be downloaded from a vendor and installed on a Web server computer. Some of these applications call for a one-time fee, and others are completely free for a merchant to use. The products that require a fee generally are geared toward enterprises that generate large monthly sales volumes. These carts typically come with premium support and upgrades for as long as the product is under license.

There exist many good licensed shopping cart software. Magento (http://www.magentocommerce.com/) is an open source, ecommerce powerhouse that includes many of the most sought after ecommerce features and has the support of a massive development and user community. The Magento platform is available in a free Community edition, but the Enterprise edition, which includes technical support and the most updated features, runs about $12,999 per year. LemonStand (http://lemonstandapp.com/) is very flexible platform that provides simplicity, ease of use, and visually appealing admin and reporting. LemonStand costs about $258. This is a great choice for tech-savvy ecommerce entrepreneurs. IBM WebSphere Commerce is a leader among large retail enterprises. This solution is very easy to integrate with social media and enables precision marketing capable of targeting offers and personalized promotions to individual shoppers. The express version starts at about $30,000, but request a quote to get specifics. Oracle ATG Web Commerce is designed for the world's largest and best-known brands. Hybris B2C Commerce is an enterprise-level, licensed ecommerce software solution. It offers a high-level of customization; is rich in features; and has an excellent record of accomplishment. OpenCart (http://www.opencart.com/) is a simple-to-use, free and open source shopping cart solution that is clearly aimed at small and startup ecommerce businesses. TomatoCart (http://www.tomatocart.com/) is an open source solution best suited for small or mid-sized commerce merchants. This licensed solution features a unique, desktop-like administration panel on top of the usual set of ecommerce features. PrestaShop (http://www.prestashop.com/en/) provides more than 240 specific ecommerce features. PrestaShop is available free and is open source.

Hosted Service Shopping Cart

A hosted service shopping cart refers to software that is never downloaded, but rather is provided by a hosted service provider and is generally paid for on a monthly/annual basis; also known as the application service provider (ASP) software model. Some of these services also charge a percentage of sales in addition to the monthly fee. This model often has predefined templates that a user can choose from to customize their look and feel. Predefined templates limit how much users can modify or customize the software with the advantage of having the vendor continuously keep the software up to date for security patches as well as adding new features. Free ecommerce carts usually are available under an open-source license that allows the user to modify the software to suit their individual needs. As long as it meets system requirements, this type of cart normally can run on any server platform. The most notable disadvantage to free carts is the lack of official support, which essentially forces the user to rely on community assistance.

There exist some good hosted shopping cart applications. Business Catalyst (http://businesscatalyst.com/) Adobe's Business Catalyst is fully customizable online store with little to no need for coding. You can insert your company logo or that of your client so that they only see the information you choose. You can use features analytics, Dreamweaver extensions, easy management and more. Flying Cart (http://flyingcart.com/) allows you to build ecommerce sites that vary in size from 50 products up to 1,000, depending on your needs. There is only a monthly fee and no transaction fees, so depending on what you are selling this could work out better for you than some of the other solutions. You can use your own domain name; accept all major credit cards, process through SSL and more. GoEmerchant (http://www.goemerchant.com/) provides you with solutions to build a full ecommerce store; you can simply add purchase buttons to your existing site if you want. The company also has a merchant gateway that can be used on an iPhone so that you can take credit card payments virtually anywhere at any time. Shopify (http://www.shopify.com/) is a template-based online store solution that provides you with an easy set up for any type of store, and then gives you numerous templates to choose from. You can also customize the site. StoreFront (http://www.storefront.net/) specializes in customized solutions for businesses that are tailored to the needs of companies of all sizes. Their systems integrate with many third-party software packages, multiple merchant gateways, and offer up customized store designs. WebStore by Amazon (http://webstore.amazon.com/) is a hosted shopping cart solution that allows anyone to use the power of Amazon's shopping technology on their own branded site. You can sell the items on your own custom WebStore, as well on the main Amazon site. Yahoo! Merchant Solutions (https://smallbusiness.yahoo.com/ecommerce) is an extremely robust hosted shopping cart that allows for cross-selling, discounted sales, creating gift certificates, wish lists and a lot more. The site features several templates you can choose from or you can have your store custom designed. Yokaboo (http://yokaboo.com/) is a simplified store system that only allows you to accept PayPal, but lets you set up a store quickly and with a look of your choosing. It is perfect for smaller merchants such as small record labels, artists, jewelry makers and more.

SHOPPING CART: BUY VS. RENT

When choosing the shopping cart software that is best for you and your customers, you have the option to choose a shopping cart supplied by a webhost or buy a shopping cart and host it yourself. There are advantages and disadvantages to both options. The biggest advantage of hosted shopping carts is technical skills are not required to set up the shopping cart as the host does the hard work. You will not be required to provide the secure certificate, and you can get straight into setting up your shop. The disadvantages of remotely hosted shopping carts are that you will need to pay a monthly fee. You also run the risk of the host closing their business, leaving you without a shopping cart. You may be given less control over your shopping cart pages, which may make it difficult to match the shopping cart design to be similar to your website.

Transaction Processing

Transaction processing starts when a shopper finishes shopping and proceeds to checkout by clicking on checkout button of the shopping cart. At that time the E-commerce software performs many calculations (such as volume discounts, sales tax, shipping costs, discounts for coupons, sales promotions etc.). The communication between shopper's web browser and seller's web server software takes place in secure mode. Transaction processing is very complex and site administrators need to make sure that all the information used in the calculations are current.

Advanced Functions of E-commerce Software

Large online businesses require advanced features in their E-commerce software. These businesses are also involved in substantial business activity that is not related to E-commerce. Thus, integration of these companies E-commerce and non-E-commerce activities become very important.

Middleware

Middleware is software that connects two otherwise separate applications (e.g. a database systems and a web server) and enables data exchange between them. Larger companies usually need middleware to link their E-commerce software and their existing information systems (e.g. accounting system). Companies can either develop their own middleware software or purchase middleware software that is customized for their businesses by the middleware vendor. Most of the cost of middleware is not the software itself, but the cost of software customization to make it work properly in a particular company. IT can cost companies up to millions of dollars to implement middleware depending on the complexity of the company's operations and its existing information systems. Major Middleware vendors include BEA Systems (Now an Oracle subsidiary), Broadvision, Digital River, and IBM Tivoli Systems.

Enterprise Application Integration

Business rules are also called business logic. An example of a business rule is: when a customer logs in, check the password entered against the password file in the database. Business logic information is usually stored in a database. Enterprise application integration refers to interconnecting this distributed business logic by using software applications to transfer this information from one application to another. For example, a program might transfer information from order entry systems in several different divisions to a single accounts receivable and sales system that integrates all enterprise-wide sales activity. In many cases, the data formats in the various applications can be different. The program transferring the data must edit and reformat the data before transferring it. XML has become a de facto standard for such data formatting. Using XML, programmers can create their own customized tags, enabling the definition, transmission, validation, and interpretation of data between applications and between organizations.

Web Services

Web services refer to a combination of software tools that can be used by application software in one organization to communicate with other application software over a network by using a specific set of standard protocols. Three protocols are used in web services: XML, SOAP, WSDL, and UDDI. We discussed XML earlier, which is designed especially for web documents. The Simple Object Access Protocol (SOAP) is a message-passing protocol that defines how to send tagged or marked up data from one software application to another across a network. Web Services Description Language (WDSL) is an XML-formatted language. Universal Description, Discovery and Integration (UDDI) is a Web-based distributed directory that can be used by businesses to list and discover each other on the Internet. It is similar to a traditional phone book's yellow and white pages. Web services allow programs written in different languages on different platforms to communicate with each other and accomplish various tasks e.g. transaction processing.

In a web service, all communication is in XML and is not tied to any one operating system or programming language. For example, programs written in Java can talk with programs written in Perl and application written for Windows applications can talk with applications written for UNIX. XML is used to tag the data, SOAP is used to transfer the data and communicate with applications, WSDL is used for describing the services available and UDDI is used for listing what services are available and their associated WSDL descriptions.

The various steps involved in actually consuming a web service can be summarized as below:

1. The client (web service consumer), queries the UDDI registry to get the details of the service and to locate it in the network (Internet).
2. The UDDI registry refers the web service requesting client to the WSDL document details.
3. The requesting client accesses the WSDL document.
4. The WSDL document gives all the relevant details of the requested web service to the client.
5. The client now sends SOAP-message request to the actual web server exposing the service.
6. The web service returns the SOAP-message response which the client receives and does the appropriate processing.

FIGURE 3-14: WEB SERVICE ARCHITECTURE

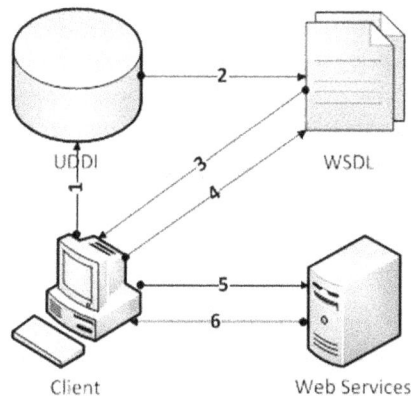

Source: www.javacodebook.com

Use of web services is increasing and some industry analysts report companies are using Web services in 25 percent of all current data integration projects (Schinder, 2009). EBay and Amazon.com are two large E-businesses that are developing their web services aggressively. Companies such as Microsoft and Sun Microsystems (now a part of Oracle) have been aggressively developing and pushing their technology initiatives (e.g. Microsoft .NET and Sun Java 2 platform, Enterprise Edition) to build and develop web services. Many companies using web services have found it a less expensive way of application integration than using middleware. For example, Merrill Lynch implemented an application integration project using web services used Web services to implement an integration project for at a cost 70% less than the cost of the same project using its older application integration approach. Amazon Web Services (AWS) are a market leader today. Many early customers of AWS were start-ups, but today the list includes most of the Fortune 500, and the US government, having received high-level security clearance from the Pentagon. Gartner believes that the total market for public cloud services is $180 billion in 2015, rising to $207 billion by next year. Gartner also estimates that AWS has five times as much computing capacity as the next 14 public cloud providers combined. Its annual revenues are estimated to be $5 billion and growing at a rate of 40 % annually. According to Gartner, AWS retains a significant lead, both in terms of scale and functionality, with Microsoft a distant second and Google an even more distant third (Gartner, 2015). Analysts at JPMorgan have assigned a $44bn valuation to AWS, suggesting it represents a quarter of the enterprise value of Amazon as a whole. High quality global journalism requires investment. The market for basic computing and data storage is a low-margin business, and AWS has increasingly been working to move up the value chain and into the more challenging area of software and enterprise services. A recent example of Amazon's shift in approach is the machine learning service it unveiled this month, to enable clients to analyze big data. It also has a data warehouse product, Redshift, which poses a potential threat to similar offerings from companies including Oracle and IBM. Amazon AWS is also targeting enterprise solutions, which has long been the domain of companies such as IBM, Microsoft and Hewlett-Packard. High quality global journalism requires investment. Amazon as a whole has even changed its mission statement, which now refers to four types of customers: "consumers, sellers, enterprises and content creators". AWS has introduced secure corporate email, secure document collaboration, and virtual desktop services. In 2014, Johnson & Johnson, the pharmaceuticals group, bought 25,000 virtual desktops, software that allows a user's PC to be stored on a remote server, instead of a physical computer (Hook, 2015).

Despite its advantages, web services does have some disadvantages. All data in web services is stored and transmitted in XML format. With so many variations of XML in use, business partners using a web services must agree on a particular XML version to use. Currently, web services management standards are developing and no history of best practices exists. Therefore, Business partners needs a detailed agreement on quality of service and service level specifications for each web service used. Security can be an issue for web services because web services feed data directly to company's internal applications, bypassing any security features used at company's network perimeter. In order to make web services simpler and more acceptable, these issues need to be worked out.

E-commerce Software for Small and Medium Enterprise

The online business activities of most small and medium businesses do not need complete coordination with the business's other activities. Following are the E-commerce software choice available for small and medium online businesses.

E-commerce Service Providers

E-commerce service providers are businesses that provide any combination of consulting, software and computer systems for E-commerce Web sites (e.g. Internet connection, E-commerce site hosting, shopping cart integration, payment processing integration etc.). These service providers are low-cost and offer free or low-cost E-commerce software for building E-commerce sites. E-commerce service provider is the fastest and easiest solution to start you online business where all that you need is needed is a business name and a list of products to be sold. Some service providers charge monthly flat rates based on the number of items in your online catalog, while a few take a percentage of your sales. These service providers take care of all the transaction processing, web servers, backups, and so forth. Some service providers may even setup a merchant account for you, if you do not already have one. E-commerce service providers, however, can be the least flexible option. Many service providers provide a limited set of options to setup E-commerce site and some service providers can also impose hidden charges or lack security and reliability. A business can check capabilities of these service providers by first trying them. Some of these services have a 30-day trial period. Yahoo! Merchant Solutions (smallbusiness.yahoo.com/ecommerce/) is example of a famous E-commerce service provider.

The HostIndex.com, TopHosts.com, and HostSearch.com are firms that provide comparisons of various web hosts and their service offerings. Major Web directories can be helpful sources; the Google Directory of Web Host Directories is especially comprehensive. When selecting a web server hosting provider, a company should ask whether the solution offered (i.e. combination of hardware and software) is scalable to meet varying levels of demand on E-commerce site. The best hosting services provide Web server hardware and software combinations that are scalable.

SOME POPULAR E-COMMERCE SERVICE PROVIDERS

Shopify and Yahoo are two large E-commerce service providers. Below, is a comparison of E-commerce platform different services provided by the two providers (Zorzini, 2015).

Both the platforms offer robust basic features and functionalities for creating a good online store. The price of starter plan provided by both is almost the same but Yahoo allows 50 products while Shopify allows only 25. With higher plan, you can sell unlimited number of products at both Yahoo and Shopify. Both providers handle calculation of shipping charges very well. There is no Provision for coupons and discount codes in Shopify starter plan. Yahoo Stores allows creation and calculation of discount/promo codes on all the plans. Shopify provides 70 supported payment gateway options as compared to just one offered by Yahoo Stores. Yahoo Store uses First Data Merchant Services (FDMS) as the payment gateway to process transactions from your store in conjunction with your merchant account. It is only possible to setup a Yahoo Store for online processing with a merchant account that uses First Data Merchant Services (FDMS. To use a payment option in Yahoo Stores, you need to have a merchant account. This is not a requirement of Shopify. Both providers provide hundreds of free as well as paid/premium themes that you can use to setup your store. Both the platforms also offer a good number of responsive themes (that display your store properly on different kinds of access devices).

Shopify provides more than a hundred apps and add-ons (both free and paid) with which you can beef up your store in various ways. Yahoo provides a limited number of apps and add-ons. Shopify offers partial multi-lingual support by allowing you to create checkout pages in multiple languages. For implementing storewide multi-lingual capabilities, you can use third-party apps. Yahoo Stores does not provide multi-lingual capabilities presently. Yahoo provides very strong SEO capabilities. Shopify holds its own with top-of-the-line SEO features and capabilities. Both platforms allow you to customize the basic elements of your store easily using menu-driven and drag-and-drop methods. For advanced levels of customization, Shopify allows raw HTML/CSS editing while with Yahoo Stores, you are forced to use RTML – their proprietary store template language that will require special programming skills, which would be difficult and costly as compared to HTML/CSS skills. Blogging functionality is a very important feature for an ecommerce store today. Blog allows you to add content to your store at will. This content is highly beneficial for SEO. Shopify provides integrated blog publishing functionality while with Yahoo Stores, you need third-party solutions. Yahoo Stores and Shopify are easy to use ecommerce platforms.

Neither Shopify nor Yahoo Stores charge any setup fees. Shopify offers a 14-day free trial while Yahoo Stores offers no free trail. Shopify offers four different plans with monthly fees ranging from $14 to $179. Yahoo Stores offers four plans, with monthly fees ranging from $13 to $249. Both platforms do not charges any bandwidth usage fees on any of their plans. Shopify does not charge any transaction fees if you use their own payment gateway. Yahoo charges transaction fee ranging from $13/month to $2409/month.

With a massive market capitalization of over $30 billion, Yahoo is definitely well ahead of Shopify. Another massive advantage of Yahoo is that it is the first and the oldest ecommerce platforms. However, Shopify outperforms Yahoo Stores in several important areas such as ease of customization, payment processing, blogging capabilities and others. While selecting either Yahoo or Shopify, businesses need careful evaluation of both options with respect to their business goals and objectives.

Another emerging competitor of Yahoo and Shopify is GoDaddy.com. GoDaddy Online Store is user-friendly eCommerce software that offers basic features and provides integration with eBay. However, this E-commerce software lacks many tools and features provided by its big competitors. In general, GoDaddy can be considered a good E-commerce platform for a modest online store. GoDaddy E-commerce software provides 800 templates to help you design an online store yourself. You can add your own logo and images. It integrates with QuickBooks and handles drop shipping orders and payments. It can manage your eBay store as well as your brick-and-mortar store, giving you extra versatility with this software. Your customers can read reviews of your products and share their finds through social media. You can also program product suggestions for upsell and cross-sell. You can create a mobile-responsive website for customers on the go. However, the software does not provide advanced and more useful features such as low-inventory alerts, abandoned cart notifications and loyalty programs. An IP country match helps you target and avoid fraud behind the scenes. GoDaddy is PCI compliant just like its competitors and backs up information every time you make a change. It employs 128-bit SSL encryption, and you can pay to have a higher level of encryption for credit card transactions. You will need to purchase an SSL certificate if you want to accept credit card payments.

Estimated Operating Expenses for a Small Online Business

The following table provides estimate of average annual costs of setting up and maintaining a small online business. The estimate does not include business-specific expenses e.g. material, salaries, utility bills etc. The estimate can vary depending on a range of fees for various services. The estimate does not include charges of payment processing (average payment processing charge is 50 cents per transaction and 2 percent of each sale's total).

Costs	Cost Estimate
Initial site setup fee	US$ 100 (one time cost)
Annual maintenance fee	US$ 1500
Domain name registration	US$ 85
Occasional HTML and site design help	US$ 400
Merchant credit card setup fee	US$ 200 (one time cost)
Total first-year cost	US$ 2285
Total Recurring Cost	US$ 1985 (excluding one-time costs)

You can also host your online business site but the proposition is costly. Consider the following table that provides estimated average annual costs of hosting a small online business yourself.

Equipment (server and networking equipment)	US$ 3000-US$ 20,000 (one-time cost)
Internet connection (A T1 or fractional T1 line)	US$ 1200 -US$ 12,000 per year
Cost of Server room	US$ 5000 per year
IT Staff costs	US$ 50,000-US $100,000 per year
Total Costs	**US$ 60,000- 137,000**
Annual Recurring Costs	**US$ 57,000- 117,000**

We can reasonably assume that the costs for the subsequent years will remain the same. Businesses should carefully evaluate the choice between self-hosting their E-commerce site and hosting with service provider. For large E-commerce sites, the costs can increase substantially and become more difficult to estimate. For large businesses typical start-up costs range from US $1 million and US$ 50 million and recurring annual costs are about 50 % of that amount.

When you build and host your own E-commerce site, you have full control including ability to customize each aspect of your site, optimize performance, and integrate legacy systems with your site. You can also change your site quickly to take advantage of emerging opportunities.

Integration in E-commerce

There are many pieces specific to the client's business model, which have to be adjusted for a successful ecommerce project. These include the website components, catalog, inventory, user system, CRM, analytics, affiliate and email marketing, payment gateways, merchant accounts, alternative payment systems, order management, accounting, dunning, reporting, and fulfilment. There is also the cart experience, recurring payments, taxes, shipping, anti-fraud tools, discounts, and PCI compliance.

It is impossible for one company or one technology to fulfill all these requirements perfectly. Your ecommerce solution is never going to have the best content management system or analytics packages. Existing companies and open source solutions focus all of their effort on perfecting these technologies and continually make improvements to them. The key is to focus on where you need flexibility. By thinking of each system as a separate service, you can customize or replace each tool independently, as long as the tools can integrate with each other effectively. If they are tightly coupled or worse, built within the same technology, you options are severely limited (Stokes, 2014).

Service-Oriented Architecture (SOA)

Service-oriented architecture (SOA) is a technology that simplifies integration. SOA provides business functionality as a service. A service is a reusable software component that can be used as a building block to form larger, more complex business-application functionality. SOA is designed not to replace but augment and extend the existing processes, where necessary.

SOA has been used for applications that provide connectivity to back-end systems inside of the enterprise. Applications using SOA are more flexible than the traditional applications developed utilizing point-to-point communication. In E-commerce, SOA can be used to provide access to service across multiple participants through a common interface that connects to multiple services.

In the complex E-commerce infrastructure, there are many service providers e.g. payment service providers, and each service provider has its own method and format of reporting data, unique security requirements, and a unique interface. Integrating multiple service providers, through a single interface, adds immediate value to the business consumer.

Cloud Computing in E-commerce

Cloud computing is a computing model in which all servers, networks, applications and other elements come from data centers and made available to end users through the internet. This approach reduces cost and complexity, while companies are able to add services when needed. The web based email services by Google and Yahoo!, the backup services of Carbonite or Dropbox, CRM applications like SalesForce.com and instant messaging (IM) and voice over IP (VoIP) by Google, Skype and others are all examples of cloud computing services.

According to a recently published article by Gartner, at least one third of all digital content will be in the cloud by 2016 (Kar, 2012). The most important reason for this exponential growth is the desire to store data, share and access it through multiple devices. With ecommerce in the cloud, online stores can manage their store wherever they are. There are also no limitations on data storage. According to analysts, within 10 years' time 80% of all computer usage worldwide, data storage and e-commerce will be in the cloud (Mearian, 2011). It is called the third phase of the internet. In the third phase, everything would be in the cloud, both data and software (bertramwelink.com, 2012).

Cloud computing provides many benefits for E-commerce. Cloud computing enables online store to use the same platform and use the same functionality. New features can be made available to everyone with a simple modification. Maintenance is central so the platform is stable. Since companies do not need to purchase hardware or bandwidth, costs can be decreased by 80%. A company can activate an ecommerce application five times faster and sell directly through a platform that is managed remotely. With cloud computing, an ecommerce business becomes more flexible and able to respond to seasonal changes or sudden increases in demand due to special promotions. Many cloud computing providers have been certified so more security can be guaranteed to customers. The explosive growth of cloud ecommerce will lead to more data exchange between the clouds. Suppliers will offer more and more possibilities to add features to their clouds for users, partners and others.

There are several types of cloud computing, of which Software-as-a-Service is probably the best. The others are Platform-as-a-Service (PaaS) and Infrastructure-as-a-Service (IaaS).

Software as a Service (SaaS)

In this model, applications are hosted by a vendor or service provider and made available to customers over a network, typically the Internet. The traditional model of software distribution, in which software is purchased and installed on personal computers, is sometimes referred to as software as a product.

SaaS is gaining popularity as the underlying technologies that support Web services and service-oriented architecture (SOA) are maturing and broadband Internet access is increasingly available around the world.

SaaS model can have two variations: hosted application management and software on demand. In hosted application management, a provider hosts commercially available software for customers and delivers it over the Web. In the software on demand model, the provider provides its customers access over the network to a single copy of software application created specifically for SaaS distribution.

SaaS provides many benefits e.g. easier administration, automatic software updates and patch management, easier collaboration, compatibility, low entry cost and risk for customers, increased customer software security, and global accessibility. Software-as-a-Service spreads software on the internet. A supplier can offer it under license, based on need and through a monthly subscription in a pay-as-you-go model. There are providers who offer the software free, when they have other ways to earn money (freemium). Examples are advertisements or making customer data available for third parties. The software is managed from a central location, where software is made available to everyone (one-to-many model). The responsibility for updates lies with the supplier, not with the customer. SaaS remains the most popular of the three cloud computing types.

Infrastructure-as-a-Service (IaaS)

Infrastructure-as-a-Service (IaaS) supplies cloud computing as an infrastructure that can be purchased based on the need of the user. Examples are the usage of servers, storage, networks and operating systems. Amazon's Elastic Cloud is an example of this. Instead of purchasing these elements, customers can purchase them as an outsourced service based on customer needs. IaaS can be offered as a public or closed infrastructure. A hybrid version is one of the possibilities. IaaS is especially useful for growing enterprises, because they can easily expand their infrastructure when more employees are hired.

Platform-as-a-Service (PaaS)

Where SaaS deals with applications, Platform-as-a-Service (PaaS) deals with development. PaaS can be defined as a platform that facilitates the creation of web applications, without the need to purchase or support software and infrastructure. PaaS is therefore the same as SaaS, but instead of making software available through the web, a platform is supplied for the creation of software.

Utility Computing

In simple words, utility computing refers to delivering information technology as a utility. Utility computing delivers, over the network, standardized processes, applications and infrastructure that provides both business and information technology functionality. Utility computing is also called e-sourcing. Customers pay only for the time they use the computing services provided by a vendor. Customers benefit from utility computing because they can transfer responsibility of non-core functions to a third-party provider and redirect the saved capital and human resources to other strategic uses. Customers can simplify the adoption of new technologies, minimize IT hiring and training obligations, and compress time-to-market for new, value-adding projects and initiatives.

Utlity computing differs from traditional outsourcing and current hosting services in three key ways.

SHARED AND SIMULTANEOUS: E-sourcing is shared, simultaneously serving multiple customers in a flexible, automated fashion.

STANDARDIZATION: E-sourcing is standardized requiring little customization or integration.

SCALABILITY: E-sourcing is scalable, providing capacity on demand in a pay-as-you-go model.

Many large and small IT companies are positioning themselves to provide utility computing. Some of them are focused on providing the hardware and software infrastructure. Some are focused on providing the content and business processes. IBM's "e-business on demand" is an example of utility computing focused on content and processes.

E-commerce Software for Mid-size to Large Businesses

There are number of E-commerce software available for midsize and large companies. These software packages differ on price, capabilities, database connectivity, software portability, software customization tools, and technical expertise required to manage the software.

Web Site Development Tools

Web page creation and site development tools e.g. recent versions of Adobe Dreamweaver and Microsoft FrontPage can be used to construct a mid-range E-commerce site. The same tools can also be used to create basic elements of dynamic web pages as easily as static web pages. The remaining elements of the dynamic pages e.g. catalog, customer service, and transaction-processing pages, can be added with development tools such as Microsoft's Visual Studio .NET software. E-commerce site developers can also use a wide variety of other tools e.g. CGI (Common Gateway Interface), Active Server Pages (ASP), Java, and JavaScript to develop interactive elements on Web pages. CGI scripts, Active Server Pages and Java Server Pages are tools that enable personalization and customization. Personalization and customization are significant marketing tools of E-commerce that could help to increase sales and revenues.

Once the site has been developed using these tools, the web designer can add purchased software modules (e.g. shopping carts and content management software) to the site. You can also consider developing your own modules to handle these functions but it is a time-consuming and costly proposition. A company should not develop its own modules unless they have some very strong reasons to do so. The final step is to either use middleware or a web services to connect the site to the company's existing product and transaction-processing databases. Most businesses can find off-the-shelf E-commerce software that meets their needs. The fact that vendors provide service, support, and even systems-integration help is a major advantage for the businesses setting up their online business.

Mid-Range E-commerce Software

Annual costs of buying and using midrange E-commerce software can range from US$ 2000 to US$ 50,000. Midrange software traditionally offers connectivity to database systems that store product catalog and provide connectivity to existing inventory and ERP systems. This can eliminate the need to run duplicate inventory systems thereby resulting in significant cost savings. Three popular midrange E-commerce software providers are Digital River, eBay Enterprise, and Intershop. Two other competing providers are IBM and Microsoft.

EBAY ENTERPRISE

The eBay Enterprise (formerly GSI Commerce) is an eBay company specializing in creating, developing and running online shopping sites for brick and mortar brands and retailers. The company also provides a variety of marketing, consumer engagement, customer care, payment processing, fulfillment, fraud detection, and technology integration services. eBay Enterprise has also consolidated together its portfolio of digital marketing technologies and acquisitions into a new suite known as the "eBay Enterprise Commerce Marketing Platform." eBay Enterprise are a partner to more than 1,000 retailers and brands, providing leading commerce technologies, marketing solutions and omnichannel operations capabilities that enabled $4 billion of ecommerce transactions in 2012.

DIGITAL RIVER

Digital River is a public company that provides global ecommerce, payments and marketing services. In 2013, Digital River processed more than $30 billion in online transactions. The Digital River's MyCommerce is a scalable, customizable, self-service e-commerce solution targeting start-ups and independent business owners worldwide. MyCommerce solutions are secured by Digital River's enterprise e-commerce infrastructure, which includes a proven payment gateway, advanced fraud prevention, 24/7 customer service and more. Customers can choose from range of solutions, each offering distinct advantages for specific markets. The tools provided are flexible and simple that can be used scale e-business on a global basis, increase efficiencies and expand online business the way customer wants.

INTERSHOP

Intershop Communications is one of the major providers of omnichannel E-Commerce solutions to large-sized companies worldwide. Intershop Commerce Suite is a modular software package that provides six specialized modules. It is the leading platform for omnichannel commerce. It provides search and catalog capabilities, electronic shopping carts, online credit card transaction processing, and the ability to connect to existing backend business systems and databases. Commerce Suite software also provides good catalog and data management tools, many built-in storefront templates, an automated e-mail facility that can send order confirmations to customers, support for secure transactions, and wide variety of site and customer reports to track Web page visits and customer activities. Management and editing of a storefront are done through a Web browser—either locally at the server or remotely through any Internet connection. Intershop's v7 platform represents a strong ecommerce solution with respect to core commerce and experience management capabilities (Businesswire.com, 2015).

IBM WEBSPHERE COMMERCE

IBM WebSphere Commerce provides an e-commerce platform that can deliver seamless and consistent omnichannel shopping experiences, including mobile, social and in-store. WebSphere Commerce helps engage your customers with immersive brand experiences through contextually relevant content, marketing and promotions, while extending your brand across customer touch points. IBM WebSphere Commerce Professional software has all the standard E-commerce features, including tools for a shopping cart, e-mail notifications upon sale completion, secure transaction support, promotions and discounting, shipment tracking, links to legacy accounting systems, and browser-based local and remote administration. WebSphere software includes catalog templates, setup wizards, and advanced catalog tools to help companies create attractive and efficient E-commerce sites. WebSphere provides a smooth connection to existing corporate systems, such as inventory databases and procurement systems. WebSphere Commerce has 3 main components: A database, an application server (Java Enterprise Edition), and a web server. IBM WebSphere Commerce only supports IBM WebSphere Application Server. It is provided with the WebSphere Commerce software. WebSphere Commerce Professional software can run on many different operating systems. You can begin by setting up a small store later more functionality can be added by executing commands and writing code. However, JavaScript, Java, or C++ expertise is required. The WebSphere Commerce Professional Edition also accommodates electronic downloadable products, such as audio tracks or software. WebSphere software can connect to existing databases and other legacy systems through IBM DB2 or Oracle databases. WebSphere Commerce Professional Edition costs $155,000 per processor. The less powerful Professional Edition of the software costs $99,000 per processor. WebSphere Commerce Express is designed for fast, easy implementation and quick creation of your online presence. IBM Commerce on Cloud enables you quickly go to market with an omnichannel commerce solution. The software helps you create an engaging brand experience across every customer touch point.

MICROSOFT COMMERCE SERVER 2009

Microsoft Commerce Server 2009 provides businesses a production-ready, out-of-the-box, contemporary E-commerce web site. The site once installed and connected to internal information systems of the company, only requires selecting skin and template to establish the look and feel of the site. Commerce Server 2009 can also be customized for content localization. Commerce server 2009 is tightly integrated with other Microsoft products.

Microsoft Commerce Server is used by six retailers among the Internet Retailer Top 500, making it the tenth most commonly deployed e-commerce platform among the Top 500 (Demery, 2011).

Microsoft Commerce Server 2009 provides functions such as user profiling and management, transaction processing, product and service management, target audience marketing, advertising and promotion, many predefined reports for analyzing site activities and product sales data, shopping cart, confirms completed sales transactions by e-mail, and support for secure transactions. The Microsoft Visual Studio .NET tools, bundled with Commerce Server 2009, allow companies to customize the sites they build. Commerce server 2009 can connect to existing accounting systems, and the administrator can oversee the site through a Web browser. Commerce Server 2009 delivers the ability to increase your business reach by selling through multiple channels through a new Default out-of-the-box shopping site based on ASP.NET Web Parts and controls deployed in SharePoint and other Microsoft SharePoint Commerce Services, and a compelling new Microsoft Multi-Channel Commerce Foundation, with a new unified run-time calling model, new extensibility points and new built-in features. Commerce Server 2009 licenses are available in a Standard Edition for U.S. $7,075 per processor and in an Enterprise Edition for U.S. $20,218 per processor. Microsoft transferred future development of Microsoft Commerce Server to Ascentium, a digital technology and services agency. Microsoft will continue to honor extended support of Commerce Server 2009 through 2019.

E-commerce Software for Large Businesses

Larger E-commerce sites deals with higher transaction volume and therefore need dedicated software to handle specific elements of their online business such as customer relationship management, supply chain management, content management, and knowledge management. Price of large-scale E-commerce software as well as its support cost is high. Large-scale E-commerce software is also called enterprise E-commerce software. Enterprise E-commerce software provides tools for both B2B and B2C commerce and can interact with a wide variety of existing systems, including database, accounting, and ERP systems. The enterprise software is also capable to make changes in the system automatically (e.g., inventory checking and order placement for items needed). In contrast, both basic and midrange E-commerce software usually require an administrator that manually makes such changes. The cost of these enterprise systems for large companies ranges from $200,000 for basic systems to $10 million and more for comprehensive solutions. Enterprise E-commerce software usually requires several dedicated computers.

Top Providers of Enterprise E-commerce Software

Examples of enterprise E-commerce software providers include Demandware, Hybris, IBM, and Oracle.

IBM

IBM's suite comprises three separate products: IBM WebSphere Commerce, IBM Sterling Order Management, and IBM InfoSphere Master Data Management Collaborative Edition. IBM WebSphere Commerce Enterprise is an omnichannel e-commerce platform that enables B2C and B2B sales to customers across channels—web, mobile, social, store or phone. It supports better marketing, selling and fulfillment with precision marketing, merchandising tools, site search, customer experience management, catalog and content management, social commerce and advanced starter stores. It dynamically optimizes content for various device types and formats including web, mobile and tablet. IBM Commerce on Cloud enables businesses quickly go to market with an omnichannel commerce solution.

HYBRIS

Hybris is an SAP company. Hybris delivers the top-rated E-Commerce software and Omni-channel solution to help enterprises across the globe innovate faster and sell more. Hybris Commerce Suite is built on robust, modular and open architecture to satisfy all omnichannel selling and fulfillment needs. Hybris Commerce Suite maximizes conversions and revenue by delivering an easy, engaging shopping experience. It deliver a cross-channel experience – whether in-store, online, via social networks, or from mobile devices. A simple ordering process and cross-sell offers are helpful to boost average order value. Businesses can reduce selling costs by extending sales to more affordable online channels. Hybris Commerce Suite supports the complete end-to-end commerce process by integrating with SAP CRM and SAP ERP. A pre-built Web shop template can be helpful for businesses to realize short time-to-value and reduce Total Cost of Ownership (TCO).

ORACLE

Oracle ATG Web Commerce enables you to deliver a personalized online buying experience for each customer by presenting relevant content and merchandizing, personalized search, customized marketing programs, and tailored websites. ATG Web Commerce enables you to deliver an engaging, consistent, and coordinated customer experience across all channels, including Web, contact center, mobile device, kiosk, or store. ATG Web Commerce helps you personalize the customer experience by creating individualized sites and relevant product content, and personalized search. Businesses can optimize the execution by tailoring recommendations, and providing assistance relevant answers at the point of need. Businesses can maximize the engagement by increasing agility through merchandising and site administration, delivering a personalized brand experience, and leveraging social data. In 2015, Oracle had plans to launch an all-new midmarket SaaS commerce offering that is intended to compete directly with Demandware.

DEMANDWARE

With Demandware Commerce, merchants can anticipate shopper's needs and syndicate relevant content so consumers can find – and ultimately buy – what they want quickly and easily. In addition, merchants can maximize profitability by featuring the optimal products based not just on consumer interest but also inventory levels. Benchmarking makes it easier for merchants to understand what is working and what is not, and adjust as needed. Personalization, search, promotions and targeting capabilities help marketers promote their brand. Demandware Commerce provides a single view of inventory – enterprise-wide – including real-time inventory counts that include product on order or in transit. Businesses can lower customer service costs by enabling self-service functions (such as order tracking, cancellations and returns management) on their site. Using a standards-based, server-side scripting language, developers can modify, configure, extend, create, debug and deploy custom business logic across any commerce channel, all to enable the best branded shopping experience for your customers. Retail intelligence highlights shopping trends, while benchmarking directs you toward growth opportunities. Predictive intelligence creates more relevance for your customers, enhancing email marketing and on-site merchandising.

Companies are also building portals to provide their customers and suppliers content useful to them. Efficient management of these portals requires software that automatically manages and rotates content on Web sites. Companies also require software to manage the knowledge that exists in their businesses. Following are the software that often used by the large E-commerce sites to achieve these goals.

Customer Relationship Management Software

The goal of CRM is to understand each customer's specific needs and then customize a product or service to meet those needs. CRM software is generally used by large companies because of its high purchase and implementation cost. Some companies create their own CRM software using outside consultants and their own IT staffs. Two major vendors of CRM software products are Oracle and MySAP CRM. Prices for these systems start around $30,000 (on average, about $1500 per user); large implementations can cost millions of dollars. We shall discuss CRM in detail in Chapter on e-CRM and e-SCM.

Supply Chain Management (SCM) Software

SCM software can help companies in a variety of tasks e.g. to provide coordinated demand forecasts using information from their partners in the industry supply chain, help in planning and operations, and warehouse and transportation management. The two major firms offering SCM software are i2 Technologies and Manugistics. Both are now part of JDA software. The cost of SCM software implementations varies tremendously depending on the number of retail locations (retail stores, wholesale warehouses, distribution centers, and manufacturing plants) in the supply chain. We shall discuss SRM in detail in Chapter on e-CRM and e-SCM, and e-SRM.

Content Management Software

Content management software can help companies manage critical business information (for example, product specifications, drawings, photographs, or lab test results) electronically and provide significant cost reduction because of less paper used. The three leading companies that provide content management software include Documentum, Vignette, and web Methods by Software AG. Content management software generally costs between $200,000 and $500,000, but it can cost three or four times that much to customize, configure, and implement.

CMS is used to manage the content of website. It consists of two elements: the Content Management Application (CMA) and Content Delivery Application (CDA). The CMA allows the content manager to manage the creation, modification, and removal of content from a Web site. The CDA uses and compiles that information to update the Web site. The features of a CMS system vary, but most include Web-based publishing, format management, revision control, and indexing, search, and retrieval. The Web-based publishing feature allows individuals to use a template and other tools to create or modify Web content. The format management feature allows documents including legacy electronic documents and scanned paper documents to be formatted into HTML or Portable Document Format (PDF) for the Web site. The revision control feature allows content to be updated to a newer version or restored to a previous version. Revision control also tracks any changes made to files by individuals. An additional feature is indexing, search, and retrieval. A CMS system indexes all data within an organization. Individuals can then search for data using keywords, which the CMS system retrieves.

Important Features of CMS

For businesses of all sizes, the important features of a CMS are very much similar. However, larger companies can need additional features. Following are some important features of a CMS.

HIGH PERFORMANCE AND SCALABILITY: A CMS should offer multiple levels of caching and each level should be easily and fully customizable. A cloud-based CMS can improve performance enhancements as required.

EXTENSIBILITY AND INTEGRATION: The CMS of your choice should easily integrate with other technology and platforms (such as different CRM systems, ERPs, Social Networks, Mobile Applications etc.) Ready-to-use connectors for most common platforms are desirable and can significantly decrease development costs.

STABILITY: Check the list of quality-assurance processes that your CMS vendor has and read reviews of their current clients, which will offer a user perspective to ensure CMS offers desired stability.

EASE OF USE: The CMS must not just be easy to use, but also intuitive, with all elements easy to find.

ADVANCED SECURITY MANAGEMENT: CMS must provide fully customizable content permissions, module permissions and user roles, and multiple prebuilt authentication options.

ADVANCED WORKFLOW AND APPROVAL PROCESS: You CMS should offer a fully customizable workflow process with configurable workflow scopes. Workflow should be followed with versioning to track the records and allow you to revert to previous versions of the document.

ONLINE MARKETING TOOLS: CMS should provide a whole range of integrated online marketing tools. These tools should be integrated to deliver and optimize real-time customer-centric marketing across multiple channels.

MULTISITE SUPPORT: Besides the corporate website, enterprise businesses now also have on-line stores, community websites, blogs, intranets and at least a few microsites to target a narrower audience and support their main website with additional links and traffic. A CMS with multisite support can provide sharing of the content/users across multiple sites and help manage content from a single point.

MULTILINGUAL SUPPORT: Today we can hardly find a language-homogenous country or market. The ability to speak to your site visitors in their native tongue is necessary. Pick a CMS that offer multiple languages and have advanced multilingual features such as Translation Management, Automatic Time Conversion or Side-by-Side Translation Comparison.

SUPPORT FOR MOBILE DEVICES: Today, your business needs either a mobile responsive Web design or a separate mobile website. As such, your CMS should be able to support both approaches. Flexible CMS should be able to hold different mobile websites or applications.

TRAINING SERVICES: On your website, you may have many editors using your CMS. Therefore, having a CMS that offers training is necessary.

HIGHLY RESPONSIVE SUPPORT: Mission-critical websites needs 24/7 quality support. Make sure your CMS offers reliable, quality 24/7 support.

REGULAR SOFTWARE UPDATES: Choose a CMS that has a bug fixing policy and provide regular updates to the CMS software.

EASY SOFTWARE UPGRADES: New features and technologies need to be adopted as soon as they become available. Look for a CMS vendor whose product is, and always will be, upgradable.

Popular Content Management Systems for E-commerce

Now a day having a powerful and dependable CMS system is becoming less of an option and more of a need. The enterprise content management (ECM) vendors are now focusing on the development and simplification of their systems as well as automation and integration of new capabilities to generate value propositions for customers. Some strong players in the market include EMC, HP Autonomy, IBM, M-Files, OpenText, Oracle, Perceptive Software and SpringCM (Roe, 2015).

EMC

EMC Documentum provides many functions focused on integration with other content management systems. Documentum is also being used to improve the functionality of other systems. It currently uses its content and process management integration to improve collaboration and search functionality SharePoint, SAP, Salesforce.com, and Microsoft Office. EMC's InfoArchive platform enables users store structured and unstructured information. Syncplicity, is an EMC product which provides document sharing, viewing and editing capabilities across platforms. It is designed for each platform individually to optimize the user experience, and supports private and public cloud, hybrid and on-premises solutions.

HP AUTONOMY

HP Autonomy is designed to deliver intelligent policy-based protection, accessibility and retention of organizational data to optimize recoverability, reduce the risks of loss and misuse, and leverage it for operational insight and value. This product is being used across a wider portfolio, especially information governance. Autonomy is capable of providing broad range of services to manage and utilize content stored in enterprise repositories in the cloud, or on premises. Autonomy also provides Records Manager for improving document capture search and disposal as well as developing integration with Office 365 and SharePoint. Autonomy also provides IDOL server that enables users search and analyze unstructured data. IDOL offers enterprises better control of the information lifecycle and access rights.

IBM

IBM has a range of ECM products that provides advanced capture, content management, and analytics and is sold as pay-as-you-go. Recently, IBM has started marketing more out-of-the-box bundled CMS. Under its Enterprise Content Management family, IBM providea multiple software for enterprise content management such as IBM Case Manager, IBM Content Foundation, IBM Content Manager OnDemand, and IBM Datacap.

M-FILES

M-Files provides a folderless document management system that also offers advanced functionality like metadata search, process and content archiving. M-Files' folderless system uses metadata, which prevents document duplication and easy retrieval, while the intuitive interface allows for easy search and the ranking of findings. Over the years M-files has collaborated with 18 hardware and software vendors including Microsoft, Salesforce and Epson to improve its functionality and end-use experience. M-Files offers document management (DMS), quality management (QMS) and enterprise asset management (EAM) solutions, while clients are able to pick whether to deploy on-premise, in the cloud, or in a hybrid environment.

OPENTEXT

OpenText offers a completely cloud-based content management platform called OpenText Core. OpenText also offers hybrid-solutions for data storage. Core also integrates social features to improve collaboration on its platform including a Facebook-style 'like' commenting feature, along with automatic synchronizing and version protection.

ORACLE

Oracle's WebCenter provides a platform that is also suited for mobile applications and integrates with the full range of Oracle middleware and business solutions. Oracle's CMS makes it easy for users to capture, manage and collaborate on content directly from the applications interface. Feature integration across Oracle's business applications drive productivity by eliminating time wasted in switching between platform interfaces.

PERCEPTIVE

Perceptive Content 7 is a fully integrated solution that is functional throughout the content lifecycle, while customization continues to be a principal reason for the selection of Perceptive over its competitors.

SPRINGCM

SpringCM integrates with Salesforce with a simple and dynamic interface. It also integrates with Oracle FusionCRM and Oracle Sales Cloud.

Mobile Content Management System (MCMS)

A mobile content management system (MCMS) is a type of content management system (CMS) capable of storing and delivering content and services to mobile devices, such as mobile phones, smart phones, and PDAs. Mobile content management systems may be discrete systems, or may exist as features, modules or add-ons of larger content management systems capable of multi-channel content delivery. Mobile content delivery has unique, specific constraints including widely variable device capacities, small screen size, limited wireless bandwidth, small storage capacity, and comparatively weak device processors.

Demand for mobile content management increased as mobile devices became increasingly ubiquitous and sophisticated. MCMS technology initially focused on the business to consumer (B2C) mobile market place with ringtones, games, text messaging, news, and other related content. Since, mobile content management systems have also taken root in business-to-business (B2B) and business to employee (B2E) situations, allowing companies to provide more timely information and functionality to business partners and mobile workforces in an increasingly efficient manner. A 2008 estimate put global revenue for mobile content management at US$8 billion. Some examples of mobile CMS are Hippo CMS and Solodev CMS.

FEATURES OF MOBILE CMS

Mobile CMSs must consider the limitations of mobile devices such as small screen sizes, low bandwidth, small storage capacity, and weaker processors. Following are some common features of mobile CMS.

MULTI-CHANNEL CONTENT DELIVERY: This feature allows users to publish from a single source document to multiple platforms and in multiple formats.

LOCATION-BASED CONTENT DELIVERY: This feature allows more targeted content delivery, based on the physical location of the user, maximizing the value of the web experience.

CONTENT ACCESS CONTROL: This feature allows authorization, authentication, and access approval to each content.

POPULAR MOBILE CMS

JOOMLA!

Joomla! is an open source CMS with an easy to use interface. It is designed for easy installation and use even for the novice user. It includes many extensions and plugins that allow extend its power.

WAPPLE

Wapple lets you build your site using the leading mobile web technology products and giving you access to reporting tools with statistics on conversions.

PLONE

Plone is an open source CMS that lets non-technical people create and maintain information for a public website or an intranet using only a web browser.

WORDPRESS

WordPress is a free and open source CMS based on PHP and MySQL. A stable product focuses on user experience and web standards. WordPress Mobile Edition is a plugin that shows an interface designed for a mobile device when visitors come to your site on a mobile device.

SITECORE

The Sitecore CMS lets you create compelling mobile apps for Android, iOS and Windows devices that connect the app with rest of your customers' experience

Knowledge Management Software (KMS)

A knowledge management solution simply indexes a wide range of information resources before filtering and prioritizing into relevant knowledge, in addition to providing an internal content management function and reports to measure knowledge base usage and knowledge gaps. The system itself indexes all the required content from all the relevant sources without the need to move any information to a central location and uses a natural language search function to allow users to access the information quickly and easily. This allows easy and timely deployment of a solution without the need to reformat or re-purpose large amounts of legacy information. Often users know how to describe what their issue is but do not know how the solution will be phrased or explained. Using a knowledge management solution, staff, partners and customers can describe the issue, problem or query in their own way and enter the phrase directly into a search. The knowledge management solution will identify solutions that are known to address similar issues and present these in order of relevance.

The process of capturing, developing, storing and sharing knowledge is known as Knowledge Management (KM). KM starts with the defining the strategy for the organization followed by Analysis, Planning, Implementation and Measurement. KM has been around for several years and with the growth of Social KM has been undergoing change and elements of Social are implemented into knowledge management. Social makes KM more effective and has created new ways of collaboration and sharing of knowledge (EditorKPS, 2014).

Knowledge management software can be utilized by companies to capture, store, protect, update, and maintain the knowledge so that it can be used anytime. Knowledge management software includes tools that can read electronic documents in a variety of formats (e.g. word and PDF) and search tools that can help users find the information required quickly. The major software vendors have Knowledge Management software offerings, including IBM Connections Content Manager and Microsoft SharePoint Technologies. Total costs for a Knowledge-management software implementation, including hardware, software licenses, and consultant fees, typically range from $50,000 to $1 million or more.

Popular Knowledge Management Software

The major software vendors have KM software offerings, including Altassian Confluence, Aptean Knova, KPS, KANA Enterprise, and Oxcyon Centralpoint. Smaller companies have also entered the market with innovative KM software and technologies. Two of the more interesting products are Entopia Quantum and Mirror Worlds Technologies Scopeware. Total costs for a KM software implementation, including hardware, software licenses, and consultant fees, typically range from $50,000 to $1 million or more.

ALTASSIAN CONFLUENCE

Some notable customers of this KMS include companies such as Groupon, Facebook, LinkedIn, Netflix, and Nike. This KMS provide features such as SaaS platform, document management, self-service portal, intelligent search, collaboration tools, user permissions, simplified versioning system, team-based knowledge and content management, automatic versioning, mobile accessibility, and integration with SharePoint. You can create content using text editor, archive and locate documents with search facility, create restrictions by user or page, consolidate all files including Word documents, PowerPoints, Excel spreadsheets, PDFs and images, sync with Active Directory or LDAP.

APTEAN KNOVA

Knova is a full-featured knowledge management software for large customer service and support organizations – especially those who need to handle complex queries across channels in industries such as high tech, telecommunications, financial services and the IT help desk. Some notable customers of this KMS include companies such as Audatex, MLB, NHL, and VMware. This KMS provide features such as SaaS platform, document management, self-service portal, intelligent search, knowledge base, and reporting. This CMS is designed for large customer service and support organizations and it can integrate websites, support forums, documents and other content. It lets users and authors search for documents and provide agent tools for assisted service. This CMS can be integrated with service desk and case notes can be saved. It includes knowledge pagelets and microsites for customized outreach and support natural language processing.

KPS

Some notable customers of this KMS include companies such as United HealthCare, Credit Suisse, British Telecom, Stanford Hospital & Clinics. This KMS provide features such as SaaS platform, document management, self-service portal, intelligent search, knowledge base, and reporting. Additional features of this KMS include natural language search, automatic FAQ's, global & individual FAQ's, guided scripting, multilingual index/search, email notifications, intelligent content routing, audit trail, quick content creation, comprehensive content curation, crowd editing, support of multiple content formats, social media ready, API's for integration, no need to rework /migrate content, interactive dashboard reports, leader board reporting, and knowledge gap reporting.

KANA ENTERPRISE

Some notable customers of this KMS include companies such as Best Buy, Chase, Geico, H&R Block, Insurance Australia Group, and JetBlue. This KMS provide features such as SaaS platform, document management, self-service portal, intelligent search, knowledge base, and reporting. Some additional features include live chat, secure messaging, web and mobile self-service, agent desktop, outbound calling campaigns, case management, guided scripting, whitemail management, experience analytics, and mobile case management

OXCYON CENTRALPOINT

Some notable customers of this KMS include companies such as Tanner Health, Barnabus Health, VCA Antech, and WholeHealthMD.com. This KMS provide features such as SaaS platform, document management, self-service portal, intelligent search, collaboration tools, and user permissions. Some additional features include centralized information, module gallery that features everything you need to deliver a comprehensive web strategy out of the box, circulation integration, CRM lead management, custom knowledge bases, email broadcasting, marketing library, and fast and painless migration.

Review Questions

1. How a client-server communications takes place?

2. Differentiate among service-oriented-architecture, software as a service, infrastructure-as-a-service, platform-as-a-service, and utility computing.

3. List some types of web content and their differences.

4. List some server-side scripting languages.

5. List some client-side scripting languages.

6. Differentiate between open-source and commercial software.

7. List some basic functionalities of web server.

8. List some important features of operating systems for web server.

9. List some operating systems for web servers.

10. List some popular web server software.

11. Define Database Management System (DBMS) and Structured Query Language (SQL).

12. List some popular DBMS software.

13. Define SPAM. List some individual and corporate level solutions to combat spam.

14. List some technical and legal solutions to combat spam.

15. List some types of content filtering.

16. List some website and Internet utility programs and their purposes.

17. List some common components of E-commerce hardware.

18. List some factors that can overload a web server.

19. Differentiate between centralized and decentralized server architecture.

20. List some basic functions of E-commerce software.

21. List some advanced features of E-commerce software.

22. What is the purpose of a product catalog?

23. What is the purpose of an online shopping cart?

24. What do you think is the role of cloud computing in E-commerce?

25. List some mid-range E-commerce software. List some E-commerce software for large businesses.

Bibliography

Allbusiness.com. (2002). *E-sourcing: Information Technology on demand: Utility computing marks a new stage in the evolution of outsourcing, and may be the first to fully deliver on the strategy's true potential. (CEO Perspectives)*. Retrieved July 23, 2009, from www.allbusiness.com: http://www.allbusiness.com/sales/internet-e-commerce/115213-1.html

Allbusiness.com. (2009). *How to Choose the Right E-commerce System*. Retrieved Sepetember 25, 2009, from www.allbusiness.com: http://www.allbusiness.com/sales/internet-e-commerce/718-1.html

antispamyellowpages. (2009). *The Definition of Spam*. Retrieved August 12, 2009, from antispamyellowpages.com: http://www.antispamyellowpages.com/what_is_spam.html

Atkearney.com. (2015, July). Global Retail E-Commerce Keeps On Clicking. Retrieved July 4, 2015, from https://www.atkearney.com/consumer-products-retail/e-commerce-index

Beal, V. (2005, January 15). *Understanding Web Services*. Retrieved July 25, 2009, from webopedia.com: http://www.webopedia.com/DidYouKnow/Computer_Science/2005/web_services.asp

bertramwelink.com. (2012, November 7). eCommerce Marketing - How cloud computing is taking over the ecommerce market. Retrieved from http://www.bertramwelink.com/index.php/cloud-computing-taking-over-ecommerce-market/

Branten, B. (2005, June 01). *High performance kernel mode web server for Windows*. Retrieved August 20, 2009, from www.acc.umu.se: http://www.acc.umu.se/~bosse/High%20performance%20kernel%20mode%20web%20server%20for%20Windows.pdf

Businesstown.com. (2009). *Optimizing Your Web Site's Performance*. Retrieved August 13, 2009, from Businesstown.com: http://www.businesstown.com/internet/design-optimizing.asp

Businesswire.com. (2015, January 22). Intershop Recognized as a Strong Performer in B2C Commerce Suites by Independent Analyst Firm | Business Wire. Retrieved July 4, 2015, from http://www.businesswire.com/news/home/20150122005130/en/Intershop-Recognized-Strong-Performer-B2C-Commerce-Suites#.VZgEhVKN1sk

Calomel.org. (2009). *"Nginx "how to" - Fast and Secure Web Server"*. Retrieved August 13, 2009, from www.calomel.org: https://calomel.org/nginx.html

Calomel.org. (2009). *Lighttpd "how to" - Fast and Secure Web Server*. Retrieved August 13, 2009, from www.calomel.org: https://calomel.org/lighttpd.html

Cauce.com. (2009). *Spam History*. Retrieved August 13, 2009, from www.cauce.com: http://www.cauce.org/history/page/2/

Demery, P. (2011, November 14). E-commerce Platforms - Microsoft sloughs off its e-commerce technology. Retrieved July 22, 2015, from https://www.internetretailer.com/2011/11/14/microsoft-sloughs-its-e-commerce-technology

Devonit.com. (2015). What is a Thin Client | Thin Client Education. Retrieved July 3, 2015, from http://www.devonit.com/thin-client-education

EditorKPS. (2014, July 28). What are Knowledge Managemnt Solutions? Retrieved July 22, 2015, from https://www.kpsol.com/what-are-knowledge-management-solutions/

eMarketer.com. (2014, December 23). Retail Sales Worldwide Will Top $22 Trillion This Year - eMarketer. Retrieved July 4, 2015, from http://www.emarketer.com/Article/Retail-Sales-Worldwide-Will-Top-22-Trillion-This-Year/1011765

FTC.gov. (2015). CAN-SPAM Act: A Compliance Guide for Business | Federal Trade Commission. Retrieved July 3, 2015, from https://www.ftc.gov/tips-advice/business-center/guidance/can-spam-act-compliance-guide-business

Gartner. (2015, May 28). Gartner: AWS Pulls Further Ahead in IaaS Cloud Market. Retrieved July 22, 2015, from http://www.datacenterknowledge.com/archives/2015/05/28/gartner-aws-pulls-further-ahead-in-iaas-cloud-market/

Hook, L. (2015, April 13). Sky's the limit for Amazon Web Services. Financial Times. Retrieved from http://www.ft.com/cms/s/0/fbff0378-e1bb-11e4-8d5b-00144feab7de.html#axzz3gc3OpiEH

HP..com. (2015). Blade Server Basics | HP® Official Site. Retrieved July 3, 2015, from http://www8.hp.com/us/en/products/servers/bladesystem/blade-server-basics.html

IBM. (2001, July 1). Choosing a database management system [CT316]. Retrieved July 3, 2015, from http://www.ibm.com/developerworks/library/ws-dbpick/

Kalczynski, P., Senecal, S., & Nantel, J. (2006). Predicting On-Line Task Completion with Clickstream Complexity Measures: A Graph-Based Approach. *International Journal of Electronic Commerce , 10* (3), 121-141.

Kar, S. (2012, July 9). Gartner Report: One Third of Digital Data will be in Cloud by 2016 | CloudTimes. Retrieved July 22, 2015, from http://cloudtimes.org/2012/07/09/gartner-digital-data-cloud-2016/

Lim, G., & Lee, J. (2002). Buyer Carts for B2B EC: The B-cart Approach. *Journal of Organizational Computing and Electronic Commerce* .

MAAWG.com. (2009, December 17). *Global Anti-Spam Organization MAAWG Liaisons With IETF and BITS/Financial Services Roundtable.* Retrieved August 14, 2010, from MAAWG.com: http://www.maawg.org/global-anti-spam-organization-maawg-liaisons-ietf-and-bitsfinancial-services-roundtable

Maribor, U. o. (2005, October). *D 4.1 – Standards and technology monitoring report.* Retrieved October 24, 2010, from mg-bl.com: www.mg-bl.com/fileadmin/downloads/deliverables/D4.1_Standards_and_technology_monitoring_report_revised_version_V1.7.pdf

Mearian, L. (2011, June 28). World's data will grow by 50X in next decade, IDC study predicts | Computerworld. Retrieved July 22, 2015, from http://www.computerworld.com/article/2509588/data-center/world-s-data-will-grow-by-50x-in-next-decade--idc-study-predicts.html

Mueller, S. (2009). *What is spam?* Retrieved August 12, 2009, from spam.abuse.net: http://spam.abuse.net/overview/whatisspam.shtml

Mullins, C. (2005, November). Database management system software: Tips for choosing the best DBMS. Retrieved July 3, 2015, from http://searchdatamanagement.techtarget.com/answer/Database-management-system-software-Tips-for-choosing-the-best-DBMS

Mywebsiteworkout.com. (2010). *What is Web Server?* Retrieved August 13, 2010, from www.mywebsiteworkout.com: http://mywebsiteworkout.com/whats-webserver.shtml

Netcraft. (2015, June). June 2015 Web Server Survey | Netcraft. Retrieved July 3, 2015, from http://news.netcraft.com/archives/2015/06/25/june-2015-web-server-survey.html

Nielsen. (2012, April 10). Newswire | Consumer Trust in Online, Social and Mobile Advertising Grows | Nielsen. Retrieved July 22, 2015, from http://www.nielsen.com/us/en/insights/news/2012/consumer-trust-in-online-social-and-mobile-advertising-grows.html

Oakley, J. (2007, June 12). *To Spam Or Not To Spam - Is That A Viable Question?* Retrieved August 12, 2009, from articleslash.net: http://www.articleslash.net/Internet-and-Businesses-Online/212752__To-Spam-Or-Not-To-Spam-Is-That-A-Viable-Question.html

Otey, M. (2014, March 18). Important New Features in SQL Server 2014. Retrieved July 3, 2015, from http://sqlmag.com/sql-server-2014/sql-server-2014-important-new-features

Payloadz.com. (2014, January 20). Key Advantages of Using a Good Online Shopping Cart | PayLoadz. Retrieved July 4, 2015, from http://talk.payloadz.com/key-advantages-of-using-a-good-online-shopping-cart/

Robert, M. (n.d.). Infrastructure as a Service options in cloud computing. Retrieved July 4, 2015, from http://www.computerweekly.com/tip/Infrastructure-as-a-Service-options-in-cloud-computing

Roe, D. (2015, January 12). 8 Companies Leading ECM Into 2015. Retrieved July 4, 2015, from http://www.cmswire.com/cms/document-management/8-companies-leading-ecm-into-2015-027691.php

Roggio, A. (2011, June). 10 Essential Shopping Cart Features. Retrieved from http://www.practicalecommerce.com/articles/2821-10-Essential-Shopping-Cart-Features

Sassen, S. (2002, October 18). *Building a High-Performance Web Server*. Retrieved July 23, 2009, from hardwareanalysis.com: http://mywebsiteworkout.com/whats-webserver.shtml

Searchexchange.techtarget.com. (2003, August 01). *What is teergrube?* Retrieved July 12, 2009, from Searchexchange.techtarget.com: http://searchexchange.techtarget.com/definition/teergrube

Service-architecture.com. (2009). *Web Services explained*. Retrieved August 12, 2009, from www.service-architecture.com: http://www.service-architecture.com/web-services/articles/web_services_explained.html

Shamsudeen, R. (2013, June 26). Oracle Database 12c review: Finally, a true cloud database | InfoWorld. Retrieved July 3, 2015, from http://www.infoworld.com/article/2611000/database/oracle-database-12c-review--finally--a-true-cloud-database.html

Spamhaus.com. (2009). *The Definition of Spam*. Retrieved July 12, 2009, from www.spamhaus.com: http://www.spamhaus.org/definition.html

Stokes, L. (2014, January 27). A Service Architecture Approach to Ecommerce. Retrieved July 4, 2015, from http://www.foxycart.com/blog/a-service-architecture-approach-to-ecommerce

Stroud, F. (2015, May 1). Top 10 Enterprise Database Systems in 2015. Retrieved July 3, 2015, from http://www.serverwatch.com/server-trends/Top-10-Enterprise-Database-Systems-in-2015.html

Symantec. (2015, July). Protection from Computer Viruses - Software Security - Latest Threats | Security Response. Retrieved July 3, 2015, from http://www.symantec.com/security_response/landing/spam/

Theregister.uk. (2005, Januray 29). *Google mystery server runs 13% of active websites And only Google runs its mystery server*. Retrieved August 13, 2009, from www.theregister.uk: http://www.theregister.co.uk/2010/01/29/google_web_server/

Traxler, D. (2012, May 18). 7 Key Ecommerce-Infrastructure Decisions. Retrieved from http://www.practicalecommerce.com/articles/3545-7-Key-Ecommerce-Infrastructure-Decisions

Unifiedmail.net. (2009). *Definition of Spam*. Retrieved October 26, 2010, from www.unifiedmail.net: http://www.unifiedemail.net/Corporate/Policies/AntiSpam/Default.aspx

van Vugt, S., & Posey, B. (2014, December). Choosing the best server OS: Linux vs. Windows comparisons. Retrieved July 2, 2015, from http://searchdatacenter.techtarget.com/tip/Choosing-the-best-server-OS-Linux-vs-Windows-comparisons

W3Schools.com. (2009). *Server-side Scripting Primer.* Retrieved July 24, 2009, from W3Schools.com: http://www.w3schools.com/web/web_scripting.asp

W3Schools.com. (2009). *Web Building Introduction.* Retrieved July 14, 2009, from W3Schools.com: http://www.w3schools.com/web/web_scripting.asp

W3Schools.com. (2009). *Web Glossary.* Retrieved June 25, 2009, from W3Schools.com: http://www.w3schools.com/site/site_glossary.asp

Wallen, J. W. in F. (2012, September 24). Five content filters suitable for both home and business. Retrieved July 3, 2015, from http://www.techrepublic.com/blog/five-apps/five-content-filters-suitable-for-both-home-and-business/

WDVL. (2009). *Web Programming 101 Part Two, Client Side Scripting.* Retrieved June 25, 2009, from www.wdvl.com: http://www.wdvl.com/Authoring/Scripting/WebWare/Client/

webdevelopersnotes. (2009). *Unix, Linux and FreeBSD operating systems for web hosting servers.* Retrieved June 25, 2009, from webdevelopersnotes.com: http://www.webdevelopersnotes.com/hosting/linux_unix_freebsd_web_hosting.php3

webdevelopersnotes. (2009). *Windows operating systems for web hosting servers.* Retrieved June 25, 2009, from webdevelopersnotes.com: http://www.webdevelopersnotes.com/hosting/windows_web_hosting.php3

Webopedia. (2009). *The Difference Between Traceroute and Ping.* Retrieved February 13, 2009, from www.webopedia.com: http://www.webopedia.com/didyouknow/Internet/2009/traceroute_ping.asp

winserverhelp. (2010, March). *IIS 7.5 and IIS 7.0 Security Best Practices – Part I.* Retrieved July 25, 2010, from www.winserverhelp.com: http://www.winserverhelp.com/2010/03/iis-7-5-and-iis-7-0-security-best-practices/

Wisegeek.com. (2015). How Do I Choose the Best Database Management System Software? Retrieved July 3, 2015, from http://www.wisegeek.com/how-do-i-choose-the-best-database-management-system-software.htm

zdnet.com, S. J. V.-N. for. (2014, August 18). As DBMS wars continue, PostgreSQL shows most momentum. Retrieved July 3, 2015, from http://www.zdnet.com/article/as-dbms-wars-continue-postgresql-shows-most-momentum/

Zorzini, C. (2015, June 2). Shopify vs. Yahoo Stores – Ecommerce Platform Comparison. Retrieved from http://ecommerce-platforms.com/compare/shopify-vs-yahoo-stores-ecommerce-platform-comparison

4

E-COMMERCE BUSINESS MODELS

Traditional Vs Electronic Business Models

A business model is defined as the methods and techniques employed by a firm to generate revenue and sustain its position in the value chain (Jalozie, Wen, & Huang, 2006). Business models are essential to understand the basics of specific businesses. Long-established business models used in the traditional businesses may or may not be useful for E-commerce. Furthermore, some business models have use in traditional business and are native to electronic business e.g. business models that focus on movement of electronic information (such as digital-delivery model and freeware model). Reliance on an established online business model does not means that the business will be successful. Business model itself is not a solution or recipe of success. The manner in which the business is run determines its success. Many other factors are responsible for the success or failure of an online business. The success of E-commerce businesses largely depends on the art of management enabled by technology.

E-commerce Business Models

E-commerce business models exploit information technology to overcome the limitations of traditional business models. E-commerce enables the creation of new business models. New business models are emerging as more and more companies adopt E-commerce. At the same time, categorization of E-commerce business models is becoming increasingly difficult. This is because the number of business models is not limited, there are overlaps in these business models, fundamentally similar business models may appear in more than one generic type, some companies may employ multiple business models and the type of E-commerce technology used can affect the classification of a business model. For example, eBay is essentially a C2C business but also acts as a B2C marketplace and also has an m-commerce business.

Benefits of E-commerce Business Model

Distribution: The use of the Web as a distribution channel substantially reduces distribution costs. Products/services can be delivered immediately. Buyers and sellers can access and contact each other directly, potentially eliminating some costs and constraints of conventional communication such as phone, letter and fax.

SELLING: Since customers order online, a significant part of the selling function is in the hands of the customers.

COMPETITIVE INTELLIGENCE: Web technology allows companies to gather market intelligence and monitor consumer choices through customers' revealed preferences when browsing and buying.

Competition in the E-commerce

Competition in E-commerce is very intense because E-commerce transactions reduce search costs for buyers, enable speedy comparisons, enables differentiation/ personalization of product and services, and offer products at lower prices, and enable provision of quality customer service. Certain other competitive factors have become less important because of E-Commerce. For example, the size of a company may no longer be a significant competitive advantage. Location of business and language are becoming less important. Product condition is unimportant for digital products, which are not subject to normal wear and tear.

It can be said that competition in E-commerce, competition between companies is being replaced by competition between networks. The company with better networks, advertising capabilities, and relationships with other Web companies (e.g., having an affiliation with Amazon.com) has a strategic advantage. It can also be said that competition is between business models. The company with a better business model will win (Turban, Lee, & Chung, 2006). In order to remain competitive a company needs to maintain its competitive advantage. Some specific ways a company can obtain a competitive advantage are by developing a global market, by obtaining favorable terms from supplier, by developing a more experienced, knowledgeable, and loyal employee base, by establishing a powerful brand name, and by any type of asymmetry that will give it more resources.

Internet and its Impact on Business

The Internet has the potential for changing business in three major areas i.e.:

- Industry Structure

- Industry Value Chain

- Firm Value Chain

The Internet can change industry structure in many ways. For example by introduction of substitute products, by increasing the bargaining power of suppliers, by increasing the power of consumers and buyers, and by changing existing barriers to entry. The increasing use of E-commerce is creating new marketing channels and size of the overall market is expanding. The global reach and universal standards of E-commerce has lowered barriers to entry and intensified competition. Reduced costs of operations, for both the industry and firms, are enabling global competition. Less steep costs for computing and communication are enabling broad-scope business strategies. Firms can use interactive tools, personalization, and customization techniques to increase richness and reduce reliance on its traditional sales force. This way companies can reduce their operational costs, develop better after-sales support strategies, and reduce threats from substitutes.

The industry value chain includes all of the value-creating activities within the whole industry while the internal value chain of a company includes all the value-creating activities within that specific firm. All the major players in an industry value chain i.e. suppliers, the manufacturers, the distributors, the delivery agents, the retailers and the customers can be impacted by Internet. Internet provides new and interactive ways of interaction for all major players. By using web-based B2B exchanges or developing direct online relationships with their customers, manufacturers can reduces their cost of goods sold. This is because the distributors and retailers are eliminated from the value chain.

Distributors can use Internet technologies to develop highly efficient inventory management systems to reduce their costs. Retailers can develop highly efficient customer relationship management (CRM) systems to strengthen their customer support services. Customers can use the Web to search for the best quality, delivery, and prices and reduce their transaction costs.

Internet can change firm value chain by changing the way business performs various business processes such as warehousing, manufacturing, sales and customer support. For example, Amazon.com uses the Internet to provide consumers access to a large inventory of items and process transactions much faster and efficiently than traditional retailers do.

Internet Value Chain

The Internet value chain includes many layers such as content rights, online services, enabling technology services, connectivity, and user interface. The content rights layer includes both commercial and user-generated content. The online services layer includes services such as e-mail, voice-over-the-Internet Protocol (VoIP), video-on-demand, gaming and e-commerce. The enabling technology services layer include services such as web hosting, billing and advertising. The connectivity layer includes both fixed and wireless network providers, Internet service providers, and content delivery network services. The user interface layer includes, for example, computers, smartphones and smart televisions.

The different layers of the Internet value chain are now beginning to merge because of increasing consolidation, with players extending their scope. New means of delivering content are being used and new business models are appearing. With advancing technological convergence, many global brands, such as Microsoft, Apple, Amazon and Google, are becoming active across various parts of the Internet value chain. Leading the front, Google has expanded from being a search engine to become a device manufacturer, operating systems developer, service provider (cloud storage, e-mail, maps, content distribution and online advertising). The telecom providers are diversifying their businesses. For example, Telefónica Digital, BT ventured into sport content and online television content via YouView. The AT&T and Verizon are offering multiple play services. Subscription television services offer premium programs, such as new release movies and live cultural or sporting events. These service providers face a direct competition from content producers that use Internet to deliver their content directly to the customers. For example, Major League Baseball and the National Basketball Association, offer live streaming of games to their subscribers. Content aggregators have started advancing to produce their own content for direct release to their subscribers. A recent example of this was the Netflix production of the hit series House of Cards. Online music streaming services (such as Spotify) are experiencing huge growth with big players, such as Google with its GooglePlay, are also entering the market. Popularity of services, such as YouTube, has generated popularity growth of user-generated content. Producers of smart television are increasingly bundling sales of their products with access to on-demand programming, such as YouTube and BBC iplayer, as well as Internet access.

FIGURE 4-1: INTERNET VALUE CHAIN

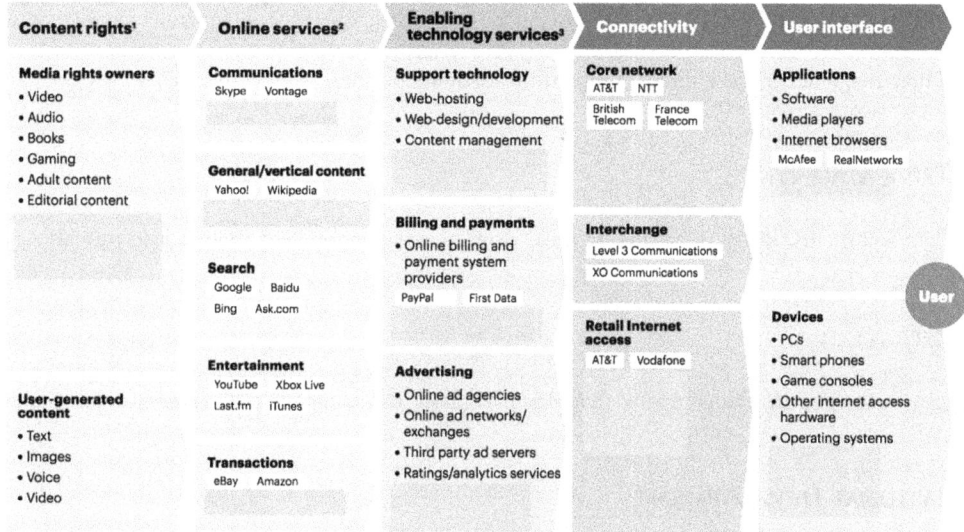

Source: A.T. Kearny Analysis

Key Elements of E-commerce Business Model

A successful business model effectively addresses eight key elements:

VALUE PROPOSITION

A value proposition refers to the benefits, including the intangible, non-quantitative ones, which a company can derive from conducting operations. According to AMIT and ZOTT (2001), e-business can create four types of value propositions. These include search and transaction cost efficiency, complementarities (i.e. bundling of goods and services together that provide more value than offering them separately), lock-in (i.e. increased switching cost that ties customers to particular suppliers), and novelty (i.e. use of innovation e.g. for creating new ways to structure transactions and connecting partners). Some of the typical value propositions provided by E-commerce include personalization, customization, convenience, reduced product search costs, reduced prices and delivery costs.

REVENUE MODEL

Revenue model describes how the company makes money from its operations. Sales model, subscription model, transaction fee model, and affiliate model are the major revenue models in E-commerce.

MARKET OPPORTUNITY

It refers to the revenue potential within a company's chosen market segment.

COMPETITIVE ENVIRONMENT

It describes how many direct and indirect competitors are doing business in the same market segment and their profitability.

COMPETITIVE ADVANTAGE

It refers to the factors that differentiate the business from its competition, and enables it to provide a superior product at a lower cost.

MARKET STRATEGY

It is the plan developed by the company that describes how the company will enter a market and attract customers.

ORGANIZATIONAL DEVELOPMENT

It is the process in which the functions within a business and skills needed to perform each function is defined. It also outlines the process of recruiting and hiring employees suitable for these functional roles.

MANAGEMENT TEAM

This is the group of individuals responsible for guiding the company's growth and expansion.

Primary Revenue Models in E-Commerce

A revenue model outlines how the organization or the E-commerce project will earn revenue. The revenue models can be first broken into two main categories: conventional revenue models adapted to the Internet and Internet-specific revenue models. The five primary revenue models used by E-commerce firms are:

1. Advertising model

2. Subscription model

3. Transaction fee model

4. Sales revenue model

5. Affiliate model

6. Internet Content Revenue Model

7. Other revenue models

In the advertising model, a business generates revenue by displaying paid advertisements on its web site. The businesses adopting this business model must be able to convince the advertisers that the business website can attract a sizeable viewership, or a viewership that meets a marketing niche sought by the advertiser. The ads can be of various types. Table 4-1 shows various types of ads.

TABLE 4-1: AD TYPES AND WEBSITES

Ad Type	Example
Display Ads	Yahoo!
Search Ads	Google
Text Ads	Google, Facebook
Video Ads	YouTube
Audio Ads	Saavn (saavn.com),
Promoted Content	Twitter, Facebook
Recruitment Ads	LinkedIn
Classifieds	JustDial, Quikr
Featured listings	Zomato, CommonFloor
Email Ads	Yahoo!, Google
Location-based offers	Foursquare

In the subscription model, a business generates revenue by offering its users content or services for a subscription fee for access to some or all of its offerings. Users generally pay a fixed fee (generally monthly) to access any type of content or service. One example of subscription model is an ISP. Users access Internet and pay the ISP a monthly fee for this service. The different types of subscription model include Software as a Service (SaaS) (e.g. freshdesk.com), Service as a Service (e.g. PayU), Content as a Service, Infrastructure/Platform as a service (e.g. Amazon Web Services, Azure), Membership Services (e.g. Amazon Prime), Support and Maintenance (e.g. Red Hat), and Paywall (e.g. ft.com, NYTimes).

In transaction fee model, a business generates revenue through commission received. The amount of commission can either be variable based on the value of transactions made or fixed for each transaction. For example, when you sell your house, you typically pay a commission to the broker, which is a percentage of the total transaction value (i.e. price of the house). The higher the value of the sale, the higher will be the commission. Commission can be fixed for each transaction irrespective of transaction amount. For example, in electronic stock trading, you pay a fixed commission per trade to your broker, regardless of the amount. Two common transaction fee based businesses are eBay and E-trade.com. When a seller successfully auctions off a product, he pays a transaction fee to eBay. Similarly, E-trade.com receives a transaction fee from the customer when it executes a trade (e.g. purchase of stock) for the customer.

In the sales revenue model, a business generates revenue through sale of its goods, information, or service to consumers.

In the affiliate model, sites receive referral fees or a percentage of the revenue from any sales that result from referring or guiding customers to other websites.

There can be other revenue models. Some companies allow you to play games for a fee or to watch a sports competition in real time for a fee (e.g., see nisn.espri.go.com). Another revenue model is licensing fees. Licensing fees can he assessed as an annual fee or a per usage fee. Microsoft, for example, charges license fee for workstation that uses Microsoft Windows operating system.

Revenue Model Issues

Some issues may arise during implementation of the revenue models discussed. Following are some of these issues and strategies companies can use to deal with them.

Channel Conflict

Technology has splintered the marketplace for manufacturers into online and offline channels, plus retail partners. In fact, most manufacturers of consumer brands rely on multiple channels to move their products and invest significant resources to cultivate them. Therefore, when they want to start selling directly online, or beef up their web store, they risk alienating their most valuable sources of sales.

A channel conflict can occur whenever online sales activities of a company interfere with its existing sales outlets. The issue can also be called cannibalization because online business consumes sales that could have been made on company's traditional sales channels. A perfect example is the computer industry. A decade ago, it was foolhardy for a computer manufacturer such as IBM, Apple, or Hewlett-Packard to sell computer products direct to the end user; by doing so, they risked losing the retailers who bought their products either directly from them or from distributors. If the manufacturers sold their wares directly to consumers, retailers would see the computer companies as direct competitors. Companies need to address channel conflict else, their retailers could go away and sell competing products. Similar issues can also arise within a company if that company has established sales channels that would compete with direct sales on the company's web site.

HANDLING CHANNEL CONFLICT IN E-COMMERCE

In order to handle channel conflict in online business several approaches can be adopted (meridianecommerce.com, 2013; Brohan, 2012).

No Competition, But Collaboration

Manufacturers can put their focus online on telling the innovation and leadership product stories that create strong desire for our products across all of their distribution channels (Love, 2013). Communicating online strategy to channel partners is a courtesy that can go a long way toward finessing potential friction. For example, it's common for manufacturers to use their web store differently from the retailers; for example, they can provide a comprehensive informational catalog of goods unavailable elsewhere. It can help reassure skittish retailers to know that the manufacturer is pursuing a top-level goal of boosting sales across all channels.

CROSS PROMOTION

Nothing will convince channel partners that you are focused on mutual success more than giving them free publicity and marketing. Remember, the key to channel harmony is collaboration. In addition, as manufacturer, you have another natural advantage here: access to those customers who come to you for information about your products, including where they can find them. This is an excellent opportunity to boost both your own brand and your channel partners' success (InternetRetailing.net, 2011).

Examples of cross-promotion include store finder, promotion of in-person store events, giving retail partners advertising space, and rewarding high-performance retailers with prominent listings on your site. Cross-promotion efforts can sometimes prove to be a delicate balance of give-and-take, but it is never necessary to shy away from online direct sales. Tailor your promotional efforts to align with the natural strengths of buying direct versus via a channel or retail partner. For example, Under Armour (www.underarmour.com) is an American sports clothing and accessories company. Under Armour integrates a robust store finder into their website, where consumers can discover retailers selling Under Armour brand garments. Given the nature of the product, many consumers want tactile access before making a purchase. However, existing customers can order more easily online, or customers seeking special deals can take advantage of Under Amour's online outlet, which can then be used to clear out inventory. Lafayette 148 (www.lafayette148ny.com) is a New York based store that offers a selection of chic and feminine designs in misses, petite and plus sizes. Lafayette 148 uses its online store to point consumers to retail locations where lead designer makes appearances. That kind of cross promotion keeps people attuned to the online store while pushing business to retail partners. Both channels can be used to stimulate interest and encourage purchases (Love, 2013). Such an approach not only pushes business to the channel best suited for it, it also serves the customers better by letting them choose where to purchase according to their own, individual priorities.

GENERATE CROSS-CHANNEL VALUE

Brand manufacturers can employ valuable business insight and intelligence to improve their products and sell them more strategically. In addition, they can deliver that insight to their offline and retail partners. For example, manufacturers might test different approaches to sales and marketing to identify the tactics that yield the most business. Brands can learn a lot about their customers by having their own online store. It opens up enormous possibilities for data mining (InternetRetailing.net, 2011). Examples of insight online stores can produce include the products customers click before they buy other items, the product details sought by the customers, the type of visuals/messaging approaches that lead to sales, and the naming/branding protocols that are successful. Additionally, the metrics generated by their online store can provide manufacturers more data that are actionable. This data can include spot trends, the most lucrative search terms, highly desired product lines, and shoppers' decision-making process.

Smart ecommerce retailers regularly run tests where they compare performance. This kind of information can be immediately implemented onsite to boost direct revenue, as well as shared with other channels to support their efforts as well. JELD-WEN (www.jeld-wen.com) manufactures building products, including windows, interior and exterior doors, and related building products. JELD-WEN leveraged their site data to show the positive impact of custom tools and guided selling. Their case convinced Home Depot to collaborate with them to try something similar on HomeDepot.com. (Lanigan, 2012). The insight gained will help manufacturers stay tuned with their customers and this, in turn, can be used to inform retailers (InternetRetailing.net, 2011).

RIGHT PRICE

You have the power to destroy your retail partners' ability to compete. They have to mark up prices to make a profit, after all. Amazon's willingness to take a loss or only break even, while plying its supply chain mastery, has helped the giant edge out competition in multiple business segments. In some cases, like e-books and bookstores, they have hastened the demise of an entire offline segment. Alternatively, you can reassure unsatisfied partners that you want to boost everyone's bottom line. Evidence suggests that price is the single issue over which the most channel conflict is generated (Kiran, Majumdar, & Kishore, 2012).

Consider the following specific pricing strategies that can yield mutual benefit.

PRICE JUST ABOVE THE COMPETITION

You could stick to list or recommended retail pricing, which would leave your prices higher than other retailers would. Consequences of this approach could be low conversion rates. Alternatively, you could match or beat retail pricing, only to start the very channel conflict you want to avoid. Many consumers want to buy from the brand source and willing pay a premium for it. You can leverage brand loyalty without threatening channel partners. The North Face, Inc. (www.thenorthface.com) is an American outdoor product company specializing in outerwear, fleece, coats, shirts, footwear, and equipment such as backpacks, tents, and sleeping bags. Outside of free shipping, The North Face does not typically discount prices. They realize that a certain core group of customers will be willing to pay a premium to purchase direct from the manufacturer and enjoy the exceptional level of service they have come to expect from the brand (Lanigan, 2012).

DYNAMIC PRICING

You can opt to price "just above" retail and offline partners. This strategy can be a problem if the prices of the retailers and offline partners are in a continuous flux. Dynamic pricing can automates the process of adjusting product prices against a pool of retailers with whom manufacturers are collaborating. Amazon.com is a good example of retailer that uses dynamic pricing.

FOCUS ON PROMOTIONS

Manufacturers may have greater freedom to compete in other areas than price. For example, a manufacturer can offer special configurations unavailable elsewhere. That makes direct Web sales compelling while still giving partners room to maneuver.

DIRECT SELLING RELATIONSHIPS

You can remove special deals from public view entirely, and focus your ecommerce efforts to serve a niche market that would other elude or be underserved by channel partners. Particularly in the case of direct selling to large organizations, you might find that you can create an agreement that leverages the advantages of both offline and online channels; giving your customers total flexibility to access your goods in whatever way is most convenient to them.

Besides the techniques discussed so far, there are some other ways to handle channel conflict

HANDLE ORDERS ONLINE WHILE PERMITTING ESTABLISHED DISTRIBUTORS HANDLE FULFILLMENT

In this scenario, a brand manufacturer refrains from direct selling. They serve as a brand portal and refer all sales traffic directly to retailers. This is the most conservative approach and tends to be used by strong brands who rely heavily on retailers (such as consumer-packaged goods). An example is Clorox (www.clorox.com). Clorox is an American multinational manufacturer and marketer of consumer and professional products. Clorox and the Clorox sub-brands provide purchase options for online retailers and offline stores. Interestingly, they also provide pricing. This model is not limited to consumer-packaged goods. JanSport, Maxsea (a software company), and JELD-WEN are other examples. JanSport (www.jansport.com) is an American brand of backpacks and collegiate apparel

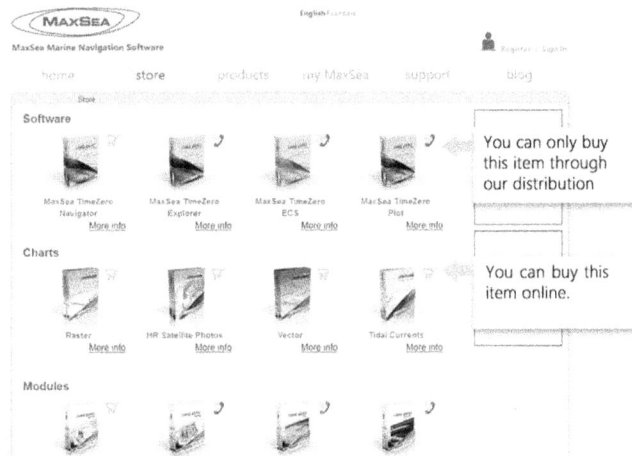

ADVERTISE ALL PRODUCTS ONLINE BUT SELL ONLINE A FEW ITEMS

Wal-Mart, QVC and HSN excel at this approach. QVC (www.qvc.com)is an American cable, satellite and broadcast television network specializing in televised home shopping. HSN (www.hsn.com) Home Shopping Network (HSN) is an American broadcast, basic cable and satellite television network. Exclusive direct selling is a big differentiating factor. Exclusives for retail distributors help maintain good established relationships with retailers. An example is Reebok. Reebok's custom shoe program gives consumers a reason to go specifically to Reebok.com. It sets them apart from their retailers without directly competing. The next iteration of this model could be Reebok offering specific custom shoes only to certain retailers.

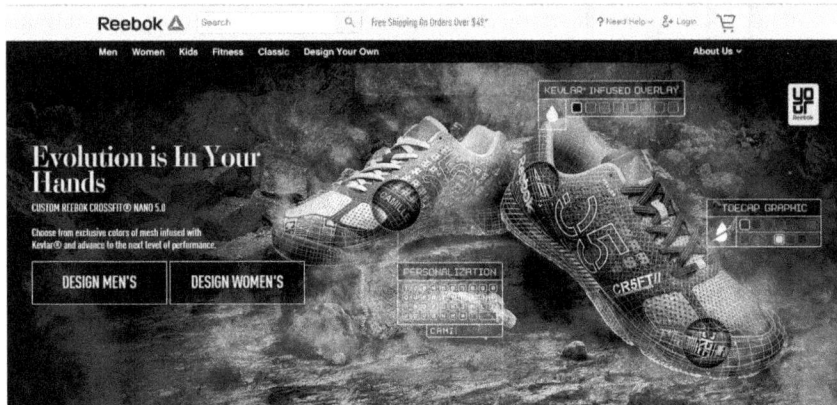

SELL EXCLUSIVELY ONLINE

Companies may also consider selling a product line that is only sold on the Internet. That way, retailers would not find them in direct competition. Another option is to sell products online under a different brand name. In any case, companies should plan carefully for possible channel conflicts when they decide to start online business.

Strategic Alliances in E-commerce

A strategic alliance is formed when two or more companies join hands together to perform an activity over a long period. Companies form strategic alliances for many reasons e.g. to make online sales. One example of strategic alliance is Levi's alliance with its retail partners in which Levi's gave its retail partners space of Levi's website to sell Levi's products. Take example of alliance between Yahoo ads ABC News. Under this alliance, ABC News provided yahoo the most updated information resource and original first handed videos of latest incidents not shown before. The major advantage that ABC News got from this alliance is to have a full time news showing on the well-known web portal. For ABC News this platform was an opportunity for going deeper. Another example is alliance between Star Airline and MasterCard. Star, an international airline decided to launch new travelling offers for its travelers in partnership with master card. The focus of this alliance is to create value for the customer as well as to generate more and more volume for the business. Under this arrangement, every order placed through master card obtained a 10% discount in travelling tickets. Both star and Master card utilize this alliance to expand their business and increase market share. Not all the alliances are successful. Take example of Cisco's alliance with Motorola. Cisco is a tech-giant that manufactures and sells consumer electronics, and networking devices. On the other hand Motorola is multinational company mainly deal in mobile sets, iPods, portable radios etc. Cisco and Motorola entered into an alliance aimed to provide a high-speed internet service on the mobile phones that enable user to visit large number of IP address in a short span of time. This alliance was advantageous for both Cisco & Motorola. They earned revenue by charging the customer for Internet access and by selling handsets used to access Internet. Some of the strategic alliances were not successful. The main reason was the inconsistent strategic business plan that lacked proper documentation to meet the legal requirements of the proposed alliance.

Mergers and Acquisitions in E-commerce

E-commerce has been the most active segment in digital media and advertising mergers and acquisitions thus far in 2011, with 67 transactions topping $6 billion. Most ecommerce deals consist of Internet retailers acquiring other operators of Internet storefronts, which sell similar or complementary product lines. Recent examples include Fanatics' acquisition of Dreams in April 2012 (both sell licensed sports gear); Shoebuy.com's acquisition of Cozy Boots (both sell footwear) in October 2011; and Shockoe Commerce Group's acquisition of Red Rock Products (both sell yoga gear). Perhaps more interesting, however, has been the recent trend of brick and mortar retailers and product manufacturers/wholesalers acquiring pure play Internet retailers in their niche categories to bolster their online presence and expertise. Recently, brick and mortar retailers and product manufacturers have been acquiring web-only Internet retailers to acquire their expertise and high growth rates. Notable examples include Walgreens (the US largest brick and mortar drug store chain with more than 7,800 stores) acquiring Drugstore.com (an Internet retailer of over-the-counter products); National Vision (a brick and mortar retailer of eyeglasses) acquiring Arlington Contact Lens Service in March 2012; and Motorsports Aftermarket Group (a manufacturer of motorcycle parts) acquisition of Motorcycle Superstore (a motorcycle part Internet retailer) in March 2012 (Kampe, 2012). In India, larger players like Flipkart and Snapdeal were on a high after having received record-breaking investments in 2014. These companies were looking to add more power to their engines through strategic acquisitions. The objective is to expand their portfolio, increase their customer base, and provide a better user experience to customers. Table 4-2 shows major Indian E-commerce acquisitions in 2015.

TABLE 4-2: MAJOR INDIAN E-COMMERCE ACQUISITIONS

Acquisition	Nature of Business
Flipkart acquired WeHive Technologies Pvt. Ltd	Flipkart is an online marketplace. WeHive is a trusted P2P marketplace with verified buyers/ sellers.
Mahindra acquired BabyOye	Mahindra group runs its offline baby care products store, Mom&Me. BabyOye is the leading online baby-care product brand.
Foodpanda acquired Just Eat India	Both are online food-ordering platform in India.
Ola Cabs acquired TaxiForSure	Both are online taxi service providers in India.
Flipkart acquired AdiQuity	AdiQuity is a mobile ad network.
Snapdeal acquired Freecharge	Snapdeal is an online marketplace. Freecharge is an online mobile recharge platform.
Flipkart acquired Appiterate	Appiterate is a mobile engagement and marketing automation company.
Snapdeal acquired MartMobi	MartMobi is a mobile technology startup.

Whether Internet retail, catalog or direct, the buyers in recent transactions have been predominantly strategic. Strategic buyers acquiring companies, titles and storefronts in similar product categories are able to leverage fulfillment, merchandising, management and house lists to meet target return on capital rates.

Fundamental Categories of E-commerce Business Models

There are various business models for E-commerce. These models can be divided into following fundamental categories.

- Business to Consumer (B2C) Models

- Consumer to Consumer (C2C) Models

- Business to Business (B2B) Models

- Other Business Models

Business to Consumer (B2C) Models

STOREFRONT MODEL

In storefront business model, an E-commerce business provides an online shopping store. Customers can use this online store to browse and shop for the products and businesses can use this store for inexpensive advertisement of their products and services. In a storefront model, an E-business can sell hard good, soft goods, and services. Customers can enter the store directly, through a portal, or through a search engine. Customers generally need to sign up with the online store before they could complete their purchase online. A storefront contains facilities for ordering and paying for products (e.g. through credit cards). Customers provide their payment information (e.g. credit card number) and shipping information. Once payment is authorized, a receipt is issued to the customer and purchased good shipped to the address supplied by the customer. The advantages to a storefront model includes extended sales and marketing, decreased marketing and promotion costs, reduced transaction costs, new product introduction and testing, and the ability to reach specialized markets.

Some forms of storefront model are:

E-SHOP: An E-shop can be an online store with a static catalogue or an online store with extended functionality e.g. interactive catalogue, mailing lists, and customers' reviews. An E-shop is setup by a single company. E-shop provides facilities whereby customers can purchase products offered at E-shop.

E-MALL: An electronic mall or e-mall is a collection of e-shops, which is often devoted to a specific service or product. For example an e-mall might be devoted to selling goods associated with a leisure activity such as fishing or toys e.g. toysrus.com. Usually e-malls are organized by a company, which charges the e-shops for administering their presence: maintaining the website, hosting the e-mall, and providing payment and transaction facilities and marketing.

E-TAILING: E-tailing or Electronic retailing refers to the practice of selling goods and services over an electronic medium like the Internet and those who conduct retail business online are called e-tailer. Books, CDs, and computer software and hardware, clothes, cosmetics, perfume, plants, toys, are the most common goods sold by e-tailers. The concept of E-tailing generally implies B2C E-commerce however, the distinction between

B2C and B2B e-commerce is not always clear. For example, Amazon.com sells books mostly to individuals (B2C), but it also sells to corporations (B2B). Barnes & Noble, has a special division that accommodates only business customers. Walmart.com sells to both individual consumers and businesses. Dell sells its computers to consumers and businesses, from dell.com, Staples sells to both markets at staples.com, and insurance sites sell to both to individuals and corporations.

PORTAL MODEL

A portal is an "infomediary". The word "infomediary" is combination of two words: information and intermediary. A portal or an infomediary" is a website that stores and organizes large amounts of data from various sources and acts as intermediary between those looking for information and those who provide information. Portals can provide services such as news, online directories, and personalized advertisements. Portals can also provide web space to other companies to host their websites. The web space is provided on the main portal site. This way a portal site can act as a major E-commerce facilitator.

A portal with information and links to websites featuring diverse subjects is a horizontal portal e.g. www.indiatimes.com and www.mantraonline.com. A portal having links and information focused on a particular subject is a vertical portal e.g. www.quicken.com, www.nuptialknots.com, or www.jguru.com. A vertical portal is also called a vortal.

Some examples of vertical portals include Espnstar.com (a site for latest soccer football information), wfn.com (women's financial network), ivillage.com (a site providing information exclusively for women), webmd.com (a source of health information), guru.com, and nfl.com. Some examples of horizontal portals include altavista.com, lycos.com, and about.com. Portals earn major share of their revenue through personalized advertising and web hosting.

DYNAMIC PRICING MODELS

COMPARISON-PRICING-MODEL

This model allows customers to poll various service merchants to get the best deal for a product/service. Revenue is generated from partnership with particular merchants. One example of such model is www.pricescan.com.

DEMAND-SENSITIVE PRICING MODEL

This model is based on the idea of selling a product to a group of customers in one transaction, thus reducing cost per person. It allows the business to reduce prices while retaining or increasing profit. One example of such model is www.mobshop.com. The MobShop Network links online buyers across the web through their partner websites to harness the power of collective volume buying. MobShop created the MobShop Network in order to make the substantial savings generated by volume purchasing available to individual buyers. MobShop's Group Buying Network, consisting of community and supplier partners, brings buyers together across the Web to transact as a group and receive discounted pricing. With MobShop's buyer aggregation service, the same values traditionally provided only to large organizations that purchase in volume are extended to individual buyers. Mobshop's network gives buyers unique access to a wide array of in-demand products at optimal prices. Conversely, suppliers now can maximize the Internet's potential to connect them with the largest pool of motivated buyers available. It is a win-win proposition that generates incredible value for both consumers and suppliers. The Mobshop Network charges its supplier partners on a per-transaction basis, so suppliers pay a small percentage of each sale price to MobShop. The community partners and customers are not charged when they participate in the MobShop Network. Another example of group-buying site is BuyWithMe. Headquartered in New York, BuyWithMe is expanding steadily and targeting a mostly female audience. Rather than specializing in impulse buys, the site offers deals for up to one week, and, like Groupon (www.groupon.com), requires a minimum number of buyers.

NAME-YOUR-PRICE-MODEL

This model allows customers to state the price they are willing to pay for a product/service. Some examples of such model are ticketsnow.com and priceline.com. Priceline.com is company that operates a commercial website that helps users obtain discount rates for travel-related items such as airline tickets and hotel stays. The company is not a direct supplier of these services; instead, it provides comparative pricing from a collection of service companies. Customers can name the price they are willing to pay for air tickets, hotel stay, car rental, and vacation packages. Priceline.com then compares this price with the prices available in their database. Customers only come to know the name of airline, rental Car Company, or hotel once the purchase is finalized. Priceline's revenue is the difference between the price an individual named and the price charged by the service establishment.

BARTERING MODEL

In this model, an E-commerce business provides an online location where various users can offer one item in exchange of another. The E-commerce business generates revenue through fees charged to users of their website. One such example is www.ubarter.com. It is an online site that helps match people up who have common needs and desires. Users can list both the products and services they need and the products and services they can provide. Any user interested in an item can be matched up and barter can be finalized. A variety of products can be bartered e.g. mobile phones, computers, PDAs etc.

REBATE MODEL

In this model, businesses offer rebates on product at leading online retailers in return for commission or advertising revenues. Such businesses provide their users with the offering of rebates from their affiliated suppliers. A user sign up with such website and go through this website when making online purchases. If the supplier is affiliated with this website, you will get a rebate (a small percentage of your purchase that is provided back to customer on purchase). One example is www.ebates.com. Rebates can be helpful for a business to attract customer attention and reward non-users of coupons. At the same time, rebates may be less appealing to consumers who want immediate gratification e.g. instant discounts.

FREE OFFERING MODEL

In this model, a business offers free products and services to generate high traffic. The revenue is generated through advertising or by receiving revenue from sites, which you have to visit before getting the free product or service. Some examples are www.startsampling.com, and www.FreeSamples.com. Typical sites, which come under this category, include gaming sites where users can play computer games using their browser, sites, which run free raffles, and sites which offer free software.

MANUFACTURER MODEL

In manufacturer model (also called direct model) a merchant utilizes Internet to sell the products and services directly to the customers. The advantages of manufacturer model are increased efficiency, improved customer service, and a better understanding of customer preferences. One example of such model is Dell Computers.

SUBSCRIPTION MODEL

In this model, users are charged a periodic (daily, monthly or annual) fee to subscribe to a service. Many sites combine free content with premium content (i.e. content which requires subscription for access) Subscription fees are incurred irrespective of actual usage rates. Subscription model can take various forms. For example, content services that provides text, audio, or video content to users who subscribe for a fee to gain access to the service. Listen.com and Netflix are examples of such model. Trust services provider charges their members a subscription fee to provide a trust seal. One example of trust service provider is Truste.

UTILITY MODEL

The utility or on-demand model is based on pay-as-you-go approach. Customers are charged for their actual usage of a product or service. Traditionally this model has been used for essential services (e.g., telephone, water, and electricity). ISPs also charge their customers for their actual use of the Internet access.

Consumer-to-Consumer (C2C) Models

AUCTION MODEL

In the auction model, participants bid for products or services over the internet. The auction sites generally do not own the products on their site, but facilitates the process of listing and displaying goods as an affiliate. Online auctions provide many benefits including no geographical or time constraints to participate in auction, large number of bidders and buyers, and network economies. Auction sites earn major share of their revenue through advertising and commissions from items auctioned on their site.

Auction Related Services

The growth of eBay and other auction sites has encouraged entrepreneurs to create businesses that provide auction-related services of various kinds. These include escrow services, auction directory and information services, auction software (for both sellers and buyers), and auction consignment services. An escrow service is designed to protect consumers along with sellers. As with traditional escrow, Internet escrow works by placing money in the control of an independent and licensed third party in order to protect both buyer and seller in a transaction. When both parties verify the transaction has been completed per terms set, the money is released. If at any point, there is a dispute between the parties in the transaction, the process moves along to dispute resolution. The outcome of the dispute resolution process will decide what happens to money in escrow. With the growth of both business and individual commerce on the web traditional escrow companies have been supplanted by new technologies.

Another service offered by some firms is a directory of auctions. Sites such as Auctionguide.com offer guidance for new auction participants and helpful hints and tips for more experienced buyers and sellers along with directories of online auction sites. Ecommercebytes.com/ is an auction information site that publishes an e-mail newsletter with articles about developments in the online auction industry.

Both auction buyers and sellers purchase software to help them manage their online e-auctions. Sellers often run many auctions at the same time. Companies such as AuctionHawk (www.auctionhawk.com), and Vendio (www.vendio.com) sell auction management software and services for both buyers and sellers. For sellers, these companies offer software and services that can help with or automate tasks such as image hosting, advertising, page design, bulk repeatable listings, feedback tracking and management, report tracking, and e-mail management. Using these tools, sellers can create attractive layouts for their pages and manage hundreds of auctions.

Several entrepreneurs have identified yet another auction-related business that meets the needs of people and small businesses who want to use an online auction, but do not have the skills or the time to become a seller. These companies, called auction consignment services, take an item and create an online auction for that item, handle the transaction, and remit the balance of the proceeds after deducting a fee that ranges from 10 percent to 40 percent of the selling price obtained. Items that do not sell are returned or donated to charity. The main auction consignment businesses include,AuctionDrop, QuickDrop, andPicture it SOLD.

Business-to-Business (B2B) Models

BUSINESS-TO-BUSINESS AUCTION MODEL

In this model, businesses bid in auction to buy items such as excess or old inventory or unwanted capital equipment. Liquidation brokers-firms are firms that find buyers for unusable inventory Items. Ingram Micro is a major distributor of computers and related equipment to value-added resellers. Often finds itself with outdated items, Ingram Micro turns over to liquidation brokers who auctions those items to their established customers. An example of a sell-side auction is AssetAuctions.com where surplus industrial equipment is sold. Sell-side auctions are sometimes referred to as forward auctions. Buy-side auctions, by contrast, reverse the traditional auction formula, which is to help the seller get the highest price for the product. Instead, the buyer initiates the auction in order to find the cheapest supplier of a product. Sellers then bid against one another, offering the lowest prices they can for their products, in order to get the buyer's business. Because the roles of the buyers and sellers are reversed in buy-side auctions, they are often referred to as reverse auctions. Not all companies use an intermediary like eBay or AssetAuctions to conduct their auctions, though. Some companies conduct their own auctions on their Web sites so they do not have to pay a fee to an intermediary. For example, General Motors auctions off reconditioned vehicles to auto dealers on its own Web site (http://www.gmonlineauctions.com).

B2B EXCHANGES

A B2B exchange is a website or collection of websites, which make the process of carrying out B2B transactions much easier. B2B exchanges enable businesses to procure services and products from each other and carry out joint activities such as marketing or project bidding. STEEL 24-7, was created in 2001 by the four largest European steel producers to provide many steel customers and suppliers one common and mutually used electronic platform. Exostar (exostar.com) is an online marketplace for the aerospace industry and was launched in October 2000. Orbitz (orbitz.com) and ChemConnect (chemconnect.com) are also two famous B2B exchanges. Terrecom (terrecom.com) is a leading B2B exchange that brings organizational buyers seeking a wide variety of goods and services together with organizational sellers seeking to find these particular products.

AFFILIATE MODEL

In affiliate model, a business offer financial incentives to affiliated partner sites to drive customers to its site. Generally, the affiliated sites are paid for their performance i.e. the affiliate is paid only if the customers transferred to the business website make a purchase. A recent example is several well-known bloggers pushing out MotoG reviews and then linking to the Flipkart exclusive sale site for commissions. Variations of affiliate model include banner exchange, pay-per-click, and revenue sharing programs. In banner exchange model, the business trades banner placement among a network of affiliated sites. In pay-per-click model, the business pays its affiliates for each customer click-through. In revenue sharing model, a business offers a percent-of-sale commission based on a customer click-through in which the customer subsequently purchases a product.

INFOMEDIARY MODEL

Data about consumers and their consumption habits are valuable, especially when that information is carefully analyzed and used to target marketing campaigns. Similarly, independently collected data about producers and their products are useful to consumers when considering a purchase. Some firms function as infomediaries (information intermediaries) and assist buyers and/or sellers understand a given market. Examples of infomediary model include advertising networks, online market research firms, incentive marketing providers, and metamediaries.

Advertising networks provide banner ads to a network of member sites. Advertisers can use these networks to launch large marketing campaigns. Advertising networks collect data about web users that can be used to analyze marketing effectiveness. DoubleClick is an example of advertising network. Online market research firms e.g. Nielsen, conducts surveys to collect data about web users. Incentive marketing providers run customer loyalty program that provides incentives to customers (e.g. redeemable points or coupons) that can be used for making purchases from associated retailers. Collected users data can be sold to other businesses for targeted advertising. Coolsavings is an example of incentive marketing provider. A metamediary facilitates transactions between buyer and sellers by providing comprehensive information and ancillary services. Metamediary does not involve in the actual exchange of goods or services between the buyer and sellers. Edmunds (Edmunds.com) is an example of metamediary.

B2B E-COMMERCE PORTALS

A B2B portal is an online market place used by manufacturers/suppliers etc. to post trade leads using their leads. Buyers and importers can find suppliers for their products. E-commerce users use it for marketing and promotion of their products and services. B2B Website Portals are most popular in countries with huge number of manufacturers to export e.g. China, India, and South Korea. B2B Website Portal benefits supplier and purchaser, exporters and importers, sellers and buyers. For global buyers, they could find their ideal suppliers and manufacturers and post their buying trade leads or threads in these B2B sites. The suppliers are of B2B portal. They use various internet-marketing tools to publicize their business partnership negotiation further with buyers. For global suppliers, they could advertise and promote their company and products by join those B2B website as member.

Alibaba.com is a Chinese owned world's largest global e-commerce platform for small and medium sized businesses around the world. ThomasNet.com, is a leading online platform for B2B seller discovery and product sourcing, and is visited by more than 1.8 million buyers every month. IndiaMART.com is India's largest B2B marketplace with a 60 percent market share. TradeIndia.com is very similar to IndiaMart and is the second largest B2B marketplace for small and medium sized companies in India. EC21.com is one of the world's top 10 largest online B2B marketplaces. Like Alibaba, it is truly global and has one million buyers and over 600,000 enquiries are exchanged every month.

ADVERTISING MODEL

In web advertising model, a web site, provides content and services (e.g. email, instant messaging, blogs) mixed with advertising messages in the form of banner ads. The website can be either the owner or just a distributor of the content/service. The advertising model works best in cases where the site receives a large number of visitors or when the visitors belong to a specific class of customers.

There can be many variations of advertising model. Some sites use Intromercials, which are animated full-screen ads placed at the entry of a site before a user reaches the intended content. Some use Ultramercials, which are interactive online ads that require the user to respond intermittently in order to close the ad and reach the intended content. Content-Targeted Advertising e.g. Google, delivers relevant ads to the customers based on the keywords searched.

COLLABORATION PLATFORM PROVIDER

In this business model, a company enables companies to collaborate with each other through its website. For example, a collaboration platform provider may provide facilities for companies who wish to come together in order to tender for a complex project in a particular market sector (such as aerospace).

E-PROCUREMENT

In this model, a company provides online procurement services. These sites generally advertise current procurement opportunities and provide various facilities for companies interested in tendering. These facilities may include the facility to fill relevant forms and track the progress of a tender. Using electronic tendering, suppliers can enjoy more tendering opportunities, lowered cost of tender submission and collaborative tendering with other companies. Companies offering tenders enjoy a substantial reduction in costs. E-procurement providers earn revenue by charging a fee for their services.

Internet (or Digital) Content Revenue Models

This refers to the written material, which is designed to share with others on the internet. As there is a global increase in competition, the originator of internet content must have skills, creative style and the ability to use relevant and concise text in order to make the content more attractive in order to ensure that it is engine optimized. Some typical examples of digital content are ebooks, blogs, podcasts, video, whitepapers, webinars, and info graphic[8].

TYPES OF DIGITAL CONTENT REVENUE MODELS

If you are a website owner or developer, the digital content revenue model you choose will play an integral role in determining the amount of revenue the website can generate. According to KPMG, an American audit, tax and advisory services firm, a second digital revolution began with the advent of social media websites such as MySpace and Facebook. This has resulted in the need for you to rethink the revenue model you choose if you operate an online business (Lewis, 2015).

The key to a business model is its revenue stream. In the converged market of content communication, the main sources of revenue are subscriptions and advertising, while strategic partnerships can help to cut costs (itunews.itu.int, 2015).

Subscription-Based Models

The delivery of digital content via the Internet has changed the nature of subscription-based models. Content providers now commonly offer both a free (basic) and a paid (premium) service. The free product is offered in an attempt to attract subscribers. For example, the Wall Street Journal provides free access to the video section of its website and non-subscribers are given a quota of free articles each month. Traditional news articles remain subject to a paywall. Video-on-demand (VoD) providers generally charge users a subscription fee for unlimited access to a content library. For example, Netflix subscribers pay a small monthly fee for unlimited access to its entire video library. Netflix's VoD service can be accessed at any time from multiple devices. YouTube, the popular online video sharing site, originated as a free service but recently announced the introduction of additional subscription channels where users can pay for access to niche or premium programming. In the United Kingdom, BSkyB has launched a new "pay as you go" model on its Internet television service. End users pay a fee for 24-hour online access to BSkyB's premium sports content.

Online Advertising Models

In online advertising models, content is given away to users at no cost, or at only a minimal cost, in an effort to generate web traffic. This traffic is then on-sold to advertisers for a profit. An estimated US$ 99 billion was spent in 2012 on Internet advertising. The most common online advertising models are based on "cost per click" and

---

[8] A visual representation of information or data, e.g. as a chart or diagram.

"pay per view". In the cost per click model, the advertiser pays each time a user clicks on a listing and is redirected to the relevant website. The pay per view model is similar, except that the advertiser pays for each click regardless of whether the user makes it to the target site or not. Another variant, the "cost per action" model, is performance based, with the advertiser paying only when a purchase is made. Google and Facebook are successful examples of the online advertising model.

Hybrid Models

The hybrid model combines subscription with online advertising. A good example of this can be found in the online music streaming industry where several large players, such as Pandora and Spotify, have emerged to rival the broadcast radio industry. Rather than apply the traditional radio business model, which relied almost exclusively on advertising revenues, online music streaming providers have tended to offer a basic service that is funded by advertisements along with the ability to subscribe to a premium service. Related business models are also being used in other online content delivery industries. For example, the New York Times has announced that its paywall has been removed for the video section of its website. Users will still be required to pay a subscription to access unlimited new articles but access to its video library will be free for users. The video library will now be funded by online advertisements.

Other Revenue Streams

Other revenue streams include product placement, expanding activities along the Internet value chain, and benefiting from business synergies. Product placement is the purposeful incorporation of commercial content into non-commercial settings in order to promote a particular product or brand. It is estimated that nearly 90% of television viewers attempt to avoid watching advertisements (Plunkett, 2010). Because product placements are directly integrated into the programs, they are more difficult to avoid, making them an attractive option for advertisers. Netflix decided to use product placements to subsidize the cost of producing House of Cards. Exclusive content deals, where premium content is made available only over certain platforms in exchange for beneficial delivery terms, is another revenue stream. Most set-top-boxes, such as AppleTV and Roku, offer their subscribers access to certain programs based on the content deals that are in place with content producers. These producers have direct access to viewers at discounted rates and set-top operators are able to increase their offering to attract subscribers. The ability to collaborate with other businesses can lead to potential cost savings and new sources of revenue.

Other Business Models

ONLINE TRADING MODEL

This business model includes the trading of financial instruments such as bonds and stocks via the internet. In this model, a business can provide various services, including financial services, to its customers. One example is www.etrade.com. E-TRADE is a financial services company based in New York City, United States that provides online services to investors. The services include buying and selling of stocks via electronic trading platforms, banking and lending products such as checking and savings accounts, money market accounts, certificates of deposit, and credit cards.

TRUST BROKERAGE MODEL

In this business model, a company provides services related to security or trust. For example, copyright is a major issue for the internet. A trust brokerage company might offer the facility for companies and individuals to register their work with them and then be able to testify to the date that the work was registered. Another trust brokerage firm may provide trust services in which it provides certification to the customers that a particular website run by a company is in fact associated with that company.

VIRTUAL COMMUNITY MODEL

A virtual community (also called a web community or an online community) fulfills social needs of people by offering them a way to connect with each other and discuss common issues and interests. Virtual communities exist in various forms, including Usenet newsgroups, chat rooms, and Web sites. Virtual communities can also help companies, their customers, and their suppliers in planning, collaborating, transacting business, and interacting in ways that benefit all of them. The success of the virtual community model is based on user loyalty. Revenue can be generated through sale of supplementary products and services, voluntary contributions, or through contextual advertising and subscriptions for premium services.

SOCIAL NETWORK PROVIDER

A social network is comprised of units (individuals or organizations). The goal of a social network is to expand relationships by finding new friends. Different types of social networks are created to serve different purposes of their members. Most of the social networks use advertising to generate revenue. Following are five common types of social networks.

General Communities: General communities are online communities of people with particular hobbies, interests, etc. These people interact with each other and share their experiences. MySpace and Facebook are two examples of general communities. General communities use advertising as their main source of revenue.

Practice Networks: These are online places where people relating to a particular area of practice (e.g. programming, photography etc.) can interact, have discussions, seek help, and obtain information and knowledge from others. For example, JustPlainFolks (justplainfolks.org) is a practice network where musicians can share ideas and experiences. The major sources of revenue of practice networks are member donation and advertising.

Interest-based Social Networks: Members of these networks share a common subject of interest e.g. travel, sports and cars. Members can exchange their ideas about the subject of interest. SailingAnarchy (sailinganarchy.com) is an interest-based social network in which members exchange ideas about sailing. The main source of revenue for this type of network is advertising and sponsorships.

Affinity Communities: These communities consist of people who share similar self- and group identification such as gender, political beliefs and religion. For example, iVillage (www.ivillage.ca) is an affinity community, which focuses on women and attracts them by topics about beauty, diet, fitness and babies. The main sources of revenue of affinity communities are advertising, sponsorship, and sales of products.

Sponsored Communities: These communities are created by organizations or governments to achieve varying objectives such as expanding brand influence by providing information, improving offline sales etc. These communities are not meant to make any profit. Westchestergov.com and Tide.com are good examples.

THIRD-PARTY MARKETPLACES

A third party marketplace is a website that offers access, through a common catalog interface, to products of a number of related companies, for example wholesalers of office supplies. The administrating company of the marketplace performs the marketing, sales, payment, and delivery of the products of participating companies. In contrast to an E-mall, the product or service providers within the marketplace are more closely integrated. Some examples of third-party marketplace include Amazon marketplace and Wal-Mart marketplace.

INFORMATION BROKERAGE

In this business model, a company offer access to information. The information provided is usually business information e.g. results of surveys of customer satisfaction for a product, financial analysis etc. The main source of revenue for an information broker is subscription fee or per-transaction charges.

CATALOG MERCHANT MODEL

In this model, a mail-order business provides customers a web-based catalog. Customers can order products either by mail, telephone, or via web.

BIT VENDOR MODEL

In this model, a business provides only deals with pure digital products and services. The business conducts both sales and distribution over the web. One example is Apple iTunes Music Store.

BUY/SELL FULFILLMENT PROVIDER

In this model, a company takes customer orders to buy or sell a product or service, including terms like price and delivery. carsdirect.com and respond.com are two examples of such businesses.

TRANSACTION BROKER

In this model, a company provides third-party services for buyers and sellers to settle a transaction. Paypal.com and escrow.com are two examples of this model.

E-LEARNING

In this model, a company or organization offers educational courses via the web. While the quality and features provided can vary, the most basic e-learning site provides its users the ability to download learning material. Sophisticated e-learning sites provide facilities to read individual lessons, attempt online quizzes, and experience simulations relevant to the subject being taught. Examples of E-learning companies include skillsoft.com, serebra.com, and kesdee.com.

E-LOAN

In this model, a business provides services to its customers to find, research, and apply for loans online. The services may include free quotes, calculators and tutorials for borrowers. Examples of E-loan providers include eloan.com and mortgagebot.com.

E-RECRUITING

In this model, a business provides services related to both employee and employers. Services for employees my include tutorial on resume preparation, help for cover letter preparation, free job searching etc. Services for employers can include facility to search suitable employees form a global pool of applicants. Examples of E-recruiting companies include guru.com, dice.com, and monster.com.

E-NEWS

In this model, a company uses Internet to offer round-the-clock access to latest news of many types e.g. entertainment, financial, sports etc. Examples of E-news providers include cnn.com, espn.com, and boston.com.

E-TRAVEL

In this model, a company provides travel-related services online to its customers. Customers enjoy the liberty to bypass a travel agent and access discounts and low fares available online. Examples of e-travel companies include expedia.com, travelocity.com, and cheaptickets.com.

E-ENTERTAINMENT

In this model, a company provides online access to various types of entertainment to its customers. Examples of e-entertainment companies include IMDB.com, universalmusic.com, and mp3.com. Internet is revolutionizing the entertainment industry by charging the ways we get entertainment. Internet offers various types of contents related to entertainment industry such as News, celebrity, interviews, music, movies, celebrity photos and gossips, social networking etc. Entertainment on Internet can be broadly classified into three general categories. Broadcast non-interactive entertainment includes real-time streaming audio, music, video and movies. Broadcast interactive entertainment is the emerging type of Internet entertainment. Typical examples include movies shown in which the viewer can change the plot lines in real-time. Non-broadcast, non-interactive entertainment includes information retrieval from the web. Information can include published text, multimedia content etc. Non – broadcast interactive entertainment includes interacting with interactive web2.0 websites such as gaming sites, adult entertainment sites, infotainment sites, and online shopping sites.

E- AUTOMOTIVE

In this model, a company provides a website providing automotive-related information. Customers can use these websites to access automobile information that can enable them to make an informed buying decision. Automotive dealers can use these websites to display vehicles. Examples of e-automotive companies include cars.com and autobytel.com.

E-ENERGY

In this model, a company provides energy exchange services. Energy providers can use these sites to services to buy on auction trade excess energy commodities, and sell materials. Examples of e-energy companies include worldenergy.com and houstonstreet.com.

E-BRAINPOWER

In this model, a company offers online services e.g. to sell unused patents and trademarks, hiring of outside contractors, consulting offer, assistance to project completion etc. Examples of e-brainpower companies include hellobrain.com and yet2.com.

E-ART DEALERS

In this model, a company provides virtual galleries that can be used by artists and filmmakers to display their work and attract new customers. Examples of e-art dealer companies include Art.com, artfulhome.com, atom.com, and art.net.

E-ADVERTISING

In this model, a company sells advertising on its website and newsletters. The advertising can come in the shape of images and banners. Forester Network (foresternetwork.com) is one example of this model. Images can be included on websites or in email newsletters. These banners come in a variety of shapes and sizes. These banners can be placed in different locations on a website.

SPONSORED DOWNLOADS

In this model, a company (such as Mequoda.com), offers free downloadable products in exchange for a user's information (such as user's email address). In the B2B segment, sponsored downloads are common. Download forms are typically longer, requesting lots of information from the user, as the user data is then provided back to the sponsor.

SPONSORED WEBINARS

In this model, a company (such as ascd.org) provides its users free access to its webinars. The company obtains sponsorship for these webinars.

SPONSORED DIRECTORIES

In this model, a company (such as metroparent.com) provides users access to various directories such as schools, party vendors, doctors, dining and other family services. Based on the popularity of these directories, companies seek sponsorship.

NATIVE ADVERTISING

Native advertising involves paid ads that align with a webpage's content. They can be articles, snippets or simply links that are assimilated into the design and flow with the platform functionality, so viewers feel the ads belong on the page. The best native advertising examples offer fantastic editorial content that does not turn off the reader as being too promotional or non-contextual. One example is Electronic House Magazine (www.electronichouse.com).

CUSTOM E-NEWSLETTERS

In this model, a company include custom sponsored newsletters and promotions as part of their advertising packages. Take example of Mr. Food (www.mrfood.com). Mr. Food specialized in practical food preparation techniques, using readily available ingredients. Subscribers of Mr. Food might get a special newsletter from one of their gourmet food sponsors. These promotions help pay for copious amounts of free content Mr. Food create for their subscribers daily.

FREEMIUM MODEL

In this model, a company sells a basic free product to as many customers as possible, but keeps the premium features exclusively for paying customers. A large number of SaaS products use this model. For instance, Dropbox offers 2GB of free cloud data storage. However, if one wants more space, one has to pay up. Other examples include Adobe Flash, Evernote, Google Docs/Drive, LinkedIn, Prezi, Slideshare, Skype, WordPress, and many mobile games like Farmville, Angry Birds etc.

SELLING DATA

If you are not paying for the product, you are not the customer; you are the product being sold. High quality, exclusive data is very valuable in the digital age. Many companies specialize in lead generation of potential customers and sell them to third parties. These companies (such as Google, Twitter and Facebook) do not charge their customers for their services. Rather they aggregate high quantities of customer data and use this data to sell contextual advertisements to companies. The data gathered can be of different types such as user data (e.g. LinkedIn), search Data (e.g. Google), benchmarking services (e.g. Comscore), and market research (e.g. MarketsandMarkets)

SPONSORSHIP/DONATIONS

Many services are sponsored by government organizations and major funds if it directly helps them or the world at large, for example, Khan Academy is funded by the Gates Foundation and Google. Then there is the Wikipedia model where the users are asked for a willing donation of small to large amounts to help support the initiative. Many browser extensions and WordPress plugins etc. also follow this model.

BUILD TO SELL

In this model, a business builds traction over time and never worried about how it is going to make money. The intention is to generate a good customer base and finally find a good large buyer of their business. Some good examples are Instagram and WhatsApp. Both were acquired by Facebook for $ 1 billion and $ 19 billion respectively.

MOBILE AND GAMING REVENUE MODELS

The Table 4-3 shows different types of such models and their examples.

TABLE 4-3: DIFFERENT TYPES OF MODELS ADOPTED BY WEBSITES

Paid App Downloads	WhatsApp
In-app purchases	Candy Crush Saga, Temple Run
In-app subscriptions	NY Times app
Advertising	Flurry
Transactions	Airtel Money
Freemium	Zynga
Subscription	World of Warcraft
Premium	XBox games
Downloadable Content	Call of Duty

Review Questions

1. Differentiate between a traditional and electronic business model.

2. List some benefits of E-commerce business models.

3. How competition in E-commerce is different from competition in traditional markets?

4. List some key components of E-commerce business model.

5. Describe a revenue model and a value proposition.

6. How E-commerce businesses can handle channel conflict?

7. List common revenue models in E-commerce.

8. Describe some revenue model issues in E-commerce.

9. List the fundamental categories of E-commerce business models.

10. Define some B2C business models.

11. Define some B2B business models.

Bibliography

AMIT, R., & ZOTT, C. (2001). VALUE CREATION IN E-BUSINESS. *Strategic Management Journal* , 493-520.

Brohan, M. (2012, June 1). June 2012 - A conflicted group - Internet Retailer. Retrieved July 5, 2015, from https://www.internetretailer.com/2012/06/01/top-500-manufacturers-need-address-channel-conflict

Chen, M. (2004). *The Value Chain of Electronic Commerce*. Retrieved August 12, 2009, from ba.yzu.edu.tw: www.ba.yzu.edu.tw/baworkshop/Dr.MinderChen/03ec_strategy.ppt

Cooper, R., & Michael, K. (2005). *The Structure and Components of E-mall Business Models*. Retrieved September 1, 2009, from University of Wollongong, Faculty of Informatics - Papers: http://ro.uow.edu.au/cgi/viewcontent.cgi?article=1377&context=infopapers

Gale Encyclopedia, o. E.-C. (2002). *Manufacturer Model - Understanding Business Models, Traditional Versus Electronic Business Models, Selling Direct*. Retrieved July 12, 2009, from ecommerce.hostip.info: http://ecommerce.hostip.info/pages/702/Manufacturer-Model.html

Gale Encyclopedia, o. E.-C. (2002). *Manufacturer Model*. Retrieved July 12, 2009, from encyclopedia.com: http://www.encyclopedia.com/doc/1G2-3405300299.html

Horsti, A., Tuunainen, V., & Tolonen, J. (2005). Evaluation of Electronic Business Model Success: Survey among Leading Finnish Companies. *Hawaii International Conference on System Sciences, p. 189c, Proceedings of the 38th Annual Hawaii International Conference on System Sciences (HICSS'05) - Track 7.*

http://ecommerceandb2b.com. (2015). Dealing with Channel Conflict in B2B E-Commerce. Retrieved from http://ecommerceandb2b.com/channel-conflict-b2b-e-commerce/

Ince, D. (2003). *Developing Distributed and E-Commerce Applications.* Pearson Addison Wesley.

insitesoft.com. (2011, March 24). Avoiding B2B Ecommerce Channel Conflict. Retrieved July 4, 2015, from http://www.insitesoft.com/blog/avoiding-b2b-ecommerce-channel-conflict/

International B2B Markets and E-commerce. (n.d.). Retrieved July 24, 2015, from http://2012books.lardbucket.org/books/marketing-principles-v2.0/s07-05-international-b2b-markets-and-.html

InternetRetailing.net. (2011, July). Blurring the Lines. Retrieved July 5, 2015, from http://internetretailing.net/magazine/archive/july-2011/blurring-the-lines/

itunews.itu.int. (2015). Business models in a converged market - Online content delivery. Retrieved July 5, 2015, from https://itunews.itu.int/En/4338-Business-models-in-a-converged-market.note.aspx

Jalozie, I., Wen, J., & Huang, L. (2006). *A Framework for Selecting E-Commerce Business Models.* Retrieved Juy 12, 2009, from swdsi.org: http://www.swdsi.org/swdsi06/Proceedings06/Papers/EC03.pdf

Kampe, C. (2012, May 22). Ecommerce Transactions Dominate Mergers and Acquisitions | Multichannel Merchant. Retrieved July 24, 2015, from http://multichannelmerchant.com/marketing/catalog/ecommerce-transactions-dominate-mergers-and-acquisitions-22052012/

Kiran, V., Majumdar, M., & Kishore, K. (2012). Distribution Channels Conflict and Management. Journal Of Business Management & Social Sciences Research, 1(1), 48–57.

Lanigan, A. (2012, October 17). Six Models for Tackling Channel Conflict. Retrieved from http://www.fluid.com/strategy/six-models-for-tackling-channel-conflict

Lewis, J. (2015). Types of Digital Content Revenue Models. Retrieved July 4, 2015, from //www.ehow.com/info_12199184_types-digital-content-revenue-models.html

Love, J. (2013, June 3). All dressed up - Internet Retailer. Retrieved July 5, 2015, from https://www.internetretailer.com/2013/06/03/all-dressed

meridianecommerce.com. (2013). Four powerful strategies for dealing with channel conflict in ecommerce: Retrieved from http://www.meridianecommerce.com/_literature_120668/Four_Strategies_for_Dealing_with_Channel_Conflict_in_Ecommerce

Onion, B. (2013, November 18). Leverage Ecommerce to Eliminate Channel Conflict | Multichannel Merchant. Retrieved July 4, 2015, from http://multichannelmerchant.com/crosschannel/leverage-ecommerce-to-kill-channel-conflict-18112013/

Philips, C. (2014, March 25). Overcoming Channel Conflict in B2B E-Commerce. Retrieved from http://www.powerretail.com.au/multichannel/channel-conflict-via-b2b-ecommerce/

Pigneur, Y. (2000). *The E-Business Model Handbook*. Retrieved July 23, 2009, from Université De Lausanne: www.hec.unil.ch/yp/Pub/00-ebmh.pdf

Plunkett, J. (2010, August 24). TV advertising skipped by 86% of viewers. Retrieved July 24, 2015, from http://www.theguardian.com/media/2010/aug/24/tv-advertising

Porter, M. ,. (1998). *On Competition*. Harvard Business School Press.

Rayport, J. (1999, July). *The Truth about Internet Business Models", Strategy-business.com*. Retrieved July 12, 2009, from strategy-business.com: http://www.strategy-business.com/article/19334?gko=6518d

Schneider, G. (2010). *Electronic Commerce*. Course Technology.

Seybold, P., & Marshak, R. (1998). *Customers.com: How to Create a Profitable Business Strategy for the Internet and Beyond*. Crown Business.

Turban, E., Lee, J., & Chung, M. (2006). Electronic Commerce: A Managerial Perspective. Prentice Hall.

5

E-COMMERCE STRATEGY

After reading this chapter, reader should be able to:

- Describe the strategic planning process for an E-commerce venture.

Understand the difference between goals, strategies, and tactics.

- Understand how E-commerce impacts the strategic planning process.

- Understand how to formulate and justify an E-commerce venture.

- Describe strategy implementation process.

- Identify common E-commerce startup mistakes.

- Understand how to peform a SWOT analysis for an E-commerce startup.

- Understand the process to build an E-commerce site.

- Understand the importance of complying with web standards when building E-commerce site.

- Understand the difficulties in measuring and justifying E-commerce investments.

- Understand various sources of funding available for E-commerce startups.

- Understand how E-commerce projects are justified.

- Understand how to research compeitors of an E-commerce business.

- Understand how small and medium-sized E-commerce businesses can compete with lare E-commerce businesses.

- Understand the components of included pages of a good E-commerce web site.

Defining E-Commerce Strategy

According to (Turban, Lee, & Chung, Electronic Commerce: A Managerial Perspective, 2006), E-Commerce strategy is defined as the formulation and execution of a vision of how a new or existing company intends to do business electronically. Building E-commerce strategy is a complex process and businesses need to invest considerable amount of time and effort for this purpose. Building an E-commerce strategy involve following steps.

- Strategy Planning

- Strategy Initiation

- Strategy Formulation

- Strategy Implementation

- Strategy Planning

Planning strategy for E-commerce is different from planning strategy for a conventional business. This is because of unique impacts of Internet on the businesses. The overall impact is high and has many dimensions. Following are some of the impacts of Internet that needs to be considered in strategy planning for E-commerce.

REACH AND RICHNESS: Internet tools (such as collaborative filtering, personalized e-mail newsletters) can enable companies to reach millions of people with rich information (Evans & Wurster, 2000).

REDUCED BARRIERS TO ENTRY: Setting up an online business is relatively easy and inexpensive and requires small amounts of venture capital.

EXPANDED REACH: Online business has no geographical boundaries. However, online businesses need to discover profitable market niches in order to succeed.

OTHER IMPACTS: Internet also produces other impacts. For example, online business alliances can reduce bargaining power of suppliers. Customer service can be improved through mass customization, personalization, and CRM.

It is therefore very important to consider the impact of Internet in E-commerce strategy planning.

E-commerce: Goals, Strategies and Tactics

Every E-commerce business needs to establish a measurable and realistic target to achieve. The goals are to be set before a company decides its strategies. Some examples of goals a) include increase sales by 30 % in next 1 year b) achieve a 10% market share by next 1 year and c) improve profit margin by 10 % within next six months. Your goals need to support your business model. To be real, a goal needs to be measurable and have a realistic chance to be successful.

In simple words, strategy is an idea on how to achieve your business goals. For example if your business goal is to achieve an increase in sales by 20% in next 1 year than you would need to develop strategies to achieve this goal. One strategy might include adding new products to grow top line revenues. Another strategy could be to improve the overall conversion rate for all visitors to your website.

Tactics are individual plans and actions to support the business strategies. For example, to improve conversion rates you could employ few tactics such as redesigning your shopping cart to reduce abandonment and adding an alternative payment method to your checkout.

E-commerce Strategy: The Phases

Strategy Initiation

In this phase, company performs a self-analysis, competitor/environment scanning, possible contribution of E-commerce to the business, and other strategy initiation issues. Many E-commerce initiatives are prepared to exploit opportunities and mitigate environmental threats. Porter's framework of four competitive strategies (cost leadership, market differentiation, innovation differentiation, market focus) can be used to compare the market strategies of both virtual firms and click-and-mortar firms. Firms can choose different competitive strategies. Virtual firms incline to use differentiation strategies based on creative marketing and innovation, whereas click-and-mortar firms prefer strategies based on market focus. Firms adopt competitive strategies to gain competitive strength but these strategies are not necessarily the best ones to improve business performance. Therefore, firms competing in electronic markets should reassess their competitive strategies and reallocate their resources to maximize the return on their investment (Koo, Koh, & Nam, 2004).

A company needs to deal with various issues about its approach to and operation of its E-Commerce strategy. Following are some of the important issues.

FIRST MOVER OR FOLLOWER

Generally, the firms decide to be first mover when they see an opportunity to make a first and lasting impression on customers, to establish strong brand recognition, to lock in strategic partners, and to create switching costs for customers. However, first movers also face risk of high cost of developing E-Commerce initiatives, making early mistakes, that market can avoid, and the risk that the move will be too early before the market is ready. Some researchers also suggest that over the long run first movers are substantially less profitable than followers (Boulding & Christen, 2001) and that switching costs and network effects are not as substantial as claimed (Porter, 2001). Research on a variety of industries showed that the lowest-priced on-line sellers do not have the highest market share. Therefore, it is important to identify the sources of differentiation advantage that can result in higher market share (Oetzel, 2004).

In determining whether a first mover succeeds or fails, (Rangan & Ron, 2001) suggest the following factors:

SIZE OF THE OPPORTUNITY: Both the size of the company and size of the opportunity for the company must be big enough if the company wants to be a first-mover. If that is not the case, the first-mover company can create an opportunity for a late mover.

COMMODITY PRODUCTS: First-movers should provide product or service that are simple enough and hard to differentiate (e.g., books, airline tickets) otherwise late-movers can differentiate themselves by offering better products and services (e.g., clothes, restaurants).

CAPITALIZE ON THE FIRST-MOVER ADVANTAGE: If a first-mover company does not capitalize on its position, late-movers can take advantage by offering a better and more innovative product or service.

EXAMPLES OF FIRST MOVERS IN E-COMMERCE

Being first to the market with a new product or process is thought to covey substantial benefits to the first mover. They are the first to launch and learn from the new product, with the ability to improve upon it quickly. Second, they can get-big-fast and build up economies of scale that render the first mover as the low cost firm. Third, they can establish their brand name in the mind of the consumer and make their brand synonymous with the identity of the product. Amazon.com is thought of a being one example of a very prominent and effective first mover.

However, not all first movers survive and come to dominate the market. Google.com was not the first search engine, nor was Orbitz.com the first e-commerce travel site. Having a strong fast second strategy of learning from the mistakes of the first mover or seeing unfilled gaps in the market may make the imitator the eventual winner. For example, AOL saw a better way to provide Internet service and soon replaced first mover Prodigy as the leading firm in the market.

STRATEGIES OF FIRST MOVERS IN E-COMMERCE

First, some have pursued a launch-and-learn strategy, where a first version of a product is introduced into the market place and then quickly modified based on the feedback from users. Second, some have followed a get-big-fast strategy where they increase their volume of sales in the hope that the extra output will lower unit costs and lead to longer-term profitability. Third, some have followed an advertising strategy, where they repeatedly place their brand name in front of the consumer, hoping to create an overwhelming awareness advantage. Such strategies are not always successful because the dynamics of the market may be so rapid that imitators have an ability to overcome any first mover advantages (Deak, 2003).

HOW FIRST MOVERS SUSTAIN COMPETITIVE ADVANTAGE

First movers can sustain their competitive advantage using their tangible and intangible assets. Some of these assets creates a potentially more lasting advantage Tangible assets involve the capital resources, such as buildings, machinery and computers, that contribute to the operating efficiency of a business. These are important in creating a competitive advantage but can be duplicated by rivals. Intangible assets include the knowledge base, brand name, experience and service quality that provides a value added shopping experience to the customer. These are harder to duplicate and tend to convey a more lasting advantage when used properly.

Probability of success also depends on whether the new entrant can truly differentiate the product (Apple iPhone) or differentiate the price (e.g. Google Docs), or the distribution (Microsoft). That is why companies with network effects are tough to beat because the costs of switching are not low. As the new entrant is inherently inferior because it lacks users and so, it is painful for new users to abandon the old product. Another overarching theme to the fast follower theory is that first movers either get an advantage and blow it or they are too early and fail too big and lose the ability to iterate until much later in the tech evolution cycle. Timing is everything and you can be too early as well as too late. For those that had first mover and lost , they are typically ones that either lost focus and stopped following core values (e.g. Friendster.com) or thought good was good enough (Yahoo search and the Google). Many search engines evolved into the portal business because it monetized well. When thinking about first entrants' lead and followers' ability to take over, it is important to bear in mind that markets are not static. Followers, fast or slow, can catch up through superior execution (for example Dropbox and AirBnB) but they can also introduce innovations that completely change the name of the game. Apple's iPod definitively falls into that category. There were great MP3 players before the iPod but they lacked the design (iRiver) and difficult to use (Sony). Elegant design combined with superior usability changed that market forever (for example iPod). Even network effects cannot protect a first mover from an innovative follower. Take example of Friendster and Facebook. Facebook became an API and is about to become a Marketplace and a payment system. This is one strong reason why no company can afford to be complacent.

WHY FIRST MOVERS FAIL IN E-COMMERCE

According to (Bresser, 1998) and (Lieberman & Montgomery, 1988), the mechanisms that benefit the first-mover may be counterbalanced by various disadvantages. These disadvantages may erode the advantages gained by first mover. These disadvantages are, in effect, advantages enjoyed by late-movers. Some of these disadvantages are as follows:

EDUCATING CUSTOMERS THE WRONG WAY: When a company brings a new concept to market, it faces a unique challenge. It must educate its customers about the new product or service. This situation poses a significant risk since would-be customer response to product cannot always be predicted, especially when perceived switching costs are high. Additionally education can be a costly process, resulting in first movers having sales and marketing costs significantly in excess of later entrants.

HIGH COST INCURRENCE: First movers often make costly mistakes that enable later entrants to penetrate the market. The internet search market is a classic example of this error. Yahoo! and many of the other search players treated search technology as a loss leader – a service designed to attract users to their revenue bearing services. As a result, they invested little in developing their search technology. This strategy provided Google with the opportunity to differentiate itself.

FREE-RIDER EFFECTS: Late movers may be able to "free ride" on a pioneering firm's investments in a number of areas including R&D, buyer education, and infrastructure development. Imitation cost are lower than innovation cost in most industries. However, innovations enjoy an initial period of monopoly that is not available to imitator firms. Nevertheless the ability of follower firms to free ride reduces the magnitude and durability of the pioneer's profit, and hence its incentive to make early investments.

MARKET UNCERTAINTY: Late movers can gain an edge through resolution of market or technological uncertainty. Early entry is more attractive when firms can influence the way that uncertainty is resolved. Firm size may also matter-in many new product markets; uncertainty is resolved over time through the emergence of a "dominant design".

Incumbent Inertia: Vulnerability of the first-movers is often enhanced by problems of "incumbent inertia." Such inertia can have several root causes: (1) the firm may be locked-in to a specific set of fixed assets, (2) the firm may be reluctant to cannibalize its existing product lines, or (3) the firm may become organizationally inflexible. These factors inhibit the ability of the firm to respond to environmental change or competitive threats (Lieberman & Montgomery, 1988).

FOLLOWERS (LATE MOVERS) IN E-COMMERCE

A second mover is a firm that sees the success of the first mover and steps in quickly to duplicate and improve upon the strategy of the first mover. The imitator saves the cost and does not bear the risk of creating the idea and the product market. They also have the ability to see unfilled gaps in the market and tailor their version of the product to fill those gaps. First movers can become fat, dumb and lazy as they stumble in their good fortune and forget about improving the product that is currently so successful.

There exist several examples of successful followers in E-commerce. Google was not the first search engine to hit the internet. Yahoo predated Google considerably. There were AltaVista and AskJeeves in the late 90-s that did a fair job of digging out data from the not-so-obscure corners of the World Wide Web. It was in 1998, with the entry of Google on the scene that search, as we know it, changed forever (Dholakiya, 2014). Facebook was not the first social network; it trumped both Friendster and MySpace. However, Mark Zuckerburg had the insight to package the tools and services in such a way that everyone fell in love with his service and there was widespread adoption of the platform (May, 2012). LinkedIn has displaced or is displacing Monster, CareerBuilder, HotJobs, and Plaxo (plaxo.com). Plaxo is an online address book that launched in 2002. YouTube was launched significantly after Google Video. After Google acquired YouTube, Google Video was shut down and replaced by Google Videos on August 20, 2012. The remaining Google Videos content was automatically moved to YouTube. Yelp (yelp.com) (a provider of user reviews and recommendations of top restaurants, shopping, nightlife, entertainment, and services) surpassed CitySearch (citysearch.com) in traffic and reviews. The Apple iPod swept all existing competitors away with superior storage, battery life and software. Apple iPad made breakthrough sales among many competitors out for over a decade with mediocre sales. Different Apple products, such as GarageBand (everything you need to play, record, and share professional-sounding music on your iPad, iPhone, or iPod touch), iMovie (lets you organize all your clips, turn them into your favorite films or trailers, and then premiere them on iMovie Theater), and iTunes (world's best digital music jukebox) were not the first product to launch in terms of solving their specific problem. Dropbox succeeded in a crowded market of online storage services. Couchsurfing (couchsurfing.com) is a hospitality exchange and social networking website. VRBO (vrbo.com) is an online vacation rental marketplace. Both of them have been around for a long time. AirBnB beat both of them thanks to a superior design and emphasizing a few critical features (such as market place, customer support, ratings, refunds, fraud, escrow) to alleviate all of users' worries. Spotify (spotify.com) is a digital music service that gives you access to millions of songs. In presence of dozens of music subscription services, Spotify was the one that won in Europe. The reason was the free service of Spotify to get users through the door and then up sell. Wesabe was a personal finance management website that analyzes a user's financial data to provide appropriate advice on how to save money. Mint (mint.com) is a free web-based personal financial management service for the US and Canada. Mint won the competition against Wesabe because of superior user experience and product management. Book Stacks Unlimited launched two years before Amazon. RIM (the maker of blackberry) was not the first to the smartphone space, and beat Palm in it. In the next phase, Apple beat RIM. PlentyOfFish (pof.com) is an online dating service, popular primarily in Canada, the United Kingdom, Ireland, Australia, Brazil, and the United States. It was a later entrant in the online dating business who bet the existing companies like Match (match.com), Lavalife (lavalife.com), and eHarmony (eharmony.com). Microsoft is another great example of a successful fast follower (Word vs. WordPerfect, Excel vs. Lotus, IE vs. Netscape, Windows vs. Mac OS)

SOLUTIONARY VS. REVOLUTIONARY

Some of the greatest advancements across industries were not huge, singular achievements but rather incremental improvements. Even bold ideas such as online music stores, departures in architecture, and new genres of music were the result of new ideas refined over time. The iPod was not the first MP3 player. Google was not the first search engine. And the list goes on. While logic should encourage us to improve what is around us, we still tend to think of innovation as creating something entirely new. Creative minds have the tendency to lose interest after the first implementation of a new idea. Marginal improvements are, frankly, less interesting for the cutting-edge creative. Nonetheless, incremental improvements often make up the difference between success and failure. Especially productive creative teams are able to find excitement in solving problems both big and small, and in varying stages. These accrued solutions make up the distance between a new idea being created, and actually being adopted. Leaders that focus on incremental progress – being "solutionary" rather than revolutionary – are the ones that truly push ideas to full fruition. Such behavior takes a tremendous amount of discipline. However, with conviction and clearly defined goals, creative energy can be channeled to refine a good idea enough to make a great impact (Belsky, 2009).

DOMESTIC OR GLOBAL?

The decision largely depends on the organizational capabilities. A company may also decide to go global selectively (i.e. in a few countries with a few products).

SEPARATE ONLINE FIRM OR INTEGRATED COMPANY?

(Venkatraman, 2000) identifies that a company should consider setting up a separate firm for its online operations when:

- Anticipated volume online business is large

- Online business requires a new business model

- An independent subsidiary can be created with no dependence on current operations and legacy systems

- Online company is to be given the independence to form new alliances, attract new talent, and raise additional funding.

Barnes and Noble, Nordstrom Shoes, and ASH Bank in New Zealand are a few examples of companies that have established separate companies or subsidiaries to form online operations.

A separate online company offer several advantages and disadvantages. Internal conflicts are reduced or eliminated; management has more freedom in pricing, advertising, and other decisions; new brands can be created quickly; new and efficient information systems can be built; external funding can be obtained if the market likes the e-business idea and buys the stock. The disadvantages include increased cost and risk of creating an independent division; and losing expertise vital to the existing company to the new online firm. Two options described above are not the only options available to a firm. A company may also form a strategic partnership (e.g., Rite Aid bought an equity stake in Drugstore.com). A company can also form joint ventures (e.g., kbkids.com was a joint venture of KB Toys BrainPlay.com).

SINGLE BRAND OR SEPARATE ONLINE BRAND?

In general, companies with strong, mature, international brands that also matches with the intent of the online business use the same brand online to capitalize on its strength and reputation. For example, BMW went online with their established brand. If that is not the case, firms may decide to create a new brand. For example, Axon Computertime (axon.co.nz) is a New Zealand based IT solutions and Services Company regarded as high-value and low-cost computer sales and configuration provider. To retain the current reputation and to take advantage of on the opportunity to deliver premium quality service in the market of computer services, Axon created Quality Direct. Quality Direct was a new division and a new brand within the parent company.

Strategy Formulation

In this phase, company evaluates specific E-Commerce initiatives and conduct cost-benefit and risk analyses for these initiatives. Company then prepares a list of approved E-commerce initiatives and decide which initiatives to implement and in what order.

Companies often make one of the three mistakes in selecting E-Commerce initiatives (Tjan, 2001). First the company may try to fund many projects indiscriminately in a hope that majority of the projects will succeed. However, company ignores the fact that organizational financial resources, time, and attention cannot support multiple initiatives. Second, a company may focus on a single high-risk and high-value E-commerce initiative. This approach is very risky. If you lose it you lose all. Third, a company may follow the new hot trend that usually ends up spending too much capital on pursuing too few opportunities. (Turban, McLean, & Wetherbe, Information Technology for Management: Transforming Organizations in the Digital Economy, 2004) identifies an additional mistake. A company may be so scared of possible loss of practicing E-Commerce or may be in a rush to practice E-Commerce because of large possible monetary benefit. In that case companies frequently jump into inappropriate E-commerce ventures.

A variety of issues arises in strategy formulation. Online and offline business integration, and pricing are some of the important issues.

For a click-and-mortar firm conflicts that arise due to the allocation of resources between off-line and online activities may cause problems when the off-line business needs to handle the logistics of the online business or when prices need to be determined (Pottruck & Pearce, 2000). Clear support by top management for both the off-line and online operations and a clear strategy of "what and how" each unit will operate are essential to address this conflict.

Using traditional pricing methods (e.g. cost-plus model) for pricing products and services of an online business has many unique aspects. First, the price comparison is easier because Internet facilitates price comparison e.g. through search engines, price comparison sites, and intelligent agents. Second, buyers can sometimes set the prices e.g. priceline.com and auction sites such as onsale.com. Third, online and off-line goods are priced differently. Pricing strategy may be especially difficult for a click-and-mortar company. Setting prices lower than those offered by the off-line business may lead to internal conflict, whereas setting prices at the same level will hurt competitiveness. Fourth, differentiated pricing can be a pricing strategy. Differentiated pricing is based on the based on the fact that some buyers are willing to pay more to receive some additional advantage. Versioning is a form of differentiated pricing which refers to selling the same good but with different selection and delivery characteristics (Shapiro & Varian, 1999). Versioning is especially effective in selling digital information goods. For example, time-critical information such as stock market prices can be sold at a higher price if delivered immediately. To remain competitive and profitable, sellers need to adopt smarter pricing strategies focused at using the Internet to optimize prices.

Strategy Implementation

In this phase, execution of the strategic plan takes place. Detailed, short-term plans are developed for carrying out the E-commerce projects agreed on in strategy formulation phase. Management evaluates options, establish specific milestones, allocate resources, and manage the projects.

The typical first step in strategy implementation is to establish a Web team, which initiates the execution of the strategic plan. This web team requires both people knowledgeable in required technology and people knowledgeable in business information/data structuring and delivery. The role and responsibilities of each individual needs to be clearly defined As the E-commerce implementation progresses, changes may be introduced in the organization therefore developing an effective change management program, including the possibility of business process reengineering, is important. According to (Plant, 2000), every Web team, also requires a project champion. The project champion is the person who ensures that the project gets the time, attention, and resources required, as well as defends the project from detractors at all times. The project champion might be the Web team leader or a more senior executive.

E-commerce project implementation often requires considerable investment of time and effort. Therefore, starting with one or few pilot E-commerce projects before initiating the big project can help company uncover problems early and make changes in the plan accordingly.

Effective allocation of resources especially the infrastructure resources that are shared by many applications (e.g. databases and the intranet) are critical for E-commerce project success. A variety of tools (e.g. Project management tools such as Microsoft Project) can be used to assist with determining project tasks, milestones, and resource requirements.

There are many strategy implementation issues, depending on the circumstances. Two important issues are application development and partner's strategy.

Implementing E-commerce requires construction of website/E-commerce applications and their integration with the existing corporate information systems (e.g., front end for order taking, back end for order processing). At this point, the company is faced with many decisions to make. Some of these decisions include the following:

- Should the website be developed internally, externally, or combination of both?

- Should the required software be built internally or use commercially available software?

- If we select commercially available software, should it be purchased from the vendor or rented from an Application Service Provider (ASP)?

- Will the website be hosted by the company or an external host (e.g. ISP)?

- If website is hosted on an external host, who will monitor and maintain the website and associated be responsible for monitoring and maintaining the website and associated information and system?

Each option has its strengths and weaknesses. Right decisions will depend on factors such as the strategic nature of the application, the skills of the company's technology workers, and the necessity to move fast or not.

Many E-Commerce applications involve many business partners. Each partners has different organizational culture, E-Commerce strategies, and profit motives. When selecting an E-commerce partner, company needs to choose a partner whose strategy aligns with or complements the company's E-commerce strategy. One popular E-Commerce partner strategy is outsourcing, in which an external vendor is used to provide all or part of the products and services that could be provided internally. Application Service Provider is an outsourcer that rents access to software applications. Using ASP, companies can save significant costs to purchase, operate, and maintain expensive applications such as an ERP system. However, companies should perform a realistic evaluation of the potential risks and rewards before selecting any option.

APPLICATION SERVICE PROVIDER (ASP)

An application service provider (ASP) delivers and manages software applications and computer services from a remote data center to multiple users. ASPs typically access these applications and services over the Internet, through a virtual private network (VPN), or through dedicated lease lines. A wide range of both applications and communications and infrastructure capabilities are available from ASPs. Among the most commonly used are enterprise applications, including enterprise resource planning (ERP), customer relationship management (CRM), supply management, human resources, and financial management. Software companies such as Microsoft Corporation are developing ASP versions of their productivity applications. When it comes to information technology (IT) and network infrastructure, ASPs can deliver network services, complex mission-critical hosting, software and hardware provisioning, infrastructure integration and support services, business continuity services, network management and administration services, and managed VPNs. ASPs also deliver network-based access to processing power and remote data storage facilities. ASPs may or may not use multi-tenancy in the deployment of software to clients; some ASPs offer an instance or license to each customer (for example using Virtualization), some deploy in a single instance multi-tenant access mode, now more frequently referred to as "SaaS".

ASPs have evolved from simply hosting Web services to building and managing e-commerce and platforms. They handle e-commerce issues such as security, registration, and payments, and provide Internet-based technologies. ASPs can provide communications platforms for messaging, voicemail, IP fax, and hosted collaboration platforms, as well as portals that offer such services as free Web e-mail, contact management, and calendaring. By hosting these services for other businesses, ASPs enable smaller businesses to benefit from high-priced software packages and systems without having to purchase them. Larger companies tend to use ASPs for outsourcing, while smaller businesses with low budgets use them to gain access to high-end enterprise computing that would be too expensive to purchase. Whatever the size of the client company, using ASPs allows businesses to focus their resources on their core competencies rather than on their information systems (IS) and information technology.

ASPs offer several benefits including quick launch for new e-commerce, supply chain, and CRM applications, less spending on buying, maintaining, and upgrading software and hardware to run basic applications, and no need of devoted IT staff and other resources to keeping up with rapid technological change. ASPs provide seamless and inexpensive upgrades; provide high levels of availability, security, backup, disaster recovery, and shadowing; provide easy up scaling and downscaling as business volumes change; and typically operate on one-to-three-year contracts with service level agreements (SLAs) that provide predictable costs to client companies. Companies can amortize their payments to ASP so that they do not have to make large capital expenditures on software and hardware.

Of course, there are tradeoffs when using an ASP. Among the factors to consider are the loss of hands-on control, the lack of a software license, and a contractual commitment lasting from one to three years. Successful ASPs must deliver on application reliability and availability. Some of the important factors considered when evaluating an ASP are guaranteed reliability and availability, faster implementation, and ability to avoid IS staffing problems. Companies are increasing their spending on ASP. Application hosting is growing due to growth of the Internet, access to larger amounts of communications bandwidth; and a widely embraced user interface in the form of Web browsers.

The ASP Industry Consortium includes more than 700 companies and includes ASP companies, software and hardware companies, network service providers, ISPs, and others. In addition to calling themselves ASPs, some also were known as managed (or management) service providers (MSPs), network service providers (NSPs), total service providers (TSPs), and software rental companies. A full-service provider (FSP) is an ASP that offers a wide range of Web-based information technology services, such as planning and creating a Web presence, software applications, and Web hosting and maintenance. Business service providers (BSPs) is a business service provider (BSP) is a company that rents third-party software application packages to their customers. A BSP is similar to an application service provider (ASP) in that it provides a cost-effective way to procure applications via networks. A management service provider (MSP) is a management service provider (MSP) is a company that manages information technology services for other companies.

Common E-commerce Startup Mistakes

According to a recent UPS study, 70% consumers want to shop their favorite retailer digitally. As a result, the category is seeing huge gains and growth annually. While handful online ventures manage to be successful over the long-term, 75% of these startups fail (Verleur, 2015). Every company starting its online operations is bound to make few mistakes. Following are some of the most common mistakes made by newly starting online businesses. It is important that businesses should learn from these mistakes and make sure not to repeat the same mistakes.

LESS EMPHASIS ON MARKETING

A brick-and-mortar business can get free traffic just by setting up a physical presence. However, the same is not true for an online business. A successful online business requires marketing and optimization of corporate website as much as possible.

OVERDUE EMPHASIS ON DESIGN REVISIONS

For a brick-and-mortar business, getting the store design right the first time is very important because it is too expensive to redo your store design once the business is operational. However, for online businesses the same is not true. Online businesses should try to build a good and acceptable website first and commit themselves to optimization after the website is launched.

MISUNDERSTANDING PRODUCT PRESENTATION

E-commerce involves face-less transactions where customers do not have the opportunity to touch or feel the products. Therefore, product presentation is extremely important for an online business. An online business should analyze the top performing websites in its niche, and pay close attention to how they describe and picture their products. Businesses should add utilize technology features (e.g. zoom-in and rotating photos) and product descriptions where needed.

NOT KNOWING WHERE TO FIND CUSTOMER ONLINE

Not every channel is going to make sense for every business. For obvious reasons, selling industrial machines on Etsy, a peer-to-peer e-commerce website focused on handmade or vintage items and supplies, as well as unique factory-manufactured item probably will not be very successful. However, if you make mittens and hats, Etsy could be a great place to make some extra cash. It is critical to do your homework beforehand and make sure you're selling something people will buy. To start, businesses should look for products similar to their products sold online. Businesses should look for marketplaces their competitors use and browse the rest of the marketplace's offerings to get a sense for its buyers and those buyers' needs and whether they could see their customers among them.

FAILURE TO GET COMMITTED CUSTOMERS

The customers of an online business can quickly go away if not satisfied. Companies should pay particular attention to remove any obstacles that may get in customers' way of online buying process. Companies should also make customer overall experience more valuable.

IMBALANCE IN USE OF PRINT AND ONLINE MEDIA

Businesses using online marketing campaign should be careful and must keep a careful balance in using print and online media. Too much reliance on any one option is not good.

IGNORING ONLINE TRUST ISSUES

Since E-Commerce is inherently a low-trust environment, an online business needs to make extra efforts to convince customers' make online purchase. This can include assuring customers of their information security and providing company policies about privacy and customer data handling.

OVER FOCUS ON HOME PAGE OF THE WEBSITE

Home page is an important part of corporate website but it is not everything. A company should not put too much emphasis on home page while ignoring other parts of the website.

OVER AMBITIOUS USE OF MARKETING TOOLS

An online business should avoid going after every latest and greatest internet-marketing tactic or E-Commerce feature. There can be many exciting new tools and marketing tactics available but companies should focus on those methods first that are sure to provide results (e.g. search engine marketing, email marketing, and website usability). Companies should perform a cost-benefit analysis of every opportunity considering the opportunity cost.

IGNORING WEB ANALYTICS

A successful ecommerce site creates a personal experience for each shopper. Fortunately, every site touch point creates the opportunity to discover more about buyers, thus providing the capability to offer personalized recommendations. Suggesting products or areas of interests makes the experience more enjoyable, while also building brand loyalty by showing the end-user that you understand them. The only way to deliver on personalization, however, is through comprehensive and exhaustive data capture. Businesses should invest in collecting site data and third-party data (e.g., historical behavior, demographics, etc.) to get a full-view of their shoppers. The data allows the company to create a bond that is often hard to accomplish with consumers online. Art startup, Artsy, does this quite well. An online business can reap significant benefits by not just monitoring but acting upon web analytics data. It is crucial that online businesses understand and interpret web analytics and be able to provide improved and enhanced products and services to their customers. Companies keep on wasting money on marketing, shuffling between agencies, hiring and firing them when they do not deliver results, neglecting the fact that a simple answer might be residing in the data. The first thing that you should do, after your website is setup is to get Google Analytics integrated properly.

FAILING TO INTEGRATE SALES CHANNELS

This mistake is especially common for click-and-mortar retailers. Failure to integrate off-line and online channels can cause channel conflict and price discrepancies. This in turn can cause brand confusion for customers.

NOT MAXIMIZING SOCIAL MEDIA

For E-commerce businesses, it is also important to master platforms like Facebook, Twitter and Google. They should make their brand accessible and pull their customer into their company's story with sneak peeks of new offerings and behind-the-scenes photos. These platforms can also help businesses create a dialogue with customers in order to know and serve customers better and survey them on which direction to take company's business.

Not Knowing Your Niche

One of the important causes of failure for online retailers could be trying to sell too many different kinds of products to too many different types of customers. Businesses need to provide a reason for people to buy just from them. Businesses also need to offer a one-of-a-kind product or serving needs that go unmet from other marketplaces can give an edge. For example, if your business was to sell pet supplies online, do not sell general supplies but focus on a category not seen anywhere else, like special toys or safety tools. Businesses should be as specific as possible, to cater to enthusiasts and hobbyists with specific product knowledge and expertise. If the products provided add value to the lives of the consumer, they will take every step to remain loyal to you.

TOO FAST TOO EARLY

E-commerce businesses, especially in the early stages, need to use their time and resources wisely. One of the mistakes many new e-commerce businesses make is building up their inventory before they even know what their demand will be, ending up with a large amounts of inventory and no one to buy it. Businesses must first get a sense of an item's demand by searching for it on different marketplaces to see how many have sold in the last few months. That will give businesses a sense of expected demand of their product.

LOST FOCUS

Before launching their business, E-commerce entrepreneurs must determine exactly what they are selling and what they are not. Many startups think they can do it all at launch and offer too many products. Too many products confuse customers, dilute your store's core value proposition, and can strain financial and management resources. At launch, the best approach is to specialize and operate in a very niche market. Once the business has established its presence in the market and achieved demonstrated success in the selected market, then further evolution of the business. Zappos (zappos.com), an online shoe and clothing shop based in Las Vegas, Nevada, is a perfect example. The early focus of Zappos was the shoe market only. Once they succeeded in this market, they branched out to a more comprehensive apparel strategy. When an online business starts, there exists a multitude of details to deal with, from shipping supplies to acquiring image photos. As to-do items add up, prioritize what keeps customers happy and your business healthy. Getting your products in the hand of customers is much more critical than fixing images that are off alignment by a few pixels.

NOT OPTIMIZING SITE NAVIGATION

A website that cannot easily be navigated is a huge turnoff for visitors. Your site visitors want an easy, convenient and intuitive experience, from homepage browsing to checkout and every page in between. Some of the most critical navigation elements in ecommerce are search (for shoppers with specific tastes), labels, and visuals (they drive navigation more than text). Checkout process must be familiar and easy to follow.

NOT USING SEO

Deploying SEO for your ecommerce site is a not a rocket science. The key is to integrate SEO right from the beginning. It is too common a mistake for companies to initiate full-scale SEO effort after the website is built, and sometimes after it's already been launched. By then the internal architecture of the site and its URL structure are already in place, making successful SEO a much bigger challenge. Businesses should integrate SEO thinking from the very start and make sure the site has the important basic title tags and metadata, and determine the most popular keywords for business category. If in-house expertise not available, the business should invest in a short-term project with an SEO consultant. Work on creating compelling product descriptions, blogs and social media that is shared or linked back to your site. That has a big impact on organic search results. Work on understanding "long tail keywords that are most relevant for your business. Long tail keywords are those three and four keyword phrases, which are very, very specific to whatever you are selling. You see, whenever a customer uses a highly specific search phrase, they tend to be looking for exactly what they are actually going to buy. Though traffic is lower for those terms, they offer higher conversion rates.

SCALING FIRST, ANALYZING AND TESTING LATER

Many of the ecommerce companies first look for high growth in the near future and leave site testing and analysis for later. Their assumption is that scaling comes first and it will not do a harm if testing is done any time later. Early analysis and testing of your site can reveal potential problems early on.

TOO MANY DISCOUNTS

Ecommerce companies are fascinated with discounts and many of them believe that it is the easiest way to acquire customers. Businesses may keep offering lucrative coupons without even realizing that they are not acquiring the right set of customers and discounting is not resulting into positive sales at all. The customers who are discount crazy might not return to your website in the future. Therefore, while discounting helps in attracting customers, it's a double edged sword and must be used with care.

NOT CHOOSING THE RIGHT ECOMMERCE PLATFORM

The importance of choosing the right ecommerce platform cannot be understated. Businesses should a very comprehensive research or consult an expert before zeroing down on a particular platform. Businesses should clearly understand their requirements, strengths etc. to ensure they make the right choice. For example, if a business is looking for a quick and easy way to get started, then using a SaaS platform can be a good option. If business want a lot of customization and control, a good choice is platforms like Magento (magento.com).

NOT OPTIMIZING FOR MOBILE

Gradually, mobile is becoming more important than desktop. Still, many ecommerce businesses do not check if their mobile websites are as responsive as their desktop version websites. In 2013, retail purchases conducted on smartphones will net a total-sales revenue of $14.59 billion. By 2017, these figures will more than double to $30.66 billion (White, 2013). Responsiveness is not only a requirement, but also a necessity for every ecommerce website. Therefore, businesses must optimize for mobile and check that their website design is compatible with small screen sizes and all major smartphone models and operating systems for a better user experience. According to a report, 78% of mobile device users claim that the "look and feel" of a company's mobile site has a huge impact on their decision to make a purchase from it" (Kansal, 2015). Ecommerce sites, which are not optimizing for these shoppers are surely missing out a lot of potential sales.

NOT OPTIMIZING FOR SPEED

Many Ecommerce businesses love to have highly decorative websites with heavy definition design and images but they overlook the fact that it might make their website slow and the user might not have a high speed Internet. According to a report, every one second of delay in site load time can reduce conversion rate by 7% (Laurence, 2014). Slow site speed not only affects the entire user experience but also the ranking of website in search engines. Friendly interface and easy navigation are important, but if your site is not loading quickly, your customer will leave your site without surfing anything further. Bounce rate is badly affected if your site load time is high.

NO FOCUS ON RETARGETING

Many visitors do not buy anything on their first visit to your website. On average, 98% of ecommerce shoppers leave without converting (Macdonald, 2013). So, retargeting becomes very important for an ecommerce business. It helps in bringing those potential customers back who once have left the site without completing the purchase, by showing them the products in their abandoned shopping cart and other similar products repeatedly. In addition, retargeting is very helpful in bringing back the users who have completed a transaction on the website.

NO BUSINESS PLAN

Decision of starting an online business is a big one that involves a good budget. Hence the decision has to be made with proper strategy and thinking. A proper business plan should be formed well before building the store, which may have within itself, the amount of marketing budget involved, the goals and aims of the company, the product which company is willing to sell online, the proposed market, which it is thinking of capturing and the viability of the products in the industry. The business plan may be broken down into various steps like processing, analyzing, evaluating, selecting, marketing, financing etc.

LACK OF EXECUTION

A lot is involved in having an E-Commerce store, which actually makes sales. Many E-commerce sites have low quality images with few lines of product information. That is a sign of bad execution. The execution plan should have product photography and a brief summary of how it helps the customers as top priority. People need to see what they are buying with clear images and read details about understand how it can add value to their lives. There is no shortcut there.

NOT HAVING SUFFICIENT BUDGET

Finance is the backbone of any of the venture. E-commerce ventures are no exception. They need good budget for marketing and getting the word out. People are not going to know about your portal on its own. You can do

SEO and hope for Google to rank your business site on first page but that takes months. Businesses need results faster and that involves expenses.

EXPECTING IMMEDIATE PROFITS

E-commerce businesses cannot expect instant return on investment. Online businesses may even lose for a certain time and there is no certain recipe of revenue generation. There are variables and they vary from business to business. In the initial stages of the site, visitor data should be collected, analyzed and processed to figure out what is going on and what to do next. With time, a business gets access to more data that can tell how the profits can be improved.

VERBAL CONTRACTS

It is especially important but neglected aspect of small online ventures. These ventures should have a proper process in place, which clearly defines about who controls what. Verbal arrangements or contracts have led to the closure of many stores as some people just so not stick to their words and cause disturbances in the business process. A proper written contract should be preferred, having defined set of terms and conditions of the relationships between the suppliers and various other peoples involved.

NO INTERACTIVE CUSTOMER SUPPORT

Customers get frustrated when they browse around an entire online shop to find someone to answer their query. In online shopping, it is easy to switch, so most people would rather leave if the help were not readily available. E-commerce businesses should try to make it as easy as possible for users to contact the business. Bad customer experience should not bring the sales down. A live chat feature can help site visitors get questions answered quickly.

NOT CREATING URGENCY TO FORCE THE SALE

Adding an urgency statement can make careful buyers take action without any further delay. You can even tempt your customers with some offers that they can use only if they order within the limited time e.g. "Order before 4pm today to receive next day delivery."

SWOT Analysis for E-commerce

A SWOT analysis takes a good deal of time. Customers can be a good source of information about what the business does well and where there is room for improvement and the business should seek their input. Ecommerce merchants should compare their businesses against both other online sellers and brick-and-mortar retailers. This complicates things because something that may be a strength when compared to a brick-and-mortar merchant (e.g. dynamic pricing) may not be a factor for ecommerce competitors if they too offer dynamic pricing. Markets change quickly and competitors may neutralize what was once an opportunity. For instance, ten years ago one of the advantages of ecommerce was the fact that websites offered 24/7 global shopping while brick-and-mortar stores did not. Now almost all traditional retailers have websites that compete with web-only sellers 24 hours a day (Kaplan, 2013). Following sections present important points that an analyst should consider when a SWOT analysis for an E-commerce business.

Strengths

Businesses should ask themselves what they do better than others in their industry do. For ecommerce merchants examples might be a larger selection of products; niche products not available elsewhere; faster or cheaper shipping than other ecommerce merchants; a wider geographical market coverage; more efficient value chain; a flexible target market; fast and accurate sharing of information among merchants and customers; a more efficient buying process; and access to niche markets; . Another important question is whether businesses make comparison-shopping easy and quick and whether they suggest other product options to their customers. These are all competitive advantages. Ecommerce vendors benefit from a structural advantage. They have lower operational costs than a brick and mortar store.

Weaknesses

Because of shipping times, there is no immediate gratification with ecommerce. Heavy, bulky and perishable goods are expensive to ship. Costs are always a headache. Security and fraud concerns mean some people are still reluctant to use their credit cards online. Allowing customers to pay using PayPal can blunt this concern. Showrooming has somewhat blunted the problem of shoppers inability of customer to touch the merchandise with shoppers looking at merchandise in stores and then using their mobile phones to place an order with an online seller, often while still in the store.

Many customer always found themselves insecure especially about the integrity of the payment process. Many online stores offer a limited number of products and do not provide personal services. Regions having limited Internet access cannot reap the benefits of E-commerce. Absence of physical and personal or direct face-to-face interaction between customer and the seller in many cases limits can limit customer trust on business.

Opportunities

Businesses should ask themselves what new technologies are available that can help grow their business and attain a competitive advantage. They should also be looking for areas where there competitors are vulnerable. Businesses should look for any new consumer trends that can benefit them. They should know not only the current statistics of Internet users in their region but also how they can use this statistics for their advantage. Ecommerce changes every day. New technologies and features have helped level the playing field with traditional retailers. For instance, improvements to shopping cart software have created a quicker, smoother, more customer-friendly shopping experience. Live chat has enhanced ecommerce customer service, blunting the traditional retailer advantage. Using Big Data analytics companies can gain insight into their customer preferences. Social media sites provide free or low-cost promotions. However, it is up to the online merchant to stay current and adopt new features.

Threats

Businesses should always be looking for broad industry shifts that might affect business growth. These include legal and regulatory changes. For instance, legislation currently before US Congress might force all Internet sellers to collect sales tax. Change in trends, fashion and fad can distress E-commerce. Low barriers to entry are a constant threat in ecommerce. It is easy to set up an ecommerce business, even in a garage. People can sell items via Facebook or eBay without setting up their own websites. Increasing privacy concerns about online transactions is still a great threat for E-commerce.

Big ecommerce merchants such as Amazon can always undercut smaller sellers on price. All ecommerce merchants are competing with Amazon and its successful $79 Prime shipping program. However, options now exist that can put online merchants on a more competitive footing with Amazon. Using such options could diminish a weakness.

In a nutshell, using SWOT analysis on a regular basis, perhaps once or twice a year, will give business a broad overview of ecommerce industry trends, show where business stand in relation to the competitors, and provide insights into mitigating the weaknesses and building strengths.

Researching your Competitors

Every E-commerce business should regularly monitor competitors' businesses (Traxler, 2013). There are various reasons for this. You may be willing to know their selling products you are not aware of or their aggressive promotion campaigns may be hard hitting your business. Researching your competitors can provide you with new ideas and keywords. You can also come to know some pricing strategies that you can adopt as well. Researching your competitors should be an ongoing exercise.

The first step to research your competitors is using a search engine such as Google. You need to put yourself into the shoes of your customer. Make sure you use Google in private browsing mode of your browser. Pick a product from your own web store and do a Google search on keywords that you think a consumer would use to find it. The private browsing mode of your browser would help ensure Google does not skew the results with your personal search history. Watch who is advertising on that keywords, the quality of the ads and whether the ads links to relevant landing pages or a home page of the advertiser. Analyze the organic search results. Organic search results are listings on search engine results pages that appear because of their relevance to the search terms, as opposed to their being advertisements. See if the results represent local stores and where do you stand in the listing. See if your competitors show up in organic search and content of the search results relevant to the keywords. See if there are any video ads appearing in the search results. These ads generally have more chances to be clicked. Repeat your search with several keywords suggested by Google. Those are the next most likely keywords that consumers use. Use a spreadsheet to track these results. This would serve as the baseline analysis that you would use to compare your results when you repeat your search next time.

The second step is to click from the search results to your competitors' sites. Watch whether you land on a product detail page, a landing page, or category list of some type. Watch the content page layout and design and compare it with your won. Look for any unique content and types of images and image features used. Look for their pricing for products in the category you chose. Put this pricing information into a spreadsheet to track prices by product for various competitors. See what the shipping policies of your competitors are and whether they promote cross-sell or up-sell items or not. Do perform random shopping at your competitor's site to see any promotions or up-sells within the cart. Browse other areas of the store to identify any promotions or seasonal offerings. Look for any new offerings and the products on clearance sale. Analyze the breadth and depth of their offerings. Think like a customer and make a note of what you liked and what you did not about their websites.

Pay particular attention to value proposition, product photography, shipping options, pricing, call to actions, site design (responsive/accessible/adaptive), and social media presence.

In the third step, make a list of the top keywords that bring you traffic, and enter them into Google. If you do not have an ecommerce store yet and are doing pre-launch research, simply search for relevant keywords. (Hayes, 2012). Now that you have a list of your top 5 competitors, and you have done basic research on how they operate, it's time to dig deeper. Use some of the tools below (some paid, most free) to gather in-depth information on what they are doing. This analysis will equip an e-commerce entrepreneur with the knowledge of trends and tactics that will affect his business.

Equipped with analysis, now assess your own ecommerce store. Try and objectively look at your online store and see how it can be improved. Use all the tactics you used on your competition and be as critical as possible. Use your learning to optimize your ecommerce store by taking advantage of your competitor's weaknesses. To remain competitive it is important to operate with flexibility and be able to pivot your direction.

Research Tools to Gather Competitive Intelligence

There are tools available that can help automate competitive intelligence gathering. Some of the popular tools are discussed below.

Alexa (http://www.alexa.com/)

Alexa is a free service that will help you analyze traffic on your competitor's ecommerce store. You can type in your competitors URL and Alexa will give you their global traffic rank, number of sites linking in, search analytics, audience insight, average site load-time, and a whole lot more.

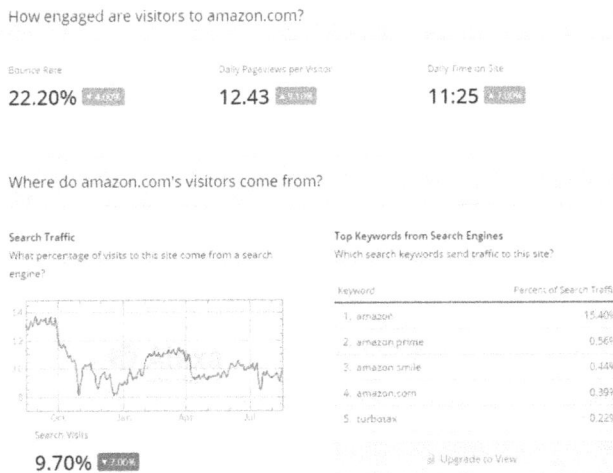

SEO Book (http://tools.seobook.com/general/website-comparison/)

This is page similarity comparison tool that you can use to easily compare page titles, meta information, and common phrases on your competitors homepages.

Google Adwords (https://adwords.google.com)

It is a keyword tool. This service will allow you easily analyze keywords and the amount of traffic generated by those keywords. It also allows you to narrow down your search by including URLs and specific categories, such as apparel, cosmetics, or whatever. You can use Google AdWords Keyword Tool to estimate how much your competition is paying per click for their ads. You can also use Google Traffic Estimator to find out the number of ad clicks and current bid prices for various keywords.

Internet Archive (http://archive.org/)

Internet Archive has been crawling the internet and taking snapshots of webpages since 1996. You can use their free Wayback Machine (http://archive.org/web/web.php) to see what a website looked like throughout the years. Analysis of the history of your competitor's sites can reveal trends in design and pricing changes.

Top Collections at the Archive

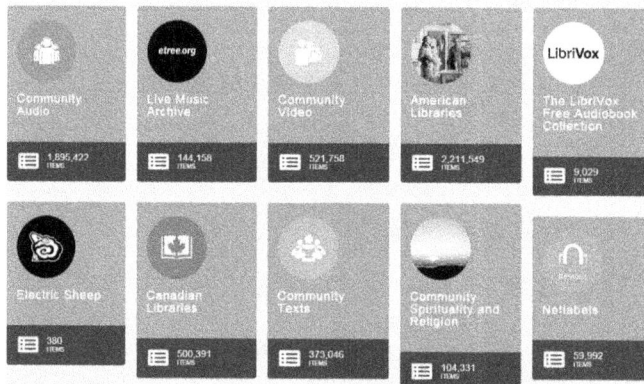

Whois (http://whois.domaintools.com/)

It is a domain tool. You can type your competitors URL in Whois to get a comprehensive record of that domain, including date registered, contact info, server stats, links in/out, and other domains the registrant owns. Following figure shows the Whois record for Amazon.com.

SpyFu (http://www.spyfu.com/)

This is a paid service ($79/month) that you can use to spy on your competitor's AdWords and keywords. You can see what worked and what did not work for your competitors AdWords campaigns. To get srated, you enter your competitor's website address on SpyFu website.

Open Site Explorer (http://www.opensiteexplorer.org/)

This service has a limited free plan and a paid service that is $99/month. You can use their service to compare your online store with up to four competitors on page authority, domain authority, linking root domains, total links, and with the pro version, you can also compare social stats. You can search your competitors' ecommerce store and see not only who is linking to them, but also what authority they have.

Google Alerts (http://www.google.com/alerts)

This is a free service from Google. You can use this service to receive email updates of the latest relevant Google results (web, news, etc.) based on your queries. Use this service to set up alerts for your competition as well as your own site. Set this service up to also get alerts for key industry terms, so you can easily monitor the broader market for new developments that could affect your ecommerce business.

Some additional tools are:

Quantcast (https://www.quantcast.com/)

You can use this tool to get demographic information on your competitor's site visitors.

Compete.com (https://www.compete.com/)

This site provides traffic estimates and trends for any website such as subdomains, search referrals, incoming traffic, outgoing traffic etc.

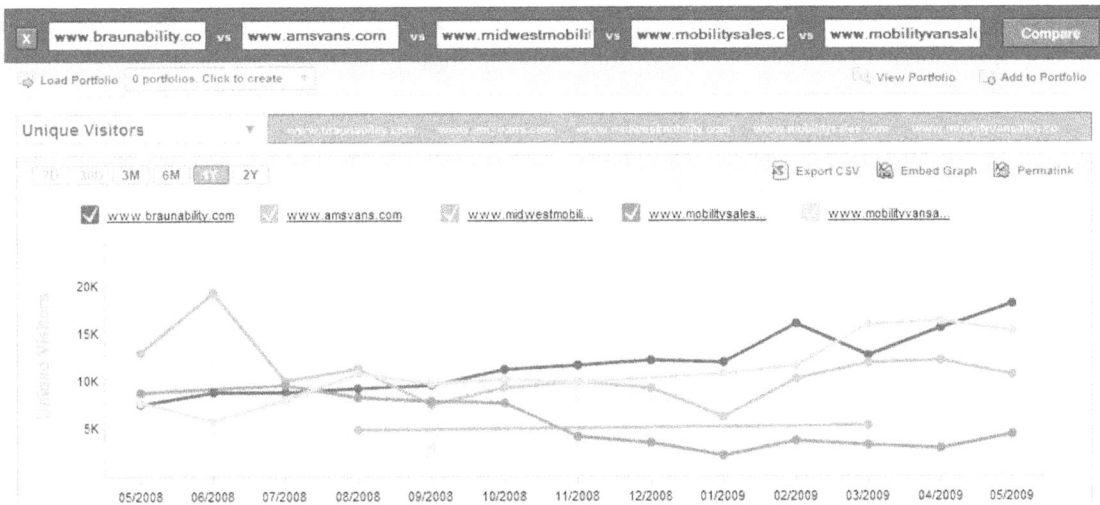

Customer Service for Ecommerce

Today's customer dynamics is entirely different as it was few years. Today's customer expects you to value them for more than the financial transaction. It is a demand and customers who do not find what they are looking for will just move on to competitors who understand how to engage them. Customer service is critical in the ongoing relationship between retailers and their customers. With the rise of social media, Ecommerce companies need to understand and embrace these new social customers, and respond to them in a personal and timely manner.

Improving Customer Service in E-commerce

This shift from customer to social customer is moving fast. However, many E-commerce businesses are far from getting this right and doing things in the wrong order. Businesses should first ways to understand the customer before investing time and money into an unknown. Following are some ways that small ecommerce businesses can provide awesome customer service for their customers (Desk.com, 2015).

OFFER MULTICHANNEL SUPPORT: Today many customers regularly use more than one support channel. In fact, 9 out of 10 customers expect to receive a consistent experience over multiple customer contact channels.

LISTEN ON SOCIAL MEDIA: Customers are craving human interaction and they are basing their buying decisions that companies care about their customers. In a study conducted by desk.com (Desk.com, 2015), 78% of customers believed that social media will be the next tier of customer service while 46% of online customers expected brands to provide customer service on Facebook. 88% of customers were less likely to buy from companies who leave social media complaints unanswered.

TRACK ALL CUSTOMER INTERACTIONS: Tracking all customer interactions eliminates wasted time for both your agents and customers; it eliminates confusion for your support agents as customers interact with a company via multiple channels, and it establishes a rapport with your company.

CREATE A SUPPORT CENTER FOR CUSTOMERS: The study conducted by desk.com (Desk.com, 2015), 90% of customers went to a company's website before calling or emailing you. Businesses should create a support center for customers to self-serve.

PROVIDE ONLINE CHAT: Chat is increasingly becoming the preferred solution for online consumers. Click-to-chat technology is gaining international momentum. Live chat is the most effective channel in terms of customer service and satisfaction (Lyoob, 2013). Internationally, the next great area of customer contact is live chat.

FOCUS ON TIMELY RESPONSES: In many ways, customer support response-time dictates how the customer perceives your company. In the study of desk.com (Desk.com, 2015), 84% of customers that reported being either likely or very likely to do business with a company if it responded via phone in less than a minute. Therefore, a prime focus of E-commerce businesses should be the timely response to customer queries.

DELIVER A WOW EXPERIENCE: A WOW experience is delivered by combining empathy and understanding for the challenge the customer is facing, identify and resolve the issue with skill, and drive and scale that learning back into the organization.

PROVIDE GREAT CUSTOMER SERVICE WITH THE RIGHT TOOLS: You can provide outstanding customer service, and there are technologies out there to help you. A business can keep track of all of its customer interactions, from any channel, all in one place.

Following is one example of great E-commerce support services. Sometimes customer is delighted even when there is a problem. Take example of ModCloth (www.modcloth.com). ModCloth is an American online retailer specializing in vintage, vintage-inspired and indie clothing, accessories and decor. They sell items from numerous different brands, and every day they offer new products. When a customer contacted them with a complaint, first, ModCloth offered a chat option for customer service. However, offering customer service alone is not enough. You should monitor it for its quality. After the chat, ModCloth contacted the customer to ask how the customer service representative handled the issue. This is where they won the customer. First, asking is great! The real excitement set in when they included a transcript of the chat. Not only did they remind the customer of the issue, how it as resolved, and whether ModCloth kept their promise, but really put themselves on the spot to do so. There is a possibility that the customer might remember the experience differently. A reminder of experience could be reminder of a terrible experience. ModCloth took the chance at their own expense, and that is where true customer delight takes shape (Keegan, 2015).

Measuring Success of E-Commerce Business

Each company measures success or failure by a different set of standards. Assessing E-Commerce is difficult because of the many configurations and impact variables involved. Sometimes the nature of what is being measured is intangible. One way to measure E-commerce project's performance is to use metrics. A metric is a specific, measurable standard against which actual performance is compared. Metrics can be both financial and non-financial. It is important to include nonfinancial measures in the measurement of strategy performance. The balanced scorecard approach is a popular strategy assessment methodology that measures organizational performance in a number of areas. (Plant, 2000) suggests seven areas for assessment in an E-commerce strategy.

These areas include financial impact, competitive leadership, brand, service, market, technology, and internal site metrics. Similarly, (Zhu & Kraemer, 2002) suggest four metric areas—information, transaction, interactions and customization, and supplier connection. [For an extensive discussion on metrics, see (Straub D. W., Hoffman, Weber, & Steinfield, 2002) and (Straub D. , Hoffman, Weber, & Steinfield, 2002)].

Important Metrics to Measure Success

Following are some important metrics, E-commerce businesses need to measure in order determine their success (Wallace, 2015; Myres, 2014; Gerber, 2013).

GROSS MARGIN: Revenue is business's bottom line. Keeping an eye on your gross margin, or the difference between your revenue and the cost of goods sold, is important as you look to reinvest profits for growth in the coming year.

COST OF ACQUIRING CUSTOMER (CAC): Traffic coming to a site is prerequisite to receive product orders. CAC reveals how much money you spend throughout the acquisition funnel, from promoting a post on Facebook, to having a visitor come to your site because of said promotion, all the way through to their finding a product they like and finally checking out. In other words, CAC is the amount of money you have to spend to get one customer. The lower the cost of acquisition, the better: i.e., you always want your cost of acquisition to go down. As a quick example, your CAC is $40 if you need to spend $200 to get five visitors to buy on your store. You may employ different techniques to bring in those visitors (e.g. SEO, paid ad campaigns, high-quality content, and social media) but all of them cost you in terms of either money or time.

CUSTOMER LIFETIME VALUE (LTV): This is the projected revenue generated by a customer during their lifetime. An easy example would be the lifetime value of online magazine subscriber who spends $20 every month for 3 years. The value of that customer would be:

$20 X 12 months X 3 years = $720 in total revenue (or $240 per year)

CONVERSION RATE (CR): Once your store gets traffic, you need to see how many visitors are buying. Conversion rate reveals just that. Conversion rate is defined as the percentage of visitors who end up buying from your store. Higher conversion rates are better. As a quick example, your conversion rate is 2% if 2 out of 100 visitors buy from your store. One way to improve conversion rate is to add video to a majority of your product pages; retailers adding video reported conversion rates close to 9%. Conversion rate is important because it directly affects your business's bottom line. Regardless of how much effort you spend on driving traffic to your store, if most visitors do not end up buying, it is all wasted.

SHOPPING CART ABANDONMENT (SCA) RATE: When your conversion rate is low, you need to understand how many visitors had an inclination to buy. To do this, you will want to examine your store's cart abandonment. SCA rate indicates the percentage of visitors who added products to their shopping cart but did not complete the checkout process. Lower SCA rates are better. As a quick example, your shopping cart abandonment is 75% if 75 out of 100 visitors with a cart leave without buying. Cart abandonment is the closest you come to earning real customers before they leave your site. Adding to the cart typically indicates an intent to purchase. The fact that they leave without buying means you lost potential customers. It gets especially bad if you paid a lot of money to get these visitors to your store. Making sure your cart abandonment is low is key to improving your conversion rate.

AVERAGE ORDER VALUE (AOV): Businesses should monitor how much money each order brings in to see how much revenue business can generate. AOV is the average size of an order on your store. The higher the average order value, the better. For example, your AOV is $35 per order if you made $140 from 4 orders. By monitoring AOV, you can figure out how much revenue you can generate from your current traffic and conversion rate. If most of your orders are small, that means you have to get a lot more people to buy in order to achieve your target. It is important to have at least a few high value orders so that your overall average is on the higher side.

CHURN RATE: If your Customer Lifetime Value (LTV) is low, it could be that many of your customers buy once and never return. This is measured by churn rate. Churn rate is the percentage of your customers who do not come back to your site. Lower churn rate is better. For example, a churn rate of 80% means 80 out of 100 customers do not come back to buy from your store. To ensure a high profit, it is important to influence your customers to keep coming back to purchase. That means you want your churn rate to be low so that once you acquire a customer, they continue to come back and purchase repeatedly. Lower churn rate means higher LTV and a healthier business overall.

PERCENTAGE OF MOBILE VISITS: Your site should be optimized for mobile and you should get traffic on your mobile site. Mobile growth continues to explode. According to a survey report (Myres, 2014), mobile sales shot up 50 percent over the holidays. Understanding the volume of visits and sales coming from this emerging channel will help ensure you are keeping a step ahead of the curve in the year ahead.

SUBSCRIBER GROWTH RATE: Many merchants use email marketing to spread news about specials, new products or just to share updates. However, mailing to a stale list of contacts will bring no benefit. If you are not seeing growth in your subscriber list, explore ways to expand your audience via outlets like Pinterest, multimedia, targeted keywords or in-person events.

VALUE PER VISIT: This data point helps you understand the value of each visitor to your website. You can calculate value per visit by dividing the revenue of your site by the number of visitors over a given period. Value per visit is especially helpful in guiding decisions around advertising and in calculating the return on your marketing investment.

NON-BRANDED TRAFFIC: This data tells you how much traffic you are getting from people who aren't looking for you specifically, but rather for something you're selling. The biggest opportunity to make more money comes from non-branded, organic traffic.

LEAD SOURCE ROI: Many online businesses start advertising on the Web without actually tracking the ROI of each particular lead source. By diligently tracking this metric, you can know which particular lead sources are profitable and which are not. On a deeper level, you can use this to split test advertisements on a granular level to find out which ones will maximize your ROI and develop the best ads.

PURCHASE FUNNEL: Understanding where and when a customer drops off the sales process is just as important as understanding the conversions coming in. Without understanding this, you cannot optimize and refine for increased conversions.

ADDITIONAL METRICS

Beyond these metrics, there are additional important metrics used by many of the large retailers. These include:

- **Cost Per Impressions (CPM):** It is calculated as (Ad Spend/(Impressions/1000)). This is most commonly used to measure the rate you would pay for a online banner ad campaign or other online advertising campaign. An Impression refers to the times the ad is served on a webpage (how many times the ad is seen).

- **Cost per Click (CPC):** It is calculated as (Ad Spend / Clicks). This is most commonly used to measure the rate you would pay for a search engine ad-campaign. CPC can also be referred to as PPC (Pay Per Click).

- **Cost per Acquisition (CPA):** It is calculated as (Ad Spend / Orders). This is most commonly used to measure the rate you would pay for a new order or customer. CPA is often used when measuring the rate you would pay an affiliate for referring an order to you.

- **Revenue per Click (RPC):** It is calculated as (Revenue / Clicks). This is the amount of revenue you can expect to get for each click on your site.

- **Cost of Sale (COS) %:** It is calculated as (Ad Spending / Revenue). This is the portion of your revenue that goes to your ad spend and is measured in percent.

E-commerce Investments

There has been a significant investment in e-business initiatives in recent years. Companies are increasingly investing in IT infrastructures to make their organizations e-business enabled. However, just like any other business projects, E-business projects must compete for funding. In many organizations, managers further invest in new technology and E-commerce in an effort to outpace competitors and to maintain or develop a competitive position. The economic justification in justifying IT expenditures is often neglected. Many IT projects are carried out without proper identification or measurement of either the benefits or costs of such investments. According to (Porter, 2001), no organization should embark on the undertaking of an E-commerce project without understanding the costs and performance of such issues. Companies are making large investments in E-commerce. At the same time, they are hard pressed to evaluate the success of their e-commerce systems (DeLone. & McLean, 2004).

Research on the business value of IT investment has provided significant insights in the context of the traditional brick-and-mortar economy [see (Bailey & Bakos, 1997); (Barua, Whinston, Shutter, Wilson, & Pinnell, 2000); (Brynjolfsson & Hitt, 1996); (Devaraj & Kohli, 2000); (Hu & Plant, 2001); (Mahmood & Mann, 1993)] but there is still a lack of tools, techniques, and approaches for evaluating e-business initiatives. E-commerce is a worldwide phenomenon, but it is usually studied at the national level. Research shows that formal evaluation of e-commerce projects provides higher level of satisfaction with the results of E-commerce initiatives (Standing & Lin, 2007).

Justifying IT projects is desirable but challenging. IT adds value by improving the competitive positioning by extending market/geographic reach and changing industry and market practices (P.Tallon, L.Kraemer, & Gurbaxani, 2000). Financial managers and other decision makers increasingly want demonstrating the value of the IT investments in the form of ROI or shareholder value format so that they can be effectively compared with alternative potential company investments. A good ROI model needs to capture IT benefits on five quantifiable dimensions relating to improvements in the production efficiency, business ecosystem benefits, customer surplus, business innovation and transaction effectiveness (Shields & Bharucha, 2003). The magnitude of benefits on these parameters is industry specific and the benefits are closely related to the critical success factors of the industry. For example, a manufacturing firm would focus on increasing production efficiency, while a financial services firm would focus on transaction effectiveness.

Funding your E-commerce Business

Arranging for the startup capital to start a new online business can be challenging. In order to avoid high interest charges you can consider to take a private loan from an individual you know or ask for a bank loan. You can also look for an angel investor. Moreover, you can also ask for grants from different businesses and government to start up your own business. Some of the sources that new E-commerce ventures can use to arrange the startup capital needed include angel investor, venture capitalists, commercial loans, and crowd funding.

Angel Investors

There are different kinds of angel investors. First are only interested in providing a start-up capital and does not take any part in operating the business. Second kinds of angel investor are ones who are likely to be involved in the business as either a sleeping partner or an active participant. For a list of angel investors, see https://angel.co/e-commerce/investors. The following is a list of some of the most active and influential early-stage angel investors (Prive, 2013).

Jeff Bezos is the Founder, President, Chief Executive Officer, and Chairman of the Board of Amazon.com. Under his leadership, Amazon.com became the largest retailer on the Web and the most widely adopted model for Internet sales. Some of his recent investments include Domo and Everfi (also The Washington Post) (as the chief investor in Bezos Expeditions: Rethink Robotics and Business Insider).

Paul Buchheit is a partner at Y Combinator, a Silicon Valley-based seed accelerator that has funded more than 550 companies in more than 30 different markets. These companies include Reddit, Dropbox, and Airbnb. In 2012, Forbes named Y Combinator as a top startup incubator and accelerator. Some of his recent investments include Lob and URX (as a partner at Y Combinator: SimplyInsured and Goldbely).

Jean-François "Jeff" Clavier is the founder and managing partner of SoftTech VC, a venture capital firm in Silicon Valley that has closed more than 150 investments since its founding in 2004. Among the successful startups that SoftTech VC has backed are Mint (sold to Intuit for $170m), Kongregate (GameStop), Milo (eBay), Wildfire (Google), and Class Dojo.

Paul Graham is a partner at Y Combinator. In 1995, Graham and Robert Morris created Viaweb, the first ASP, which in 1998 became Yahoo! Store. In 2002, he devised a spam-filtering algorithm that has inspired the current generation of filters. Businessinsider.com named him to their Top 50 Early-Stage Investors in Silicon Valley in July 2012, and February 2010, Businessweek.com listed him as number 11 Top Angel in Tech.

Sharad SharmaSharad was the CEO of Yahoo! India R&D before founding BrandSigma and was responsible for emerging markets engineering and several key global products. Sharad is an evangelist for developing technology product businesses in India and leads in efforts to nurture the ecosystem. Some of the startups he invested in include Frrole, Mobilewalla, HashCube, Druva Software, Kwench Library Solution, Vayavya Labs, Unbxd, Consure Medical, Aurus Network Infotech Pvt. Ltd

Kunal Bahl is the co-founder and CEO of Snapdeal.com, which is among India's leading online marketplaces. He started at Jasper Infotech Ltd in 2007 with a seed amount of INR four million INR. After tweaking the business model half a dozen times, he came up with the successful marketplace model for Snapdeal.com.

Sachin Bansal is the co-founder and CEO of Flipkart.com. He founded the company in 2007, which started as a humble online bookstore and later ventured into other product categories that proved to be a huge success in India. Some of the startups he invested in include Ather, TouchTalent, Roposo, NewsInShorts, MadRat Games, Spoonjoy

There are many advantages of angel investing. Angel investors can provide a small amount of funds at the starting stage of the company by utilizing their personal capital for the investment unlike the venture capitalists. As compared to the traditional financial lender, angel investors have more informal investment criteria. Therefore, there can be a negotiation in deals. Due to this elasticity, they can be a perfect source of capital for initial stage businesses. A part from the funds, angel investors also provide the required expertise, support and contacts in order to help business to grow. By being optimistic, angel investors invest into the business against a huge amount of returns to offset the risk. Another advantage of obtaining capital from angel investor is that there are no accrual payment rates therefore; entrepreneurs do not have to worry about high monthly payments and fees. The angel investors are nowadays located all over the place. They tend to invest in nearly all markets worldwide. Regardless of the market sector, angel investors are more attracted in businesses, which have the potential of growth and profitability.

There are many drawbacks of angel investors as well. Many angel investors are less probable to make follow-on investment because of the chances of losing more capital by reinvesting into the business when the business fails. While most of the angel investors look further than the monetary gains, there are few investors who are usually impatient and do not give any mentoring or guidance to the new entrepreneurs. In return, the angel investors would expect a little proportion of stake in a firm. Moreover, investors may employ skilled professionals for everyday activities. Angel investors generally have certain amount of control in the business. Most of the time investors lack the knowledge of industry experience. Therefore, entrepreneurs must only seek investors who have enough industry experience. While for the venture capital (VC) firms proper well documented directories are being maintained but, no proper national register is maintained for angel investors. Due to such dissimilarity, there is no national recognition of angel investors as their VC counterparts.

Venture Capitalists (VC)

Venture capitalists firms have an aim to make a profit by investing in those companies that have potential to grow. Their work is to pool their resources in small businesses as an investment to reduce their risk. Businesses need to have a solid plan in order to secure funds from a venture capital company. Venture investors may want to have a say in important business decisions and involvement in day-to-day business operations. Both parties should clearly state in writing, the level of participation expected by both parties before a financing agreement is finalized. Venture companies may give a mixture of start-up capital and the funds required to nourish an online business. However, entrepreneurs who do not want to give sole decision-making power to may not find this option of financing very attractive. It is not very easy to obtain venture capital funds. Firms should demonstrate that they can give high rates of return within a five-year period to obtain their finds but in most of the cases, they may reject the offer. Dundee Venture Capital (http://www.dundeeventurecapital.com) invests $50,000- $2 million in startups with an ecommerce and web services focus. The e.ventures (www.eventures.vc) is a global venture capital fund with investment focus on early stage opportunities in consumer Internet, media, and mobile internet.

The venture capital firms can provide businesses with many advantages. Startup firm can be benefited by the consultants on their staff provided by the venture capital firms that are experienced in specific markets. Not all entrepreneurs are good managers. When venture capitals provide funds to the business they demand little percentage of equity and a little say in business management. This could help the entrepreneurs in managing their business. Venture capital firms also provide consultant who are specialized in hiring right people for the business and this would avoid making the wrong decision and hiring an inappropriate person for the company. For a startup business, venture capital firms provide them with further help such as in legal matters, payroll matters and tax issues.

There are some drawbacks of venture capitalists as well. The venture capital firms may want to add their member of a team, in a management of the startup company to ensure that company is successful and growing. However, this may result in some internal disputes. Sometimes startup companies are asked to give the majority share of equity to the venture capital firms in exchange of their funds. This means, the venture capital firm has now more right over the business and mainly it is controlled by them.

Business Line of Credit or Commercial Bank Loans

When planning to open a new business, many entrepreneurs turn to their trusted banks for financing. However, it is very challenging for a new business owner to secure a bank loan or line of credit. Banks are usually supportive towards established businesses because, they tend to dislike risk. Financing depends on the entrepreneur's personal credit and loans that may be needed on the security of assets. Repayments are usually inflexible dependent on the whether the business is thriving or not.

Due to these reasons, banks are often not preferred for financing when other choices are available. Start-up capital should prepare a professional e-commerce business plan especially if a bank loan or bank credit financing is only source available to increase the likelihood of funding approval.

There are many advantages of obtaining commercial loans. Besides a standard business loan, commercial lenders can propose noncommercial loans. As the conditions of repayment of every commercial banks differ, a business loan offers a low interest rate as compared to other funding options. After reviewing the business plan banks rarely have any direct power over how the money is spent. Along with the original loan, only interest fees are given to the lender. The lender will not have a right over the gains. Most commercial loans often permits you to claim a subtraction on business taxes related to the payments of interest that businesses are getting on business loan.

There are many drawbacks of acquiring commercial loans as well. Like other sources of firm financing, commercial lenders are usually stricter in their rules and therefore, for reviewing process they require more information. A standard business loan is usually restricted to pre-existing business that has a successful financial history. Lenders tend to do this in order to increase the probability that a loan will be repaid. There is stringent procedure to make sure a high possibility of the loan being reimbursed which has made the commercial loan procedure very complicated than other alternatives. A commercial lender usually persists on collateral while others do not.

Crowd Funding

Crowd funding can help you find a community of small investors to fund your business, without the risks of traditional financing. Following is a list of crowd funding sources. Some sources focus on funding creative projects, others focus on meeting specific needs in the marketplace or community.

33needs (http://33needs.com/): 33needs provides crowd funding for social entrepreneurs, social enterprises and companies with a social mission. Investment dollars are exchanged for rewards offered by crowdfunded companies, as well as points to redeem for special offers.

Appbackr (http://www.appbackr.com/): Appbackr is a wholesale marketplace for mobile phone apps. The developer posts an app or app-in-development to the appbackr marketplace. Backers can purchase a bulk of apps wholesale, and the developer receives immediate payment.

ChipIn (http://www.chipin.com/): ChipIn is a web-based service that simplifies the process of collecting money from groups of people.

Cofundos (http://cofundos.com/): Cofundos crowdfunds open-source software projects.

IndieGoGo (http://www.indiegogo.com/): IndieGoGo offers anyone with an idea the tools to build a campaign and raise money. Unlike many crowdfunding sites, you keep all the money you raise, even if you do not meet your goal.

Kickstarter (http://www.kickstarter.com/): Kickstarter is a funding platform for creative projects in the world. Project creators are required to offer rewards (such as products, benefits, and experiences) to project backers.

MicroVentures (http://www.microventures.com/): MicroVentures targets companies that are creating technologies, products and services in core areas, such as business products, consumer products, electronics, online technology, and more.

Peerbackers (http://peerbackers.com/): Peerbackers is for business owners to raise capital from their peers in exchange for tangible rewards.

Pozible (http://www.pozible.com.au/): Pozible is an Australian crowd-funding website. Each project has a funding goal and a time limit set by the creator.

ProFounder (https://www.profounder.com/): ProFounder is a crowd-funding platform for entrepreneurs to raise investment capital from their communities. ProFounder helps you calculate payments that each investor is due and manage the distribution of funds owed.

Quirky (http://www.quirky.com/): Quirky offers product designers and inventors the chance to bring their products to market. If your idea is selected and brought to market, you will earn a share of the revenues.

RocketHub (http://rockethub.com/): RocketHub is a community for those with projects and those who contribute. Submit your project to the "Launchpad" for fueling.

In summary while deciding for the right option for financing companies should rely on their personal and business financial status, including the past as well as the present, and on their own specific wishes and requirements. The best way is to evaluate the pros and cons of the option according to your business requirements and only then, you will be able to make the right and profitable decision.

Strategies of Successful E-commerce Businesses

The ecommerce market is growing, and it is growing fast. In 2013, the growth of ecommerce significantly outpaced brick-and-mortar both domestically and worldwide. 2014 kept up this pace of growth, and according to recent projections from eMarketer (eMarketer.com, 2014), this trend is going to continue in the coming years. Worldwide total retail will continue to grow between 5-6% through 2018 and worldwide ecommerce will grow at a rate between 13-25%. On average, ecommerce retailers are making over $1 million in monthly revenue by the end of year three. There are a few areas where top-performing E-commerce businesses are creating competitive advantage (rjmetrics.com, 2015).

BUILDING A NEW BUYING EXPERIENCE

From solving the inconvenience of mattress shopping (Casper.com) to delivering all the ingredients for a healthy home-cooked meal (Plated.com), ecommerce companies are building lasting consumer brands by blurring the line between product and service.

Buying a mattress is one of the worst consumer experiences in the world. Yet, few startups have really tried to improve it. Buying a mattress will never be like buying an iPhone. We spend a third of our life sleeping on a mattress, which is probably also in line with the amount of time we spend on our smartphone. All the while, we drop hundreds of dollars on a phone every two years, but mattresses are a considered to be a pricey purchase, even when we only buy one every five or more years. Casper thought that this was because the mattress stores were downright terrible. Setting aside Casper's supportive beds, convincing the average customer to splurge on a web-based, one-size-fits-all model was challenging.

100 NIGHT TRIAL, 10 YEAR WARRANTY

Sleep on it, lounge on it, dream on it — if you don't love it, we'll pick it up
for free and give you a full refund. No springs attached.

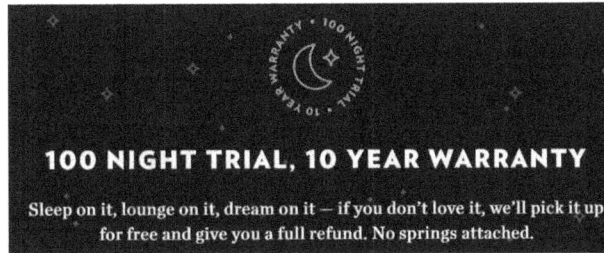

Fortunately, some of the work had already been done there. Tuft & Needle (https://www.tuftandneedle.com/), another mattress startup that offers a similar to Casper, had the No 1-rated mattress on Amazon. Both companies offer free delivery and returns on all mattresses within a 30-day period. Tuft & Needle foam beds are a few hundred dollars cheaper than Casper's, but also do not include latex foam (Hamburger, 2014). This is what Tuft & Needle say on their website.

WHY WE STARTED

With high sticker prices and questionable
quality, shopping for a mattress didn't make
any sense. We set out to fix that.

Plated is like a do-it-yourself food-delivery service, much like Blue Apron. Customers go to the website and enter his/her ZIP code. Customer can also let Plated know whether he/she have any dietary restrictions, or if he/she simply just does not like something. On which day food is delivered depends on where customer lives. Each week, there are many menu items created by Plated chefs. It costs around $12-$15 per plate, and customer needs to order a minimum of four plates. If customer spends $50, however, customer gets free shipping (shipping costs around $6). Customer can order a la carte, or sign up for a weekly subscription, which comes with its own perks, such as 20% off all plates. Customer can even get a couple bonus plates if he/she refers a friend. Once customer is done with his/her order, a box full of portioned ingredients is delivered on customer's doorstep and customer can start cooking (Levy, 2014).

YOU CUSTOMIZE

Tell us what you like to eat and we'll get started on your first box. Once you sign up, you can add, swap, or remove recipes.

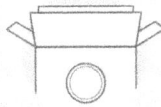

WE DELIVER

Each week, we'll send you fresh and seasonal ingredients and step-by-step recipe cards. Skip or change deliveries anytime.

YOU COOK

Explore new cooking techniques, discover unique ingredients, and cook delicious meals at home.

Blue Apron is a Plated competitor. The basic idea behind Blue Apron is that people who are strapped for time but want to make their own home-cooked food can sign up to receive three meals a week that will come in either two, four, or six-person portions. Big boxes of pre-measured ingredients will arrive once a week, with simple recipe cards that will instruct customers how to cook three fresh, out-of-the-ordinary meals that supposedly only take around 35 minutes to prepare. Each week, the company lists six recipes to choose from that suit a wide range of different tastes, catering to vegetarians as well as meat lovers. The most basic plan costs $60 a week, or roughly $10 per person per meal (D'Onfro, 2014). See what Blue Apron says on their website.

BUILDING PASSIONATE COMMUNITIES

Amazon owns commodities, but best-in-class companies build competitive advantage by engaging with vibrant communities. Customer Community pages provide a home on Amazon.com for thousands of topics, with new ones being added every day by customers. Bark & Co. (bark.co) has built a strong community around shared interests of dog lovers, just as Chubbies Shorts (chubbiesshorts.com) has done with shorts for "bros." In these companies, customer loyalty fuels growth. With the Chubbies loyalty program, after a second purchase, customers get a beer cozy. After a fourth, Chubbies sends a tank top (which resell on eBay for up to $70).

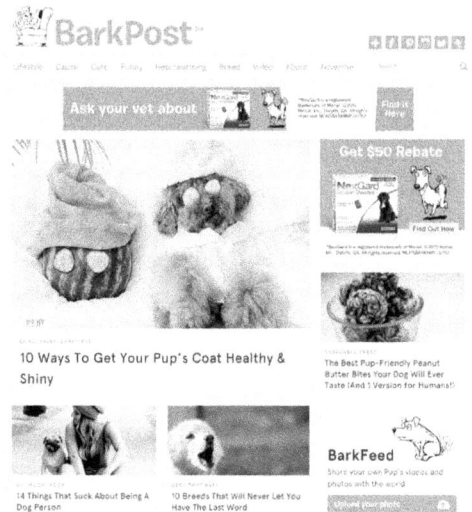

BLENDING CONTENT AND COMMERCE

Few companies have re-imagined the catalogue and turned it into an editorial experience. Examples of this include Thrillist (Thrillist.com) and Glossier (glossier.com). Thrillist is a digital media company that offers city guides for metro hubs like New York City, Miami and San Francisco. The company recommends restaurants, products and events via free, daily emails. Also operates JackThreads (jackthreads.com), a flash sales site for men's clothing. Glossier is a new way of thinking about and shopping for beauty products.

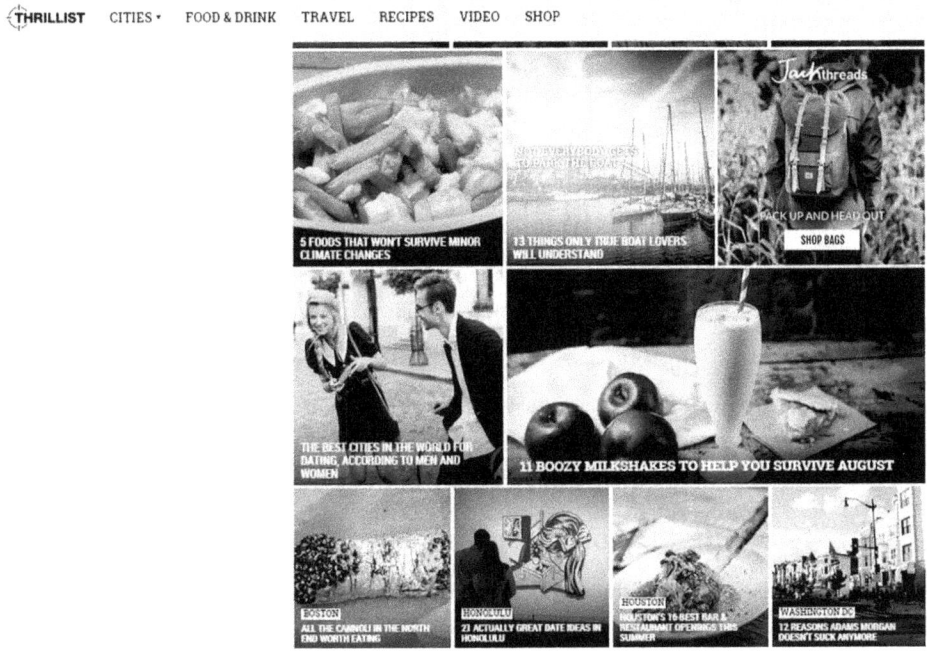

Want a Glossier inbox?

We do this thing where we send an occasional email with everything new and good that you'll probably want to know about: new products, g.IRL posts, promos, and parties. Sign up below.

OFFERING A "TRY BEFORE YOU BUY" EXPERIENCE

Many buyers refrain from online shopping due to the fear they may not actually get what they saw in the picture on the website. This hurdle needs to be overcome by any ecommerce retailer. However, some of today's biggest success stories come from companies who have made overcoming this hurdle a core part of their experience. Warby Parker's home (warbyparker.com) try on kits for prescription glasses builds returns into the experience.

Birchbox (birchbox.com) reimagined the cosmetic sample, creating a unique buying experience. In the wake of Warby Parker's success, loads of imitators sprung into being. There is Rivet and Sway, the online retailer that specialized in eyeglasses for women, and Lookmatic (http://lookmatic.com/) which lets you swap out colors, and Classic Specs (http://www.classicspecs.com/). Rivet & Sway unexpectedly closed their business 16 months after raising a $2 million financing round. Made Eyewear, another competitor, has a huge advantage over Waby Parker, one that allows it to beat Warby on both price and personalization. Customers can engrave whatever words they choose into their eyeglass stems and swap out various colors for frames and stems. Even better, customers can test out three pairs for free through Made's Home Try-On program (www.madeeyewear.com/about-us), and all three pairs come with the a prescription lens. In other words, you immediately get to keep the one you like because it is ready to go. If you do not like any of them, you send them all back and it costs you nothing.

The bottom line is this that top performing E-commerce businesses excel at two things. First, they have a product and brand experience that customers love and keep coming back for, and secondly they have a keen focus on their KPIs, particularly around customer acquisition.

Strategies to Compete against E-commerce Giants

It is natural that top competitors also use each other's products and services. One example is eBay and Google. The eBay helped users find hard-to-find, unique products. Google's goal of organizing the world's information also helped users find hard-to-find, unique products. The mechanisms and models were different, but the overlap was clear and eBay came to view Google as our top competitive threat. At the same time, eBay was one of Google's top advertisers at the time, and paid Google a lot of money (Jordan, 2013). Today, Google's most challenging competitor is not Bing, Baidu, or Yahoo!. It's Amazon. Amazon is a vertical search engine focused on helping users find products. The overwhelmingly dominant way to find things on their site is the search box. Users enter a keyword phrase and are presented with results that match his or her query. The order of the search results is determined by algorithms that seek to optimize relevance and monetization. Shopping on Amazon is a superior user experience with an increasingly comprehensive product assortment coupled with their ever-expanding direct sales supplemented by third-party merchants who sell on the platform. Prices are usually extremely competitive that many users would not need to use Google to comparison-shop at different merchants. Amazon offers the fastest and most cost effective shipping solutions.

Shopping experience on Google is different. Users have an infinite selection provided they can manage to find what they are looking for amidst the forest of search results. Users need to work to find the best price, typically by scanning different search results to check both prices and shipping costs. If the users find a product they want to buy from a new merchant, they need to enter all the payment and shipping information from scratch.

While shopping on Google can take considerable time at Amazon its very quick. E-commerce merchants now also have a very viable advertising alternative to Google. They can list their products for sale on Amazon through the Amazon Marketplace program. Amazon is currently generating billions of dollars in sales for these merchants, and these third-party sales are growing significantly faster than Amazon's direct business. This is a great challenge for Google because virtually all of Google's revenue comes from advertising ($44 of $46 billion in 2012, excluding the Motorola acquisition), and the majority of that comes from search. Presumably, Google's largest advertising category is shopping. A report found that four of Google's largest 10 categories were different segments of retail in which Amazon competes, and that many of Google's largest advertisers were retailers (Jordan, 2013).

It is clear that Amazon is a brutal competitor for brick and mortar merchants due to their large and growing cost advantages and a strong commitment to having the lowest prices anywhere. These same drivers also make Amazon a heavyweight competitor for e-commerce companies. Amazon is a monster competitor to online merchants as well. Amazon enjoys economies of scale far beyond their online competition, and they can use that power to offer hyper-aggressive prices and fast, cheap shipping. Amazon is larger than the next dozen largest e-tailers combined (Jordan, 2013).

While giants like Google have cash flows to support their big-ticket efforts to compete against another giant (i.e. Amazon), small and medium businesses do not. Here are some strategies that both offline and online retailers are using to compete against companies like Amazon. Not all are mutually exclusive i.e., many companies deploy multiple strategies (Jordan, 2013).

SELL DIFFERENTIATED PRODUCTS

Most sales at Amazon comprises of things like media, electronics, home and garden, and toys. Most best-selling products at Amazon are produced by large manufacturers that market them heavily and distribute them broadly through multiple retail channels. These products are essentially commodities and for most commodities, price is the key differentiator. Consumers know that Amazon usually has the lowest prices, along with free and fast shipping. It is extremely important to narrow your ecommerce niche and dominate it. E-commerce businesses try to sell in categories where Amazon is less dominant. One such category is soft-line products. The term "soft-line product" primarily refers to items that are literally soft, such as clothing and linens. Online companies like NastyGal and Zappos (before their acquisition by Amazon) and offline companies like Nordstrom and Neiman Marcus have successfully pursued soft-line strategies and have managed to weather competition from Amazon. Home improvement retailers have also used this strategy successfully because bulky or heavy items are less suited to online distribution. A related strategy is to feature products from companies that typically are not distributed or searched for on Amazon. One example is companies that primarily sell goods from design firms that lack extensive national distribution. These design firms lack broad awareness, so users do not typically find these products when searching on Amazon.

DEVELOP YOUR OWN PRODUCTS

Many retailers compete with Amazon by developing their own products. These products can be largely insulated from direct price comparison as they are proprietary and the producing company can elect not to have them sold by other online retailers. A number of the best performing offline chains pursue this strategy, including

Lululemon and Victoria's Secret. Online retailers like Chloe & Isabel in jewelry, Julep in cosmetics, ShoeDazzle in women's shoes, and Poppin in office goods are pursuing this strategy as well. While it's clearly much more work to design and source your own products, retailers that do are often rewarded with higher gross margins as they both cut out expensive middlemen and avoid head-to-head price competition.

MERCHANDISE PRODUCT DIFFERENTLY

Amazon.com, at its core, is a search engine for products. This type of search engine works best when consumers know exactly what they are looking for. Many consumers use Amazon's ubiquitous search box. Merchandising on Amazon is almost completely algorithmic (things like "others searching for 'x' also looked at 'y' and 'z.'"). A number of companies are trying to compete with Amazon by building a great browse experience, showing consumers a targeted assortment of attractively displayed products. Offline retailers have done this through beautiful window displays and in-store end caps. In addition, a new breed of online merchants is doing this too.

DEPLOY ALTERNATIVE DISTRIBUTION STRATEGIES

A number of online retailers are trying to put themselves directly in front of consumers before they think to consider searching for a product on Amazon. "Flash sales" companies like One Kings Lane and The Clymb send daily emails that display an assortment of goods at attractive prices. Other companies like Birch Box or Trunk Club use a subscription model that sends you a highly curated selection of products, typically on a monthly basis.

LEVERAGE UNIQUE STRENGTHS

Compared to Amazon, brick and mortar retailers are at a disadvantage because of their higher real estate, labor, and inventory costs. However, a number of merchants are trying to flip this disadvantage on its head and put their network of local stores to use. Wal-Mart has allowed consumers to pick up online orders at their local store on the day it was ordered. Both take advantage of Wal-Mart's massive inventory in geographically spread out locations. Wal-Mart is considering crowdsourcing its local, same-day delivery to Wal-Mart customers, who would receive discounts on their shopping bill in exchange for their efforts. Alternatively, Williams-Sonoma (www.williams-sonoma.com) has used both its store locations and catalogs to build its online business. Williams-Sonoma, Inc. is a high-end American consumer retail company that sells kitchenware, furniture and linens, as well as other housewares and home furnishings, along with a variety of specialty foods, soaps and lotions. They have been willing to cannibalize themselves, believing that someone else will do it if they do not. Over 40% of its revenue now comes through the online channel.

FIND AND DOMINATE A NARROW NICHE

It is extremely important that E-commerce business narrow their ecommerce niche and dominate it. They can do it through content marketing. Amazon.com does not do content marketing. They buy PPC, they do conversion optimization, they do SEO, they release products, they claim more verticals, and they do many other things. However, they do not do content marketing very well. They do not even do email marketing that great. This leaves E-commerce businesses with a huge opportunity to go into their niche, do great content marketing, and start to rank for all kinds of awesome keywords (Patel, 2014).

CREATE A SHIPPING INNOVATION

One area that Amazon has completely dominated is the area of shipping. It is, in fact, one of the company's greatest successes. Members of Amazon's Prime service can get two-day shipping free and next-day shipping for just a few dollars on each order. In addition, they offer same-day shipping. A business does not win merely by offering something. The business wins by giving it without being asked. Take example of Zappos. Their shipping policy surprised their customers. Zappos was delivering the customer's packages before they anticipated them. Businesses do not have to offer free shipping or some other extraordinary same-day service, but they can create a shipping service that exceeds expectations and provides satisfaction.

DEVELOP A SUBSCRIPTION SERVICE

Amazon has also ramped up their revenue by creating subscription services such as Subscribe and Save (recurring shipments), Amazon Prime (annual fees), and Amazon Simple Storage Service (AWS). Amazon launched a new service called Amazon Payments so they can help other companies keep charging customers on a recurring basis. The subscription service is one of the smartest ways to sell a product. You do not just get a one-and-done transactional experience with your customer. Instead, you get a relationship, and revenue every month or year. You do not have to be a software provider to make the subscription model work. Any form of recurring deliverables warrants recurring payments.

BOAST CUSTOMER SERVICE EXPERIENCE

One of the best ways to compete with Amazon is by providing personalized, one-on-one service to customers. Such personalized service is currently a weak area of Amazon. Your brand can achieve viral spreadability through passionate brand evangelists. Customer service is the idea that you have personal interaction with and relationships with your customers. There is consensus around the fact that customer retention via customer service and relationships has a better ROI and cost-effectiveness than other methods of marketing. If your customer knows you care, then you can hang on to most of your customers. Overdelivering to your customer can powerfully reshape your identity as a brand, and compel your customers to stay with you

BUILD A LOYAL FAN BASE

Building a loyal fan base is not easy, but it is possible. Everything you do as a brand, from social media outreach to content marketing, must have a personal and close-connected feel to it. It is about cohesion, connection, and knockout service. For example, you can buy a GoPro camera from Amazon but you cannot connect with GoPro as a brand. This is one strong reason why loyal GoPro fans stick to the GoPro site to buy their GoPro swag. They love this brand.

In summary, to compete with Amazon, E-commerce retailers need to outdate Amazon at their own game. E-commerce retailers can succeed based on their product and the way they market that product.

Review Questions

1. What is E-commerce strategy?

2. What steps are involved in building E-commerce strategy?

3. List some important issues to be considered in E-commerce strategy planning.

4. List some important issues to be considered in E-commerce strategy implementation.

5. List some common E-commerce startup mistakes.

6. List steps involved in system development life cycle for an E-commerce site.

7. List some components of a good E-commerce site.

8. List six web pages of a good E-commerce site.

9. How compliance with web standards is important for an E-commerce site?

10. Why justifying E-commerce investments has been difficult?

11. How do you perform SWOT analysis for an E-commerce business?

12. How small and medium-sized E-commerce businesses can compete with large E-commerce businesses?

13. How would you you research competitors of your E-commerce business?

Bibliography

Bailey, J. P., & Bakos, Y. (1997). An exploratory study of the emerging role of electronic intermediaries. *International Journal of Electronic Commerce* , 7-20.

Barua, A., Whinston, A., Shutter, J., Wilson, B., & Pinnell, J. (2000). *Measuring the Internet economy*. Retrieved July 26, 2009, from CISCO systems, University of Texas: http://ai.kaist.ac.kr/~jkim/cs492a/internet_economy-UT.pdf

Belsky, S. (2009). Be Solutionary, Not Revolutionary. Retrieved from http://99u.com/articles/6033/be-solutionary-not-revolutionary

Boulding, W., & Christen, M. (2001, October). First-Mover Disadvantage. *Harvard Business Review* , p. 20.

Bresser, R. K. F. (1998). Strategische Managementtheorie. Walter de Gruyter.

Brynjolfsson, E., & Hitt, L. (1996). Paradox Lost? Firm-level Evidence on the Returns to Information Systems Spending. *Management Science* , *42* (4), 541-558.

D'Onfro, J. J. (2014, June 1). Here's What It's Really Like Cooking With Blue Apron — The NYC Food Startup That's Worth Half-A-Billion Dollars. Retrieved July 11, 2015, from http://www.businessinsider.com/blue-apron-review-cooking-startup-2014-6

Deak, E. J. (2003). Economics of E-Commerce and the Internet With Economic Applications Card. South-Western Pub.

DeLone., W. H., & McLean, E. R. (2004). Measuring e-Commerce Success: Applying the DeLone & McLean Information Systems Success Model. *International Journal of Electronic Commerce , 9* (1), 31–47.

Desk.com. (2015). 7 ways to provide exceptional customer service for ecommerce | Infographic | Desk.com. Retrieved July 12, 2015, from https://www.desk.com/success-center/customer-service-tips-ecommerce-infographic

Devaraj, S., & Kohli, R. (2000). Information Technology Payoff in the Health-Care Industry: A Longitudinal Study. *Journal of Management Information Systems , 16* (4), 41.

Dholakiya, P. (2014). 4 Lessons from Google to Improve Your Site Search. Retrieved from http://www.advancedwebranking.com/blog/how-to-improve-your-site-search/

eMarketer.com. (2014, July 23). Worldwide Ecommerce Sales to Increase Nearly 20% in 2014 - eMarketer. Retrieved June 27, 2015, from http://www.emarketer.com/Article/Worldwide-Ecommerce-Sales-Increase-Nearly-20-2014/1011039

Epstein, M. J. (2004). *Start to Measure Your E-commerce Success*. Harvard Business School Press.

Evans, P., & Wurster, T. S. (2000). *Blown to bits: How the new economics of information transforms strategy*. Boston: Harvard Business School Press.

Gerber, S. (2013, August 26). 9 Data Sets Every Ecommerce Company Should Measure. Retrieved July 5, 2015, from http://mashable.com/2013/08/26/9-data-sets-every-e-commerce-company-should-measure/

Goliath.com. (2004, December). *Measuring the payoffs of IT investments*. Retrieved December 13, 2009, from Goliath.com: http://goliath.ecnext.com/coms2/gi_0198-188834/Measuring-the-payoffs-of-IT.html

Gupta, S. (2009, August). *Key Strategies for Maximizing E-Commerce Investments*. Retrieved March 15, 2010, from risnews.edgl.com: http://risnews.edgl.com/retail-best-practices/Key-Strategies-for-Maximizing-E-Commerce-Investments35858

Hamburger, E. (2014, April 22). "Sleep startup" Casper dreams of overturning the mattress racket. Retrieved July 11, 2015, from http://www.theverge.com/2014/4/22/5638400/casper-dreams-of-overturning-the-mattress-racket

Hayes, M. (2012, June 7). 8 Tools to Research Your Competition – Shopify. Retrieved July 17, 2015, from http://www.shopify.com/blog/6128722-8-tools-to-research-your-competition

Hu, Q., & Plant, R. (2001, Jul-Sep). An empirical study of the Casual Relationship between IT investment and Firm Performance. *Information Resources Management Journal* , 15-25.

Jordan, J. (2013a, May 9). Godzilla vs. Mothra, The Sequel | Jeff Jordan. Retrieved from http://jeff.a16z.com/2013/05/09/godzilla-vs-mothra-the-sequel/

Jordan, J. (2013b, October 24). How to compete with Amazon - Fortune. Retrieved July 11, 2015, from http://fortune.com/2013/10/24/how-to-compete-with-amazon/

Kansal, R. (2015, May 16). 7 Critical Business Mistakes that eCommerce Startups Must Avoid. Retrieved from http://www.iamwire.com/2015/05/7-business-critical-mistakes-ecommerce-start-ups-avoid/116347

Kaplan, M. (2013, April 3). SWOT Analysis for Ecommerce Companies | Practical Ecommerce. Retrieved July 4, 2015, from http://www.practicalecommerce.com/articles/3971-SWOT-Analysis-for-Ecommerce-Companies

Keegan, C. (2015, May 20). 3 Amazing Examples of Ecommerce Companies Ensuring Customer Delight. Retrieved July 5, 2015, from http://blog.hubspot.com/marketing/amazing-examples-ecommerce-customer-delight

Koo, C. M., Koh, C. E., & Nam, K. (2004). An Examination of Porter's Competitive Strategies in Electronic Virtual Markets: A Comparison of Two On-line Business Models. *International Journal of Electronic Commerce* , *9* (1), 163-180.

Laudon, K., & Traver, C. G. (2009). *E-Commerce 2010 (6th Edition)*. Prentice Hall.

Laurence. (2014, November 4). E-commerce website tips for Christmas. Retrieved from http://www.intouchmarketing.co.uk/6-tips-to-get-your-ecommerce-site-ready-for-christmas-january-sales/

Levy, K. (2014, November 3). I Tried Plated, The DIY Food-Delivery Site Started By A Couple Of Wall Street Guys Who Didn't Want To Get Fat. Retrieved July 11, 2015, from http://www.businessinsider.com/how-to-cook-using-plated-2014-11

Lieberman, M. B., & Montgomery, D. B. (1988). First-mover advantages. Strategic Management Journal, 9(S1), 41–58. http://doi.org/10.1002/smj.4250090706

Lyoob, J. (2013, September 3). 7 Reasons Why Customers use Web Chat for Customer Service| Stratus Contact Solutions | Multichannel Contact Center in USA. Retrieved from http://stratuscontactsolutions.com/7-reasons-why-customers-use-web-chat-for-customer-service/

Macdonald, M. (2013, August 6). Shopping Cart Abandonment: Why Online Retailers Are Losing 67.45% of Sales and What to Do About It. Retrieved July 12, 2015, from https://www.shopify.com/blog/8484093-why-online-retailers-are-losing-67-45-of-sales-and-what-to-do-about-it

Mahmood, M., & Mann, G. (1993). Measuring the Organizational Impact of Information Technology Investment: An Exploratory Study. *Journal of Management information System*, *10* (1), 97-122.

May, B. (2012). The Content Graph. Lulu.com.

Myres, A. (2014, January 12). 7 key analytics metrics to measure year end performance. Retrieved July 5, 2015, from http://blog.bigcommerce.com/7-key-ecommerce-metrics/

Oetzel, J. M. (2004). Differentiation Advantages in the On-line Brokerage Industry. *International Journal of Electronic Commerce*, *9* (1), 105.

P.Tallon, P., L.Kraemer, K., & Gurbaxani, V. (2000). Executives' Perceptions of the Business Value of Information Technology: A Process-Oriented Approach. *Journal of Management Information Systems*, *16* (4), 145-173.

Patel, N. (2014, December 4). A Five-Step Guide to Competing with Amazon. Retrieved July 11, 2015, from https://blog.kissmetrics.com/competing-with-amazon/

Peters, C. (2009). *A Nonprofit's Guide to Building Simple, Low-Cost Websites*. Retrieved December 15, 2009, from techsoup.org: http://www.techsoup.org/learningcenter/webbuilding/page11890.cfm

Plant, R. T. (2000). *Ecommerce: Formulation of Strategy*. Prentice Hall PTR.

Porter, M. (2001, March). Strategy and the Internet. *Harvard Business Review*, pp. 63-78.

Pottruck, D. S., & Pearce, T. (2000). *Clicks and Mortar: Passion Driven Growth in an Internet Driven World*. John Wiley & Sons, Inc.

Prive, T. (2013, December 16). 20 Most Active Angel Investors. Retrieved July 5, 2015, from http://www.forbes.com/sites/tanyaprive/2013/12/16/20-most-active-angel-investors/

Rangan, S., & Ron, A. (2001). Profits and the Internet: Seven Misconceptions. *Sloan Management Review*, *42* (4), pp. 44-53.

rjmetrics.com. (2015, April 2). The Five Indicators of Breakout Ecommerce Growth. Retrieved July 1, 2015, from https://blog.rjmetrics.com/2015/02/04/the-five-indicators-of-breakout-ecommerce-growth/

Shapiro, C., & Varian, H. (1999). *Information Rules: A Strategic Guide to the Network Economy*. Boston: Harvard Business School Press.

Shields, J., & Bharucha, J. (2003). *Measuring IT Returns: Are we Counting Chickens for Eggs?* Retrieved June 26, 2009, from Infosys.com: www.infosys.com/research/publications/Documents/CE-06-03.pdf

Standing, C., & Lin, C. (2007). Organizational Evaluation of the Benefits, Constraints, and Satisfaction of Business-to-Business Electronic Commerce. *International Journal of Electronic Commerce*, *11* (3), 107.

Straub, D. W., Hoffman, D. L., Weber, B. W., & Steinfield, C. (2002). Measuring e-Commerce in Net-Enabled Organizations: An Introduction to the Special Issue. *Information Systems Research* , *13* (2), 115-124.

Straub, D., Hoffman, D., Weber, B., & Steinfield, C. (2002). Towards New Metrics for Net-Enhanced Organizations. *Information Systems Research* , *13* (3), 227-238.

Tjan, A. K. (2001, February). Finally, A Way to Put Your Internet Portfolio in Order. *Harvard Business Review* , pp. 76-95.

Traxler, D. (2013, March 5). How to Research your Ecommerce Competitors. Retrieved from http://www.practicalecommerce.com/articles/3933-How-to-Research-your-Ecommerce-Competitors

Turban, E., Lee, J., & Chung, M. (2006). *Electronic Commerce: A Managerial Perspective*. Prentice Hall.

Turban, E., McLean, E., & Wetherbe, J. (2004). *Information Technology for Management: Transforming Organizations in the Digital Economy*. Wiley.

Ueland, S. (2011, June 15). 13 Crowdfunding Websites to Fund Your Business. Retrieved from http://www.practicalecommerce.com/articles/2853-13-Crowdfunding-Websites-to-Fund-Your-Business

Venkatraman, N. (2000). Five Steps to a Dot-Com Strategy: How to Find Your Footing on the Web. *Sloan Management Review* , *41* (3), pp. 15-28.

Verleur, J. (2015, January 9). 4 Startlingly Basic Mistakes That Doom Most Ecommerce Startups. Retrieved July 5, 2015, from http://www.entrepreneur.com/article/241447

Wallace, T. (2015, January 22). 11 Need to Know Ecommerce Metrics. Retrieved July 5, 2015, from http://blog.bigcommerce.com/6-vital-ecommerce-metrics/

White, L. (2013, November 7). Mobile Trends By The Numbers, Just In Time For The Holidays. Retrieved July 12, 2015, from http://marketingland.com/a-look-at-mobile-trends-in-numbers-63815

Zhu, K., & Kraemer, K. L. (2002). e-Commerce Metrics for Net-Enhanced Organizations: Assessing the Value of e-Commerce to Firm Performance in the Manufacturing Sector. Information Systems Research , 13 (3), 275-295.

6

WEBSITE USABILITY

Web Presence of E-commerce Business

Importance of Effective Web Presence

A business's presence in the physical world conveys its public image to its stakeholders. Such presence can be created in the form of stores, factories, warehouses, and office buildings. Importance of a company's public image increases as organization's size increases. On the web, its importance increases significantly as the customers of an online business know the company only through its web presence. Therefore, creating an effective Web presence can be critical even for the smallest and newest firms operating on the Web.

Effective web presence is important for both For-profit organizations and not-for-profit organizations. For-profit organizations can enhance their images by providing information on their websites. For not-for-profit organizations, this image-enhancement capability is a key goal of their Web presence efforts. They can use their Web sites as a central resource for communications with their varied and often geographically dispersed constituencies. Their Web sites allows these constituencies to integrate information dissemination with fund-raising in one location. Visitors who become engaged in the issues presented can be taken to a page offering memberships or other opportunities to donate using a credit card. Their web sites also provides a two-way contact channel for people who are engaged in the organization's efforts but who do not work directly for the organization. This combination of information dissemination and a two-way contact channel is a key element on any successful electronic commerce Web site. Interestingly, in accomplishing this combination of elements in their Web presences not-for-profit organizations are ahead of many businesses.

Pillars of Effective Web Presence

There are four pillars of effective web presence namely value, trust, usability and presentation. If a business get these four elements right, they will have an effective and engaging website that attracts more visitors (Marketing Donut, 2015).

VALUE

Value means whether your website provide information that is of real value to the customer. Customers want value and want to know how the site they visit will help them solve their problems and achieve their goals. Tell your customers, in language they understand, exactly how you help clients in their position. For each group, describe their business problems and say how you will solve them. Show the benefits you will bring. Serve your customers with valuable content that will help solve their business problems (such as educational articles, papers, resources, eBooks, video clips, audio files, cartoons etc.) Show that current customers have had success. Provide case studies and testimonials that show the real benefit of what you do.

TRUST

Trust and credibility are big issues on the web. Not all company websites present on Internet are reputable. There are few ways businesses uses to improve their sites to gain customer trust. First, provide information on your people i.e. your management team and key customer contacts. Show photos of real people so customers know whom they are dealing with. Enable your customers to make contact with your team directly. Second, use social media and provide links from your website. One of the major benefits of getting your company into social networking is the ability to show that your company is made up of real people with opinions, passion and expertise in their marketplace. Social media enables you to connect with your customers. Whether it is via a company blog, LinkedIn, Twitter, another platform or a combination of the lot, social media makes good business sense. Third, keep your website up to date by providing provide fresh content and regular update. Fourth, provide testimonials from customers and case studies that tell the story of their success thanks to your services or products. Fifth, be approachable and genuine. People like to do business with people. Genuinely communicate through your site and you will form a connection.

USABILITY

People visit websites for their utility. If the site is convenient, people will use it; if not, they will not. Today, users are far less tolerant of difficult sites. Usability is more important than ever. Make your website easy to use, so your customers can get to the information they want, fast. Pay close attention to navigation and plan and organize your content carefully. If you are redesigning your site, build a wireframe first. A wireframe is a visualization tool for presenting proposed functions, structure and content of a Web page or Web site. Test your navigation with real customers. Give them a task and see how easy it is for them to achieve this. Tweak the navigation accordingly. Follow web conventions such as recognizable page names. Web layout has become standardized. Write for the web properly. Poor writing makes websites fail. Design your home page carefully. This is where web usability usually succeeds or fails. Make contact easy. Make your contact details very, very obvious. Make sure your site is accessible to everyone, including the disabled. For this purpose, follow WC3 guidelines.

PRESENTATION

For your website, color schemes, branding and imagery are important of course, but must not be prioritized at the expense of usability and content. Hire a professional web designer to make the site visually appealing to your customers. Bad design can frighten customers away while good design adds interest and will help to draw them in. Do not overcomplicate things. Make your website interesting but simple, consistent and free of clutter. Pay attention to typography as well as graphics. Make sure your content is easy to read. Avoid overstuffed design and splash pages, as they will detract from your content.

Web Presence Goals

Managers of businesses operating in a physical world focus on very few objectives that are image driven. For such companies, the presence of a physical business results from satisfying many objectives e.g. finding a suitable physical space and balance the inventory and storage needs. This physical presence rarely involves designing the physical space.

On the other hand, online businesses need to create websites for building their presence. A well-designed website can provide many opportunities to create and enhance a business image effectively. For example, it can serve as a sales brochure, a product showroom, a financial report, an employment ad, and a customer contact point. Each entity that establishes a Web presence should decide which features the Web site can provide and which of those features are the most important to include.

Achieving Web Presence Goals

An effective site is one that creates an attractive presence that meets the objectives of the business or organization. These objectives may include, attracting visitors to the Web site, making the site interesting enough that visitors stay and explore, convincing visitors to follow the site's links to obtain information, creating an impression consistent with the organization's desired image, building a trusting relationship with visitors, reinforcing positive images that the visitor might already have about the organization, and encouraging visitors to return to the site

Web Presence and Brand Image

It is important that a business's web presence is consistent with its brand image. Web presence goals may be different for different firms.. Take example of Coca Cola and Pepsi. Both companies have developed powerful brand images in the same industry but they have developed significantly different Web presences. Both companies make frequent changes to their Web pages, but the Coca Cola usually includes a trusted corporate image such as the Coke bottle on its pages. On the other hand, Pepsi usually fills its web pages with hyperlinks to a variety of activities and product-related promotions. These Web presences convey the images each company wishes to project. Each company's web presence is consistent with other elements of their marketing efforts. Coca Cola projects itself as a trusted classic, and Pepsi projects itself as the upstart product favored by a younger generation.

Website Usability

Research shows that current web presence of few businesses accomplishes all of their goals for their Web sites. Even sites that succeed in achieving most of these goals often fail to provide sufficient interactive contact opportunities for site visitors. Many large firms do not acknowledge and use the Web's capability for two-way, meaningful communication with their customers. Industry consultants argue that use of this communication process is not optional; companies that fail to communicate effectively through this channel will lose customers to competitors that do.

Usability is an important element of creating an effective Web presence. Usability provides a measure of quality of user experience with a product or system (e.g. a website, software, or a mobile device). Usability also includes the ease with which a user can use a product or service to achieve his objectives and his satisfaction with the process.

Two international standards further define usability and human-centered design (IBM, 2008).

1. Usability refers to the extent to which a product can be used by specified users to achieve specified goals with effectiveness, efficiency and satisfaction in a specified context of user.

2. Human-centered design is characterized by: the active involvement of users and a clear understanding of user and task requirements; an appropriate allocation of function between users and technology; the iteration of design solutions; multi-disciplinary design.

What does usability measure?

Usability is a combination of factors related to user interface. These factors include:

- **Ease of learning**: How fast a new user can learn the user interface sufficiently well to accomplish basic tasks?

- **Efficiency of use**: How fast an experienced user can accomplish tasks?

- **Memorability**: If a user has used the system before, can he or she remember enough to use it effectively the next time or does the user have to start over again learning everything?

- **Error frequency and severity:** How often do users make errors while using the system, how serious are these errors, and how do users recover from these errors?

- **Subjective satisfaction:** How much does the user like using the system?

Customer-Centric Web Site Design

An important part of a successful online business is a Web site that meets the needs of potential customers. When constructing the website the focus is on meeting the needs of all site visitors. A customer-centric approach of designing a website puts customer at the center of all design activities. This approach is a key methodology to carry out usability. It involves users throughout all stages of Web site development, in order to create a Web site that meets user's needs. This approach considers an organization's business objectives and the user's needs, limitations, and preferences. The objectives for your website must balance the needs of users and the needs of your organization. If that is not the case, you will not be able to meet your organization's objectives because users avoid using websites, which are not helpful for them. On the business side, you can lower operating and redevelopment costs by developing a Web site correctly the first time.

A customer-centric approach leads to some guidelines that Web designers can follow when creating a Web site that is intended to meet the specific needs of customers, as opposed to all Web site visitors. These guidelines include the following:

- The website design should be based not on the organizational structure but on the anticipated user navigation of links on your site.

- Provide visitors with quick access to information

- Avoid using inflated marketing statements in product or service descriptions.

- Avoid using difficult to understand terms and business jargon

- Your web site should work for visitors who are using the oldest browser software on the oldest computer connected through the lowest bandwidth connection. A company may need to create multiple versions of Web pages to do that.

- Design features and colors should be consistent.

- Navigation controls should be clearly labeled or otherwise recognizable

- Text should be visible on smaller monitors

- Color combinations should not impair viewing clarity for color-blind visitors

- Conduct usability tests by having potential site users navigate through several versions of the site.

- Cultural orientations, especially high-context communication, are important in website design. Not paying attention to this orientation can make websites less clear, less attractive, and less interactive (Usunier, Roulin, & Ivens, 2009).

- Quality of Web site content, Web site appearance, and extent to which the Web site provides convenient services to facilitate users' on-line activities influence Web site usability (Tung, Xu, & Tan, 2009).

- The links that lead the on-line consumer to a firm's Web site affect the perception of the firm's trustworthiness. Firms should carefully consider the navigation routes through which their Web sites are reached (Stewart & Malaga, 2009).

- Benchmarking is a vital activity in Web site development. A comparison of their websites with that of competitors can provide managers an opportunity to improve the quality of their websites (Pang, Suh, Kim, & Lee, 2009).

- The ability to evaluate business Web sites from the consumer's point of view is of obvious importance (Loiacono, Watson, & Goodhue, 2007).

- Web site appeal is significant in building initial trust of customer, which in turn has a significant effect on intention to use the Web site in the future.

- Success retailing on Internet requires web site traffic. Web site traffic is a prerequisite to transactions in e-tail. Publicity and product assortment increases traffic to your website (Nikolaeva, 2005).

- Trust and economic conditions have significant positive impact on on-line shopping behavior. In fact, they explain more than 80 percent of the variability in on-line shopping behavior (Mahmood, Kohli, & Devaraj, 2004).

These guidelines can help make visitors' Web experiences more efficient, effective, and memorable.

Creating a User-Centric Web Site

The process of creating a customer-centric website includes the following steps.

1. Planning

2. Analysis

3. Design

4. Testing and Reinforcement

5. Implementation and Retest

PLANNING

The first step is to build a plan for your website. To develop a plan you need to think and set agreement on the website scope, website audience, and objectives your firm wants to achieve through this website. You can begin by collecting information about your firm's primary business objectives and their relation to the Web, users of your website, tasks and goals of your users, information needs of users, website functions needed by users, users experience level with the web, and the hardware and software used by users to access your website. You should create a site name and write a short description of your website. List all possible major groups of audiences that you intend to serve through your website (e.g. public, researchers, advocates, students etc.) and think about their needs from your website. Think about your objectives in business terms and set measureable objectives (e.g. reduce emails, increase customer satisfaction etc.). By setting meaningful and measurable objectives, you can measure your success once the website is launched. Satisfaction, accuracy, and time are some typical usability goals. For example, you can set a usability goal for the overall time the user will take to carry out a task on your site. To set measurable usability goals, one technique is to observe users performing similar tasks. For example, if you are developing a Web application, your application must allow users to do their tasks at least as fast with as few errors and as much success and satisfaction as their current way of working.

The second step in planning is to assemble a project team. For a successful project, you need a project team with the right mix of skills. Number of people required will depend on the scope of your project. There are many roles to perform in a project team. One person may have the skills to fill more than one of the roles or several people may be needed for some of these roles. Some of the roles you need to fill include Project Manager, Usability Specialists, Web Content Writers, Information Architects (responsible for building site navigation structure), Graphic Designers, Developers and Programmers, and Subject Matter Specialists. A Usability Specialist can be an in-house consultant. When hiring a Usability Specialist you must review the individual specialist's skills, their past performance, list of clients, variety of project work, education, and references to ensure that you select a qualified expert. There are resources available on the web to look for usability specialists e.g. HCI Consultants (Special Interest Group for Computer Human Interaction), Human Factors and Ergonomics Society (HFES), and Usability Professionals Association (UPA). You can also ask others who have hired usability consultants or you can look at lists from some other professional societies.

The third step in planning is to develop a statement of work (SOW). A statement of work is prepared when you need outside help with the user-centered design process or with specific usability activities. SOW describes the work to be performed and usually includes location of the work, period of performance and timeline and deliverable schedule, level of effort ,and any special requirements (such as security clearances, travel required, special skills or knowledge). Vendors are invited to bid for performing work described in SOW. Vendors respond with a proposal and cost estimate. The evaluation criteria to select a vendor may include the vendor's plan for performing the work, the skills and experience of the individuals who are proposed for the work, the past performance and other projects that the company has completed, and the price.

The fourth step in planning is to hold kick-off meeting. A kick-off meeting is held to develop an understanding of how your internal team views the current Web site and the project to develop a new site. It is important to consider the differences between what you find and what your team thinks. It is critical to address the differences between reality and thinking of your team so that the team is able to learn and adjusts its plans to match the realities.

ANALYSIS

You should start analysis phase by reviewing your current website. This review can provide you with an understanding of what works and what areas need improvement. The objectives of site review are: to see how well your current site meeting your organization's objectives and the usability goals you have set for the site, to understand how well your site is meeting your users' needs, and to find out whether your site complies with the basic web standards and practices.

For achieving these objectives, determine the types of users who visit your site, the reasons they come to the Web site, how well users can find information on your web site, users level of understanding with the content of your web site, how much users enjoy using your web site, users expectations about your site, their levels of experience with the web and similar types of sites, the ways of working with information, and the technology available to them (e.g. broadband or dial-up). There can be several ways to gather this information.

- By reviewing the emails that users send to the web master

- By reviewing the questions users call your organization about

- By reviewing posted questions to the website

- By analyzing website logs and looking for most popular pages and most searched items

- By conducting usability testing which involve users and you can observe actual behaviors and listen to users' comments.

- By using contextual interviews that allow users to do their own work in their own home or work site and let you observe and listen to actual behaviors in the user's environment with the technology they use.

- By using online surveys to ask users how they use the Web site. However, these surveys do not collect actual behavior

- By individual interviews of users that can provide a lot of information about users. However, these interviews do not collect actual behavior.

- By using focus groups that involve small group discussions involving eight to 12 people and moderated by a trained facilitator. This technique is a self-reporting technique and may not be relied upon as indicator of actual user behaviors.

User analysis is complemented by task analysis. Task analysis is used to learn various aspects of your site users e.g. users goals, their expectations from your site, specific tasks users must do to meet their goals, what steps they take to accomplish those tasks etc. Through task analysis, you can determine what your web site must supports, the appropriate content, and applications to be included in your Web site etc.

Developing personas and scenarios can further assist your understanding your users. A fictional representation of a user group on your site is called persona. A persona has all the characteristics, which are most representative of that user group. Using personas your team can focus on building a site with features that users will actually use rather than what the users ask for. A scenario refers to user tasks, questions, and stories that website must satisfy. There can be different levels of details in scenarios. Using scenarios, you can determine what content the site must have and which pieces of content must be easiest to find and understand.

To determine your site compliance with basic web standards and practices, you can perform a heuristic evaluation (performed by one or more usability experts) to review your goals as well as your users scenarios to identify potential usability issues on your site. You can also review the available usability and web design guidelines.

DESIGN

This phase starts with determining website requirements. Web site requirements tell you what the Web site must have and what it must allow users to do. These requirements describe the features, functions, and content of the site. These requirements might be general features and functions (such as search or contact us) or these might be specific features and functions for your site (such as some specialized applications e.g. e-mail alerts). Initially, requirements can be as simple as single line statements of what a website must allow a user to do. Over the times detailed requirements can be developed.

The next step is to conduct a content inventory. A content inventory is a list of all the content (the information) on your site. It categorizes and describes the information on every page of your site. You can use the content inventory to determine what is on the site and what needs to be done to different parts of the site. An updated content inventory is very valuable for growing websites. A spreadsheet or a database application can be used to create a content inventory. The content inventory should include the overall topic or area to which the page belongs, the page title and URL, a short description of the information on the page, the date page was created, date of last revision and when the next page review is due, other pages that this page links to, and page status. A content inventory has to be dynamic in nature.

To better organize your website in way that makes sense to users, card-sorting technique can be used. Card sorting is a technique in which users are asked to organize the content from your web site in a way that makes sense to them. Users review items from your web site and then group these items into categories. Card sorting can help you in many aspects of website design e.g. what content should be placed on a web page, what should be the structure of the website, and how the home page categories should be labeled.

Results of card sorting can help you develop the information architecture of your website. Information Architecture informs your users what categories of content you provide on your web site. Information architecture consists of the main categories for your website navigation menu. Navigation links on a home page offer users the opportunity to quickly learn about the site itself and the information offered. Information gathered in user analysis, personas and scenario development, and task analysis should be reviewed when developing information architecture. Home page of your site reflects the reasons users come to your site. You should make sure that the most important and most frequently performed tasks are represented on the home page. You can use information gathered from card sorting to perform grouping of content. The key is to look at the categories to see which groups of information have the most agreement among users and for which types of content users faced difficulty in categorizing. You should than develop a plan to deal with such content which was difficult to categorize.

A site map is a visual representation of the information architecture of your site that can help you organize your site structure. Before you create a site map, you should review the information gathered during website requirements analysis and content inventory to help you identify all of the features and content on the site. You should review the users' priorities from the Task Analysis and Scenarios to determine the items that need to appear on the home page. You should use the data from the Card Sorting activity to identify the main categories and the labels for the home page categories. Once you identify the links, features, and content that needs to appear on the home page, you can define the content that will appear on second-level and content pages.

Once you have developed the information architecture of your website, the next step is to create content for your website. When you start creating content pay attention to how users read online information. You need to present information in a form that makes sense to them. When users read online thy typically scan the site looking for relevant words that match the information they are seeking. Most users scan the web pages for keywords and sentences. According to a Nielsen study of 2008, 79% of users scanned Web pages; they read only 20-28% of the words on the page. Therefore, writing for web should be in a style that accommodates that behavior. Too much text on web pages can overwhelm users that are unable to pay attention for longer times. One technique is to break your text into small manageable sections. You should use short sentences and words that are easy to understand and include examples because users prefer examples and will often read them instead of the text.

It is also a good idea to design your site in parallel. In parallel design, several independent teams create an initial design from the same set of requirements. The design team then considers each design. Through parallel design, you can generate many different, diverse ideas in a short period and ensures that the best ideas from each design are integrated into the final concept.

Before you actually start full-scale development of your website, you should first build a prototype of your website. A prototype is a draft version of a Web site that can be anything from few pages, or a fully functioning Web site. Prototypes allow you to explore your ideas before investing time and money into website development and implementation. Although prototype can be built anytime ideally, it should be created as soon as possible. You can begin with a prototype, consisting of home page and few secondary pages, to see the information architecture of your site is working. Using prototypes, you can get users feedback while still planning and designing your website. According to a Nielson study of 2003, gathering usability data as early as possible can bring significant improvements in user experience. Nielsen estimates that it is 100 times cheaper to make changes before any code has been written than to wait until after the implementation is complete.

You will also need to be concerned about accessibility when programming your website. Accessibility is an important aspect of the web site design process that should be considered throughout the entire development process. An accessible website not only facilitates information access by disabled but also improve usability for all the users of the website. Once you have your page designs, information architecture, navigation menu, and content ready you can begin programming your site.

TESTING AND REINFORCEMENT

Once programming or coding of your site is done, you are ready to test it for usability. There are two types of testing techniques.

1.   Usability evaluations, which typically do not include users working with the product

2.   Usability tests, which focus on users working with the product

Usability evaluation techniques can include surveys/questionnaires, observational evaluations, guideline based reviews, and expert reviews. These techniques require a lot of judgment by evaluators and usually do not include representative users. A usability evaluation can be performed as soon as you have a prototype website. Usability evaluation results can become the basis of usability testing by developing hypotheses about what could be serious problems. You can then develop the usability test around those hypotheses.

Usability tests always include test participants. It is the only way to know if the web site actually has problems that keep people from having a successful and satisfying experience. A formal usability lab is not required for performing usability tests and these tests should be performed early and often. This test can be performed even at user's home. Cost of usability tests depends on many factors including website size, the extent of tests, types or participants etc. Generally, it is advisable to budget for more than one usability test. Building usability into a Web site is an iterative process. Usability tests should be planned carefully. Usability specialists and test team should spend time to be familiarized with the site. They should perform some dry runs using scenarios before performing actual tests.

The results of usability evaluation and usability test may or may not be consistent with each other. Usability evaluation predicts the problems or successes that users will have with the Web site while usability test with representative users tells you whether your predictions are valid.

IMPLEMENTATION AND RETEST

It may not be possible to implement all the recommendations of a usability test. A workable approach is to develop priorities based on fixing the most global and serious problems. Fixing most serious and widespread problems first can help you achieve your site usability goals quickly. You should, on priority basis, find out the changes that users require in the website. Remember that it costs more to support users of a poorly designed site than to fix a site, which is still under development. The most successful strategy would be to start an iterative site design process. You should begin with a partial prototype of your site, perform early usability tests, fix the problems, expand the prototype, again perform usability tests and repeat this process.

Usability Improvement Guidelines/Checklist

Following are some guidelines that can be used to improve the usability of a website.

MAKE YOUR WEBSITE SPEAK FOR ITSELF

Many users enter a website through a link and not from the home page. A good tag line explaining what the website is about can help the users. Such tag line can be included next to the website logo or somewhere on the page where it can be noticed. An example is cnn.com website.

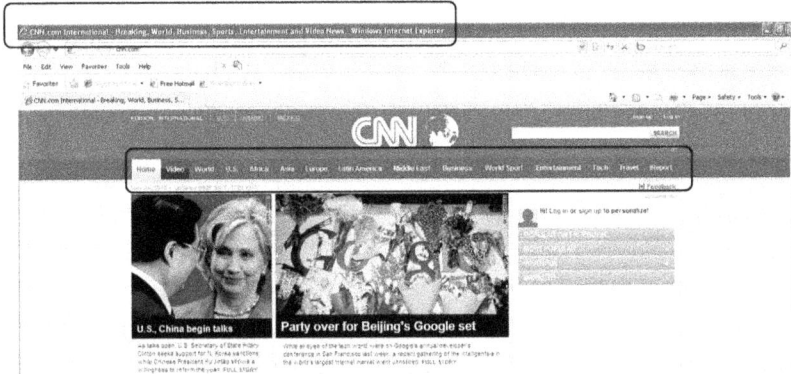

KEEP HOMEPAGES CLUTTER FREE

Homepage of a website should include only necessary information required for visitors to decide the purpose of the site and the next page they want to visit. For example, consider the home page of ning.com, a social networking website.

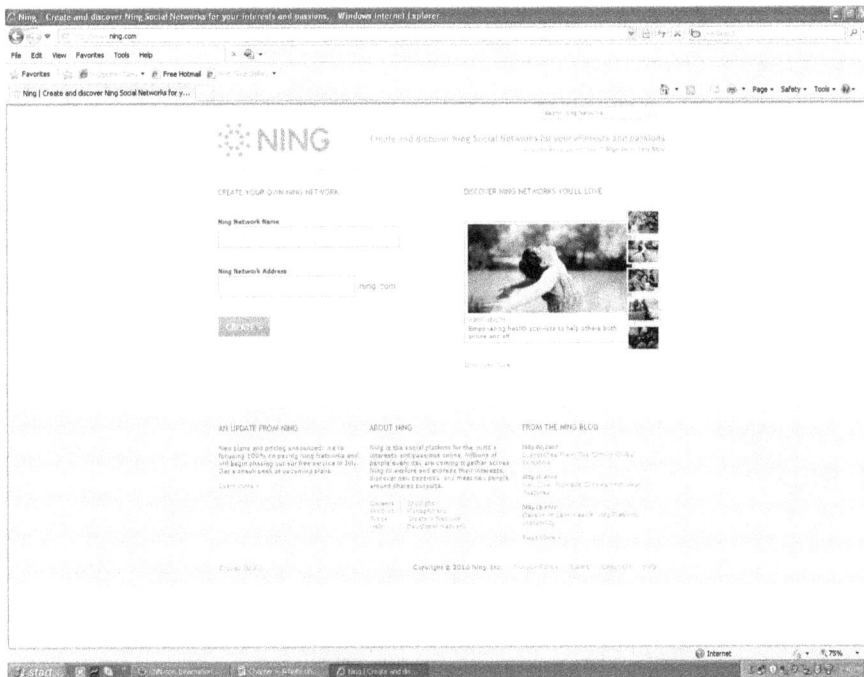

USE A CONSISTENT DESIGN WITH CONVENTIONS

The website should provide users with a consistent design that keeps navigation elements in the same format and position at all times. This design consistency, or uniformity among the sections, is important. Good examples of well-designed E-commerce sites include Amazon.com and eBay.com. These sites exemplify strong branding; their logos and color schemes appear on every Web page. Contact information is easily located.

PAY CLOSE ATTENTION TO YOUR WEBSITE NAVIGATION

Primary form of navigation on each website can be different e.g. tabs or menus. However, irrespective of the primary method of navigation chosen, it must be consistent across the site. For example, Amazon.com uses menus while cnn.com uses tabs as primary method of navigation. The three-click rule is an unofficial web design rule concerning the design of website navigation. It suggests that a user of a website should be able to find any information with no more than three mouse clicks. It is based on the belief that users of a site will become frustrated and often leave if they cannot find the information within the three clicks. Although there is little analytical evidence that this is the case, it is a commonly held belief amongst designers that the three-click rule is part of a good system of navigation. Critics of the rule suggest that the number of clicks is not as important as the success of the clicks (Porter, 2003).

FIGURE: 6-1: THREE-CLICK NAVIGATION

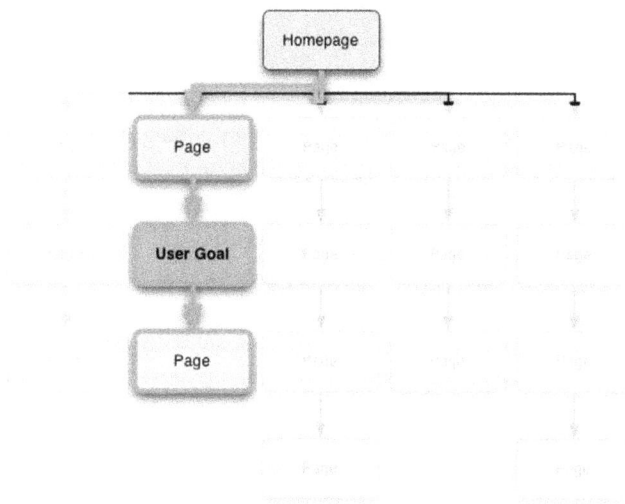

Your sitemap is an excellent navigation tool. The good thing about a sitemap is that it provides your visitor with direct access to anything on the website. However, in order for your sitemap to function as a navigational tool it is necessary for your visitors to be able to access it easily. Provide a link to your sitemap on every page on your website. The sitemap should be your secondary means of navigation, not the primary one.

There are no certain rules about how and where to put your navigation bar. Many webmasters put it along the left side of their webpage. Some put it on the top. Occasionally a website will put it on the right side, forcing the visitor to look in that direction. When deciding on the placement of your navigation bar you should consider the effect it will have on the space you will have available for your content. Navigation bars can take up a lot of space that could otherwise be used for content. This is why it is advisable to use links only to your main topics and branch out from them on following pages. Every website should have some type of theme - a way of giving your site an identifying look. You can do it by using a unique logo, a particular color combination or a recognizable idea such as gardening or pets. Your theme should be considered when designing your navigation scheme. For a professional look, it is very important that you blend your navigation components into the layout of your webpage. Develop a theme and work it into the way that you display your navigation links. (i.e. icons, bullets, colors, etc.). It is also important to note that you should always include a text-only copy of your navigation links, as some people surf with images turned off or even with text-only browsers. Page headers and footers are a logical place to include your text-only links.

DESIGN FOR QUICK RESPONSE AND LOAD TIMES

Faster pages make a better user experience. The user notices page load times, either consciously or subconsciously and it is possible that search engine ranking of pages is directly related to the page load times. As a result, sites offering pages with long load times cannot only downgrade their search engine rankings but also lose users. It is estimated that the E-commerce industry loses $1.1 to $1.3 billion yearly on slow load times. It is estimated that a web page should ideally load in one tenth of a second however pages loaded within a second are also considered well-designed.

HAVE A SEARCH OPTION ON YOUR SITE

Some of the problems associated with site layout can be solved by adding a search engine but surprisingly many sites either don't provide a search box or hide it away somewhere on the site. Search feature can make it easy for customers to required information on a website quickly and many users use the search feature as a primary method of navigating a site. By not having a search box, a website will risk losing the potential customers. A website does not need to code its own search feature. A website can use famous search engines such as Google to get their site indexed and then use code provided from Google and plug it in a search box. You can add a search engine to your site free through www.freefind.com. You submit your e-mail address and URL, and the site then sends you the HTML code via e-mail for implementing the search engine feature on your site.

PERFORM USABILITY TESTING EARLY AND OFTEN

Usability testing does not have to be expensive and you can start with selective groups of customers and ask them to perform certain tasks (like signing up or posting a comment) and take notes. When you identify a problem, fix it and test again.

MAKE HYPERLINKS EASY TO DETECT AND CLICKED

Hyperlinks are designed to be clicked and they should be easily detectable. The figure below shows a portion of Hacker News (http://news.ycombinator.com/). Look at the small hyperlinks created for comments. They are too small and clicking them is harder than it should be.

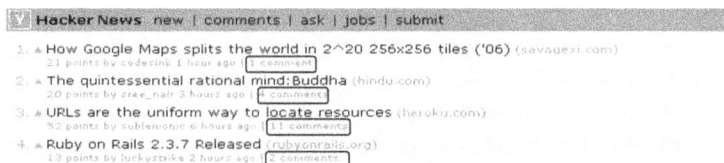

Hand movements with the mouse are not very precise and large clickable area can make it easy to move the mouse cursor over the link and click. The figure below shows a portion of the comments link, from newspond.com. Notice that the comments links are much larger.

AVOID INADEQUATE AND EXCESSIVE PAGINATION

Pagination refers to splitting up content onto several pages. Pagination is often used on websites to display long lists of items that cannot be displayed on a single page (see example of Google search results below). Pagination is adequate here because displaying too many items on one page would increase the page load time.

Pagination may also be used to increase page views. Many blogs and magazines get their revenue through advertising. They may use excessive and un-necessary pagination of content to get more page views and thus charge more for each ad. However, this approach has two problems. First, having to load several pages just to read one article can annoy your site visitors. Second, it is not good from SEO point of view. Search engines use the content on the page to make sense of what it is about and then index it accordingly. Content spread on several pages can dilute content on individual page (i.e. less keywords and less relevance to the topic) and downgrade the search engine ranking of the pages.

AVOID DUPLICATE PAGE TITLES

Web page titles are the pieces of text we write between the <title> tags in the <head> section of HTML code of a web page. The title of each Web page is important and creating a generic title (e.g. by using name of the website, in the website's template and then re-use the same title across the whole website) should be avoided. This approach has two problems. A generic title cannot communicate effectively to users what this page is all about and user will become confused whether they are at the right place or not. The second problem is SEO. Search engines use page titles as one of the important information to gauge relevancy of a particular page to a particular search term. When people look at the search results of a search engine they usually look at the title of the page to figure what this page is all about. Unique page titles succinctly describe the content of the page, including a couple of words searched by people. Look at the cnn.com Europe News pages below. See how the title of the page changes compared with the title of the home page.

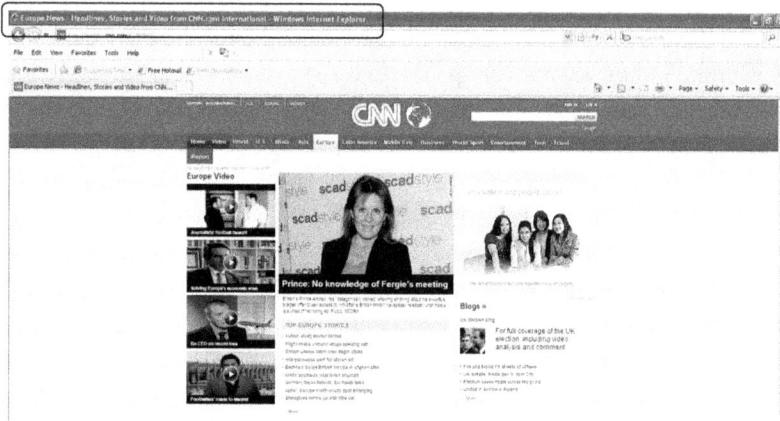

And here's how the page is displayed in a Google search result:

Europe News - Headlines, Stories and Video from CNN.com International
Provides continuous coverage and analysis of issues impacting the world. Available in English
www.cnn.com/EUROPE/ - Cached - Similar

MAKE CONTENT EASY TO SCAN

A usable site is the one that has good design and good content. Good site design guides the visitors around the website, focus their attention on the things that matter and help them make sense of information provided. However, visitors also need to read content to process information. When presenting this content it is necessary to understand how visitors read information online. Presented with a lot of information of a page, visitors try to consume as much of this content as possible. Doing so, visitors jump from one piece of content to another and from one website to the next. Visitors tend not to read websites top to bottom rather they start reading whatever pops out at them first, and then move to the next thing that captures their interest. To take advantage of this browsing pattern, designers should structure the content on the website a certain way. Each page should have a few points of focus i.e. parts of the page that could draw users attention. This can be achieved by using stronger, higher-contrast colors and larger fonts, images, such as icons, next to text to give these areas visual appeal. The focus points should be informative yet concise. Without reading any further, visitors should be able to understand what this bit of content is all about. Look at the snapshot from culturedcode.com. See how the features are split up into little bite-sized segments, each with its own icon and heading so that users can easily scan the list.

Your own agenda
A smart Today list automatically gathers all you need to look at, well... today!

Tags
With tags, you have them all: contexts, priorities, energy, time, and more. It's up to you.

Repeating To-Dos
Every week on Tuesday and Saturday? On the last day each month? We've got you covered.

Projects
You can never have too many. Avoid project overload by setting less pressing ones to inactive.

Get started
Have a task that cannot be started until a certain date? Use the Scheduled list to get it out of your head and be reminded when the time is right.

Areas of Responsibility
Things helps you organize life and work according to your responsibilities.

Due Dates
Every list can easily be filtered and sorted by due date.

People
Delegate to-dos to your coworkers and teammates.

Notes & References
Include notes and links to files, web sites or email messages. (email links require Mac OS X 10.5)

iCal Sync
Your to-dos, systemwide. (requires Mac OS X 10.5)

Search
Quickly search across your entire library.

iPhone Sync
Take your to-dos with you using the most popular paid task manager in the entire App Store.

At your fingertips
Convenient keyboard shortcuts make you even more efficient.

Dock Badge
See what tasks still lie ahead of you. (requires Mac OS X 10.5)

Multilingual
Available in English, French, German, Japanese, Russian, and Spanish.

Spotlight and Quick Look
Find your to-dos with Spotlight. Instantly preview them with Quick Look.

Fully scriptable
Unleash the power of Things with custom AppleScripts.

PROVIDE WAY FOR VISITORS TO GET IN TOUCH

It is natural for online buyers to be doubtful about buying products online. With many articles providing tips on shopping from a reliable and professional website, the informed online shopper becomes selective while selecting a website to purchase from. Displaying contact information on your website makes a great difference to the sales on your website. If your targeted visitors are able to contact you to get instant information, it is a great aid to increase your sales. Create a separate "Contact us" page with details of your contacts such as phone number, email address, etc. You can also provide a toll free contact number to increase number of phone calls and it will significantly increase sales on your site. Keeping your customers in contact is also important to build a successful community of loyal customers. Such community is important to build successful websites and social web applications. Still many websites lack ways through which their website users can get in touch with them. Some websites do it quite efficiently. Look at the website of Coca-Cola. When someone clicks on the Contact Us link on the official Coca-Cola website, he or she is presented with the following page.

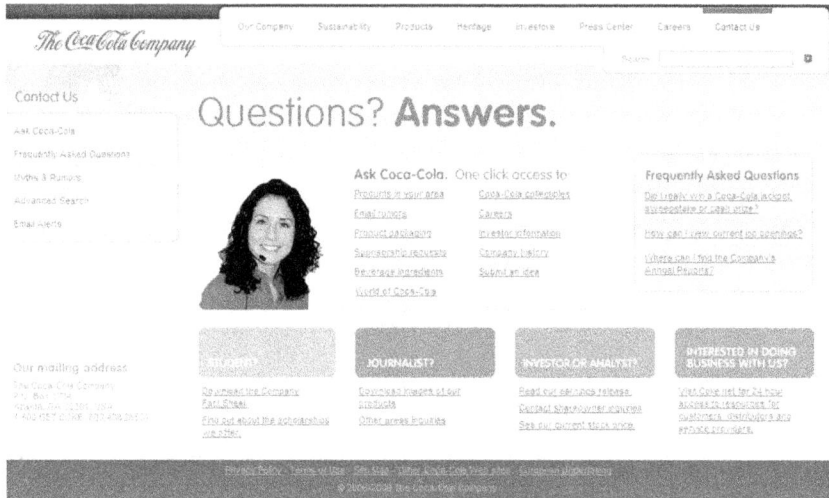

The page does not provide an email address or phone number and many links take you to the automated FAQs. The "Submit an idea" link opens up a form with terms and conditions attached. Coca-Cola uses contact forms to bypass the problem of showing the business email address on a page. In any case, the business is still likely to receive spam unless website uses some good CAPTCHA[9] or other spam protection mechanism. However, things like CAPTCHA are barriers to user interaction and likely to degrade the user experience. Therefore, designers should use these features carefully. Another way to get in touch with customers or visitors is online forums. One big advantage is that if the business is not able to respond to a customer query another helpful customer may help that person out, solving his or her problem. One such example is support forums operated by Dell.com.

AVOID EXCESSIVE USE OF USER REGISTRATION REQUIREMENT

A website may have content and features that require visitors to register before using. However, websites should be cautious and must not over use the registration feature. Features that require registration must not be mixed with those that do not. Excessive use of registration feature can act as a barrier to using the website. For example,

---

[9] A CAPTCHA is a program that can generate and grade tests that humans can pass but current computer programs cannot. For example, humans can read distorted text as the one shown below, but current computer programs can't.

the home page of pixlr.com, an online graphic editing application, has a link titled "Jump in n' get started!" When visitor clicks on this link it opens up the application. No registration is required.

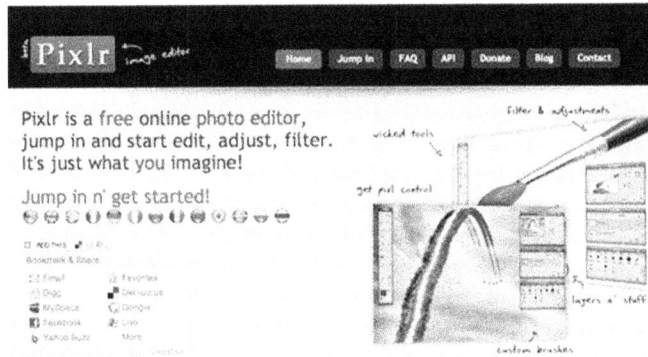

Another example is posterous.com, a blog hosting network. It does not even require registration to start using it. Visitors just send an email with their post and a new blog is created for them.

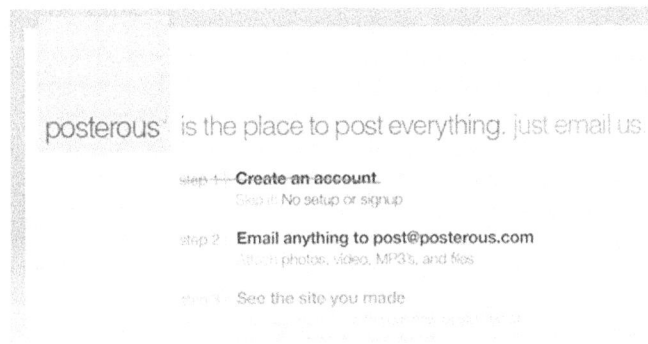

USE REDIRECTS FOR PERMALINKS

A permalink is a link to a page that does not change. Problem occurs when a website moved to another domain or its structure is reorganized. In that case, if a user uses the old link it will see a 404 error-message (Page not found). Designers should setup web pages using 301 redirects (instructions stored on the web server that redirect visitors to appropriate pages).

AVOID LONG REGISTRATION FORMS

Filling registration forms can act as a barrier to using a website because filling these forms is time and effort consuming. Remembering the user names and passwords used on registration forms is another hassle. The sign up forms used on the website should be kept as short as possible and show only necessary information required. For example, look at the signup form of readoz.com. The form is long with many optional fields.

New User Registration

Select Your Status

☐ I represent a business/organization whose publication(s) I would like to publish on ReadOz and I would like to 🔲
apply for the Official Publisher designation.

Login details

Login Name: [_____] ✱

Email: [_____] ✱

Password: [_____] ✱

Confirm Password: [_____] ✱

Personal info

First Name: [_____] ✱

Last Name: [_____] ✱

Company Name: [_____]

Job Title: [_____]

Address: [_____]

City: [_____]

Country: [Please Select ... ▼]

State: [Please Select ... ▼]

Zip Code: [_____]

Enter Code: [_____] KEELND

Another example is signup form of tumblr.com. The form is very concise and provides just three fields.

tumblr. [Sign up] [Log in] [Explore ▾] [_____]

Sign up

E-mail address

[_____]

Password

[_____]

URL

[_____] .tumblr.com

[Sign up and start posting!]

MAKE SIGN-IN/LOG-IN LINKS VISIBLE

The Log-in/Sign-in link on your web page should be visible. A greater font-size or use of icons would do the job. For example, hotmail.com uses large font-size and buttons for the Sign-in links.

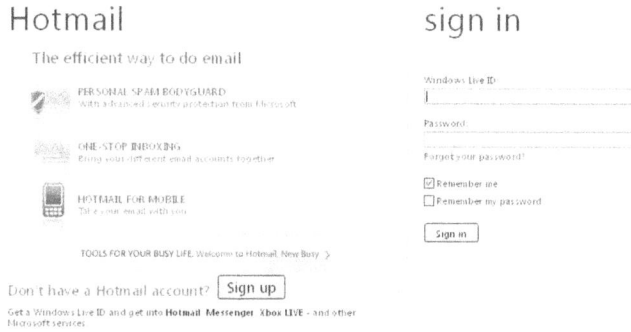

AVOID POP-UPS FOR CONTENT PRESENTATION.

Most modern web browsers blocks the pop-ups. Therefore using pop-ups to present the important content is not probably a good idea.

AVOID INVISIBLE LINKS

Site visitors need to know where they are, where they have been and where they can go next. Look at the following snapshot from real.com.

Solid circles show links while dotted circles are not. To avoid confusing site visitors in this way, some basic rules can be used. For example, non-link text should be presented in a different way (e.g. by making them bold or enlarged). If graphics are used a link than they should either be placed adjacent to the text-based links or some text should be embedded in the graphics.

AVOID AD CLUTTER

Ad clutter is not good and in many cases it can cause the important content on the site go unnoticed by the visitors. Look at the following snapshot from overstock.com. The page is not filled with ads. Visitors can easily differentiate what is a link and which is not.

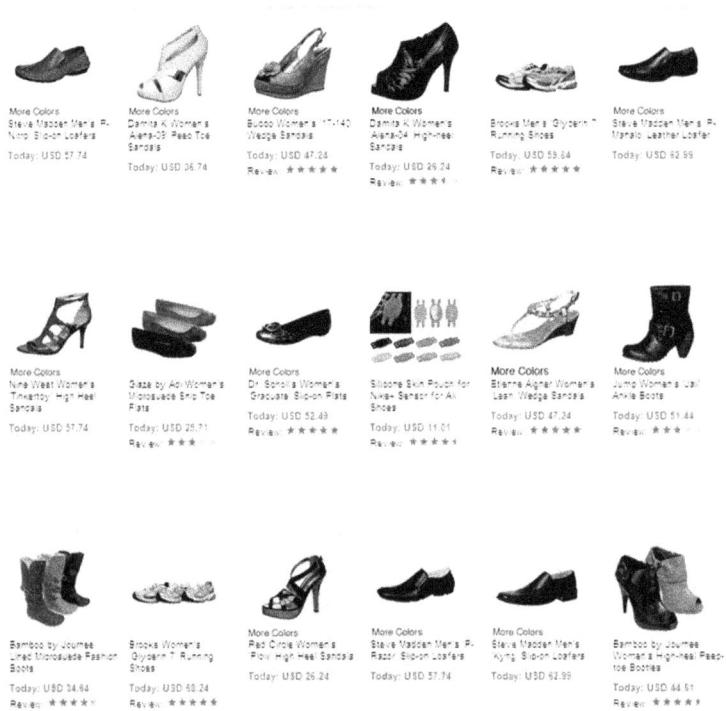

DROP-DOWN MENUS

Drop-Down menus are useful and they can save a large amount of vertical space on a web page. However such menus requires precise movement of mouse cursor over different sections and if the distance between different levels of navigation is too large (for instance because some navigation items have more text) users have to move the mouse horizontally. Look at the following snapshot from brita.net. In case of a change in vertical position of a mouse focus, user will have to start navigation with the menu again.

NO CLICKS

A trend is emerging in which the websites requires the visitors to just point to the content he or she is willing to use. Visitors do not have to click at anything. One example is Dontclick.it. Within this website, you will not find any buttons. Instead, visitors navigate the contents of website by pointing the mouse to the areas of the site they are interested in. However, designers should use this feature with caution as many visitors may still want to use the same button-style navigation.

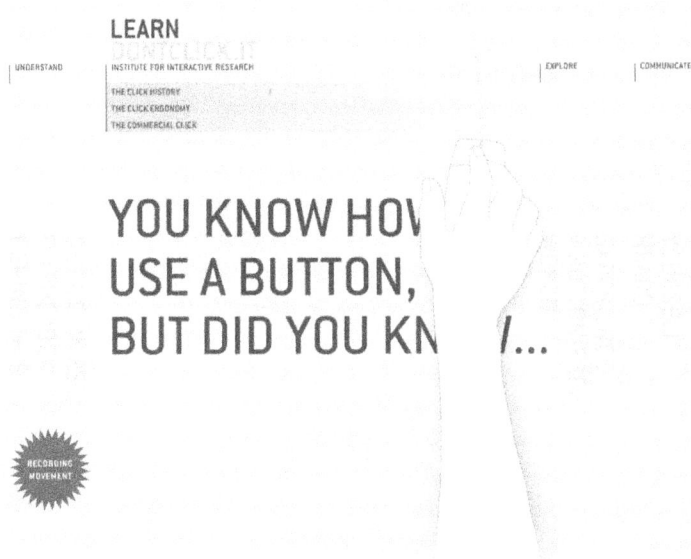

HIGHLIGHT IMPORTANT CHANGES

Visibility of the web site status is one of the most significant elements of a usable site. Visitors must notice immediately what is going on the website and changes occurring. One way to do this is with animation. The animation can be noticed by visitors fairly well if the rest of the page is static. For example playing an animation, when the user add an item into their shopping cart can be noticed by the user and it will also be an indication for the user that the action he performed actually worked. However, animations should be used carefully. Use of too many animations can slow down the speed of user interaction with the website, which in turn can frustrate the user. Web services such as www.eprise.com help businesses update their content regularly. This is particularly important for e-businesses offering timely information.

ENABLE KEYBOARD SHORTCUTS

To offer visitors more responsive and interactive user interfaces, a website can integrate keyboard shortcuts or navigation. This little feature can significantly improve the workflow of your site visitors and make it easier for them to get their tasks done. Look at the following snapshot from boston.com. See that you can navigate the web page using keyboard using letters "j" and "k".

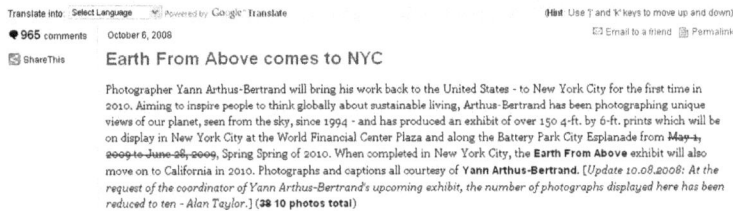

The shortcuts should be intuitive and self-explanatory. For example, it would not make sense to make the shortcut letters for "Up" and "Down" navigation too far apart from each other on the keyboard; rather, they should be close together.

AVOID USING DARK BACKGROUNDS

Avoid using dark backgrounds - in particular black - for text, which will then be hard to read. Do not use a small font that is difficult to read. Backgrounds and pictures will also slow down the downloading of pages.

USE A COMMON LANGUAGE

Websites are written in the HTML, in one of its versions, or in similar languages like XML. You want to write your website in a language that most browsers can read, like the basic HTML.

HAVE ADEQUATE EMAIL FACILITY

If you select a free or low fee website and/or e-mail server, the ISP may restrict the sizes of your e-mail and the number of website hits. Make sure that you agree to the ISP's rules. You can also have your own e-mail and website servers.

USE FEATURES TO ENHANCE USER EXPERIENCE

There is variety of features you can add to your site to enhance the consumer's shopping experience. One example is using image-enhancing software. For example, Magic ToolBox (www.magictoolbox.com) offers a variety of software tools that allows consumers to take a closer look at your merchandise. Customers can focus on a particular area, zoom in, zoom out, rotate the image, and see texture and quality. Community-building tools can also enhance your visitors' experience and increases the possibility they will visit the site frequently. For example, your e-business can provide a place where people can ask questions and find answers or locate an event near them. MyEvents.com is an application service provider that offers both Internet and wireless access to common files, including calendars, reminders and bulletin boards.

CREATE A SEPARATE LANDING PAGE FOR YOUR VISITORS

Landing pages are usually small, standalone web pages designed for a single purpose, usually a marketing campaign. They are called "landing" pages because they are most often used as a destination for a link. When someone clicks a link in an email blast or an online ad or something else, they "land" on a page with information and a call to action specific to the campaign. This is usually much more effective than simply linking them to your website homepage or some other generic site. Clear and precise information has the potential to convince your prospective customers. You need to make it a point that the landing page for your visitors should be separated from the home page of your website. Home page is often overloaded with information and takes longer loading time, which may confuse your visitors and discourage them from going any further. A unique landing page designed for a specific purpose makes it easier for your visitors to understand and take interest.

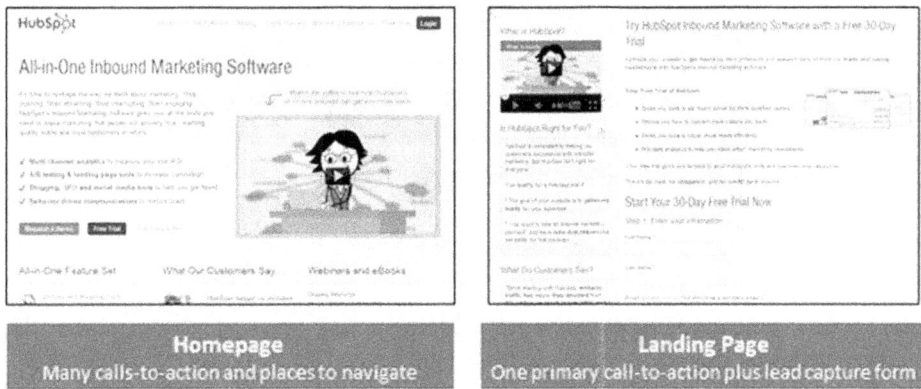

Homepage
Many calls-to-action and places to navigate

Landing Page
One primary call-to-action plus lead capture form

FOCUSES ON A SINGLE THEME ON A SINGLE PAGE

You should let your targeted visitors get a clear idea about your products or services from the landing page. Whatever products or services you provide, it is essential that the related items should be displayed on a single page. If you try to sell items of different themes on the same page, chances are that your prospective customers get distracted and lose interest in your products. Focusing on a single theme per page also makes it easier to select targeted keywords for the page that ensures relevant traffic.

CREATE MULTIPLE WEBSITES FOR DIVERSE THEMES

It is not recommended to sell products from very different industries on a single website. First, it will be difficult to select the right kind of keywords to promote your website. Second, you may find it difficult to target your audience from a completely different market if your products do not belong to the same industry. Therefore, you should create multiple websites with each site serving an entirely different purpose with entirely different content.

SHORTEN SHOPPING CART CHECK OUT

One of the effective ways of attracting your visitors is to introduce an improved shopping cart system. Often, visitors lose interest when they have to click through different options to finish their buying processes. With an easy buying process, your visitors are more likely to accept the buying process. Try to provide them simple ways by asking only the necessary information to fill in.

DON'T USE SPLASH PAGES BEFORE YOUR HOME PAGE

Splash pages are those "click here to enter" pages that you sometimes see when you go to a website's homepage. Several years ago, it was in vogue to add these intro pages to a website, but today they are not very common, probably because they are really annoying.

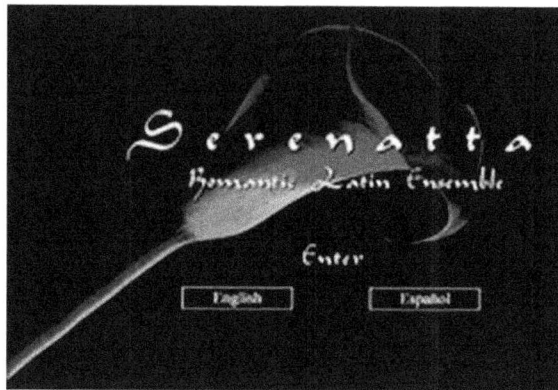

As web-designing techniques have matured over the years, splash pages have all but disappeared, because they are unnecessary and can cause people to avoid your website.

DON'T USE AUDIO FILES AS CUTE TRICKS

Many websites use music track or sound effect looping on and on in the background. In many cases, it is unnecessary. If you feel it is necessary, give the visitor control over it from the page with a well-marked, obvious set of controls. Remember that navigation is navigation, animation is animation, and sound is sound. For mp3 sites or gaming sites, audio and music is a critical part but for reading your engaging text, audio and music is not that much important.

USE ADVANCED PERSONALIZATION TECHNIQUES

In addition to the standard personalization technologies, there is technology called rules-based filtering. Using this method customer need to answer a set of question to get the appropriate information presented. The rules can range from the country the user's comer from to the specific need of each customer. A site may ask for zip code or the name of your hometown, for example, to find out which local news you may be interested in. it can also involve question about personnel preferences. Another approach is collaborating filtering. Collaborative filtering (CF) is a technique commonly used to build personalized recommendations on the Web. Some popular websites that make use of the collaborative filtering technology include Amazon, Netflix, iTunes, IMDB, LastFM, Delicious and StumbleUpon. In collaborative filtering, algorithms are used to make automatic predictions about a user's interests by compiling preferences from several users. It helps in refining the process of customizing a particular website to the customer. This technology tries to serve relevant material that may be of interest to customer by combining the preference of customer with the preferences of a group with similar interest. In this way, customer automatically recommended product, service, & information to other with similar interest. See figure below.

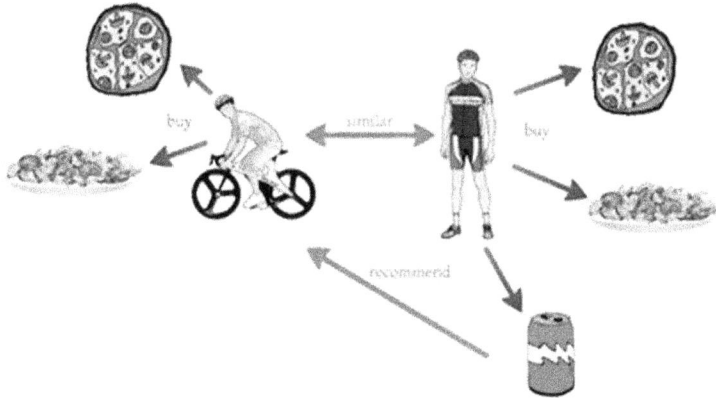

Review Questions

1. Why an effective web presence is important for a company?

2. How a company's web presence can affect its brand?

3. What is website usability? What does it measure?

4. What is customer-centric website design? What steps are involved in creating a user-centric website?

5. List some tips to improve website usability.

Bibliography

Charlton, G. (2014, May 14). Adaptive web design: pros and cons. Retrieved July 7, 2015, from https://econsultancy.com/blog/64833-adaptive-web-design-pros-and-cons/?utm_campaign=bloglikes&utm_medium=socialnetwork&utm_source=facebook

IBM. (2008, March 11). *Information architecture essentials, Part 4: Improving usability of information systems.* Retrieved December 24, 2009, from ibm.com: http://www.ibm.com/developerworks/library/ar-infoarch4/

Lake, C. (2014, January 13). 18 pivotal web design trends for 2014. Retrieved July 7, 2015, from http://econsultancy.com/blog/64096-18-pivotal-web-design-trends-for-2014/?utm_campaign=bloglikes&utm_medium=socialnetwork&utm_source=facebook

Loiacono, E., Watson, R., & Goodhue, D. (2007). WebQual: An Instrument for Consumer Evaluation of Web Sites. *International Journal of Electronic Commerce , 11* (3), 51-87.

Mahmood, M. A., Kohli, R., & Devaraj, S. (2004). Introduction to the Special Issue: Measuring the Business Value of Information Technology in e-Business Environments. *International Journal of Electronic Commerce , 9* (1), 5-8.

Marketing Donut. (2015). Websites: the four pillars of wisdom | Marketing Donut. Retrieved July 13, 2015, from http://www.marketingdonut.co.uk/marketing/online-marketing/your-website/websites-the-four-pillars-of-wisdom

Nikolaeva, R. (2005). Strategic Determinants of Web Site Traffic in On-Line Retailing. *International Journal of Electronic Commerce , 9* (4), 113.

Pang, M., Suh, W., Kim, J., & Lee, H. (2009). A Benchmarking-Based Requirement Analysis Methodology for Improving Web Sites. *International Journal of Electronic Commerce , 13* (3), 119-162.

Porter, J. (2003, April 6). Testing the Three-Click Rule. Retrieved July 13, 2015, from http://www.uie.com/articles/three_click_rule/

Scott, D. (2014, May 13). The five golden rules of responsive web design. Retrieved July 7, 2015, from https://econsultancy.com/blog/64823-the-five-golden-rules-of-responsive-web-design/?utm_campaign=bloglikes&utm_medium=socialnetwork&utm_source=facebook

smashingmagazine.com. (2007, September 27). *10 Usability Nightmares You Should Be Aware Of.* Retrieved January 26, 2010, from www.smashingmagazine.com: http://www.smashingmagazine.com/2007/09/27/10-usability-nightmares-you-should-be-aware-of/

Stewart, K. J., & Malaga, R. A. (2009). Contrast and Assimilation Effects on Consumers' Trust in Internet Companies. *International Journal of Electronic Commerce , 13* (3), 71.

Tung, L., Xu, Y., & Tan, F. (2009). Attributes of Web Site Usability: A Study of Web Users with the Repertory Grid Technique. *International Journal of Electronic Commerce , 13* (4), 97-126.

Usability.gov. (2009). www.usability.gov. Retrieved January 2010

Usunier, J., Roulin, N., & Ivens, B. (2009). Cultural, National, and Industry-Level Differences in B2B Web Site Design and Content. Internatioal Journal of ELectronic Commerce , 14 (2), 41-88.

7

E-COMMERCE MARKETING

Learning Objectives

After reading this chapter, reader should be able to:

- Understand significance of Internet marketing.

- Understand the difference between Internet marketing and E-marketing.

- Understand the common Internet marketing mistakes made by companies.

- Understand the advantages and disadvantages of Internet marketing.

- Understand 7 P's of Internet marketing.

- Understand various customer segmentation and analysis techniques

- Understand how to create and maintain brands on the web.

- Understand applications of various Internet tools in Internet marketing.

- Discuss Internet marketing communication and its various tools.

- Discuss the ways in which a Web site can be used as a marketing communications tool.

- Understand web positioning and its strategies including Search Engine Optimization (SEO) and Search Engine Marketing.

- Discuss the role of social networking sites in E-commerce.

- Understand various web and social analytics used in E-commerce

- Recognize and describe important Web 2.0/3.0 tools and their applications in E-commerce.

- Understand virtual worlds and their use in E-commerce.

- Discuss Internet users and their Internet usage patterns.

- Discuss the basic concepts of online consumer behavior and purchasing decisions.

- Understand how consumers behave online.

- Discuss various online consumer aids, including comparison-shopping aids.

- Describe Internet marketing in B2B, including organizational buyer behavior, marketing, and advertising.

- Discuss the issues of trust in E-commerce and factors affecting consumer trust building in E-commerce.

- Describe consumer market research in E-commerce including its methods.

- Describe partner relationship management (PRM).

Introduction

Companies initially used Internet merely as an alternative for traditional advertising but today many companies are using Internet to create innovative online marketing strategies. Internet marketing today is affecting all businesses though its significance to businesses varies for different products and markets. For electronics equipment companies (e.g. CISCO) Internet marketing is very significant because Cisco gains majority of its global revenue through online business. For digital products (e.g. software, music, and information), Internet marketing is extremely significant because for most such businesses 100% sales occurs online. For FMCG brands, Internet marketing is less significant because most of the sales revenue is generated through conventional sales channels and not through online business.

With consumers spending increasing amounts of time on Internet, Internet is becoming increasingly important in influencing consumers' purchases decisions. Internet has become an effective way to reach target markets, increase consumers' frequency and depth of interactions with the brand, support full range of organizational functions and processes that deliver products and services to customers and other key stakeholders, support the entire marketing process, and as a powerful communications tool to integrate various functional parts of the organization. Internet marketing is continuing to impact various businesses and many new Internet-based business models that currently exist only within particular sectors will eventually spread through other sectors.

Impact of E-commerce on Marketing

Stockdale and Standing (2004) state that e-commerce facilities the business and the basic principles of marketing are the same. The effect of e-commerce on consumers and marketing practitioners brings about a new experience for both groups as they try move from an offline-based kind of relationship to one that is fundamentally online based. While marketing practitioners try to invent new strategies to meet the demands of e-commerce and by implication attract and win the hearts of customers, consumers also seek marketers to help them make the right choices in their purchases (Yazdanifard & Samuel, 2012). According to Allen and Fjermestad (2001) e-commerce offers a new way to interact with customers and by connecting with the traditional advertising techniques the effect can be doubled. Chaffey et al. (2000) mention that companies can add additional products and services around their core products. Allen & Fjermestad (2001) state that e-commerce can lead to price standardization. According to the literature, the effect of e-commerce on the "place" should be more visible than the other three elements of the marketing mix. Chaffey (2002) explains the big opportunities for covering large customer area and international expansion. According to the literature, the increased ability to exchange information and cover larger customer area is among the main benefits (Xu & Quaddus 2009; Turban et al. 2008).

Internet Marketing and E-marketing

(SMITH & CHAFFEY, 2005) defines Internet marketing as "Achieving marketing objectives through applying digital technologies". These digital technologies include Internet media (such as web sites, e-mail) as well as other digital media (such as wireless, mobile, cable and satellite). In practice, Internet marketing also uses online promotional techniques (such as search engines), banner advertising, and links or services from other web sites.

An alternative term is e-marketing or electronic marketing. (Strauss & Frost, 2001) define e-marketing as the use of electronic data and applications for planning and executing the conception, distribution and pricing of ideas, goods and services to create exchanges that satisfy individual and organizational objectives". The definition by Straus and Frost is the official definition for E-marketing adopted by the E-Marketing Association. This term has a broader scope and considers improving both internal and external marketing processes and communications by using information and communications technology.

Common Internet Marketing Mistakes

Theodore Levitt, in his article "Marketing Myopia" identified some factors that form the basis of the downfall of many organizations and at best seriously weaken their longer-term competitiveness. According Levitt, any organization that defines its business by what it produces is said to be suffering from marketing myopia. The factors identified by Levitt are still relevant and can be used by organizations as reminder to avoid costly mistakes when starting their Internet marketing. The factors include:

1. Organization's misunderstanding of the business they are in

2. Absence of focus on customer needs and market opportunities

3. Unwillingness to innovate

4. Narrow vision in terms of strategic thinking

5. The lack of a strong and visionary CEO

6. Not giving marketing its true importance.

According to Levitt, businesses need to be customer-centric rather product-centric and businesses using technology should not ignore the fact that technology is only an enabler, not an objective. Customers on Internet seek information to make informed purchase decisions and companies should use strong exhortations to influence customer purchase decisions. The need for market orientation is a critical aspect of Internet marketing strategy and companies should explore using Internet to create new opportunities for value addition. Businesses that fail to understand Internet marketing broader applications within the total marketing process and focus on just using it as a communication and selling tool cannot integrate and fully establish the Internet marketing as strategic marketing management tool. Internet as a media is capable to communicate and sell, but this is only one important aspect of the marketing process to which the Internet can contribute. When developing an Internet marketing strategy, businesses need to decide which marketing functions can be assisted by the Internet. Businesses tend to first use the Internet to restrict applications to promotion and selling rather they should use Internet as a relationship building and service delivery tool.

Advantages and Disadvantages of Internet Marketing

There are many advantages of Internet marketing for businesses (Gangeshwer, 2013).

ADDITIONAL SOURCE OF REVENUE: Additional source of revenue made possible by an alternative marketing and distribution channel.

MARKETING PENETRATION: The Internet can be used to increase sale of existing products into existing markets by utilizing the power of Internet advertising to increase product and company awareness amongst potential customers in an existing market.

MARKET DEVELOPMENT: Internet can be used to develop new markets for a company's products by using Internet advertising that is cheap, has an international reach, and does not require a supporting sales infrastructure in the customers' country.

NEW PRODUCT AND SERVICE DEVELOPMENT: New products (e.g. information products such as market reports) or services (e.g. stock trade) can be developed which can be delivered using Internet.

DIVERSIFICATION: Using Internet, new products developed can be sold into new markets. Using search engines, you can promote your product or service directly to people who are actively looking for it.

Ease of Tracking Advertising, Promotion, and Sales: Using Internet marketing allows you to easily monitor your advertising and identify/target specific markets individually. Doing so, you could and focus your efforts on the most effective advertising medium thereby increasing effectiveness of your marketing. Internet marketing also allows use of rich media advertisements including text, audio, graphics, and animation in web ads. Internet advertising is capable of providing fresh and up-to-the-minute information and you can track statistics such as people who actually saw the ad or even opened the page featuring the advertisement. Advertisers can even record time consumers spent looking at the advertisement. Companies using Internet advertising can provide bonus offers for purchases made online as they spend less overall on their marketing. They can use low cost e-newsletters to keep their old and new customers informed and updated on new products.

Methods of E-commerce Marketing

There are many different types of online marketing methods available to promote products or services (Huang & Benyoucef, 2013). Following we discuss some of the popular ones.

ARTICLE MARKETING: It is composing articles, uploading them, and driving traffic to websites expecting to sell products, services, or opportunities. You can get two benefits from article marketing. First, if you create articles with truly valuable content and upload and promote the article, you have a great likelihood that readers will follow the links that you include in that article, especially if you have a strong "call to action" near the link. Second, the links that you place in your articles can become incoming links to your site, increasing the value of your site in the eyes of the search engines. The most powerful and elusive part of SEO is creating a web of incoming links to your site, and article marketing is a fundamental part of an effective link building strategy.

FORUM MARKETING: Forum Marketing is sharing the knowledge about the area of your expertise on online public forums. People ask questions and you answer their questions based on your expertise. You create your online reputation so that people may trust you for your answers. One example is http://quora.com. This is an online forum where people share ideas and views. Most questions are technical. Most of the forums allow a link back to your website in your signature text, which would appear below the post you make.

SEARCH ENGINE MARKETING: Search engine marketing (SEM) is a form of Internet marketing that involves the promotion of websites by increasing their visibility in search engine results pages (SERPs) primarily through paid advertising.

PAY PER CLICK ADVERTISING: Pay per click (PPC), also called cost per click, is an internet advertising model used to direct traffic to websites, in which advertisers pay the publisher (typically a website owner) when the ad is clicked. It is defined simply as the amount spent to get an advertisement clicked.

LINK EXCHANGE: The practice of exchanging links with other websites. You place another site's link on your site, usually on a links page, and in return, the other site places a link on their site back to you. A link exchange is a confederation of websites that operates similarly to a web ring. Webmasters register their web sites with a central organization, that runs the exchange, and in turn receive from the exchange HTML code, which they insert into their web pages.

CLASSIFIED ADVERTISING: It refers to Small messages grouped under a specific heading (classification) such as automobiles, employment, real estate, in a separate section of a newspaper or magazine. These relatively inexpensive ads are usually a column wide, do not include any graphics, and are typeset (see typesetting) by the printer or publisher of the publication.

EZINE MARKETING: Ezine is a newsletter that is sent via email that members subscribe, request information from a particular market niche. Ezines are also classified as an electronic magazine. Ezines sent weekly or bi-weekly, but they can also be delivered daily when using ezine solo ads posted. Ezine marketing is a strategy of using ezines to market your business online.

LIST BUILDING: This refers to building a subscribers database who are entered into the list of that database when they opt in for your ezine or agree to a free download against providing you their email address. You can contact these subscribers with your offers and promos.

VIRAL MARKETING: viral marketing is any marketing technique that induces Web sites or users to pass on a marketing message to other sites or users, creating a potentially exponential growth in the message's visibility and effect

ONLINE PRESS RELEASES: Here, you build a press release for your website and submit it to one or more press release sites like prweb.com. Press releases are done to create awareness among the web visitors and are displayed on various news channels or sites on the net. Press releases also build increased back links. Both free and paid submissions are available.

JOINT VENTURES: Here two or more marketers come together and promote a product or service in a way that it will benefit them all. Joint ventures are a great way to build your business because the marketing efforts are combined and results are always more than individual efforts.

AFFILIATE MARKETING: Affiliate marketing is internet advertising that allows any online business to affiliate themselves with web site owners (known as affiliates or publishers) using affiliate programs. Affiliates make money by generating sales, leads and traffic for the Merchants business

RESELL RIGHTS MARKETING: You can offer resell rights to your product where people would be able to sell it and keep all the money. It increases your product value and helps in building in name because more sales are made than when you sell it just on your own.

RSS MARKETING: RSS stands for Really simple Syndication and it may be difficult to conceptualize in the beginning. RSS works by RSS feeds which needs to be generated by website. People can subscribe to these RSS feeds and can view the content of the website via their RSS reader. By this, people are enabled to receive the content directly on their desktop.

BLOG MARKETING: A weblog, also called a blog, is a journal that a blogger maintains and it contains information that is instantly published. Blogging is a very popular activity. Blog marketing is any process that publicizes or advertises a website, business, brand or service via the medium of blogs. This includes, but is not limited to marketing via ads placed on blogs, recommendations and reviews by the blogger, promotion via entries on third party blogs and cross-syndication of information across multiple blogs

SOCIAL BOOKMARKING: Social bookmarking is a way for people to store, organize, search, and manage "bookmarks" of web pages. Users save links to web pages that they like or want to share, using a social bookmarking site to store these links. Social bookmarking can introduce sites to others with relevant tastes, drive traffic to your site, and valuable backlinks.

VIDEO MARKETING: Video marketing is a new type of internet marketing and advertising in which business create 2-5 minute short videos about specific topics using content from articles and other text sources. The videos are then uploaded to various video sharing websites like YouTube for distribution and exposure.

The 7 P's of Internet Marketing

Elements of the marketing mix, the four P's, include product, price, place, and promotion. Research shows that the 7 Ps' model of marketing is perfectly able to adapt to the changes in the society and the market produced by the diffusion of digital technologies, particularly Internet [see (Aldridge, Forcht, & Pierson, 1997); (Peattie, 1997); (O'Connor & Galvin, 1997); (Bhatt & Emdad, 2001); (Allen & Fjermestad, 2001); (Möller, 2006); (Jain, 2013)]

PRODUCT: Clear facts about the product or service being sold online can be readily made available to the customers so the customer knows the facts and not sales persons' assumptions. Buying process can be customized for returning customers to make the repeat purchases easier.

PRICE: The internet has made pricing very competitive and many costs associated with traditional business have disappeared. Many places to shop from and easy access to information, have increased customers bargaining power, made prices competitive, and placing price pressures on traditional retailers. With technology to track repeat customers, companies are able to provide targeted incentives.

PLACE: Online customers can purchase directly from manufacturers without needing retailers. The challenge for online retailers is to ensure that the product is delivered to the customer within a reasonable time. In this context, the choice of location becomes important.

PROMOTION: New forms of product promotion are available with Internet e.g. banner ads, e-newsletters, e-mail, Web public relations (e.g. publishing stories based on product or service launches on the company's webpage) etc. These new methods of promotion are less expensive and can reach a large number of audiences in very less time. Having a recognizable domain name is first stage towards promoting products and services online. However, sometimes companies may find it difficult to get a domain name of their choice.

PEOPLE: When you are doing Internet marketing, you have to be able to outsource the work and hire the people to do things like write effective content. You will want the people who can provide answers to visitors' questions and provide them with the information they are looking for. Lastly it is very important to ask if you have the people to share information on the product, service, or company on blogs, forums etc. by posting about it.

PROCESS: It is very important for Internet marketing too. You need to ask if you have resiliency in your site, the ability to handle large number of customers, the proper support at all times, and a system to answer FAQs. There will be a time when an internet marketing company scales campaigns according to each client's needs. There are times when clients tell the internet marketing company to scale back the efforts since they cannot handle the volume, and have to expand resources, and the marketing company adjusts campaigns accordingly.

PHYSICAL EVIDENCE: Online, it is difficult for the customer to know how the product is going to benefit them. So it will be necessary to communicate in a way that the customer will be able to feel confident in purchasing a product or service. You can do reports and articles that will excite the customers about the product and the service. Video and images also help the customer feel comfortable. If you talk to any Online Marketing Company representative, you can learn about how they take care of video, images, content, reviews, articles, and synchronize all of your offline and other online marketing campaigns for accountability.

Marketing Strategies for E-commerce

There are two broad strategies of E-commerce marketing.

Product-based Marketing Strategy

Product-based marketing is built around the idea that if you build a better mousetrap the world will beat a path to your door (Dutt & Kashyap, 2014). Building a cheaper mousetrap will work, too. Product differentiation and low-cost based strategies are classic examples of product-based strategies. Some strategies are better suited for some products than others are. For example, high-tech companies can command high premiums for products if no competitor can offer the same capabilities, while products for which little differentiation is possible, such as table salt, is more likely to compete on price or placement. When creating a marketing strategy, managers must consider both the nature of their products and the nature of their potential customers. Many office supply stores on the Web believe customers organize their needs into product categories. Best Buy (www.bestbuy.com) is an example of product-based marketing strategy. On their home page, you will notice that more detailed product category links are at the left-hand side. Staples designed its page to meet the needs of the customers who has a specific product category in mind.

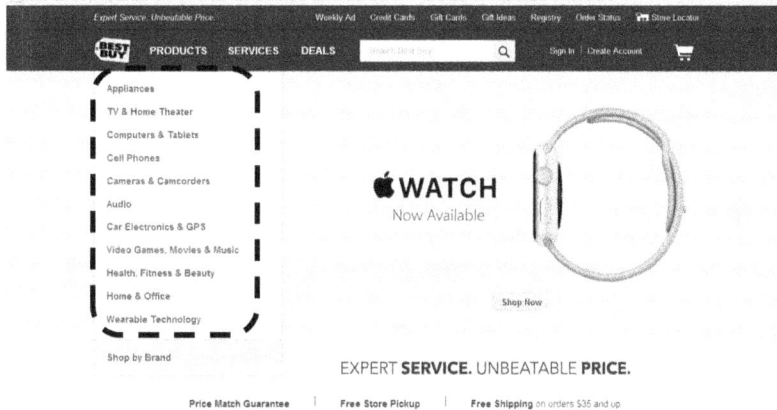

Another example is strawberry.net, a cosmetics store. You can see all the product categories listed on the top.

Liberty Books (www.libertybooks.com) sells all kind of books, magazines, novels etc. On their websites, we can see categories of different items sold that any customer can select.

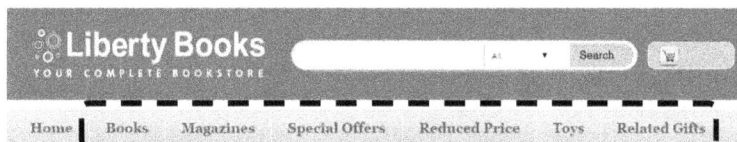

If customer arrives at these websites looking for a specific type of product, this approach works well. Many marketing researchers and consultants advise companies to think as if they were their own customers and to design their web sites so that customers find them to be enabling experiences that can help customers meet their individual needs.

Customer-based Marketing Strategy

Customer-based strategies are built on the understanding that it is often easier and more profitable to keep an existing business relationship instead of acquiring a new customer every time you have to make a sale (Virginia Phelan, Chen, & Haney, 2013). Customer-based strategies are also motivated by the realization that some customer segments are more profitable than others are. Good first step in building a customer-based marketing strategy is to identify groups of customers who share common characteristics. B2B sellers are more aware of the need to customize product and service offerings to match their customers' needs

Macy's is an American super store. Macy's sells products classified into different categories and different product lines. The home page of this company includes links to separate sections of its site that are designed to meet the needs of each major customer groups. By following the links, the Macy's different customers can find specific products and services targeted to each of their needs.

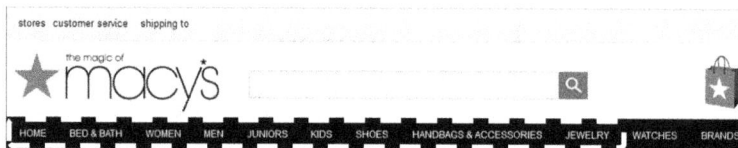

United Colors of Benetton is a company that sells clothing for men, women, and kids. Their website has different categories to serve their customers according to their needs.

The use of customer based marketing strategy approaches was pioneered on B2B sites. B2B sellers were more advertising on the web are likely to fail for the same reasons--- many people ignore or resist messages that lack content of any specific personal interest to them. Companies can use the web to capture some of the benefits of personal contact. Thus, the internet provided a new vehicle for achieving high levels of customer-focused marketing strategies.

Customer Segmentation in E-commerce

A segment is a subset of people who you can identify within your audience, that you can describe via their attributes and behaviors, and that you can reach with precision across channels. Personalization platforms come with many "out-of-the-box" segments and, while it is great to be able to distinguish your new visitors from your returning visitors these targets only address part of the interactions you are having with a customer. As a marketer, you need a way to reconcile their behavior across channels, which is where segmentation becomes more powerful (Wilmoth, 2015).

To create better ecommerce customer segments, you need to use three categories of information:

Context: This is real-time data (e.g. location, weather, and device type) and demographic data (e.g. age and gender). All of these data points impact the mindset of your customer, and how they interact with your brand.

Behavior: This is observed data i.e. what your customer is doing on your site, what products, brands, categories they express interest in, and what actions they take, like adding items to carts, wish lists etc. All of this information gets you closer to understanding a customer's buying intent.

Relationship: This historical and cross-channel data makes it possible for you to recognize and relate to your customer. Taking a backward look at the customer will also help you determine stability of your segment. Are they tied to a particular purchase cycle? Do they only shop with you during a particular season or time of day? Do they only make a purchase when presented with a promotional offer?

These categories progress in a certain way. They start broad, then narrow in scope. Not all of your customers are going to fall into a segment. That is why there are natural breaking points. Context provides you with the break points to form groups (e.g. female, college educated, urban etc.) while behavior will show you how people differ within the group. This is also, where you start to form better segments. Companies need to examine what types of behavior are taking place within the group, how often, and when. This will help to zero in on your most valuable customers. The next step is to analyze the relationship, which will bring forward the attributes, behavioral patterns, and even changes in context over time that help you understand when, and with what urgency a customer is ready to convert.

Software-based Customer Segmentation

With marketing budgets being squeezed, marketers see improved targeting as critical. That means improved customer segmentation based on the ability to score customers' propensity to buy, churn or default – and react accordingly, at speed. Software-based segmentation can provide companies such capabilities. For example, Optimizely (www.optimizely.com) is a software that you can use easily target the right audience for your tests and personalized experiences. You can build your own segments and audiences using advanced targeting capabilities (geography, URL, browser etc.) and third-party integrations, or by using existing lists. Optimizely's visitor segments allow you to categorize your visitors for the purposes of filtering your results and even adding personalization to the user experience. Once you have created a visitor segment, you can also target your experiments to run only for that visitor segment.

Analytics in E-commerce

The online marketing space is constantly shifting as new technologies, services, and marketing tactics gain popularity and become the new standard. E-commerce businesses are being significantly affected by these constant evolutions. In order for these businesses to survive and thrive, they need to be able to make better decisions faster. This is where web analytics comes into play (Butilon, 2015). The resources of many small-to-medium E-commerce businesses are finite. Having access to statistical information from all areas of your online marketing and sales activities gives you a competitive advantage and, understanding shifts in consumer behavior gives you insights into the demands of your market.

Web Analytics

According to Web Analytics Association, Web Analytics is the measurement, collection, analysis and reporting of Internet data for the purposes of understanding and optimizing Web usage. Web analytics can be used as a tool for business and market research to measure the results of traditional print media advertising campaigns. Some important industry bodies involved in developing definitions within web analytics include JICWEBS (Joint Industry Committee for Web Standards), Web Analytics Association, and IAB (Interactive Advertising Bureau).

There are two categories of web analytics. Off-site web analytics refers to web measurement and analysis regardless of whether you own or maintain a website. It includes the measurement of a website's potential audience (opportunity), share of voice (visibility), and buzz (comments) that is happening on the internet as a whole. On-site web analytics measure a visitor's journey once on your website. This includes its drivers and conversions; for example, which landing pages encourage people to make a purchase. On-site web analytics measures the performance of your website in a commercial context. This data is typically compared against key performance indicators for performance, and used to improve a web site or marketing campaign's audience response (Marz & Warren, 2015).

The analysis can be used for likelihood of customers purchase based on previous purchases, website personalization for customers, determining sales volume, observing visitor's geography, visit pattern, and purchase pattern, and designing web advertising campaign.

Following are some key definitions related to web analytics.

Click or Click-Through or Ad Click: If the visitor clicks the banner ad to open the advertiser's page, that action is called a click or click-through.

Ad Views: Refers to the number of times (during a specific period e.g. a day) users call up a page that has a banner on it. Ad view is also called impression.

Button: It is a small banner linked to a Web site. It may contain downloadable software.

Hit: A hit occurs each time a user requests for data from a Web page or file.

Visit: A visit occurs when a visitor requests a page from the Web site. When the visitor loads further pages from the same site they are counted as part of the visit for a specified period. This period is chosen by the site administrators and is dependent on the type of site. A site featuring stock quotes might use a short time period because visitors may reload the page quickly to check the change in stock price. A government site might use a long period because visitor would be expected to load multiple pages over a longer period during a visit and would use a longer visit time window.

Unique Visit: A unique visit is a count of the number of visitors to a site, regardless of how many pages they view per visit.

Trial Visit: Trial Visit occurs when a particular visitor loads a page from a web site very first time. Subsequent pages loaded are called repeat visits.

Click-through Rate or CTR: CRT can be used to measure the success of an online advertising campaign. CRT is calculated:

CRT = number of users who clicked on an ad on a web page / number of times the ad was delivered (i.e. impressions).

For example, if a banner ad was delivered 100 times (impressions = 100) and one person clicked on it then the CTR would be 1 percent.

Bounce Rate: It is the percentage of visits where the visitor enters and exits at the same page without visiting any other pages on the site in between.

Run of Site (ROS): Refers to displaying a banner ad throughout a website or a banner network with no targeting by keyword or site category. Run of site advertising costs substantially less than more targeted advertising.

Popular Tools for Web Analytics

GOOGLE ANALYTICS: It is one of the best free tools that any website owner can use to track and analyze data about web traffic. Google Analytics helps marketers and website owners understand traffic patterns, sources, conversions, bounce rates, paid search statistics, and more. It can be integrated with the Internet's largest paid search platform, Google Adwords. This helps marketers to effectively track and improve landing pages.

MOZ ANALYTICS (/moz.com/products/analytics): This tool is more appropriate if you are looking for more qualitative data about your website traffic. It can be used by marketing teams to monitor SEO-related issues and map social media campaigns in conjunction with Google Analytics data. It provides data on issues that can hurt organic SEO results and provides relevant keyword recommendations that hold immense value to websites especially if your business is inbound-marketing focused.

CRAZY EGG (www.crazyegg.com): Using this tool, businesses can get a visual picture of how site visitors navigate their sites, along with features such as heatmaps and scrollmaps. Businesses can track what made users click and how far they scroll down pages.

KISSMETRICS (/kissmetrics.com/): This analytics tool tracks user behavior on your website. Businesses can access easy-to-understand visual charts detailing user engagement patterns and habits before and after making purchases. This information allows business to identify website flaws and bugs, improve and fix them and then increase conversions.

ADOBE ANALYTICS: Adobe Analytics is the industry-leading solution for applying real-time analytics and detailed segmentation across all of your marketing channels. There are different versions of this product. For example, Analytics for mobile app improve mobile app discovery and engagement with integrated app analytics, messaging and intelligent location marketing. Analytics for Video standardize content and advertising measurement of viewer engagement across digital devices and platforms.

The Marketing Funnel

One of the basic concepts of web analytics is funnels. All marketing activities can and should be seen in terms of funnels. The idea of funnel analytics is that your target audience will go through a systematic flow or funnel until they make a purchase on your site (F. Nunes, Bellin, Lee, & Schunck, 2013). A typical marketing funnel may look like this:

1. A fan on your business's Facebook page sees one of your posts.

2. The fan clicks on the post.

3. The fan arrives on a landing page advertising a specific product and clicks on "add to cart."

4. The fan clicks on checkout.

5. The fan enters their personal information and finalizes the purchase.

At each step of the process, a certain percentage of people will drop out of the funnel. Knowing these percentages will help you determine the barriers and psychology behind your customers.

Another classic example of a marketing funnel is that of an email campaign; let us look at an example:

1. You send 1,000 of your past customers an email promoting your summer sale.

2. Out of the 1,000, 970 are delivered by your email service.

3. Out of the 970, 350 are opened by past customers.

4. Out of the 350, 50 click on one of the product links in the email.

5. Out of the 50 that clicked, 45 add the product to their shopping cart.

6. Out of the 45 that added one or more products to their shopping cart, 10 finalize their purchase.

When looking at this funnel, we can see that the campaign resulted in a 1% conversion rate. This number, however, does not tell the whole story. We can see that there were significant drops at the "open email," "click on product link," and "finalize purchase" stages. This tells us where our focus should be. You can setup funnel statistics using Google Analytics. Google provides great support for e-commerce analytics natively within Google Analytics and provide both the high-level and drilled-down information needed to make important choices about how you conduct your online business. Figure 7-1 shows a marketing funnel.

FIGURE 7-1: MARKETING FUNNEL

Traffic Tracking

Web analytics (such as Google Analytics) can track four categories of traffic: search, referral, campaign, and direct. Search traffic is traffic that comes directly from search engines like Google, Bing, and Yahoo. Referral traffic is traffic that comes directly from a link on a different website. Campaign traffic is traffic that has been tagged by the marketer. Direct traffic is traffic that does not have a known source. A good practice is manual tagging of your marketing campaigns so they are better segmented within Google Analytics. Let us explain this with an example. Let us say you publish two posts to Facebook. The first is a regular post about your industry, and it links to a blog post on your site. The second is a special post announcing a new promotion. If these two posts are not manually tagged, all the traffic from both posts will appear in Google Analytics as Referral traffic with Facebook being the referrer. It would be more useful to tag each of these posts as different campaigns. You can tag links using UTM tags. A UTM code is a simple code that you can attach to a custom URL in order to track a source, medium, and campaign name. This enables Google Analytics to tell you where searchers came from as well as what campaign directed them to you. UTM tags are small snippets of text, which appear at the end of a link.

FIGURE 7-2: USE OF UTM TAGS

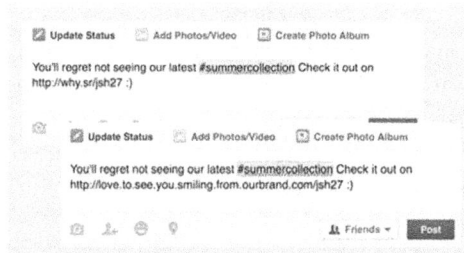

E-commerce businesses need to use UTMs on every single link that drives traffic to any of their web page. The more segmented your traffic the more you can learn about the different channels.

Setting up Web Analytics

Google Analytics has become a standard tool when it comes to web analytics because of its ease of use, informative reports, and the fact that it is free. Google Analytics is a very powerful tool for e-commerce sites because Google allows you to send all your sales data to your Google Analytics account. Once Google Analytics is integrated to your site, all your sales will be tied to actual sessions, allowing you to connect sales to specific marketing channels.

Setting up e-commerce tracking in Google Analytics is a multi-step process, which requires that you first enable e-commerce tracking in your Google Analytics admin and then make some changes to your code. Making changes to code is tricky and a programmer's help should be sought for doing this. Google Analytics can provide both the high-level and low-level data on their customers. Without easy access to the information that can help you make necessary business decisions, you will be blind to movements in the market and potentially miss thousands of dollars in additional revenues.

There are some other important web analytics metrics. You can setup E-mail marketing funnel. Social media has become a popular marketing channel for online stores over the last few years. As with email marketing, it is possible to map out entire funnels for your social media (i.e. Facebook, Twitter etc.) marketing activities.

Social Media Tools for E-commerce Analytics

Like any quality marketing endeavor, yours should be based on quality market research. It should be done before you try to acquire traffic and truly engage people who are looking for your products and services. This can be done easily if you have a number of tools handy, have a good process in place, and know what to do with the insights you will gather. (Smirnova, 2014). Most of the marketing activities of E-commerce businesses today revolve around social media. Below we discuss some social listening tools you can use for gathering marketing analytics. These tools can be used for competitive analysis, content/influencer discovery, campaign management, publishing automation and conversation listening.

COMPETITIVE ANALYSIS TOOLS

Rival IQ (http://www.rivaliq.com/)

Rival IQ is perfect for getting full intelligence coverage on your target market. You can setup landscapes (up to 30 sites) in your category or content publishing niche, find the hottest and most engaged topics and channels. You can spot gaps where your competition is not present. You can mine the top 100 SEO keywords for search volume, traffic share and search rank positions. This is a great tool if you are new to a customer segment. Using this tool, you can get a clear picture of what works for your competition in terms of engaging and converting online audiences on social channels.

Site Alerts (sitealerts.com)

Site Alerts is great tool to can learn where the traffic comes from for the sites you are targeting. You also gauge what technology these sites use on their backend such as ecommerce platforms, marketing tools, analytics etc. You can also determine when those technologies were implemented.

BuzzSumo (http://www.buzzsumo.com/)

BuzzSumo is tools you can use to learn what people are talking about and who your influencers are. It also tells you which top bloggers to watch.

Keyhole.co (http://keyhole.co/)

Keyhole.co is a tool you can use to see what topics of interest are a high priority for your audience, who engages best with which hashtags (from your set of competitors) and even get some sense of placement where this traffic originates from.

Compete (http://compete.com/)

Compete is a tool you can use to know who is who in terms of size of traffic, types of traffic and what the site is all about.

CONTENT DISCOVERY TOOLS

Scoop.it

Scoop.it allows you to have topics tracked online. It is a good, easy way to create a content library that starts building itself daily or weekly. In addition, you can also select some topics of your choice from the pre-set categories or again name your own.

Similarweb (http://www.similarweb.com/)

Similarweb is perfect for discovering upcoming sites or apps that cover your topics for future partnerships, seeing topics that are trending and building your library of creative ideas from other people's posts.

Feedly (http://feedly.com/)

Feedly is more for searching and finding sites on the topics of your interest and creating a bookmarked page with all of them in one place. It is very handy to discover sites that cover your stories and its easy user interface allows you to create your own collections of stories to spin your content from. In addition, it is free and accessible from iOS or android devices via apps.

CAMPAIGN MANAGEMENT TOOLS

Tailwind (http://www.tailwindapp.com/)

Tailwind is a Pinterest analytics tool that you can use to manage your Pinterest marketing. If you have an account, you can see how your brand page performs (how many followers, pins, repins and likes you have). It gives you data on your boards' performance, shows likes, hashtags and comments.

Piqora (https://www.piqora.com/)

Piqora is a tool you can use to optimize your Pinterest, Tumbler and Instagram campaigns.

Curalate (http://www.curalate.com/)

Curalate is very similar to Piqora, yet it also bring image tagging intelligence (that track images across the web), which results in more concrete & different kind of customer specific insights.

Shoutlet (http://shoutlet.com/)

Shoutlet provides a platform for large E-commerce businesses with thousands or millions of products to create sales-driven campaigns at scale, monitor conversations, and provide third party data from interests, groups and activities to measure your campaigns across other social channels. It also integrates with Kenshoo, Google Analytics and your CRM system.

Qwaya (http://www.qwaya.com/)

Qwaya is a Facebook marketing platform for smaller online stores and allows you to manage Facebook campaigns with more ease and at a scale.

SocialApps HQ (http://www.socialappshq.com/)

SocialApps HQ is a tool for running contests or lead-generation campaigns on Facebook. It allows you to set campaigns up through forms, tabs, and features within the channel.

Ritetag (https://ritetag.com/)

Ritetag is a great tool to gain insights on what hashtags to use (which ones are visible, high, moderate and overused). This makes a big difference if you sell via Twitter posts, as the right tags bring more retweets, follows and sales.

Tagboard (https://tagboard.com/)

Tagboard can be used to create custom boards or landing pages for special events and launches that you do via engaging your social audiences. You can organize the entire campaign around a specific hashtag, pulling posts from Facebook, Twitter, and Instagram.

Trendsmap (http://trendsmap.com/)

Trendsmap is a good tool to see what tags are trending around the globe with top users, cities, videos and links. It is also perfect to get a sense what works on Twitter in other countries or how your campaign performs across regions.

Topsy (http://topsy.com/)

Topsy is a tool you can use to bring you photos, links, videos, and influencers for your specific topic in many languages.

PUBLISHING AUTOMATION TOOLS

HooteSuite (http://hootsuite.com/)

HooteSuite works best to automate your social posts on Twitter, Google+, and LinkedIn for new campaigns. It has great analytics reports and many extra apps as add-on features.

Sprout Social (http://sproutsocial.com/)

Sprout Social is a tool to be used when you have a team or outsourced people do your social posts and you are doing many times a week.

Social Annex (http://socialannex.com/)

Social Annex is more of an enterprise level platform that offers scalable publishing of your social and user generated content on your site, which could be Q&A, reviews, ratings and comments.

CONVERSATION LISTENING TOOLS

Trackur (http://www.trackur.com/)

Trackur is good tool for keywords, brand mention monitoring and tracking across the web.

Social Mention (http://www.socialmention.com/)

Social Mention is a real time search tool for your brand, or keywords of your choice across blogs, questions, videos, bookmarks and images.

IceRocket (http://www.icerocket.com/)

IceRocket is a search tool for content across social channels like Twitter, Facebook and blogs that generates plenty of searches to look through and can be used as a site discovery tool. You can see trends for specific terms like seasonality and subscribe to your searches via RSS.

TalkWalker (http://talkwalker.com/)

TalkWalker is a tool to see what people are saying on the web about your brand and it even does sentiment analysis marking. You can find influencers, segment conversations by themes, demographics and results.

Newsblur (http://www.newsblur.com/)

Newsblur is a tool that you can use to create a big content library in one place for all the subscriptions that you have on the web. You can subscribe to a new publication to see the content (topics titles, kind) the site is doing and let you see quickly if this could be a potential partner.

Personalization in E-commerce Marketing

In retail, there is a sales tactic known as suggestive selling. In essence, suggestive selling is an attempt by a sales person to sell the customer an accessory for a product that they are already buying. Therefore, if you walk into a shopping mall to purchase a phone, the sales person may suggest a car charger to go along with the phone. This tactic has now spread across industries. Dropbox offers Packrat, which gives users unlimited deletion recovery. It is an add-on to their core product. (Patel, 2014). For e-commerce companies, suggested selling is a tried and true tactic for increasing average order size. Take example of Amazon. Amazon relies heavily on suggested selling that you can notice while browsing their product pages. This personalization has provided Amazon great success and it is now a common practice of E-commerce businesses (Agarwal, 2015).

Optimizing your Product Pages for E-commerce Marketing

To buy or not to buy is the question. In addition, a product page is where this crucial decision is made (Chawla, 2014). Optimizing your product pages for a great customer experience can increase the conversion rate of your e-commerce store. There are many ways you can optimize your product pages. It is assumed that social proof helps to improve conversions. This is why many E-commerce businesses have social sharing buttons on their product pages. However, evidence exists that it may not be the case for every company. In a study, a company tested their original product page with social sharing buttons against the new version with no sharing buttons, their "add to cart" click-through rate improved by 11.9%. A couple of arguments can be made to justify why social sharing buttons are most likely a bad idea for product pages. First, these buttons compete with the main conversion goal of the page (i.e., adding items to cart) and distract visitors from it. Second, unless it is something exceptionally cool or exclusive, not many people like sharing a product page or landing page on their social profiles. Therefore, the negative social proof actually can backfire in such cases. Third, these buttons sometimes increase your page load times. An experiment found that adding a Facebook "Like" button to E-commerce sites added 1.3 seconds page load time (Ogborne, 2012). Note that every second of delay in page load time can reduce conversion by 7%. (Work, 2011). E-commerce businesses should run some tests to see if adding social buttons work for them.

Many E-commerce businesses feel that authenticity of the product can be a more important concern for brand-loyal shoppers. In an experiment, Express Watches (a UK based online retailer of Seiko watches) found that the product page with a badge of authenticity increased online sales by 107% (Deswal, 2012). E-commerce businesses selling branded products should test whether their customer price or authenticity and design their product pages accordingly.

The product page should only provide necessary information to keep the page crisp. Underwater Audio tested two versions of their product page for waterproof headphone sets. The focused new page with no distractions engaged more visitors and increased Underwater Audio's online sales by 40.81% (Chawla, 2013). E-commerce businesses should check their product pages to ensure these pages do not have any irrelevant links and do not offer too much information that the majority of visitors do not need to make their purchase decision. Such unnecessary information can sometimes distract visitors from your conversion goal or might even overwhelm them. Establish a clear eye flow on the page to help visitors easily scan the page.

When you are competing with established brands, it is important to pay close attention to what makes you a relatively better choice. In addition, it is important that your prospects do not overlook it. Paperstone (paperstone.co.uk) is an e-commerce store of office supplies in the UK that competes with some big names, like Viking and Staples, in its industry. Paperstone displayed a small price comparison table on product pages, comparing their prices with two known brands in the industry. The table was placed right below the "add to cart" button so that visitors would notice it. This assured visitors they were getting the best deal. It is important to note that these comparison tables were shown only for products where Paperstone was offering relatively lower prices. This move by the company increased their online sales by 10.67% (Deswal, 2014).

Small and medium E-commerce businesses competing with big names in the industry, should find the one thing that should make their prospects buy from them instead of their competitors. They should place this information strategically so that their visitors do not overlook it.

Optimizing User Interfaces for E-commerce Marketing

A good user interface is about more than just pretty graphics and stylish fonts. It is about all the "little things" that add up to a professional, compelling brand presence. It is about creating websites that users actually look forward to browsing and using (Basińska, Dąbrowski, & Sikorski, 2013). There are some user interface tweaks to improve conversion rates (Jacob, 2014).

Calls to Action

Oftentimes, particularly with the influx of mobile users, having a large clickable (or tappable) area on your call to action buttons is crucial to getting them clicked. These buttons should provide a more responsive and user-friendly option. Small links may look slim and sleek on a desktop, but are squint-worthy exercises in frustration on a smartphone. To reduce lead times, these images can be swapped with a more sophisticated CSS button (Tzuaan, Sivaji, Yong, Zanegenh, & Shan, 2014).

Error Messages

Error messages should be friendly. Figure 7-3 shows some phrases to be avoided in your error messages.

FIGURE 7-3: ERROR MESSAGES TO BE AVOIDED

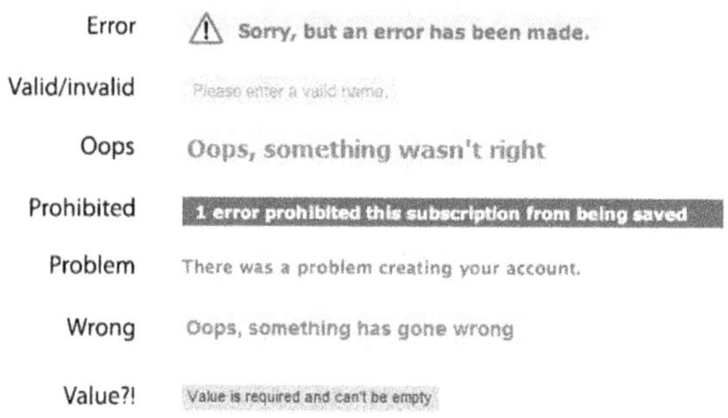

Error	⚠ Sorry, but an error has been made.
Valid/invalid	Please enter a valid name.
Oops	Oops, something wasn't right
Prohibited	1 error prohibited this subscription from being saved
Problem	There was a problem creating your account.
Wrong	Oops, something has gone wrong
Value?!	Value is required and can't be empty

Build Around Results

A common focus in the user interface designing process is asking yourself "What does the user want?" or "What does the user need?" and then constructing around those wants or needs. Fact of the matter is people want results, and they want to know they are on the right track to achieving those results. Including things like forward-focused systematic navigation can help them get in, get out, and get on with their lives.

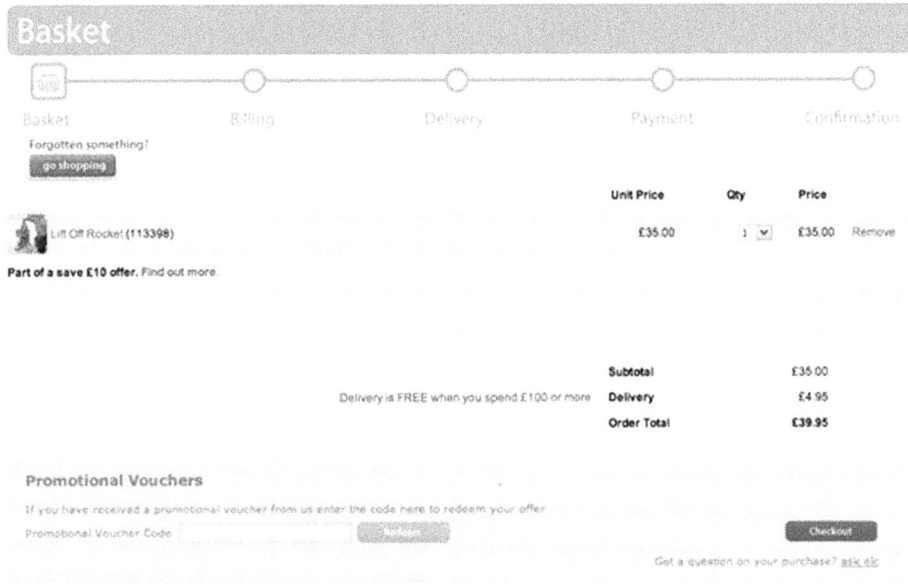

Design process does not matter much to the end user and they are only interested in getting things done with as little frustration as possible. The easier you can make the process, the more appreciative the end user will be and the more likely they will come back and tell their friends about their positive shopping experience.

Use Icons

Icons help give users a point of reference – particularly when navigating a site or when scrolling and scanning over a page. Common, everyday icons might be the best choice especially if you are serving an international audience where icons that are more creative might not be immediately clear. If you are going to use esoteric icons, make sure to label them clearly, so that there is no misunderstanding exactly where clicking them will take you. Icons are meant to supplement your navigation, not replace it.

Use Sliders Sparingly

Sliders are very popular with many modern websites for several reasons. They are trendy. They let you pack a lot of information above the fold and let you highlight key areas of focus such as specials or sales. However, many times they scroll so fast and try so hard to fit everything in that they end up defeating the original purpose of sharing information that is meant to get attention. Make it simple to navigate. Decrease the number of slides and use text concisely to communicate your ideas clearly with fewer words. Consider if you actually need to use a slider or if a static image would be better suited to your target audience. Run a split test between both ideas to determine what works best for you.

White Space

White space is often an afterthought in user interface design. When you are working with limited screen space (such as designing for mobile), it can be tempting to fit as much as possible into that small space. Proper use of white space increases focus and lead the eye to key visual elements, headlines and other areas of your site that you want visitors to focus on. It presents your site as more welcoming, open and accessible – all perceptions that you want to encourage. White space combined with clear images and short, to-the-point text gets your point across far better than trying to pack everything into a tight, closed space. A good example is website of Metta.

Persona Analysis

A buyer persona is a fictional customer who closely resembles your ideal customer. Constructing a believable, 'human' persona makes it much easier to understand and empathize with your customers, so building personas should be your first step. Personas should be ready at the planning stage of content marketing or social media campaigns (Rashid et al., 2013). A persona replaces a lot of statistical data about customer with a picture of customer that can be internalized with the customer you're trying to attract and relate to them as human beings (Campbell, 2015).

A persona is different from a definition of an ideal customer. The ideal definition of customer consists of dry statistics. For example, a definition of a customer might say, "they're usually between 30 and 45, they normally reside in South London and they're mostly men". It is hard to picture these customers while creating content. Now let us look at persona that may describe a customer, as "George is a 39-year-old man from South London who's interested in your product". With such basic description, it is easier to put a face to a name. Now, instead of asking, "Will this perform well with x demographic?" you can ask yourself, "Will George like this?"

Having just one customer persona is probably too limiting. Many experts recommend making between three and five customer personas, a number that is big enough to represent the variety inherent in a large audience, but small enough to retain the value of specificity. To construct a persona, begin with a business card template and fill in the blanks for your ideal customer. You will want to determine name, job title, and location. After that, add a resume with age, gender, family, education, and career. Next, try to imagine your customer's hopes, fears and values. It is even better if you can go beyond your imagination. A great way to gain this kind of data is to use quizzes and polls on Facebook and on your site. People love these and they consistently generate engagement. In addition, they give you the details necessary to create true-to-life buyer personas. Finally, add an 'elevator pitch'. This pitch should be tailored specifically to this customer. It will give you a clear idea of what you want to say to this customer, and that clarity will come through in other messages. As you develop your personas, you need to answer other questions that would bring these personas to life. Some of these might include computer literacy, hobbies, and news sources, and real quotes from actual customers, drawn from interviews, polls, comments online or other sources are a goldmine of useful information. Nothing can humanize your persona more than some actual human input.

To build a persona you need to get information from various sources. You can check your site's existing traffic analytics. This set of data shows you who your audience is, what they want and how they go about getting it. You can use social media listening skills to build an image of your customers via their Facebook and Twitter interactions with your brand. You can use surveys, interviews and polls to gather data that will allow you to create effective marketing personas. A persona is a fiction, but to be useful it has to be rooted in truth. Real data will reveal that truth and the more of it that comes from customers. When you are making decisions about how to orient your marketing efforts, you should almost feel like your customer persona is in the room with you. You should know if your customer persona would like, hate, or love everything about it. Below, is a sample persona for a customer named Toby.

Toby – "Fashion Phone Upgrader"

"One year in phones is a long time"

Toby loves technology and has to be seen with the newest and coolest digital gadgets. His phone is not just about making calls; he loves using its wealth of features for everything he can: surfing the web, writing emails, social networking and using it as a personal organiser.

Because he gets bored quickly with his phones, Toby is always looking for the latest toy and pays attention to new releases. He frequently upgrades part way through his contract and is willing to pay the upgrade fee to get the best phone. To him, a contract is a mere inconvenience, but something he endures to get a bigger discount off his new phone

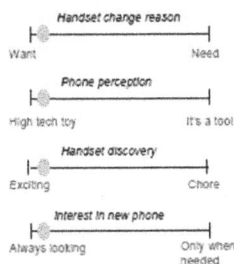

Key Characterics
- Age 20-35
- Is tech savvy
- Loves showing off his new phone to friends
- Would find a way to get out of his current contract for the latest phone
- Keeps up to date with the latest phones online
- Gets bored with phones quickly

Goals
- Have the latest, coolest phone
- Be up to date with the newest phones on the market
- Use as many features on his phone as possible

Behaviours

Handset change reason
Want ——————— Need

Phone perception
High tech toy ——————— It's a tool

Handset discovery
Exciting ——————— Chore

Interest in new phone
Always looking ——————— Only when needed

With your entire buyer personas defined, you should now have a much better idea of who, how, and where you should be marketing to achieve the greatest level of success. The demographic information of each persona will help you target your ads more effectively, and the psychographic information can help you write copy that is more engaging for each particular persona. For example, we know from creating our persona above that advertising a sale or discount to Alex will likely be more effective than trying to advertise quality since we know Alex is price conscious.

Cohort Analysis

Cohort Analysis enables you easily compare how different groups, or cohorts, of customers behave over time. This gives you quick and clear insight into customer retention trends and the health of your business. You can see how your latest customers compare to those from several years ago, or compare users who joined over the holiday season with another group that joined in the summer and see if those holiday shoppers really stuck around. There could be many use cases for cohort analysis. You can use cohort analysis to examine where cash flow is coming from and understand the health of your business; see how much monthly or quarterly revenue is driven from newer and older cohorts; study customer retention patterns to see if they are getting better or worse; and to compare cohorts of users from different segments (Mason & Fienberg, 2012).

Customer Life Time Value (LTV)

The lifetime value (or LTV) of customer is probably one of the easiest metrics to calculate, but many overlook its importance. Many people throw around the term "ROI". ROI is an oversimplified concept. The key concept missing from most marketers' perspectives on ROI is that, although we know intuitively that the goal is to develop lasting relationships with consumers, it's not a metric that most ecommerce marketers track and therefore isn't a metric that they can work on improving. If you are trying to calculate your ROI but just modeling the individual sales you get from your investment instead of looking at the value of customers you acquire and nurture, you're missing out on the big picture (Ammon, 2013).

LTV calculations vary wildly based on methodology. This is because the metric is a complex one to calculate and includes both profits obtained from a customer as well as expected profits in the future as well. There can be various use cases of LTV analysis. We can use LTV analysis to compare LTV across channels and ads within channels. We can also use LTV analysis to find lower value customers and work on improving their value (Ovchinnikov, Boulu-Reshef, & Pfeifer, 2013).

When calculated correctly, you will know how much your customers spend, how often they spend it, and what programs and rewards inspire those buyers to become regular customers. Calculated correctly, knowing what kinds of customers have the highest LTV can help you determine where your company should invest in growth. In addition, tracking this metric and the influencing variables helps you figure out how to affect those variables and improve this key metric.

LTV Calculation: An Example

Companies can use multiple way to calculate LTV. On a most basic level, the E-commerce business should focus on a formula with variables they know how to influence so that this metric is actionable for them. One basic formula is:

$$LTV = (Average\ Order\ Value)\ x\ (Number\ of\ Repeat\ Sales)\ x\ (Average\ Retention\ Time)$$

Let us assume a business offers a subscription service at the rate of $10 per month for women and $20 per month for men. In return for the $10, customers receive a box of goodies from various companies hoping to find lifelong customers. If we apply this to formula above, we get:

$$LTV = (\$10)\ x\ (12\ months)\ x\ (3\ years) = \$360$$

Each female subscriber has a value of $120 per year. If the average customer is around for 3 years, that gives business an average LTV of $360. However, this is the LTV for a basic service only. What if the company also offers a free service for a month in order attract more subscribers? Would losing that $10 per customer be worth the additional revenue generated by the increased subscriptions? Would investing in identifying people who are at-risk of canceling and nurturing them increase this? Or would providing more opportunities for people to upgrade be able to increase the average order value? These are the types of questions answered by LTV. LTV helps companies determine the points of leverage to improve business.

Trend Analysis

Trend Analysis lets you quickly compare month-over-month and year-over-year metrics. Even better, you can view these metrics either for your entire customer base or for specific segments. With this analysis, you can quickly spot and act on new trends in revenue, profit, customer acquisition, CLV, average order size, and more. There are many use cases of trend analysis (Barr, Taylor-Robinson, Scott-Samuel, McKee, & Stuckler, 2012). We can use trend analysis to quickly get a sense on how our key ecommerce metrics (total revenue, total profit, customer lifetime value, and number of new customers) are performing compared with last month and last year. Trend analysis makes it easy to get a quick overview of any customer segment. You can zoom in on customers from different regions, customers from different acquisition sources, and more.

Techniques to Increase Customer Retention

Customer retention is the act of getting more of your customers to stay loyal to your brand or business. A successful customer retention strategy turns one-time shoppers into loyal, repeat purchasers that buy more, more often. A customer retention strategy will boost your profitability while encouraging repeat business that drives a sustainable long-term business model. (McEachern, 2015).

There is ample statistical evidence that supports the idea of customer retention. Increasing customer retention rates by just 5% can increase profits by 25% to 95% (Reichheld & Schefter, 2000). Attracting a new customer costs five times as much as keeping an existing one (Lee Resources, 2010). Globally, the average value of a lost customer is $243 (Kickassmetrics.com, 2012). 71% of consumers have ended their relationship with a company due to poor customer service (Kickassmetrics.com, 2012). The probability of selling to an existing customer is 60–70%. The probability of selling to a new prospect is 5-20% (Kickassmetrics.com, 2012). 82% of companies agree that retention is in fact cheaper than acquisition and marketers are more focused on acquisition than retention (Econsultancy, 2014).

Following we discuss some ways E-commerce firms can use to increase customer retention rates (Charlton, 2015; McEachern, 2015; Kwon & Kim, 2012; Polo, Sese, & Verhoef, 2011; Cho & Fjermestad, 2015)

DELIVERY: Delivery can be a pain for online retailers. They may sell great products, provide an excellent online experience, yet the final step in the process is often in the hands of third parties who do not necessarily share the company's values. Here, a reliable courier and close monitoring of service levels helps, but you can also keep customers informed on the progress of their delivery and make the process as convenient as possible. Retailers should never underestimate the 'want it now' mentality. If customers know that they will receive goods quickly when they order, they will keep coming back.

PACKAGING: Not all retailers have complete control over the delivery process, but they can ensure that the packaging provides a 'WOW' experience for customers. You could also add little extras and surprises.

CUSTOMER EXPECTATIONS: This is vitally important, as delivery issues are guaranteed to deter repeat purchases. Focus on under-promising and over-delivering. Take example of Zappos. Zappos promises delivery within five business days but the majority of orders is shipped overnight.

PERSONALIZATION: Personalization allows you to increase customer retention through more relevant and tailored experiences with the brand. Personalization requires you to collect data about your customers that you can later use to make recommendations and tailor promotions to each customer's individual needs. There are a few ways to use personalization. An example of this is Bodybuilding.com; they take all your previous purchase information and make product recommendations based on what you like and the product you are currently looking at. Another example is U.S. Patriot Tactical (uspatriottactical.com). Their homepage provides relevant offers and click through opportunities based on the interests of their contacts. A simple way to use personalization to increase customer retention is through email. Many email service providers allow you to mark emails with personalization such as name and company. When a shopper is addressed by name with content they care about, they will be much more loyal to you and your store.

REGISTRATION: If retailers can persuade customers to register without making it a barrier to purchase, then there are huge benefits in terms of retention. They can track orders, receive special offers and, most importantly, repeat purchases are easier if delivery and payment details are saved.

EASY REPEAT PURCHASES: Amazon's one-click payments is a very important role in its success online, as it makes purchases incredibly simple so encourages shoppers to keep coming back. In combination with next day delivery via Prime, it makes it almost too easy to buy from the site. It works by saving the customers card details and delivery address so they only have to enter a username and password. It is also especially valuable on mobile devices, as consumers do not want to waste time trying to enter credit card details on a smartphone.

EASY ACCOUNT INFORMATION RETRIEVAL: Most web users have so many passwords that remembering them all is almost impossible. This means that, if they have not purchased from a site for a while, then a forgotten password can be a real barrier. E-commerce businesses should make it convenient for customers to retrieve their account information. Many E-commerce sites allow just submitting an email address and send a link to rest the password.

SOCIAL MEDIA CUSTOMER SERVICE: Offering great customer service via social media can help customers to avoid the pain of the call center queue, and offer a more personal touch.

IMPROVE EMAIL CUSTOMER SERVICE: A 2011 study found that email is the preferred customer service channel for 44% of consumers. However, email customer service is often poor, or non-existent (Charlton, 2011). Problems include the sheer length of time it takes many companies to respond and no reply email addresses, which prevent a conversation.

CONVENIENT PRODUCT RETURNS: Offering free and convenient returns is a great way to persuade first-time customers to buy, but is also a great retention tactic (Charlton, 2012). If customers know they can return items easily if they change their minds, they are more likely to come back again. Stone & Strand (www.stoneandstrand.com) is a jewelry store. They offer free product returns to satisfy customers with their purchase.

On the flipside, charging for returns, though retailers have costs to cover, can deter customers from returning to a website. The cost of the return needs to be weighed against the risk of losing repeat business.

POST-PURCHASE CONTACT: If a customer has just made their first purchase, this is a good time to follow up with a welcome email and some up and cross sell suggestions.

BIRTHDAY / EVENT EMAILS: Emails triggered by specific events, such as a customer's birthday, abandoned checkouts etc. can be a very effective retention tactic.

REMINDER EMAILS: If a customer has not made a purchase for a while, then a gentle nudge may be enough to tempt them back. It also helps to sweeten the email with a discount.

REFER A FRIEND FOR GIFT VOUCHERS / DISCOUNTS: This is a common tactic for financial sites. For example, first direct will credit your account with £100 for every friend you refer.

OFFER EXCLUSIVE DEALS FOR SOCIAL FOLLOWERS: Luxury flash-sale site Gilt (www.gilt.com) has been offering exclusive sales to Facebook fans. This gives people a real reason to keep coming back, and to use the brand's Facebook store.

LOYALTY PROGRAMS: A loyalty program is a fantastic way to boost customer retention. When a shopper is given additional value (like points) for shopping at your store, it becomes much more difficult to choose a competitor for their next purchase. Points create a switching cost. If a customer moves to a competitor, they will be leaving money (points) on the table. Ecommerce loyalty programs allow you to reward more than just loyalty. You can reward points to drive other profitable actions like reviews, referrals, and social sharing. Points are often used to mask the true value of what you are rewarding. For example: you may be willing to pay 3 cents for a share on Facebook, but 30 points is a stronger motivator for your customer than 3 cents.

GAMIFICATION: You can get the power of gamification working for your customer experience and purchase process. With gamification you are encouraging shoppers to complete actions by making it more enjoyable and adding a sense of competition. Sites that incorporate gamification often have leaderboards, status, and badges so shoppers can display where they stand relative to others. Gamification is great because it can be incorporated into other things like promotions and even loyalty programs.

SUPPORT SYSTEMS: Support systems are anything added to your site to improve customer service and satisfaction. These systems could be help desk software like Zendesk or Freshdesk or live chat software like Olark or Zopim. All of these support systems allow you to resolve customer issues and conflicts quickly and efficiently, which provide a few key benefits. The first benefit of support systems is that it provides a one on one experience. You can easily address customer issues and quickly get things resolved. A quickly resolved conflict can actually create a lifelong customer. Secondly, live chat software allows you to engage with customers in real time. Shoppers do not want to wait; live chat is a direct and real time connection to your customer. Customer connections lead to retention.

CUSTOMER RELATIONSHIP MANAGEMENT (CRM): CRM is a tool used to increase satisfaction by keeping track of a customer's entire journey. A CRM tool is used in tandem with other tools on this list to provide a total customer retention strategy. You can use your CRM to track which customers have received which badges with your gamification tactics or which customers have earned points in your loyalty program. A CRM has many operational benefits, but it also can help with customer retention. When all customer information and interactions are stored in one place, it becomes much easier to provide an amazing customer experience (the backbone of a total retention strategy).

Pricing Strategies

The price of the product is generally the first thing noticed by a customer visiting a product page on your site. Your store's pricing position will determine how many of each type of customers you attract. In general, E-commerce businesses adopt one of the four common pricing strategies: cost-plus pricing, target return pricing, value-based pricing, and psychological pricing. Cost-plus pricing, known in retail as keystone pricing, is the wholesale cost of a product plus a mark-up percentage. Target return pricing is a price that sets a goal for its return on cost. Value-based pricing is pricing an item to appeal to customers over alternative products or competitor prices. Psychology pricing is based largely on name recognition and a customer's perceived value of a product regardless of its true value.

Many individuals use the Web to find the lowest price, and price-comparison tools and sites even help them to find the best price or substitute. However, many more use and shop the Web to look for unique or hard-to-find items, or to benefit from the convenience of round the clock shopping facility and home delivery. So we can say that no matter business is physical or online pricing is of same importance. Every pricing decision is based on two values: ceiling and the floor. The price ceiling is the limit of what the market will bear, that is, how much customer will spend to purchase a particular item. The price floor is the cost of the goods plus all the expenses of doing business. The spread between the ceiling and the floor is the business's profit margin. To develop an appropriate pricing strategy, an E-commerce business need to take into consideration many factors (Elizabeth, 2014).

CHEAP PRICE OR GOOD VALUE

Every E-commerce business needs to decide about its Unique Selling Proposition (USP). Whether you can make the lowest price offered as USP or not, it depends. A research by Stanford University (LaPlante, 2009) revealed that a marketing message that emphasizes the experience associated with a product is more powerful than a message that focuses on money.

You may or may not make the lowest price in the market for your products as your USP (Unique Selling Proposition).It depends. A research by Stanford University (LaPlante, 2009) revealed that a marketing message that emphasizes the experience associated with a product is more powerful than a message that focuses on money. Another research by Wharton business school (Wharton, 2009) tested the following marketing messages for a lemonade stand:

1. "Spend a little money and enjoy C&D's lemonade"

2. "Spend a little time and enjoy C&D's lemonade" (focused on time savings)

3. "Enjoy C&D's lemonade" (neutral approach)

The research found that second message attracted double the number of customers and these customers were willing to pay twice as much for the lemonade. We can see a subtle word play can drastically change the perceived value of the product. It is clear that a focus on cheapest price may not be a good idea for every business. There are cases where customers are only interested in getting the authentic product irrespective of the seller they are getting it from. One good example is computer software. For example, customer wants to ensure he is getting genuine Microsoft Office DVD. In that case, price becomes important. Here a seller could lead by price.

OPTIMIZE PRODUCT PAGES

A research (Knutson, Rick, Wimmer, Prelec, & Loewenstein, 2007) found that consumers could be divided into three groups based on their brain activity at the time of spending: Spendthrifts, Tightwads, and Unconflicted. Tightwads felt a sharp pain before purchasing while Spendthrifts did not feel much pain. The unconflicted were in the middle. Another research (Rick, Cryder, & Loewenstein, 2008) developed a Spendthrift-to-Tightwad scale and categorized consumers according to the pain experienced by them at the time of purchasing. The research found that 24% of the customers were Tightwads. It clearly shows that a significant percentage of customers hold hard onto their money. As such, E-commerce businesses need to optimize their offers to appeal to these careful spenders. To attract these customers, your product page design must highlight the utility of the product. For example, the following product picture shown on a product page is a good fit for spendthrifts and unconflicted customers.

However, for tightwad customers, the following product picture showing the practical value might be a more appropriate choice.

SUBTLE WORD PLAY

The Wharton study (Wharton, 2009) also tested customer behavior when a hypothetical offer was presented to them. The subjects had to pay $5 fee for overnight shipping of a free DVD box set. This fee was described in two ways: "A small $5 fee" and "a $5 fee." The version with the addition of the word "small" increased offer conversions by 20% for tightwad customers. Think about the number of syllables in the price you are quoting. In a 2012 study, researchers found that "consumers non-consciously perceive that there is a positive relationship between syllabic length and numerical magnitude." That means, if a price takes longer to say aloud, people think of it as more expensive. Therefore, when you're writing prices on your website or marketing materials, imagine your customers saying the price out loud. If you write a price as "$1,599", for example, people will say, "one thousand, five hundred and ninety-nine" in their heads. Remove the comma, so that it's "$1599", and they are more likely to read it as "fifteen ninety-nine". You might just make some more sales as a result (Blackman, 2014).

OFFER BUNDLING

Bundling related product into a kit is the most common pricing strategy visible in the bricks and mortar business. The same strategy can be used by online stores. By bundling the products, online customers can offer the products at relatively low prices. This can offer savings to both the buyer and to the seller, who saves the cost of marketing both products separately. Justifying one purchase is so much easier than making several small purchases.

Frequently Bought Together

Price for all three: **$103.97**

Add all three to Cart

Add all three to Wish List

Show availability and shipping details

☑ **This item:** ASUS XONAR DG Headphone Amp & PCI 5.1 Audio Card $29.99

☑ WD Blue 1TB Desktop 3.5 Inch SATA 6Gb/s 7200rpm Internal Hard Drive $52.99

☑ Asus 24x DVD-RW Serial-ATA Internal OEM Optical Drive DRW-24B1ST (Black) $20.99

OFFER LIMITED-TIME DISCOUNTS

To attract, tightwad customers, you can use a hello bar to show them a limited-time discount offer on your product page. A Hello Bar is a bright horizontal bar that appears on the top of the page. It contains one simple statement and a call-to-action.

Place your order in the next 3 hours and get 35% OFF! Place Order NOW!

While the discounts will usually increase the absolute number of units sold but the Average Order Value (AOV) can go down. E-commerce businesses need to check their revenue figures carefully to see if they are able to achieve the right balance between the number of conversions and the discount offered by them.

SMALL OR BIG: FREE IS FREE

Marketers use the free products and services for the promotion of new products. It create awareness in the eyes of customers. Free is such a tempting term that it can make the demand very high. There is a sensible economic logic for giving away things free and that is to gain customer's attention. Take an example from Amazon. The company noted that their sales in France were drastically lower than other European countries. The reason was that they were charging a 20-cent shipping fee in France, while providing free shipping in other countries (Dan, 2008). Note the power of the word "Free". A 20-cent shipping fee is nothing but it cannot match the charm of the word "Free."

TEST DIFFERENT PRICING BRACKETS

You should carefully set pricing options for your product. Too many options can confuse visitors and too few options can limit your profit potential (Poundstone, 2010). Take example of a case study of purchase patterns of consumers for a beer. (Poundstone, 2010). In the first test, two options were made available – regular beer and premium beer. Figure below shows the price and percentage of customers that opted for this product price.

$1.80 $2.50
20%   80%

A majority of the people opted for premium beer. However, to cater to those looking for the cheaper option, a third option was added to the mix:

$1.60 $1.80 $2.50
0%   80%   20%

Surprisingly, no one opted for the cheaper option but the ratio of the regular and the premium beer changed drastically. Therefore, a more expensive option than the premium beer was added.

$1.80 $2.50 $3.40
5%   85%   10%

The same principle was used with travel packages. When people were offered to choose a trip to Paris (option A) vs a trip to Rome (option B), they had a hard time choosing. Both places were great, it was hard to compare them. Now they were offered three choices instead of 2: trip to Paris with free breakfast (option A), trip to Paris without breakfast (option A-), trip to Rome with free breakfast (option B). Now overwhelming majority chose option A, trip to Paris with free breakfast. The rationale is that it is easier to compare the two options for Paris than it is to compare Paris and Rome (Laja, 2014). Another research from Yale found that if two similar items are priced the same, consumers are much less likely to buy one than if their prices are even slightly different.

Researchers asked participants to choose or 'pass' (keep their money) on two different packs of gum. When the packs of gum were priced the same at 63 cents, only 46% made a purchase. Conversely, when the packs of gum were priced differently (at 62 cents and 64 cents) more than 77% of consumers chose to buy a pack (Kim, Novemsky, & Dhar, 2013). This example is a clear indication that testing different price brackets for your product is very important.

PURCHASING POWER-BASED CUSTOMER SEGMENTATION

A 2006 study of New York's Fulton fish market found that the dealers constantly charged less prices from Asian buyers than White buyers (The Economist, 2012). Over time, the dealers learnt that most Asian buyers readily reject higher prices and group against dealers who offer high prices. Hence, the dealers preferred this method of staying profitable. In E-commerce, this trend is catching up. Orbitz is a travel website that uses this technique as their software determines if the visitor is a Mac user or a Windows PC user. They found that Mac users prefer pricier hotels than Windows PC users. Hence, Orbitz recommended expensive deals to Mac users, whereas Windows PC users were shown relatively affordable deals. This variable pricing strategy could also backfire through. Take example of Amazon where the variable prices affected more than half of the company's top-100 best-selling DVDs, or digital versatile discs (Cnet.com, 2002). Therefore, it is important that E-commerce businesses should test such segmentation-based pricing strategies before implementation.

PSYCHOLOGICAL PRICING

It is a human psyche that people consider $499.95 is less than $500. Many studies show that products sell best at certain price points, such as $197, $297, $397. One use of psychological pricing is in price-ending numbers. Buyers consider prices ending in uneven numbers as a better deal. Many studies published between 1987 and 2004 confirmed higher sales for prices ending with the number 9 ($1.79, $79, $49, and more). The average increase in sales was reported at 24% (Poundstone, 2010). The challenge with this strategy is that products ending in an odd number are also often perceived as being lower in value. Businesses need to ensure that they chose the right price and the right strategy for your specific product or service. In another study, MIT and the University of Chicago ran an experiment in which they printed three versions of a mail order catalog. One women's clothing item was sold for $39, while other versions of the catalog carrying the same item were priced at $34 and $44. All three catalogs were sent to an identical sample size. $39 catalogs had more sales than the other two and had a higher profit per sale. The '$39' price surpassed even its cheaper counterpart ($34) (Anderson & Simester, 2003). The Figure below shows how Zivame (http://www.zivame.com/), a lingerie store, implements this psychological pricing strategy.

Penny My Caribbean Cruise
Liquid Drape Maxi Dress
Rs. 2295

Penny My Bohemian Chic
Liquid Drape Maxi Dress
Rs. 2295

Penny My Weekend Movie
Date Liquid Drape Dress
Rs. 1995

Penny My Sun N Sand Liquid
Drape Dress
Rs. 1995

COMMUNICATE VALUE BEFORE THE PRICE

If you are selling a scented candle for $150, it might seem expensive. However, if you explain that this scented candle does not melt quickly and can last for one week at a stretch, you justify the price well.

Pricing strategies are very important for your E-commerce business and A/B testing them is usually very complicated. But the risks definitely have their rewards too.

Dynamic Pricing

Dynamic pricing is a pricing strategy in which prices change in response to real-time supply and demand. The dynamic pricing strategy is a popular strategy for E-commerce businesses that allows retailers to remain competitive with round-the-clock price monitoring and changes, boosting profits by 25% on average. Dynamic pricing gives retailers the flexibility to decrease prices to increase sales when they are sluggish and increase prices to generate more profit when they are booming. Many retailers opt for pricing intelligence software that has the ability to scan thousands of products. Dynamic pricing can be difficult to manage. Because of this, many retailers have turned to technology to automate what was once a manual process. Pricing intelligence software has already caught on as 22% of retailers have chosen to implement it. An additional 7% retailers plan to start using it within the next six months and 36% in the next year (Shpanya, 2014). Research has found that price optimization software improves gross margins by 10% (Shpanya, 2015). Retailers using an in-house or third party dynamic pricing software can set repricing rules to ensure that their pricing always matches their brand identity and never goes below cost. As an added bonus, there is no possibility of unwillingly engaging in a margin depleting price war. Dynamic pricing also provides retailers with additional insights on market trends. Retailers can implement different price levels and observe price elasticity before finding the optimal market price.

Amazon, one of the largest retailers that use dynamic pricing, changes its prices every 10 minutes on average. The company saw a 27.2% increase in sales from 2012 to 2013 and generated over $44 billion in sales in 2013. This resulted in Amazon being named one of the top 10 retailers in the US for the first time. Walmart, also changes its prices roughly 50,000 times a month. In 2013, its global online sales grew by 30%, a growth rate that topped Amazon's for the first time in 5 years. Walmart's success has continued with a 27% increase in global web sales in first quarter alone. Best Buy and Sears also incorporate dynamic pricing into their pricing strategies. Best Buy's online sales increased by 25% in 2013 and has already seen a 20% increase in 2014. Sears also implemented price changes to about 25% of its products during the holiday season. It experienced a 17% increase in online sales in 2012 (Shpanya, 2015).

There are some dynamic pricing techniques that E-commerce businesses can utilize.

SEGMENTED PRICING: This technique is used to appeal to a larger market. Using this strategy, retailers have tiered prices from value to premium in order to capture as much of the market as they can. Apple has recently started using this strategy for its iPhones, creating a value product to complement its premium product (5C and 5S, respectively). Apple also uses segmented pricing for other products, fluctuating its prices according to how much memory each item holds.

PEAK PRICING: Peak pricing allows retailers to take advantage of fluctuations in demand, increasing prices when demand is high or when competitors have low inventory. Retailers could utilize this strategy during the holidays when consumers have high demand for various products while shopping for gifts for their loved ones. This could also be used when there is a special event such as a championship game approaching. Consumers would want to properly represent their teams with shirts, caps and more. Retailers could capitalize on this by slightly increasing prices to accompany the increase in demand.

TIME-BASED PRICING: Time-based pricing allows retailers to adjust prices according to the time of day or the time a product has been on the market. Retailers can increase the demand for an older product by marking it down. Microsoft used this strategy when pricing its game console, the Xbox. It originally had the Xbox 360 priced at $399, however when the Xbox One was released, with motion sensing technology, the Xbox 360 was marked down to $299.

PENETRATION PRICING: Penetration pricing allows retailers to set a lower price than the eventual market price in order to persuade consumers to try their product. Once customers are drawn to their store, retailers can then gradually increase prices as the item becomes more and more popular.

Online Branding

Consumers purchasing products over the Internet generally have incomplete information about the retailer's credibility and this is why retailer's brand becomes very important. Price, objective product information, and perceptions of retailer credibility are the three important attributes when consumers select retailers for online shopping (Su, Consumer E-Tailer Choice Strategies at On-Line Shopping Comparison Sites, 2007).

Rapidly changing field of online marketing is presenting brand managers with many challenges e.g. choice between traditional and electronic media. Many firms are firmly shifting their brand emphasis online. One example is Philips deal with CNN in 1996. In this deal Philips became sponsor of all CNN channel's online delivery services, including mobile news bulletins, streaming video and RSS alerts. In 2002, AMD (a technology company) launched an extensive, integrated global print and online branding campaign called "AMD me" that was ran on major media across key global markets. Industry analysts are still debating whether Internet will be able to offer powerful branding results comparable with traditional media such as TV and radio.

Online content with its flexibility, speed and relative low cost is an attractive choice for two purposes. First, it can be used to promote a brand's image and performance. Online content can supplement ongoing traditional media campaigns to create a tangible consumer response. Second it can be used to respond damaging information – whether real or perceived – that can inflict lasting impact on a product or service. Apple Computer's 2002 Switch campaign is a great example of a campaign where a billboard or television ad alone was unlikely to produce a direct consumer response. In this campaign, various ads were shown and each ad guided users to Apple's web site, where the campaign included content that is more substantial and information reinforcing the benefits of moving to a Mac[10].However not all products and services are well suited to this kind of integrated campaign to promote consumer response. Brand managers need to decide carefully before launching a campaign to minimize risk of losing customers.

A brand community is a specialized community, which has no geographical boundaries and is based on a structured set of relationships among brand lovers (Muniz & O'Guinn, 2001). A branded good or service is at the centre of brand community. A brand community might exist around many things. e.g. rock bands (Tom Petty and Lifehouse), TV shows (Star Trek and Xena Warrior Princess), cars (Ford Broncos and Saab), artists (Van Gogh and Jackson Pollock), authors (Shakespeare and Jack Kerouac), computers (Apple Macintosh), movies (Star Wars), and beverages (Coca-Cola). Brand communities may go beyond geographical boundaries and include various consumer groups. Brand communities have potential power both for repeat purchases and for valuable word-of-mouth recommendations.

B2C Branding

To build a brand for your E-commerce business, you first need to uncover what makes your brand remarkable. Ask yourself some questions such as: Does your ecommerce business have a mission? Is there a problem your product solves? What do your current customers think of your business? What do potential customers think of your business? What standards do you want customers to associate with your business?

Next, do some market research and try to get inside of your customer's heads to know what they think about your business. You can do this by emailing a quick survey to your current customers, or, implement a survey functionality within your store. To get immediate better branding, get a great, recognizable logo designed by someone with branding experience; integrate your brand into every social media outlet you use, by using the same profile picture on all platforms; and be consistent with your tone and style across all platforms.

B2C Branding Strategies

Your brand strategy will help define how your customers see your business and product. This strategy will help you stand out from the competition. Building a brand strategy for your ecommerce store is also a powerful marketing, customer retention, and loyalty technique and is necessary for any new or established ecommerce

---

[10] For details, see http://www.apple.com/pr/library/2002/jun/10switch.html

business. It's important to determine your brand strategy and positioning before branching out into any other aspects of business. With declining brand loyalty (Mochari, 2013), building a strong customer base is more important than ever before. The Internet has turned people into savvy online shoppers, and the power has shifted from business to consumer. Often, shoppers bypass the idea of brand loyalty if the product is available at a lower price somewhere else. Here we will discuss some ways to build a powerful brand strategy for ecommerce businesses (Schreiber, 2015).

YOUR UNIQUE SELLING PROPOSITION

Determining a unique selling proposition is a foundational step in any good brand strategy. The goal is to differentiate your brand from your competitors, and give your customers a reason to purchase a product from you. It should highlight a story, philosophy unique aspect or goal of your business and product. Let us take the example of Mast Brothers Chocolate (http://mastbrothers.com/).

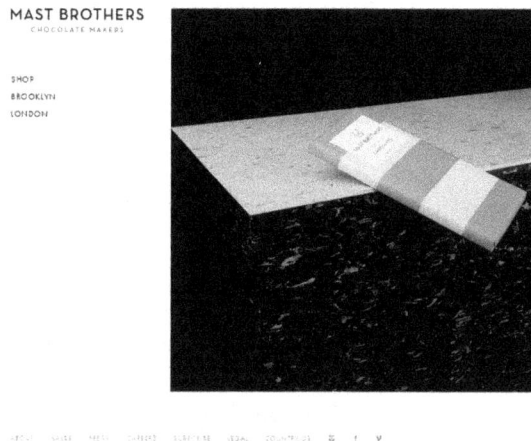

The Mast brothers hand make chocolate in New York City, and their shop is one of the few shops where bean-bar chocolate is produced in house. Not only is their chocolate delicious, but the way they make it is very unique – especially in a mass-produced world. They use that to leverage their unique selling proposition. Rather than using the industrial methods of larger companies like Nestle or Hershey, the Mast Brothers produce all their chocolate bars by hand using old school techniques like a stone grinder. A recent study published in Harvard Business Review (Nunes & Downes, 2014) showed that nostalgic, artisanal and disrupted technology goods are making a comeback. If you are producing a handmade product, much like the brothers are, consider using this story as an initial way to uniquely position your brand.

DISTINGUISH YOUR BRAND ON PRODUCT QUALITY

Using your brand's product quality as a marketing tool is an extremely effective way to build brand loyalty. Nowadays, if the product you are selling is not built or designed with quality in mind, your customers will not return and you may find them express their dissatisfaction across the many social media platforms available to them. To get lifetime customers, you need to consistently over deliver on the quality of product. Take example of Casper (casper.com), a mattress manufacturer. Casper has leveraged this brand positioning strategy. They make their product quality and features the most important part of their product page by listing it right under the title. It is immediately visible to a potential customer.

OUR MATTRESS MEETS THE COVETED GOLDILOCKS STANDARD OF 'JUST RIGHT.'

We made the Casper from responsive materials to ensure great sleep for nearly everyone, from 40 to 400 pounds. So whatever your weight or sleeping position (back, side, starfish), you'll love your bed.

The Casper promise is a fantastic way to promote the quality of their product and a risk removal tactic, which helps build consumer trust. A study (Patel, 2013) showed that when stores added a money back guarantee to their products, a 21% increase in sales was seen. Of that increase in sales, only 12% of customers asked for their money back. Another outstanding way for Casper to use quality to position their brand, is by including Instagram pictures right on their product page. The pictures are from happy customers who have given the mattress an outstanding review.

CHANGE THE BRANDING RULES

Another strategy is to change the rules of the branding game. Taking risks can sometimes bring new attention to your ecommerce store. If your product is quirky or unique, you can position your marketing efforts around that. By finding a custom that exists in your consumers and completely shattering it, you might just see a loyal customer base start to develop. The goal here is to make space for your brand in an unconventional way to get people talking. Take example of Cards Against Humanity (http://cardsagainsthumanity.com/).

Cards Against Humanity is not only a ridiculously fun game, but part of the reason they have gained so much attention is through their brand positioning. *"Cards Against Humanity is a party game for horrible people. Unlike most of the party games you've played before, Cards Against Humanity is as despicable and awkward as you and your friends."* It is not often that we will see a brand insult its customers. Cards Against Humanity is notorious for having fun with their customers. In fact, on a Black Friday they sent 30,000 customers cow poop. Yes, 30,000 people are so loyal to Cards Against Humanity that they bought poop at $6 a piece (Machkovech, 2014). This was not their first stunt, as in previous years their Black Friday sales included things like increasing the cost of the game by $5. Cards Against Humanity are going against the grain by doing something so unique and unheard of, and it has only worked in their favor.

PERSONAL CUSTOMER EXPERIENCE

Let us assume you are someone in his early 30's, wear plaid shirts, appreciate craftsmanship and the qualities that go into making something by hand. You look online to buy a plaid shirt and come across a brand that hand sews all of their shirts locally. After purchasing the product, it arrives beautifully packaged, customized with a handwritten note on the inside. That is a personalized customer experience. A great example of a company that does this with user experience in mind is Loot Crate (https://www.lootcrate.com/) a monthly subscription box for gamers. Not only are they using custom packaging, but also they include a fun little activity for their customers to do that actually promotes their brand.

GIVE SOMETHING BACK

Brands are able to position themselves well in a busy online space by getting in the spirit of giving back to their customers. No matter how many sales you have for your ecommerce store, there is always an opportunity to share your gratitude with your customers. You will want to make sure that you take some time to go beyond to thank your customers for their business. Not only does this position your brand as being thankful and recognizing the support of your customers, but it is a surefire way to get a customer for life.

Why Online Brands Fail?

Usually, the problem behind brand failure lies in the fundamental approach they take to online retail. Too many brands have built a digital presence with a "waiter" mentality, relying on a website that acts as a glorified order form and delivering product to consumers who have come looking for it. Instead, e-commerce should be thought of as a "salesperson," a workhorse increasing the value of current customers and actively acquiring new ones. Following are some reasons why E-commerce brands and suggestions that E-commerce businesses can follow to avert the downfall of their brands (Rekuc, 2015).

IN-APPROPRIATE SITE DESIGN: A site that converts well your existing customer base is not enough to convince brand new users to buy. Your product detail pages must sell the product and your policy pages must instill trust as if your site visitor has never heard of you. This starts with compelling copy, a place for customer reviews, 21st century shipping/return policies, and professional photography.

FLAWED DIGITAL MARKETING: For enterprising brand marketers, paid search is a customer acquisition vehicle. A site using SEO to acquire new business is focusing on driving new customers with non-brand terms. A highly competitive brand can typically support multiple comparison-shopping engines, including eBay shopping network, Google Shopping, and Nextag, among others. Brands can sustain these channels simply because the products they carry are competitive in the marketplace, their site effectively sells them, and their digital marketing team goes after new customers ruthlessly. The salesperson uses social media to encourage customer conversations create an engaging culture associated with the brand, and drive micro-conversions (email sign ups, shares, referrals, etc.) that can lead to increased sales.

NOT GOOD ENOUGH OFFERS: By investing in aggressive promotion offers, you can make rest of your marketing endeavors much more efficient. Pure play e-commerce companies are a great place to look for inspiration. Let us take online retailer Warby Parker (https://www.warbyparker.com/) as an example. The company allows new visitors to try on 5 pairs of glasses completely for free. Warby Parker also pays shipping and returns with no commitment on the part of the shopper. That is a convincing way to get customers to try their product. Another example is Audible.com (http://www.audible.com/) (an Amazon company), which lets shoppers try their first audio book completely free. Both of these companies use a loss leader to act as a salesperson on their site. They remove the traditional e-commerce barriers (such as shipping costs and product uncertainty) to drastically improve initial conversion rates. To sustain this strategy, they have to be diligent about making sure these loss leaders result in paying customers and making sure lifetime customer values exceed customer acquisition costs.

INADEQUATE USE OF EMAIL: A good salesperson always gets a phone number (or in recent years an email address) to actively follow up after a potential customer has gone home. Smart brand marketers know that even when they drive a ton of new traffic to their site, they are only going to get orders from a very small percentage of those visitors. Typical conversion rates hover around 2% or 3%. The next step is to convince the other 97% of visitors to return. Both Warby Parker and Audible.com collect email addresses in exchange for their free trials. Not only are they giving the customer a taste of their product, they have made it easy to follow up on that sample later on. In the same vein, Blue Nile (http://www.bluenile.com/), an online jeweler, uses a giveaway pop-up to collect emails as soon as a customer "walks in the door." Given how many visitors land on their site on a daily basis, Blue Nile is probably subscribing tens of thousands of email subscriptions for the cost of a single giveaway. In addition, considering the only cost

WIN A $5,000
SIGNATURE DIAMOND

Enter to win a Blue Nile Signature Diamond, cut to the most exacting standards they are the top diamonds in the world.

to run this giveaway is the $5,000 diamond itself this is an incredibly affordable campaign. Email may not be as attractive and new as other digital marketing channels, but once you start building a database of email subscribers, you can begin reaching out to them with marketing emails and eventually increase the efficiency of your list with clever A/B tests.

B2B Branding

If you buy the wrong brand of peanut butter, you can always buy another brand when you finish the jar. However, you will be in great trouble if you buy the wrong computer software or hardware solution for your business and you have a major problem. Brands matter in the B2B world, sometimes more than B2C. A strong brand means the perception of reduced risk and makes doing business easier. As with B2C, buyers of B2B products value vendors' reputation because it makes the purchase decision safer (Regensburg, 2014). In fact, a Forbes survey (McKinsey & Company, 2013) found that B2B purchasing decision-makers consider the brand as a central rather than a marginal element of a supplier's value proposition. Their survey found that decision-makers say the brand is almost as important as the efforts of sales teams in encouraging them to make out a purchase order. Creating a brand and adding value plus differentiating one's brand from another in the mind of the potential customer is of great importance. Differentiating one's brand, companies can use various strategies, leveraging on the origin of the goods or the processes to manufacturing them. Ultimately, a strong B2B brand will reduce the perceived risk for the buyer and help sell the brand (Qulech,2007).

In B2B branding, there are three dimensions of benefits upon which B2B firms should build positioning platforms (Anne Maarit Jalkala & Joona Keränen, 2014): Functional (what the product does), Economic (what the brand means to the customer in time and money), and Emotional (how the brand makes the customer feel). Brands that deliver beyond the functional and economic levels with emotional benefits will command an incremental price premium and create strong competitive advantage and customer brand loyalty. B2B sellers tend to talk in terms of product features, performance, and ROI. B2B buyers may be looking for trust, confidence, ease, and security.

Salesforce.com is a company is a great example of B2B branding principles. Salesforce.com started with a seemingly simplistic brand vision, "No Software." Rather than focusing on its myriad features, they leveraged "Emotion". Recognizing that sales teams were incredibly frustrated with the complexity of using, maintaining and updating sales and marketing software, they built a brand around relieving that pain, and in the process, accelerated the revolution in business known as Software-as-a-Service (SaaS). Since then, Salesforce.com never lost sight of the simplicity of the emotionally driven "No software" message.

B2B Branding: Best Practices

A new logo and tagline may be several of the deliverables of a B2B branding process. To create a new brand or refreshing an existing brand, there are is a variety of factors to take into consideration. Following are some best practices to help make the entire process smoother for B2B businesses (Jensen, 2015).

KNOW YOUR AUDIENCE: Clearly define your audience and their pain points. Knowing your target market enables you to appeal to them in your strategy.

FOCUS ON YOUR STRENGTHS: Determine what your firm's strengths are and how they add value for clients. Determine your few significant strengths and make sure they are all relevant to your clients.

CREATE A STRONG MISSION STATEMENT: Once you know your strengths and values write and refine your mission statement. Your mission statement must be concise, specific, and written down. This is very important because all your future decisions must be aligned with this mission statement.

BE CONSISTENT: Pick a business focus area and stick with it. For example, if you are focusing on "efficiency," make sure you carry that message throughout all your brand-related pieces. The very nature and layout of all your marketing and sales material should portray the feeling of "efficiency."

DIFFERENTIATION: The most effective brands take extra efforts to differentiate themselves. Make it clear for your clients what makes your firm and your services unique and valuable.

DO NOT COMPLICATE DECISION MAKING: Seeking input of everybody is good for innovation and ideation. However, in the process of branding you should include only those individuals in the decision making whose opinions really matter. Identify the main stakeholders in the branding process and ensure they have final decision ability.

OFFER WHAT YOU CAN: Do not promise what you cannot deliver.

AVOID COPYING THE COMPETITION: Even if your competition has great ideas and strategies, never copy them. Your firm is unique and different. Copying a competitor makes it look like you do not bring anything new or different to the market and does nothing to make you standout.

USE SOCIAL MEDIA: There are three ways to improve your B2B social media presence. First, you should use Twitter. Top B2B companies post to twitter many times a week. Contrary to our established beliefs, these tweets should be more over the weekends (Brandwatch, 2015). A report found that business audience were actually more engaged on weekends, retweeting and mentioning twice as often per post. Second, you should create a Twitter account and use it for conversations (Brandwatch, 2015). Third, post visuals to Facebook. More than half of the Facebook posts created by top B2B brands contain a photo and the tactic works. This is because these posts drive the most engagement. Photos may receive the most likes of any content format. However, videos foster the most comments and shares (Brandwatch, 2015).

B2B Brand Promotion

B2B brand promotion works in different formats as compared to B2C brand promotion. B2B brands avoid mass-market broadcasts and generally use media that can be targeted at a specific business audience, such as direct marketing both on and offline. B2B companies are present where their potential customers are: Trade Shows, Exhibitions are increasing in online communities (Qulech,2007).

Cost of Branding

Branding has always been considered a key to retail success and consumers are thought to be more willing to search products with strong brand recognition, as well as pay a little more for them. According to (DAYAL, LANDESBERG, & ZEISSER, 2000), Internet sites such as Amazon.com are putting established brands (e.g., traditional brick-and-mortar booksellers) at risk by creating quick brand recognition. However, businesses must exercise caution and avoid excessive spending in their drive to establish brand recognition quickly. The e-tailers experience suggest that most customers, especially long-term loyal customers, come to a Web site from affiliate links, search engines, or personal recommendations (Carton, 2001).

Emotional Vs. Rational Branding

Emotional branding is an effort to establish and maintain brands using some emotional appeal in advertising and promotion of the product. Apple, automaker Lexus, retailer Target and outdoor clothing line Patagonia are examples of rational brands. Emotional branding works well when target audience of advertisement is in a passive mode of information acceptance e.g. print media, radio, television, and billboards. Emotional branding on the web becomes difficult because web by nature is an active medium of communication controlled largely by the customer. Customers on the web can quickly click away from emotional brands.

Rational branding is an effort to establish and maintain brands using cognitive appeal of the product or service in its advertising and promotion. Web-based e-mail services e.g. Yahoo and Hotmail are examples of rational brands. They provide their users a valuable service i.e. an e-mail account and in exchange, the users see advertisement on the web pages that provide them e-mail service.

Online Brand Leveraging

Using the strategy of brand leveraging, a company can extend its dominant position to other products and services and strengthen its established brands. One good example is Amazon.com. Amazon.com expanded its original book business into CDs, videos, and auctions thus leveraging its dominant position by adding features useful to existing customers. Yahoo! is another example. Yahoo has continued to strengthen its leading position by acquiring other Web businesses (such as GeoCities and Broadcast.com) and expanding its existing offerings.

Online Brand Consolidation

New on-line brands face difficulties in attracting consumers because they are relatively unknown compared with their established and known counterparts. Using brand consolidation, a business can leverage its established brand of existing Web site. New online businesses can also establish their own brands and persuade consumers by associating themselves with well-known brands (Eschenbrenner, Nah, & Siau, 2008). A good example of brand consolidation is Time Warner's CNNMoney.com. Time Warner created CNNMoney.com as a one-stop-shop for Fortune Magazine, Money Magazine and a host of other information-hungry money-topic audiences. Another example is WeddingChannel.com, an online bridal registry providing wedding planning services and access to every item that a bride and groom might need. WeddingChannel.com connects to several local and national department and gift stores (e.g. Neiman Marcus and Tiffany & Co.). Brand and logo of each participating store is displayed prominently on the WeddingChannel.com site.

Affiliate Marketing

Brand leveraging and consolidation strategies work only for firms that already have Web sites that dominate a particular market. New entrants can use affiliate marketing as a useful tool to generate revenue. Affiliate marketing is gaining popularity among low-budget Web sites. In affiliate marketing, the website of one firm (i.e. the affiliate firm) includes descriptions, reviews, ratings, or other information about a product that is linked to another firm's site that offers the item for sale. For every visitor who follows a link from the affiliate's site to the seller's site, the affiliate site receives a commission.

Besides generating revenue through commissions, the affiliate site also obtains the benefit of the selling site's brand in exchange for the referral. CDnow and Amazon.com were two of the first companies to create successful affiliate programs on the Web. Sellers of other products and services also use affiliate-marketing programs to attract new customers to their Web sites. Affiliate commissions can be based on several variables. In the pay-per-click model, the affiliate earns a commission each time a site visitor clicks the link and loads the seller's page. In the pay-per-conversion (or pay-per-performance) model, the affiliate earns a commission each time a site visitor is converted from a visitor into either a qualified prospect or a customer. Such site usually pays a percentage of the sale amount rather than a fixed amount per conversion. Some sites use a combination of these methods to pay their affiliates. On each completed sale, affiliate can earn a commission ranging from 5 % to 20 % of the sale amount. The commission earned depends on many variables such as the type of product, the strength of the product's brand, how profitable the product is, and the size of an average order. Pay-per-performance affiliate marketing can be further classified into two popular types: Pay-per-sales (PPS) and Pay-per-lead (PPL).

In a pay-per-sale (PPS) type of affiliate marketing, the merchants pay the affiliate a certain fee whenever the visitor he has referred to the merchant's site actually buys something from the merchant's site. Affiliates are often paid on commission basis, although other merchants would opt to pay a fixed fee. However, no matter what the basis of the fee is, it is generally higher than the fee paid to affiliates in a pay-per-click affiliate program.

The pay-per-lead (PPL) type of affiliate marketing is a slight variation of the PPS type and is often used by insurance and finance companies and other companies who rely on leads for their company to grow. In this type of affiliate marketing, the affiliate is paid whenever the visitor he referred to the merchant's site fills up an application form or any similar form related to the business of the company. Compensation for this type of affiliate marketing is based on a fixed fee whose rates approximate that of the fixed fee in the PPS type.

Affiliate Marketing Tiers and Residual Income

Affiliate marketing can have different levels or tiers in the affiliate network by which payments are made. In a single-tier affiliate-marketing program, the affiliates are only paid based on the direct sales or traffic he has referred to the merchant. All the previously mentioned affiliate marketing types (i.e. PPS< PPL, and PPC) fall under the single-tier classification. In two-tier affiliate marketing programs, the affiliate is not only paid for the direct traffic or sales that he refers to the merchant's site, but also on every traffic or sales referred by various other affiliates who joined the affiliate program through his recommendation. Multi-tier affiliate marketing works the same way, although the affiliate gets additional commission for a wider number of affiliates in different tiers in the affiliate network.

In residual income affiliate marketing, the affiliate is paid not only once for every customer he has referred to the merchant's site. Rather, the affiliate is also paid whenever the customer he has referred returns to the site and purchase another product. Compensation for such type of affiliate marketing is based on either sales percentage commission or fixed fee basis.

Types of Affiliate Marketing Websites

Affiliate marketing sites can be of different types. Informational Website or Blog is a type of website that exists to provide information first, and sell products second. Many of the world's most popular websites fall into this category (e.g. Tom's Hardware).A product review/comparison site specializes in providing detailed reviews of products in a specific niche, or many different niches. Kayak is an example of a brilliantly executed comparison website in the travel niche. A virtual storefront is a website that looks like an ecommerce website but transfers the reader to a merchant's website when they click the "Buy" button. This type of website can work well with a merchant that sells everything, such as Amazon. A mini-site is a very small website, usually consisting of around 5-10 total pages. This type of website provides information on an extremely narrow topic (such as battery-powered lamps). A reward website gives visitors an incentive to buy through affiliate links by giving them a portion of the proceeds in the form of points, discounts or other bonuses.

You can visit an affiliate-program broker site that offers affiliate program opportunities for a number of Web sites. An affiliate program broker is a company that serves as a clearinghouse or marketplace for sites that run affiliate programs and sites that want to become affiliates. Commission Junction and LinkShare are two popular affiliate program brokers.

To join an affiliate network or program, generally you need to complete their online application form. Some programs provide instant approval while some require a human check of application before approval. Once approved, you will be provided with your specific affiliate HTML code that you can use on your site. Most of the affiliate networks and programs require that you own a domain name before they approve your affiliate application. It is a good idea a get a domain name registered under your name before filing an affiliate application.

One very important factor in your selection of an affiliate program is the target audience of your website. For example, if you are targeting sports fans on your site, links to affiliates with sports news, sports goods and the like may generate more revenue than other banners you might put on your sit. Another important factor is the number of affiliate networks or programs you might join. You may join as many programs or networks you want but it is a good idea to join affiliate networks and programs that are relevant to your site. Advertising for many different affiliate networks on your site may earn you smaller amounts from each network. In addition, most affiliate networks and programs have a minimum payout amount so it can take you a long time before you actually get paid. A good strategy is to choose advertisements relevant to your target audience because these advertisements are more likely to be picked up by your audience than general advertisements.

There is another approach in affiliate advertisement called Automated Context-Sensitive Advertising. For example, when you sign up with Google AdSense (a banner advertising network of Google) it automatically checks the pages of your site and determines the most relevant advertisement for the various pages. This approach results in advertisements targeted at the interests of your site visitors. Such targeted advertisement tends to provide better performance and returns.

The Future of Affiliate Marketing

The affiliate marketing industry has, over the years, undergone numerous changes that have variably affected the key players. The future of affiliate marketing has raised quite a stir with market players predicting different trends. The future of affiliate marketing depends on individuals who can adapt to a fast changing industry. The future of affiliate marketing looks even more favorable for affiliates through the advent of marketing data analytics. Affiliates keen to position their affiliate links on niche coupon sites will be able to capitalize on the currently underused traffic. SMS-centered Pay per Call will ensure that affiliates maximize on commissions and merchants on conversions. Pinterest will still be a great platform for affiliates to do promotions through creative images and semi-offsite techniques for boosting image-based links. Similarly, how-to videos on YouTube will be instrumental for raking in more landings through embedded in-video clickable links. The future of affiliate marketing will benefit from the lengthened consumer demand chain, which will only grow bigger with increased competition. Apart from direct benefits, the opportunity for advertisers and publishers to profit will be further good news for affiliate marketers. Research shows that total affiliate market spending is bound to increase by more than a billion dollars to around $4.5 billion in the next two years. Ultimately, the future of affiliate marketing relies on the actions of the key players. For affiliates, complementing typical marketing techniques like PPC and blog marketing with alternative promotional avenues is key (Rogers, 2014).

Applications of Internet tools in Internet Marketing

The applications of the Internet for marketing are not limited to the use of e-mail and the World Wide Web-most commonly used tools by businesses for Internet marketing. Many of the other tools such as e-mail, IRC and newsgroups, that formerly needed special software to access them, are now available from the web. Table 7-1 lists applications of different Internet tools.

TABLE 7-1: APPLICATIONS OF INTERNET TOOLS

Internet Tool	Application
E-mail	Used to send messages and documents (e.g. promotional messages)
Internet Relay Chat (IRC)	A communication channel used by marketers for text-based chat with customers.
Usenet News Groups	An electronic bulletin board widely used by users for discussing a particular topic.
File Transfer Protocol (FTP)	Marketers can use FTP to allows customers of downloading various files (e.g. promotional materials, price lists, and product specifications)
Telnet	Remote access software. Marketers could use it, for example, to provide retailers remote access to corporate system so that retailer could check his order status.
World Wide Web (WWW)	A widely used tool to disseminate information and run business applications over the Internet.

Internet, Intranets, Extranets, and Internet Marketing

Intranets and Extranets provide two types of opportunities for marketing. Intranets and extranets can support the marketing process first by facilitating communication and control between staff, suppliers and distributors and second by providing better control through their application at various levels of management within a company. Table 7-2 illustrates potential marketing applications of Internet, intranet, and extranet for supporting marketing at different levels of managerial decision-making (Chaffey, Mayer, Johnston, & Ellis-Chadwick, 2002).

TABLE 7-2: MARKETING APPLICATIONS OF INTERNET, INTRANET, AND EXTRANET

Type	Internet	Intranets and Extranets
Strategic Application	- Environmental Scan - Competition Analysis - Market/Customer Analysis - Decision Making	- Internal data analysis and reporting - Business Intelligence - Marketing Information - Monitoring and Control
Operational and Tactical Applications	- Advertising and Promotion - Direct Marketing - Public Relations - Marketing Research - Virtual Teams - E-mail	- Training/Education - Product/Service Information - Collaboration - Customer Support/Service - Datawarehousing

Internet Marketing Communication

Internet, as a new marketing medium, is developing a revolution in marketing that is significantly changing the traditional view of advertising and communication media. Internet is the least expensive and most cost-effective marketing communication tool, which is by far the fastest growing (Turban, Lee, & Chung, 2006). Internet is the least expensive marketing communication tool as well as the most cost-effective. People from around the world can create a business relationship for a fraction of the cost of any other marketing method (Janal, 1997).

Marketers use the Internet to gather data for marketing planning and interacting with customers [(Strauss & Frost, 1999); (Heinen, 1996)]. Internet allows an interactive multimedia and many-to-many communication network to interact with the customers and/or firms. Using many-to-many communication, Internet supports discussion groups, multiplayer games, chat, file transfer, e-mail and global information access and retrieval systems. Using Internet, consumers can obtain product information as well as put product-related content on Internet. Marketers can use Internet to provide customers with media rich product catalogues, online customer support, and even distribute some products. In comparison, traditional media follows a passive one-to-many communication model where content is provided to a mass market of customers with limited forms of feedback from the customer. Markets ability to interact with customers is limited somewhat. Internet has become an established global channel through which both existing and potential customers can be targeted, and through which organizations can both publicize and present commercial offerings.

Individuals and businesses share some characteristics but are obviously different in purpose, function, motivation, and decision-making processes. Creating communications to appeal to the buying power of these entities should be done in recognition of their differences, with special attention paid to the decision-making processes. All communications should always remain professional, in line with branding and marketing strategies, and clearly demonstrate value, but these are achieved differently in business-to-business (B2B) and business-to-consumer (B2C) efforts (Naiduk, 2013).

Internet Marketing Communication Strategy: B2C Vs B2B

EMOTION VS LOGIC: Understanding why people buy can help you influence what they buy. General consumers differentiate between small ticket and big ticket purchases. However, emotions often play a role in their buying decisions. These consumers want items that make them feel good and that they can picture themselves using with a level of enjoyment. Businesses have very different incentives for buying goods and services. For a business, all purchases are investment and they stand to gain a profit from them. Business purchase decisions are based on logic. Businesses want to buy items because they feel they can provide them some benefits, such as cost and time.

IMMEDIACY VS. RELATIONSHIP: Consumer purchases may or may not be made with the expectation of an ongoing relationship. While many retailers to the consumer market do want to establish relationships and build trusting reputations, consumers sometimes are more interested in how a product will help them now. When communicating with businesses with the intent of creating a sale, the process is typically much longer. There is more often a relationship-building period. There is a series of informational and rapport-building calls and meetings before a purchase is made. Good B2B communicators acknowledge and respect that element of the process and they make it a priority to build a strong and trusted professional relationship. Depending on the nature of the good or service at hand and the ticket price, the relationship will continue long after the decision to buy. It is important that you provide your customers the right content at the right time so that your conversion rates can be increased. Figure 7-5shows how the content is provided to the B2B customers' right from the beginning till they make the purchase.

FIGURE 7-5: B2B PROMOTION CONTENT

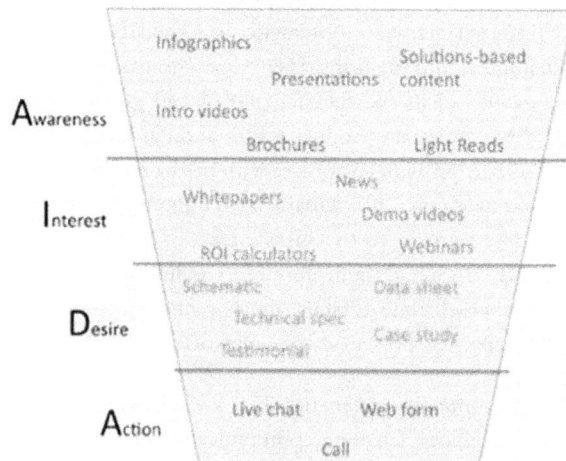

BROAD VS. SPECIFIC TARGET MARKETS: B2C communications are generally aimed at a larger target market. Businesses have unique wants and needs specific to their industries, operations, and goals. Effective B2B communications are often more specific and direct, backed by proper research. Customized communications, proposals, reports, and so on are common in B2B efforts for that reason.

SINGLE VS. MULTIPLE DECISION MAKERS: When designing marketing communications, it is essential to identify the decision-maker(s). For most B2C efforts, it can be reasonably assumed that the decision-maker is one individual, or at most a few key members of a family or group. The purchase decision is considered a low-risk decision. The decision-making process for businesses is often much more involved and there can be many people involved. The purchase decision is considered a risky decision. For that reason, B2B communications must clearly reveal the benefits of going with one company over another.

Internet Marketing Communication Tools

There are several Internet marketing communication tools available: web site, e-mail, newsgroup, and search engines etc. Marketers can use these tools to move closer to their customers and add value to their products. The six main tools of web marketing communications are:

1. **Online Advertising** (banner ads, pop-up and pop-under ads, interstitial ads, Newspaper like standard ads, Uniform Resource Locator (URL), Chat room advertising, online sponsorships, other forms of advertising)

2. **Online Personal Selling** (video conferencing, Webcasting and Web Conferencing, virtual exhibitions)

3. **Online Sales Promotion** (coupons, sampling, contests, sweepstakes etc.)

4. **Online Public Relations** (forums, newsletters, newsgroups, online communities, , cause marketing, viral marketing)

5. **Online Direct Marketing** (E-mail, computerized home shopping)

6. **Corporate Web Site**

Online Advertising

Advertising is defined as non-personal communication of information by an identified sponsor through various media. It is usually paid for and persuasive in nature about products (goods and services) or ideas (Smith, 1998).

Today companies can use many of Internet's tools such as banner advertising, pop-up boxes, and links from other sites to communicate with the target audience easily and efficiently. Many companies are using web to advertise their products or services. With web advertising, consumers have more control over advertising exposure because they can select how much content they wish to view. Web advertising may be either text based or multimedia based. Text-based advertising works through e-mail and bulletin boards while multimedia advertising uses banners, buttons, interstitial and sponsorships.

According to Interactive Advertising Bureau, online advertising represented 10% of the advertising industry (over USD 51 billion) in 2009 (Lynch, 2010). The role of online advertising is growing rapidly. Since 2003, online advertising is growing at a double-digit rate outperforming the overall advertising growth significantly. It is predicted that in 2010 Internet advertising will account for about 15% of global advertising spending (American Association, 2009). In the United States, online advertising contributed to US$ 300 billion and directly employed more than 1.2 million people (InteractiveAdvertisingBureau, Economic Value of the Advertising-Supported Internet Ecosystem, 2009).

High-consideration and information-intensive products are well suited for Internet advertising. For these products, consumers typically want to search prior to purchase. Some examples of such products are computer hardware, automotive services, travel services, and financial services.

FUTURE OF OFF-LINE ADVERTISING

The major traditional advertising media are television (about 42%), newspapers (about 31%), magazines (about 16%), and radio (about 11%) (Shimp, 2008). However, offline advertising is still important. So far, the most successful Internet marketing communication campaigns are the ones that incorporated consistent imagery with ads and combined both offline and online communication medium in the campaign. Furthermore, in most developed countries offline media such as television and radio have nearly 100% market penetration and a significant majority of adults read a newspaper every day. Therefore, for Internet marketing to be successful online communication media needs to be integrated with traditional media such as Print, TV (Chaffey, Mayer, Johnston, & Ellis-Chadwick, 2002).

ONLINE ADVERTISING: ISSUES AND PROBLEMS

While on-line advertising provides opportunities to serve consumers more effectively, at the same time it faces many objections against its use. Most of the objections are raised because of concerns about privacy and the use of personal information. A recent report shows that a significant number of businesses provided their consumers inadequate information on their transactions, and their rights. There were irregularity in the information provided 55% of websites selling consumer electronic goods found to have irregularities in the information provide about consumer rights and total cost of the product (Europa, 2009).

One of the key issues related to online advertising relates to behavioral advertising. Behavioral advertising tailors advertising to individual consumers, based on available tracking data about their online activities and related information collected from the Internet or elsewhere over time. Behavioral advertising can be beneficial for consumers because it is personalized and provide the consumer with useful information (e.g. availability, price and characteristics of products that may be of great interest), and filter out advertising for products for which a consumer has no interest. Critics argue that collection and storage of consumer's personal data for advertising and marketing purposes may undermine consumer privacy. Critics are also concerned about the invisibility of the data collection to consumers, irregularities in current disclosures about the data collection practice, and the risk of this behavioral advertising data (including sensitive data) falling into wrong hands or used for unanticipated purposes (FTC, Self-Regulatory Principles for Online Behavioural Advertising, Behavioral Advertising - Tracking, Targeting, & Technology, US FTC Staff Report, 2009). There are reports of large data breaches and unexpected transfers of personal data to third parties (OECD, 2006). Many consumers do not want tracking of their online activities for advertising purposes, even if it may provide them some discounts or other benefits (Turow, King, Hoofnagle, Jay, Amy, & Hennessy, 2009). Policymakers in various countries have begun to explore this issue and develop guidance for advertisers. United States FTC, for example, issued a set of proposed principles to encourage and guide industry self-regulation (FTC, Self-Regulatory Principles for Online Behavioural Advertising, Behavioral Advertising - Tracking, Targeting, & Technology, US FTC Staff Report, 2009).

Another critical issue for organizations is to determine the best way to utilize on-line media for advertising purposes. Effectiveness of web advertising strategies in influencing the behavior of Web users varies. (Wang, Wang, & Farn, 2009) suggests that to achieve Web advertising effectiveness, appropriate Web advertising strategies should be implemented in accordance with Web users' goal-directedness and product involvement.

Industry groups have also started to develop self-regulatory principles on behavioral advertising. In July 2009, a set of principles guiding the use of behavioral ads was proposed by an alliance of US advertising groups (InteractiveAdvertisingBureau, Self-Regulatory Principles for Online Behavioral Advertising, 2009). A set of self-regulatory guidelines, called *Good Practice Principles*, was issued in the United Kingdom by Internet Advertising Bureau (IAB). These principles aimed to provide good practices for companies and increase consumer awareness about the collection and use of data for online behavioral advertising purposes. A website www.youronlinechoices.co.uk was launched to help consumers understand online behavioral advertising. Office of Fair Trading (OFT) of United Kingdom recently launched two market studies the extent of impact of behavioral advertising on consumers shopping online and misleading pricing techniques (OFT, 2009).

The Interactive Advertising Bureau (IAB) empowers the media and marketing industries to thrive in the digital economy. It is comprised of more than 650 leading media and technology companies that are responsible for selling, delivering, and optimizing digital advertising or marketing campaigns. Together, they account for 86 percent of online advertising in the United States. Working with its member companies, the IAB evaluates and recommends standards and practices and fields critical research on interactive advertising. The organization is committed to professional development, elevating the knowledge, skills, and expertise of individuals across the digital marketing industry. The IAB also educates marketers, agencies, media companies and the wider business community about the value of interactive advertising. Founded in 1996, the IAB is headquartered in New York City (IAB.net, 2015).

ONLINE ADVERTISING REGULATIONS

The FTC (Federal Trade Commission) is an American agency that provides consumer protection and competition jurisdiction in broad sense of economy. The FTC has set rules and regulations to prevent unfair commerce practices, just like advertising in different mediums is governed by rules and regulations, online advertising too has rules that control and govern it. These rules and guidelines protect businesses and consumers - and help maintain the credibility of the Internet as an advertising medium. There are many regulations that the FTC act enforces upon businesses. The FTC Act prohibits unfair or deceptive advertising. That is, advertising must tell the truth and not mislead consumers. A claim can be misleading if relevant information is left out or if the claim implies something that is not true. Sellers are responsible for claims they make about their products and services. The third parties, (websites in this case) may too be liable for making or disseminating deceptive representations if they participate in the preparation or distribution of the advertising, or know about the deceptive claims. Website designers are responsible for reviewing the information provided by the advertisers in order to substantiate the claims. They may not rely on the advertisers when they say that the claims are substantiated. FTC Act requires that disclaimers and disclosures must be clear and conspicuous.

The ASA (Advertising Standard Authority) is a UK based independent organization. It aims to ensure that adverts are honest and truthful and do not mislead customers. The ASA governs and overlooks all mediums of advertising including advertising on websites. They build trust by enforcing the Advertising Codes written by the Committee of Advertising Practice and acting swiftly when marketing communications break the rules. The committee that sets these advertising codes is called CAP- the Committee of Advertising Practices.

Online Advertising: Tools and Techniques

Banner Ads

Banner ads are small advertisement (typically a rectangular graphic) bought by companies and placed on other companies' advertising vehicles (such as search engines, chat rooms, online magazines, and Web pages) for purposes of brand building or driving traffic to a site. The most common form of web advertisement is banner advertisements. Banner ads range in size from few inches to full-page width. They can contain several colors. A well-designed banner ad could leave the visitors with some knowledge about the product or with a positive image of the company.

Most Web sites have voluntarily agreed to use standard banner sizes called interactive marketing unit (IMU) ad formats. The Interactive Advertising Bureau (IAB) has established voluntary standards for IMUs. IAB is a not-for profit organization that promotes the use of Internet advertising and encourages effective Internet advertising. IMU ad formats include a large number of standard ad formats, but many advertisers continue to use the four standard formats. The reason is that most advertisers believe that almost every Web site will be able to display ads in those formats properly. Table 7-3 lists some of the most commonly used standard banner sizes with their suggested maximum file sizes.

TABLE 7-3: COMMONLY USED STANDARD BANNER SIZES

Banner size (pixels)	File Size	Template Name
728 long x 90 high	25 KB	Leaderboard Banner
468 long x 60 high	20 KB	Full Horizontal Banner
392 long x 72 high	20 KB	Full Horizontal Banner with navigation bar
234 long x 60 high	15 KB	Half Banner
120 long x 240 high	20 KB	Vertical Banner
120 long x 90 high	10 KB	Button 1
120 long x 60 high	10 KB	Button 2
125 long x 125 high	15 KB	Square Button
88 long x 31 high	5 KB	Micro Button

Courtesy: www.cmykreative.com

Banners can be of many types. A keyword banner is the one that appears when a web user uses the search engine to query a predetermined. Keyword banners are used by companies to narrow their target audience. A random banner is the one that appears randomly and used by companies to introduce new products. Banners can be static or interactive. In a dynamic banner, the advertisement changes with each refresh of the page and enable the display of a numerous ads on a web site without using too many banners. In contrast, Static banners contain one advertisement that does not change.

Advertising agencies and web site design firms can create banner ads. Charges for creating banner ads range from about $100 to more than $1500, depending on the complexity of the ad. Companies can make their own banner ads by using a graphics program or the tools provided by some Web sites. AdDesigner.com is an advertising-supported Web site that offers free downloadable graphics.

BENEFITS AND LIMITATIONS OF BANNER ADS

Clicking on banner ads, a user can be taken to advertiser's site and many times directly to the shopping page of that web site. Banner ads can also be customized to the targeted individual or market segment of web users. Companies can also force customers to see banner ads before their required page is loaded or before they could access the free information or entertainment that they want to see. Finally, banners may include attention-grabbing rich media e.g. a movie clip or animation.

However, banners can be costly. Running a successful marketing campaign requires placing banner ads on high traffic websites and that can cost a large percentage of the advertising budget. Another disadvantage is the limited amount of information that can be placed on the banner. Hence, advertisers need to come up with creative but short message to attract viewers.

Pop-Up and Pop-under Ads

A pop-up ad launches automatically in a new browser window when a visitor visits or leaves a site, on delay, or on other triggers. A pop-up ad appears in front of the active window. A pop-under ad is an ad that also appears in a new browser window when a visitor visits or leaves a site, on delay, or on other triggers. The difference is that the pop-under ad window underneath the current browser window; when users close the active window, they see the ad. Pop up ads are likely to generate more profits than the banner ads on the websites and so pop up ads seem to appear on the major websites. The pop-up windows can contain anything: text, graphics, a form to collect information or email addresses, even a little game. Most pop-up windows can be minimized with relatives ease, so if the pop-up window is being used to rotate ads on a time basis, your advertisement may not even be visible but still be charged for it!.

Interstitial Ads

An interstitial ad is a full-page ad that appears before the actual webpage appears in the web browser. For example, the code of an interstitial ad can be attached to a target page and when visitors try to load this page in their browser by typing its address in the address bar they will see a full-page with ad (the interstitial ad) before continue to see the target page. The template of interstitial ad can be customized. If visitor clicks on the ad, a new window to the advertiser's website will be opened without interfering with the navigation within your website. These ads remain while content is loading. An interstitial may be an initial Web page or a portion of one that is used to capture the user's attention for a short time. Interstitial ads are literally "in between" other screens of information.

BANNER AD PLACEMENT

Companies can utilize three different ways to have their banner ads displayed on various web sites. The first is to use a banner exchange network e.g. 1800banners.com. Banner exchanges are a group of websites that exchange banner advertising with each other. After joining a banner exchange network a business will be provided with an html code to insert into business website. This html code will automatically display banner ads on business website and in return, business will generate advertising credits for its own banner to be displayed on the free banner exchange. Most banner exchanges give a 2:1 exchange ratio, which means a business get one free banner display per every two-banner ads displayed on its site. Both businesses make their profit by selling the extra ad space to other businesses. Banner exchanges are not free and businesses may find it difficult to find a group of other web sites that have formed an exchange or that belong to an exchange that are not direct competitors. This limitation prevents many businesses from using banner exchange networks. Not-for-profit information Web sites are more likely to find a banner exchange network suitable. Some banner exchanges will not allow certain types of banners.

Businesses can also place their ads on sites that appeal to one of the company's market segments. Businesses pay these sites to host their ads. This technique is called banner swapping. Banner swapping is probably the least expensive but difficult to arrange from of banner advertising. A business must first locate sites that could generate a sufficient amount of relevant traffic and then contact owners of the site to inquire if they would be interested in a banner swap. That can involve considerable time and effort. First, because smaller sites may not have an established pricing policy for advertising and second larger sites usually have high standard rates that they usually discount for larger customers only. Companies can seek services of advertising agencies to negotiate lower rates and help with ad placement as because the agencies can consolidate their clients' budgets and buy large blocks of advertising space at one time. Advertising agencies can also help businesses in design of the ads, banner creation, and identification of appropriate web sites.

Companies can also use a banner-advertising network that acts as a broker between advertisers and Web sites that carry ads. The larger banner advertising networks, such as DoubleClick and ValueClick offer internet-advertising solutions for publishers of web sites and online advertisers. These networks often broker space primarily on larger Web sites (e.g. Yahoo!) that have high traffic rates and are more expensive.

Google AdSense

Google AdSense is an ad serving application run by Google Inc. Website owners can enroll in this program to enable text, image, and video advertisements on their websites. These advertisements are administered by Google. These advertisements generate revenue on either a per-click or a per-impression basis. Google uses its Internet search technology to serve advertisements based on website content, the user's geographical location, and other factors. Those wanting to advertise with Google's targeted advertisement system may enroll through AdWords. AdSense has become a popular method of placing advertising on a website because the advertisements are less intrusive than most banners, and the content of the advertisements is often relevant to the website.

Many websites use AdSense to monetize their content; it is the most popular advertising network. AdSense has been particularly important for delivering advertising revenue to small websites that do not have the resources for developing advertising sales programs and sales people. To fill a website with advertisements that are relevant to the topics discussed, webmasters implement a brief script on the websites' pages. Websites that are content-rich have been very successful with this advertising program, as noted in a number of publisher case studies on the AdSense website.

There are different types of Google AdSense.

AdSense for Feeds: AdSense for Feeds works by inserting images into a feed. When the image is displayed by a RSS reader or Web browser, Google writes the advertising content into the image that it returns.

AdSense for search: AdSense for search, allows website owners to place Google search boxes on their websites.

AdSense for mobile: AdSense for mobile content allows publishers to generate earnings from their mobile websites using targeted Google advertisements.

AdSense for domains: AdSense for domains allows advertisements to be placed on domain names that have not been developed.

AdSense for video: AdSense for video allows publishers with video content to generate revenue using ad placements from Google.

To use Google AdSense, the webmaster first needs to insert the AdSense code into a webpage. Each time this page is visited, the code display content fetched from Google's servers. For contextual advertisements, Google's servers use a cache of the page to determine a set of high-value keywords. If keywords have been cached already, advertisements are served for those keywords based on the AdWords bidding system. For site-targeted advertisements, the advertiser chooses the page(s) on which to display advertisements, and pays based on cost per mile (CPM), or the price advertisers choose to pay for every thousand advertisements displayed. Search advertisements are added to the list of results after the visitor performs a search.

Bing Ads

Bing Ads (formerly Microsoft adCenter and MSN adCenter) is a service that provides pay per click advertising on both the Bing and Yahoo! search engines. As of the fourth quarter 2013, Bing Ads has 15.6% market share in the United States. You also have the ability for your ads to show on their syndicated partners such as Facebook, Amazon, Monster, WebMD, CNBC, and Viacom.

FIGURE 7-6: BING ADS

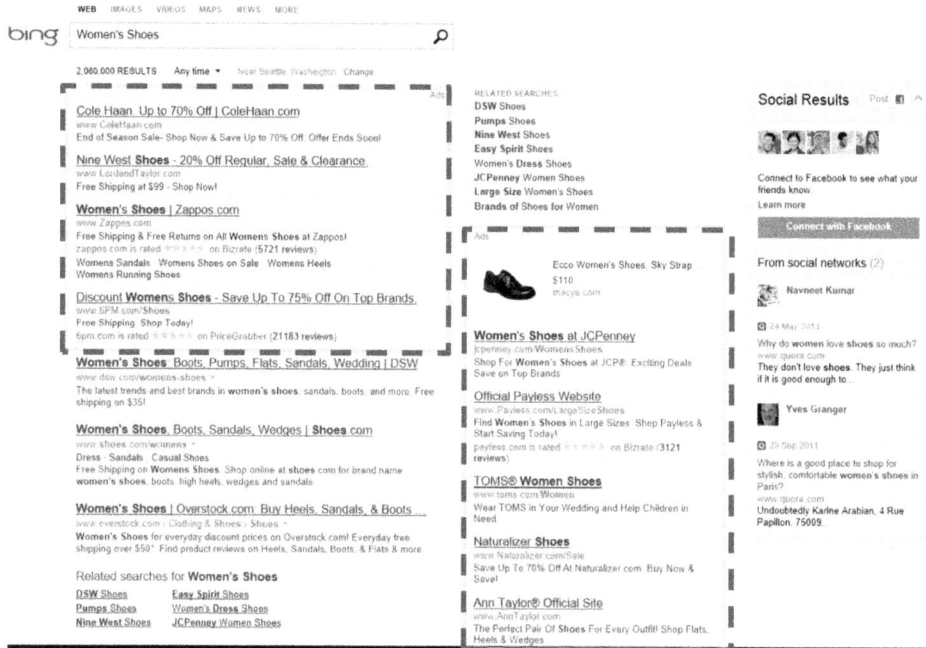

Courtesy: fp.advertisning.micrsoft.com

By bidding on the right keywords, a business can shows their ads only to people who are interested in their products. On top of that, they only have to pay for the ad when someone clicks on it and visits their website. Less people advertise using Bing Ads than Google AdWords, so the competition, and the amount you have to pay to reach your target customer, can be much lower. In fact, research (Pretorius, 2013) shows that your cost per click when using the Bing Ad network can be up to 70% lower than when advertising using AdWords. To

382

setup your Bing Ads account, you first sign up for a free $50 Bing Ads coupon. You then setup your Bing Ads account and import or create your first campaign and first Ad. You then choose the Keywords against which you want to shows your ads. You then choose match type and set the bids. At the end, you choose your campaign name, budget, and location targeting.

Yahoo! Advertising

Yahoo advertising (also called Yahoo! Search Marketing) is a keyword-based "Pay per click" or "Sponsored search" Internet advertising service provided by Yahoo.

BANNER AD COSTING

Use of online advertising is increasing. Therefore, the issue of measuring web site effectiveness has become important. One can measure mass media efforts by estimating audience size, circulation, or number of addressees. The price of a mass media advertisement can be calculated in terms of dollar amount paid for every thousand people in the estimated audience. Measuring Web audiences is more complicated because of two reasons. First, the web is interactive and second the value of a visitor to an advertiser depends on how much information the site gathers from the visitor (for example, name, address, e-mail address, telephone number, and other demographic data).

Costing Internet advertising is more difficult than doing so for conventional advertising for two reasons: (1) it is difficult to measure the effectiveness of Internet advertising (2) There is disagreements on pricing methods.

Several methods are available for measuring advertising effectiveness, cost-benefit analyses, and pricing. There are three general methods for calculating cost of banner ads. These are Cost per Click (CPC), Cost per Thousand (CPM), and Cost per Action (CPA).

Cost per Thousand (CPM)

The "M" in CPM derives from the Latin word "mille" which means "thousand". CPM is not an absolute cost and it estimates the cost per 1000 views of the ad. It is calculated as:

$$\text{Total budget} / (\text{Impressions}/1000) = \text{CPM}$$

In a CPM campaign, a business pay for each how many times the ad appears. If ad performs well and lots of users click through business's site, then each click-through costs the business less. On the other hand, the lower the click-through rate, each click costs you more. For example, in a CPM-based campaign, a business may pay $3 for 1,000 impressions. If the ad receives a click-through ratio of just 1 percent (i.e. 10 clicks) business is paying $3 for 10 clicks. Rates vary greatly depending on how much demographic information the web site obtains about its visitors and what kinds of visitors the site attracts. Most rates range between $1 and $50 CPM.

Cost per Click (CPC)

Banner ads can also be sold on CPC basis. The manner in which ads are displayed is same for both CPM and CPC. The difference is in the payment for the advertising. In a CPC campaign, a business pays a fixed amount for each click and sets a ceiling payment. Ceiling payment means that business can set a limit on the amount of times the ad can be clicked and cap the total amount spent on the campaign. This way business can reduce risks and avoid unexpected advertising bills. By paying only for actual clicks, business save money and reach only targeted customers. Using CPC, business can control their advertising budget and gauge response to their advertising campaigns. For example, if a business decides to go for CPC campaign, the ad will display each time a user enters in a keyword that matches predetermined keyword list. Business will not pay each time the ad displays; only when it's actually clicked. Let us say a business is spending $1 for each click.

There are several places where businesses can find CPC-based advertising. Google AdWords is one of the most well-known CPC-based advertising programs. With AdWords, businesses pick the keywords they want to use and only pay for as many clicks as they can afford. There are also many companies offering CPC-based banner ads in addition to the traditional CPM advertising method.

Cost-Per-Action (CPA)

Under this model, payment is based solely on qualifying actions such as sales or registrations. The actions are defined in a cost-per-action agreement and relate directly to some type of conversion. This model is somewhere in between CPC and CPM model where the publisher takes most of the advertising risk as their commissions are dependent on good conversion rates.

Similarly, CPL (Cost per Lead) advertising is identical to CPA advertising and is based on the user completing a form, registering for a newsletter or some other action that the merchant feels will lead to a sale. Also common, CPO (Cost per Order) advertising is based on each time an order is transacted. CPE (Cost per Engagement) is a form of Cost per Action pricing first introduced in March 2008. Differing from cost-per-impression or cost-per-click models, a CPE model means advertising impressions are free and advertisers pay only when a user engages with their specific ad unit. Engagement is defined as a user interacting with an ad in any number of ways.

Cost per conversion describes the cost of acquiring a customer, typically calculated by dividing the total cost of an ad campaign by the number of conversions. The definition of "Conversion" varies depending on the situation: it is sometimes considered a lead, a sale, or a purchase.

Some other methods of banner ad costing are:

CPV (Cost per Visitor)

 In this method, advertisers pay for the delivery of a Targeted Visitor to the advertiser's website.

CPV (Cost per View)

In this method, the price paid by an advertiser for each unique user view of an advertisement or website (usually used with pop-ups, pop-under and interstitial ads).

Which Method is Appropriate?

It is a good idea to try, evaluate, and track the results of different methods of advertising at the same time. This way a business can determine what works best for them.

If the desired keywords are very popular and expensive then CPM-based advertising campaign may be a more suitable alternative for a business. Although click-through ratios for CPM-based advertising are low, businesses should remember that a well-design un-clicked (but viewed) ad could grab a viewer's attention and help in promoting brand recognition and play an important role in promoting their business. Businesses can also use websites that offer targeted CPM-based advertising that can greatly increase overall effectiveness of ads. Targeted CPM-based advertising can be very beneficial for small companies.

CPM-based advertising do have some disadvantages. First, there is a requirement of a minimum purchase of impressions that could be more than the business requires. The minimum purchase can amount to one million impressions though there are companies that offer low minimums. Businesses should plan carefully before deciding to buy the impressions. Even if budget allows buying a large number of impressions, businesses should check various companies before making a purchase decision.

CPA model can work only on sites where actual purchases can be made (e.g. cattoys.com). At the Ritchey Design (ritcheylogic.com) (Maker of handcrafted road and mountain bicycle components) or Coca-Cola's website, users only get information and brand awareness, thus this method would be inappropriate for these types of merchants.

BANNER ADS EFFECTIVENESS

There are dramatic differences in the effectiveness of various forms of advertising. If you pay for advertising, then it is probably important for you see some results. However, if you waste money on inefficient advertising, you are missing better opportunities and the results may not come at all. The main objective in measuring advertising effectiveness is to determine the effect of each advertising campaign from the results of our measuring and compare it with its price. Then we can decide which campaigns bring the best value for the money spent. It is also important to realize the various factors influencing advertising. The medium, ad copy (exact wording), the format, audience all affects the final success of the campaign. Therefore, it is necessary to judge the effectiveness in context.

Measuring the costs and benefits of advertising on the Web is the most difficult thing for businesses using banner advertisement. Most of the benefits of advertising are intangible. Metrics have been developed by companies to evaluate the number of desired outcomes their advertising yields. For example, rather than comparing the number of click-through obtained per dollar of advertising, businesses can measure and compare the number of new visitors to their site who buy for the first time after arriving at the site by way of a click-through. Businesses can then calculate the advertising cost of acquiring one customer on the Web and compare that to how much it costs them to acquire one customer through traditional channels. Another approach to measure banner ad effectiveness is by using brand measures (e.g. brand awareness and attitudes). One other approach is to use actual transactions (i.e. conversion rates).

Monitoring the effect of your internet advertising can be easy and cheap when you use quality tools. Google Analytics is a great free website analysis tool. When you implement it in your website and set up your campaigns properly, you will have all the information you need to decide which advertisement you should drop and which brings you good return on your investments.

Besides basic functions like monitoring number of visits and page views, it offers you a variety of statistics regarding visitor segmentation, traffic sources etc. Google Analytics also allows setting up goals on your website and tracking conversions – goals accomplished by visitors. Then you can mine for data and see for example what the average transaction value is for visitors from London, how many pages they view per visit etc. If you set value for your goals, you can also see return on investments (ROI) for the money spent on advertising. This is possible because of tags, which are added to the links pointing to your website. Another advantage of this tool is that it helps you to test the different versions of either website or ad copy. Thus, you can create much better land page or more effective ad this is called A/B testing.

TECHNIQUES FOR EFFECTIVE BANNER ADS DESIGN

Researching banner ads is integral to finding the formula that works best for you. The second rule of creating effective banner ads is to keep it short and simple. Some Techniques to build effective banner ad are as follows.

Produce a Visual impact

Banner ads are displayed amongst a large amount of other information, competing with several other factors on the page for the user's attention. Unlike print media, where an advertisement can grab the entire page or appear as a two-page spread, banner ads get limited display area. So it is imperative to keep the banner ads simple, easy to understand and to the point. Brief is good. Clutter is not. A study (Reyna, 2014) found that banner ads that have too much going on result in lesser clicks and lower brand recall. Banner ads should be simple and any graphics should be used only when it enhances the message. If graphics is to be used than it should enhance, not distract from the message. An effective banner should have the right colors that can attract the eyes, font styles that are not too fancy and readable, and uses large text for call to action and smaller ones for the contents. Such banner ad has a touch of creativity but still looks simple. Also, check out the other banner ads on the website to ensure that your ad uses colors that stand out from the crowd. Make sure that the color scheme that anyone is selecting for his or her banner ad must be humane enough for a person to look at.

Use Good Call to Action

The banner ad should include an attention-grabbing element, a call for action and a reason to click through (unless goal is branding). Words, such as 'Free' and 'This is Your Last Chance', create a feeling of urgency. In making your banner, you need to choose the right words that could give an immediate message and sense of urgency to the audience. It should also attract the users in a way that their attention could be diverted towards the banner even if they are reading something in the website.

Use Dynamic Content

Rich media usually enhances click-through rate. It is best if animation supports the message to be delivered through advertisement but use of animation should not slow down the download. A study (Education B.V., 2013) found that animated ads generated 15% higher click thru rates than static banners did. Coupled with an intelligent but witty message and the appropriate design this form of advertising can work very well. Be very careful when using banner animation as if it takes too long to download. Keep the file size low. About 45-70kb is typically a good size for a banner ad. Most websites you advertise on will have limitations regarding the size. In any case, it benefits to have fast loading banner ads to appeal to all audience groups. A good rule of thumb is to examine your target market when determining the file size. For example, if you are selling advanced computer software- chances are you do not have to worry about an audience with slow load times. If you are selling computers, you may want to appeal to those with slow loading times. Check your banners loading time (on average a surfer will move on after 6 seconds if your page does not load efficiently) (Education B.V., 2013).

Involve the Audience

Invite your target market to participate in some activity carried on at your site. Perhaps a Game, Quiz, Give-away, work on the premise that people like to play games. Simple games have been found to increase click-through rates.

Provide Something Free

Viewers will only respond to a compelling proposition. Your banners must be attractive or interesting enough to be successful in generating response. People react favorably to banners that provide them a chance to win something. A banner where they can get something free or a special discount also gets high click rates. Develop your banner's message around the most persuasive reason why people would want to go to your site, be it the information you provide, special offers and promotions, or products that can make their life much better. E-mail newsletters and product updates can work very well with a targeted group. Screen savers can produce good results as they can be used with subliminal advertising. Free downloads beta software can and do create effective responses.

Don't Include Your Brand Name

Adding your brand name, even if it is popular, can be counterproductive. Viewers often assume they already know everything they need to know about the brand and totally ignore the ad. Users have a high level of recall of online ads coupled with increased brand awareness if the banner announces a new product or service.

Change Your Banners Frequently

Remember that after third or fourth impression, response rates drop dramatically. Therefore, you should rotate your ads often. You should avoid putting too many ads on the website. As a benchmark, if number of ads is greater than three it is more likely that customers will leave the site. One way to avoid too many ads is to use fewer and larger ads per page or use dynamic banners. Use of bold colors in banner ads is effective and banners at top of page about twice as likely to be noticed. Bigger banners get more attention and you should fit ad message to content of page. Last but not the least you should test your banner ads continually.

Use Relevant Keywords

Attach your banner to relevant keywords for an increased chance of being located through search engine. Consider sponsoring keywords and phrases for different search engines. For example, if you would like to promote your website for business cards; you could enable users to easily find you every time someone enters the word "business".

Use Attention-grabbing Content

Use an image that will surely get the attention of the users. The image depends on you and the type of product or service you will be advertising. Use the appropriate image that could get the interest of the users. You may also include a sound like a drum roll or a music hit or whatever. The important thing there is, you should catch their attention. Have one clear message on your banner ad. It can be something as simple as "Save 25% when you order today!" Always design the banner ads while keeping in mind the context in which it will be displayed. While the message can be the same across different websites, try to choose your colors based on the background of the website on which the ad is to be displayed. Keep your banner size small with regard to bytes say under 10k-15k this will speed download, and then use wider banners - either 468 or 500 pixels wide.

Choose a Strategic Location

Choose the right place to display your banner ads. You can create a fantastic banner ad with crisp copy and appealing graphics and still find that it is failing miserably. Check to see if you are targeting the right audience. If you are advertising expensive gadgets to customers with low-income, chances are you won't achieve higher ROI on your ads. Study the audience on the website where you intend to display the banner ads to determine if that audience is right for the products you are looking to advertise. Look for a location where in the user could immediately see it. It would be wise to place it above the fold. Most people place it at the top or at the left side. Actually, wherever you place it, as long as it looks attractive, the users will still notice it.

OTHER BANNER AD TYPES

Disguised Banner Ads

Recently, banner ads are losing their ability to grab viewers' attention mainly because of increasing proliferation of banner ads on the web. As a result, developers are turning towards new approaches in banner ads design to regain customer attention. The first approach is to use animated GIFs with moving elements in banner ads rather than using stationary graphics. The second technique is to introduce rich media effects (such as movie clips) in banner ads. The third technique is to use interactive effects in banner ads. Interactive effects in banner ads are introduced by writing programs (such as Java programs) that could respond to a user's click with some action (other than simply loading the advertiser's page into the browser). The fourth technique is building banner ads that appear to be dialog boxes. These ads are designed to induce users to click a button in the ad to perform an action (e.g. fix the error) but actually by clicking on the button, user is taken to the ad sponsor sites or begin installing a program on the user's computer.

Newspaper-Like Standardized Ads

These ads are online and full-column-deep. They are also called skyscraper ads. These ads are very tall with heights often ranging from 500 to 800 pixels (and widths often ranging from 120 to 160 pixels). With standards in place, skyscraper ads have become more common at advertising networks. Some of these ads are interactive. Users can click on a link inside the ad to get more information about a product or service.

Text Ads

Text ads are probably the dominant from of online advertising today. A text ad consists of a few lines of copy together with a link or an email address for action. These ads are popular in newsletters and RSS feeds. A well-known example of text ads is Google's ad sense program, which delivers contextually relevant text ads (ads targeted based on the content of a page).

HTML Ads

HTML Ads combine graphics and text with other HTML elements such as pull-down list, check boxes or forms.

Rich Media Ads

Rich media ads make use of multimedia elements such as sound, animation. Rich media ads capture more clicks than any other advertisement medium as they tends to be more eye catching and attractive to the viewers.

Email Ads

These ads are distributed by a publisher through email to opt-in audiences. Advertisers can individually sponsor a publisher's email newsletter or they can purchase classified ad space.

Advertorials

These are advertisements in editorial form that appear to contain objectively written opinions. Online advertorials are typically featured on publisher's websites. These advertorials promote products and services related to the website's content.

RSS Advertising

RSS stands for Rich Site Summary or Really Simple Syndication. RSS advertising is the integration of RSS into online advertising. RSS advertising can avoid Spam, email filtering and pop-up blocking system.

Superstitial Ads

This is an animated ad. It uses video, 3D content or Flash to provide a TV-like advertisement.

Unicast Advertisement

It is an advertisement that consists of a video played like a TV commercial, usually in a pop-up or pop-under advertisement. Unicast ads have the same influence and power as a regular TV commercial, except they allow the viewer to click on the ad in order to go to the company's website, or get more information

Display Ads

Display ads are often available in many standard shapes and sizes, including banners, leader boards, skyscrapers, large boxes, and other sized graphical ads. Display ads use eye-catching visuals that quickly grab the attention of website visitors browsing the pages on which they are featured.

Sticky Ads

This form of integrated advertising overlays an ad of any form in a fixed position in the browser window, in such a way that it is not affected by scrolling and has to be manually closed by pressing a small close button that is often hard to find.

Pushdown Ads

These ads expand to almost a full screen upon loading a page. Some publishers require the ad to collapse after a certain time. The viewer may then click to re-expand the Banner and read the contents.

Fixed Panel Banner

This ad appears set in the browser. The panel (measuring 336 x 860 pixels) rolls to the top and bottom of the page as the user scrolls up and down.

Floating Ads

These ads appear when you first go to a webpage, and they "float" over the page for a certain time. While they are on the screen, they obscure your view of the page you are trying to read, and they often block mouse input as well. These ads appear each time that page is refreshed.

Homepage Takeover Ads

A display ad implementation that involves creative from a single advertiser appearing in all available units on a site's homepage.

XXL Box Banner

A colossal sized (468 x 648 pixels or more) Banner used by prominent advertisers for brand effectiveness. These massive advertisements can comprise several pages and video.

Hybrid Ads

Hybrid ads combine other advertising types, such as text and banners, to make a more effective pitch to visitors.

Ad Targeting Methods

CONTEXTUAL TARGETING

When ads are served based on related content a user is currently reading or browsing online, it is known as contextual targeting. For example, if you are reading an article on a news website about sports, you may see contextual ads for sports gear, memorabilia, or game tickets. Contextual ads are purchased through major search properties like Google, Yahoo, MSN, and through many other contextual ad networks. Ad relevancy is typically determined by algorithms that will assess the appropriateness of the ad in relation to the displayed content.

FIGURE 7-7: CONTEXTUAL TARGETING

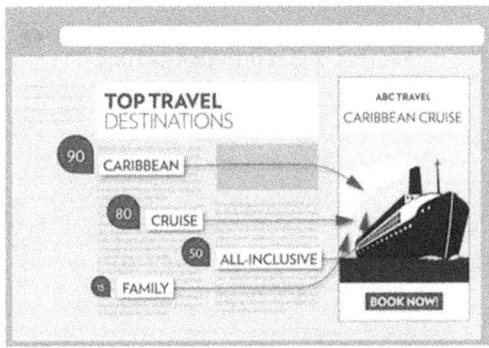

BEHAVIORAL TARGETING

When ads are served based on user behavior, it is known as behavioral targeting. Behavioral targeting is based on a variety of online factors such as recent online purchases, searches, and browsing history, as well as demographic details such age or gender. For example, if you recently visited a real estate website, you may see behavioral ads selling mortgages.

FIGURE 7-7: BEHAVIORAL TARGETING

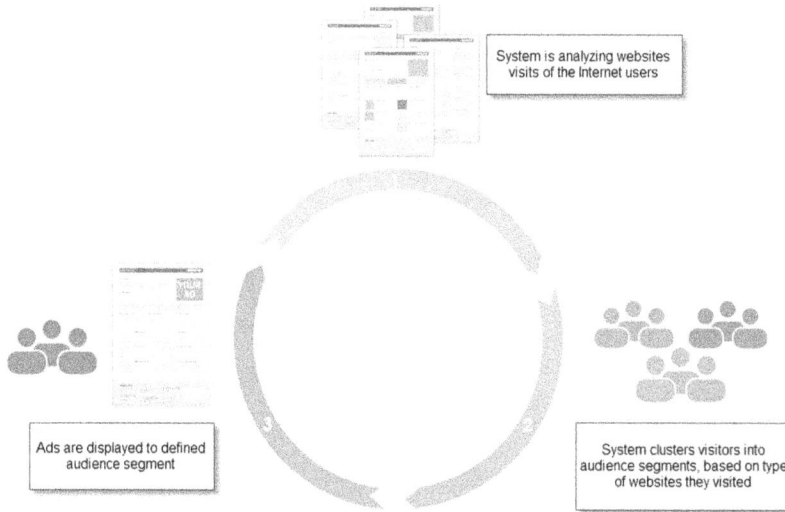

System is analyzing websites visits of the Internet users

Ads are displayed to defined audience segment

System clusters visitors into audience segments, based on type of websites they visited

GEO-TARGETING / LOCAL ADVERTISING

When ads are served based on a user's geographical location, it is known as geo-targeting. For advertisers interested in targeting users within a specific locality or region, geo-targeted online advertising is an effective solution. An increasing number of websites now offer geo-targeting capabilities. Local advertising also includes network buys through radio, television and newspaper websites, as well as localized search engines and directories such as Yahoo! Local, Google Local, and AOL City Guide.

FIGURE 7-8: GEO-TARGETING

Chinese restaurant near karachi

Rating

Yuan Tung
★★★★ Chinese
Tariq Rd

Little China
Chinese
Tariq Rd

Imperial Court Chinese Restaurant
Chinese

More chinese restaurant

Uniform Resource Locator (URL)

A URL (Uniform Resource Locator) is the unique address for a resource that is accessible on the Internet. URL can be thought of as a straightforward to find the name of a site similar to postal or ZIP codes. The major advantage URL as an advertising tool is that it is free. The major disadvantage of URLs is that the chance that a specific site will be placed at the top of a search engine's display list is minimal. Second, different search engines index their listings differently; therefore, it is difficult to make the top of several lists.

Advertising in Chat Rooms

A chat room can be used for advertising (Gelb & Sundaram, 2002). The advertising in a chat room combines with the activity in the room, and the user is aware of what is being presented. The main advantage of advertising in chat rooms is that it allows advertisers to cycle through messages and targets the chatters repeatedly. There are other ways your business can use chat rooms for advertising. Host a free online seminar in your own chat room about a subject of your expertise. Use your chat room to meet with your current customers and answer any questions or address and problems they may have. Regularly schedule free events in your chat room and be certain that your customers are made aware of when they will occur. For example, you might have an expert in the field available to answer questions on a certain day and between certain hours. You might also host other people's chat rooms as an expert yourself. You, of course, could charge for this but it might be wise to do it free to gain publicity.[5]

Online sponsorships

With increasing use of banner ads, increasing partnerships among firms to provide users with useful content, online sponsorships are increasing. Sponsorships on the web are important because in many ways the firms on the web are squealing after similar targets. Rates of sponsorships depend on many factors e.g. web site's circulation and the number of impressions purchased. Sponsorship of content sites is similar to advertising, but generally, it is low price with a smaller link to the site and lower placement on a page than a full advertisement. Sponsors can provide sponsorships for a variety of reasons e.g. sporting events and charity. Sponsorship enables the marketers to target their consumers in an efficient & relevant manner. Therefore, if Mercedes Benz wants to reach CEOs, they can do so more efficiently by sponsoring a golf tournament than by advertising on TV.

Other forms of advertising

There are many other forms of online advertising including ads in newsgroups, ads in computer kiosks, advertising on Internet radio and TV, and advertising to members of online communities. Some use an advertorial, which is material that looks like editorial content or general information but is really an advertisement. There are also ads that link users to other sites that might be of interest to community members and targeted ads that can also go to the members' portals. In addition, the use of domain name for brand recognition and advertisement on cell phones and other mobile devices is growing.

Online Personal Selling

In personal selling, firm's sales force makes personal presentation to the customers for the purpose of making sales and building customer relationships. Internet technology can take over some of the activities offered by the personal sales person (e.g. accepting purchase orders) but some activities cannot be replaced. These include non-transaction activities (e.g. limited service/maintenance functions at the customer's site) pre-transaction activities (e.g. site visits), and post-purchase activities (e.g. training).

Similar to advertising and direct mail, the web provides excellent opportunities for personal selling. Some of the commonly used personal selling tolls used online include Webcasting and Web Conferencing, and virtual exhibitions.

Video Conferencing

Video conferencing applications allow for audio and video communication across the Internet. Generally, such connections are person-to-person between anyone on the Internet with appropriate hardware, though forums can be set up by using a reflector, where everyone connected can be seen by anyone else connected to the same reflector. Unfortunately if such a group gets too large, the video can become excessively slow to update, and voice communication can break up. The bandwidth and synchronization required by video is significant, and it can often be difficult to maintain an efficient person-to-person communication on the asynchronous packet-based Internet.

Webcasting and Web Conferencing

Webcasting and Web conferencing enhance business communication. Web conferencing is a low-cost solution that enables marketers to create a relationship with a customer, provide support, and distribute sales information at the customer's convenience. Web conferencing allows businesses to meet and collaborate online, in real time from anywhere in the world. Another advantage of online video conference is that all people can participate in at the same time. One disadvantage of video conferencing is that due to lack of physical interaction the passion in a person's voice as well as the subtlety of a comment might be lost.

Webcasting is essentially broadcasting over the Web. Using webcasting on their websites, businesses can stream seminars, conferences, shows, sales presentations, conduct live Webcasting events, and store prerecorded Webcasts on their site to be viewed at any time. One of the most famous Webcasts is the annual Victoria's Secret lingerie fashion show. Live webcasting transmits content live or continuously as it is being presented there and then. The added benefit of live webcasting is that a user can chat with friends or family across the world. On-demand webcasting refers to viewing or listening to past content. On-demand webcasting is generally popular because most of the people are not somehow able to view live webcasts.

Evoke Communications is an Internet communications service provider specializing in Webcasting and Web conferencing services (www.evoke.com). Evoke Webcasting allows you to stream live or recorded audio, video and other visual media over the Internet or a corporate Intranet.

Virtual Exhibitions

With virtual exhibitions, suppliers, competitors and customers all come under one virtual roof. Customers can access these exhibitions day and night from anywhere they are. Customers can reduce their expense by eliminating the need to visit specific shows physically. Customers can go to the virtual stands, examine products and services and leave their details for future quotations.

Online Sales Promotion

Sales promotion is a short-term incentive to encourage the purchase or sales of a product or service. Online Sales promotion stems from the premise that any brand or service has an established perceived price or value, the "regular" price or some other reference value. Sales promotion is believed to change this accepted price-value relationship by increasing the value and/or lowering the price. Sales promotion activities include coupons, rebates, product sampling, contests, sweepstakes, and premiums (free or low-cost gifts). Sales promotional activities are aimed at the consumer and may be difficult to replicate on Internet. However, the use of coupons that can be downloaded and printed from the Internet is now common and is a measurable demonstration of a website's effectiveness. Product sampling, contests and sweepstakes are also widely used on the Internet.

Sales promotions offer a number of advantages to the owner of a small business. By offering a reduced price on a popular item, you can lure customers away from competitors, which may ultimately help turn them into regular shoppers. Businesses like airlines and hotels successfully use rewards programs to encourage customer loyalty. Giving away free products or services is a good way to get people to try them for the first time, which may lead to a purchase. A sales promotion can help you provide information to potential customers that aid them in making a decision. This can be beneficial for products or services that are complicated or are unfamiliar to consumers. A notable disadvantage of doing promoting businesses online is that it may be difficult for the businesspersons and consumers to thoroughly evaluate the legitimacy of a transaction. Another disadvantage of promotion via the Internet is that the customers and businesspersons are isolated. There is little personal contact between customer and salesperson prior to and after the sales is closed. Thus, the prospect for repeat sales may thus be diminished.

Coupons

Both online and traditional print media coupons can add value to a business. Many firms offer electronic coupons. Many firms allow the customers to search their database of coupons by zip code to find the ones available in their locality. They also send e-mail notification as new coupons become available, attempting to build brand loyalty. When retailers drive customers to the Web site through point-of-purchase or traditional advertising, coupon redemption generally increases substantially.

Like other sales promotion tools, coupons have their advantages as well as their problems. On the plus side, they have the advantage of passing along savings directly to consumers, as opposed to trade allowances given to retailers by producers. Consumers perceive coupons as a temporary special offer rather than a price reduction, so the withdrawal of coupons usually does not have an adverse effect on sales. In addition, coupons often create added traffic for retailers, who have the option of doubling or even tripling the value of manufacturers-coupons at their own expense to create even more store traffic. Moreover, retailers often receive additional compensation from manufacturers for handling the coupons. On the other hand, the increased quantity of distributed coupons has been paralleled by falling redemption rates. In addition, excessive coupon distribution also increases the likelihood of fraud and misredemption. Coupons that are issued for established brands tend to be redeemed primarily by loyal users who would have purchased the product without a coupon. Coupons can be beneficial if used as part of an overall marketing strategy. Companies should set expiration dates and attribution metrics and ensure that affiliate commission rates are in line with new and existing customer mixes. (referenceforbusiness.com, 2015).

Product Sampling

Some sites allow users to sample product prior to purchase. Many software companies allow free download of fully functional demo versions of their software. The demo normally expires in 30 to 60 days, after which time the user can choose to purchase the software or remove it from the system. Online music stores similarly allow customers to sample music before ordering the CD. Market research firms often offer survey results as a sampling to entice businesses to purchase reports.

The advantage of sampling is that it gives the recipient a firsthand look at the product and encourages future sales if the recipient is pleased with the services offered. Online Sample Sales have become popular over the past few years because they allow everyone to take advantage of savings. Sample sales are events held to sell items, usually designer clothing and accessories, at a reduced price. The items are usually samples that were created before the item went into mass production, or extra items created during design process. The price cuts can be very big, in some cases up to 70 or 80% off of retail prices.

Contests and Sweepstakes

In order to attract visitors and get returning visitors, many sites hold contests and sweepstakes. Contests require skill whereas sweepstakes involve only a pure chance drawing for the winners. Just as in the traditional brick-and-mortar business, these sales promotion activities create excitement about brands and entice customers to visit the sites. *Online sweepstakes* and *contests* are an effective way to create excitement about your brands and products and increase traffic and sales. The number of *sweepstakes* and *contests* conducted over the Internet has grown exponentially over the last decade. Internet marketing enables companies to conduct *sweepstakes* and *contests* in a timely, relevant, personal, and cost-effective manner. Sweepstakes and contests conducted online, in particular those that are open to children under age 13, create even more obligations for marketers.

The distinction between a sweepstakes and a contest is that in a sweepstakes the element of consideration must be eliminated because the winner is selected by chance; the entrant does not have to pay or do anything substantial to enter. In a contest, the element of chance is eliminated; the winner is selected based on skill. In a contest, it is permissible to require a purchase in order to enter except in certain states.

Online Public Relations

Public Relations is a set of management, supervisory, and technical functions that foster an organization's ability to strategically listen to, appreciate and respond to those persons whose mutually beneficial relationships with the organization are necessary if it is to achieve its missions and values. Public relations are used for many purposes e.g. to build good relations with various key constituencies by obtaining favorable publicity, to build a good corporate image, and handling or heading off unfavorable rumors, stories and events. Internet facilitates the traditional methods of public relations and enables public relations professionals to communicate efficiently with various key constituencies directly and to expand the depth and breadth of public relations. For example, using Internet, news releases can be distributed widely. Word of mouth can be accelerated across the Internet. Newsgroups and influential discussion groups can be targeted with information. Software tools are available that can crawl the Internet and search for defamatory comments about customers or their products.

Use of web in imparting information is both economical and effective, if people are persuaded to visit the site and understand (and believe) the information contained therein. Web offers a variety of methods for companies to boost their corporate image and sales through publicity. For example, companies can use forums, bulletin boards, newsletters, newsgroups, e-mail, and other methods for publicity.

Forums

The forum is a formal mechanism used by the community to exchange information, generally through posted messages that are organized into threads (threads are subject headers for a discussion). Online forums support communities formed around a preferences, satisfaction or dissatisfaction, and use of products specific interest. Using online forums, marketers can gather a range of consumer marketing data on such communities.

Newsletters

Online or electronic, like print counterparts, may appear either regularly or irregularly but changes their content (Zimmerman, 2001). Their basic purpose is to disseminate information quickly in a field of interest. Some items that are typically found in newsletters include research updates, preprints and interviews.

Newsgroups

Newsgroups are a way to share discussions with specialized audiences. They are essentially electronic bulletin boards that are available for Internet users to post their questions and replies without charge. Newsgroups offer several different methods of sending messages. A user can post a message for everyone in the group or respond to someone else's comments on a particular topic.

Online Communities

An online or virtual community is a place that organizes and brings together individuals, groups, and businesses around common interests or purposes. Online communities can range from identity-and interest-based communities to industry-based B2B exchanges and may include Internet message boards, online chat rooms, and virtual worlds. Each community has its own rules, culture, interface, and features, depending on the purposes to which it is dedicated and the community it serves.

To have a valuable community, businesses first need to consider how the community will provide value not only for the customers, but for the business as well. Secondly, businesses need to assemble and implement the appropriate technical infrastructure. Finally, businesses must decide on the social infrastructure, which includes the basic ground rules for the content to be covered, the ethical rules for community members and so on.

Cause Marketing

Cause marketing is a marketing strategy that involves using affiliate marketing. Cause marketing generally benefits a charitable organization thus supporting its cause. In cause marketing, when visitors click a link on the affiliate's Web page, a donation is made by a sponsoring company. The page that loads after the visitor clicks the donation link carries advertising for the sponsoring companies. One example of cause marketing is of Yoplait's "Save Lids to Save Lives" campaign in support of the Susan G. Komen for the cure. The company packages specific products with a pink lid that consumers turn in, and in turn, Yoplait donates 10 cents for each lid. Another example is the American Heart Association's stamp of approval on Cheerios, the popular breakfast cereal. The American Heart food certification program grants use of its "Heart Check" icon and name to dozens of cereals and juices meaning that products meets the Associations' low fat, low-cholesterol standard. In 2007, Singapore Airline launched a cause marketing campaign to bring awareness to Doctors without Borders. Product Red is another example of one of the largest cause-related marketing campaigns. Product Red was created to support The Global Fund to Fight Aids, Tuberculosis & Malaria (aka" The Global Fund") and includes companies such as Apple computer, Motorola, Giorgio Armani, and The Gap as participants.

Viral Marketing

Viral marketing is a marketing strategy that relies on word of mouth publicity by company's existing customers about the company's products or services they have used. Similar to affiliate marketing that uses web sites to spread the word about a company, viral marketing uses individual customers to do the same. Blue Mountain Arts, an electronic greeting card company, is an example of viral marketing. Electronic greeting cards are e-mail messages that include a link to the greeting card site. When people receive Blue Mountain Arts electronic greeting cards in their inbox, they click a link in the e-mail message that opens the Blue Mountain Arts web site. A greeting card recipient is likely to search for cards that he/she might like to send to other friends. When a greeting card recipient sends electronic greeting card to his/her friends, their friends could then also send greeting cards to their friends. Each new visitor to the site could spread the information i.e. the knowledge of Blue Mountain Arts. Viral Marketing has become a very trendy way to promote and brand products/services of a company or organization. Many groups/companies are now using this strategy to spread their message. One of the most effective ways of implementing a viral marketing campaign is through videos. Many major companies are using videos to attract a wide audience. These videos can be seen on various social media sites such as YouTube, Face book, MySpace, Dig, among others.

Mobile viral marketing is a marketing communication or distribution concept that relies on consumer to transmit mobile viral content(products, services or ideas) using mobile communication techniques to other potential consumers in their social sphere and to animate these contacts to also transmit the mobile viral content. The explosive growth of mobile devices from niche to mainstream in recent years has resulted in increasing interest by marketers in this lucrative market. While the most common mobile marketing campaigns use SMS, in Europe and Asia numerous brands are also known to provide mobile content and applications to customers. This encourages customers to interact directly with traditional marketing campaigns and extends brand reach. SMS and MMS coupons, logos, ringtones and branded games and applications are all used to various degrees depending on the campaign and desired result. Viral marketing has considerable value in the mobile market as it takes advantage of the inherent nature of cell phones as communication vehicles, enabling people to share information and content within circles of friends and colleagues. By adding a viral component to branded applications, marketers can significantly expand campaign reach.

Online Direct Marketing

Internet is an excellent channel for direct communication with customers on an individual basis because the communication is immediate and can involve direct interaction. Using Internet, customers responses can also be solicited in real time. Online direct marketing includes E-mail, direct response advertising[11] (e.g. phone now or fill in the coupon ads), computerized home shopping and home shopping networks. The most important goal of direct marketing is to gain a response. Companies need to choose carefully targeted individual consumers in order to obtain an immediate response. One-to-one online direct marketing can develop strong and lasting relationships with loyal customers.

Bloch, Pigneur, Segev (1996) suggested that B2C direct marketing can impact E-commerce in many ways. Direct marketing can increased product and service promotion and contact with customers becomes more information-rich and interactive. Direct reach to customers and the bidirectional nature communication can create new distribution channels for existing products. Delivering information and products over Internet reduces companies' costs and administrative overhead substantially. Customer service becomes quick and more efficient. Customization of products and services is possible and online order reduces both processing time and mistakes.

Mobile Advertisement

Mobile advertising is a form of advertising via mobile (wireless) phones or other mobile devices. It is a subset of mobile marketing. Mobile advertising targets users according to specified demographics. Mobile networks identify related mobile profiles, preferences, and displays corresponding advertisements when consumers download and uses data services like games, applications (apps) or ring tones. The Mobile Marketing Association (MMA) is a non-profit global trade association that fosters mobile marketing and advertising technologies. It regulates associated terms, specifications and best practices. MMA also oversees global mobile advertising units in messaging, applications, video, and television and on the Web.

Mobile advertising can be done in the following ways:

---

[11] Direct response advertising is a type of advertising that can be web and press advertisements and solicits an immediate response from customers.

Mobile Web: Text tagline ads, mobile Web banner ads, WAP 1.0 banner ads, rich media mobile ads

Multimedia Messaging Service: Short text ads, long text ads, banner ads, rectangle ads, audio ads, video ads, full ads

Mobile Video and TV Advertising Units: Ad breaks, linear ad breaks, nonlinear ad breaks, interactive mobile video and TV ads

Mobile Applications: In-app display advertising units, integrated ads, branded mobile applications, sponsored mobile applications

According to Gartner, the mobile advertising market will continue to be driven by smartphones and tablet devices, which will enhance growth to $19 billion by 2015 (Sullivan, 2015). The inevitable result of the mobile phone explosion is that mobile phone advertising is increasing at a rapid pace. While only a short time ago only big advertisers could afford and use mobile advertising, that trend is ending and small to mid-sized business owners are enjoying the benefits of mobile advertising as well. One of the most popular mobile advertising avenues is when customers "subscribe" to receive SMS text messages from vendors they choose. Since the customer has already expressed an interest in the vendor and his offerings, the ads sent by the vendor typically have much higher conversion rates than other advertising methods. Not only are most small business owners able to find a mobile advertising program that is within their budget, the cost to set up a phone advertising campaign is typically very affordable as well. Because most mobile phone ad campaigns are based on consumer's opting in, it is very easy for a small business owner to target their ideal customer at the onset. The small business owner can advertise based on location, age, even gender. This ability to zero in on the consumer that is most likely to respond to your mobile ad makes it even more beneficial to advertise via mobile phone ads. Mobile advertising is not the same as other forms of advertising, however, and it is important for the small business owner to keep the disadvantages of mobile phone ads in mind when deciding on their mobile advertising strategy. The first challenge with mobile advertising is the slow data transfer speed of mobile devices. In an environment where most mobile phone users have grown up with super high speed Internet, they can get very frustrated with the delayed reactions still experienced with most mobile devices. While technology is advancing rapidly, this is still a challenge for the mobile phone advertisers.

Computerized Home Shopping

In computerized home shopping, the home computer is linked with a store. Users can browse the store, select the items required, and inspect it (e.g. by using image magnification and rotation technology). Home shopping Networks are now carried into millions of homes worldwide. For example, the Internet Shopping Network (ISN), recently purchased by the Home Shopping Network, is extremely accessible to users.

QVC.com (an initialism for "Quality, Value, Convenience") is an American cable, satellite and broadcast television network, and multinational corporation specializing in televised home shopping that is owned by Liberty Interactive. QVC broadcasts to 235 million households in six countries as QVC US, QVC UK, QVC Germany, QVC Japan, QVC Italy and QVC/CNR (China) - with QVC France due on air in mid-2015. QVC broadcasts live in the United States 24/7 (apart from the Christmas show which is pre-recorded) to more than

100 million households, and ranks as the number two television network in terms of revenue (#1 in home shopping networks), with sales in 2010 giving a net revenue of $7.8 billion.

Home Shopping Network (HSN.com) is an American broadcast, basic cable and satellite television network that is owned by HSN, Inc., which also owns catalog company Cornerstone Brands. Based in St. Petersburg, Florida, United States, the home shopping channel has former and current sister channels in several other countries. HSN also has an online outlet at HSN.com.

E-mail Advertisement

E-mail advertising is the online equivalent of direct mail advertising in which businesses uses e-mail to send their product information to people or companies listed in their mailing lists. E-mail ads often direct visitors to Web sites using hyperlinks. E-mail is a significant communication medium since it is widely used and total cost of sending one e-mail message to a customer costs considerably less than sending a message through traditional means.

To facilitate the process, businesses can also purchase mailing lists that contain e-mail addresses of highly targeted audiences that have asked to receive specific kinds of e-mail messages. Purchase of mailing list can add a little cost (from few cents to a dollar) to the total cost of each e-mail message sent. The exact cost of mailing list can be calculated. If you purchase a list of 10,000 names at $300 per thousand, the total cost would be $3,000. With an assumed response rate of 5%, you could expect 500 responses, at a cost of $6 each.

E-mails advertising provide many advantages. It can reach a wide variety of targeted audience and conversion rates[12] on requested e-mail messages range from 10 percent to more than 30 percent (Click-through rates on banner ads are currently under 0.5 percent and decreasing). Another advantage is that e-mail is an interactive medium, and can combine advertising and customer service. Compared to other media investments such as direct mail or printed newsletters, e-mail is less expensive. The delivery time for an e-mail message is short (i.e., seconds or minutes) as compared to a mailed advertisement (i.e., one or more days). An advertiser is able to "push" the message to its audience, as opposed to website-based advertising, which relies on a customer to visit that website. E-mail messages are easy to track. An advertiser can track users via auto responders, web bugs, bounce messages, unsubscribe requests, read receipts, click-through etc. These mechanisms can be used to measure open rates, positive or negative responses, and to correlate sales with marketing.

Email advertising has disadvantages as well. Firewalls, spam filters and email blacklists are common features of the modern Internet landscape, so emails do not always get through to every address. While opt-ins and white-list requests are viable options, they are only effective if the customer is aware of your business and agrees to allow the email through the filtering gauntlet. Most people enjoy an attractive, colorful advertisement, but not every recipient has the computer power or connection speed to quickly load and display email photos or graphics. In traditional print advertising, repetitive ads are necessary for the eventual notice and response of the prospect. With email advertising, the same email sent repeatedly is likely to be ignored or deleted. It is common practice

---

[12] The conversion rate of an advertising method is the percentage of recipients who respond to an ad or promotion.

for other companies to opt-in to competitor's email lists for purposes of obtaining pricing, special offers and information. Within minutes of sending your exclusive offer email, it may not stay exclusive for long and may result in pricing wars.

Opt-in or Permission Marketing

Opt-in marketing refers to sending e-mail messages, containing specific information, only to specific people. These people agree voluntarily to be marketed to by indicating an interest in receiving information about the product or service being promoted. Two companies that offer opt-in e-mail services are PostMasterDirect and yesmail.com. These companies provide the e-mail addresses to advertisers. The rates of this service vary depending on the type and price of the product being promoted. Typically, the price of this service can range from $1 to 25-30 % of the selling price of the product.

Marketers need to be careful in executing Opt-in marketing and should attempt for gradually gaining increased customer trust and a stronger relationship. First, the consumers should be offered an incentive to participate (e.g. a prize, airline mileage points, free information, etc.). Second, the marketers should reinforce this incentive to guarantee that the consumers maintain the permission. Third, the marketers should offer additional incentives to get even more permission from the consumer. Over time, marketers should leverage the permission to change consumer behavior toward profits. What is unique about this form of marketing is that it holds distinct advantages for both consumers and marketers.

Consumer Advantages	Marketer Advantages
Messages are anticipated, personalized and relevant.	Cost of technology and infrastructure is low and feedback is instantaneous.
Has ability to opt-out or unsubscribe at any time.	Customer loyalty and pass-along marketing
Privacy is preserved.	Marketers can segment their audience better by knowing how many and the types of people who visit their sites

The Corporate Web Site

A Web site is the primary tool used by every online business to advertise its products or services and attracts customers. It is an advertising medium as well as a bi-directional and quick customer communications forum and distribution channel without geographical and temporal limits. Web sites provide benefits for both companies and consumers. A company can display its identity and advertise its product and services to many people. Companies can get feedback directly from customers. A web site can enhance the company's image and provides tangible benefits both to the company and to its leadership. A web site can also improve communications with other companies, thus improving the efficiency of business process by increasing direct sales and reducing cost.

Web sites exist in all kinds, shapes and sizes. The most critical decision facing businesses is what function the Web site should serve. The primary functions a business Web site can provide are provide information, showcase advertising and marketing, sales support, public relations, customer service communication and feedback, and E-commerce. If multiple functions are to be served, it may become necessary to create separate Web site. For example, a company may use one web site to promote individual products as well as the overall company while using another web site to improve the performance of sales and post-sales.

Web site is the key tool in Internet marketing communication. Using web site, a business can improve their communication performances by building consistency in communication. A web site can also become one of the most important factors in judging a company because web site is the key interface between businesses and consumers. Well-designed web sites can provide electronic distribution of information that can match the effectiveness of personal selling.

Web site is an excellent advertising media on Internet. Companies can completely control all the aspects of advertisement e.g. its look and feel, its content, and interaction with prospective customers. With web advertisement, consumers can have increased control over advertising exposure because of their ability to select how much content they wish to view. Mainstream traditional sales promotions can be leveraged across the website. In a sense, a URL address listed on packaging of a good containing a sales promotion can invite the audience to check out the website.

Many Web sites are designed to serve as public relations vehicles. For public relation professionals, the World Wide Web is an excellent platform because it gives them the ability to build an interactive relationship with stakeholders through media relations, employee communication, government relations and customer relation.

Websites and Their Types

Business websites are created for a wide variety of reasons and in a wide variety of forms. Each has a different purpose, requires different computer hardware and software, and requires different monetary and personnel resources. Decisions about server hardware and software should be driven by the volume and type of Web activities expected, how many visitors will be connecting to the Web site, and what types of files (graphics, multimedia, or text) will be delivered through the site. The company must also assess its existing information technology staff. Some companies have a large staff with a depth of experience, while others have a small or relatively inexperienced staff.

Types of sites include:

Development Sites: These simple sites are used by companies to evaluate different Web designs with little initial investment. A development site can reside on existing PC running Web server software. Multiple testers access the site through their client computers on an existing LAN.

Corporate Intranet Sites: These sites reside on corporate networks that house internal memos, corporate policy handbooks, expense account worksheets, budgets, newsletters, and a variety of other corporate documents.

Corporate Extranet Sites: These Intranets allow certain authorized parties outside the company (such as suppliers) to access certain parts of the information stored in the system.

Transaction-processing Sites: These sites are available 24 hours a day, seven days a week and require fast and reliable hardware and software for handling high traffic volumes that occur periodically.

Content-delivery Sites: These sites deliver content such as news, histories, summaries, and other digital information and generally available 24 hours a day, seven days a week. Visitors must be able to locate articles quickly with a fast and precise search engine.

Website Domain Names Issues

Choosing Website Domain Names

Most people will think of your website by its name. If your company's name is also your URL, people will automatically know where to go. For example when people think of delta (an airline of USA) they do not have to wonder what URL to type into their browser to get there. The name of the company is also the URL i.e. www.delta.com. An important part of establishing a Web presence that is consistent with the company's existing image in the physical world is to obtain identifiable names to use on the Web. New startups may first obtain the domain name and then name their business after the domain name that they have acquired. Companies with strong brand name or reputation generally prefer the domain names for their Web sites that reflect that brand name or reputation. This is because it is the easiest thing for customers to remember and is the first thing that customer will try in their browser when looking for your brand. It also takes a lot of time and money to establish the business reputation. An unidentifiable name becomes a problem for these businesses when they start their business online. One example is delta airline of USA. The original domain name of Delta Air Line was www.delta-air.com. After several years of complaints from confused customers who could never remember to include the hyphen, the company purchased the domain name www.delta.com. If a suitable domain name is not available, these companies try to buy over the domain name from the current owner. Table provides the name and price of some famous website names. Another example is Proctor & Gamble and Reckitt Benckiser. Proctor & Gamble owns Pampers.com and Reckitt Benckiser owns Pimples.com.

Domain names can be long or short. While short domain names are easy to remember and less likely to be misspelled long domain names are usually easier to recall. With the proliferation of websites, it is much more difficult to obtain and register a short meaningful domain name. A trick is to get a long domain name with your site keywords. Today many people use search engines to look for a particular product or service they want. Such domain names are search engine friendly. Search engines can distinguish keywords better and make your site prominent in search results for those keywords. For example, if you have a website providing skin treatment products and you get a domain name www.skinacnetreatment.com it can perform better in search for "skin treatment" or "acne treatment" or just "acne".

Rich (2006), suggests some useful approaches to choosing a thriving domain name. One approach is to keep the company name as the domain name. This will help people relate to the website and will be able to reach the desired website. Some popular examples are:

Company name	Domain Name
Dell Computers	www.dell.com
Nike	www.nike.com
Sony Corporation	www.sony.com

If the company sells product or service, the domain name can also be on the name of product or service the company offers. Companies can also come up with nonspecific domain names but only if they can afford extensive marketing and promotion campaigns to make people aware of the company existence online and what it deals in. For example, www.amazon.com is a nonspecific domain name, as people cannot figure out the nature of business this entity conducts by looking at the name but due to comprehensive marketing campaigns, Amazon has created an image and now people are well aware that it deals in books, videos, music cds etc.

While registering a domain name with a web host, many a times the web host recommends a name containing plural form of the domain name. For example if you wanted to register a domain name bestbook.com the web host may suggest you e.g. thebestbook.com or bestbooks.com. This may occur because the required domain name is not available. However, such plural forms of domain names can be a problem. Your prospective customers may fail to the "s" in the domain name. Therefore, while marketing your domain name you must advertise your site with the full domain name.

After coming up with a creative domain name for the website, one need to check immediately whether it is available or someone else has already registered it. To check, one can use www.whois.net, a domain based research service. It keeps record of all domain names taken, those available, even those that were registered previously but are now available. If the domain name is already registered then one can get the help from registrars. These registrars like www.godaddy.com offer alternative domain name suggestions for negotiable money.

The fully qualified host name of a computer on the Internet, such as www.delta.com, has three major parts: the host name, the domain name and the top-level-domain (TLD). Most Web servers use www as the host name. The TLD usually describes the type of organization that owns the domain name. For example, the .com TLD usually refers to a commercial venture, .org TLD usually refers to a nonprofit organization, .net is usually used for network-related sites, and .org normally denotes a non-profit organization. There are country specific TLD as well, such as .us for the United States, .ca for Canada and .uk for the United Kingdom. Choosing your domain name also requires deciding which top-level-domain (i.e. .com, .org, .edu etc.) or country-specific top-level-domains (i.e. .us, .in, .nz etc.) should be chosen. Selection is not straightforward. If your business serves local population, getting a country-specific TDL may benefit your business if the people in your country prefer to deal with a local business. If your business operates internationally, many prefer to get a domain name with ".com" extension. Two arguments are given in favor of this approach. Fist, many people assume that a particular domain name will have a ".com" extension. Missing a .com extension in your domain name, you may lose your customers. The second argument is the algorithm a web browser use to locate a website when a user simply types a name like "Alba" into the browser. When web browsers search the domain name "Alba" they will first look for Alba.com before looking for Alba.net or Alba.org etc. In that case people who type only "Alba" will be delivered to Alba's competitor's site if Alba doesn't also own the ".com" extension. Many companies buy domain names with other extensions besides the ".com" extension.

Potential site visitors may misspell the URL of a site. As a result, visitors could be taken to another website. To protect their brand and reputation, companies often buy more than one domain name to redirect such visitors to the intended site. For example, Yahoo! owns the name Yahow.com and Google owns the name Gogole.com. Other companies may own many URLs because they have many different names or forms of names associated with them. For example, Barnes and Noble (www.barnesandnoble.com) owns the name books.com and Bank of America (www.bankofamerica.com) owns the name loans.com. General Motors (www.gm.com) owns the names generalMotors.com, chevrolet.com, chevy.com, gmc.com.

Buying, Selling, and Leasing Domain Names

There are certain steps need to be followed when buying a domain name. The first step is to determine whether the domain name required is for sale. WhoIs (ww.whois.com) is online directory of registered domain names. WhoIs database can be checked to find out whether the intended domain name is available or not. If the domain name is available, the next thing to do is to create a budget for the purchase of domain name, and getting a reliable estimate of its value. Value of a domain name depends on many factors e.g. domain name length, popularity of word sued etc. Many sites e.g. www.cubestat.com offer free calculators to calculate the value of the domain name. You can also calculate the budget by estimating how much the domain name worth to you. Usually, ownership of a domain name requires a one-time registration fee followed by recurring annual fees. These generally fall between $30 and $200 per name. Once budget is decided, you need to decide from where you want to purchase it. There are several methods of purchasing a domain name, including private purchase.

Determining the worth of Domain name

There are many companies providing the service of assessing the value of domain names but all are faced with one common flaw that all domain names are 'unique'. Hence, which domain name is worth more and which is worth less is difficult to gauge. However, there are some basic guidelines available, which can help determining the potential or worth of domain names. An individual may receive email from people willing to buy the domain name of his website without it being listed anywhere for sale. It is a biggest indicator that this domain name has some worth and is readily acquirable. If the domain name's TLD is .com, it means it is worthy because .com still occupies the position of king of all TLDs. If the domain name is shorter, it is better unless the alphabets are replaced by numbers for the sake of keeping it short. For example, www.goforit.com is a good short domain name but www.go4it.com is less worthy and less attractive.

Selling Domain Names

There are few approaches when it comes to selling a domain name. First, you can sell your domain name independently. Under this option, you can opt for just waiting to see if potential buyers approach you for the domain name you own. But this can only be possible if the seller owns the most wanted domain name. Another better idea is to generate a webpage with that domain name and putting content on it, which says "Following domain names are for sale, please contact <e-mailaddress@domain.com> for more details." This requires little time and effort. A third option is to create an active site. This is for those who are willing to put strenuous effort on marketing and promoting the domain name. They can create an active and attractive site and keep valuable content on it to attract greater number of visitors. Also put somewhere on the site that this domain name is "for sale". When considerable number of people visits this website, among them can be potential buyers as well.

The second approach to sell your domain involves selling the domain name via broker. Various companies work as a broker between the sellers and buyers of domain names also called "domain name brokers". Broker simply acts as a representative of buyer when approached by buyer and as a representative of seller when approached by him. He tries to maximize the good of the party he represents. When he is dealing with a seller on buyer's behalf, he tries to quote the lowest price and highest when dealing with buyer on seller's behalf. Selling the domain name through a broker requires some simple steps as suggested by Monas & Hooker (2008). First, you should explore broker's website to get the idea of how the broker operates. Then send him an exploratory email asking for some relevant information and informing about seller's interests. Then asses broker on how quickly he responds and how appropriate the response is.

Domain Name Leasing

The term "lease" has literal meaning of rent and the terminology "domain leasing" has the same application of renting the domain name to some other company's website for money. A domain name can be put to lease in two ways. You can put your domain name in a directory for example www.noktadomains.com/domainnameleasing.html or you can independently market you domain to customers.

Such companies try to match advertisers with the owners of domain names. Domain name owners however, set the leasing price themselves so that they get the desired amount of money than parking the domain. The major advantage of leasing domain name to advertisers is that after acquiring the domain name, advertisers will put relevant material on the site, which will enhance the search engine value of the domain and then if anytime in the future; lessee withdraws, leaving behind the high valued traffic that other potential renters or even buyers would readily want. Moreover, if the renter starts up a business online on the website operating with the leased domain name, it is highly probable that the renter buys the domain name as it has become the integral part of the business structure.

Domain Brokers and Registrars

Domain brokers are legitimate online businesses that sell, lease, or auction domain names. Some example of URL brokers are domainholdings.com and Guta.com. Companies can also obtain domain names that have never been issued, or that are currently unused, from a domain name registrar. ICANN (the Internet Corporation for Assigned Names and Numbers) maintains a list of accredited registrars. Many of these registrars offer domain-name search tools on their Web sites. A company can use these tools to search for available domain names that might meet their needs. Another service offered by domain name registrars such as DirectNIC.com and GoDaddy.com is domain name parking. Domain name parking, also called domain name hosting, is a service that permits the purchaser of a domain name to maintain a temporary simple Web site (usually one page) setup for you so that the domain name remains in use. The fees charged for this service are usually much lower than those for hosting an active Website are.

To register the domain name, choice of a good registrar is important. Most famous registrars are www.godaddy.com and www.networksolutions.com. Once on the registrar's website, a person is asked to fill out the form with relevant information for example: full name, the name of company, person's email address, address of the company (street, city, state and Zip), phone number and fax number. Whether the website will be used for personal or business purposes will be asked along with the nature of business (sale of products or services etc.). The term for registration will also be asked which can be minimum one year but if prepaid for a longer period the annual registration becomes cheaper. When the form is filled, payment needs to be made either using credit card or through the PayPal account. The annual fee for domain name registration is as low as $8.95 when using www.godaddy.com or it can be as high as $34.99 per year with some additional services. As the registration fee per domain name is cheap, companies in order to make sure their customers do not navigate towards competitors' websites, register more than one domain names especially they try to acquire three most famous TLD's .com, .net and .org. They may not put them to use instead "park" them.

Web Positioning

World Wide Web provides a more level playing field for smaller firms to compete with their larger counterparts [(Berthon, Pitt, & Watson, 1996);(Laflin, 2001); (Paul, 1996)]. Web positioning is the marketing of the website itself and it is crucial to the success of online businesses. It involves making potential visitors aware of the existence of the website and making it easy for visitors to arrive there. Website is considered as the face of an online business. Today, web users have limited time available to explore thousands of sites available on the web. Just building an E-commerce website is not enough and developers must work continuously to prominently position the website among millions of other competing websites to seek attention of busy web users.

Web Positioning Strategies

Companies can adopt two basic strategies to position their websites. Under first strategy, the company seeks a high ranking on major search engines, such as Google and Web directories such as Yahoo. Under second strategy, the company advertises the website's presence, address, and benefits in as many appropriate places and ways as possible (Berthon, Pitt, & Watson, 1996).

The strategies of organization's web positioning should be implemented continuously. A company should make continuous efforts to gain and maintain high search engine placement through appropriate key terms and seek every opportunity to promulgate the fact that the company's website exists and the site's address.

WORD OF MOUTH PUBLICITY

Word of mouth is one potentially effective way of informing potential visitors about the website's existence and address. This can be done, for example, at meetings with customers, clients, vendors and other stakeholders. The website's address can also be distributed in written and electronic communications e.g. text material distributed by the organization, including letters, business cards, invoices, bills, posters, E-mail, and marketing materials.

Legal Issues Affecting You and Your Website

There are many legal issues and statutory requirements that may impact your online marketing. Ignorance of the laws will not protect you if you get in trouble. Following are some of the important legal issues and statutory requirements that every online marketer should know of.

Compliance and Your Website

By law, you have to include your company information on your website and in your emails. For example, if you're a company in the UK Companies Act requires that you update both your website and your email signature files.

You are required to put many types of information on both your website and business emails. This includes your company registration number, place of registration, registered office address. In addition, on your website, you also need to include, your company name, postal address and company email address; the name of any trade bodies or professional associations that the business is part of, including membership or registration details; your VAT number, even if the website is not being used for e-commerce transactions; and any prices on the website must clearly state whether they are inclusive or exclusive of tax and delivery costs. All of this information only needs to be on your website once, for instance on your Contact Us page, or your Terms and Conditions. The contact form or newsletter sign up on your website must have a data protection statement.

Email Marketing

In email marketing there are many legal requirements. Your marketing message must include an "unsubscribe" option and your company information. You cannot share or sell emails and other personal information without a consumer's explicit consent. You must make it clear from the subject line that your email is in fact a form of direct marketing. UK office of Fair Trading (OFT) imposes heavy penalties for not fulfilling any of these requirements.

Blogging and Social Media

Blogging and social media tools are excellent ways to connect with your customers. However, there are some stringent legal requirements for their use in business. There are tough penalties for pretending to be a consumer to promote your company via social media. If you are a blogger who receives money, gifts or anything else in return for reviews, you are required by laws to tell your readers if you are sent a product to review.

Search Engine

A search engine is a computer program that can access a database of Internet resources, search for specific information or keywords, and report the results. Google, AltaVista, and Lycos are popular search engines. Portals such as AOL, Netscape, and MSN have their own search engines. There are special search engines, organized to answer certain questions or search in specified areas e.g. Ask.com and allmovie.com. Thousands of different public search engines are available e.g. zabasearch.com (for people search in USA). In addition, thousands of companies have search engines on their portals or storefronts.

The term "search engine" is often used generically to describe both crawler-based search engines and human-powered directories. These two types of search engines gather their listings in radically different ways.

Crawler-Based Search Engines

Search engines, such as Google and Yahoo!, use special software, called crawlers, to find pages for their search results. Crawling refers to the automated browsing of a website pages. Not every page is indexed by the search engines and crawlers look at a number of different factors to decide whether to crawl the page on a site (e.g. distance of pages from the root directory of website).

Upon its visit to a website, the search engine first crawl the robots.txt file. This file is placed in the root directory of a website. This file instructs the crawlers as to which pages of the site are not to be crawled. Webmasters can use this file to avoid selected web pages not to be included in search results of a search engine. Crawler may keep a cached copy of the files found on web pages visited. Web pages can also be excluded from crawling by using a meta tag specific to crawlers. Web Pages typically prevented from appearing in search results may include login specific pages such as shopping carts. If you make any changes in the web pages, crawler will eventually detect those changes. These changes can affect your website ranking by the search engine.

Human-Powered Directories

A human-powered directory, such as the Open Directory, depends on humans for its listings. You submit a short description to the directory for your entire site, or editors write one for sites they review. A search looks for matches only in the descriptions submitted. Changing your web pages has no effect on your listing. Things that are useful for improving a listing with a search engine have nothing to do with improving a listing in a directory. The only exception is that a good site, with good content, might be more likely to be reviewed free than a poor site.

Components of a Crawler-based Search Engine

Search engines consist of three main parts. Search engine spiders follow links on the web to request pages that are either not yet indexed or have been updated since they were last indexed. These pages are crawled and are added to the search engine index (also known as the catalog). When you search using a major search engine, you are not actually searching the web, but are searching a slightly outdated index of content, which roughly represents the content of the web. The third part of a search engine is the search interface and relevancy software. Figure 7-9 shows working of a crawler-based search engine.

FIGURE 7-9: WORKING OF CRAWLER-BASED SEARCH ENGINE

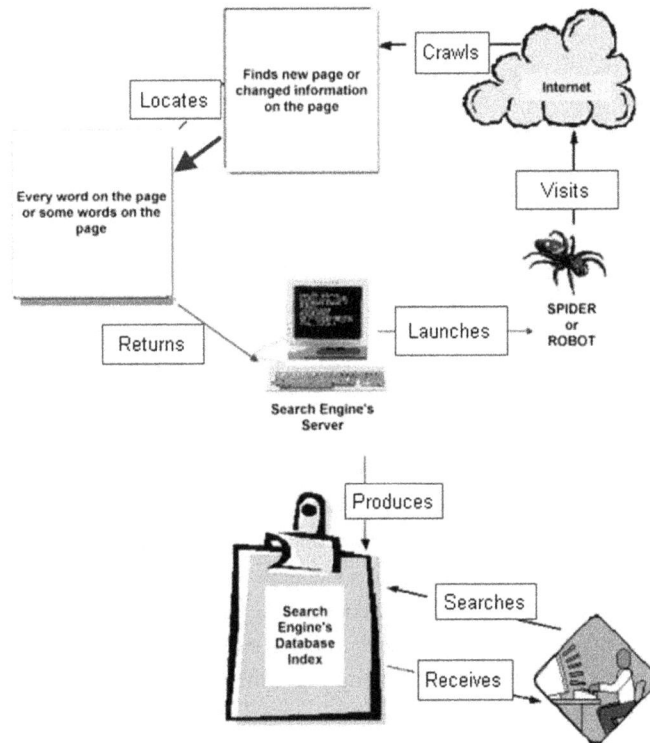

Search Engine Optimization (SEO)

SEO refers to the process of improving the rank of website in "natural" or un-paid search results of search engines. Search engines are not paid for natural search traffic and don't guarantee continued referrals. In general, the earlier a website appears in the search results, the more visitors it will receive from the search engine. SEO primarily involves editing website content (including HTML and associated coding) to increase the relevance of content to specific keywords and to make the website easier to search by the search engines. SEO may be incorporated into web site design and development. A search engine friendly website is the one whose various elements (design, menus, content etc.) are optimized for search engine. Special types of search engines are now available e.g. audio search engine (e.g. volumo.com) and video search engine (e.g. blinkx.com).

Search engines rank websites according to the keywords that a potential visitor inserts into a search box. Google Page Rank is a number from 0 to 10 and it measures Website popularity in its databases. You can see the Page Rank of the site you visit, if you downloaded a Google Toolbar for your browser. Page Rank is one of the most important factors when Google is calculating SERP (Search Engine Result Page) to decide where to position your links among all search results. Search Engines want to place links from the most authoritative source of information first. Company's goal should be to obtain highest possible ranking in major search engines, under the common words people use to search for the kind of goods and services the company sells. For example, a

411

fitness club in New York might seek to gain a high ranking on search engines under key terms such as "fitness," "fitness club" "exercise," "strength," and "New York".

Determining Web Page Relevance and Ranking

To a search engine, relevance means more than finding a page with the right words. Hundreds of factors influence relevance. Search engines typically assume that the more popular a site, page, or document, the more valuable the information it contains must be. This assumption has proven successful in terms of user satisfaction with search results. Popularity and relevance are not determined manually. Instead, the engines employ mathematical equations (algorithms) to determine relevance and then to rank the pages. These algorithms often comprise hundreds of variables. In the search marketing field, these variables are called "ranking factors."

Web Pages with keywords appearing in the title are assumed more relevant than other pages on the same topic. Search engine will also check to see of the keywords appear the top of a web pages, such as in the heading or in the first few paragraphs of text. Frequency is the other major factor in how search engines determine relevancy. A search engines will analyze how often keywords appear in the relation to other words in a web page. Those with a higher frequency are deemed are relevant than other web pages. By analyzing how pages link to each other, a search engine can both determine what a page is about and whether that page is deemed "important" and thus deserving of a ranking boost. In addition, sophisticated techniques are used to screen out attempts by webmasters to build "artificial" links designed to boost their rankings. Using click through measurements, a search engine may watch what results someone selects for a particular search, and then eventually drop high-ranking pages that are not attracting clicks, while promoting lower-ranking pages that do pull in visitors.

Building Search Engine Optimized Website

Creating search engine optimized web page pages that rank highly under specific key word(s) is a complicated task. Several issues need to be addressed e.g. what actual words are on the page, different search engines' criteria for gaining a high listing and how many hyperlinks to the website there are from other websites. Due to the complication and importance of search engine optimization many organizations have introduced devoted individuals to develop sophisticated search engine placement strategies and techniques. Criteria used by search engines for ranking the search results change often. A website which was ranked, under a key term, on the first page of a major search engine one month may find itself downgraded to fifth or sixth page the next month. Downgraded search engine rankings should be actively investigated and appropriate actions be taken to regain the ranking. For multinational companies, successful search optimization may require translation of web pages in local languages, domain names with local TLD, and web hosting with a local IP addresses.

Search Engine Optimization Focus Areas

When optimizing a website for search engines, we can focus on two areas: on-page optimization and off-site optimization.

ON-PAGE OPTIMIZATION

On-page optimization refers to keyword optimizations made in the code of the web page. This type of optimization is easy to control. Eye tracking studies have shown that searchers scan a search results page from top to bottom and left to right (an F-shaped pattern), looking for a relevant result. To increase the number of searchers who will visit the site results placement should be near the top of the search result page. The price of the banner should depend on where it is located on the page. Figure 7-10 shows screenshot of Eye-tracking Heat-maps of three websites. The areas where users are colored red; the yellow areas indicate fewer viewers, followed by the least-view Gray areas didn't attract any fixations.

FIGURE 7-10: EYE-TRACKING HEAT MAP

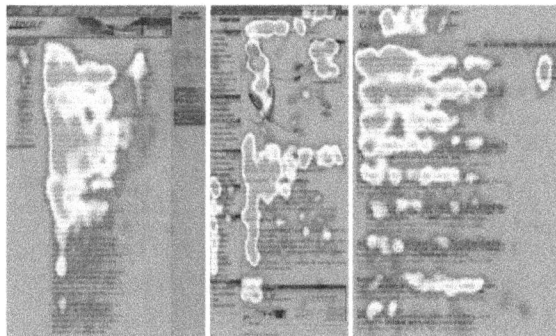

Courtesy: www.useit.com

OFF-SITE OPTIMIZATION

This involves having links coming into your web site. This task is difficult to control. You need to inform the world about your presence and the content you have so that you could get quality keyword links coming in. The social components of today's website i.e. RSS feeds, Blogs, Forums, Comments, Polls etc. have made this task easier. These social components are becoming a commodity and are now becoming part of a large number of websites. Following are the methods that can be adopted to perform off-site optimization of your website.

- Create & submit XML Sitemap to search engines

- Create & promote RSS Feeds

- Submit web site to web directories

You can use software tools to create your XML site map e.g. Sitemap Writer Pro. Once you have generated the XML site map you can submit it to the major search engines such as Google. XML site map informs a search engine what, when and how to spider. You will need to submit the XML site map to search engines from time to time in order to keep track of changes on your web site.

RSS (Really Simple Syndication) provides pull style of interaction with your audience. Your site audience can easily get your site updates when they want to, in either online of offline mode. Promote your RSS feeds with the RSS syndicators (examples) and directories whenever there is a new content.

There are many directory systems on the Internet that carry listing of large number of website. One example is DMOZ.org. DMOZ.org (The Open Directory Project) is the Internet's largest directory system containing over 4 million web sites. Registering your web site with DMOZ can lead to more visits to your website and more links to your website because other directory systems also copy DMOZ content. Google takes URL's from DMOZ and then matches these URL's with Google Page Rank number. Google than copies these URLs into Google Directory sorted by Google Page Ranks. There are other free and for-fee directories to submit your sites to e.g. Yahoo Directory (http://dir.yahoo.com/). Businesses should only submit URLs to directories that are relevant to their industry and have a good page rank. Every directory has submission guidelines and in order to be accepted, businesses need to abide by these guidelines. Site submission to a directory system is a time consuming process. Both Yahoo Directory and the DMOZ require manual submission and human editorial review. It can easily take six months before you could see any benefits of directory submission.

Search Engine Optimization Techniques

Search Engine Optimization (SEO) techniques can be classified into two broad categories. White hat SEO techniques are those that search engines recommend, and black hat SEO techniques, which search engines, do not approve.

WHITE HAT SEO TECHNIQUES

White hat SEO techniques conform to the search engines' guidelines. Use of these techniques ensures that the content indexed and ranked by a search engine is the same content the users will see. White Hat SEO techniques may include providing quality content on the site, Providing pages with proper titles and meta data , effective use of keywords and key phrases throughout the website, and having good links on the website (Good links are those links on your website that are from other web pages. These links are highly regarded by the search engines and are contextually relevant to the content of the website).

BLACK HAT SEO TECHNIQUES

Black hat SEO techniques attempts to improve search engine rankings by attempting to trick the algorithm used by search engines. These techniques degrade both the relevance of search results and the user-experience of search engines. Common black hat techniques are use of hidden content inside the code, Mea tag stuffing (i.e. loading meta tags with too much text), improper use of Meta keywords (i.e. inserting too many keywords in meta tags), and link farming (i.e. having too many links on your website from other unrelated sites).

Search engines do not favor black hat SEO techniques and may penalize sites they discover using black hat techniques. Search engines either reduce rankings of such sites or remove them from their listings from their databases altogether. Such actions can be taken automatically by search engines or by a manual site review. In 2006, Google removed BMW Germany, Ricoh Germany, and PPC Agency BigMouthMedia for use of deceptive practices. However, these companies quickly fixed the problem pages and were able to regain their status in Google's list.

SEO for Secure Websites

If you have pages on your Web site where users provide sensitive data, such as a credit card number or other type of account information, you can make these pages both secure and SEO friendly. The Internet solution for protecting sensitive information is to put those Web pages on a *secure server*. Technically, this means that the Web page is on a secure port on the server, where all data is encrypted. Secure servers can cause duplicate content problems if a site has both a secure and non-secure version of a Web page. Two versions of the same page end up competing against each other for search engine rankings, and the search engines pick which one to show in search results. In addition, because people link to both versions of the page, neither page can rank well because they have split their link equity (Clay & Esparza, 2011). There are some practices available for making secure servers SEO-friendly.

First, webmasters must avoid creating two versions of their site, or of any page on their site. Even if they exclude their secure pages from being indexed, people link to them at some point and the search engines find the secure versions through those links. Second, you should only secure those pages that need to be secured. Third, you should not to try to rank for pages on a secure server. Fourth, you should secure your pages by putting them behind a logon. Search engine spiders cannot crawl pages that require a logon to access, so they definitely will not be indexed. You also raise the user-friendliness of your site by including a logon because users will clearly understand why they have moved into a secure server environment and feel more comfortable entering their account information there.

Search Engine Marketing

Search engine marketing (SEM) is the process of improving the volume or quality of traffic to a web site using paid placement, contextual advertising, and paid inclusion in search results. Through SEM, vendor gets an advantageous position in the paid-placement, promotional section of the search-results page. The largest SEM vendors are Google AdWords, Bing Ads, and Yahoo! Search Marketing. Yahoo! provides a service that guarantee either crawling for a fixed fee or cost per click. Such programs usually guarantee inclusion of the pages in the database, but do not guarantee specific ranking within the search results. However, the use of SEM has generated significant controversy because originally search engines produced unbiased results from searching the Web's vast collection of pages. Search engines may or may not inform the user that the results of a query have been paid for by participating firms. Consumers object to this because the impact of fees being paid for search inclusion might distort the result list and exclude otherwise valid links.

Search engine marketing has become a principal means of reaching potential customers just when they appear to search for a product. SEM looks attractive to the seller yet the buyer is inclined to trust the natural search results and research shows that customers fundamentally mistrust the paid placement of search results [(Feng, Bhargava, & Pennock, 2003); (Jansen & Resnick, 2006); (Sen, 2005)]. Buyers using search engines to look for information tend to trust and follow links shown in natural search results section. Still, most on-line sellers prefer paid placements to search engine optimization. Sellers believe that SEO is more expensive; the results do not justify its cost, and does not consistently results in high search-results rankings (Sen, 2005).

Paid Placement

It is also called Paid for Placement (PFP). In paid placement, website owners pay to have their site included and ranked higher than other results in an index of similar or competitive content. The web site owner typically will bid on certain keywords or phrases. Paid placements are usually marked as "Features Listings," "Sponsored Links," etc. For PFP engines, the challenge is to monitor the key phrases that marketers are bidding on and the abstracts that are submitted. There is also the risk of "Perceived Relevancy" where the users are shown non-relevant results, both geographical and content, because the bid-system is allowing too many low-quality/non-brand-name customers into their index. One good thing about paid placement is that you can control the costs. You can buy as many or as few keywords as you want, and you decide how much you are willing to pay for each click. This way you can control your budget.

Paid Inclusion

Paid Inclusion, or pay for inclusion (PFI) allows website owners pay to have guaranteed inclusion and refresh (in the case of crawler-based services) in the index. The services are usually priced on a per-URL basis and guarantee one year of inclusion. For volume submissions, greater than 1,000 URL's, for example, the marketer often pays on a cost-per-click basis. Paid inclusion only guarantees scheduled index inclusion; it has no effect on ranking. However, most of the crawler-based engines do consider refresh rate in the ranking algorithms. For PFI engines, the "Perceived Relevancy" problem is less of an issue because we use static rank (link analysis) and dynamic rank (keywords, etc.) to determine relevancy. The challenge, of course, is to remain very strict on spam detection and elimination. Paid inclusion can provide increase targeted search traffic and improve site brand awareness. You can maximize number of indexed pages and ranking and optimize ranking on highly targeted keywords and geographies.

Contextual Advertising

Contextual advertising is targeted advertising that typically occurs on a banner or pop-up ad on a website. Contextual ads target ads to a specific user based on the keywords on the page he or she is visiting (hence, the context of the ad comes into play). Search engine advertising is also a form of contextual advertising, since the ads that appear in the results are based on the keywords the searcher uses. Publishers and advertisers bid for the use of keywords in an auction system run by a contextual advertising company. Publishers and advertisers only profit from the advertisement when a user clicks on the ad.

Use of contextual advertising involves some potential privacy issues. This is because users occasionally need to install software onto their computer through a process known as third-party hyperlinking. In this system, the software searches the context of the page, site, or the user's history and then displays advertisements related to the content in the form of hyperlinked PPC keywords instead of regular banners. Google AdSense is a contextual advertising program, as is Google Content-Targeted Advertising. This program was one of the first of its kind. The ads appear in search results and on websites within Google's Content Network. Developers can accomplish this by adding lines of JavaScript code to their web pages. When the code is integrated, AdSense searches for relevant ads using Google's search algorithm. Contextual ads will be displayed when the algorithm finds a match. The program also considers the location and languages of the visitors. Bing Ads and Yahoo! Advertising Network are alternative contextual advertising companies that provide these types of services. These sites give small business owners the opportunity to advertise without dealing with each individual sponsor. Large sponsors often will not advertise on small sites on an individual level (Kim, 2010).

Beyond Search Engines

While search engines *are* a primary way people look for web sites, they are not the *only* way. People also find sites through word-of-mouth, traditional advertising, traditional media, blog posts, web directories, and links from other sites. Since the advent of Web 2.0 applications, people are finding sites through feeds, blogs, podcasts, vlogs (a blog in which the postings are primarily in video form) and many other means. Sometimes, these alternative forms can be more effective than search engines. The most effective marketing strategy is to combine search marketing with other online and offline media.

Using Google to Improve Your Site Search

Google is a very popular search engine. Nearly 75% of all search traffic today is routed via Google (Dholakiya, 2015). Internet users mostly have come to expect a similar level of excellence in all forms of search that they encounter online. One of the key components of other forms of search is 'site search'. The way Google performs search provide many important techniques for improving your on-site search function

Autocomplete: The Google's autocomplete or autosuggest feature offers suggestions for your complete search term. Based on the millions of search requests Google receives every single day, it has developed sophisticated predictive models that help users speed up their search process. Very often, results begin to be populated even before the user hits the Search button.

A site search engine that offers autocomplete helps users in speeding up their search process. It also offers users clues on all related data or products that it carries. This is helpful while searching on a new site where the user limited knowledge about its contents. One good tool to achieve this is Algolia (www.algolia.com). Algolia is a hosted cloud search as a service. Using Algolia, site owners can index their database with an API and then relevant search results show up on-site in real-time. The auto-suggest option can even correct typos.

Semantic Search: The idea behind semantic search is to mimic and understand natural language instead of relying purely on keywords for securing results. Semantic search is based on two key concepts: user intent and context. Use these two key insights into your site search to offer users more useful and clearly filtered search results. Configure your site search to see the context behind the phrase entered by the user. For example a search for "bags for gym" on Amazon and Wal-Mart, provides clearly different results. The Amazon search results clearly show the search engine understands the meaning behind search term and shows user exactly what he/she has in mind i.e. gym bag. On the other hand, search results on Wal-Mart clearly show that search engine is unable to determine the real user intent behind search. As such, the search engine uses the words "gym" and "bag" in completely unrelated contexts and provides mixture of irrelevant results.

To introduce semantic search on your site, one good tool is Celebros. Celebros (www.celebros.com) is the global leader is semantic site-search, merchandizing and navigation conversion technologies.

Personalization: Personalization ranges from a customized salutation to tailor-made search results for each user. In general, Google personalizes search results for every user based on past searches. If a user is logged in with a Google account, the search results are more personalized. This is because Google can dig deep into your past search history, picks up what you have liked or shared and ranks such results higher than other results. Site owners can use cookies to help their internal search engine remember past searches and learn about the user's browsing habits. From price differentiation to personalized merchandise options, personalized search helps not just the user, but also the site owners. Unbxd (unbxd.com) is a custom search tool that enables marketers to define and fine tune business rules and regulate exactly what is displayed in a search results page. You can control the degree of personalization and the exact amount to which specific items need to be merchandised on your site.

Advanced Search: Advanced search option enabled a use to zero in on results of a particular search. For example, a user can first search a laptop and then refine its search to find laptops manufactured by a particular manufacturer with price tag below $500. There can be many more such criteria. This feature has been adopted by e-commerce sites in the avatar of faceted search. Faceted search allows shoppers to fine-tune search results and zero in on the exact item that they have in mind. See Figure for an example from BestBuy:

Swiftype (swiftype.com) is a site search tool that provides a good faceted search solution, which allows you quickly refine your results by date, author, location and content types.

Social Networking Sites and E-commerce

A social network is made up of individuals (or organizations) which are tied by one or more specific types of interdependency (e.g. friendship, common interest, like/dislike, relationships etc.). Social networking sites provide a list of contacts for each user and allow users to share ideas, activities, events, and interests within their individual networks. A variety of social networking sites have emerged in recent years. Some popular social networking sites include Facebook, Linkedin, Twitter, and MySpace.

Most social networking sites invite users to provide personal data to build their profiles and users may post their own material (such as music, photos, or videos). The number of social networking sites users is growing at an exponential rate. Worldwide visitors to Twitter.com rose to 19.1 million in the month of March 2009: an increase of 95% compared to the month of February 2009 (Techcrunch.com, 2009). In the United States, Facebook.com is a very famous site that reaches over 91 million American users per month. Yahoo! Shopping (shopping.yahoo.com), Yub.com and Tribe (tribe.com) are three innovative retail sites that are active in capitalizing on the benefits of community content. The growth in social networking sites has raised privacy concerns particularly with respect to the collection and use of consumer's personal information.

Having recognized the importance of social networking sites, online merchants are increasingly shifting a considerable portion of their marketing efforts from to social networking sites to gain market share or retain customers in the economic downturn (Nusca, 2009). A recent survey indicated that more than 90% of companies intended to have a presence on social networking sites (E-comMag, 2009). In the United Kingdom, the number of adults signing up to social networking sites has almost doubled since 2007 (OFCOM, 2009). Online merchants are increasing the use of expensive search marketing programs, which includes paying for appearances in top Web search results. E-commerce sites are increasing their R&D investments into social features (e.g. message books, electronic notifications, photo albums etc.). 73.3% of companies used social networks to hire employees (EcommJournal, 2010).

Another impact of social networking websites in E-commerce is sharing of product recommendations, notable by young consumers. Such a medium for sharing opinions online about products and brands is fluid and informal in contrast with the structured mechanism of product recommendations found on E-commerce sites such as Amazon.com.

E-tailers can benefit from social networking sites in several important ways. First, the users of social networking sites can be a good source of feedback on product/service design, marketing and advertising campaigns, and customer service. Second, visibility of niche retailers and products can be increased by using these users in viral marketing campaign. A common effect of viral marketing is increased website traffic.

Facebook in E-commerce Marketing

With more than 500 million registered users. Facebook is a great opportunity for e-commerce marketers. The user of Facebook involved in many activities such as socializing, writing on each other's walls and playing online games. Facebook is a very important internet-marketing tool because there is a growing trend of socialization and people are becoming blind to traditional advertising.

They key to Facebook marketing is to build up your friends' network and spread word about your business periodically. While businesses should create their profile on Facebook, they should not explicitly sell on their profile but include a link to their website and information about their business. You create a fan page for your business. A fan page is a profile page for a business or brand. On this page, you can be explicit about selling your products and services. You can include pictures, talk about your services, and link to your business website etc. You can create a group about an issue, problem or need that your potential prospects face. For example, if you sell audio-visual equipment for handicapped people, you might create a group related to handicapped people. The Facebook marketplace is a place to buy or sell products. Posting of products/services is free, and your products are searchable on and off Facebook. Your products and services must comply with Facebook's terms of use). The Facebook ads are pay-per-click ads. These ads provide a picture of products/service. They are easy to set up.

FACEBOOK IN E-COMMERCE MARKETING: EXAMPLE CASES

Following are some use cases that serve as examples of Facebook to enhance individual brands (Vahl, 2014).

Qatar Airways

Qatar Airways post about things their fans care about: Football. On its Facebook page, Qatar Airways posts a number of pictures about football (soccer in America). Fans are very responsive to questions and posts on the company's page, as well. Qatar also changes their cover photo often, with a new message for their fans. This is a great way to promote something new.

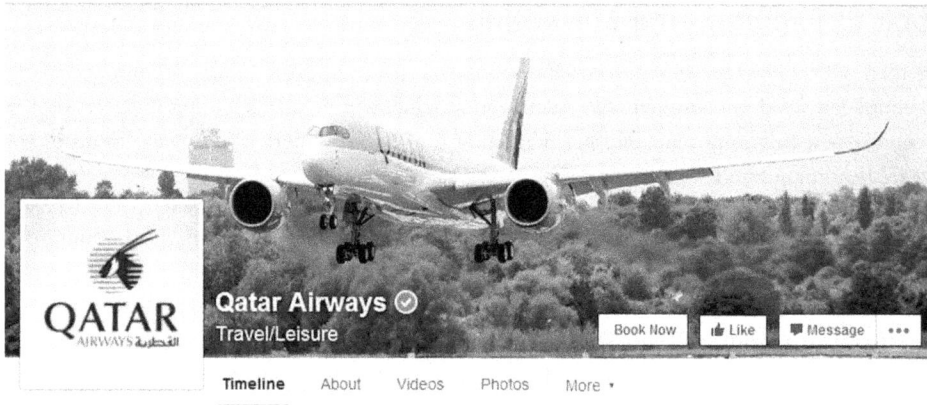

Dove

Dove's Facebook page inspires their fans to tell a story. Dove also encourages tagging by asking their fans to tell a story about someone who means something to them. Tagging someone else in a post can also be a great way to help get the word out about a timeline contest.

Oreo

Oreo, the cookie maker, fills its Facebook page with beautiful images combined with recipes. Oreo also uses Facebook hashtags to provide some fun to their fans.

Humans of New York

Humans of New York have developed itself into a book and a movement to connect on a deeper level with individual people who live in New York. Each day, they share several posts that tell people's stories, along with their image. It is a great how they tell their fans a story about their brand directly in a Facebook post.

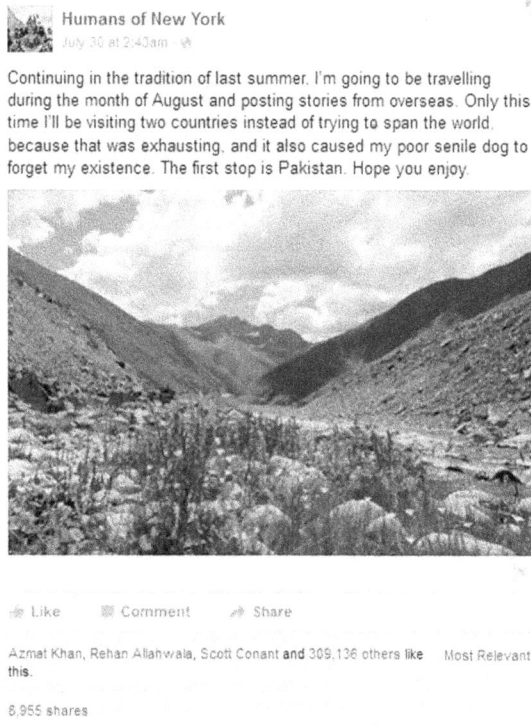

Agilent Technologies

Agilent is a B2B company that uses their Facebook page to tell where their instruments are being used. They brand their images and make them eye-catching. Agilent also connects people to the brand by sharing personal information. For example, they feature interesting stories about inventors a link to the story. They also post pictures of their happy customers.

Agilent Technologies
June 2

He invented the blood bank... then wasn't allowed to donate. Read more @ http://bit.ly/1Q3tERp

HE INVENTED THE
BLOOD BANK...
THEN WASN'T ALLOWED
TO DONATE

Agilent Technologies

Like   Comment   Share

863 people like this.

32 shares

Agilent Technologies shared Praveen G Warner's post
July 29 at 2:33am

Another happy Agilent customer!

Girls Who Code

This non-profit does a great job of mixing up the posts on their Facebook page. They keep fans up to date about what is happening with their mission. Girls Who Code also entertains their fans with games. People can play directly from a post on Facebook. Showing celebrity photos help endorse the mission of the site.

Girls Who Code
August 5 at 8:44pm

"My dreams, in creating this brand and in it coming to life, would not have been possible without technology."

Jessica Alba's Honest Company Gets Behind Girls Who Code, Hosts Summer Program
Eighty young women participated in Girls Who Code's first summer immersion program in greater LA.

Like   Comment   Share

914 people like this.   Most Relevant

90 shares

425

Stella and Dot

Stella and Dot is a network marketing jewelry company using their page to promote their products and share information. One thing they do very well is cross-promote to their other channels, such as Instagram. When you connect to your audience on multiple channels, it improves your chances of being seen. The company also provides shareable images to inspire their audience. Quotes are a great source of inspiration.

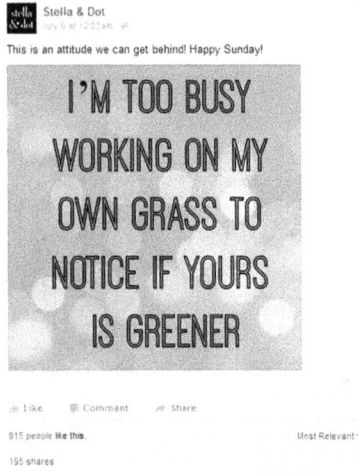

Illegal Pete's

This restaurant has a small fan base, but gets good engagement by running contests. Their fans can tag their friends in the comments.

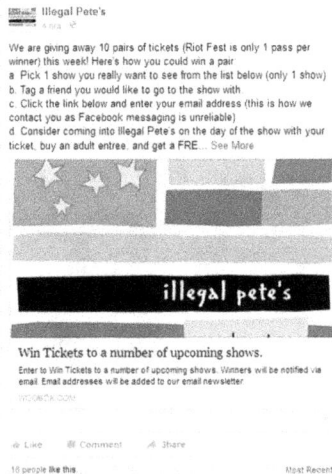

Starbucks

With more than 37 million fans, the Starbucks Facebook page is one of the top brand pages. Starbucks give people tips to help them use their products. Starbucks also shows holiday-themed images that are a little different and fun.

To get more customer engagement, likes and shares, you should inspire your audience, use great graphics and video; make your information shareable, useful and fun; and cross-promote to other channels. More important is that business shows its personality throughout Facebook marketing.

Twitter in E-commerce Marketing

Twitter can offer great help to get traffic to blogs, articles, improves your SEO, and build an online presence. Twitter provides a platform where you can not only socialize but also get information, get opinions, ask your twitter followers for their opinions, for favors etc. Following people on twitter not only provides insight about them but you can also create potential audience for yourself. You can post links to blog posts and articles you have written, comment on other peoples to follow who might fall under your target market.

Twitter can be a cost-effective way to reach customers and prospects around the world. There is no fee to open and operate a twitter account. You can send and receive tweets via mobile devices such as cell phones at no cost.

Twitter can be used to locate people who share similar interest. This can help you target potential customers. You can also build your own following and develop long-term relationships that can lead to sales. Twitter can be used to establish your field expertise, which in turn can help build your credibility and gain customers or clients. For example, a real estate business can send frequent tweets containing brief tips as how to prepare your home for sale or impact of change in mortgage bites. A link to your website can be included in these tweets for additional information. Using twitter, you can send out message at any time immediately. If a quick tweet can be used to inform your followers that can offer your business a competitive advantage over other businesses of information that can be of immediate benefit to them. Your followers can check your link and biography anytime. It can help you build your presence on twitter and twitter offers demonstrate your ethics and personality to make you an appealing person for them to buy from. Twitter is a good promotional tool to announce various activities such as upcoming online events (webinars, virtual tradeshows etc.). As an internal communication tool within your business, twitter can be used for effective communication within the organization.

TWITTER IN E-COMMERCE MARKETING: EXAMPLE CASES

Following are some use cases of Twitter in E-commerce marketing. These use cases serve as an example of some good practices of using Twitter for E-commerce (Smarty, 2015).

Diet Coke #ShowYourHeart Campaign

Coca-Cola made ran a marketing campaign for their Diet Coke brand to raise awareness of heart health, particularly related to women. For every #ShowYourHeart photo submission they had on Twitter, they donated $1 to women's heart health programs around the world. They also gave away five trips to selected winners of a drawing to attend the Red Dress Collection Fashion Show in New York City. This campaign helped Coca Cola bring awareness of their brand and engage with users in a way that boosted their visibility through sharing alone.

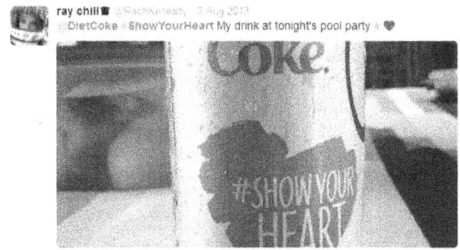

Zaggora's Christmas Pic Contest

Zaggora is a fitness wear brand that offers clothing to help burn additional calories during workouts, while providing a comfortable fit. In 2013, Zaggora held a contest for Christmas that asked followers to post a funny photo of them showing off their Christmas gifts. To enter, a fan used the #zaggora hashtag. This was a simple and great way to get the tag trending and reach more people. The winner of a Zaggora kit of useful products was the person who posted the most "hilarious" Christmas gift pic, as chosen by a panel of judges.

This both increased visibility, and made people aware of what they sold, and why they would want it.

@MaybellineIndia's #ILovePinkBecause Contest

The ILovePinkBecause Contest asked users of Maybelline why they loved the color pink. Random winners selected through the day would win items from the new product line they were promoting, 30 Shades Of Pink. This campaign brought a huge amount of attention to the new products.

Taco Bell's Personable Tweeting Campaign

Taco Bell target people with high follower counts, in order to become visible to those who have them on their feed. This effective tactic caused a sharp increase in Taco Bell followers. They also offer gift cards for retweets.

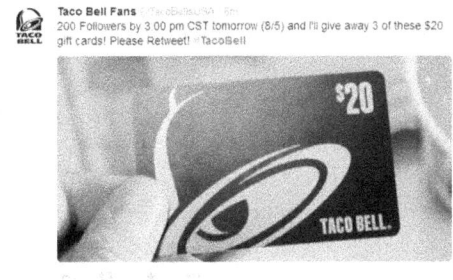

LinkedIn in E-commerce Marketing

LinkedIn is the leading online professional directory of individuals and companies. Individuals use LinkedIn for professional networking, connecting, and job searching. Companies use LinkedIn for recruiting and for providing company information to perspective employees. LinkedIn members can search for jobs; join groups, research companies, and network with members of their network. Companies can post information about the company and job listings on company pages. Companies can also contact LinkedIn members for recruiting purposes.

With more than 120 million business professionals, LinkedIn is an online marketing must for B2B marketers. Besides personal and business profiles, LinkedIn has several other valuable features for Internet marketers. LinkedIn allows you to interact with business associates, past employers, and other in your field. It brings the power of a networking event online. You can create a customizable business page. This page can provide your logo and a summary, as well as the staff of the company. You can also add a tab displaying your products and services. Using apps, you can extend other functionality to your page, like polls. Customers or clients can log on and recommend your product/services they used. That would help show others the quality of whatever you are selling.

LINKEDIN IN E-COMMERCE MARKETING: EXAMPLE CASES

The 71 % of consumer marketers in North America are using LinkedIn. This is because some of the most influential consumers are on LinkedIn. Following are examples of some great use of LinkedIn by strong consumer brands (Lazzaroni, 2014).

Delta Air Lines

Delta is running an "Innovation Class" campaign. In this campaign, Delta gives its current and future customers an opportunity to sit next to a leader in their field on a Delta flight to an industry event. To facilitate engagement, Delta uses a custom LinkedIn API to invite consumers to sign up using their LinkedIn account. Professionals are able to nominate themselves and a LinkedIn connection for a chance to learn, share ideas, talk about goals or even collaborate on a project with a specifically selected mentor. Delta also uses a suite of LinkedIn products, including Sponsored Updates, to promote the program in members' personal feeds. Nominated connections will receive an InMail inviting them to engage with "Innovation Class," where they can follow Delta, watch videos, discover upcoming events and read about previous events.

DELTA INNOVATION CLASS
POWERED BY [in]

What is Delta Innovation Class?

A mentoring program at 35,000 ft

Fruit of the Loom

Fruit of the Loom (FOTL) is a global manufacturer of quality products including underwear, panties, bras, t-shirts, sweats, sweatshirts, socks, thermals and more. Fruit of the Loom started a campaign in which they provided a free underwear to new hires on new LinkedIn. To date, FOTL has provided 25,000 new hires on LinkedIn with "fresh fruit for your fresh gig" as part of their quest to help America "Start Happy." The landing page of Fresh Gigs provide many interesting information such as where the Fresh Gigs are (broken down by state); who's getting the most Fresh Gigs (broken down by industry); the gender breakdown of Fresh Gigs; and the specific styles the top Fresh Gigs industries are choosing.

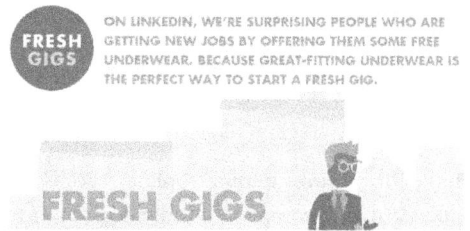

FRESH GIGS

ON LINKEDIN, WE'RE SURPRISING PEOPLE WHO ARE GETTING NEW JOBS BY OFFERING THEM SOME FREE UNDERWEAR. BECAUSE GREAT-FITTING UNDERWEAR IS THE PERFECT WAY TO START A FRESH GIG.

FRESH GIGS

Secret Deodorant

Proctor & Gamble started "Secret Deodorant Showcase" page on LinkedIn to highlight its brand. This page features stories, videos and other content that helps its audience stay "100% Fearless at Work." The page allows Procter & Gamble to drive engagement with a specific audience through messaging that aligns with the aspirational mindset of the LinkedIn platform.

Mercedes-Benz USA

Mercedes-Benz ran "Driven to Perform" campaign that awarded a LinkedIn member with a two-year lease of a Mercedes C-300. To win, members must be nominated by someone in their LinkedIn network. Mercedes was asking LinkedIn members to nominate colleagues by highlighting their professional accomplishments and attributes, along with how the nominee's personality qualities meet the sophistication of the Mercedes-Benz brand. The campaign was aligned with the LinkedIn mindset. This was clearly shown by the campaign message what stated *"Recommending a colleague or peer on LinkedIn is nothing new. Winning them a two-year lease on a Mercedes-Benz is. We believe that great performance deserves great performance. So tell us about someone who exemplifies the teamwork and innovation that went into the game-changing 2015 C-Class. You might earn them the performance bonus of a lifetime."*

Driven to Perform | Who should be one of the first drivers of the all-new 2015 C-Class?

in Recommend Someone >

Recommending a colleague or peer on LinkedIn is nothing new. Winning them a two-year lease on a Mercedes-Benz is.

We believe that great performance deserves great performance. So tell us about someone who exemplifies the 'the best or nothing' work ethic that went into the all-new 2015 C-Class. You might earn them the performance bonus of a lifetime.

* a two-year lease

Generating Sales through Social Networking Sites

Besides being used as online marketing platforms for generating web traffic, social networking sites can be used to generate direct sales, if used properly and effectively. To succeed, a business needs the combination of right product and a proper approach to communicate with social media audiences. However, online businesses must be watchful for few things to increase their probability of success.

Avoid Aggressive Marketing: Users of social networking sites generally dislike spam even if it is of some real value. Marketing products aggressively increases the probability that the marketing message will either be ignored or face rejection and criticism. The marketing strategy should approach these users as future customers that deserve long-term care and attention and avoid pushy marketing efforts and short-term sales tactics.

Gain Feedback and Create Hype: Users of social networking sites can be used to get feedback on products under development. By doing this, businesses can expect to have much greater sales success upon product launch. Another advantage is that by using social networking sites, businesses can create hype about the products.

Ongoing Relationship Efforts: Compared with any other marketing platform, social media requires more input to get measurable output. To generate sales, companies need establish two-way communication and real relationships with their consumers. Company can increase its probability of sales increase with ongoing relationships. This is because, with ongoing relationships, market becomes more aware, interested, and starts trusting the business.

Social Commerce

Social commerce is a subset of E-commerce that involves using social media, online media that supports social interaction and user contributions, to assist in the online buying and selling of products and services. The concept of social commerce can include user-generated advertorial content on e-commerce sites, or collaborative e-commerce tools that enable shoppers to get advice from trusted individuals, find goods and services and then purchase them. Today, the social commerce has expanded to include range of social media tools and content used in the context of e-commerce, such as customer ratings and reviews, user recommendations and referrals, social shopping tools (sharing the act of shopping online), forums and communities, social media optimization, social applications and social advertising.

Example Cases of Social Commerce

As per current estimates, social commerce accounts for between 5 – 20% of sales for nearly 60% of all businesses that market themselves on social media. However, this number is about to take a sharp upturn, with social commerce set to account for $30 billion in sales in the US alone. Catching on to the fact that Pinterest is a fabulous social network for retail curation and user wishlists, Nordstrom started highlighting items that were popular on Pinterest with a "Popular on Pinterest" tag on the physical item in stores.

The experiment was so successful that every Nordstrom outlet across the US started showcasing its most popular items on Pinterest with a 'Top Pinned' section inside physical stores. Shop assistants were equipped with an in-house iPad app that showed trending items for the day and helped them tag these items appropriately in-store.

Coke started the first step of its social marketing campaign with product personalization. For the first time in history, 250 of the most popular first names in each country were shortlisted and printed on the iconic red and white Coke labels, instead of the Coke logo. Coke then used mass media channels like television, outdoors and radio to communicate to users that their favorite drink just might have their name on it. Each bottle also carried a hashtag #ShareACoke to remind users to post pictures of their personalized Coke bottles on social media using the hashtag.

The experience of seeing one's own name on Coke bottles was so novel and addictive that people actually paid premium prices just to lay their hands on their 'own' bottles of Coke and shared them on social media like wildfire. Images of Coke bottles shared on Instagram, Twitter and Facebook with the #ShareACoke hashtag were then plastered across digital billboards, across the country (Dholakiya, 2014).

SOCIAL SHOPPING

Shopping is a very social activity. With friends, people can exchange ideas to help make wise shopping decisions. Online social shopping is a technology-based solution provided by e-commerce businesses to help shoppers involve their friends in shopping experience. Online social shopping activity attempt to mimic the same social interactions a shopper can have in a physical shopping mall/stores. Social shopping sites provide a range of products e.g. electronics to apparel to home décor. Shopper can interact, chat and share information in new ways. Social shopping sites also provide online retailers with an opportunity to enhance their web presence.

Social shopping is the new frontier for retailers and businesses alike. With consumers becoming increasingly tech and shopping savvy you need to distribute your financial resources. It is not to say you need to invest and sacrifice your business model, but remain open to this untapped revenue stream. Things have radically changed and buying habits have drastically shifted. E-commerce businesses need to careful watch the buying patterns and target what sort of interactions generate leads via these social sites.

Social shopping sites can provide product reviews, price comparisons, polls (e.g. which shoe is best with this dress?), wish lists and shopping lists that customers can create, and blogs or message boards that users can use to chat with friends and other customers and share section of products with deals/discounts. Social shopping sites can help customers find hard-to-locate items, find the best price, get products at their doorstep, easily compare products and their prices, receive product updates, and feel more confident in their online shopping because they can get friends recommendation. Social shopping sites provide many advantages. Sites such as styleHive, provide their users opportunity to interact with brand representatives through community blogs & profiles. Users can use this interaction as a source of expert advice. Users can exchange and view each other's wish lists. That makes it *much* easier to shop for appropriate for your friends/family.

Some popular social shopping sites can be divided into different categories. Group shopping sites (Half off Depot, Plum district, Groupon, Kactoos, Living social, and Buy with Me) encourage groups of people to buy together at wholesale prices. Online shopping communities (e.g. Listia) allow users to communicate and aggregate information about products, prices, and deals. Recommendation engine sites (shop socially, Shopow, Blippy, Left of trend and swipe) allow their uses to ask a fellow shopper for advice. Social Shopping Marketplaces brings together independent sellers and creates forum for them to display and sell their wares to buyers. Shared shopping sites: allow shoppers to form ad-hoc collaborative shopping groups in which one person can drive an online shopping experience for one or more other people, using real-time communicate among themselves and with the retailer. Some popular social shopping sites are discussed below.

Wanelo.com: Wanelo is an abbreviation for "Want, Need, Love," and is an online community-based e-commerce site that brings together products from a vast array of stores into one pinboard-style platform.

Fancy.com: A social shopping site offers trendy products ranging from fashion items to useful gadgets and novelty gifts.

Pinterest: Pinterest is not organized like traditional e-commerce sites. Products are mixed in among a variety of other pins, such as product reviews, articles, quotes, and much more. Pinterest allows you to make pin boards of your favorite pins, and follow your favorite pinners.

Fab.com : This site provides product offerings in department store categories, many available at discount prices through flash sales.

Polyvore: Polyvore is a fashion-focused community site for visual collection creation, product discovery, and purchasing.

Luvocracy: Luvocracy encourages shopping based on other members' recommendations. The site also offers rewards for using its service and acts as a third-party price negotiator for product purchases.

Opensky.com: This site is a gamified social shopping with a variety of products and deals. The site features mainstream products that you can browse from category sections or from your custom news feed.

Faveable: This site is comprised of various items, which are organized by categories or can be discovered via a simple keyword search.

Ownza.com: This site provides a monetary incentive for online browsing and shopping, by giving you cash back percentages for the products you purchase.

Etsy: It is the famous online shopping site of handmade products. It provides sellers an online storefront for listing goods for a fee.

Web 2.0 Tools and E-commerce

The Web 2.0 tools are interactive in nature. These tools are redefining today's online experience. Web 2.0 technologies improve customer experiences, boost sales productivity, and help CRM efforts. With increasing reliance of product development and sales strategies on web, the importance of web 2.0 technologies will continue to increase. Following are the important web 2.0 technologies.

Really Simple Syndication (RSS): It is a means of alerting users about news and other Web content generated by an online publisher.

Blogs: A blog can be a website or part of a website. Blogs are maintained both individuals and businesses. Blog allows its users to reflect, share opinions, and discuss various topics. User may comment on blog posts.

Wikis: These websites allow visitors to contribute content.

Podcasts: A podcast is also called non-streamed webcast. It is a series of digital media files (either audio or video) that are distributed through websites for playback on mobile devices or personal computers.

Social Networking: These web sites enable users to build online social communities and interact with their contacts.

Peer-to-Peer Applications: These applications enable peer-to-peer sharing e.g. of music, video and text files.

Asynchronous JavaScript and XML (Ajax): It is a method that combines different programming languages to create real-time, interactive Web applications.

Mashups: It refers to a web page of application that combines two or more data sources under a unified interface, to provide some unique information or service.

By 2008, few sites took advantage of Web 2.0 technologies to improve usability, interactivity and navigation. However, successful e-commerce operations invested continually in upgrades to their sites (Gartner.com, 2008).

Use of interactive Web 2.0 technologies is also raising privacy and security and businesses are cautious in adopting these technologies. A good approach for businesses is to get started by using simple technologies (such as blogs and wikis) internally before moving for full deployment of advance web 2.0 technologies.

Examples of Web 2.0 Tools Usage

Burpee, a very old family-run gardening business, managed to implement some of the most basic Web 2.0 technologies (a blog and an RSS feed) with impressive results. Burpee launched its own feed to distribute a blog that featured customer's gardens, alerted subscribers to new offerings and special sales, and provided gardening tips. It also added a tool that enabled customers to post user reviews on products. Burpee's sales increased after launching the feed.

TAC Worldwide, an IT staffing firm, cultivated a database of 150,000 résumés after 38 years of operations. They launched a new Web-based recruitment toolkit called TACSource. Within three months, the TAC résumé count jumped to 3 million. TACSource, using mashup technology, provided three functions. First, it allowed TAC Worldwide to connect directly to clients via the Web so that a company can submit a job description to a recruiter online in real time. Second, it allowed TAC Worldwide to search external job postings in an automated way. Third, its algorithms allowed recruiters to see automated displays of résumés that are good matches for job openings. TAC also piloted an RSS feed so that a client firm could, for example, opt to receive an RSS feed that would tell it which sort of talent and skill sets are available in the area. Meanwhile, potential job candidates would opt to receive ticker-like updates of employment news in specific industries or regions.

Web 3.0 Tools and E-commerce

Web 1.0 was all about driving online commerce and trying to find "anything" on the Web. It produced companies like Yahoo, Amazon.com, eBay, Netflix and Blue Nile. Web 2.0 has been primarily focused on social networking through online communities. Facebook, Twitter and LinkedIn, have been the most notable companies to emerge. Web 2.0 has brought an onslaught of user-generated content in the form of blogs, podcasts, appending comments at the bottom of articles, posting reviews of restaurants, movies, stores, and hotels. Media has become truly interactive, as opposed to the one-way world we were used to. Social Semantic Web (Web 3.0) will organize itself around two different elements: context and the user. By "context," we mean the intent that brings us to the Web, our reason for surfing. Looking for a job is "context," as is planning a trip or shopping for clothes. Fundamental to context is the user. In addition, when we fuse a specific user with genuine context, we end up with truly personalized service. Take example of travel website. In a Web 3.0 world,, a personalized travel agent will help you find and book a highly customized itinerary, leveraging all the power of previous generations of Web technology –searching (both generic and vertical), community building, content and commerce (Mitra, 2014).

Web 3.0 technologies have been ignored in the past by many Ecommerce developers, online retailers and Search Engine Optimizers alike. Google is pushing the use of structured data for web and expanding Rich Snippets to more search results and collecting user feedback along the way. Developers increasingly use microformats or RDFa to markup content on websites involving review or people/social networking content. This ensures that Google can better understand the content and generate useful Rich Snippets. Rich snippets are a few lines of text that appear under every search result of web pages that have implemented that feature. They are designed to give users a sense for what is on the page they will visit, and why it's relevant to their search query. Depending on the nature of the products or services you offer, your rich snippet could appear similar to the one in the Figure.

Wise **Registry Cleaner** Free - Download.com - **CNET** Download
download.cnet.com/ Registry-Cleaner.../3000-18512_4-10606508
★ ★ ★ ★ ★ Review by CNET staff - Free - Windows - Utilities/Tools
5 Jan 2012 – **Registry cleaners** are fine when used with care but can cause problems when used carelessly

In mid-2011, Google, Bing and Yahoo joined forces in supporting Schema.org, to create and support a common vocabulary for structured data markup on web pages. The final frontier is personalization. Baynote is a company that offers personalized recommendations. By applying principles of neuroscience, Baynote can predict what shoppers would be looking for.

Virtual Worlds for E-commerce

Virtual worlds are three-dimensional, computer-generated worlds containing interactive objects (Park, Nah, DeWester, Eschenbrenner, & Jeon, 2008). Virtual worlds are the next stage of the Internet and they present opportunities for businesses to find new ways to provide value to customers. Examples of virtual worlds include Second Life (secondlife.com), There (there.com), Active Worlds(activeworlds.com), and Kaneva(kaneva.com). In virtual worlds, people are represented through their digital representations, called avatar. These avatars can move, walk, and interact with other avatars (Pratt, 2008). Some virtual worlds provide communication capabilities that include textual, visual, and auditory to facilitate involvement and learning, which can help to create brand awareness or enhance the value of a brand (Eschenbrenner, Nah, & Siau, 2008). Such learning can occur by interacting with others through their avatars or with other objects in the environment.

Almost any business activity that can be conducted on the Web can be conducted in virtual worlds. Users can look at virtual products, participate in designing or customizing products, and can have text or audio conversations with the avatars of business representatives. Virtual world increases the richness of conducting business activities (Brandon, 2007).

There exists significant potential and opportunities for businesses in virtual worlds. With exponentially increasing user based of virtual worlds, businesses are increasingly participating in these environments (Schwarz, 2006). Social virtual worlds have become an industry with an investment of more than $1 billion and more than 100 virtual worlds either in development or live. Many businesses have begun using virtual worlds as a new channel to reach consumers (Richardson, 2008).

Market Segmentation

Marketers segment the markets to formulate effective marketing strategies that appeal to specific consumer groups. t Market segments are logical groups of customers that can be created in many ways e.g. by geography, demographics, psycho- graphics, and benefits sought. Data modeling and data warehousing are two tools that are commonly used for segmentation. Using data mining and Web mining, businesses can look at consumer buying patterns to create finer segments. For examples Royal Bank of Canada segments its millions of customers at least once a month to determine credit risk, profitability, and so on (Turban, Lee, & Chung, 2006). Segmentation can be very effective in the Web environment. (Hutt, Lebrun, & Mannhardt, 2001) found that segmentation by age, employment status, family role, and household structure produces the best result for customer acquisition.

Quantitative Techniques of Market Segmentation

The two quantitative techniques that can be used for market segmentation on the web are:

1.)   CHAID   2.)   RFM

CHI-SQUARED AUTOMATIC INTERACTION DETECTION (CHAID)

CHAID primarily serves as a market segmentation technique. As an example, consider a response tree shown in Figure 7-11. This tree represents a market segmentation of the population under consideration. The five bottom branch boxes are the segments. The segments are prioritized for targeting based on first their level of responsiveness, and second on their size. The upper segments, defined by response rates larger than the overall response rate are the "low-hanging" fruits, which are high yielding and require little effort to obtain. The lower segments, defined by response smaller than the average, are "high-floating" fruits, which are not high yielding and require extra effort to acquire. The middle segments, defined by response about equal to the average, offer the marketer a choice either to use the current business-as-usual strategy to yield average results (10%), or implement an unexpected forceful strategy to efficiently stimulate these segments to produce greater than average results.

FIGURE 7-11: CHAID TREE

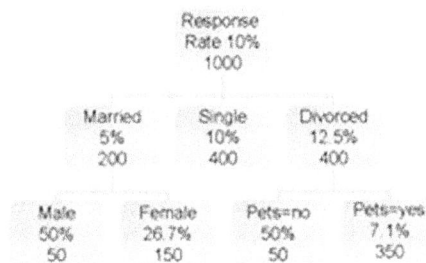

RECENCY, FREQUENCY, AND MONETARY (RFM)

RFM is a predictive modeling and segmentation technique that can dramatically improve email click-troughs and conversions. It can be used to identify best customers and to send them special, tailored offers. RFM stands for Recency (date of the last purchase or the date of the last email click-through or the date of the last lead-generation conversion), Frequency (number of times each recipient purchased over a set period of time or number of times each recipient clicked over a set period of time or number of lead-generation conversions over a set period of time) and Monetary (total amount spent over a set period of time or total estimated value based on factors like cost-per-lead and revenue-per-lead over a period of time or cumulative engagement score derived from different metrics over a set period of time). It relies on the premise that someone who recently bought something, who shopped often and who spent a lot is more likely to respond to your next campaign than someone who bought something a long time ago, shopped infrequently and spent next to nothing. Coupled with customer lifetime-value analysis, RFM enables organizations significantly increase response rates by sending offers only to the subset that is most likely to buy. In the direct-mail world, where every name comes with set production and shipping costs, there is a strong motivation to pare lists down. As a result, direct marketers have used RFM for decades. Once you have decided which RFM metrics make the most sense for your business, you will need to tie your email database to the system that contains purchase or conversion history, such as your CRM or Web-analytics tool. After that, you are ready to perform RFM segmentation. Many popular data-mining and statistical analysis tools generate ready-made RFM-classification reports. Unlike other predictive-modeling techniques, RFM is based on past customer behavior and does not require heavy statistics-driven analysis. In traditional direct mail, you perform a break-even analysis to determine which mailing recipients are profitable. You look at the test-group response rate for each RFM cell, and then stop mailing to cells whose response rates are less than the rate required to break even on mailing costs.

The Internet Users: Population and Usage Trends

Internet Users Population

Both developed and less-developed countries still have relatively low percentages of Internet users. Even countries with high levels of education and employment, long histories of Internet use, and high percentages of broadband installation have large number of non-users of Internet (CentrefortheDigitalFuture, 2010). According to (InternetWorldStats, 2009), in 2009, Asia accounted for the largest number of Internet users, followed by Europe and then North America. As a percentage of the population, North America led the world, with almost 75% of the population having access to the Internet. Twenty countries of the world account for around 77% of the total Internet users. China has the highest percentage of the Internet users (20%) followed by United States (13%). The statistics reveals that, even in technologically advanced developed countries, Internet has not yet become part of life for millions of people around the globe.

According to Statista.com (2015), Current internet user figures suggest that there are more than 2.5 billion users worldwide with a global penetration rate of 35 percent with North America accounting for 14 percent of global internet users. Facebook has more than 1.2 billion monthly active users, presenting a global reach of almost 50 percent of internet users worldwide. Following are some important Internet usage statistics.

In 2014, global mobile data traffic amounted to 2.5 exabytes per month. In 2019, mobile data traffic worldwide is expected to reach 24.3 exabytes per month at a compound annual growth rate of 57 percent. In 2014, global mobile video traffic amounted to 1,377,497 terabytes per month and is expected to more than double to 4,104,719 TB per month in 2016. As of the second quarter of 2015, it was found that mobile devices accounted for 45 percent of organic search engine visits. As of 4th quarter 2013, it was found that 34 percent of internet users worldwide used their mobile devices to research products or services they intended to purchase.

Internet Usage Trends

(CentrefortheDigitalFuture, 2010) and (Center for Digital Future, 2014) identified following trends in Internet usage by users worldwide.

Gender Differences: There exist notable differences between men and women in their use of Internet. In general, Men use Internet more than woman does.

Age: Age is related to level of usage of Internet. It is a general observation that Internet use increases as age decreases.

Non-Usage: Surprisingly the report found that the cost of accessing Internet was not the primary reason for not using Internet. Instead, most users did not Internet because they thought it was not useful or they had no interest in using Internet.

Online Purchasing: In general, online purchase is not yet a frequent activity of Internet users. One reason for this trend is high levels of concern about the security of credit card information during online purchasing.

Contact with Family and Friends: In general, Internet use has a positive effect on contact with family and friends. Contact with friends and family is increases because of Internet use.

Productivity at Work: In general, using the Internet at work has improved workers work performance or productivity.

Information on the Internet: There exist widely divergent views on reliability of online information. While many users think that only half or less of online information is reliable the Internet is still considered as an important source of information by large majority of users.

Using the Internet to Look for News: Large percentage of Internet users goes online to get local, national, or international news.

The Internet and School-Related Work: Large number of student Internet users is going online to seek information for their school-related work.

Customers Switching to Internet: In US, television viewers are migrating to the Internet and trend will likely continue especially with the new technology that enables mobile Internet access. With increasing experience with the Internet, users spend less time watching TV.

Mobile Internet: The most popular online activities of mobile Internet users and other Internet users include using search engines, reading news and sports information, downloading music and videos, and sending/receiving email and instant messages (IDC, 2009).

Online Consumer Behavior

Online consumer behavior can be defined as any Internet-based activity related with the user's consumption of goods, services, and information. Further, Internet consumption includes (1) information gathering through passive exposure to advertising, (2) shopping (including browsing and information search), and (3) the selection and buying of goods, services, or information (Ahuja, Gupta, & Raman, 2003).

Research suggests that many factors can contribute to understanding of the consumer behavior online. Session characteristics and the click stream behavior are two most important factors to predict the consumer behavior online. Trust, economic conditions, and technological understanding of on-line shoppers are some other important factors [see (Mahmood, Kohli, & Devaraj, 2004)].

Stages of the consumer decision process are the same in both online and offline consumer behavior but there exist some differences between the two. The general model of consumer behavior requires modification to adapt to consumer behavior online. Many factors can influence consumer behavior online including website features. Web site features include the content, design and functionality of the site that can influence the customers. Same concept applies to a physical store where consumer behavior can be influenced by store design by arranging goods and promotions along the tracks consumers most like visits. Proper store design and precision tracking of consumers is not a new concept. The challenge is its technical implementation on the web.

Online Consumer Purchase Decision

A consumer go through a decision making process before making a purchase decision. Many people can play role in this decision making influencing the ultimate decision made by the customer. (Kotler & Armstrong, 2002) suggest the following major roles are played by people in the consumer decision-making process.

Initiator: This is the person who actually initiates the idea of a particular product or service.

Influencer: This is the person who is in a position to influence a final purchase decision.

Decider: This is the person making the ultimate purchase decision.

Buyer: This is the person who makes an actual purchase.

User: This is the person who consumes or uses a product or service.

These roles can be played by a single individual or many people. Marketers need to analyze the roles carefully to properly target advertising and marketing.

Several models have been developed to describe the details of the purchase decision- making process. One general model, used by (O'Keefe & McEachern, 1998), consists of five major phases. Each phase can have several activities and many involve one or more decisions. The five phases are:

Need Identification: In this phase marketers get the consumer recognize a gap between their actual and desired need and convince them that a particular product or service the seller offers can fill this gap.

Information Search: Once the need is identified, the consumer searches for information on the various alternatives available to satisfy the need. Catalogs, advertising, promotions, and reference groups are the sources that can influence consumer's decision-making. Information search is a critical stage in the Internet purchase process (Su, Characteristics of Consumer Search On-Line: How Much Do We Search?, 2008).

Evaluation of Alternatives: After searching and analyzing information, consumer will come up with a small set of preferred alternatives. Consumers perform further evaluation and possibly negotiate terms with sellers to develop purchase criteria. Consumer then uses this criterion to evaluate the set of alternatives.

Purchase and Delivery: In this phase, the consumer makes the purchase decision, arrange payment and delivery.

After-purchase Evaluation: This phase consists of customer service and evaluation of the usefulness of the product.

Remember that these phases are a general model and it is not necessary that all consumer's decision-making will necessarily proceed in this order. Some consumers may proceed to a phase, revert to a previous phase, or skip a phase altogether.

Online Consumer Purchase Decision Aids

Rapid growth of E-commerce has provided consumers with ability to make purchase decisions in computer-mediated environments. Availability of many alternatives, multiple decision criteria and a real-times flow of information is overwhelming consumers. Finding products that meet consumer needs has become an increasingly difficult task. Many electronic aids are now available to facilitate customers purchase decisions (Solomon, Dahl, White, Zaichkowsky, & Polegato, 2014). Following are the common electronic decision aids available to consumers.

INTELLIGENT SOFTWARE AGENT

An intelligent agent is a software program that can perform routine tasks that require intelligence e.g. searching products and services, comparing prices, interpreting information, monitoring activities, and providing assistance. Users can even chat or collaborate with intelligent agents (Bhamra, Verma, & Patel, 2014). One example of intelligent agents is called shopbot that that specialize in shopping for items like books, music, consumer electronics products, and computer hardware and software. A shopbot query merchant'ss databases simultaneously when a user presents his request. Pricefinder (www.pricefinder.com) and pricewatch (www.pricewatch.com) are examples of a shopbot.

BUSINESS RATING SITES

These are sites that rate e-tailers e.g. Bizrate.com. Bizrate.com is an independent, third-party online shopping guide that surveys online buyers on an ongoing basis. They compile their findings and provide them in the form of store ratings. Customer ratings and reviews increase site traffic, conversion rate and average order size. They also improve customer loyalty, retention and improve search engine optimization. As we can see, adding customer reviews and ratings to e-commerce site is not just a helpful tool for customers to interact with site - they lead to purchase decisions (Gay & Pho, 2013).

TRUST VERIFICATION SITES

Trust verification sites e.g. TRUSTe (truste.com) and Verisign (verisign.com) evaluate a businesse's trustworthiness (Özpolat & Jank, 2015). TRUSTe operates a privacy seal program. It certifies more than 3,500 websites including leading online portals and brands like Yahoo, Facebook, Microsoft, and IBM.

FTC of USA has launched an online tool called Online Shopping Assistant. This tool, developed for econsumer.gov by Canada's Office of Consumer Affairs, will help you evaluate online businesses and decide whether to proceed with ordering from them. The Shopping Assistant is designed to remain on your screen while you review a merchant's web site. Users can click "Yes" or "No" in response to the questions in the Shopping Assistant against information they find on the business's website. User can bookmark the shopping assistant and easily pull it up when you are considering an online transaction. The tool is available at http://www.econsumer.gov/english/resources/resources.shtm.

FIGURE 7-12: FTC SHOPPING ASSISTANT TOOL

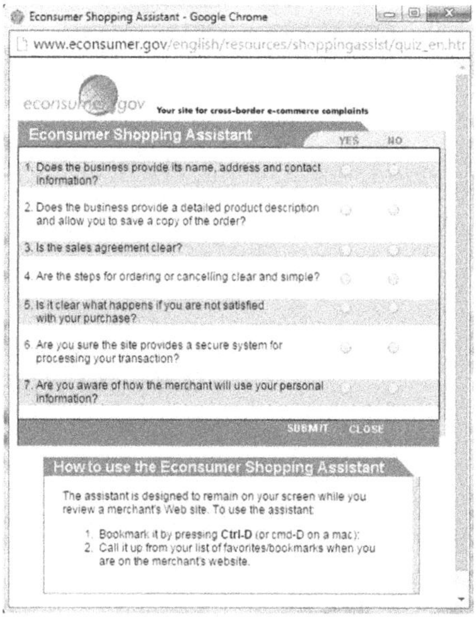

CONSUMER COMMUNITIES

There are also communities of consumers available who offer advice and opinions about various businesses. One example is www.complaintsboard.com (Héder, 2014).

RECOMMENDER SYSTEMS

Recommender systems are intelligent applications that assist users in a decision-making process. They are useful in situations where users do not have sufficient personal experience make a product selection from an overwhelming set of alternative products or services. The scope of recommender systems is expanding and they are becoming more and more popular even in simple E-commerce Web sites (Jansen & Resnick, 2006). Recommender systems have been used for recommending many things including books, CDs, movies, news, electronics, travel, financial services, and many other products and services (Adomavicius & Tuzhilin, 2005; Bobadilla, Ortega, Hernando, & Gutiérrez, 2013).

Recommender systems can use several criteria to recommend products e.g. top overall sellers on a site, demographics of the customer, and past buying behavior of the customer. Recommender systems can help enhance E-commerce sales in three ways. First they can convert browsers into buyers by helping customers find products they wish to purchase. Second, they can improve cross-sell by suggesting additional products for the customer to purchase. Third, they can improve customer loyalty by creating a value-added relationship between the site and the customer (Schafer, Konstan, & Riedi, 1999).

Few examples of recommender systems include Amazon.com "Customers Who Bought This Item Also Bought" feature and eBay's Feedback Profile. The "Customers Who Bought This Item Also Bought " feature is found on the information page for each item in Amazon product catalog. The Feedback Profile feature at eBay allows both buyers and sellers to contribute to feedback profiles of other customers with whom they have done business.

FIGURE 7-13: WORKING OF RECOMMENDER SYSTEM

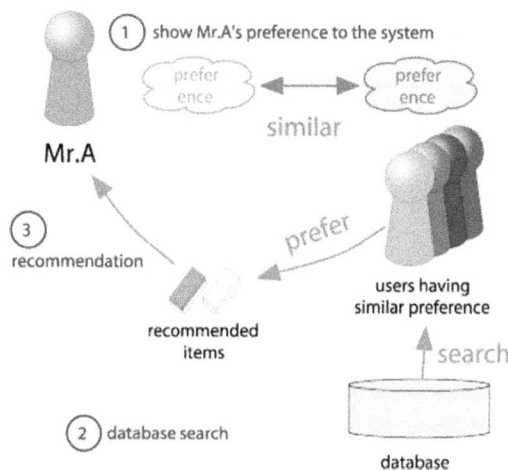

Online shopping Trends

Figure shows a ranking of countries in E-commerce Index. In the following sections, we shall discuss online shopping trends in different regions of the world. We shall take some examples of countries leading E-commerce world and some emerging countries on the E-commerce map. For ranking, we shall use the 2015 Global Retail E-commerce Index developed by atkearney.com (a global management consulting firm that focuses on strategic and operational CEO-agenda issues facing businesses).

FIGURE 7-14: THE 2015 GLOBAL RETAIL E-COMMERCE INDEX

Rank	Change in rank	Country	Online market size (40%)	Consumer behavior (20%)	Growth potential (20%)	Infra-structure (20%)	Online market attractiveness score
1	+2	United States	100.0	83.2	22.0	91.5	79.3
2	-1	China	100.0	59.4	86.1	43.6	77.8
3	+1	United Kingdom	87.9	98.6	11.3	86.4	74.4
4	-2	Japan	77.6	87.8	10.1	97.7	70.1
5	+1	Germany	63.9	92.6	29.5	83.1	66.6
6	+1	France	51.9	89.5	21.0	82.1	59.3
7	-2	South Korea	44.9	98.4	11.3	95.0	58.9
8	+5	Russia	29.6	66.4	51.8	66.2	48.7
9	+15	Belgium	8.3	82.0	48.3	81.1	45.6
10	-1	Australia	11.9	80.8	28.6	84.8	43.6
11	-1	Canada	10.6	81.4	23.6	88.9	43.1
12	+2	Hong Kong	2.3	93.6	13.0	100.0	42.2
13	+6	Netherlands	8.9	98.8	8.1	84.6	41.8
14	-3	Singapore	1.3	89.4	15.7	100.0	41.5
15	+13	Denmark	8.1	100.0	15.1	75.5	41.4
16	0	Sweden	8.8	97.2	11.8	77.7	40.9
17	Not ranked	Mexico	10.0	53.3	58.6	68.0	40.0
18	Not ranked	Spain	13.2	73.1	20.2	80.1	39.9
19	+1	Chile	2.7	71.8	49.3	73.2	39.9
20	+6	Norway	8.2	99.4	5.6	76.3	39.5
21	-13	Brazil	19.6	57.4	28.0	72.4	39.4
22	-7	Italy	12.3	71.6	27.8	70.7	38.9
23	+6	Switzerland	7.1	89.6	7.4	82.5	38.8
24	-1	Venezuela	1.7	54.1	79.4	55.7	38.5
25	-4	Finland	6.4	98.3	3.8	77.3	38.4
26	-8	New Zealand	1.7	86.4	25.9	75.4	38.2
27	Not ranked	Austria	5.9	85.3	19.0	74.8	38.1
28	Not ranked	Saudi Arabia	1.1	46.6	67.3	74.6	38.1
29	-17	Argentina	5.7	70.3	43.9	64.3	38.0
30	-3	Ireland	4.9	74.4	27.6	74.1	37.2

Online Shopping Trends in Developed World

China

China is the leading country in E-commerce world with annual e-commerce sales of $426.26 billion. Chinese consumers spend the most on travel products when shopping online, forking out an average of RMB3,750 per year, according to report by McKinsey (ChinaTravelNews, 2015). Although the internet coverage in smaller cities is less extensive, internet users in those cities are just as active in online shopping as their urban counterparts. China is one of the most socially active countries in the online world and Chinese netizens spend an average of 78 minutes a day per person on social media compared with 67 minutes for Americans. Referrals contributed to 50% of new interactions among Chinese users, compared with 40% for Americans. These trends are showing signs of gaining momentum. Social media is stimulating online shopping and the time spent on online shopping, watching films and reading news and articles is increasing significantly. Chinese e-commerce businesses are also taking their direct sales models to popular mobile platforms through semi-private groups of 50 to 100 users on SNS app Wechat. Brick-and-mortar retail venues are increasing becoming "showrooms", for digital savvy shoppers. In the case of digital products – about 16% are sold online, compared to only 1% five years ago. O2O shopping is already used by 71% of all e-commerce consumers in China and 97% of these shoppers plan to increase O2O shopping in the next six months. A third of the respondents who have not used O2O services yet are willing to try it within the next six months. Below are the types of O2O services consumers are interested to try. While a high percentage of consumers in tier three and four cities shop online, at the rate of 68% and 60% respectively, consumers in rural areas are even more keen online shoppers having 25% more frequent online shoppers among users than in major cities. Demand for food products online has shot up and the category now has the highest purchase rate of all online products. Chinese users made 34 food purchases per year compared with 22 clothing purchases per year in 2014. Food safety is the biggest concern among online shoppers, with 65% percent of respondents being "very concerned" about the issue compared with 36% and 26% in the USA and the UK respectively.

Europe, UK, USA, and Canada

E-commerce is the fastest growing retail market in Europe. Sales in the UK, Germany, France, Sweden, The Netherlands, Italy, Poland and Spain are expected to grow from €156.28 billion in 2014 to €185.39 billion in 2015 (+18.4%), reaching €219.44 billion in 2016. In 2015, overall online sales are expected to grow by 18.4% (same as 2014), but 13.8% in the U.S. on a much larger total. The recession has induced many shoppers to buy online rather from traditional stores, whilst above-average growth in countries with smaller ecommerce sectors shows there has been an element of catch up. Retail focus on the growing use of mobile technology is an additional factor in making online retailing attractive and convenient. European online market is dominated by the UK, Germany and France, which together are responsible for 81.3% of European sales in these eight countries. Apart from the UK and Germany, market shares are comparatively low in most European countries. The countries with the highest online shares of their internal markets are: the UK (15.2%); Germany (11.6%); and France (8.0%). Other countries with high market shares are Sweden and The Netherlands. At present Germany has the fastest-growing online sector.

US online spending were $306.85 billion in 2014 and expected to reach $349.20 billion in 2015, an increase of 13.8%. The US share of retail in 2014 was 11.6%. The European market share was 7.2% in 2014 and 8.4% is projected for 2015. The US is still the leader in online retailing compared to Europe. With a similar population to the eight countries surveyed, 57.4% of the US public was e-shoppers compared to 46.7% in Europe. Every online shopper in Europe is expected to spend $1329.54 in 2015 compared to $1815.52 in the US. Canada's online sector is comparatively small, but is forecast to grow from US$20.82 billion in 2014, to reach $23.56 billion in 2015 and $26.99 billion in 2016.

Many retailers already report that up to one-half of website browsing occurs through customers using mobile devices, both smartphones and tablets. However, a much small proportion actually uses their mobile device to make the final purchase. In 2014, total ecommerce via mobiles in Europe was €23.77 billon, which is expected to grow by 88.7% to €44.87 billion in 2015. The UK figures are £8.41 billion in 2014 rising to £14.95 billion in 2015. The U.S. mobile share is predicted to grow from 18.7% of all online retail spending to 26.8%, or the equivalent of $57.38 billion to $93.58 billion.

The growth of online sales at such a rate will inevitably reduce the market for traditional shops. By the time that online sales represent 5% or more of domestic retailing then the continued growth of online retailers is expected to occur at the expense of conventional stores. In Europe as a whole, online retailers in 2015 are expanding 14.2 times faster than conventional outlets creating major strategic issues for store-based retailers.

In general, there exist three stages in online market development and business strategy: mature, mid-range, and immature. Mature markets are those where market share is 9.5% or above, 55%+ of the population are internet shoppers, rapidly developing mobile use (15%+ of all online in 2014), multiple online providers throughout each sector and 12+ purchases pa by each shopper. In mid-range markets, market shares range between 6.5%-9.5%, there is a wide range of suppliers, more than ten purchases pa per shopper, 45% are online shoppers and a smaller mobile use. In immature markets, online market share is below 6.5%, a patchy takeup (regionally or demographically) of online retailing, fewer than ten purchases pa, and some trade sectors are comparatively less developed. Mature markets, such as the US, the UK and Germany, are expected to grow more slowly, recruiting a percentage of non-users but mainly growing because existing eshoppers place more orders or buy more expensive items. However online growth in Germany is continuing at a very high rate, so maturity is a tendency rather than a scientific law. Mid range markets, such as France, The Netherlands and Sweden, will grow by recruiting more users as well as persuading shoppers to buy more frequently. Immature markets, such as Italy, Spain and Poland, have to overcome structural issues in the quality of their telecommunication networks, but can be expected to develop rapidly by increasing the number of eshoppers in their population and then inducing them to purchase more regularly (Retailersearch.org, 2015).

Online Shopping Trends in Developing World

Brazil

Brazil has one of the world's most connected populations. More than half of the Brazilian population uses the Internet, and those who do are connected nearly all the time. The country's online market size was about $13 billion in 2014. Online retailers still find Brazil to be a growing e-commerce market that is impossible to ignore.

In particular, Brazil is highly connected, with heavy use of mobile phones and broadband Internet, attributes that will drive future growth. Some structural challenges have also stymied Brazil's e-commerce growth, starting with burdensome regulations and taxes. The country's shipping network is not as developed as in other markets; it is well developed in the regions with the highest demand, but there is still progress to be made in the more remote areas. Additionally, Brazil's history of payment by installments can create working capital challenges for e-commerce retailers.

Government and business are trying to take steps to improve this. The government plans to invest $30 billion in infrastructure. eBay has begun allowing shoppers to fund their PayPal accounts using Boleto Bancario (a payment method in Brazil) ; shoppers are invoiced at checkout and can pay bills at banks, post offices, lottery agents, and some supermarkets. Meanwhile, Rio de Janeiro-based B2W, Brazil's largest online retailer, acquired Direct, a local logistics provider, part of a year's worth of investment in logistics and technology. Walmart, one of the most active international players in Brazilian e-commerce, is investing in distribution, part of a push to capture a larger share of Brazil's online retail market. In 2014 the company redesigned its Brazilian website; it also has plans to increase inventory and hire more employees. Social media is particularly popular in Brazil, whose population ranks third in the world in terms of total users of Facebook. Kanui.com is one e-commerce site that has been particularly adept at tapping into social media; its Facebook page (with more than 2.8 million followers) includes not only promotions, but also content marketing such as educational videos for using sports gear and apparel (atkearney.com, 2015).

Mexico

Mexico's e-commerce market grows rapidly thanks to a young, connected population that is increasingly willing to shop online. Almost half of Mexicans are connected to the Internet, and of these users, roughly two-thirds make purchases online. Total online sales was $6.6 billion in 2014. Research has shown that Mexican online shoppers today are looking primarily for bargains or doing information gathering. Services such as travel make up a large portion of online sales in Mexico, a sign of a market early in its development as consumers appear more comfortable spending money online for intangible items rather than physical goods. Online shoppers are still wary of delivery, as most Mexican retailers still use independent contractors for last-mile delivery and product return and refund policies are limited. These signs and others, however, indicate that this market is primed for growth. Only 17 percent of mobile phone users were shopping on their smartphones as of 2012, a percentage that is expected to increase in coming years. Higher-income consumers in Mexico's three largest cities accounted for more than half of all online sales in 2013, so as the rest of the country catches up there will be an opportunity for online retailers.

Mexico's proximity to the United States makes it a prime location for cross-border e-commerce. Roughly, half of the country's online shoppers make purchases on foreign websites; the e-commerce leaders in Mexico are international giants such as U.S.-based Amazon and Walmart, Argentine leader MercadoLibre, German-owned Linio, and Brazil-based Dafiti and Netshoes. Walmart, under a variety of banners, has led the way for international retailers, thanks to an assortment that is much larger than that of Amazon or any other local retailer and a localized channel that offers same-day delivery. Among Mexican brands, Soriana offers same-day delivery, which appeals to online shoppers specifically interested in purchasing fresh food from their devices. Mexican department store chains Liverpool and Coppel offer free shipping for price-conscious Mexican consumers; other websites typically base delivery rates on product weight and distance (atkearney.com, 2015).

Internet Marketing in B2B

Internet marketing in B2B and B2C is entirely different with major differences with respect to the nature of demand and supply and the trading process [see (Coupey, 2001) for details]. Next sections discuss the differences between B2B and B2C consumers' buying behavior and the marketing and advertising methods used in B2B.

Characteristics of B2B and B2C Online Buyers

B2B buyers are usually in a group and the factors that affect individual consumer behavior and organizational buying behavior are quite different. B2B online customers also consider situational factors besides product attributes. Further, individual B2B buyers' demographic characteristics, cultural backgrounds, attitudes toward technology, and economic factors influence the decision-making process (Gattiker, Perlusz, & Bohmann, 2000; Zhang & Wang, 2015).

B2B purchases are different from B2C purchases in many respects. Businesses buy large quantities of direct materials. They also buy indirect materials (such as office supplies) to support their production and operations processes. B2B buyers are more concerned about obtaining specific information, such as delivery conditions and pricing options. The B2B terms of negotiations and purchasing processes are more complex. In many cases, B2B buyers pose higher requirements for quality and after-sale support because of company policy. Although the number of B2B buyers is much smaller than the B2C buyers, their transaction volumes are far larger.

Differences may also occur between online and off-line shopping for both B2B and B2C customers (Gattiker, Perlusz, & Bohmann, 2000). While price, brand name, word of mouth, warranty etc. may be more important in online shopping, in off-line shopping other product features may be more important.

Marketing and Advertising in B2B E-commerce

The marketing and advertising processes for B2B E-commerce have major differences as compared with B2C E-commerce. Traditional B2B marketers, for example, may use trade shows, advertisements in industry magazines, e-mail and paper catalogs, and sales force. In online world, businesses use a variety of methods to reach business customers because traditional marketing methods may not be effective, feasible, or economical. Some of the popular online marketing techniques include online directory services, matching services to find business partners, marketing and advertising services of exchanges, co-branding or alliances, affiliate programs, online marketing services (e.g., see digitalcement.com), and e-communities (see b2bcommunities.com).

B2B Online Marketing Methods

In B2C E-commerce, a seller may advertise in traditional media (such as magazines or television shows) to targeted to its intended audience. It is also true for B2B sellers. They can use trade magazines and directories targeted to their intended audience. B2B vendors can also contact all of its targeted customers individually when they are part of a well-defined group. For example, to attract companies to an exchange for auto supplies, one might use information from industry trade- association records or industry magazines to identify potential customers. Another technique used to bring new customers is through an affiliate program, which operates just as a B2C affiliate programs operates. A company pays a small commission every time the affiliate company brings traffic to its site.

Many of the advertising methods used for B2C E-commerce re applicable to B2B E-commerce. For example, a B2B seller can use an advertising network provider, such as DoubleClick (doubleclick.com), to target its customers.

MARKETING SERVICES

There exist many B2B marketing services. Digital Cement (digitalcement.com) is a service that provides corporate marketing portals to help companies market their products to business customers. National Systems (nationalsystems.com) is a company that provides tailored marketing and advertising services based on its competitive intelligence on pricing, product mix, promotions, and content. Business Town (businesstown.com) is a company that provides information and services to small businesses, including start-ups. DHCommunications (dhcommunications.com) is a company that offers tools that help increase traffic to a B2B company's Web site.

AFFILIATE PROGRAMS

Affiliate programs work almost the same way in both B2B and B2C E-commerce. However, with B2B, there are additional types of affiliate programs available. For example, Schaeffer Research (schaeffersresearch.com) is a leading provider of research, education, recommendations, and strategies for stock and options traders. Schaeffer Research offers financial institutions a content alliance program in which content is exchanged so that all can obtain some free content.

INFOMEDIARIES AND DATA MINING

While data about consumers, suppliers, and other businesses is valuable for companies, data about companies and their products is useful to consumers when considering a purchase. The term infomediary (a composite of information and intermediary) refers to web site that helps both consumers and sellers understand a given market. It does so analyzing the data and converting it to information useful for consumers and sellers. Infomediaries can be of two types: those intended for consumers and those intended for businesses. Generally, we find infomediaries of the later type. Infomediaries analyze the clickstream data and mine it to produce useful knowledge that B2B sellers can use to increase sales and reduce marketing expenses. Infomediaries provide this service because many consumers and businesses do not have the specialized knowledge and systems to perform such data analysis on their own. SAS Institute (sas.com) and WebTrends (www.webtrends.com) are examples of Infomediaries and data mining specialists.

B2B internet Marketing Strategies

Educate Your Prospects

Help your prospect feel like an expert to make this more likely to buy than if they feel confused (Doligalski, 2015). Never talking down to clients or using a hard sell. Business professionals want to feel smart, competent and in charge of the transaction. They like to collect all the data to make informed decisions, and enjoy complicated products (like PCs) that demand a learning curve and are difficult to master. In addition, if the prospect feels like they are being pushed, they are likely to walk away, because they view a purchase as a rational decision based on information. One of the most precious commodities for the business professional is their time. So, embrace timesaving technologies like email or web meetings.

Create a New Framework

Focus on creating a favorable framework through which people can evaluate your product or service. The traditional approach to selling a product or service is to focus on the particular features and benefits of the solution, and why it is better/faster/cheaper than the competition. Instead, create a new framework for business professionals to apply their specific industry experience, customer feedback and financial analysis to your solution. Rather than focus on price, your conceptual framework allows prospective clients better understand the focuses that shape their business and how your solution is an integral part of their success.

Create Opportunities to Communicate

When you first meet a potential customer, you typically ask for a business card and try to learn more about their company. Implied in the exchange of business cards is the opportunity to follow-up with more information. Create reasons to communicate high-value ideas and information to your business prospects, and balance persistence against overload. Create industry-specific email distribution lists and forward relevant articles/ upcoming events to both your clients and sales prospects. Forward the latest whitepaper or research study that may help them do their job more effectively. Reinforce through this communication that part of the 'value-add' of purchasing from your company is the added attention you will bring to the relationship.

Speak Their Language

When it comes to communication to prospects that have different functional positions, a one-size-fits-all approach will not work. The CFO of a company will often focus on price, capital expenditures and measurable return on investment (ROI).To the CTO, however, the pressing issue may be how the product may abscond scarce IT resources, or how well the product integrates with the company's existing technology. And the chief marketing Officer may have altogether different issues-concerns with getting timely reporting from the IT staff on the effectiveness of the last marketing campaign may be top of mind. Begin by creating a laundry list of bullet point that examines your product or service from the viewpoint of each of your constituents. Then personalize your email communications, segmented by job function. Make sure toy also give the marketer the financial and technical ammunition necessary to convince their respective counterparts.

Email training series

One of the best ways to educate your prospects and stay top-of mind is with an email-based training series. Allow the experts within prospects sign up, they receive a series of email messages at a regular interval. Inside the email is a mini-summary of the individual topic and a short case study that examines the practical application of your solution. A quiz after each lesson can provide immediate feedback on what they have learned.

Product Discussion list

An effective way to get feedback on your product or service is to engage your customers in the product development process by using an email discussion list. Allow your engineer, designer and customer service representatives to interact with your customers in discussion form to shape the future of your offering. A moderated list allows control over what information is posted and who is able to send to the list.

Additional information by email

Give your prospects way to receive additional information from your company. An email-based approach can work better because many professionals use their email inbox (and subfolders) to organize their content. Email is a way to reach out to prospect's inbox and deliver the information directly to them. In addition, you have just created an additional reason to contact them with a personalized message.

Building Trust in E-Commerce

Trust is defined as the psychological status of involved parties who are willing to pursue further interactions to achieve a planned goal (Turban, Lee, & Chung, 2006). When people trust each other, they have confidence that involved parties will keep their promises. E-Commerce researchers have developed several models to explain the E-Commerce–trust relationship [e.g. (Lee & Turban, 2001); (Birkhofer, Schoumlgel, & Tomczak, 2000); (McKnight & Chervany, 2001)]. Trust is particularly important in global E-Commerce because of two reasons: the difficulty in taking legal action in cases of a dispute and the potential for conflicts caused by differences in culture and business environments. The speed and extent of both B2B and B2C E-commerce largely depends on the level of trust and confidence that consumers have in online shopping.

However, there are challenges to be addressed to build and maintain consumer confidence. In 2007, half of the cross-border complaints and disputes filed with European Consumer Centre Network were related to Internet (Commission, 2009). Leading reasons for complaints were delivery problems and dissatisfaction with the products purchased. Data gathered by econsumer.gov (an initiative of the International Consumer Protection and Enforcement Network that can be used by consumers to file complaints involving cross-border transactions including E-commerce) reveals similar pattern. Figure 7-16 shows top products or services for E-consumer complaints. Note that the percentages are based on the data gathered between January 1st to December 31, 2008.

Figure 7-15: Top Law Violations for E-commerce Complaints)

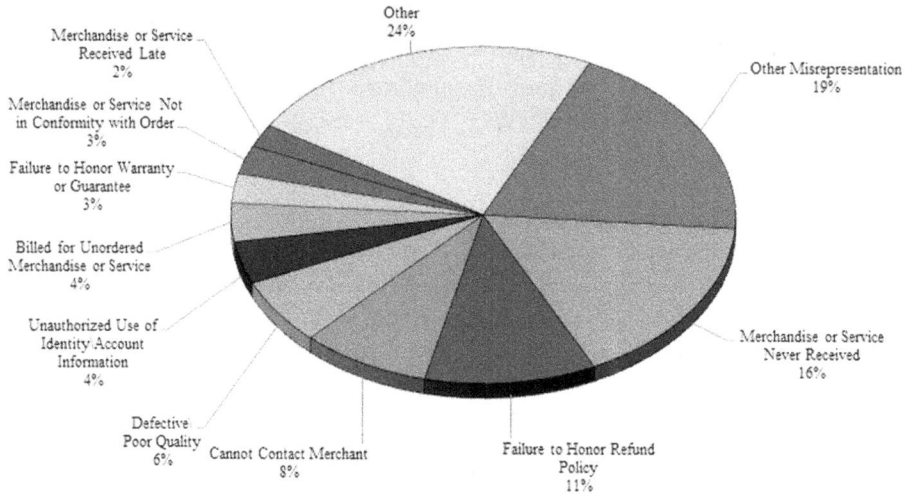

[1]Percentages are based on the 29,621 econsumer law violations reported from January 1 to December 31, 2013. One complaint may have multiple law violations.

Source: (FTC, Cross-Border Fraud Complaints, January – December 2013)

Figure 7-16: Top Products or Services for E-commerce Complaints

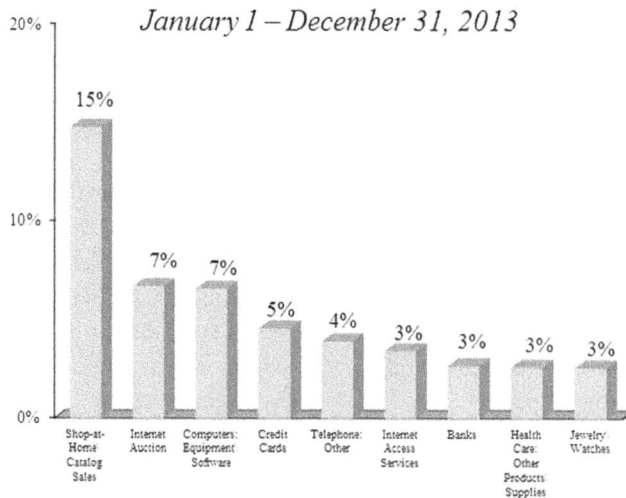

[1]Percentages are based on the 23,437 econsumer complaints received from January 1 to December 31, 2013.

452

Source: (FTC, Cross-Border Fraud Complaints, January – December 2013)

Establishing the necessary level of trust for E-Commerce depends on the type of trust. For example, (Shapiro, Sheppard, & Cheraskin, 1992) shows that desired level of trust between sellers and buyers is determined by the following factors: the extent of initial success experienced by each party with E-commerce and each other, well-defined roles and procedures for all parties involved, and realistic expectations about outcomes from E-Commerce. Conversely, trust decreases with the following: user uncertainty about the E-commerce technology, lack of initial face-to-face interactions, and lack of enthusiasm among the buyers and sellers.

Many third parties operate services aimed to increase trust. Some important companies are BBBOnLine (bbbonline.org) and TRUSTe (truste.com) and. Some other useful resources include escrow providers and reputation finders (see cyberalert.com and cymfony.com). These sites provide intelligence and research about spying on businesses.

Factors That Influence Trust Building in E-commerce

Many factors can influence trust building in E-commerce. Following are some common factors.

Brand Recognition: Brand recognition plays a very important role in building E-commerce trust. But that means customers need to be assured that they are dealing with the actual brand manufacturer. E-Commerce security mechanisms can help solidify trust by ensuring the authenticity of the trading parties. To build transaction integrity into the system, online sellers must do two things: disclose and update their latest business status and practices to potential customers and guarantee information and privacy protection.

Security: Sellers and buyers must trust the E-Commerce computing environment and the E-Commerce infrastructure. If people do not trust the security of the E-Commerce infrastructure, they will not feel comfortable about using credit cards to make E-Commerce purchases.

Uniqueness: Inclusion of key differentiators, culture, and values in an E-commerce site design can help enhance the overall customer experience and make the site unique. This uniqueness can help the company gain the competitive advantage and increased customer trust.

Social Media: Using social media (e.g. social networking sites), online businesses can give themselves a unique personality and identity. This identity can be very helpful in creating trust through relationship building.

Visibility: An online business should strive to gain as mush positive exposure as possible. Positive exposure can be gained for example by running blogs, by listing in search engines, press releases, by organizing contests, By sponsoring events, etc. The more positive exposure your business has more the customers trust you.

Industry Self-Regulation: While individual businesses can make efforts to increase customer trust on E-commerce, industry level initiatives are also required to do the same. Private sector took many initiatives in this regard. In 2000, United States Better Business Bureau developed a Code of Online Business Practices for E-commerce firms and established an online reliability seal program that can be used by consumers to check the credibility of the business they are dealing with (FTC, U.S. Implementation of the OECD E-Commerce

Guidelines, Mozelle Thomson, 2000). To increase consumer trust in behavioral advertising, leading US marketing and advertising industry associations issued their self-regulatory program for online behavioral advertising (InteractiveAdvertisingBureau, Self-Regulatory Principles for Online Behavioral Advertising, 2009). BeCommerce initiative is an initiative of Belgian direct marketing companies aimed to stimulate confidence in distance shopping. Secure Online Shopping Association and the Net Consumer Association in China established a website certification system that provides trust mark. Euro-Label is an initiative of European business community that provides electronic shopping trust mark. Global Trustmark Alliance is an international alliance aimed to promote safe e-commerce.

Public-Private Partnerships: Public-private initiatives have taken many forms. United Kingdom collaborated with online auction and price comparison sites to develop messages for consumers about their rights online. A Code of Practice for Consumer Protection in Electronic Commerce was developed in Canada. The code was developed by business, consumer groups and government to establish benchmarks for good online business practice (IndustryCanada, 2004). Microsoft, Yahoo, Western Union, and African Development Bank developed Advance Fee Fraud (AFF) Coalition in 2008 to raise global awareness among consumers of fraudulent activities online.

Consumer Education: Consumer education is important because it enables them to make informed decisions and increases their confidence in E-commerce. Many governments have been actively involved in educating consumers. UK government established two websites to educate consumers about on-line purchases: Consumer Direct (consumerdirect.gov.uk), which provides tips for safe shopping online, and GetSafeOnline (getsafeonline.org) which warns consumers about scams. United States FTC maintains a website OnGuardOnline.gov that provides education about a number of issues e.g. identity theft, children's privacy, online shopping etc. The eYouGuide, was released by the European Commission in 2009. This guide explained consumer rights in online shopping.

Ways to Build Customer Trust and Loyalty

The increasing number of online transactions has not diminished the value of relationship building with customers. Building trust is necessary as consumers become more wary of cyber theft and customer loyalty is essential for any business that plans to survive into the future. E-commerce business can use many tactics to build customer trust and loyalty online (Roesler, 2014).

Use Social Media to Connect with Fans: A 2014 study by Deloitte (Deloitte, 2014) found that 75% of online Americans say product information found on social channels influences their shopping behavior and enhances brand loyalty. A large number of people use social media to quick analyze a company and to see what other people have to say about it. For a business, it is essential to have an effective social media presence to build customer loyalty.

Offer Customer Service Assistance Online: A study in 2014 (Tierney, 2014) found that for US mobile shoppers, 75 percent would prefer to use live online chat versus calling to speak with an agent and 60 percent will abandon their online shopping carts and never make purchases from an online retailer again if they experience poor customer service. The internet is often the first place people turn for customer service assistance. A business can reduce the number of customer calls they get by having a Frequently Asked Questions section on the site. Similarly, a customer service chat on the website can reassure customers who are trying to buy something that they can get help if they need it.

Use Loyalty Programs: A study by Nielsen in 2013 (Nielsen, 2013) found that 76 % of North American consumers are more likely to choose retailers that offer loyalty programs. The percentages are higher in Asia (92%) and Latin America (82%); and slightly lower in Europe (72%). Three out of five US consumers say that there is loyalty programs offered where they shop. E-commerce businesses should consider implementing loyalty programs. There are many ways e.g. email coupons, loyalty points for doing certain actions, free shipping perks, contests, etc. E-commerce businesses need to ensure that loyalty programs are easy to use. Tierney (2014) found 48 % of US loyalty program members experienced frustration when attempting to redeem rewards. Frustrating rewards programs would be the opposite of building trust and fostering loyalty.

Encourage Customer Reviews and Display Them Prominently: Customer reviews are important for online consumers since they are not able to hold the product in their hand or test it before they buy it. Nielsen (2013) found that 70 % of global consumers trusted online reviews. According to Roesler (2013), consumers trust reviews as if they were coming from friends and family. Business owners can use this to their advantage and use the words of their satisfied customers to boost confidence in online shoppers. Encouraging reviews also lets current customers know that a business cares about their opinion. This builds loyalty, and often, these customers will become brand ambassadors to their friends, family and people they meet online.

Offer Multiple Payment Options: With what seems like a never-ending stream of breaches and hacks, consumers are increasingly wary of how and where they pay for things online. Business owners will encounter more customers who only use a certain payment method and the sites that have it are at an advantage over those that do not. One study (Capgemini, 2014) found that 73 percent of online consumers want several payment options when making purchases.

Professional Web Design: Make sure your site looks professional, with high-definition graphics and user-friendly navigation so that visitors stick around to explore the site further. While first impressions are important in any type of business, they are even more important for online stores. It is crucial to take the time to create a compelling and aesthetically pleasing design. Make sure there are no broken links on your site. This conveys a sense of unprofessionalism and failure to pay attention to detail. It can also lead people to wonder whether you are even still in business.

Brand Humanization: Once prospective customers have a certain level of interest in your products or services, many will want to know more about your business and company culture. Essentially, they want to know who they're potentially doing business with. One of the most effective ways to answer this question is by creating an inviting "About Us" page. Good "About Us" pages include a brief history of your business, what types of products or services you offer, and what your unique philosophy is. Add any awards you have received, major accomplishments, and organizations of which you are a part. Adding some images of team members along with brief biographies can further develop trust by humanizing your brand. The tone is also important. Wherever possible, allow prospects to see the people that power your brand.

Contact Information: Due to the number of scam artists and business schemes out there, you need to prove that you run a legitimate business. This can be done easily by including a "Contact" page with a telephone number, email and mailing addresses, and other information. Many online businesses offer only a generic email address. You should go beyond and include employees' email addresses, Twitter account user names, or LinkedIn profile URLs. This helps establish extra credibility. It also streamlines the networking process and assists in acquiring new customer and business relationships.

Privacy Policy: In a time where identity theft and the leaking of sensitive information have become all too common, it is smart to prove that customers' private data will remain secure when they do business with you. This can be done by creating a privacy policy and inserting it alongside the About Us and Contact page underneath your website's header or on the sidebar. Privacy policies provide a brief statement saying that you will not share any private information like customer name, address, financial records, credit information, etc.

Certified Seal: Getting a certified seal from Trust Guard or the Better Business Bureau is another way to quickly increase your credibility. Both companies have the same purpose, but slightly different features. Although they require a financial investment and that you meet certain guidelines for approval, increased conversions will often cover the cost of these efforts and expenses.

Secure Checkout: Once a customer is ready to make a purchase, there is one final detail that can influence their decision to follow through with it: perceived security of the transaction. Ensuring that the third-party payment vendor you choose is reliable and remains current on the latest security technology is extremely important. PayPal and Stripe are recognized brands that most customers will feel comfortable with. Use SSL encryption to prevent unauthorized parties from accessing financial information. Adding an icon or brief text at the checkout area stating that all transactions are secure can be helpful, too.

Frequently Updated Blog with Great Content: A blog that is updated often is a sign that your website is well managed and cared for; it's an indication of how a prospective buyer or client will be treated by your company. Make sure your blog content is authoritative, trustworthy, and written by an expert in your field. Not only does this increase conversion rates, but it also will net you higher search engine rankings.

Building customer trust and loyalty is vital for online businesses that want to succeed. Customers are just a click away from E-commerce business. Consumers will not stay with a business that does not make them feel valued or where they have ongoing concerns about privacy and security. Use of tactics described above can help an E-commerce business make a lasting rapport with customers that can last a lifetime.

Electronic Word of Mouth (eMOW)

Over the years, the Internet has become a participative platform in which users can contribute to developing, rating, commenting on, and distributing digital content and customizing Internet applications (OECD, Participative Web and User-Created Content, 2007). Electronic Word-of-Mouth (eWOM) is a phenomenon that occurs when consumers use World Wide Web to talk to other people when looking for opinions on a particular brand, product or company. The communication can take place through a variety of methods e.g. through blogs and social networks). Businesses need to understand many things in order to develop successful electronic communities, do effective web advertising, or increase web site usability. In this regard, critical is the understanding of consumption behavior of web users and their motivation for participation in Internet-related activities. Electronic word-of-mouth is bringing fundamental changes the way Web users get information to decide on their consumption behaviors. Electronic word-of-mouth reviews play an important role in consumption behaviors of online users. Persuasive impact of these reviews depends on both their quality and their quantity. Many factors influence web users' determination of perceived credibility of Electronic word-of-mouth review and subsequently adoption of these reviews. These factors include strength of arguments in review, credibility of the source providing review, confirmation with web users' prior beliefs, and consistency of recommendation in the reviews (Cheung, Luo, Sia, & Chen, 2009). (Park, Lee, & Han, 2007) found that quality of on-line reviews had positive effect on consumers' purchasing intention and purchasing intention increased with increase in number of reviews.

Electronic word-of-mouth reviews are becoming increasingly important in making purchase decisions and for product sales. According to (Bizreport.com, 2007), 62% of US Internet users read product reviews written by other consumers. More than 8 out of 10 respondents who read these said that the reviews had affected their buying intentions (they either became determined to buy the products or bought different products altogether).

E-Commerce Market Research

Market research is an important tool for any organization seeking a commercial Opportunity, particularly for small firms looking to grow their business. From discovering a gap in the market to ensuring customer satisfaction and planning effective marketing campaigns, research can provide the market intelligence needed to encourage success, enhance competitiveness and maximize profits. Online research has revolutionized research, providing both opportunities and challenges to researchers and users of research. On the positive side, the growth of online technology has enabled researchers to offer clients a fast and cost effective method for reaching their target audiences. Online research as a proportion of the whole market is growing rapidly, with the establishment of specialist online agencies and new methodologies coming into the fore ever year that passes. Internet research offers many opportunities, most of which are currently unexplored. This is an exciting time for companies to consider the possibility of online surveys and a definite growth area for the future. However, it is important that its significant potential harnessed correctly. Without harnessing this potential, online research may not be used to achieve desired benefits.

E-Commerce market research attempts to answer some typical questions e.g. what are the purchase patterns for individuals and groups (segmentation)? What factors encourage online purchasing? How can we differentiate between real buyers and browsers? How does an individual navigate? What is the optimal Web page design? Answers to these questions help E-commerce sellers to advertise properly, to price items, to design the Web site, and to provide appropriate customer service. E-Commerce market research can be conducted by conventional methods, such as telephone or shopping mall surveys, or it can be done with the assistance of the Internet. Interest in Internet research methods is on the rise. Online market research can provide such data about individuals, about groups, and even about the entire Internet. For a review of online market research techniques, technologies, and issues see (Miller & Dickson, 2001).

An online marketing research study can be conducted by the company itself, or the company may choose to hire an outside company to manage this. How and by what method a company decides to conduct a market research plan is dependent upon how objective or anonymous the company wishes to remain. An online marketing research program can include any number of methodologies. Some companies may use a direct method, by which a survey is sent out to existing customers to determine satisfaction levels and gather feedback. A company may also contact consumers and gather information from a specific demographic, such as an age group, or a specific geographic region. Companies may also use a "blind" method, by which a random sample of the population is contacted for opinions online.

Advantages of Online Market Research

Market research on the web is generally faster, more efficient and allows access to a more diverse and extended audience base. Also market researchers can conduct a very large study much more cheaply than with other methods. The larger sample size used in online research can provide increased accuracy and the predictive capabilities of the results. Online surveys cost a lot less than similar-sized offline surveys. Online market research is often interactive allowing personal contact with customers, and businesses can gain better and increased understanding of the customer, the market, and the competition.

Disadvantages of Online Market Research

Online market research requires many issues to address. Accuracy of the information provided by customers, and appropriate design of web questionnaires are two important issues. Companies also need to provide incentives for true completion of web surveys, which is critical for the validity of the search results. Customers may also have concerns about the security of the information provided and it can impact the truthfulness of the responses provided. Analysis of questionnaires can take long time and become costly if the data is large. Some respondents may not be able to respond to the online questionnaires because of lack of suitable computer or Internet connection. Ethics and legality of Web tracking is another issue. Another important issue is to have truly representative samples, because the researcher does not who is participating in the research. Some large companies establish their own market research department while most companies use third-parry research companies such as AC Nielsen.

Methods of Online Market Research

Online market research can be done using a variety of methods e.g. E-mail questionnaires (one-to-one communication with specific customers), moderated focus groups conducted in chat rooms, and questionnaires placed on Web sites. Professional marketing research companies frequently conduct online voting polls (e.g. see cnn.com, nielsen.com). Companies offer incentives such as games, prizes, or free software to draw customers.

Online Surveys: These surveys can either be passive (that only allow users to fill in a questionnaire) or interactive (one in which respondents can also add their comments to questionnaire, ask questions, and discuss issues).

Online Focus Groups: Research firms can create panels of consumers recruited online and verified by telephone. These panels can be used to conduct moderated online focus groups e.g. See NPD (www.npd.com) and Toluna (us.toluna.com). Companies maintain database of consumers and then calls them periodically to verify that they are who they say they are.

Feedback from Customers: Companies can also seek direct feedback from their customers. Companies can ask customers directly what they think about a product or service. Software tools provided by companies such kampyle.com can be used for this purpose.

Customer Scenarios: According to (Seybold, 2001), customer scenarios are situations that describe the customer's needs and the manner in which the product fulfills the needs. For example, one customer may buy a car because he needs to replace his old card whereas another customer may buy a similar car to give someone as a gift. The information gathered is used to design products and advertising. For example a company can use customer scenarios to offer Web tools to help engineers design electronic devices.

Tracking Customer Movements: Companies can also track customers' web movements using web server transaction logs or cookies. A transaction log is a record of users' activities at a website. Internet Profile Corporation (IPC) (ipro.com) is a company which collects data from a company's client/server logs and provides the company with periodic reports that include demographic data such as where customers come from. A cookie is a small file that can be placed on the hard drive of the user computer when they visit a website. When the customer returns to the site, the cookies can be used to find what the customer did in the past

Spyware: Spyware is software that stays on the user's hard drive and continually tracks the user's actions It periodically send gathered information to its instigator concerning the user's activities. It is typically used to gather user information, through information for advertising purposes. Most of the time users cannot control what data is sent via the spyware, and unless using special tools, cannot uninstall the spyware.

Conducting Market Research for Your Online Business

Conducting market research for your small business idea is essential (Ferenzi, 2015). A thorough research will help an entrepreneur uncover whether or not there is a healthy demand around your business idea. When completing market research around your business idea there are several different methods you can use. It is a good idea to use a combination of methods to obtain an accurate picture of your market and make the best business decision possible.

Keyword Research: Keyword search is an easy, fast and free way to get a feel for your product or business demand. The Google Keyword Tool helps you visualize how many people are searching for your business or product idea during any given period. This tool can also provide you with various related keywords, which can help you generate ideas around which words to use in your product name, descriptions, blog posts and your website as a whole. Conducting keyword research and then optimizing your store for search engines will help drive quality traffic to your online store.

Trend Search: Google Trends is a tool you can use to do a search for a product or business idea and then view the search volume for that keyword throughout the past few years. You can also compare the volume of searches between a few terms, see geographic information and even view how particular events affect search popularity.

Social Media Search: Examining social media is a great way to start understanding the volume of conversations and mentions around your business idea. Social media also helps to uncover aspects of your target market you can use within your marketing tactics later on. By probing into all of the different social media channels available, you can identify how potential customers talk about your product or industry as a whole. This will help you learn their language so you can leverage it later via your product descriptions, blog posts, ads, social media and promotions. Using the language of your targeted customer will help to drive quality traffic to your site, build a loyal customer base and increase conversions. Topsy (topsy.com) is a Social Media Analytics tool for Twitter. It allows you to search by time & place, set alerts, and analyzes sentiment for every tweet ever made. Several social media analytics tools have been discussed in the beginning of this chapter.

Data Warehouse and Data Mining

In simple language, a warehouse is a place where something is stored. A Data Warehouse is a central location where consolidated data from multiple locations are stored. Data Warehouse is not loaded every time when a new data is generated but the end-user can assess it whenever he needs some information. There are certain timelines determined by the business as to when Data Warehouse needs to be loaded whether on a daily, monthly or once in a quarter basis.

In a data warehouse, the data is stored and procured from the transaction system. In this case, there are two concepts-OLTP and OLAP. The former deals with recording transactions, while the latter analyses the data and this is where the data warehouse is utilized. Every transaction made through an ATM is recorded in an OLTP system, and so are various other activities. Implementing data warehouse could help a company avoid various challenges.

In an era of intense competition, it is not sufficient to just take decisions alone. It must be taken on time. With a data warehouse, a business can analyze what products are sold, what is not selling, when the sale goes up, what is the age group of customers who are buying a particular product and several other queries.

Generally, data mining (sometimes called data or knowledge discovery) is the process of analyzing data from different perspectives and summarizing it into useful information - information that can be used to increase revenue, cuts costs, or both. Data mining is the analysis of data for relationships that have not previously been discovered. For example, the sales records for a particular brand of tennis racket might, if sufficiently analyzed and related to other market data, reveal a seasonal demand. Data mining parameters include Association (looking for patterns where one event is connected to another event), Sequence or path analysis (looking for patterns where one event leads to another later event), Classification (looking for new patterns), Clustering (finding and visually documenting groups of facts not previously known), and Forecasting (discovering patterns in data that can lead to reasonable predictions about the future also called predictive analytics) (Rouse, 2008). Web mining, a type of data mining used in customer relationship management (CRM), takes advantage of the huge amount of information gathered by a Web site to look for patterns in user behavior.

Applications of Data Warehousing and Data Mining in E-commerce

Data mining is important in creating a great user experience at an E-commerce site. Data mining is a systematic way of extracting information from data. Techniques include pattern mining, trend discovery, and prediction. For E-commerce sites, data mining plays an important role in many areas of their business (Hu, 2010).

Product Search: When the users search for a product, they typically query a few keywords that can match many products. For example, "Verizon Cell phones" is a popular query. However, a query like this can match thousands of listed items. E-commerce sites can use user click-through rates or product sell-through rate to rank products. Both indicate a facet of the popularity of a product page. In addition, user behavioral data can provide the link from a query, to a product page view, and all the way to the purchase event. Through large-scale data analysis of query logs, an E-commerce site can create graphs between queries and products, and between different products. E-commerce sites can also mine data to understand user query intent. When a user searches for "Mercedes Benz", are they searching for a new car, or just repair parts of the car? Query intent detection comes from understanding the user, other users' searches, and the semantics of query terms.

Product Recommendation: Recommending similar products is an important aspect of E-commerce businesses. A good product recommendation can save hours of search time and delight our users. Typical recommendation systems are built upon the principle of "collaborative filtering", where the aggregated choices of similar, past users can be used to provide insights for the current user. Discovering item similarity requires understanding product attributes, price ranges, user purchase patterns, and product categories. Given the hundreds of millions of items sold on large E-commerce sites (such as eBay), and the diversity of merchandise, this is a challenging computational task. Data mining provides possible tools to tackle this problem, and we are always actively improving our approach to the problem.

Fraud Detection: A problem faced by all e-commerce companies is misuse of systems and, in some cases, fraud. On a C2C sites (such as eBay), sellers may deliberately list a product in the wrong category to attract user attention, or the item sold is not as the seller described it. On a typical storefront site (such as Amazon), retailer face problems with users using stolen credit cards to make purchases or register new user accounts. Fraud detection involves constant monitoring of online activities, and automatic triggering of internal alarms. Data mining uses statistical analysis and machine learning for the technique of "anomaly detection" that is, detecting abnormal patterns in a data sequence. Detecting seller fraud requires mining data on seller profile, item category, listing price and auction activities. By combining all of this data, we can have a complete picture and fast detection in real time.

Business Intelligence: Every company needs to understand its business operation, inventory and sales pattern. Large E-commerce businesses have large and diverse inventory. These stores (such as eBay) have no complete product catalog that can cover all items sold on their website. For such businesses, determining the number of a particular product sold is challenging. Such item (e.g. sunglasses) can be listed under different categories, with different titles and descriptions, or even offered as part of a bundle with other items. Using inventory intelligence, E-commerce sites can use data mining to process items and map them to the correct product category. This involves text mining, natural language understanding, and machine learning techniques. Successful inventory classification also helps E-commerce site to provide a better search experience and gives a user the most relevant product.

There is a growing need for data mining and its huge potential for e-commerce sites. The success of an e-commerce company is determined by the experience it offers its users, which these days is linked to data understanding. Data mining can provide great potential benefits to E-commerce sites in this regard.

Partner Relationship Management in E-commerce

The channel is most critical component of marketing and distribution strategies. Channel partner relationships are complex and partner's potential impact on revenue can be high. In many cases, companies lack adequate knowledge about any given channel partner and its importance in the overall marketing strategy. Internet and web technologies can help companies enhance their understanding of partners and build effective and strategic relationships with them.

Partner relationship management can be thought of as a method for delivering the right information to the right individual at the right time for the purpose of growing revenue and market share. Partner relationship management involves collecting detailed information about the partners through their contacts across all the touch points in the company. These touch points can include human contacts as well as self-service applications such as interactive voice response (IVR) systems, Web-based applications and a variety of enterprise and legacy systems.

This detailed information is then used to create unique and detailed partner profiles. These profiles can then be used for multiple purposes. For example, they can be used to segment, micro-segment or create a specific individual segment that is unique to one partner. They can also be used to develop partner strategies specific to each channel partner.

Subsequent information gathered from partners can be stored in a central partner profile database. Such database can provide consistent full-spectrum view of each partner, leveraged knowledge for decision-making, complete history of past and current contacts, segmentation and targeted strategies, complete customized communication, and measurement of channel strategy initiatives.

Partner relationship management (PRM) is different from traditional Customer relationship management (CRM). While CRM has a distinctly consumer focus PRM has a partner focus. The volume of contacts in PRM is low and contacts are targeted and relevant. PRM is both communication and information intensive and information is collected from diverse touch points.

oftware solutions are available to manage PRM. For example, salesforce.com provides Sales Cloud application that also has the capability to integrate CRM with PRM.

Review Questions

1. Differentiate between Internet marketing and E-marketing.

2. List some common Internet marketing mistakes.

3. List some advantages and disadvantages of Internet marketing.

4. How online branding is different from conventional branding?

5. Differentiate between emotional and rational branding.

6. Differentiate between online brand consolidation and leveraging.

7. What is the purpose of affiliate marketing?

8. List some Internet tools used in Internet marketing and their use.

9. List some uses of Internet, Intranet, and Extranet in Internet marketing.

10. List some common Internet marketing communication tools.

11. List some issues and problems of online advertising.

12. What are some benefits and limitation of online banner ads?

13. List some methods of banner ads placement.

14. Differentiate among Cost-per-click (CPC), Cost-per-Thousand (CPM), and Cost-per-Action (CPA).

15. List some tips for effective banner ad design.

16. Differentiate between webcasting and web conferencing.

17. List some tools of online public relations.

18. Differentiate between cause marketing and viral marketing.

19. Define opt-in or permission marketing.

20. What is the importance of corporate website in Internet marketing?

21. List some types of corporate websites.

22. List some properties of a good domain name.

23. List some web positioning strategies.

24. List focus areas of search engine optimization.

25. List search engine optimization techniques and their differences.

26. Differentiate between search engine optimization and search engine marketing.

27. List some common Web 2.0/3.0 tools and their use in E-commerce.

28. List some current Internet usage trends.

29. List some online consumer purchase decision aids.

30. How Internet marketing for B2B is different from Internet marketing for B2C?

31. List some factors that influence customer trust building in E-commerce.

32. List some advantages and disadvantages of online market research.

33. List some customer segmentation and analysis techniques used in E-commerce.

34. List some web and social analytics techniques used in E-commerce.

Bibliography

Adomavicius, G., & Tuzhilin, A. (2005). Toward the Next Generation of Recommender Systems: A Survey of the State-of-the-Art and Possible Extensions. *IEEE Transactions on Knowledge and Data Engineering*, *17* (6), 734-749.

Agarwal, D. (2015). WEB PERSONALIZATION AND RECOMMENDATION MODEL FOR TRUST IN ECOMMERCE WEBSITES FROM AN INDIAN PERSPECTIVE.

Ahuja, M., Gupta, B., & Raman, P. (2003). An empirical investigation of online consumer purchasing behavior. *Communications of the ACM, 46*, pp. 145-151.

Aldridge, A., Forcht, K., & Pierson, J. (1997). Get linked or get lost: Marketing Strategy for the Internet. *Internet Research: Electronic Networking Applications and Policy*, *7* (3), 161-169.

Allen, E., & Fjermestad, J. (2001). E-commerce marketing strategies: an integrated framework and case analysis. *Logistics Information Management*, *14* (1/2), 14-23.

Allen, E., & Fjermestad, J. (2001). E-commerce marketing strategies: an integrated framework and case analysis. Logistics Information Management, 14(1/2), 14–23.

American Association, o. A. (2009). *Group M study says Internet advertising spending will reach 15 % of total in 2010.* Retrieved August 12, 2010, from www2.aaaa.org: http://www2.aaaa.org/news/agency/Pages/092409_GroupMSTudy.aspx

Ammon, T. (2013, October 18). How to Calculate the Lifetime Value of Ecommerce Customers. Retrieved July 7, 2015, from http://blog.hubspot.com/ecommerce/how-to-calculate-the-lifetime-value-of-ecommerce-customers

Andam, Z. (2003, May). *e-Commerce and e-Business*. Retrieved October 15, 2009, from apdip.net: www.apdip.net/publications/iespprimers/eprimer-ecom.pdf

Anderson, E. T., & Simester, D. I. (2003). Effects of $9 price endings on retail sales: Evidence from field experiments. Quantitative Marketing and Economics, 1(1), 93–110.

Anne Maarit Jalkala, & Joona Keränen. (2014). Brand positioning strategies for industrial firms providing customer solutions. Journal of Business & Industrial Marketing, 29(3), 253–264. http://doi.org/10.1108/JBIM-10-2011-0138

Ascenetworks.com. (2008). *Mitigating eBusiness Growing Pains*. Retrieved June 12, 2009, from www.ascenetworks.com: http://www.ascenetworks.com/solutions/e-business.htm

Barden, A. (2009, May 13). 10 Proven B2B Marketing Communication Strategies to Drive Revenue Growth. Retrieved from https://andrewbarden.wordpress.com/2009/05/13/10-proven-b2b-marketing-communication-strategies-to-drive-revenue-growth/

Barr, B., Taylor-Robinson, D., Scott-Samuel, A., McKee, M., & Stuckler, D. (2012). Suicides associated with the 2008-10 economic recession in England: time trend analysis. Bmj, 345, e5142.

Basińska, B., Dąbrowski, D., & Sikorski, M. (2013). Usability and relational factors in user-perceived quality of online services.

Berthon, P., Pitt, L., & Watson, R. T. (1996). Marketing communication and the world wide web. *Business Horizons , 39* (5), 24-32.

Bhamra, G. S., Verma, A. K., & Patel, R. B. (2014). Intelligent Software Agent Technology: An Overview. International Journal of Computer Applications, 89(2).

Bhatt, G. D., & Emdad, A. F. (2001). An analysis of the virtual value chain in electronic commerce. *Logistics Information Management , 14* (1/2), 78-84.

Birkhofer, B., Schoumlgel, M., & Tomczak, T. (2000). Transaction - and Trust-Based Strategies in E-Commerce - a Conceptual Approach. *Electronic Markets , 10* (3), 169-175.

Bizreport.com. (2007). *Online Consumers Turn to User Reviews*. Retrieved June 24, 2010, from bizreport.com: www.bizreport.com/2007/10/online_consumers_turn_to_user_reviews.html

Blackman, A. (2014, October 13). The Psychology of Pricing - Tuts+ Business Tutorial. Retrieved July 28, 2015, from http://business.tutsplus.com/tutorials/the-psychology-of-pricing--cms-22206

Bloch, M., Pigneur, Y., & Segev, A. (1996). Leveraging Electronic Commerce for Competitive Advantage: A business value framework. *Proceedings 9th International Conference on EDI-IOS, "Electronic Commerce for Trade Efficiency and Effectiveness"*, (pp. 91-112).

Bobadilla, J., Ortega, F., Hernando, A., & Gutiérrez, A. (2013). Recommender systems survey. Knowledge-Based Systems, 46, 109–132.

Brandon, J. (2007, May 2). *The top eight corporate sites in Second Life*. Retrieved July 24, 2010, from www.computerworld.com: http://www.computerworld.com/s/article/9018238/The_top_eight_corporate_sites_in_Second_Life

Brandwatch. (2015, May 11). 24% of top B2B brands lack social media presence. Retrieved July 16, 2015, from http://www.brafton.com/news/social-media-news/24-top-b2b-brands-lack-social-media-presence

Bruceclay.com. (2009). *The Customer Life Cycle*. Retrieved July 24, 2009, from www.bruceclay.com: http://www.bruceclay.com/analytics/customerlifecycle.htm

Brugueras, J. (2009). *Internet Users Demographics*. Retrieved September 12, 2010, from ezinearticles.com: http://ezinearticles.com/?Internet-Users-Demographics&id=2189063

Bryson, S. (1996). Virtual reality in scientific visualization. *Communications of the ACM , 39* (5), 62-71.

Butilon, J. (2015, March 2). An Introduction to Analytics for Ecommerce Websites. Retrieved July 7, 2015, from https://blog.kissmetrics.com/intro-to-ecommerce-analytics/

Campbell, K. (2015, January 6). How to Develop Buyer Personas for B2C Ecommerce Marketing. Retrieved from https://www.prestashop.com/blog/en/how-to-develop-buyer-personas-for-b2c-ecommerce-marketing/

Carton, S. (2001). *The Dot.Bomb Survival Guide: Surviving (and Thriving) in the Dot.Com Implosion*. McGraw-Hill Companies.

Center For Digital Future. (2014). 2014 Digital Future Report. Retrieved August 14, 2015, from http://www.digitalcenter.org/wp-content/uploads/2014/12/2014-Digital-Future-Report.pdf

CentrefortheDigitalFuture. (2010). *World Internet Report 2010*. Retrieved September 15, 2010, from www.digitalcenter.org: http://www.digitalcenter.org/pages/site_content.asp?intGlobalId=42

Chaffey, D., Mayer, R., Johnston, K., & Ellis-Chadwick, F. (2002). *Internet Marketing: Strategy, Implementation and Practice*. Financial Times/ Prentice Hall.

Charlton, G. (2011, September 14). 44% prefer email for customer service: survey | Econsultancy. Retrieved July 25, 2015, from https://econsultancy.com/blog/7999-44-prefer-email-for-customer-service

Charlton, G. (2012, October 31). Why online retailers need to offer free returns this Christmas | Econsultancy. Retrieved July 25, 2015, from https://econsultancy.com/blog/10998-why-online-retailers-need-to-offer-free-returns-this-christmas

Charlton, G. (2015, July 3). 21 ways online retailers can improve customer retention rates. Retrieved July 7, 2015, from http://econsultancy.com/blog/11051-21-ways-online-retailers-can-improve-customer-retention-rates/?utm_campaign=bloglikes&utm_medium=socialnetwork&utm_source=facebook

Chawla, S. (2013, August 30). Compelling Copy + Clean Design + Fewer Distractions = 40.81% Increase in Sales. Retrieved from https://vwo.com/blog/split-test-increased-website-sales/

Chawla, S. (2014, August 21). 4 Simple eCommerce Product Page Tests That Take Less Than an Hour. Retrieved July 7, 2015, from https://blog.kissmetrics.com/simple-ecommerce-page-tests/

Cheung, M., Luo, C., Sia, C., & Chen, H. (2009). Credibility of Electronic Word-of-Mouth: Informational and Normative Determinants of On-line Consumer Recommendations. *International Journal of Electronic Commerce , 13* (4), 9-38.

ChinaTravelNews. (2015, February 25). 2014 online shopping and O2O trends for Chinese netizens - ChinaTravelNews. Retrieved July 30, 2015, from http://www.chinatravelnews.com/article/89418

Cho, Y., & Fjermestad, J. (2015). USING ELECTRONIC CUSTOMER RELATIONSHIP MANAGEMENT TO MAXIMIZE/MINIMIZE CUSTOMER SATISFACTION/DISSATISFACTION. Electronic Customer Relationship Management, 34.

Ciotti, G. (2014, February 6). 5 Ecommerce Pricing Experiments that Will Make You Want to Run an A/B Test Today – Shopify. Retrieved July 7, 2015, from http://www.shopify.com/blog/12109933-5-ecommerce-pricing-experiments-that-will-make-you-want-to-run-an-a-b-test-today

Clay, B., & Esparza, S. (2011). Search Engine Optimization All-in-One For Dummies. John Wiley & Sons.

Cnet.com. (2002, January 2). Now showing: random DVD prices on Amazon. Retrieved July 28, 2015, from http://www.cnet.com/news/now-showing-random-dvd-prices-on-amazon/

Commission, E. (2009). *The Consumer Markets Scoreboard, Second Edition.* Retrieved July 10, 2010, from ec.europa.eu: http://ec.europa.eu/consumers/strategy/docs/2nd_edition_scoreboard_en.pdf

Coupey, E. (2001). *Marketing and The Internet.* Prentice Hall.

crm.businessdecision.be. (2009). *A Customer based Marketing Approach.* Retrieved August 25, 2009, from crm.businessdecision.be: http://crm.businessdecision.be/1826-customer-based-approach.htm

Dan, A. (2008). Predictably irrational: the hidden forces that shape our decisions. New York. NY, Etats-Unis: HarperCollins Publishers.

DAYAL, S., LANDESBERG, H., & ZEISSER, M. (2000, May). Building digital brands. *McKinsey Quarterly* , pp. 42-51.

Dayrit, K. (2010). *How to Effectively Use Social Networking Sites to Generate Sales.* Retrieved August 12, 2010, from e-commerce-marketing.suite101.com: http://e-commerce-marketing.suite101.com/article.cfm/how-to-effectively-use-social-networking-sites-to-generate-sales

Deswal, S. (2012, June 7). 107% increase in sales shows that customers care more for authenticity than low prices. Retrieved from https://vwo.com/blog/increase-in-sales/

Deswal, S. (2014, February 17). Price Comparison Information on Ecommerce Store Increased Conversions by 10%. Retrieved from https://vwo.com/blog/ecommerce-price-comparison-increased-conversions/

Dholakiya, P. (2015, April). 4 Lessons from Google to Improve Your Site Search. Retrieved July 17, 2015, from http://www.advancedwebranking.com/blog/how-to-improve-your-site-search/

Doligalski, T. (2015). Internet-Based Customer Value Management. Springer.

Dutt, R., & Kashyap, A. (2014). A Study of Brand Preferences and Consumer Behavior Aspects in Durable Goods with special reference to Washing Machines in India. Journal of Marketing Vistas, 4(2), 46.

E-comMag. (2009). *Les entreprises du Net parient sur le contenu multimédia et le participatif.* Retrieved July 14, 2010, from ecommercemag.fr: www.ecommercemag.fr/xml/Breves/2009/02/24/28374/Les-entreprises-du-Net-parient-sur-le-contenu-multimedia-et-le-participatif/

EcommJournal. (2010). *73.3% of companies use social networks to hire employees.* Retrieved August 10, 2010, from ecommerce-journal.com: http://ecommerce-journal.com/news/28835_733-companies-use-social-networks-hire-employees

Econsultancy. (2014, August). Cross-Channel Marketing Report 2014 | Econsultancy. Retrieved July 25, 2015, from https://econsultancy.com/reports/cross-channel-marketing-report

Education B.V., S. (2013). Webvertising: The Ultimate Internet Advertising Guide (Softcover reprint of the original 1st ed. 2000 edition). Wiesbaden: Vieweg+Teubner Verlag.

Elizabeth, V. (2014, February 14). Ecommerce Pricing Strategy: How to Price Your Products Online. Retrieved July 7, 2015, from https://www.ometria.com/blog/ecommerce-pricing-strategy-how-to-price-your-products-online

Encyclopedia, E.-c. (2009). *Buyer Behavior of Online Consumers - Attracting and Retaining Online Buyers: Comparing B2B and B2C Customers, Introduction, Who Buys Online?* Retrieved July 15, 2009, from encyclopedia.jrank.org: http://encyclopedia.jrank.org/articles/pages/1331/Buyer-Behavior-of-Online-Consumers.html

Eschenbrenner, B., Nah, F., & Siau, K. (2008). 3-D virtual worlds in education: Applications, benefits, issues, and opportunities. *Journal of Database Management* , 91-110.

Europa. (2009, September 9). *Consumers: EU crackdown on websites selling consumer electronic goods*. Retrieved August 10, 2010, from europa.eu: http://europa.eu/rapid/pressReleasesAction.do?reference=IP/09/1292&format=HTML&aged=0&langu age=EN&guiLanguage=en

F. Nunes, P., Bellin, J., Lee, I., & Schunck, O. (2013). Converting the nonstop customer into a loyal customer. Strategy & Leadership, 41(5), 48–53.

Feng, J., Bhargava, H. K., & Pennock, D. (2003). Comparison of Allocation Rules for Paid Placement Advertising in Search Engines. *ACM International Conference Proceeding Series; Vol. 50, Proceedings of the 5th international conference on Electronic commerce*, (pp. 249-299). Pittsburgh, Pennsylvania.

FTC. (2000, February). *U.S. Implementation of the OECD E-Commerce Guidelines, Mozelle Thomson*. Retrieved April 19, 2009, from www.ftc.gov: www.ftc.gov/speeches/thompson/thomtacdremarks.shtm

FTC. (2009, April). *Cross-Border Fraud Complaints, January – December 2008, Consumer Sentinel Network*. Retrieved April 12, 2010, from www.ftc.gov: www.ftc.gov/sentinel/reports/annual-crossborder-reports/crossborder-cy2008.pdf

FTC. (2009, February). *Self-Regulatory Principles for Online Behavioural Advertising, Behavioral Advertising - Tracking, Targeting, & Technology, US FTC Staff Report*. Retrieved April 12, 2010, from www.ftc.gov: www.ftc.gov/os/2009/02/P085400behavadreport.pdf

Gangeshwer, D. K. (2013). E-Commerce or Internet Marketing: A Business Review from Indian Context. International Journal of U-and E-Service, Science and Technology, 6(6), 187–194.

Gartner.com. (2008). *Press Release*. Retrieved July 15, 2009, from www.gartner.com: http://www.gartner.com/it/page.jsp?id=681608

Gattiker, U., Perlusz, S., & Bohmann, K. (2000). Using the Internet for B2B Activities: A Review and Future Directions for Research. *Internet Research*, *10* (2), 126-140.

Gay, S., & Pho, K. (2013). Online reputation management: the first steps. The Journal of Medical Practice Management: MPM, 29(2), 81.

Gelb, B., & Sundaram, S. (2002). Adapting to word of mouse. *Business Horizons*, *45* (4), 21-25.

Héder, M. (2014). The machine's role in human's service automation and knowledge sharing. AI & Society, 29(2), 185–192.

Heinen, J. (1996). Internet marketing practices. *Information Management & Computer Security*, *4* (5), 7-14.

Helpscout.net. (2015). 25 Fun, Quirky and Memorable Customer Appreciation Ideas. Retrieved July 16, 2015, from http://www.helpscout.net/25-ways-to-thank-your-customers/

Hosford, C. (2009, May 8). Study: Referring links critical in search query rankings. Retrieved July 16, 2015, from http://adage.com/article/btob/study-referring-links-critical-search-query-rankings/275459/

Huang, Z., & Benyoucef, M. (2013). From e-commerce to social commerce: A close look at design features. Electronic Commerce Research and Applications, 12(4), 246–259.

Hutt, E., Lebrun, R., & Mannhardt, T. (2001). Simplifying Web segmentation. *The McKinsey Quarterly* (3).

IAB.net. (2015). IAB - About the IAB. Retrieved July 16, 2015, from http://www.iab.net/about_the_iab

IDC. (2009). *Press Release*. Retrieved August 15, 2010, from www.idc.com: http://www.idc.com/getdoc.jsp?containerId=prUS22110509

Imediaconnection.com. (2006). *Ecommerce and Social Networks*. Retrieved July 16, 2009, from www.imediaconnection.com: http://www.imediaconnection.com/content/7744.asp

IndustryCanada. (2004, January). *Canadian Code of Practice for Consumer Protection in Electronic Commerce*. Retrieved August 24, 2010, from cmcweb.ca: http://cmcweb.ca/eic/site/cmc-cmc.nsf/vwapj/EcommPrinciples2003_e.pdf/$FILE/EcommPrinciples2003_e.pdf

InteractiveAdvertisingBureau. (2009). *Economic Value of the Advertising-Supported Internet Ecosystem*. Retrieved June 10, 2010, from www.iab.net: www.iab.net/media/file/Economic-Value-Report.pdf

InteractiveAdvertisingBureau. (2009, July). *Self-Regulatory Principles for Online Behavioral Advertising*. Retrieved July 10, 2010, from www.iab.net: www.iab.net/media/file/ven-principles-07-01-09.pdf

InternetWorldStats. (2009, December 31). *Internet Usage in Asia*. Retrieved June 26, 2010, from internetworldstats.com: http://www.internetworldstats.com/stats3.htm

Jacob, S. (2014a, October 23). Using Interactive Content to Increase Conversions: 4 Examples from Top Companies (And How You Can Do It Too!). Retrieved July 7, 2015, from https://blog.kissmetrics.com/interactive-content/

Jacob, S. (2014b, December 5). 6 User Interface Fixes That Can Greatly Increase Conversion Rates. Retrieved July 7, 2015, from https://blog.kissmetrics.com/ui-ux-conversion-rates/

Jain, M. K. (2013). An Analysis of Marketing Mix: 7Ps or More. Asian Journal of Multidisciplinary Studies, 1(4).

Janal, S. (1997). *Online Marketing Handbook: How to Promote, Advertise, and Sell your Products and Services on the Internet*. NY: John Wiley & Co.

Jansen, B. J., & Resnick, M. (2006). An Examination of Searcher's Perceptions of Nonsponsored and Sponsored Links during Ecommerce Web Searching. *Journal of the American Society of Information Science and Technology*, 57 (14), 1949-1961.

Jensen, K. (2015, March 10). 10 Do's and Don'ts of B2B Branding. Retrieved July 16, 2015, from https://www.bopdesign.com/bop-blog/2015/03/dos-and-donts-of-b2b-branding/

Kickassmetrics.com. (2012, May 12). Customer Retention Statistics. Retrieved July 25, 2015, from https://blog.kissmetrics.com/retaining-customers/

Kim, J., Novemsky, N., & Dhar, R. (2013). Adding small differences can increase similarity and choice. Psychological Science, 24(2), 225–229.

Kim, L. (2010, September 15). 10 Facts and Trends about Contextual Advertising. Retrieved from http://www.searchenginejournal.com/10-facts-and-trends-about-contextual-advertising/24098/

Knutson, B., Rick, S., Wimmer, G. E., Prelec, D., & Loewenstein, G. (2007). Neural Predictors of Purchases. Neuron, 53(1), 147–156. http://doi.org/10.1016/j.neuron.2006.11.010

Kotler, P., & Armstrong, G. (2002). *Principles of Marketing*. New Jursey: Prentice Hall.

Kwon, K., & Kim, C. (2012). How to design personalization in a context of customer retention: Who personalizes what and to what extent? Electronic Commerce Research and Applications, 11(2), 101–116.

Laflin, M. (2001). *Sport and the Internet—The impact and the future*. Retrieved July 12, 2009, from New Technology in Sports Information and Sports Information Management. International Association for Sports Information: http://multimedia.olympic.org/pdf/en_report_60.pdf

Laja, P. (2014). Pricing Experiments You Might Not Know, But Can Learn From. Retrieved from http://conversionxl.com/pricing-experiments-you-might-not-know-but-can-learn-from/

LaPlante, A. (2009, March 1). Jennifer Aaker: The Happiness-Time Connection | Stanford Graduate School of Business. Retrieved July 28, 2015, from http://www.gsb.stanford.edu/insights/jennifer-aaker-happiness-time-connection

Lazazzera, R. (2014). How To Build Buyer Personas For Better Marketing – Shopify. Retrieved July 7, 2015, from http://www.shopify.com/blog/15275657-how-to-build-buyer-personas-for-better-marketing

Lee, M., & Turban, E. (2001). Trust in B2C Electronic Commerce: A Proposed Research Model and its Applications. *International Journal of Electronic Commerce* , 6 (1).

Lynch, S. (2010). *Online ads worth €100m in 2009*. Retrieved September 13, 2010, from www.irishtimes.com: http://www.irishtimes.com/newspaper/breaking/2010/0727/breaking34.html

Machkovech, S. (2014, November 28). Cards Against Humanity calls bull**** on Black Friday, sells cow feces. Retrieved July 16, 2015, from http://arstechnica.com/gaming/2014/11/cards-against-humanity-calls-bull-on-black-friday-sells-cow-feces/

Macinnes, I., Li, Y., & Yurcik, W. (2005). Reputation and Dispute in eBay Transactions. *International Journal of Electronic Commerce*, *10* (1), 27-54.

Mahmood, M. A., Kohli, R., & Devaraj. (2004). Introduction to the Special Issue: Measuring the Business Value of Information Technology in e-Business Environments. *International Journal of ELectronic Commerce*, *9* (1), 5-8.

Mallikarjunan, S. (2014, June 9). 10 Ecommerce Buyer Persona Questions Every Marketer Should Ask. Retrieved July 7, 2015, from http://blog.hubspot.com/ecommerce/ecommerce-buyer-persona-questions-to-ask

Marks, K. Z. (2009). *Marketing Specifics - Product Based Marketing.* Retrieved June 25, 2009, from ezinearticles.com: http://ezinearticles.com/?Marketing-Specifics---Product-Based-Marketing&id=1574180

Marz, N., & Warren, J. (2015). Big Data: Principles and best practices of scalable realtime data systems. Manning Publications Co.

Mason, W. M., & Fienberg, S. (2012). Cohort analysis in social research: Beyond the identification problem. Springer Science & Business Media.

McEachern, A. (2015, February 19). 5 Customer Retention Tools for Long Term Ecommerce Success. Retrieved July 7, 2015, from http://blog.hubspot.com/marketing/customer-retention-tools-long-term-ecommerce-success

McKinsey & Company. (2013, June 24). Why B-To-B Branding Matters More Than You Think. Retrieved July 16, 2015, from http://www.forbes.com/sites/mckinsey/2013/06/24/why-b-to-b-branding-matters-more-than-you-think/

McKnight, D. H., & Chervany, N. L. (2001). What Trust Means in E-Commerce Customer Relationships: An Interdisciplinary Conceptual Typology. *International Journal of Electronic Commerce*, *6* (2), 35-59.

Miller, T. W., & Dickson, P. R. (2001). On-line Market Research. *International Journal of Electronic Commerce*, 139-167.

Mitra, S. (2014, September 16). From E-Commerce to Web 3.0: Let the Bots Do the Shopping. Retrieved July 29, 2015, from http://www.wired.com/2014/09/e-commerce-to-web-3-0/

Mochari, I. (2013, December 13). As Brand Loyalty Declines, 2 Trends You Should Know for 2014 | Inc.com. Retrieved July 16, 2015, from http://www.inc.com/ilan-mochari/2-trends-consumer-products-2014.html

Möller, K. E. (2006). Comment on: The Marketing Mix Revisited: Towards the 21st Century Marketing? by E.Constantinides. *Journal of Marketing Management*, 439-450.

Muniz, A. M., & O'Guinn, T. (2001, March). Brand Community. *Journal of Consumer Research*, 412-32.

Naiduk, C. (2013, October 9). B2B vs B2C Communications - YourVelocity.com. Retrieved July 28, 2015, from https://www.yourvelocity.com/b2b-vs-b2c-communications/

Nusca, A. (2009, May 5). *Forrester: e-commerce better suited to withstand economic downturn.* Retrieved July 15, 2010, from zdnet.com: http://www.zdnet.com/blog/btl/forrester-e-commerce-better-suited-to-withstand-economic-downturn/17470

O'Connor, J., & Galvin, E. (1997). *Marketing and Information Technology.* London: Pitman Publishing.

OECD. (2006). *Report on the Cross-border Enforcement of Privacy Laws.* Retrieved July 14, 2009, from oecd.org: www.oecd.org/dataoecd/17/43/37558845.pdf

OECD. (2007). *Participative Web and User-Created Content.* Retrieved July 12, 2010, from http://213.253.134.43: http://213.253.134.43/oecd/pdfs/browseit/9307031E.PDF

OFCOM. (2009, October). *Social networking surge.* Retrieved September 12, 2010, from www.ofcom.org.uk: www.ofcom.org.uk/consumer/2009/10/social-networking-surge/

OFT. (2009, October). *OFT launches market studies into advertising and pricing practices, Press release.* Retrieved July 12, 2010, from www.oft.gov.uk: www.oft.gov.uk/news/press/2009/126-09

Ogborne, M. (2012, September 12). The True Cost of Adding Social Buttons. Retrieved July 25, 2015, from http://lastdropofink.co.uk/market-places/the-true-cost-of-adding-social-buttons/

O'Keefe, R. M., & McEachern, T. (1998). Web-based customer decision support systems. *Communications of the ACM, 41,* pp. 71-78.

Ovchinnikov, A., Boulu-Reshef, B., & Pfeifer, P. E. (2013). Revenue management with lifetime value considerations: balancing customer acquisition and retention spending for firms with limited capacity. Forthcoming in Management Science.

Özpolat, K., & Jank, W. (2015). Getting the most out of third party trust seals: An empirical analysis. Decision Support Systems, 73, 47–56.

Park, D.-H., Lee, J., & Han, I. (2007). The Effect of On-Line Consumer Reviews on Consumer Purchasing Intention: The Moderating Role of Involvement. *International Journal of Electronic Commerce , 11* (4), 125-148.

Park, S. R., Nah, F. F.-H., DeWester, D., Eschenbrenner, B., & Jeon, S. (2008). Virtual World Affordances: Enhancing Brand Value. *Journal of Virtual Worlds Research , 1* (2).

Patel, N. (2013, June 27). What Converts Better: Free Trial or Money Back Guarantee? Retrieved July 16, 2015, from http://www.quicksprout.com/2013/06/27/what-converts-better-free-trial-versus-money-back-guarantee/

Patel, N. (2014, July 2). Infographic: Personalization Drives eCommerce Revenue. Retrieved July 7, 2015, from https://blog.kissmetrics.com/personalization-improves-ecommerce-revenue/

Paul, P. (1996). Marketing on the Internet. *Journal of Consumer Marketing* , 27-39.

Peattie, K. (1997). The marketing mix in the third age of computing. *Marketing Intelligence & Planning* , *15* (3), 142-150.

Polo, Y., Sese, F. J., & Verhoef, P. C. (2011). The effect of pricing and advertising on customer retention in a liberalizing market. Journal of Interactive Marketing, 25(4), 201–214.

Poundstone, W. (2010). Priceless: The myth of fair value (and how to take advantage of it). Macmillan.

Pratt, M. K. (2008). *Avatars get down to business*. Retrieved July 14, 2010, from www.computerworld.com: http://www.computerworld.com/s/article/318544/Avatars_Get_Down_to_Business

Pretorius, K. (2013, June 18). Why Using Bing Advertising Is Becoming a Must | Paid Search. Retrieved July 28, 2015, from http://www.mediavisioninteractive.com/blog/paid-search/why-using-bing-advertising-is-becoming-a-must/

Rashid, A., Baron, A., Rayson, P., May-Chahal, C., Greenwood, P., & Walkerdine, J. (2013). Who am I? Analyzing digital personas in cybercrime investigations. Computer, (4), 54–61.

referenceforbusiness.com. (2015). Coupons - advantage, disadvantages. Retrieved July 26, 2015, from http://www.referenceforbusiness.com/small/Co-Di/Coupons.html

Regensburg, P. (2014, September 9). Why B2B Branding Works - And Why It's More Important Than Ever. Retrieved July 16, 2015, from http://blog.raincastle.com/why-b2b-branding-works-and-why-its-more-important-than-ever

Reichheld, F. F., & Schefter, P. (2000, October 7). The Economics of E-Loyalty - HBS Working Knowledge. Retrieved July 25, 2015, from http://hbswk.hbs.edu/archive/1590.html

Rekuc, D. (2015, March 24). Why Most Brands Fail At E-Commerce (And 4 Keys To How To Fix It). Retrieved July 16, 2015, from http://marketingland.com/brands-fail-e-commerce-fix-122172

Retailersearch.org. (2015). Online Retailing Research - Centre for Retail Research, Nottingham UK. Retrieved July 31, 2015, from http://www.retailresearch.org/onlineretailing.php

Reyna, S. M. (2014). How To Create bannerAds That Works. Retrieved July 28, 2015, from http://www.powerhomebiz.com/vol35/bannerads.htm

Richardson, B. (2008). *Tips on Using Virtual Worlds for E-commerce*. Retrieved August 12, 2010, from www.emarketingandcommerce.com: http://www.emarketingandcommerce.com/article/tips-using-virtual-worlds-e-commerce/1

Rick, S. I., Cryder, C. E., & Loewenstein, G. (2008). Tightwads and spendthrifts. Journal of Consumer Research, 34(6), 767–782.

Rogers, M. (2014, August). What Is The Future Of Affiliate Marketing? Retrieved July 28, 2015, from http://www.affiliatemarketertraining.com/what-is-the-future-of-affiliate-marketing/

Schafer, J. B., Konstan, J., & Riedi, J. (1999). Recommender systems in e-commerce. *Proceedings of the 1st ACM conference on Electronic commerce*, (pp. 158-166).

Schreiber, T. (2015, January 15). 5 Brand Strategies to Uniquely Position Your Ecommerce Business Above the Competition – Shopify. Retrieved July 16, 2015, from http://www.shopify.com/blog/16692816-5-brand-strategies-to-uniquely-position-your-ecommerce-business-above-the-competition

Schwarz, J. (2006). *Bold new opportunities in virtual worlds*. Retrieved August 15, 2009, from www.imediaconnection.com: http://www.imediaconnection.com/content/8605.asp

Sen, R. (2005). Optimal Search Engine Marketing Strategy. *International Journal of Electronic Commerce , 10* (1), 9-25.

Seybold, P. (2001, May). Get inside the lives of your customers. *Harvard Business Review* , pp. 81-88.

Shapiro, D. L., Sheppard, B. H., & Cheraskin, L. (1992). Business on a Handshake. *Negotiation Journal* , 365-377.

Shimp, T. (2008). *Advertising, Promotion, and Other Aspects of Integrated Marketing Communications*. South-Western College Pub.

Shpanya, A. (2014, August 18). Why dynamic pricing is a must for ecommerce retailers. Retrieved July 7, 2015, from https://econsultancy.com/blog/65327-why-dynamic-pricing-is-a-must-for-ecommerce-retailers/?utm_campaign=bloglikes&utm_medium=socialnetwork&utm_source=facebook

Shpanya, A. (2015, January). The Retail TouchPoints Blog • Dynamic Pricing: Why The Future Of Retail Is Now. Retrieved July 28, 2015, from http://retailtouchpoints.tumblr.com/post/111857413867/dynamic-pricing-why-the-future-of-retail-is-now

Sid. (2014, October 30). 3 Ways to Use Analytics to Increase Your E-commerce Conversion Rates. Retrieved July 7, 2015, from https://blog.kissmetrics.com/analytics-ecommerce-conversion-rates/

Smirnova, Y. V. (2014, August 19). 26 Social Listening Tools to Infuse Your eCommerce Marketing with Awesome Intelligence. Retrieved July 7, 2015, from https://blog.kissmetrics.com/26-social-listening-tools/

Smith, P. R. (1998). *Marketing Communications: an integrated approach*. London: Bell and Bain Ltd.

SMITH, P., & CHAFFEY, D. (2005). *e-Marketing excellence: at the heart of e-Business*. Oxford: Butterworth Heinemann, UK.

Solomon, M. (2009). *Sold on Web 2.0 Tools*. Retrieved July 12, 2010, from Searchcio.com: http://searchcio-midmarket.techtarget.com/magItem/0,291266,sid183_gci1265841,00.html

Solomon, M. R., Dahl, D. W., White, K., Zaichkowsky, J. L., & Polegato, R. (2014). Consumer behavior: buying, having, and being. Prentice Hall Upper Saddle River, NJ.

Stockdale, R., & Standing, C. (2004). Benefits and barriers of electronic marketplace participation: an SME perspective. Journal of Enterprise Information Management, 17(4), 301–311.

Strauss, J., & Frost, R. (1999). *Marketing on the Internet: principles of online marketing*. New Jersey: Prentice Hall.

Strauss, L., & Frost, R. (2001). *Internet Marketing*. NJ: Prentice Hall.

Su, B.-c. (2007). Consumer E-Tailer Choice Strategies at On-Line Shopping Comparison Sites. *International Journal of Electronic Commerce , 11* (3), 135.

Su, B.-c. (2008). Characteristics of Consumer Search On-Line: How Much Do We Search? *International Journal of Electronic Commerce , 13* (1), 109.

Sullivan, M. (2015, February 26). Programmatic Advertising on Mobile with Choozle & Upcoming Webinar | Choozle SimplaMATIC. Retrieved July 26, 2015, from http://blog.choozle.com/programmatic-advertising-on-mobile/

Tanoury, D., & O'Leary, T. (2000, June 5). *Partner Relationship Management: Emerging Vision In The E-Commerce Marketplace*. Retrieved July 14, 2009, from crn.com: http://www.crn.com/it-channel/18834080;jsessionid=3UCZREQMZJEMJQE1GHPSKHWATMY32JVN

Techcrunch.com. (2009, April). *Twitter Eats World: Global Visitors Shoot up To 19 Million*. Retrieved August 13, 2010, from www.techcrunch.com: http://www.techcrunch.com/2009/04/24/twitter-eats-world-global-visitors-shoot-up-to-19-million/

The Economist. (2012, June 30). How deep are your pockets? The Economist. Retrieved from http://www.economist.com/node/21557798?fsrc=scn/tw_ec/how_deep_are_your_pockets_

Thefreelibrary.com. (2009). *Advantages and Challenges of Online Marketing*. Retrieved July 12, 2010, from www.thefreelibrary.com:
http://www.thefreelibrary.com/Advantages+and+Challenges+of+Online+Marketing-a01073786649

Turban, E., Lee, J., & Chung, M. (2006). *Electronic Commerce: A Managerial Perspective*. Prentice Hall.

Turow, J., King, J., Hoofnagle, C., Jay, B., Amy, & Hennessy, M. (2009, September). *Americans Reject Tailored Advertising and Three Activities that Enable It*. Retrieved July 12, 2010, from ssrn.com: http://ssrn.com/abstract=1478214

Tzuaan, S. S., Sivaji, A., Yong, L. T., Zanegenh, M. H. T., & Shan, L. (2014). Measuring Malaysian M-commerce user behaviour. In Computer and Information Sciences (ICCOINS), 2014 International Conference on (pp. 1–6). IEEE.

Virginia Phelan, K., Chen, H.-T., & Haney, M. (2013). "Like" and "Check-in": how hotels utilize Facebook as an effective marketing tool. Journal of Hospitality and Tourism Technology, 4(2), 134–154.

Wang, K., Wang, E., & Farn, C. (2009). Influence of Web Advertising Strategies, Consumer Goal-Directedness, and Consumer Involvement on Web Advertising Effectiveness. *International Journal of Electronic Commerce , 13* (4), 67-96.

WATABE, K., & IWASAKI, K. (2007). Factors Affecting Consumer Decisions about Purchases at Online Shops and Stores. *9th IEEE International Conference on E-Commerce Technology and The 4th IEEE International Conference on Enterprise Computing, E-Commerce and E-Services*, (pp. 249-299).

Wharton. (2009, September 16). Time vs. Money: Analyzing Which One Rules Consumer Choices. Retrieved July 28, 2015, from http://knowledge.wharton.upenn.edu/article/time-vs-money-analyzing-which-one-rules-consumer-choices/

Wilmoth, H. (2015). 3 Criteria for Creating Better Ecommerce Customer Segments. Retrieved from http://www.monetate.com/blog/3-criteria-for-creating-better-ecommerce-customer-segments/

Work, S. (2011, April 28). How Loading Time Affects Your Bottom Line. Retrieved July 25, 2015, from https://blog.kissmetrics.com/loading-time/

Xu, J., & Quaddus, M. (2010). E-business in the 21st Century: Realities, Challenges and Outlook. World Scientific Publishing Co., Inc.

Yazdanifard, R., & Samuel, O. (2012). The Influence of E-Commerce on Marketing Practitioners and Consumers. Computer Technology and Application, 3(3).

zenithoptimedia.com. (2006). *Press Release*. Retrieved April 12, 2009, from www.zenithoptimedia.com: http://www.zenithoptimedia.com/gff/pdf/Adspend%20forecasts%20December%202006.pdf

Zhao, X., Fang, F., & Whinston, A. B. (2006). Designing On-Line Mediation Services for C2C Markets. *International Journal of Electronic Commrce , 71-93.

Zimmerman, J. (2001). Marketing on the Internet: seven steps to building the Internet into your business. Canada: Maximum Press.

Amir Manzoor

8

E-COMMERCE PAYMENT SYSTEMS

Chapter Outline

ELECTRONIC PAYMENTS (E-PAYMENTS)

METHODS OF ELECTRONIC PAYMENTS

ELECTRONIC PAYMENT TECHNOLOGIES

TRANSACTION SECURITY PROTOCOLS

TRANSACTION SECURITY STANDARDS

MOBILE TRANSACTIONS

MOBILE PAYMENTS

E-FINANCE

PHISHING ATTACKS

FRAUD IN E-COMMERCE

MICROPAYMENTS

BUSINESS-TO-BUSINESS (B2B) TRANSACTIONS

ELECTRONIC FUNDS TRANSFER (EFT)

E-PAYMENT STANDARDS

OPEN AND CLOSED LOOP PAYMENT SYSTEMS

PAYMENT CARDS

ELECTRONIC WALLETS

STORED VALUE CARDS

PAYMENT GATEWAYS

MERCHANT ACCOUNTs

PAYMENT AGGREGATORS

ALTERNATIVE PAYMENT SYSTEMS

INTERNET TECHNOLOGIES AND BANKING INDUSTRY

AUDIT OF E-COMMERCE TRANSACTIONS

Learning Objectives

After reading this chapter, reader should be able to:

- Describe the traditional payment systems.

- Understand concepts of electronic payments.

- Understand terminology associated with electronic payments.

- Describe methods of electronic payments.

- Understand electronic cash and its working.

- Describe various electronic cash systems.

- Understand how to secure electronic cash.

- Describe transaction security protocols.

- Describe payment methods in B2B E-Commerce, including payments for global trade.

- Describe the features and functionality of electronic billing presentment and payment systems.

- Describe electronic funds transfer (EFT) and its major systems.

- Describe e-payments standards.

- Describe various types of payment cards.

- Describe electronic wallets and their types.

- Discuss the different categories and potential uses of smart cards.

- Describe a payment gateway, its working, and selection process.

- Describe alternative payment systems.

- Describe impact of Internet technologies on banking.

- Describe the processes and parties involved in e-checking.

- Dsecribe how E-commerce transactions are audited.

- Describe how to get a merchant account.

Payment Methods

Today, four basic ways to pay for purchases dominate both traditional and B2C E-commerce: Cash, checks, credit cards, and debit cards. A small but growing percentage of consumer payments are made by electronic transfer. Credit cards are by far the most popular method that consumers use to pay for online purchases. Recent surveys have found that more than 85 percent of worldwide consumer Internet purchases are paid for with credit cards. Increasing number of payment solutions offerings from merchants is providing them opportunities to expand their customer base (OECD, 2006).

Cash as Method of Payment

Cash, as a method of payment, offer many advantages e.g. instant convertibility without the need of intermediation and very low to none transaction costs for small purchases. Merchants are not exposed to any financial risk and both merchant and consumer can remain anonymous.

Cash is a tamper-proof payment system, which does not require any expensive special hardware or authentication to complete the sales. Sales made on cash cannot be repudiated.

However, use of cash does have some disadvantages. For example, transaction costs for using cash as a payment method are significant and consumers are exposed to financial risk by carrying cash for purchases. There is no time lag between actual purchase of item and actual payment. Without an agreed upon return policy between the parties, purchases made on cash are tend to be final and cannot be reversed. There is no security against unauthorized use of cash.

Introduction to Electronic Payments

Electronic payment refers to exchange of money electronically. Handling of electronic payments is an important function of E-commerce sites. Irrespective of their format, electronic payments are far cheaper, faster, convenient, and economical than conventional payment methods e.g. paper checks. The estimated transaction cost of billing one person by mail range between $1 and $1.50 while billing the same person over Internet can cost an average of 50 cents per bill (Schneider, 2010). Total savings in costs can be very large keeping in view the volume of electronic payments that could take place. Savings of paper can produce a significant impact on environment by reducing the energy consumed and wastes generated in the papermaking process.

Domestic and International Payments Mechanisms

Domestic (within country) and international payment mechanisms differ significantly. Domestic payments are normally settled using either a private clearinghouse, or a public one. Credits and debits of banks are normally netted and each bank subsequently pays only the difference between all its credits and all its debits to the clearinghouse. These payments are considered final because these payments are made to the banks through their accounts with the central bank. Domestically, banks can accept each other's liabilities, for several reasons. One reason is a better understanding of each other on domestic level, faster payments, time lag before final regulation is short, and banks know that the Central Bank overlooks the payment system, supplying liquidity when necessary, and in some countries, there are insurance schemes, which guarantee settlement. Internationally, banks normally handle payments through their foreign correspondents. This process can be fast for large value payments but for small value payments, it is usually very slow, requiring sometimes weeks to complete.

With respect to electronic payments, there can be several scenarios with different issues. For domestic payments, electronic payments can be processed quickly and could be settled and regulated in the same way as checks, or payments can be made via ATMs.

Automatic Teller Machine (ATM)

Automatic Teller Machine (ATM) is an electronic banking outlet, which allows customers to complete basic transactions without the aid of a branch representative or teller. There are two primary types of automated teller machines, or ATMs. The basic units allow the customer to only withdraw cash and receive a report of the account's balance. The more complex machines will accept deposits, facilitate credit card payments and report account information. An ATM identifies a customer using a card that contains a magnetic strip or smart card chip. The user provides a personal identification number or PIN to access account information. Banking is networked, which allows customers to access account information from any ATM in the world. Sometimes a fee is charged for using an ATM outside of the customer's banking system. ATM Processing is similar to credit card processing, except with ATM machines the processing center uses ATM networks instead of credit networks. To begin with, we will have to program or re-programmed your ATM machine with a TID (terminal ID number) assigned to your ATM at location. This is the number that identifies your ATM machine on our ATM processing system.

For international payments, the process is complicated. Banks are hesitant to accept electronic payments for several reasons. They are unaware of credit rating of foreign bank as well as financial and political stability of the country of foreign bank. Another reason is that that internationally, there are no loss-sharing agreements or insurance schemes. In this situation, banks have two options. They can collect electronic payment from foreign bank and wait until settlement before crediting beneficiaries account. If beneficiary were a merchant, he would be faced with the question of whether to deliver the goods immediately, or wait settlement. The banks can also refuse to accept the electronic payment. Recipient of electronic payment would then have to verify the electronic payment himself with the bank that sent electronic payment. Practically that means a merchant will need to open an account with all the banks.

Basic Problems with Electronic Payments

Electronic payments have some basic problems. First, they are not direct and simple. A complex system is required to ensure that sellers will be paid, buyers only give what is due, and payment mechanism is secure, reliable and sufficiently simple-to-use. To become popular as a means of handling money transfer, online payments need to address the basic trust issue. To have wide applicability, electronic payment systems need to be integrated with other organizations where they have a large number of users and have a demand of an electronic payment system (e.g. financial institutions and electronic marketplaces such as eBay). An electronic payment system needs to balance the convenience of user with the provision of anonymity and privacy. Payment security is another important factor for consumers' decision to make online purchase or not. According to a survey, more than 40% of users indicated they would not transact online because of the fear that their personal information could be stolen (ITU, 2006). A 2006 report of Office of Fair Trading, UK showed that 79% of Internet users were very concerned about the risk to the security of their payment information when shopping online (OFT, 2007). Recent large scale hacks into companies' customer records have also contributed to undermine consumer confidence in providing their personal details on-line. Take example of the hacking of Yahoo's HotJobs.com. In this case, hackers gained control of every Yahoo! service victims used. For details, see http://www.xiom.com/whid-2008-32. Acceptance of the idea of using technology to bypass actual money meets opposition from people at all levels. Wristbands are a part of wearable contactless payment devices meant especially for low-level purchases. In the UK, Barclaycard's bPay wristband can be used at more than 300,000 locations including shops, bars, cafes and public transport. Here is an electronic payment system, which has caught on at a popular level, but only for transactions less than $50. Big transactions are a problem. Once again, though a viable methodology the problem seems to be general acceptance of the electronic payment system itself. Starbucks has an app specific to their store, which is scanned at the counter. However, the PayPal Beacon moves that idea up an order of two. It uses Bluetooth Low Energy (BLE) that connects to a customer's PayPal app when the customer enters the store. The smart device notifies with a vibration or sound that check-in is complete. When paying, the customer simply notifies the store they are paying with PayPal and the purchase is completed. Though a growing technology, BLE and PayPal Beacon are not available everywhere. Large retail stores are connected to older legacy technology that represents a sizeable investment for the chain and a problem for electronic payment systems of the newer sort (Adams, 2014).

Governments and industry both are making efforts to improve consumer protection and confidence in online payment systems. Technological solutions and secure payment methods are being developed for consumers and card issuers have adopted many additional protections. However, laws vary from country to country and in many countries, they tend to be generic with no specific regulations for E-commerce. The consumer liability for an unauthorized charge varies from country to country. Some countries may hold consumer liable for a portion of an unauthorized charge while in other countries liability may depend on the degree of negligence on the part of the consumer or the payment mechanism used (e.g. credit or debit card), or the nature of the transaction (domestic or international).

Another important concern is fairness of electronic transactions. Because of lack of face-to-face communication in E-commerce, fairness in transactions becomes very important. Several fair transaction protocols have been proposed e.g. see [(Fan & Liang, 2008); (Wang & Li, 2007); (Liu, Luo, & Si, 2007)].

Electronic Payments Terminology

Before starting a detailed discussion of electronic payments, let us first get familiarized with a few terms that are commonly used in electronic payment process.

INVOICING: It refers to the process of notifying the buyer how much he has to pay and when the payment is due.

CLEARANCE: It refers to the transmission, reconciliation or confirmation of payment orders matching the invoice. Clearance is necessary before payment actually takes place.

SETTLEMENT: It refers to recording of debit or credit status of both buyer and seller. Actual payment may take place later. Either gross settlement (in which transaction is settled individually every time) or net settlement (in which transactions can be offset and only net transaction is settled) takes place.

COLLECTION: It refers to the process of crediting funds to collecting party and debiting from paying party. Collection involves authorization of the financial service provider (e.g., bank) to confirm the credit or debit.

PAYMENT GATEWAY: A payment gateway is the service that automates the payment transaction between the shopper and merchant. A merchant gateway can be thought of as equivalent of a physical point-of-sale terminal in a retail store.

MERCHANT ACCOUNT: A merchant account refers to a business banking relationship in which the bank and business agree to accept credit card payments. The payment gateways delivers information to merchant accounts. Businesses receive funds through their merchant account provider. Without merchant account, a business cannot directly accept payments by any of the major credit card brands such as VISA and Master Card.

Electronic Payments: Major Stakeholder and Requirements

In Electronic payments, many stakeholders are involved e.g., buyers, merchants, banks, credit card companies, and network providers. The interests of each are often different and conflict of interest may arise. Buyers prefer payment systems, which are low-cost, low-risk, reliable, and can be repudiated. Merchants prefer payment systems, which are low-cost, low-risk, reliable and low probability of repudiation. Currently most of the risk of checking and credit card fraud, any charges for repudiation, and expense of hardware required to process the payments are borne by the merchant. Merchants also prefer all sales to be final but government regulations and protection provided by the credit and debit card providers limit the risks to individual consumers. The financial intermediaries (banks, credit card companies, and network providers), concerned about the transactions security, opt for maximizing the transaction fees they receive in order to transfer the risk of fraud or abuse on to either the merchants or the buyers. Following are some major requirements of e-payments. The importance of each varies for each stakeholder involved in e-payments.

APPLICABILITY: For buyers it refers to the number of customers using a particular electronic payment system e.g. credit card. For merchants it refers to the number of merchants that accept a particular electronic payment system e.g. credit cards. For credit card companies it refers to the number of financial institutions, which issues its credit cards. For banks, it refers to the number of financial institutions which issue the same type of credit cards e.g. VISA. For network providers it refers to the number of financial institutions, which issue a particular type of credit card. Wider applicability is preferred by all players.

BUYER COST: It refers to the cost per transaction paid by the customer to the payment system provider.

CONVERTIBILITY: For buyers it refers to the ability of electronic cash to make payment in other currencies. For merchants it refers to the ability to transfer received money directly to merchant account.

PRIVACY: It refers to private information exchange when consumers provide credit card and personal information to merchants for the transaction.

SECURITY: It refers to various security measures (e.g. use of SSL, and requiring both the card number and password) adopted by the players to provide secure payment processing.

CONVENIENCE: For all players it refers to the convenience in using the system for processing payments.

ANONYMITY: For all parties it refers to the fact that one is not identifiable to persons other than those one communicates with or third parties one has revealed to. For banks and Credit card companies, it refers to their ability of being able to trace the source of money.

REGULATORY ASPECT: For all players it refers to the concern as to who regulates the various players in payment system.

FINANCIAL RISK: For merchants it refers to the risk of charge backs and frauds.

CREDIT COST: For credit card company it refers to various charges e.g. charge from the buyers and merchants, charge for the late payments, and membership fees.

MERCHANT COST: For merchants it refers to the fees charged by financial institutions to provide merchants account.

NON-REPUDIATION: For merchants it is the ability of payment system to ensure that a party to a transaction cannot deny the authenticity of their signature that they originated.

AUTHORIZATION TYPE: For merchants it refers to both online and offline authorization of payment mechanism.

ATOMIC EXCHANGE: For merchants it refers to automatic exchange of good and payment.

INTEROPERABILITY: For banks it refers to different parties joining the network

FLEXIBILITY: For network providers it refers to the independence of payment method with a particular network provider e.g. credit card is independent with a particular network provider.

Methods of Electronic Payments

METHOD 1: INSECURE TRANSMITTAL OF CREDIT CARD INFORMATION

It is the simplest way of paying with credit cards online. In this method credit card holder's details is transferred via the Internet (usually through a web form) without security. After receiving the details, the merchant performs manual processing of authorization, clearance, etc. This method is simple but it lacks security and requires manual work.

METHOD 2: SECURE TRANSMITTAL OF CREDIT CARD INFORMATION

In this method, cardholder's details can be securely transmitted to the merchant using a secure encrypted link (e.g. using Secure Socket Layer). After receiving the details, the merchant performs manual processing of authorization, clearance, etc. The method is secure but still requires manual work.

METHOD 3: USING AUTOMATED PAYMENT GATEWAY

Large companies with thousands of transactions occurring at given time need automation of payment process. Payment process automation can be achieved by replacing the manual extraction of information from the web server by a payment gateway. There are two scenarios. In first case, the payment gateway software, residing in the merchant server, collects the cardholder's details and process it automatically. In the second case, the merchant can redirect the customer to pay at some payment site e.g. PayPal. Here payment gateway is provided by a payment service provider. The method of using automated card gateways is secure and requires no manual work. However, it is more costly and required much more technical work.

Electronic Payment Technologies

The major electronic payment technologies include electronic cash, payment cards, electronic wallets, and smart cards. Another less used payment technology is called scrip.

Scrip

Scrip is digital cash minted by a company instead of by a government. Scrip is often a form of credit and generally used for company payment of employees and for making payment in situations where use of regular money is unavailable e.g. during war times. Most scrip cannot be exchanged for cash and must be exchanged for goods or services by the company that issued the scrip. One example of scrip is eScrip (escrip.com) which focuses on the not-for-profit fundraising market. Other forms of scrip include subway tickets, gift certificates and gift cards.

A given type of scrip might have served one or more functions: as a stimulant to business, relief for unemployment, a weapon against chain stores, and/or a means of municipal finance (Gatch, 2012). The many types of scrip issued can be reduced to five basic categories defined by what gave people the confidence to use it as a money substitute. The first category, "reputational scrip:' comprises private currency issues by corporations, organizations, and even individuals, and has numerous antecedents in American financial history. The scrip was issued by individual companies to meet payrolls and which was redeemable in the company store. Such scrip circulated because of the economic power and reputation of its issuers. During the 1930s, corporations with steady receivables could issue scrip good for purchases of their products and services. The second category is the bank and financial scrip such as clearing house certificate, an emergency currency that private-banking settlement-association (clearinghouses) issued to meet the liquidity crises. The third category was stamp scrip. Between 1932 and 1934, hundreds of communities across the country issued scrip that required the placing of some type of stamp as a condition of its further circulation. The fourth type of scrip was barter and self-help scrip. Chronic mass unemployment in the early 1930s encouraged the establishment of barter exchange and self-help cooperatives. In these arrangements, the unemployed joined either to barter their labor with willing employers or to produce their own goods, which they subsequently traded for other necessities. The fifth type of scrip was tax anticipation notes.

Some companies still issue scrip notes and token coin, good for use at company points of sale. Among these are the Canadian Tire money for the Canadian Tire stores and gasbars in Canada, and the Disney dollars, used at Disney resorts. In the retail and fundraising industries, scrip is now issued in the form of gift cards, eCards, or less commonly paper gift certificates. Physical gift cards often have a magnetic strip or optically readable bar code to facilitate redemption at point of sale. Great Lakes Scrip Center, the US's largest broker of scrip gift cards, reports more than 17,000 non-profit organizations have raised more than $500 million with scrip fundraising (shopwithscrip.com, 2015). Canada's largest scrip fundraising company, Fundscrip, demonstrated scrip's popularity by boasting a 75% repeat purchase rate among groups seeking to easily support their respective causes, and $12 million in fundraising as of 2015 (fundscrip.com, 2015). Gift cards are increasingly being offered as consumer incentives at no cost or for a substantial discount. For example, reward points given for the use of credit cards can be exchanged for a variety of gift cards and some companies will offer high value gift cards at a discount. Gift cards are readily available at a discount using online trading services where users can trade, donate, buy and sell gift cards. Sellers use these websites because the gift card is not to a store of their liking, and buyers because this provides opportunities to buy these cards for less than they are nominally worth at the business. Other websites sell a collection of third party gift cards. Businesses offer credits to these websites for employee incentives, usually in return for a discount; however, discounts for consumers are rarely offered (giftcardswapping.com, 2015).

Electronic Cash (E-cash)

Electronic cash (also known as e-money, electronic currency, digital money, digital cash or digital currency) refers to any value storage and exchange mechanism created by a private (non-governmental) entity that does not use paper documents or coins and that can serve as a substitute for government-issued physical currency. Digital cash can be readily exchanged for physical cash on demand. Digital cash represent value because it is backed by a trusted third party, usually a bank. To date, use of digital cash has been relatively low-scale and success stories are rare. One example is Hong Kong's Octopus card system that started as a transit payment system and has now become a widely used electronic cash system. Another example is Chipknip. Chipknip is the electronic cash system used in the Netherland. Value can be loaded to your smart card via Chipknip loading station. Chipknip can be used to make payments at many places including payment for parking, grocery shops etc. In addition, there are other sophisticated forms of electronic cash, such as Mondex, E-Cash, and Visa cash.

Electronic cash has very low fixed costs. It is particularly useful in sale of low-price goods/services and facilitate payments for customers who do not have credit cards. Customers may not have credit cards for many reasons e.g. because they do not qualify for it (due to minimum income requirements or past debt problems or age requirement) or they traditionally preferred to make purchases using cash. In 1999, online teen buyers represented a consumer market worth an estimated $141 billion (Sandoval, 1999). They are not eligible to obtain credit cards because they are too young. All these people represent a good market for electronic cash. However, there is still lack of common standards among all electronic cash issuers. As a result, electronic cash is not accepted universally and there are issues in accepting one issuer's electronic cash by another issuer.

WORKING OF ELECTRONIC CASH

To use electronic cash, a consumer first opens an account with an electronic cash issuer (such as a bank or a private vendor of electronic cash, such as PayPal) and presents proof of identity. User deposits money into its account by sending an encrypted message to the bank. This message informs the bank to deduct a certain amount from the user's account. The bank receives the message and verifies its authenticity by using digital signatures. The bank deducts the amount from user's account, generates electronic cash coins, encrypts the message, signs it with its digital signature and returns it to user. This encrypted message contains the electronic cash that user can spend or store in an electronic wallet or on a stored-value card. When merchants receive electronic-cash from a consumer during a transaction they contact the bank that issued the electronic cash to verify the electronic cash has not been double spent and to credit the amount to the merchant's account.

HOLDING ELECTRONIC CASH

There are two generally accepted approaches to holding e-cash: online storage and offline storage.

ONLINE CASH STORAGE

In Online cash storage, customer does not hold e-cash. A trusted third party (usually an online bank) is involved in all transactions and this third party also consumers' cash accounts. Merchants contact the consumer's bank to receive payment for consumer purchases. No additional hardware is required and online cash storage can be implemented with existing Point-of-Sale devices with a software upgrade. However, real time verification of electronic cash increases the costs of keeping the transaction secure and anonymous. The electronic cash issuer bank also needs to build and maintain an ever-growing database of serial numbers of used electronic cash coins.

OFFLINE CASH STORAGE

In offline cash storage customer, holds e-cash and no third party is involved in the transaction. Merchants collect electronic cash coins from users, store them in a storage media and later go to the bank that issued this electronic cash exchange e-cash for physical cash. To protect against fraud either hardware or software safeguards are utilized. These safeguards ensure merchants that bank will either accept the user's electronic cash or identify double spending and punish the guilty party.

ADVANTAGES OF ELECTRONIC CASH

E-cash is software in nature and can be programmed to provide many functions that physical cash can never provide. This ability opens up a whole range of exciting functionality that digital money may offer. Besides this, digital money offer may more advantages as shown in the following table.

TABLE 8-1: ADVANTAGES OF ELECTRONIC CASH

Advantage	For Users	For Banks	For Retailers
Convenience	Funds can be accessed from home as well as payments for items. Smart cards can be used to initiate financial transactions wherever user may be. Mobile payments enable users make payments on the go.	Digital cash greatly reduces processing effort, resulting in less time and staff required.	E-cash can be immediately credited to retailers' accounts for immediate use.
Security	Current security standards are competitive .Strong encryption techniques, passwords, and biometric information of the user can be used to protect the digital cash.	Costs of fraud and abuse could be reduced.	Transaction fees for electronic cash are lower because of smaller operating costs for the issuer. Costs for counting, storing and transporting cash are also reduced.
Intractability	True anonymous digital cash makes it impossible for anyone to link payment to payer and provides unconditional intractability.	Handling costs for e-cash are minimal.	There is no need to hold the cash. Processing and holding costs of cash are reduced.

DISADVANTAGES OF ELECTRONIC CASH

ALGORITHM: The algorithms being used in most electronic cash systems, though very secure, are still prone to security breaches.

USER UNDERSTANDING: Use of electronic cash requires a basic technical understanding of its technology in order to use it and inspire confidence.

LEGAL PROBLEMS: The anonymity of electronic ash makes it an attractive choice for money laundering and tax evasion purposes.

ELECTRONIC CASH SYSTEMS

Following are some of the service providers that currently offer electronic cash services and bill presentment and payment systems.

CLICKSHARE

Clickshare (clickshare.com) is an electronic cash system that users can use to buy items (e.g. information, music, video, software etc.) with simplicity and security. Users first open an account with a Clickshare membership service provider (e.g. an ISP), and can then make single-click purchases throughout the web without providing their personal or financial information each time they make an online purchase. Clickshare keeps track of transactions and bills the user's Clickshare membership service provider. The membership service provider then bills the user for his or her purchases. Users get a single bill for all the purchase they made online from various vendors. Clickshare also tracks user movement on the Internet using the standard HTTP protocol and does not use cookies or software wallets.

PAYPAL

PayPal (paypal.com) is an electronic cash system that provides payment-processing services to businesses and to individuals. To use PayPal, online merchants and consumers first must register for a PayPal account. PayPal does not charge a fee to open a PayPal account. No minimum amount limit is applied to PayPal account and customers can add money to their PayPal accounts by authorizing a transfer from their checking accounts or by using a credit card. Money can be sent instantly and securely to anyone with an e-mail address, including an online merchant. PayPal earns a profit on the float (money that is deposited in Pay-Pal accounts and not used immediately). PayPal also charges a transaction fee to businesses that use PayPal services to collect payments. Transaction fee is not charged from individuals who use PayPal to send money to other individuals. Fees apply only if the sender uses a credit or debit card, or if you receive any payment for goods or services. Anyone with a PayPal account (individuals, online merchants etc.) can withdraw cash from their PayPal accounts at any time by first registering with PayPal and then requesting that PayPal send them a check or make a direct deposit to their checking accounts. The standard rate for receiving payments for goods and services is 3.4%. If you receive more than €2,500.00 EUR per month, you are eligible to apply for PayPal's Merchant Rate - which lowers your fees as your sales volume increases. Your fees can be as low as 1.9%, based on your previous month's sales volume. There's no fee to send an invoice or money request. You pay a fee when a customer sends you money. Seller Protection is a free service provided to our members. The Table 8-2 presents transaction fees charged by PayPal.

TABLE 8-2: PAYPAL FEES

Sending money – Personal payments	It is free within the Canada to send money to family and friends when you use only your PayPal balance or bank account, or a combination of their PayPal balance and bank account. There is a fee to send money as a personal payment using a credit card. The fee in Canada is 2.9% plus $0.30 CAD of the amount you send. For example, if you send $100.00 CAD by credit card, the fee would be $3.20 CAD ($2.90 + $0.30). For personal payments, the sender is responsible for the fee.
Receiving money – Personal payments	It is free to receive money from friends or family in Canada when they send the money from the PayPal website or mobile app, using only their PayPal balance or their bank account, or a combination of their PayPal balance and bank account. The fee of 2.9% + $.30 also applies if you click Request Money on the PayPal website or mobile app.
Goods and services – Purchase payments	There is no fee to use PayPal to purchase goods or services. However, if you receive money for goods or services (such as from selling an item on eBay), the fee for each transaction is 2.9% plus $0.30 CAD of the amount you receive.
International payments	There is a fee when you send a payment to someone in another country or if you receive a payment from someone in another country. Please note that exchange rate fees also apply if there is a currency conversion. The Fees section of the PayPal User Agreement has specific information. You can find the User Agreement by clicking Legal Agreements at the bottom of any PayPal page.
Withdrawing money	You can withdraw money from your PayPal account to your bank account at no cost. PayPal does not charge you a fee for withdrawing Canadian currency to your local bank account. However, there may be a Currency Conversion fee for withdrawing other currencies to your local bank. Some banks may charge fees for electronic funds transfers. You can contact your bank directly for more information.

There are two types of PayPal accounts: Personal and Business. To open a personal PayPal account, you will need to provide your name, address, phone number, and email address. All PayPal accounts allow you to send and receive payments.

- **Personal**: Recommended for individuals who shop and pay online, or wish to send or receive personal payments for shared expenses such as splitting of dinner bills or rental charges.

- **Premier**: Recommended for casual sellers or non-businesses who wish to be paid online, and who make online purchases.

- **Business**: It is recommended for merchants who operate under a company/group name. It offers additional features such as allowing up to 200 employees limited access to your account and customer service email alias for customer issues to be routed for faster follow-ups.

You can have one Personal account and one Premier or Business account. You can add more email addresses, debit or credit cards, and bank accounts to your account, but each account must have its own email address and financial information. You can also upgrade your Personal account to a Premier or Business account.

PayPal's micropayments price is 5% + $.05 and is designed for merchants who process low-value transactions (typically under $10 in value). The micropayments rate is available to all merchants and in all countries where Business accounts are available. If you sign up for micropayments, you will be charged the micropayments rate on all transactions regardless of payment size.

As soon as you receive the payment, it goes into your PayPal balance. When you have a PayPal balance, you can use those funds to either send a payment or make a withdrawal. You have the option of holding multiple currencies in one account. You can open a balance in any of our supported currencies, make or accept payments in multiple currencies, convert your balance from one currency to another, or close out your balance in any currency. You can designate a primary currency. A primary currency is the one you use most often for sending and requesting payments. The currency defines your sending and withdrawal limits. When you buy something that is for sale in a currency you hold, PayPal will take money from that currency balance first. If enough money is not available in that currency, PayPal will debit your primary currency balance and apply our exchange rate.

PayPal holds money in reserve just in case you receive payment reversals or chargebacks and your PayPal balance is not enough to cover them. The amount being held in reserve can be found on your Pending Balance page. From time to time, PayPal may need to adjust your reserve amount. If that happens, PayPal will email you about the changes. In deciding whether to apply payment holds, PayPal reviews many factors including account tenure, transaction activity, business type, past customer disputes, and overall customer satisfaction. Holds are typically placed on new sellers or sellers who have limited selling activity - until they build up a history of successful transactions, merchants selling in higher-risk categories like electronics or tickets, where we see higher levels of chargebacks and disputes, and sellers who have performance issues, or a high rate of buyer dissatisfaction or disputes.

You can cancel only payments that have a status of "Unclaimed." If the payment is completed, you will not be able to cancel the payment yourself, as the recipient will already have received the money. To reverse a completed payment, you will need to contact the recipient and request a refund. You can issue a full or partial refund up to 60 days after you receive a payment. PayPal offers buyers purchase protection for unauthorized account access and use, items not received from the seller, and items significantly not as described.

The PayPal Mobile app is available for iPhone, Android, and Windows Mobile. You can download the app to your computer from our website. PayPal Here™ is a mobile payment solution that includes a free app and a thumb-sized card reader for your iPhone. It lets you simply and securely accepts multiple forms of payment wherever your business takes you within Canada. PayPal Here™ helps you increase your customer base, and make more sales. At this time, PayPal Here™ is only available to a select group of Canadian businesses. We are working closely with these customers to learn and determine the best possible offline payment solutions for Canadian merchants.

PAYZA

Payza (payza.com) (formerly Alertpay) is the global online payment platform that specializes in e-commerce processing, corporate disbursements, and remittances for individuals and businesses around the world. Payza members will be able to load funds to and from their accounts via bank transfer, bank wire, credit/debit card and check, according to the specific regulations for each country.

GOOGLE CHECKOUT/GOOGLE WALLET

Google Checkout was an online payment processing service provided by Google aimed at simplifying the process of paying for online purchases. Users store their credit or debit card and shipping information in their Google Account, so that they can purchase at participating stores by clicking an on-screen button. Google Checkout was discontinued on November 20, 2013. The company offers a new solution, called Google Wallet, for certain payments. Google Wallet makes it easy to pay - in stores, online or to anyone in the US with a Gmail address. It works with any debit or credit card, on every mobile carrier. Tap and Pay feature allows you to pay with your Android phone everywhere you see this symbol . You can use your Google Wallet Card to pay with your Google Wallet Balance in-store at millions of Debit MasterCard locations. You can also use it to withdraw cash at an ATM, and you will get instant notifications for every transaction, right on your phone. Google Wallet allows you to store your gift cards/loyalty cards/coupons and easily redeem them in stores at checkout. You can also instantly view your gift card balance.

2CHECKOUT

2Checkout (http://www.2checkout.com/) is an online payment processing service that bundles a payment gateway and merchant account into a single package for simple integration with your site. 2CheckOut provides a reliable, dependable, affordable and highly secure payment option for many small to medium online business enterprises. 2Checkout is available in 197 countries.

SKRILL

Skrill (skrill.com) (formerly Moneybookers) is an e-commerce business that allows payments and money transfers to be made through the Internet. Moneybookers allows for sending and receiving payments to/from 200 countries and territories in 41 currencies, supporting major credit/debit cards and around 100 local payment options.

WORLDPAY

WorldPay (http://www.worldpay.com/) is an online payment gateway. WorldPay is a leading, single-source provider of electronic payment processing services – including credit, debit, EBT, checks, gift cards, e-commerce, customer loyalty cards, fleet cards, ATM processing and cash management services.

AUTHORIZE.NET

The Authorize.Net Payment Gateway can help you accept credit card and electronic check payments quickly and affordably. More than 350,000 merchants trust us for payment processing and online fraud prevention solutions.

SQUARE

Square (https://squareup.com/)supplies a free card-reader dongle that you plug into your smartphone. Link it to your bank account, and then simply swipe a credit card to collect money. It even e-mails receipts. Square charges a fee of 2.75 percent per swipe.

AMAZON PAYMENTS

Amazon Payments (https://payments.amazon.com/) is a way for customers to purchase goods and services at websites across the internet using the payment methods in their Amazon.com accounts, such as their Visa or MasterCard. Popular crowdfunding site Kickstarter uses Amazon Payments.

MYCHECKFREE.COM

MyCheckFree.com is a popular electronic payment system that allows you to receive your bills directly to your email and pay any billers who are affiliates, including major department stores, clothing chains and even commercial banks.

WEPAY

WePay (https://www.wepay.com/) is a payment processor that allows Internet merchants to accept credit cards and bank account payments online.

PROPAY

ProPay electronic payment system (http://www.propay.com/) that processes credit cards through the Internet via touch-tone phones; used to make purchases at mobile businesses, trade shows/fairs and taxis.

DWOLLA

Dwolla (https://www.dwolla.com/) is a direct competitor to PayPal. One of the newcomers in the third-party payments space, the company is processing over $1 million per day.

STRIPE

Stripe (http://stripe.com/)is a simple way to accept payments online. Stripe provides an excellent payment solution for web developers who would like to integrate a payment system into their projects using Stripe's robust API. Stripe powers commerce for thousands of sites across the web.

CLEAREXCHANGE

Clearxchange (https://www.clearxchange.com/payments/) is the first network in the U.S. created by banks that lets customers send and receive person-to-person (P2P) payments easily and securely using an email address or mobile number. The clearXchange service enables customers of their member banks to send and receive person-to-person (P2P) payments safely and easily through their online or mobile banking service. To send payments or payment requests, customers of their member banks only need to know the recipient's email address or mobile number. To receive payments and payment requests from customers of member banks, recipients just need an email address or mobile number and a checking or savings account with any U.S. bank. The clearXchange service does not move funds. Instead, they provide the information needed by our member banks to transfer the funds and complete the transaction.

AMERICAN EXPRESS SERVE

Serve (https://www.serve.com) is a Full Service Reloadable Prepaid Card with online Account access and a Mobile App.

VISA CHECKOUT

VISA Checkout (https://checkout.visa.com) is an instant-buy option, which aims to speed up online payments on both phone and PC. After setting up an account, you can register any debit or credit card for instant access. Once you are on a website that supports Visa Checkout, you simply tap or click the Visa Checkout logo in order to pay for your products instantly.

BITCOIN

Bitcoin is a virtual currency that enables direct payment over the Internet between two individuals by skipping the intermediary, such as a bank or credit card company. Bitcoin transaction, with fees that are much lower than what financial institutions charge, rely on cryptography to prevent double spending, counterfeit, or theft.

In 2008, "Satoshi Nakamoto" (a pseudonym for a person or group whose identity is a mystery) wrote a paper outlining bitcoin's design. The rules state that the number of bitcoins in circulation will grow at a decreasing rate toward a maximum of 21 million coins. The currency is now maintained by an open-source community that has no centralized authority that regulators can target. Currently, nearly 11 million bitcoins are in circulation (Lee, 2013).

Before a bitcoin can be purchased, a user must install a virtual wallet onto a personal computer or mobile device. The wallet, which is similar to personal finance software, keeps track of your bitcoin balances and all transactions. To buy a bitcoin, real money must either be deposited through an online payment company or transferred directly from a bank account into an account on a third-party web site that connects bitcoin buyers and sellers. Once the funds are available, a buyer can place an order for a bitcoin, similar to trading stocks, through an exchange such as Bitstamp. Bitcoins can also be purchased from third parties such as BitInstant, which sends the coins directly into the virtual wallet. Bitcoins can be used to buy from online vendors, such as George's Famous Baklava in

New Hampshire, which sold a dark chocolate pastry for 14 bitcoins in 2011 (worth about $1900 today) (Lee, 2013). Bitcoins are also being bought and traded as investments.

SECURING ELECTRONIC CASH

Electronic cash has unique security problems. This is because E-cash need to have two important characteristics. First, e-cash must be able to avoid double spending (i.e. to duplicate and spend the digital cash twice) and second, e-cash should be anonymous (i.e. entire e-cash transaction occurs only between two parties).

The user anonymity in e-cash is maintained by using Blind Signature Protocol. Using a blind signature protocol a user can obtain from a bank a digital coin properly signed by the bank. With blind signature protocols, bank (the signer of e-cash) does not learn information about the coin it signed and user cannot obtain more than one valid signature after one interaction with the bank (Boldyreva, 2003). Blind signature protocol eliminates the association between the user identification and the serial number of the coin. However, this raises the problem of double spending. To tackle this problem, the merchant needs to verify the digital coin with the bank at the point of sale in each of the transaction. This verification requires extra bandwidth, causes a potential bottleneck of the system, and requires synchronization between bank servers (i.e. synchronization of coin's serial numbers every time a coin is deposited into the bank).

A temper-resistant device, trusted by the bank, is used to verify the authenticity of the digital coin but it does not check whether the digital coin has been double spent. Secret splitting is a commonly used method to protect the privacy of the user as long as the user does not double spend the E-cash. Secret splitting splits a secret (in this case the value of digital coin) into multiple pieces and each secret piece is valueless on its own. None of the parties to the secret can learn the secret (the value of digital coin) unless all parties cooperate with one another to combine their secret pieces. This way buyer is protected from unauthorized redemptions of digital coins by seller. When digital coin serial numbers are added to the digital coin pieces, both bank and buyer are protected from double spending (Tewari, O'Mahony, & Peirce, 1998). The system could be implemented so that the digital coin itself can be reusable. Merchant can use another temper-resistant device to spend the digital coin at another place before final deposit of the digital coin to the bank for verification.

FUTURE PROSPECTS AND CHALLENGES FOR E-CASH

Electronic cash has been successful though there is a long way ahead. To make e-cash a global commercial success and a popular alternative payment system requires wide acceptance and establishing some common standards for electronic cash disbursement and acceptance. Electronic cash need to be independent and portable. When electronic cash is independent, it is unrelated to any network or storage device i.e. its existence does not depend on a particular proprietary storage mechanism that is specially designed to hold one type of electronic cash. Electronic cash should ideally be able to pass transparently across international borders and be automatically converted to the recipient country's currency.

Transaction Security Protocols

SSL and 3-D Secure are two standards that protect the integrity of online transactions. SET is another protocol but it is outdated and mot used anymore.

3-D Secure

The 3-D Secure is a collective of Verified by VISA (VBV) and MasterCard Secure Code (MSC). It is the most recent fraud prevention initiative available now. 3D Secure is also the only fraud prevention scheme that is available that offers companies liability cover for transactions that are verified by the checks. This provides additional protection to companies using the scheme as opposed to those that do not. The following cards can be used with 3-D Secure: Visa, MasterCard, International Maestro, and Laser.

3-D Secure provides an additional security layer for online credit and debit card transactions by adding an authentication step for online payments (americanexpress.com, 2010). Analysis of the protocol by academia has shown it to have many security issues that affect the consumer, including greater surface area for phishing and a shift of liability in the case of fraudulent payments (Murdoch & Anderson, 2010). The basic concept of the protocol is to tie the financial authorization process with an online authentication. This authentication is based on a three-domain model (hence the 3-D in the name). The three domains involved in the 3D-Secure process are Issuer Domain, Interoperability Domain and Acquirer Domain. The issuer domain is where the cardholder and issuing bank act. The acquirer domain is where the gateway and acquiring bank act. The interoperability domain is where all the "connecting" services act (Directory Server & ACS). Below we is the description how 3D Secure works by contrasting a regular (non 3D Secured) credit card payment with a 3D Secure credit card payment and providing detail on how international buyers are affected by 3D Secure for payments to merchants.

WORKING OF A REGULAR CREDIT CARD PAYMENT

In a regular credit card (non-3D Secure) authorization process, there are four primary parties involved: The card holder, the payment gateway, the acquiring bank, and issuing bank. Figure 8-1 shows the simplified diagram of the process of a regular credit card transaction without 3D Secure.

FIGURE 8-1: NON 3D-SECURE TRANSACTION

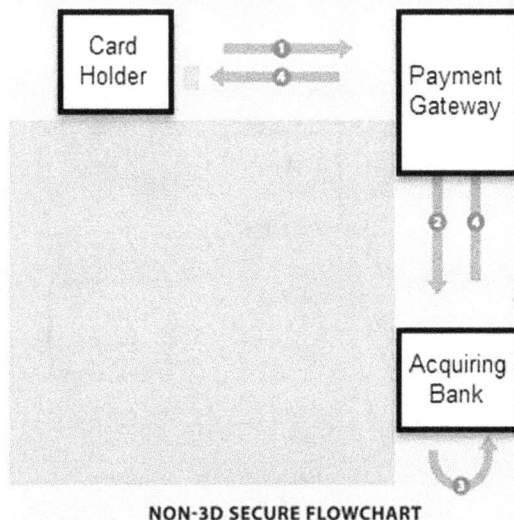

NON-3D SECURE FLOWCHART

1. The card holder enters their card information (16 digit card number, expiry date etc.) on the payment gateway
2. Payment Gateway submits the data to our acquiring bank
3. The acquiring bank authorizes the transaction (by communicating with the credit card network and issuing bank)
4. The response (success or failure) is passed back up the chain to the card holder

When comparing this process with 3D Secure, there is no communication between the issuing bank and the cardholder. As part of the acquiring bank's authorization process, the issuing bank looks purely at the card account and authorizes or denies the transaction based on "regular" account parameters (card active, funds available, card not expired etc.).

AUTHORIZATION VS. AUTHENTICATION

When discussing credit card transactions, the terms authorization and authentication are distinct. Authorization is the act of the issuer verifying the validity of the card details provided and consenting to the charge based on internal rules (e-commerce allowed, acquiring country allowed, funds available etc.). Authentication refers to the cardholder providing confirmation to the issuing bank, that it is indeed them performing a transaction. They are "authenticating" themselves in a manner similar to providing a known password to login to a website.

WORKING OF 3D SECURE TRANSACTION

With 3D Secure, a number of additional steps are added to the credit card process with the aim of authenticating the cardholder performing the transaction. Figure shows a very simplified diagram of 3D-Secure transaction process.

FIGURE 8-2: 3D SECURE TRANSACTION PROCESSING

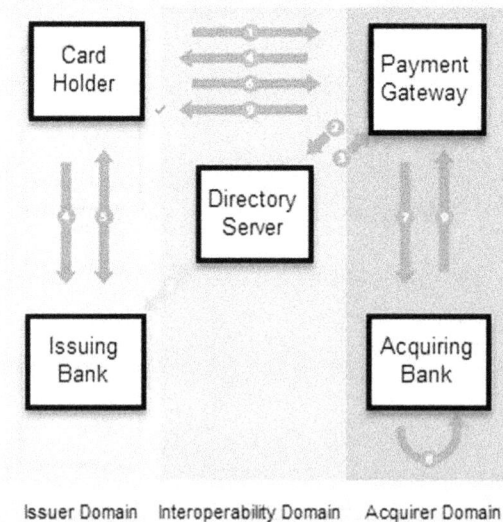

| Issuer Domain | Interoperability Domain | Acquirer Domain |

1. The card holder enters their card information (16 digit card number, expiry date etc.) on the Payment Gateway
2. Payment Gateway contacts a directory server to ascertain whether the card is enrolled in 3D Secure
3. The directory server responds with a message indicating that the card is registered
4. Payment Gateway uses the message to redirect the cardholder to a "3D Secure" page served by the issuing bank
5. The cardholder authenticates themselves to the issuing bank on the 3D Secure page (One Time PIN, known password etc.)
6. The result of this authentication is returned to Payment Gateway
7. Payment Gateway submits the card information and the 3D Secure authentication result to our acquiring bank
8. Our acquiring bank authorises the transaction (by communicating with the credit card network and issuing bank)
9. The response (success or failure) is passed back up the chain to the card holder

Use of 3-D Secure offers many advantages.

Liability Shift: The main benefit to companies using the 3D-Secure scheme is the availability of a liability shift for a successfully verified transaction. This offers protection by the card issuers against chargebacks as the liability is assumed.

No extra Cost: There are no extra costs to add 3D Secure onto your merchant account. Your acquiring bank may charge to add this onto your merchant number however; you may also find that your transaction charges lower because of using 3D Secure.

Easy to set up and Control: The set-up of the 3D-Secure scheme is controlled within your merchant account administration area. You can also set custom rules that could automatically validate cards registered with 3D Secure.

There are drawbacks of using 3-D Secure as well (sagepay.co.uk, 2015).

Chargebacks: Fully authenticated 3D-Secure transactions do not guarantee a liability shift; this is decided on the discretion of your merchant bank.

Not all Cards are Included: Right now, there are no similar initiatives for JCB or Diner's club cards.

SET (Secure Electronic Transactions)

SET is a set of standards for customers, merchants and banks to perform credit card transaction online. SET allows automatic processing of transactions in which all parties identify one another. SET was introduced jointly by VISA and MASTERCARD with technical designed by GTE, IBM, Microsoft, Netscape, RSA, SAIC, Terisa and VeriSign. SET utilize the digital certificates to authenticate the parties involved in transaction to minimize the possibility of repudiation charges. In contrast, SSL only provides server-side authentication and cannot protect against repudiation.

Here is how a transaction using SET works. First, the customer opens a credit card account with a bank that provides electronic payments and supports SET. The customer receives a digital certificate signed by the bank. The merchants have their own certificates. When customer places an order, the merchant sends a copy of its certificate that can be used by the customer to verify the identity of the merchant. On order placement and payment by the customer, the merchant requests the payment authorization from customer's bank. Once the payment is authorized, the merchant confirms the order and ships the goods or provides the service to the customer. Figure 8-3 illustrates working of SET.

FIGURE 8-3: SET OPERATION

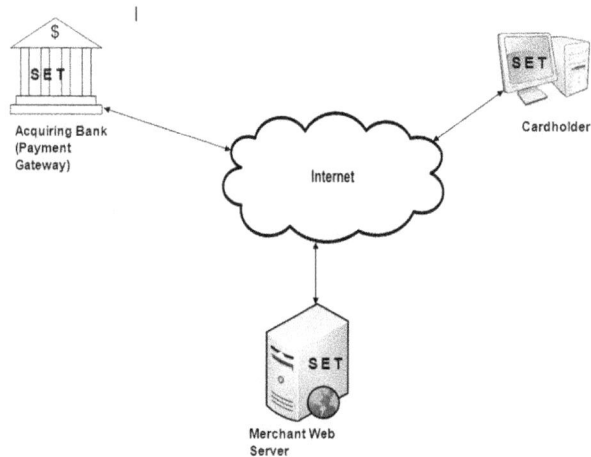

SET was intended to become the de facto standard payment method on the Internet between the merchants, the buyers, and the credit-card companies. Despite heavy publicity to win market share, it failed to gain widespread use. Reasons for this included network effect (need to install client software), cost and complexity for merchants to offer support, contrasted with the comparatively low cost and simplicity of the existing SSL based alternative, an client-side certificate distribution logistics.

Secure Socket Layer (SSL)

The Secure Sockets Layer (SSL) protocol uses encryption to provide privacy and integrity for data communication through a reliable end-to-end secure connection between two points over a network (IBM, 2015). SSL uses digital certificates which are electronic ID cards issued by a trusted party, to exchange keys for encryption and server authentication. The trusted entity that issues a digital certificate is known as a Certificate Authority (CA). The CA issues a digital certificate based on a limited period. When the expiration date passes, another digital certificate must be acquired. With SSL, the data that moves between a client and server is encrypted by a symmetric (private) key algorithm. An asymmetric (public) key algorithm is used for the exchange of the keys in the symmetric algorithm.

When a client attempts to connect to a secure server, an SSL handshake occurs. The handshake involves the following events:

- The server sends its digital certificate to the client.
- The client verifies the validity of the server digital certificate. For this to occur, the client must possess the digital certificate of the CA that issued the server digital certificate.
- If the handshake succeeds, these events occur:
    1. The client generates a random symmetric key and sends it to the server, in an encrypted form, by using the asymmetric key in the server digital certificate.
    2. The server retrieves the symmetric key by decrypting it.
    3. Because the server and the client now know and can use the symmetric key, the server and client encrypt data for the duration of the session.

When a client browser sends a request message to a server's secure Website, the server sends a hello request to the browser (client). The browser responds with a client hello. The exchange of these greetings, or the handshake, allows the two computers to determine the compression and encryption standards that they both support. Next, the browser asks the server for a digital certificate (i.e. proof of identity). In response, the server sends to the browser a certificate signed by a recognized certification authority. The browser checks the serial number and certificate fingerprint on the server certificate against the public key of the CA stored within the browser. Once the CA's public key is verified, the endorsement is verified. That action authenticates the Web server. Both the client and server agree that their exchanges should be kept secure because these exchanges involve transmitting sensitive private data (such as credit card numbers) over the Internet. To implement secrecy, SSL uses public key (asymmetric) encryption and private key (symmetric) encryption. Although public-key encryption is handy, it is slow compared to private-key encryption. That is why SSL uses private-key encryption for nearly all its secure communications. Because it uses private-key encryption, SSL must have a way to get the key to both the client and server without exposing it to an eavesdropper. SSL accomplishes this by having the browser generate a private key for both to share. Then the browser encrypts the private key it has generated using the server's public key. The server's public key is stored in the digital certificate that the server sent to the browser during the authentication step. Once the key is encrypted, the browser sends it to the server. The server, in turn, decrypts the message with its private key and exposes the shared private key. From this point on, public-key encryption is no longer used. Instead, only private-key encryption is used. All messages sent between the client and the server, are encrypted with the shared private key, and also known as the session key. When the session ends, the session key is discarded. A new connection between a client and a secure server starts the entire process all over again, beginning with the handshake between the client browser and the server.

FIGURE 8-4: WORKING OF SSL

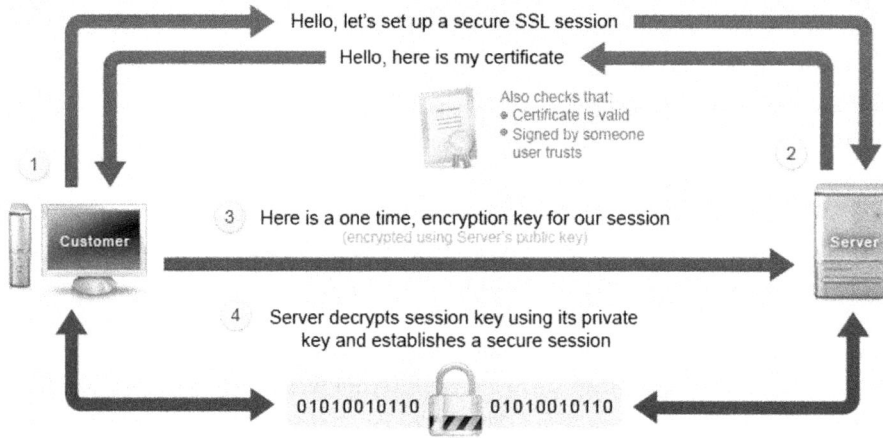

Hello, let's set up a secure SSL session

Hello, here is my certificate

Also checks that:
- Certificate is valid
- Signed by someone user trusts

1

2

3 Here is a one time, encryption key for our session
(encrypted using Server's public key)

Customer

Server

4 Server decrypts session key using its private key and establishes a secure session

01010010110 01010010110

The primary reason why SSL is used is to keep sensitive information sent across the Internet encrypted so that only the intended recipient can understand it. This is important because the information we send on the Internet is passed from computer to computer to get to the destination server. Any computer in between can see the credit card numbers, usernames and passwords, and other sensitive information if it is not encrypted with an SSL certificate. When an SSL certificate is used, the information becomes unreadable to everyone except for the server you are sending the information to. This protects it from hackers and identity thieves.

SET VS., SSL: A COMPARISON

The main difference between SET and SSL is that SET uses digital certificates for all involved parties, unlike SSL, which has only recently introduced this feature to its newer versions. As a result, SET provides for better authentication. As well, SET has better overall security. Unfortunately, it does have its drawbacks including complex implementation and higher costs than SSL.

SSL protocol secures data transmissions with a combination of public-key and symmetric-key encryption. SSL protocol always begins with a handshake that allows the server to authenticate itself to the client. If the server is not authenticated, the connection cannot be established. Whereas, SET uses dual signatures to secure a transaction. SET requires the purchase of software to use for an e-commerce site. The design of SET protocol requires the installation of an e-wallet on the client.

Unlike SSL, SET does solve the issue of identifying the cardholder in electronic transactions. However, there is a real cost involved in doing that, and is it really the area of greatest risk. It is generally agreed that the greatest risk with e-commerce is catastrophic losses from large-scale theft of credit card numbers. If that occurred, it would more likely happen with break-ins to file servers, and not during the transmission of transactions through a secure pipe. SET does not protect against that-SET begins and ends with the individual credit card transaction.

SET and SSL share some common characteristics too. Both SET and SSL offer:

- Confidentiality of information, and protection against; hacking; or other interception during transmission across a public network, through the encryption of data.

- Integrity of data (like SET, SSL has the capability to determine if messages have been altered).

- Verification that the merchant has been certified by a trusted Certificate Authority. This capability exists with SSL for the cardholder to verify that the merchant site is legitimate. If merchants want to conduct transactions in a secure environment, they must provide business licenses and other notarized proof of ownership to be certified by the CA.

LIMITATIONS OF SET AND SSL

A downside of both SSL and SET protocols is that they both require use of cryptographic algorithms that place significant load on the computer systems involved in commerce transactions. For the low and medium e-commerce applications, there is no additional server cost to support SET over SSL. For the large and medium term e-commerce server application, support of SET requires additional hardware acceleration resulting in 5-6% difference in server cost.

DISADVANTAGES OF SSL

Use of SSL does involve some disadvantages. Cost is an obvious disadvantage. SSL providers need to set up a trusted infrastructure and validate your identity so there is a cost involved. Because some providers are so well known, their prices can be overwhelmingly high. Performance is another disadvantage to SSL. Because the information that you send has to be encrypted by the server, it takes more server resources than if the information were not encrypted. The performance difference is only noticeable for web sites with very large numbers of visitors and can be minimized with special hardware. According to tests run by IBM, SSL can reduce the time it takes for data to travel between the user and website by two to ten times. It also increases network traffic by as much as 300 percent. This should not be a problem for broadband users, but dial-up and some mobile Internet users will experience greater load times on SSL websites. Regular Renewal, like a website domain and hosting plan, an SSL certificate expires after a short period of time---usually one to five years. You have to renew the SSL protection regularly and pay the subscription price again forever in order to keep the protection. If you forget to renew the SSL protection, your website will display an error on the user's computer stating that the certificate is not valid. Handling SSL from the client side is not always trivial, and languages may require additional extensions installed (which may or may not be available in a shared host environment).

Hyper Text Transfer Protocol Secure (HTTPS)

Hypertext Transfer Protocol Secure (HTTPS) is the secure version of HTTP, the protocol over which data is sent between your browser and the website that you are connected to. The 'S' at the end of HTTPS stands for 'Secure'. It means all communications between your browser and the website are encrypted. HTTPS is often used to protect highly confidential online transactions like online banking and online shopping order forms. Web browsers such as Internet Explorer, Firefox and Chrome also display a padlock icon in the address bar that visually indicates that a HTTPS connection is in effect.

FIGURE 8-5: COMPARISON OF HTTP AND HTTPS

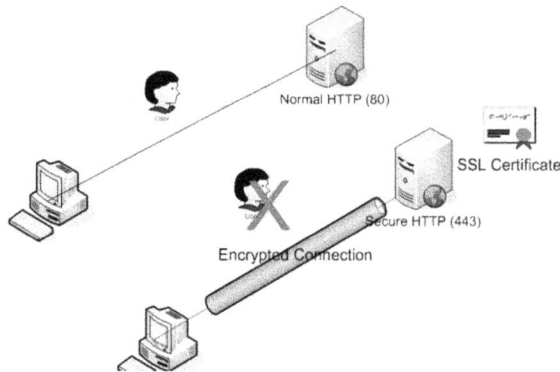

HTTPS pages typically use one of two secure protocols to encrypt communications - SSL (Secure Sockets Layer) or TLS (Transport Layer Security). Both the TLS and SSL protocols use what is known as an 'asymmetric' Public Key Infrastructure (PKI) system. An asymmetric system uses two 'keys' to encrypt communications, a 'public' key and a 'private' key. Anything encrypted with the public key can only be decrypted by the private key and vice-versa.

As the names suggest, the 'private' key should be kept strictly protected and should only be accessible the owner of the private key. In the case of a website, the private key remains securely ensconced on the web server. Conversely, the public key is intended to be distributed to anybody and everybody that needs to be able to decrypt information that was encrypted with the private key.

When you request a HTTPS connection to a webpage, the website will initially send its SSL certificate to your browser. This certificate contains the public key needed to begin the secure session. Based on this initial exchange, your browser and the website then initiate the 'SSL handshake'. The SSL handshake involves the generation of shared secrets to establish a uniquely secure connection between yourself and the website. When a trusted SSL Digital Certificate is used during a HTTPS connection, users will see a padlock icon in the browser address bar. When an Extended Validation Certificate is installed on a web site, the address bar will turn green.

🔒 Allied Bank Limited [PK] https://allied.direct.abl.com.pk/allied.proxy/

All communications sent over regular HTTP connections are in 'plain text' and can be read by any hacker that manages to break into the connection between your browser and the website. This presents a clear danger if the 'communication' is on an order form and includes your credit card details or social security number. With a HTTPS connection, all communications are securely encrypted. This means that even if somebody managed to break into the connection, they would not be able decrypt any of the data, which passes between you and the website.

The major benefits of a HTTPS certificate are:

- Customer information, like credit card numbers, is encrypted and cannot be intercepted
- Visitors can verify you are a registered business and that you own the domain
- Customers are more likely to trust and complete purchases from sites that use HTTPS

Transaction Security Standards

EMV

EMV stands for Europay, MasterCard, and Visa, the three companies that originally created the standard. EMV is a technical standard for smart payment cards and for payment terminals and automated teller machines, which can accept them. EMV cards are smart cards (also called chip cards or IC cards) which store their data on integrated circuits rather than magnetic stripes, although many EMV cards also have stripes for backward compatibility. They can be contact cards, which must be physically inserted into a reader, or contactless cards, which can be read over a short distance using radio-frequency identification technology. Payment cards, which comply with the EMV standard, are often called chip-and-PIN or chip-and-signature cards, depending on the exact authentication methods required to use them. The standard is now managed by EMVCo, a consortium with control split equally among Visa, Mastercard, JCB, American Express, China UnionPay, and Discover. The most widely known chip card implementations of EMV standard are VIS (Visa), M/Chip (MasterCard), AEIPS (American Express), CUP (China Union Pay), J Smart (JCB), and D-PAS (Discover/Diners Club International). Visa and MasterCard have also developed standards for using EMV cards in devices to support card-not-present transactions over the telephone and Internet. MasterCard has the Chip Authentication Program (CAP) for secure e-commerce. Its implementation is known as EMV-CAP that supports a number of modes. Visa has the Dynamic Passcode Authentication (DPA) scheme, which is their implementation of CAP using different default values. The EMVCo standards have been integrated into the broader electronic payment security standards being developed by the Secure POS Vendor Alliance, with a specific effort to develop a common interpretation of EMVCo's place relative to, and interactions with, other existing security standards, such as PCI-DSS.

Payment Card Industry Data Security Standard (PCI DSS)

The Payment Card Industry Data Security Standard (PCI DSS) is a proprietary information security standard for organizations that handle branded credit cards from the major card schemes including Visa, MasterCard, American Express, Discover, and JCB. Those cards, which are not part of a major card scheme, are not included in the scope of the PCI DSS. The PCI Standard is mandated by the card brands and administered by the Payment Card Industry Security Standards Council. The standard was created to increase controls around cardholder data to reduce credit card fraud via its exposure. Although the PCI DSS must be implemented by all entities that process, store or transmit cardholder data, formal validation of PCI DSS compliance is not mandatory for all entities. Currently both Visa and MasterCard require merchants and service providers to be validated according to the PCI DSS. Validation of compliance is performed annually, either by an external Qualified Security Assessor (QSA) that creates a Report on Compliance (ROC) for organizations handling large volumes of transactions, or by Self-Assessment Questionnaire (SAQ) for companies handling smaller volumes. Issuing banks are not required to go through PCI DSS validation although they still have to secure the sensitive data in a PCI DSS compliant manner. Acquiring banks are required to comply with PCI DSS. In the event of a security breach, any compromised entity, which was not PCI DSS compliant at the time of breach, will be subject to additional card scheme penalties, such as fines.

Mobile Transactions

Mobile devices and smartphones are still new ideas in the tech community. Compared to home computers and laptops, the common mobile smartphone has only been around for a couple of years now. However, having quickly grown to popularity, there are more early adopters now than we have ever seen before.

Working of Mobile Transactions

Customer visits an online shopping website and after checking out of the website, customer inputs his mobile account number into his mobile payment app and click pay. Customer will receive a USSD[13] alert asking to input 4-digit MPIN (mobile pin code), also informing him about the amount being charged, the online shopping website that is asking for the amount, date, time, and unique transaction ID. After customer enters the correct MPIN, customer receives an SMS informing him about the successful transaction. On the screen, the retailer will inform the customer of a successful transaction. Figure 8-6 shows how a mobile transaction works.

FIGURE 8-6: MOBILE TRANSACTION

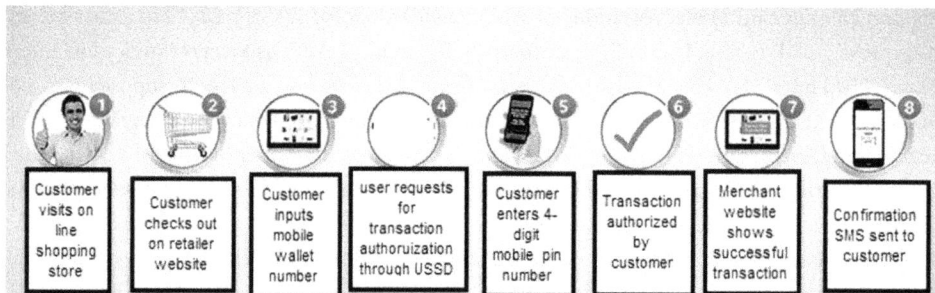

| Customer visits on line shopping store | Customer checks out on retailer website | Customer inputs mobile wallet number | user requests for transaction authoruization through USSD | Customer enters 4-digit mobile pin number | Transaction authorized by customer | Merchant website shows successful transaction | Confirmation SMS sent to customer |

---

[13] Unstructured Supplementary Service Data (USSD) is a protocol used by GSM cellular telephones to communicate with the service provider's computers.

Securing Mobile Transactions

Mobile devices can often be less secure than computers within your own home network. There is also the possibility of running into bugs in the mobile app or website. Security for mobile phones has been advancing tremendously, but has to hit a peak. Following are ways that consumers can use to secure their mobile transactions (Rocheleau, 2015).

DOWNLOAD APPS FROM TRUSTWORTHY SOURCES: Downloading third party applications from other areas outside the App Store/Play Store is very risky. Always download apps from trustworthy sources.

CHECK APP REVIEWS AND RATINGS: In the App Store/Play Store listings page you will see rating mar for each application. The chart ranges from half star all the way up to five full stars. Besides the actual user ratings, you can also see how many people have voted, along with some of their reviews. Check the app reviews before downloading and possibly purchasing items through their software.

PROTECT YOUR MOBILE DEVICES WITH PASSWORDS: Always use a device password to protect your mobile device. You may also lock specific applications as well, but this may prove more convoluted than useful.

SEND DATA OVER A SECURE INTERNET CONNECTION: Whenever you are purchasing things or transferring money online, always do this over a secure connection.

CHECK MOBILE WEBSITES FOR HTTPS: When you first load up any website in your browser always check for the secure HTTPS connection. This will guarantee that any data passed between your device and the server is only shared between those two machines.

Mobile Payments

The acceleration of smartphone adoption has resulted in an eruption of the mobile payment market. Online payments using mobile devices continue to rise globally and now accounting for 27.2% of the total online payments made. The average transaction value (ATV) of digital goods purchased via tablets has surpassed the figure for desktops/laptops. (Adyen.com, 2015). The transactional value of global mobile payments has skyrocketed from only $240 billion in 2011 to nearly $1 trillion forecasted by 2015. Additionally, the shift from PC to mobile is more and more apparent, with smartphones being the most common starting place for online activities (versus PC only a few years ago). Interestingly, it the internet and technology companies, not the banks, will lead the mobile payments market share over the next two to four years. There are three primary methods for mobile payments: **O2O (online-to-offline), enforced security, and partnerships with banks**.

O2O has become very prevalent in the mobile world, including making purchases through mobile apps, QR code scanning for commercial purposes and deal-of-the-day websites such as Groupon, Gilt Group and Living Social. There are various benefits of O2O. First, it will allow businesses to track every transaction that is done and monitor the effects of campaigns in each channel. Second, O2O will benefit from the diversity of online advertising methods, while providing customers with the real-life experience with products offline. Finally yet importantly, it will help businesses track and study consumer behavior by pushing advertisements on a real-time basis. Despite the countless security measures that have been implemented through e-commerce transaction phases, e-commerce providers are still facing various security threats give the nature of the industry. As the mobile payments industry matures, we will see increase self-imposed security measures and regulatory requirements governing the industry. Value-added services (such as mobile wallet and microfinance) are emerging within mobile payments. For example, China Merchants Bank began cooperating with Tencent's WeChat in May 2013 to provide social media banking. The service initially used to notify customers about things like their account balances and credit card purchases. Today, the service has expanded to an offering range of banking services, including money transferring and credit card payment. The Kotak Mahindra Bank of India offers hashtag banking. Customers can bank with Kotak using Twitter. Using Twitter, customers can access various services such as account balance, mobile & DTH recharge, order book, transaction history, cheque book request and many more. To avail this service, customers need to link their twitter handle to their bank account. They can then get instant updates simply by tweeting to @KotakBankLtd. Kotak bank maintains a hashtag list that customers can use to know what to tweet. The responses come to customers as a DM so only customer can view them. The growing mobile payments will have a variety of impacts on businesses such as retail category, geographic presence, price range, and the targeted demographic population.

Paytm is India's largest mobile commerce platform. Paytm started by offering mobile recharge and utility bill payments and today it offers a full marketplace to consumers on its mobile apps. Paytm has over 80 million registered users. In a short span of time, Paytm has scaled to more than 60 Million orders per month. Paytm entered India's e-commerce market in 2014, providing facilities and products similar to businesses such as Flipkart, Amazon.com, Snapdeal. In 2015, it added booking bus travel.

Instamojo Inc. is a web-based Indian startup that primarily provides customers a platform to sell digital goods like ebooks, reports, comic books, music, software, templates, photos, tutorials etc. by listing the item on the website and sharing the web link with others. The company has over 4,300 sellersand has received funding from angel investors like Rajan Anandan and Sunil Kalra. Instamojo pays the seller 95 percent of every successful transaction for digital goods and 98 percent of every successful transaction for physical goods and products not hosted on their servers.

E-Finance

E-finance, loosely defined as the use of internet in the provision of financial services. From online brokerage services and e-commerce to credit management and online payments, many academics and industry professionals believe that the financial services industry as a whole is undergoing a fundamental transformation because of this phenomenon (CKGSB, 2014).

Alibaba has been a first mover by combining payment with commerce, with Google, Amazon and Apple trying to follow suite to develop comparable payment services. As the internet increasingly shifts from PC to mobile, there will be a rapid migration and emphasis on mobile payments, and companies who are best positioned for this transition to mobile will reap the benefits of this transformation. Mobile payments will be especially critical in the O2O world, where it will effectively link the real world with the online world and serve as a source of information flow, capital flow and logistic flow. The continuous growth and innovation in e-finance will fundamentally alter financial systems. In China, Baidu, Alibaba and Tencent ("BAT") were the first wave of innovative businesses to leverage e-finance.

Rapid progress in information and communication technology (ICT) is a key factor changing the financial sector in many countries and its effects are predominantly strong in finance. Financial services are intangible. Progress in ICT has significantly reduced the cost of providing financial services. Low operating costs are influencing both the performance of e-finance providers and the structure of the finance industry namely the "e-broking". E-finance is the most promising areas of e-commerce as financial services are information- intensive and often involve no physical delivery. E-finance is a driving force that is changing the landscape of the finance industry vitally, specifically towards a more competitive industry and has distorted the boundaries between different financial institutions, enabled new financial products and services, and made existing financial services available in different packages (Burger, 2008).

Two emerging areas of e-finance -- peer-to-peer lending and microfinance -- are disrupting the status quo even further and moving their markets into a new stage of development. Facilitators of peer-to-peer borrowing and lending, such as Prosper.com, GlobeFunder (http://www.globefunder.com/) and Lending Club (http://www.lendingclub.com/) are taking advantage of the networking and online application processing power of the Internet. People's Bank of China was planning to regulate e-finance platforms after several investment products launched by e-commerce firms have become popular. The central bank was targeting online investment products sold by Alibaba, its rival Tencent and Baidu, which offer annual yields that top 7% (Wong, 2014).

Phishing Attacks

Phishing is a technique for committing fraud against the customers of any type of online businesses. Phishing attacks are particularly important to financial institutions because customers of financial institutions expect a high degree of security to be maintained over their personal and financial information. The basic structure of a phishing attack is simple. The attacker sends e-mail messages to a large number of recipients who might have an account at the targeted Web site (New Domino Bank) is the targeted site in the example shown in the figure 8-7. The e-mail message tells the recipient that someone has attempted to gain access to his account and as a result, the account has been locked. It is necessary for the recipient to go to bank website and unlock his account. The e-mail message includes a link that appears to be a link to the home page of the New Domino Ban Web site. However, the link actually leads the recipient to the phishing attack perpetrator's Web site, which is disguised to look like the targeted Web site. The unsuspecting recipient enters his or her login name and password, which the perpetrator captures and then uses to access the recipient's account. Once inside the victim's account, the perpetrator can access personal information, make purchases, or withdraw funds at will.

FIGURE 8-7: A PHISHING MESSAGE

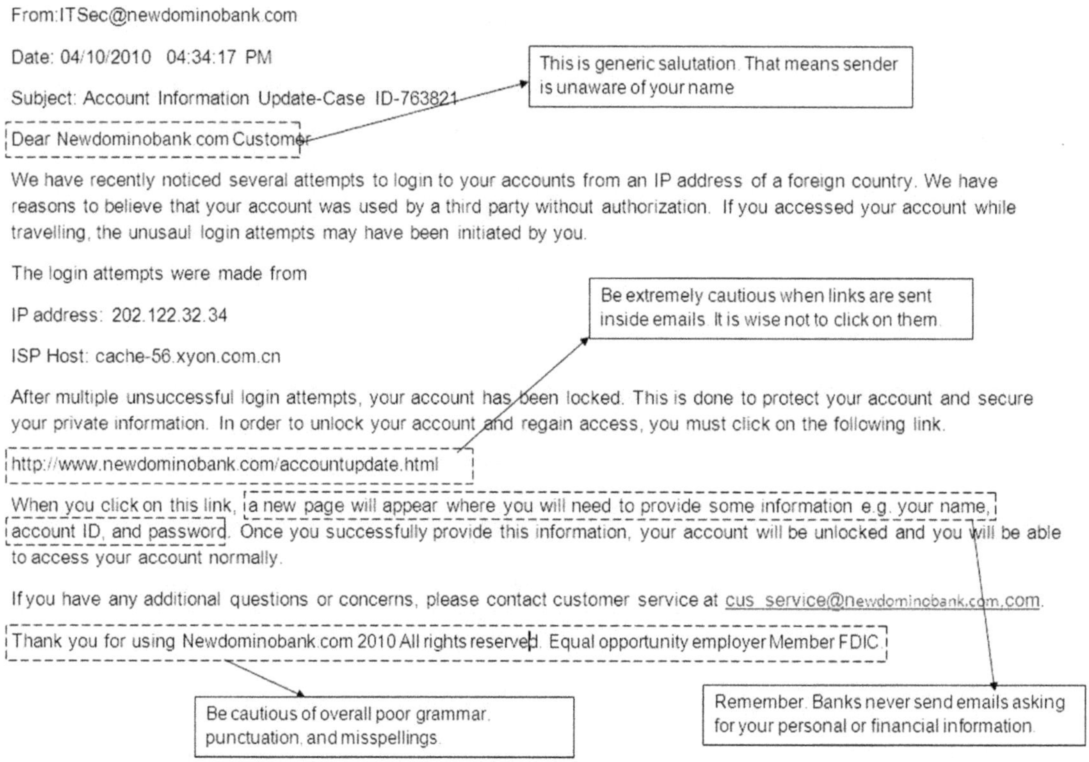

From:ITSec@newdominobank.com

Date: 04/10/2010 04:34:17 PM

Subject: Account Information Update-Case ID-763821

> This is generic salutation. That means sender is unaware of your name

Dear Newdominobank.com Customer

We have recently noticed several attempts to login to your accounts from an IP address of a foreign country. We have reasons to believe that your account was used by a third party without authorization. If you accessed your account while travelling, the unusaul login attempts may have been initiated by you.

The login attempts were made from

IP address: 202.122.32.34

ISP Host: cache-56.xyon.com.cn

> Be extremely cautious when links are sent inside emails. It is wise not to click on them.

After multiple unsuccessful login attempts, your account has been locked. This is done to protect your account and secure your private information. In order to unlock your account and regain access, you must click on the following link.

http://www.newdominobank.com/accountupdate.html

When you click on this link, a new page will appear where you will need to provide some information e.g. your name, account ID, and password. Once you successfully provide this information, your account will be unlocked and you will be able to access your account normally.

If you have any additional questions or concerns, please contact customer service at cus_service@newdominobank.com.com.

Thank you for using Newdominobank.com 2010 All rights reserved. Equal opportunity employer Member FDIC.

> Be cautious of overall poor grammar, punctuation, and misspellings.

> Remember. Banks never send emails asking for your personal or financial information.

The links shown in phishing e-mails are frequently disguised. One common technique to hide real URL is to use "@" sign after domain name (Drake and Oliver, 2004). This instructs the Web server to ignore all characters that precede the "@" and only use the characters that follow it. For example, a link that displays:

https://Newdominobank.com@218.25.28.127/login.html

looks like it is an address at old National Bank. However, the "@" sign causes the Web server to ignore the "Newdominobank.com" and takes the user to the login.html page at the IP address "218.25.28.127"

The e-mail sent to the customer by phishing attack perpetrator contains a link that appears as follows:

http://www. NewdominoBank.com/index.html

But when the victim clicks the link, the browser opens a completely different URL:

http://fishboom.com/fish/webscr.dll

Instead of the URL it shows in the e-mail client, the link in the phishing e-mail actually includes following JavaScript code:

<A   onmouseover="window.status='http://www.Newdominobank.com/cgi-bin/webscr?cmd=_login-run'; return true" onmouseout="window.status='http://www.Newdominobank.com/cgi-bin/webscr?cmd=_login-run"href="http://   fishboom.com.com/fish/webscr.dll">http://www.Newdominobank.com/cgi-bin/webscr?cmd=_login-run</A>

Many e-mail client software does not show this code and victims may never know they opened a phishing website. Phishing attack executors can use a variety of other techniques to hide the URLs e.g. using code that pops up windows that look exactly like a browser address bar. The window is coded to pop up over the browser's address bar.

Phishing Attack Countermeasures

An effective way to combat phishing is to improve users' ability to identify SPAM or source of an email message. The key to achieve this ability is the improved mail transfer protocols used on the Internet. There are several ongoing efforts to improve the Internet's mail transport protocols so that e-mail recipients' could identify the source of an e-mail message. The most effective technique for companies to counter phishing attacks however is to educate their Web site users e.g. by continuously updating them about phishing attacks and their prevention. Another option for the companies is to obtain services of consulting firms that specialize in anti-phishing work. These consulting firms monitor the Internet for new Web sites that use the company's name or logo and monitor the chat rooms used by criminals to learn about possible phishing attacks.

Software are available that can notify you when you are being directed to a Web site known to be fraudulent. These products continually update a list of known fraudulent Web sites and allow your browser to access this list. Of course, many phishers constantly change the sites they use to get around this software and to escape detection. One product of this kind is Microsoft's Phishing Filter.

Some software tries to detect phishing by looking for characteristics of previously detected attacks and guessing whether a given site is likely to be fraudulent. One free product that does this is Spoof guard (crypto.stanford.edu/SpoofGuard/).

Automatic display of domain name: Other software fights phishing by displaying information such as the real (as opposed to the spoofed) domain name of any Web site you visit. One free product of this kind is SpoofStick.

Some Web-based email providers offer built-in privacy features. However, in order to reap the benefits, both the sender and recipient must use the service. Hush mail is one provider of secure Web email.

Fraud in E-commerce

Quite often fraud is not considered when selling online and only when you get a letter from your bank saying a transaction is about to be reversed, or you get an email from PayPal saying your account is suspended pending an investigation is the small business website owner introduced to credit card fraud online. The front line of defense against the Internet fraudsters is a proactive approach on the part of anyone who collects, possesses, uses or transmits sensitive data.

The Uniform Commercial Code (UCC)

UCC regulates and defines the responsibilities of counterparties in business and banking transactions. The code states that liability and monetary loss in a fraudulent transaction is split between the counterparties in a transaction based on each party's due diligence and negligence. Consequently, to reduce liability in the event of a fraudulent transaction, it is important to have proper controls in place. Advances in technology have reduced the effectiveness of traditional fraud prevention techniques and have even enabled new forms of fraud. For example, in the past, many governments relied upon physical security features embedded in check stock to prevent check fraud. These included watermarks, unique colors, and graphical designs. Advanced duplication technology and remote deposit capture have reduced the effectiveness of these physical measures to prevent fraud. The 2014 was not a good year for companies and data security. There were a large number of cases in which personal and financial information (such as login credentials, personal details, and credit card numbers) was stolen from consumers and companies. Sometimes it happens by way of a security loophole other times, a rogue employee.

In January 2014 Target reported a massive data breach that affected somewhere around 70 million users. On top of credit card numbers and PIN information, the thieves got away with full names, physical addresses, and phone numbers. The Target incident is one of the highest profile breaches in the past five years. In February, hackers were able to get into Kickstarter's databases and pull customer information, including usernames, passwords, email addresses, and phone numbers. In March 2014, eBay experienced a big setback when a significant portion of their customer database was hacked into and stolen. To pull this off, the hackers used compromised eBay employee login credentials to navigate the corporate network. Stolen data included full names, passwords, physical addresses, and phone numbers. In April 2014, AT&T reported being hacked by three of their contractors. These contractors gained access to a database of personal records, which included social security numbers. (Lee, 2014).

For online merchants and their customers, the growing threat of online theft is a real concern. With every step retailers take to tighten security and prevent malicious activity, fraudsters seem to up their game and outwit them. This is especially true as brick-and-mortar retailers prepare to switch to EMV later in 2015, at which point online merchants become easier targets than traditional POS systems. Online fraud cost companies billions of dollars each year, cutting dramatically into earnings. Both online and off, the true cost of dealing with online fraud is growing, with retailers losing $3.08 for every dollar of fraud they incurred in 2014. This is up from $2.79 in 2013. Mobile fraud is driving these costs upward, since the cost of online fraud is higher on mobile platforms than through other forms of payment.

When a customer conducts a fraudulent transaction, a retailer loses the merchandise, as well as the cost to prepare and ship that merchandise. This does not even include the cost to secure systems in order to prevent theft in the first place. As fraud becomes more prevalent, online merchants are tasked with trying to keep up with the latest techniques, forcing them constantly spend money into fraud detection and prevention. Online shopping has brought "friendly fraud" to an all-new level, as customers seem to find it easier than ever to reverse transactions. Chargebacks are designed to protect customers from fraudsters, but in recent years, customers have begun using it in the place of refunds. The perception that "the customer is always right" seems to extend to credit card companies, who put the burden of proof on retailers when making a decision on a dispute.

Unfortunately, an estimated 86% of chargebacks are fraudulent, which is bad news for both retailers and customers. With friendly fraud, a customer receives an item and disputes it, getting a full refund while also being able to keep the item. Whether a chargeback claim is decided in a retailer's favor or not, the retailer is charged a fee and, over time, multiple chargebacks can leave a retailer without a payment-processing provider (Rampton, 2015).

Fraud Protection

There are services that work in the favor of retailers, including Authorize.net's Advanced Fraud Detection Suite. This fraud protection can be customized to a business's needs but it generally includes multiple filters that catch fraud before it happens. Chargeback prevention is generally in the hands of the e-tailer, however. For e-commerce businesses, it's vital to have generous return policies and communicate those policies clearly to customers. Instead of pushing your policies on every customer who requests a return, try to work with each customer to find an amicable solution to every issue.

Online fraud is an ongoing problem for retailers, but with the right precautions in place, a business can reduce the damage significantly. Companies should learn to spot suspicious transactions and protect against them, preferably using processing software that stops it before it comes through (Rampton, 2015). There are some good business practices that can be used to minimize fraud in E-commerce.

NO FREE EMAIL ADDRESSES. By not allowing free email addresses, you will immediately eliminate over 90% of credit card fraud. This is the single biggest protection method.

BIN COUNTRY MATCHING: This method applies only to website owners who have their own merchant facility. The BIN or Bank Identification Number refers to the first four digits on a credit card, which is used to identify the bank that issues that credit card. BIN's are stored on publicly available databases so a well-designed ecommerce payment page will run a check on the customers BIN and report to you that the order was placed with a card from another country or an odd bank. This is important; it allows you to identify behaviour that is at odds with the majority of your customers. For example, it is very common for a fraudster to buy a stolen credit card number online and use it. If you only trade in USA, you can limit the use of credit cards that are not from USA.

IP ADDRESSES COUNTRY MATCHING: Every internet user has an IP address assigned to them in order to use the internet. IP addresses are assigned regionally which means your IP address reveals what city you live in. A well-designed shopping cart payment portal will run a check on the users IP address to determine what country that person lives in. A smart way to secure you is to run an IP address check in conjunction with a BIN check. If the customer is in another country and the credit card is from the US, you know the order is likely fraud and you can cancel the order and inform the bank that the owner's card is used for suspicious activity.

USE THE AVS: Use your credit card processor's AVS (Address Verification System). This system typically captures the numeric portion of the ZIP code, and sometimes the address line of the billing address on your order form. Then it compares these with the address and ZIP code listed on the customer's credit card account. If they do not match, the charge is not authorized. Nor will it help if the credit card owner's ID has been stolen or is being used without authorization, for example, by a family member. However, used in conjunction with other methods, it will cut down your rate of fraudulent orders.

Micropayments

Merchants must pay a transaction fee for each credit card transaction. For inexpensive items, this fee may actually be higher than item's price. Micropayments are payments that generally do not exceed $10.These payments are used for nominally priced products and services (music, pictures, text or video) to be sold over the Web. Some examples of micropayment service providers include Cartio Micropayments (cartio.com), Digicash (digicashinc.com), and Cybercash (paypal.com/cybercash). These companies aggregate the micropayments and charge the merchant at the end of of the billing period in order to justify the transaction fees.

PayPal defines a micropayment as a transaction of less than five British pounds while Visa prefers transactions under $20, and though micropayments were originally envisioned to involve much smaller sums of money, practical systems to allow transactions of less than 1 USD have seen little success. One problem that has prevented their emergence is a need to keep costs for individual transactions low, which is impractical when transacting such small sums even if the transaction fee is just a few cents.

Micropayments have to be suitable for the sale of non-tangible goods over the Internet. This imposes requirements on speed and cost of processing of the payments: delivery occurs nearly instantaneously on the Internet, and often in arbitrarily small pieces. On the other hand, the bottleneck in sales of tangible goods, handling and shipping, sets a lower bound particularly for costs to remain economical. With the rising importance of intangible (e.g. information) goods in global economies and their instantaneous delivery at negligible cost, "conventional" payment methods tend to be more expensive than the actual product. On the other hand, billing for small portions of a product or service reduces the need of security.

Advantages and Disadvantages of Micropayment Systems

Using micropayments provides many advantages

ANONYMITY: Setting up an online account with a micropayment service provider allows one to conduct financial transactions online with some anonymity.

SPEED: Micropayment accounts allow for quick and convenient purchase of real and virtual goods and services.

SCALABILITY: Micropayment systems can grow easily to accommodate additional trades, and new products, or services.

SECURITY: Fewer online transfers of actual payment lead to fewer opportunities for actual theft or abuse. Further, it is much easier to contain the scope of theft or abuse using a micropayment system.

There are also some disadvantages of using micropayments.

INSECURE DATA: If sensitive account information is compromised, the account holder is left vulnerable to more than just the losses from the investment in the account, often secondary or tertiary accounts may be compromised as a result.

MICROPAYMENT VENDOR/PROCESSOR DISHONESTY: Account holders may lose their investment in the micropayment system if the payment processing company is dishonest, or otherwise deceptive.

EXCESSIVE, TAXES, FEES, AND CHARGES: Individual transactions end up costing the buyer more over the long term as individual taxes, fees, and charges, when combined and compared with a single larger purchase, reveal that the purchases actually cost more than if a single large purchase was made.

EXCESSIVE MAINTENANCE COSTS: With the explosion in the sheer number of micro transactions, actually auditing or reviewing such transactions quickly becomes extraordinarily expensive. Proportionally the number of customer disputes over failed or undesired individual purchases increase as well.

Early Micropayments systems

The micro-payment idea originated with banks in the early 90's. It was a pretty solid idea; load around 150 dollars on a card to buy things you need, which is safer than using a credit card and more convenient than using cash. Around 1995, micropayments were put into practice in the US and Belgium. One of the most famous micropayment systems in Belgium was called Proton. It grew in popularity after 1996 but around 1998 there were some questions about the security of the card. If the amount that you wanted to charge up was small enough, you could do that without giving your pin number. The banks corrected that security flaw immediately. Nevertheless, the damage had been done to Proton's image (bitcoinist.net, 2015). After 2001, Proton's popularity began to decrease, so banks began putting fees on the system. The card users were hit by fees as well as the surcharge for merchants and retailers that used proton terminals in their shops. This eventually led to the slow but certain death of Proton at the end of 2014. Though banks gave up on micropayments, other sectors did not abandon that idea so easily. The gaming sector picked up where banks left off and modified micropayments to suit their needs. These micro transactions were successfully introduced in Massively multiplayer online role-playing games (MMORPGS) and then spilled over to other games, like Assassin's Creed Unity. Steam and PlayStation Network have integrated micro transactions as well. Some game reviewers and the gaming community in general, have commented on this recent development. In the late 1990s, established companies like IBM and Compaq had microtransaction divisions, and research on micropayments and micropayment standards was performed at Carnegie Mellon and by the World Wide Web Consortium.

MILLICENT

Millicent, originally a project of Digital Equipment Corporation, was a micropayment system that was to support transactions from as small as 1/10 of a cent up to $5.00. It grew out of The Millicent Protocol for Inexpensive Electronic Commerce, which was presented at the 1995 World Wide Web Conference in Boston, but became associated with Compaq after that company purchased Digital Equipment Corporation. The payment system utilized symmetric cryptography.

NETBILL

The NetBill electronic commerce project at Carnegie Mellon University researched distributed transaction processing systems and developed protocols and software to support payment for goods and services over the Internet. It featured pre-paid accounts from which micropayment charges could be drawn. Initiated in 1997, NetBill seems to have died completely sometime after 2005.

Current Micropayments Systems

KYASH

It enables you to collect cash payments as less as INR 50/ from a countrywide network of small shops. If a customer has a mobile and cash to pay then he can transact using Kyash (http://www.kyash.com).

M-PEASA

M-Pesa is the world's most successful money transfer service. It enables millions of people who have access to a mobile phone, but do not have or have only limited access to a bank account, to send and receive money, top-up airtime and make bill payments. Customers register for the service at an authorized agent, often this is a small mobile phone store or retailer, and then deposit cash in exchange for electronic money, which they can send to their family or friends. Once they have registered all transactions are completed securely by entering a PIN number and both parties receive an SMS confirming the amount that has been transferred. The recipient, who does not have to use the same network, receives the electronic money in real-time and then redeems it for cash by visiting another agent. In addition to the millions of person-to-person (P2P) and customer-to-business (C2B) transactions, M-Pesa has made a big difference to people's lives in other ways. In Tanzania for example, where the cost of travel prevents many people from getting the medical care they need, one non-governmental organisation (NGO), Comprehensive Community Based Rehabilitation in Tanzania (CCBRT), has used M-Pesa to send patients the money to pay for their travel to its hospital. In Kenya, water company customers use M-Pesa to buy credits, which is then used to pay for fresh, clean water when they need it.

ITELEBILL

iTelebill (www.itelebill.com) are the mobile and phone payment specialists for online micro billing requirements. They specialize in the Online Dating, Social Networking, Web Streaming, and Online Publishing industries.

YIPPSTER

Customers can use this service at (Yippster.com) to pay for premium digital content with their phone talk time. This means if a customer has a prepaid phone, the charge is deducted from his/her talk time balance. If the customer has a post paid phone, the charge is added to the monthly phone bill.

HEARTLAND PAYMENT SYSTEM

Heartland Payment Systems, Inc. (www.heartlandpaymentsystems.com) is a Fortune 1000 company that provides debit, prepaid, and credit card processing, mobile commerce, e-commerce, check processing, payroll services, billing services, marketing services, security technology, lending services and a growing line of industry-specific business facilitation solutions for small to mid-sized merchants and enterprises. The easier it is for consumers to spend the more they buy. That's why, at Heartland Payment Systems, we're changing the landscape of cashless payments by creating state-of-the-art solutions that manage the small-ticket payments cycle from start to finish. Enabling your customers to pay for purchases using credit/debit cards, hotel room keys and contactless tags that can be affixed to mobile devices has a positive effect on your bottom-line.

PAYPAL MICROPAYMENTS

PayPal offers support for Micropayments to merchants for US to US, UK to UK, Australia-to-Australia, and EU-to-EU transactions for business and premier accounts. PayPal's micropayments price is 5% + $.05 and is designed for merchants who process low-value transactions (typically under $10 in value). The micropayments rate is available to all merchants and in all countries where Business accounts are available. If you sign up for micropayments, you will be charged the micropayments rate on all transactions regardless of payment size. Each PayPal account is associated with only one merchant processing rate. That rate determines the fee that is applied to funds received into that account. For example: if your Premier/Business account rate for receiving funds is 2.9% + $0.30, using PayPal's 5% + $0.05 micropayments rate would reduce the total transaction fee charged to payments received below the value of $12 (per payment). However, if you accept payments that are greater than $12, you would pay a lower processing charge by accepting the payment into the account set with the 2.9% + $0.30 rate.

FLATTR

Flattr is a micropayment system - more specifically, a micro donation system - that launched publicly in March 2010 on an invite-only basis, and then opened up to the public on 12 August 2010. Flattr is a project started by Peter Sunde and Linus Olsson. Users are able to pay a small amount every month (minimum 2 euro's) and then click Flattr buttons on sites to share out the money they paid in among those sites, kind of like an Internet tip jar. (The word "flattr" is used as a verb, to indicate payments through the Flattr system - so when a user clicks a Flattr button and they are logged in to the Flattr site, they are said to be "flattring" the page they are on.) Sunde said, "We want to encourage people to share money as well as content.

PAYCLICK

A micropayment system set up by Visa Inc in Australia, Payclick allows users to fund an account that is then drawn from when purchases at participating online retailers are made.

EASYPAISA

With Easypaisa (www.easypaisa.com.pk), customers have access to the easiest way to conduct their financial transactions, whether they are related to paying bills, sending/receiving money within Pakistan, receiving money from abroad, purchasing airtime (easyload) for their mobile phones or giving donations etc. Any person can use Easypaisa services by visiting their nearest Easypaisa shop, Telenor Franchise, Telenor Sales & Service Center or a Tameer Microfinance Bank branch. As of June 2011, there are thousands of Easypaisa shops operating all over Pakistan. Approved by the State Bank of Pakistan, everyone can enjoy secure branchless banking service through Easypaisa with instant transactions.

Zong mobile payments (www.timepey.com) are a micropayment system that charges payments to users' mobile phone bills. This service can be used to purchase virtual goods in online games and social networks.

Uses of Micropayments

ONLINE MAGAZINES AND NEWSPAPERS: A micropayment solution could be used to pay subscription fee to online magazines and newspapers.

ONLINE MUSIC AND VIDEOS: A micropayment solution could be used to charge €0.01 to listen to a song directly online or €1 to download it.

SOFTWARE: Independent software developers could charge small amounts of money for their software components. A micropayment solution would make it possible to charge for example $0.1 and thereby hopefully find more buyers.

GAMES: An online car game might charge $0.1 for a new set of tires, or a multiplayer boxing game might charge $0.05 per match. Gambling and betting games are strong candidates for these types of payment solutions.

MESSAGE BRIDGING: Email is usually free but people would probably be willing to pay for services such as sending fax, paper invoices, physical postcards, or SMS.

ALERTS: It is possible to charge for services such as stock price alerts or weather alerts sent to a cell phone while sailing. These could be initiated through the Internet in which case the proposed micropayment solution could be used.

Business-to-Business (B2B) Transactions

The overall size of the B2B payment market and size of the average payment are much larger than B2C. B2B payment systems are complex and require many documents to complete the transaction (e.g. bill of lading, financial documents, regulatory documents etc.). These systems also need to be integrated with existing ERP (Enterprise Resource Planning) systems of the company. B2B payment systems can be used to expedite both transactions conducted on exchanges and those conducted between existing trading partners.

Desirable Characteristics of B2B Payment Systems

B2B payment system should possess few desirable characteristics.

- **PAYMENT GUARANTEE:** It should provide assessment of creditworthiness of buyer to seller (to guarantees the payment by the buyer to the seller)

- **ESCROW SERVICE:** It should provide some guarantee to buyer that goods would be of the quality specified and will be delivered.

- **NON-REPUDIATION:** It should permit sellers to conduct business with unknown parties without fear of repudiation or fraud.

- **MULTIPLE PAYMENT METHODS:** It should support multiple payment methods that buyers can use to pay with (such as credit card, debit card, ACH check, EFT etc.)

- **FASTER COLLECTION:** It should provide faster collection of funds to the seller

- **REPORTING SERVICES:** It should provide reporting of delivery of goods, approval, invoicing, payment, collections, etc.

B2B Electronic Payment Solutions

With the increased use of consumer online bill pay and online banking, more and more payments are going electronic, yet most business payments are made via paper check. A 2010 study by the Federal Reserve found that between 2000 and 2009, B2B remittance checks were up 53 percent from 3.9 billion to 6 billion. There could be multiple reasons for this slow adoption. The accounting systems most businesses use are not tied into their banking system, so business owners find it easier to simply print out checks than re-enter payment information into their bank website. Potential issues of online payments also deters some business owners. There is a perceived concern about security of online payments. However, paper checks are ultimately less secure. Forged signatures, hacked bank accounts, changed amount, and payee information are all common problems businesses can face with check payments. When proper security measures such as two-step authentication and payment verification are taken, online payment solutions are more convenient, secure and cost-effective than checks. Most electronic payment software also allows direct integration with banking and accounting systems to make keeping track of payments even easier.

Government programs and policies are certainly a useful method for pushing the adoption of e-invoicing and B2B electronic payment processes. Chile mandated e-invoicing, specifically, in 2014, estimating that it would save the country $600 million annually and help reduce tax fraud. The U.S. Department of the Treasury implemented an e-invoicing requirement in 2011, resulting in a savings of $450 million every year. Governments could simply invoke similar legislation in an attempt to encourage B2B electronic payment processes, citing a reduction of fraud and spending as key factors.

However, despite the push for electronic B2B transactions by the government and business networks, enterprises are not going to forget about paper unless new technologies support payment processing and tasks associated with e-invoicing and supply procurement. Cloud is one technology streamlining electronic B2B transactions. Cloud technology can provide real-time collaboration capabilities and a simpler way to share information and documents and businesses can make procurement more transparent and easy to manage. However, such digital solutions will need to seamlessly become a part of B2B transactions or businesses will risk having those programs go unused. For example, SAP is a popular platform in the data centers of businesses across industries because it integrates with everything from enterprise resource planning software to analytics tools to customer management suites. It is important that B2B payment processing technologies work with those existing tools, as operational efficiencies directly impact the bottom line.

Another factor driving the digitization of B2B payment processing is the Internet of Things. In the near future, manufacturers and suppliers will have all of their equipment, machinery, transportation, goods and other devices connected to one another in a machine-to-machine network. The data from the IoT will improve production and inventory tracking, but when layering that information and integrating it with payment data, businesses will achieve a level of visibility and management unmatched in the modern day. However, in order to prepare for this future, manufacturers and suppliers will need to start digitizing their B2B processes now. The last and most important requirement for a B2B electronic payments future lies in cyber security. To protect the corporate data of clients, B2B suppliers will need cutting-edge cyber security solutions that offer simplicity and control. Until those systems are in place, wholly relying on electronic B2B transactional methods will be dangerous and potentially illegal in an era when the government is becoming more involved in setting data protection standards. B2B payments have a long road ahead of their improvement. It is not an impossible end goal to reach, but it will be critical to start the journey sooner rather than later (Delegosoftware.com, 2015).

A variety of companies provides B2B payment services that are secure and fast.

TRADECARD (TRADECARD.COM)

TradeCard provides a web-based platform that all parties (buyers, suppliers and service providers) can access through the Internet. TradeCard services can securely support both cross-border and domestic trade transactions. To use services of TradeCard, buyers and sellers must first become TradeCard members. TradeCard assess potential members to determine their credit worthiness and perform various other checks including anti-money laundering. The buyer can then create and send TradeCard a purchase order. TradeCard informs seller that a purchase order is pending. The purchase order can either be approved by the customer or renegotiated with the buyer. Once approved by the seller, the purchase order is stored in TradeCard and seller can select payment protection, if available. Seller will create an invoice and packing list on TradeCard as soon as the goods are ready to be shipped. The TradeCard then matches the invoice and other optional documents (such as the packing list) against the purchase order to make sure all required documents have been provided. TradeCard will inform both buyer and seller in case any discrepancies are found. Buyer and seller will then have the option to renegotiate. Once documentary compliance has been achieved, TradeCard automatically sends payment instructions to a participating financial institution, which debits the buyer's account and credits the seller's account. On January 7, 2013, GT Nexus and Trade Card merged to create World's Biggest Cloud-based Business Network for Global Trade and Supply Chain Management

Paymentech

Paymentech (chasepaymentech.com) is one of the largest payment solutions providers for point-of-sale transactions on the Internet. Paymentech supports all types of credit and debit cards provides secure processing of all transactions.

SKRILL

Skrill Payouts (**Skrill.com**) makes it simple and fast. All you need is the recipients' email addresses and payment amounts. There is no fee for receiving money. You can use this service to pay customers, freelancers and suppliers worldwide easily.

BOOSTB2B

Boost's B2B payment platform (boostb2b.com) streamlines commercial payments through automated transaction processing, expedited reconciliation, elimination of potential security risk and lowering cost of acceptance. As the only acquirer in the United States exclusively focused on B2B payments, Boost combines specialized knowledge, proprietary relationships, innovative technologies and unique solutions to help its clients transform their procurement activities from costly operations into profit centers.

AMEX SUPPLIER PAYMENT SOLUTIONS

American Express Supplier Payment solutions are B2B payment solutions available to both enterprises ($250 million + revenue) and mid-size businesses ($4-250 million revenue). Some of the B2B payment services offered include corporate purchasing card, Business travel account, Vpayment, and Buyer initiated payment (BIP).

PAYSCAPE

Payscape (https://payscape.com/) provides B2B payment gateway solutions. It simplifies transactions for B2B customers with business-specific payment methods, including corporate credit cards and checks.

ING

Dutch bank ING partnered with Basware for a new capital management service for business-to-business firms. The purpose of this partnership was to facilitate payments of its clients. The new services were based on the Basware+Mastercard Pay solution and focused on streamlining companies' cash flow management, funding, and payment to suppliers through strengthening ING's electronic invoicing offerings. ING will integrate the Basware Commerce Network to allow companies to use Basware Pay to purchase items from suppliers and settle invoices. In addition to getting access to working capital, ING customers will also benefit from Basware B2B management resources so suppliers can get paid faster. The new service will also facilitate companies looking to pay their suppliers early to take advantage of early payment discounts. According to ING, buyers could save up to 1 percent when suppliers are paid within three-to-five days (Pymnts.com, 2015).

BASWARE

Basware (www.basware.com) is the global leader in providing purchase to pay and e-invoicing solutions. We empower companies of all sizes to unlock value across their financial operations and business networks. Services provided include Purchase to Pay, Accounts Payable Automation, Receiving e-invoices, Sending e-invoices, e-Procurement etc.

B2B PURCHASE CARDS

The traditional B2B payment methods (checks, financial EDI, letters of credit, and EFT) may be too expensive for SMEs. B2B purchasing cards are special-purpose payment cards that re maintained with a merchant and issues to a company's employees. These cards are used solely for the purpose of limited payment for non-strategic materials and services (e.g., stationery, office supplies, computer supplies etc.).Purchase cards are similar to charge cards except that the balance at the end of the month need to paid in full to the merchant. B2B purchase cards can be used both for off-line and online purchases.

The U.S. federal government remains the largest buyer of products and services anywhere, spending over $600 billion annually. Most B2B catalogs and multichannel merchants have had success with government sales via the Federal SmartPay charge card. Since the program inception in 1989, there has been about $200 billion in sales that have gone through the federal purchase charge cards. The charge card program began for several reasons, among them to eliminate costly paper-based purchasing practices for small purchases. There was a significant cost-savings for the government in avoiding the paper-based procurement method. Another reason was to give front-line managers faster access to the products and services they required to get things done on a timely basis. Most state governments and many local governments also use purchase cards for official business (Amtower, 2015).

ELECTRONIC LETTERS OF CREDIT

A letter of credit (LC) is a written agreement by a bank to pay to the beneficiary (the seller) on account of the applicant (the buyer) a sum of money upon presentation of certain documents (Chung, Lee, & Turban, 2006). In general, five steps are required in a LC: issuance, credit advising, confirmation, transfer, and negotiation. When LCs issued online, these steps can be conducted much faster and conveniently. For example, Royal Bank of Canada issues electronic LCs. A business can submit a LC request electronically to the Royal Bank. Royal bank can review and process this request immediately and provide instant payments. Customers can also view export letters of credit online.

E-BILLING

E-billing systems, also called Electronic Billing Presentment and Payment (EBPP) systems, are essentially physical check free systems. Using EBPP, a company can display a bill online process its actual payment. Payments are generally electronic transfers from consumer checking accounts that are conducted through the ACH (Automated Clearing House). EBPP systems provide substantial cost savings (e.g. savings in postage and processing of paper documents). These systems are becoming increasingly popular in both the B2C and B2B markets. Furthermore, online bills can be used for marketing and promotion such as offering rebates, and savings offers.

MYCHECKFREE

MyCheckFree (mycheckfree.com) is an international online bill processor that provides online payment processing services to both large corporations and individual Internet users. For billers, online billing provides convenience and lower costs. While it costs anywhere between 50 cents and $2 to process a paper bill, e-bills cost 35 to 50 cents each. Billing companies can partner with CheckFree to provide their customers online billing and payments services. CheckFree works directly with those companies and provides customers the facility to view and pay their bills instantly using one convenient, secure, online place. Customers can pay their online bills using a checking account, major credit cards, or electronic checks where accepted by the billing company.

The advantage for the consumer for using such a service is the ease of bill viewing and payment with a click of a button. No need for handling paper bills, writing checks, or mailing payments. Bills of different companies are found electronically at the checkfree site and are saved up to six months for further viewing by the consumer. Payment method through a bank account, for instance, needs to be setup only once and used repeatedly for all bills received at checkfree. Consumers do not have to go from one site to another using different account names and passwords to make payments. The service guarantees payment to the billing company within two days of the payment being scheduled by the consumer. If the payment is not made by CheckFree on time, they handle any late charges associated with the delay in payment. Additionally, this service is offered free of charge to consumers, with the cost passed on the billing companies themselves.

CHECKFREEPAY

CheckFreePay (CheckFreePay.com) is the largest processor of walk-in bill payments in the United States. At a CheckFreePay agent location, you can make payments to many of the companies that send you a monthly bill including electric, gas, wireless, cable / satellite, healthcare, insurance, credit cards, and car loans and leases. A large agent network of nearly 25,000 locations makes it easy to find a convenient and secure location to make bill payment. Agent locations include large national retailers, convenience stores, grocery stores, and neighborhood bodegas.

MY MASTERCARD BILLERS

This is a service offered by Master Card. Using My MasterCard billers, a user can search for billers that accept MasterCard, save them to his custom list, and set up e-mail reminders to automatically inform him when approaching payments are due. A user can also access the billers' Web sites to make payments directly from his saved list. Using your MasterCard card to pay your bills gives you enhanced convenience, flexibility, and can help you earn rewards associated with your MasterCard account. My MasterCard billers allows you to search for billers that accept MasterCard, save them to your custom list, and set up e-mail reminders to automatically inform you when approaching payments are due. You can also access your billers' Web sites to make payments directly from your saved list. In addition, remember, if you cannot find some companies, check back often as new billers are added periodically.

PAYTRUST

Paytrust (paytrust.intuit.com) is a provider of online bill delivery, payment and management services. Paytrust offers two services. Total Bill Management service allows members to receive, review, pay and organize all their bills online. PayAnyone service allows members to make payments online, but they will continue to receive paper bills at home. You can use PayTrust to pay anyone with a U.S. mailing address. To issue a payment from the PayTrust Bill Center, you will need to set up a Payee Profile. Most payees are Business Payees, and many of them can be found in our Common Payees database.

Electronic funds transfer (EFT)

EFT is a system in which transfer of money takes place from one bank account directly to another account without any exchange of physical money. EFT can be initiated through an electronic terminal, including Credit card, ATM, Fedwire and point-of-sale transactions. EFT can be used for credit and debit transfers, direct deposit payroll payments, electronic bill payment, electronic Benefit Transfer etc. Banks process the transactions the ACH network. ACH (Automated Clearing House) is the secure transfer system that connects all U.S. financial institutions. ETF provides many benefits including reduced administrative costs, increased efficiency, simplified bookkeeping, and greater security. Three major EFT systems are:

1. SWIFT: (Society for Worldwide Interbank Financial Telecommunication)

2. Fedwire: U.S Federal Reserve's system providing a real-time gross settlement system

3. CHIPS: U.S-based private sector system

SWIFT

Banks transmit messages that contain highly sensitive information. They need a highly reliable and highly secure network through which they can communicate to avoid intercepted or lost messages. SWIFT (Society for Worldwide Interbank Financial Telecommunication) allows for the quick and secure exchange of standardized financial messages between financial institutions throughout the world. With message standardization, both banks and customers can benefit from uniform policies and procedures across different banks. SWIFT is a network of over 8,300 banks, securities, and corporations located in over 208 countries, controlled by its shareholders. Formatted messages created by members organizations are forwarded to SWIFT. SWIFT delivers these messages to the recipient member organizations. These formatted SWIFT messages are preset brief documents that provide standardized conditions for the transfer of funds between banks. These messages contain the name and code of the originating and receiving bank, the amount of transfer, and one of many preset codes that can be used to provide a message to the receiving bank. The main use of SWIFT messages is for payment. SWIFT also provides its members with a secure email messaging system that members can use to pass highly confidential email messages between member institutions. The SWIFT network is comprised of leased lines and dial-up lines and most members connect to the SWIFT network's access points through leased lines or dial-up lines.

Russia and China are discussing setting up a system of interbank transactions which will become an analogue to International banking transaction system SWIFT, First Deputy Prime Minister Igor Shuvalov" announced back in September of this year. "According to Shuvalov, Russia has been also discussing establishment of an independent ratings agency with China. Concrete proposals will be made by the end of 2014, he said." (tass.ru, 2014) The Central Bank of Russia (CBR) has launched a new SWIFT-style payment service aimed at moving away from Western financial dominance. The system is already operating, and will be fully functional within six months. According to Russian regulators, the new service will allow credit institutions to transmit messages in a SWIFT format through CBR to all Russia's regions without restrictions (emergingequity.org, 2014).

Fedwire

FedWire (Federal Reserve Wire Network) is a high-speed electronic communications network that links the U.S. Federal Reserve Banks, the U.S. Treasury Department, and other federal agencies. FedWire processes high-value payments in the US e.g. Reserve Banks and the Treasury use FedWire for high-value time-sensitive payments, such as funds transfers between reserve banks, purchases or sales of Fed Funds transfers between correspondent banks, and sales of U.S. Government securities. Fed Wire transfers are immediate and are effective usually within minutes of the time a payment is initiated. The Federal Reverse guarantees payments to repository institutions, and absorbs credit risk if the payment banks have insufficient funds.

Clearing House Interbank Payments System (CHIPS)

The Clearing House Interbank Payments System (CHIPS), together with the Fedwire, forms the primary U.S. network for large-value domestic and international US doallar payments. Financial institutions own the CHIPS and any banking organization, present and regulated in United States, may become an owner and participate in the network. CHIPS processes over 95% of the US$ cross-border payments. Only the largest banks dealing in U.S. dollars participate in CHIPS while many smaller banks have accounts at CHIPS-participating banks to send and receive payments. CHIPS are operated by The Clearing House, which also provides ACH (Automated Clearing House), an electronic network for financial transactions in the U.S, which processes large volumes of credit and debit transactions.

The automated clearinghouse (ACH) system is a nationwide network through which depository institutions send each other batches of electronic credit and debit transfers. The direct deposit of payroll, social security benefits, and tax refunds are typical examples of ACH credit transfers. The direct debiting of mortgages and utility bills are typical examples of ACH debit transfers. While the ACH network was originally used to process mostly recurring payments, the network is today being used extensively to process one-time debit transfers, such as converted check payments and payments made over the telephone and Internet.

The Reserve Banks and Electronic Payments Network (EPN) are the two national ACH operators. As an ACH operator, the Reserve Banks receive files of ACH payments from originating depository financial institutions, edit and sort the payments, deliver the payments to receiving depository financial institutions, and settle the payments by crediting and debiting the depository financial institutions' settlement accounts. The Reserve Banks and EPN rely on each other to process interoperator ACH payments--that is, payments in which the originating depository financial institution and the receiving depository financial institution are served by different operators. These interoperator payments are settled by the Reserve Banks (federalreserve.gov, 2015).

E-Payment Standards

The standardization of E-payment mechanisms is essential to the success of E-commerce. The PCI Security Standards Council (pcisecuritystandards.org) is an open, global forum whose goal is to enhance data security during online transaction. The council's founding members include American Express, Discover Financial Services, JCB International, MasterCard Worldwide, and Visa, Inc. The council has developed a set of standards called PCI Data Security Standards (PCI DSS). These standards have been developed to help merchants ensure the safe handling of sensitive consumer information. To become PCI-DSS compliant, merchants need to complete a self-assessment security questionnaire and perform a quarterly vulnerability scanning of their servers and network connections. PCI DSS compliance is now required for all businesses.

OFX

Transaction information also needs to be standardized so that all parties are able to accept and read the information. The Open Financial Exchange (OFX) (www.ofx.net) is a widely adopted, unified and open standard for the secure electronic exchange of financial data between financial institutions, businesses and consumers via the Internet. OFX was Developed and presented by Intuit, Microsoft and Checkfree in 1997 and it is not a financial institution. OFX is based on widely accepted networking, Internet and security standards. OFX is used by more than 5,500 banks and brokerages as well as major payroll processing companies around the world.

Jalda

Jalda is an open and global payment method owned by EHPT, a company jointly owned by Ericsson and Hewlett Packard (LBI.com, 2010).Using Jalda, payments can be made from a PC, mobile phone or any wireless device with Internet access. Jalda can handle very small to large amount transactions. To use Jalda, both consumers and retailers are connected to a special account and the system administers all transactions for both parties. Jalda offers digital certificates for authentication and communication is SSL encrypted. Jalda's Application Programming Interface (API) presents the payment standard on the online merchant's site. This is connected to a server managed by an Internet Payment Provider. When a transaction is made, it travels from the consumer to the online merchant's API and then to the Internet Payment Provider.

EMV

EMV standard, developed in 1999 by Europay International, MasterCard International and Visa International, is a global standard for chip-based credit card and debit cards. Current membership of EMVCo includes MasterCard, American Express, Visa, and JCB. Each member owns equal share of EMVCo. By 2009, more than 944 million EMV compliant chip-based payment cards were in use. For more details, see (emvco.com).

Open and Closed Loop Payment Systems

In closed loop payment system, the card issuer pays the merchants that accept the card directly and does not use an intermediary, such as a bank or clearinghouse system. It is called a closed loop systems because no other institution is involved in the transaction. American Express and Discover Card are examples of closed loop payment systems. Both American Express and Discover issues cards directly to consumers.

In an open loop payment system, three or more parties (including one or more intermediary banks) are involved in the transaction. Intermediary banks coordinate the transfer of funds from the customer's bank to the merchant's bank. Visa and MasterCard are examples of open loop payment systems. Neither Visa nor MasterCard issues cards directly to consumers. Visa and MasterCard are credit card associations that operated by the member banks. These member banks, which are also called customer issuing banks, issue credit cards to individual consumers and are responsible for processing transaction.

In a closed-loop system, the platform is managed by a single company, which signs all contracts directly with cardholders and merchants. Amex, Diners Club, and private label cards like the "Pass" card issued by the French multinational retailer Carrefour are often referred to as closed-loop retail payment systems. Amex issues cards that can only be accepted by merchants affiliated with its platform and charges both consumers and retailers directly. The system used by Carrefour for its "Pass" card is very similar, except that its acceptance network is limited to Carrefour stores. Those systems are not necessarily specific to payment cards: for example, the issuance of gift checks accepted by a group of shops corresponds to a three party system.

FIGURE 8-8: CLOSED-LOOP PAYMENT SYSTEM

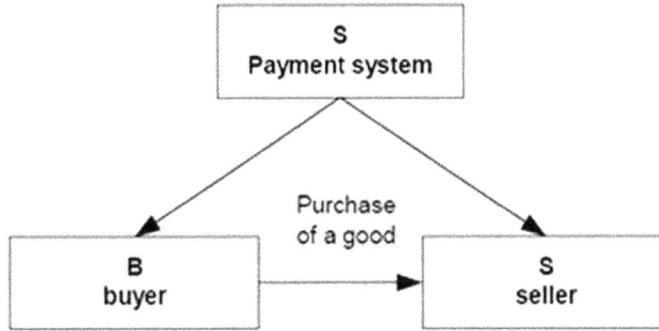

The organization of open-loop payment systems is more complex, for its members act as intermediaries between the platform and its end-users, consumers and merchants. Two levels of pricing must be taken into account the pricing of the services provided by the platform to banks, and the pricing of services provided by banks to end-users. In this case, the impact of the prices chosen by the platform on end-users depends on the degree of competition between banks. The Visa and MasterCard payment card systems are examples of open-loop systems. Banks pay fees to become members, but remain free to choose their pricing policy as regards cardholders and merchants.

FIGURE 8-9: OPEN-LOOP PAYMENT SYSTEM

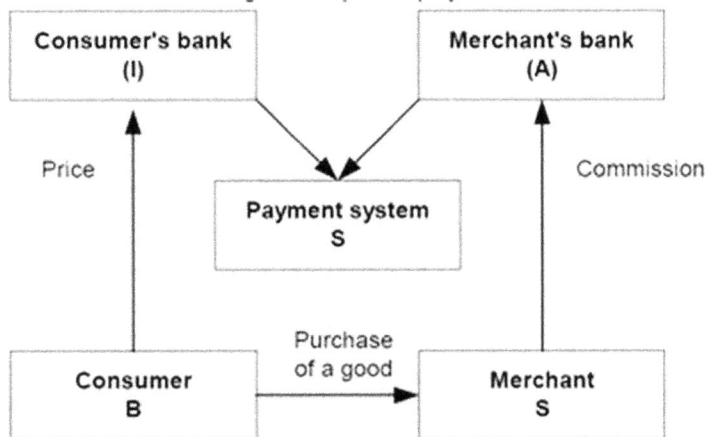

Payment Cards

In general, the term payment card refers to all types of plastic cards that consumers (and some businesses) use to make purchases. The main types of payment cards are credit cards, debit cards, and charge cards. Increasing use of payment cards online (whether debit cards, credit cards or some other type) is increasing the volume of online B2C transactions.

Credit Card

A credit card is a card issued by a financial company that gives its holder an option to borrow funds (primarily short-term financing). Credit cards charge interest on any unpaid balance and have a spending limit based on the user's credit history. At the end of each billing period, user has the option to either pay off the entire credit card balance or pay a minimum amount. Credit cards are widely accepted by merchants around the world and provide assurances for both the consumer and the merchant. Among global consumers, credit cards have become preferred method of payment online (Data Monitor, 2008).

Credit cards, as the form of online payment, are disadvantageous for merchants. Merchants must pay a significant transaction fee of (generally 3%- 5% of the sale amount plus a fee of 20 to 30 cents per transaction and other set-up fees) and the risks of a transaction are largely born by the merchant. To mitigate this, most merchants that accept credit cards online require a minimum purchase amount (often $10 or $15) to cover these fees and earn profits.

Credit card associations (such as Visa and MasterCard) are non-profit organizations that set credit card standards while banks are the organizations that issue the credit cards, process credit card payments, and receive interest. Third-party clearinghouse performs the verification of credit card accounts and their balance.

To accept credit card payment online or offline, a merchant needs to apply for a merchant account which is used to accept credit card payment. If merchant accepts offline payment, he needs to install the card reader. For online payments, merchant needs to install a piece of software to accept credit card in his web site.

ADVANTAGES OF CREDIT CARDS

Power and Ease of Purchase: With credit cards, you do not have to carry physical cash and can buy many things for which cash payments are not accepted.

PROTECTION OF PURCHASES: Credit card users are protected against unauthorized purchases made against their credit card provided they report their card lost, damaged, or stolen.

CREDIT HISTORY: Building a good credit history can help users both to get credit cards and to get loans.

WORLDWIDE ACCEPTANCE: Major credit cards can be used anywhere in the world. Any currency conversion required is handled by the card issuer.

ADDITIONAL BENEFITS: Credit cards can offer additional benefits e.g. discounts available from particular stores, bonus offers etc.

DISADVANTAGES OF CREDIT CARDS

IMPULSE BUYING: Credit cards can make a user impulsive buyer spending more than he should.

ACCRUED INTEREST: Credit card companies charge users on unpaid balance. If users make minimal payments, the interest can accumulate to large amounts.

FINANCIAL RISK: Lost or stolen cards may result in some unwanted expense and inconvenience.

TYPES OF CREDIT CARDS

Credit cards can be of several types. The cards differ in way the rates charged and how a merchant is responsible for processing them.

CONSUMER CARD: It is a type of credit card, which is neither linked to a checking account nor associated with any business.

CHECK CARDS: This card is similar to debit cards but you do not need a PIN code to complete the transaction. Same equipment used to accept credit cards could also accept check cards. The funds for check card transactions are immediately withdrawn from your checking account.

REWARDS CARDS: A rewards card offers its cardholder special rewards (e.g. airline miles) for using that card to make purchases. Processing transactions involving reward cards cost more than a traditional credit card.

BUSINESS CARDS: These cards, also called corporate cards, are carried by business owners and typically used for business-related purchases. Processing transactions involving business cards cost more than a traditional credit card.

PURCHASE CARDS: These cards are very similar to business cards except that these cards can be used to make purchase at predetermined merchants. The predetermined merchants are determined by acceptable Standard Industrial Classification list. This list maintained by the customer and kept at card issuing bank.

VIRTUAL CREDIT CARDS: With a virtual credit card, a credit card issuer issues a special number that can be used in place of regular credit card numbers to make online purchases. This allows users to use a credit card without having to disclose the actual credit card number.

MAJOR CREDIT CARD COMPANIES

VISA AND MASTER CARD

Visa and Master Card both are two leading companies in the world. Both operate at very similar lines. From consumer point of view, there is no real difference between the two. Both are very widely accepted in over one hundred and fifty countries and it is very rare o find a location that will accept one but not the other. The Visa card consumers can register their purchases and file their claims online of charge. The consumer purchases are protected in case of theft or damage for the first 90 days from the day of purchase.

Consumers can over-spend their Visa/MasterCard i.e. spending over the available credit limit, which can cause overdraft fees. Many banks are now charging over-limit fees or non-sufficient funds fees based upon pre-authorizations, and even attempted but refused transactions by the merchant. In some countries, Visa/Master Cards, debit cards offer lower levels of security protection than credit cards. There are many hidden charges present while using VISA/MasterCard.

The distinguished features of MasterCard include the hologram security device, enhanced clearing system that uses advanced technology to enhance transaction-processing operation. MasterCard is also the online PIN-based global debit card.

AMERICAN EXPRESS & DISCOVER CARD

Discover and American Express are two cards that are not used as much. However, they still have benefits. The main downside to using either Discover or American Express is that neither card is as widely accepted as Visa or MasterCard (although Discover is accepted at slightly more places than American Express). The Discover card is a major credit card, issued primarily in the United States. Most cards with the Discover brand are issued by Discover Bank. Discover Card transactions are processed through the Discover Network payment network. As of February 2006, the company announced that it would begin offering Discover Debit cards to banks, made possible by the Pulse payment system, which Discover acquired in 2005. American Express Company or AmEx is an American multinational financial services corporation headquartered in New York, United States. The company is best known for its credit card, charge card, and travelers cheque businesses. Amex cards account for approximately 24% of the total dollar volume of credit card transactions in the US, the highest of any card issuers.

METHODS OF PROCESSING CREDIT CARDS

Today a majority of credit card transactions are processed electronically using various processing methods. In all cases, either the card is swiped through a credit card terminal/reader to read card's magnetic strip or the credit card information is manually entered into a credit card terminal, a computer or website.

CREDIT CARD TERMINAL

A credit card terminal is stand-alone dedicated electronic equipment (most terminals consist of a modem, keypad, printer, magnetic stripe reader, power supply, and memory card) that allows a merchant to swipe or key-enter a credit card's information as well as additional information required to process a credit card transaction. When a credit card is processed, it contacts the network to verify if the credit card can be authorized. The machine will either upload the electronic funds directly to the merchant bank, or use a polling service provider to process and submit the required data to the merchant bank.

AUTOMATED RESPONSE UNIT (ARU)

ARU is designed for merchants who wish to accept transactions without having to buy or lease any processing equipment. ARU can be utilized by any size business, but is usually employed by smaller merchants with fewer transactions. Merchants can call a toll-free number from any phone (cell phone, pay phone, landline, etc.) to get authorization of credit card transaction.

CREDIT-CARD FRAUD ON THE INTERNET

Credit-card fraud is a significant problem for many e-businesses. A chargeback results whenever a credit-card holder claims an unauthorized purchase or when a purchase was not received. In online businesses, merchants neither have a scan of the card nor a customer signature. Therefore, the cost of chargeback, if any, is born by the merchant. VISA has developed a list of "best practices" to be used by merchants when conducting credit-card transactions. The list includes implementing internal fraud protection systems, using encryption, using VISA tools, and incorporating intercompany security practice. Merchants failing to meet the requirements may not be able to accept Visa credit cards as a method of payment. A type of online credit card fraud, known as electronic shoplifting (a practice used by cardholders who take advantage of the chargeback rules to fraudulently revoke purchases on their credit cards) is becoming a serious issue for e-businesses. For example, a customer may place an order using his credit card and later dispute the transaction claiming he never got the item while actually he did. Alternatively, a customer may place order for two of the same product and later claim that he got only one while actually he got both. Electronic shoplifting results in charge backs and duplicate orders.

There are techniques available to combat electronic shoplifting. One way is to call the credit card company to authorize all credit card purchases. Secondly, companies can wait for online checks to clear before shipping orders. Third, companies can do shipping only via courier companies that offers instantaneous tracking and delivery confirmation. Many businesses e.g. Amazon.com, have setup internal fraud departments to perform further verification of credit card transactions that is beyond electronic authorization.

Debit Card

A debit card (also known as a bankcard or check card) is a plastic card, which is linked to customer's checking account. When used in a purchase, a debit card deducts the amount of the sale from the cardholder's bank account and transfers it to the seller's bank account. Debit cards can also carry the name of a major credit card issuer, such as Visa or MasterCard, by agreement between the issuing bank and the credit card issuer. Through this arrangement, banks ensure that their debit cards will be accepted by merchants who recognize the credit card brand names. To use a true debit card, a user will need to enter a PIN number in order to complete the purchase.

Use of debit card offer many advantages. Non-credit worthy customers may find it difficult or impossible to obtain a credit card. These customers can easily obtain a debit card. Use of a debit card is limited to the existing funds in the linked bank account. Therefore, customers do not pay any debt because of its use, or charged interest, late fees, or fees exclusive to credit cards. A debit card can be used at an ATM to obtain cash. The disadvantages of debit cards are that banks may charge over-limit fees or non-sufficient funds fees based upon pre-authorizations, and even attempted but refused transactions by the merchant. While the holder of a credit card is provided with certain protection against the unauthorized or fraudulent use, the debit cardholder may be held liable for hundreds of dollars in fraudulent debit transactions.

Charge Card

A charge card is similar to a credit card except that it carries no spending limit and the balance must be paid in full when the statement is received at the end of the billing period i.e. there is no "minimum payment". These cards neither involve lines of credit nor do they accumulate interest charges. A partial payment (or no payment) results in a severe late fee (as much as 5% of the balance) and the possible restriction of future transactions and risk of potential cancellation of the card.

Single-Use Payment Cards

Single-use cards give consumers a unique card number that is valid for one transaction only. Therefore, this card cannot be used to complete unauthorized transactions on the consumer's account or selling the card number to others. In 2000, American Express was the first to offer single-use cards but stopped offering it in 2004.Two major banks issuing these cards are MBNA and Citigroup.

Electronic Wallets

An Electronic Wallet, also called digital wallet, is a piece of software that allows a user to make an electronic payment with a financial instrument. An electronic wallet serves many purposes. First, it provides security and encryption for the user personal information and for the actual transaction and second, it keeps a database of user-inputted information (shipping address, billing address, payment methods, and other information) and third, it authenticates the consumer through the use of digital certificates and other encryption methods. Consumers need to enter their information just once in the electronic wallet, instead of having to enter their information at every site with which they want to do business. Digital wallets provide convenience for the consumer and lower transaction costs because order entry can be expedited. With electronic wallets, online shopping becomes more efficient. When consumers select items to purchase, they can then click their electronic wallet to order the items quickly. Some electronic wallets can also hold e-cash from various providers and provide at an E-commerce site's checkout page. Using digital wallets merchants can lower their transaction costs and increase branding and marketing opportunities for their products. Financial intermediaries benefit from the use of digital wallets because they receive processing fee for each transaction. Some major electronic wallet providers are Microsoft .NET Passport, Yahoo, CyberCash, AOL, and Google.

In the future, electronic wallet owners could track the purchases and maintain receipts for those purchases. The electronic wallets can then use a Web robot to suggest where the consumer might find a lower price on an item that he or she purchases regularly. For a wallet to be useful at many online sites, it should be able to populate the data fields in any merchant's forms at any site that the consumer visits. This accessibility requires compatibility between electronic wallet software and merchant's websites so that a wallet can recognize what consumer information goes into each field of a given merchant's forms and populate it correctly. This problem has been eliminated by sites and wallet software that use ECML (Electronic Commerce Modeling Language) technology. ECML is a protocol that dictates how online retailers structure and setup their checkout forms. Microsoft, Discover, IBM, and Dell Computers are examples of vendors who incorporate both electronic wallet technology and ECML.

Electronic Wallet Characteristics

Every good electronic wallet should possess some desirable characteristics.

EXTENSIBLE: A wallet should be able to accommodate and inter-operate with different payment mechanisms used by users. For example, if a client uses both credit card and digital cash than the digital wallet should be able to hold both and make payments using either credit card or digital cash.

CLIENT-DRIVEN: Any action by digital wallet should be driven by client and not the merchant e.g. merchant should not be capable of automatically launching a client's digital wallet every time the client visits a merchant's web page that offers a product for sale.

SYMMETRIC INTERFACE: A digital wallet should re-use, whenever possible, the same infrastructure and interfaces within wallets, vendors, and banks.

GENERALIZED INTERFACE: Wallet interface should be similar regardless of what type of device or computer it is running on. A digital wallet running on a personal digital assistant (PDA) should have substantial functionality in common with a digital wallet running on a PC.

Electronic Wallet Categories Based on Purpose

In general, there are three kinds of e-wallets namely closed, semi-closed and open.

CLOSED WALLET: A closed wallet is one that a company issues to its consumers for in-house goods and services only. These instruments do not carry the advantage of cash withdrawal or redemption. Several online shopping portals such as Flipkart, Jabong and MakeMyTrip offer such closed wallets. It is an account where money is credited in case of a refund due to cancellation or return.

SEMI-CLOSED WALLET: In the payments space, companies such as MobiKwik, PayU and Paytm offer semi-closed wallets. As per the RBI, a semi-closed wallet can be used for goods and services, including financial services, at select merchant locations or establishments that have a contract with the issuing company to accept these payment instruments. Semi-closed wallets do not permit cash withdrawal or redemption by the holder as well.

OPEN WALLET: Such wallets can be used for purchase of goods and services, including financial services such as funds transfer at merchant locations or point-of-sale terminals that accept cards, and cash withdrawals at automated teller machines or business correspondents. These kind of wallets can only be issued by banks.

Electronic Wallet Categories Based on Data Storage

Electronic wallets fall into two categories based on where they are stored.

CLIENT-SIDE ELECTRONIC WALLET

A client-side electronic wallet stores a consumer's information on his or her own computer. In general, these wallets are easy to maintain and fully compatible with most E-commerce Web sites. In order to use it, users need to download and install the wallet software onto every computer used to make purchases online. Once installed, the user begins by entering all the pertinent information. Storing sensitive information (such as credit card numbers) on users' computers provides stronger security. This is because attackers will have to launch many attacks on user computers, which are more difficult to identify.

CyberCash's wallet is a client-side wallet. It is now part of PayPal. CyberCash's wallet software program, which buyers use when making a purchase, must be downloaded and installed on the buyer's machine before they can make a purchase. The wallet software handles communication between buyer and merchant and secures the communication by encrypting it. Before start using the wallet software, users need to create a wallet ID and password. This ID and passwords are registered with CyberCash. Buyers can create multiple wallet IDs, each with its own password. After creating the ID and password, users must bind at least one credit card to the wallet. Binding a credit card includes entering credit card information (such as credit card number, expiration date, shipping address and phone number) to the wallet. Multiple credit cards can be bound to the wallet. Once the wallet ID is established, and at least one card has been bound, the buyer is ready to start using the wallet for online purchases.

SERVER-SIDE ELECTRONIC WALLET

A server-side electronic wallet stores a customer's information on a remote server owned by a particular merchant or wallet provider. Server-side wallets remain on a server and thus require no download or installation on a user's computer. They must be enabled before a consumer can use it on merchant's site. These wallets provide security, efficiency, and added utility to the end-user. Server-side electronic wallets employ strong security measures but still vulnerable to security breaches. These security breaches could reveal users' personal information (including credit card numbers) to unauthorized parties. Server-side wallets require a high-level of acceptance by merchants' users before their acceptance by consumers. For the same reason, only a few server-side wallet vendors will be able to succeed in the market. Digital wallets are called the future of real-world payment technologies (Clark, 2015).

Microsoft .NET Passport is a server-side electronic wallet operated by Microsoft. Anyone who obtains a Hotmail account is signed up automatically for a Passport account. The Passport Wallet service provides standard electronic wallet functions, such as secure storage and form completion of credit card and address information. When requested by a participating merchant, a consumer's secure information is released to the merchant so that the consumer does not need to enter data into a form.

Yahoo! Wallet is a server-side electronic wallet offered by Yahoo! Yahoo! Wallet lets users store information about several major credit and charge cards, along with Visa and MasterCard debit cards. Yahoo! hosts a number of services and shops that it can be certain accommodate its own wallet. Therefore, Yahoo! Wallet is certain to have a large number of merchant acceptability.

Google provides an online payment processing service called Google Checkout. Besides wallet services, Google Checkout also offers fraud protection, as well a unified page for tracking purchases and their status. Google

Checkout is advantageous for merchants because Google Checkout button can be displayed on their AdWords ads and Product Search listings. The Google Checkout button can highlight their store and informs potential customers that shopping with these merchants will be safe and secure. Google claims to protect 98% of Google Checkout orders (on average) from chargebacks resulting from claims of unauthorized purchases and non-receipt of goods.

The consumer payments industry will process about $2.7 trillion of card-based payments in the US in 2015. This explains why there is a massive amount of mobile-centric innovation occurring in the payments space and why competition for the consumer's wallet is so intense (Annamalai, 2015). The recent changes in the payments competitive landscape, there exist many strong mobile wallet players and new card-based payment devices. Mobile wallet solutions enable paying through the phone at the POS as the primary modality of payment. Below we discuss some of the contemporary mobile wallet solutions.

APPLE PAY

Apple Pay was introduced in late 2014 and appears to have grown in acceptance by merchants and users ever since. Apple Pay is now accepted at over 700,000 locations. This compares to just 220,000 points of sale at introduction, reflecting both the acceptance of Apple Pay and the quickening conversion to NFC terminals by merchants. Apple Pay offers the best user experience for making mobile payments currently in the marketplace. Apple Pay pioneered the concept of tokenization for mobile payments with the help of Visa and MasterCard. This authentication has the potential to effectively eliminate traditional card fraud associated with magnetic stripe based credit cards. When the customer presents the phone near the NFC enabled contactless payment terminal, the wallet loads automatically without having to launch an app thereby providing a smooth user experience. With Apple Pay, payment credentials are stored in the cloud, while the token is stored in a secure element (does not need an internet connection to use Apple Pay). Card enrollment and provisioning is automatic via iTunes or by taking a photo of the card (Green Path). Apple Pay can be used in NFC enabled POS as well as in-app. Apple managed to bring most of the payment networks and banks/credit unions to accept Apple Pay. More financial institutions are signing up to attract the affluent Apple customers as part of their customer base. However, there are some drawbacks of Apple Pay. A fair number of businesses still do not have support for NFC contactless terminals limiting the use of Apple Pay. No rewards program is another limitation with Apple Pay. Apple Pay is proprietary to the iOS ecosystem, effectively ruling out the Android user base which has a significant number of users in the US and worldwide.

As more of the newer iPhones are sold, Apple Pay will gain traction and people will try Apple Pay. Apple Pay compatibility with older iPhones (5S, 5C) may add additional users who have not upgraded to the newer iPhones. As we see more merchants adopting NFC enabled terminals to support EMV liability shift in 2015, the number of locations accepting Apple Pay will grow.

ANDROID PAY/GOOGLE WALLET WITH NFC

Google Wallet was the mobile wallet pioneer in the US in using NFC technology. However due to lack of partnerships and the telecom providers unwillingness to install the wallet as part of the new phones being sold, it did not see early mainstream adoption. After the launch of Apple Pay, Google did a major reboot to their wallet initiative and is planning to re-launch the wallet and the Payments API platform with a new name "Android

Pay". Being an API, Android Pay will not be proprietary. This allows Google to open up the API to device manufacturers and developers to build wallets that may result in mainstream adoption. Android has a very healthy user base in the US and the rest of the world. Its user base also tends to be younger and more of a digital native. By separating the Android Pay mobile payments API and the Google wallet the proprietary offering, Google is opening up the playing field for more innovation. This may allow financial institutions to finally integrate payments into their mobile banking offerings, which has always been a longstanding pipe dream. There is a very good chance host card emulation (HCE) being integrated into the Android Pay API to support phones which do not have an NFC radio/Secure Element built in. The Android OS provides built-in support for HCE since its Kit Kat release. With Apple Pay already launched and gaining momentum, Android Pay's late entry may impact market share. Android has an 80% worldwide market share. With an open API approach, Google is betting on sheer volume of its user base to make this payments solution a success. After the API launch, financial institutions may be able to integrate this into their mobile banking offerings, providing a more seamless payment experience and increasing loyalty when combined with their rewards programs. In developing nations (like India and Africa), the notion of direct carrier billing is very frequently used in payments. It would be interesting to see how the new Android Pay API may work with that model. With Android's open model and fragmented ecosystem, the success of Android Pay may not be imminent.

SAMSUNG PAY

Samsung acquired the mobile payment startup LoopPay. Samsung is one of the first device manufacturers to provide a wallet offering directly competing with Apple Pay. Samsung Pay is directly aimed at Apple Pay in most of the capabilities being offered. It supports tokenization and biometrics and has relationships with the major card networks and banks as launch partners. Samsung Pay boasts a much broader adoption rate (close to 90% of POS) for the terminals in US due to its inclusion of LoopPay's Magnetic Secure Transmission (MST) technology, allowing it to be used in traditional non-NFC, magstripe only POS readers. Samsung Pay stores credentials in the cloud; however, the token is transmitted from the cloud to the device at the time of purchase. This requires cellular or data connection, which prohibits usage in areas, which have weak cellular signals, or no data connection. However, Samsung Pay may fallback to LoopPay's MST to complete the payment. Unlike Apple Pay, the customer must launch the app to use Samsung Pay to make a payment near the terminal. Samsung Pay only supports POS purchases, (both NFC and existing magstripe). There are no in-app purchases. Samsung may decide to make the LoopPay technology a hardware chip, which could be resold to device manufacturers, making MST available to most of the Android family of phones. Samsung Pay also supports Private Label Credit Cards (Apple Pay does not). The slower growth of NFC terminals in US may result in a timing advantage for Samsung Pay with Loop MST technology. Most other mobile wallets rely on NFC capabilities. With the EMV Liability shift, most of the merchants who decide to upgrade may move to an NFC card reader to future-proof their investments. This puts into serious question how long Samsung will be able to maintain the benefit and advantage offered by the LoopPay acquisition.

MCX/CURRENTC/PAYDIANT WALLET

MCX is the powerful merchant conglomeration that is building their own mobile wallet to bypass the interchange fees paid by merchants to the payments network. MCX is basing its mobile wallet using Paydiant's QR code / cloud wallet technology to their in store payments. MCX is backed by a powerful conglomeration of merchants.

This wallet allows payments from bank accounts, gift cards and select merchant branded private label credit cards. The customer transaction (SKU level) data is kept internal to merchants. The use of QR-code technology does not provide a fluid POS experience as Apple Pay and other mobile wallets. This wallet does not accept major credit cards backed by payment networks like Visa/MasterCard/Amex/Discover. One of the primary funding sources is a customer's bank checking account, which the customers may be unwilling to share due to security reasons. MCX promises to pass along significant savings to customers by eliminating the swipe fees paid to card networks. This may be a powerful behavioral change, which customers may support in the long-run when they see savings in action.

PayPal Wallet

PayPal does not provide a mobile wallet in the traditional sense. The PayPal app allows you to tap into a list of merchants who accept PayPal in a geographic area. By checking into the merchant when you are in the location, PayPal allows to make a payment using your PayPal funding sources. By using your picture taken at the time of application enrollment, PayPal provides secure payments at POS. Check in can be made manually or automatically if the PayPal app senses a BLE enabled PayPal Beacon at the merchant store. PayPal has a rich portfolio of 'cards on file' for moving funds in and out. PayPal mostly uses the bank ACH funding to minimize the swipe fees paid to card networks. Low incidence and effective fraud fighting methods are one of the core strengths of PayPal. It also has already tokenized millions of deposit bank accounts. By acquiring Paydiant, PayPal now also can support QR code-based payments at a POS terminal. This provides many synergies with the MCX merchants. The Paydiant acquisition may help PayPal to crack the POS payment issue.

While the vast majority of consumers are still using plastic cards to make debit and credit card payments, a revolution is on the horizon. Digital alternatives will soon replace traditional POS payments. The end game will be making purchases using mobile devices and NFC. Until then, some believe all-in-one payment cards may lighten consumer wallets.

UNIVERSAL CARD PAYMENTS

A universal card payment refers to a standalone plastic device along with mobile phone as a supporting accessory to conduct payment at the POS. Following are some notable universal card payment devices. These devices allow consumers to store more than one credit/debit/loyalty/ID card on a single piece of plastic and allow switching between cards using electronics integrated on the card.

Coin Card

Coin (http://www.onlycoin.com/) is a secure, connected device that can hold and behave like the cards you already carry. A single Coin can help lighten your wallet by consolidating your debit cards, credit cards, gift cards, loyalty cards, and membership cards. Multiple accounts and information - all in one place. Coin works by letting you add all of your debit, credit, and loyalty cards onto one piece of technology, the Coin. After signing into the Coin app with the same credentials used to order the Coin, users are asked to create a unique six-digit tap code. It uses a combination of long taps and short taps,

of your choosing, to ensure no one can get into the Coin app or the Coin itself unless they know the code, or have control of the user's smartphone. Once the app is set up, users can pair their Coin and add new cards by manually entering information, swiping the card through an included card reader that goes into the headphone jack of the phone, or by taking a picture of the card as you would with Apple Pay. Coin has built a custom 128-bit encryption layer for bluetooth that secures sensitive information and prevents man-in-the-middle attacks. The Lock-and-Find feature provides a real-time validation that the owner of Coin is present at the time of the transaction. If owner is not there, Coin will lock itself and the owner can find Coin's last known location in the mobile app. The Coin remains locked when not in use. As soon as you're ready to make a transaction, a single tap on the Coin's solitary button will wake the device, do a quick search for your specific smartphone, and after a couple of seconds it will unlock. If your phone is turned off, on Airplane mode, or otherwise unavailable, you can unlock the Coin by entering the same six-digit Morse-style pin code that you will use each time you access the Coin app. The Coin stays alive for seven minutes once it is unlocked (so that a waiter can have the time to swipe), and then automatically locks and goes to sleep. It also remembers its last-known location and alerts the user as soon as it thinks that the smartphone has been separated from the Coin. Users can save up to eight cards on the Coin at a single time, and they can re-sync different cards stored within the app as long as they are within reach of their smartphone (Crook, 2015).

PLASTC CARD

Plastc (https://www.plastc.com/card) manages to bundle twice as many features into the same svelte form factor. Like Coin, Plastc synchronizes with your phone using Bluetooth. It lets you swipe between 20 cards or barcodes using an E-ink touchscreen, and offers NFC (like the iPhone 6) for contactless transactions, an EMV "chip" for the latest card readers, and RFID so you can replace your office building's ID card. Plastc includes various security measures, like a PIN code you must enter on its E-ink screen, and a requirement that you can only add cards with your name on them, which makes it more difficult to use for card skimming. Plastc's "remote wipe" mode can wipe your card when it is lost. Plastc can show live card balances on its E-ink screen so you can check before you pay. You can pre-order Plastc. By October 2015, Coin will already be antiquated as American retailers are forced to buy new card readers to support EMV chips, which Coin does not work with. These new readers will also likely support NFC, which is a good news for Plastc because it does work with NFC and EMV (Hamburger, 2014).

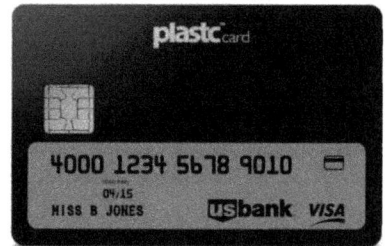

SWYP CARD

SWYP (http://www.swypcard.com/) is a next-generation secure electronic wallet. It is an electronic device that looks, feels and works just like a conventional plastic card such as a credit card. However, unlike a regular plastic card, SWYP can transform itself into any of your cards at the click of a button. In short, instead of carrying 25 different cards, you carry one SWYP. SWYP can replace any magnetic stripe based card. This includes credit cards, debit cards, loyalty cards, gift cards and frequent flyer cards, to name a few. SWYP is

designed to work at any Point of Sale (POS) where you can swipe your regular plastic card. That is more than 10 million locations in the US alone. On your SWYP card, you can store any credit card issued to you in much the same way you can store them in your physical wallet. SWYP has a chip that stores account information for each of your cards. You can select any card using buttons on the SWYP card. Information including name, account, CVV, and expiry date is displayed on the card, and the magnetic stripe on the back is programmed accordingly to transform your SWYP into the chosen card. If your credit card has a magnetic stripe in the back, then you should be able to add it to your SWYP. SWYP does not have an annual fee or per-transaction fee associated with it.

STRATOS CARD

The Stratos Cards (https://stratoscard.com/) is one credit card that stores all your cards (i.e. credit, debit, gift cards, rewards cards etc.) Like the competition, Stratos stores up to three main cards on the physical card at one time, but with the Stratos app it is able to suggest other cards in your inventory to use at certain locations, like a Starbucks gift card and Starbucks for instance. You swipe in cards using a card swipe attachment, also similar to Coin. Stratos is able to discern your location through the app via GPS, and the app lets you easily switch out cards whenever you need them. You can also set a security timer so if the card does not detect your phone for a certain amount of time it locks down. One disadvantage of Stratos is that it does not use chip and PIN security, at least not yet. As part of the Stratos Card's $95 annual membership ($145 for two years), you will get a new card every year, including upgraded ones when a new chip and PIN version does arrive (Orf, 2015).

The universal payment cards may be considered as a mobile payment bridge technology for consumers who are not willing to pay through phone for various reasons. Consumers can be ambivalent in terms of their phone choices and these offerings provide a way to not lock them into a specific iOS or Android ecosystem. These cards promise to work with both these major operating systems. The only hurdle, which may hinder the adoption for these cards, is the initial purchase price. Companies are not sure whether consumers be willing to pay an initial purchase price (anywhere from $50 – $150) to use these cards.

Stored Value Cards

Stored-value cards are capable of storing monetary value. These cards can be either smart cards or plastic cards with a magnetic strip. Common stored-value cards includes prepaid phone, copy, subway, and bus cards. Many a times the "stored-value card" and "smart card" are used interchangeably.

There are many features that distinguishing stored-value card from a credit card. First, the stored value card contains monetary value and not just the cardholder's details. Second, the stored value card can be used by any one, with no identity of owners. Third, the merchant takes the money instantly from the card. Fourth, the stored value card caters for low value transactions. Therefore, it is a form of micropayment with prepayment at the client side.

Magnetic Strip Cards

Most magnetic strip cards are stored-value cards in which the card's magnetic strip stores the monetary value added to the card. These cards are passive i.e. they cannot send or receive information, nor can they increment or decrement the value of cash stored on the card. The processing must be done on a special reader device.

Smart Cards

A smart card is a stored-value card. It is a plastic card with an embedded microchip to store information. A smart card can store about 100 times the amount of information that a magnetic strip plastic card can store. A smart card can hold private user data, such as financial information, health insurance information, medical records, and so on. Smart cards are safer because the information stored on a smart card is encrypted and can be designated as "read only" or as "no access". Smart cards can require the user to have a password to use them. Smart card can have a picture on its face to identify the user. Smart cards can provide strong defense against skimming (a fraud technique in which criminals copy information from the magnetic stripe of a credit or debit card) because the information stored on smart card is encrypted. According to Eurosmart, over 6.1 billion smart cards were shipped worldwide in 2011, with smart card shipments forecast to grow 13% to 6.9 billion in 2012(United States Securities and Exchange Commission, 2012).

Smart cards can be of two types: contact and contactless smart cards. Contact smart cards require a smart card reader to read and update the information on the smart card. A contactless smart card can transfer information without needing a smart card reader. The information transfer speed is faster than the contact smart cards. One example of contactless smart cards is the card used for automatic toll payment at tollbooths. The card is placed in a device in the car charge your account as you drive through tollbooths.

TYPES OF SMART CARDS (MECHANISM WISE)

- Contact Smart Card
- Contactless Smart Card
- Combination Smart Card
- Hybrid Smart Card
- Proximity Smart Card

CONTACT SMART CARDS

The contact smart cards are named so because they come in contact with the reader. These smart cards are the size of a credit card. A metallic chip is embedded inside the plastic card with a microprocessor and a memory or only with a memory. The contact smart cards are widely used in network security, access control, e-commerce, and electronic cash and as health cards.

CONTACTLESS SMART CARDS

The contactless cards do not directly come in contact with the card. These cards have an antenna built in the card. The antenna of the contactless cards is used to communicate to the card reader for reading and writing data on the card. The working of these cards is based on radio frequency identification technology. These cards are used as parking cards, student identification and electronics passports.

COMBINATION SMART CARDS

The combination smart cards are a combination of the contact smart cards and contactless smart cards. These cards can be read and written with contact or without contact with the reader. The antenna of the card is used or the contact pads are used to manipulate data on the smart card. The combination smart cards are used as vending passes, meal passes, access control and network security.

PROXIMITY SMART CARDS

The proximity cards are contactless cards and they have an antenna embedded in the card. However, the proximity cards are read-only cards and the information on these cards cannot be manipulated. The proximity cards also use the radio frequency identification (RFID) technology. The applications of these cards include access control, identification and security.

HYBRID SMART CARDS

The hybrid cards have more than two technologies embedded inside a single smart card. These cards use any two of the features of the smart cards in a single chip. Some applications of smart card require more than two technologies like the proximity card and the contact card integrated in a single chip.

TYPES OF SMART CARDS: (CAPABILITIES WISE)

- Microprocessor-Based Smart Cards
- Memory-Based Smart Cards

MICROPROCESSOR-BASED SMART CARDS

The microprocessor-based smart cards have greater memory storage as compared to those cards without microprocessor. The security of data on the microprocessor cards is greater than any other storage device because it has the microprocessor embedded in the plastic card along with the memory. The microprocessor-based smart cards available today have an eight bit processor and 512 bytes random access memory (RAM) and 16 KB read-only memory. Some of the microprocessor-based smart cards use cryptography for securing a digital identity. The microprocessor-based smart cards have a card operating system (COS) that manages the data on the card. The card operating system makes it possible to make the smart cards multi-functional.

MEMORY-BASED SMART CARDS

The memory-based smart cards are used for applications in which the function of the card is fixed. These cards need a card reader to manipulate the data on the card. Memory-based smart cards communicate to the reader using some synchronous protocols. Memory-based smart cards have no processing power and cannot manage the data stored in them. These cards are widely used as prepaid phone cards.

APPLICATIONS OF SMART CARDS

Currently, disposable, prepaid phone cards are the most widely used smart cards in the United States. Transit cards are quite popular in a few large metropolitan areas e.g. Washington, D.C. and New York. Applications of smart cards are increasing. The following are some of the more important applications.

LOYALTY CARDS: Retailers use these cards to identify their loyal customers and reward them. Boots advantage card) (boots.com) and Shell Company's plus points (shell.com) are examples of loyalty cards.

IT CARDS: Many modern PCs are equipped with smart card readers. Users can use smart cards to protect their privacy.

TRANSPORTATION CARDS: Smart cards are being used to pay for transportation. One such card that can be used for other payments, such as in vending machines, restaurants, and gas stations, is Octopus Card in Hong Kong.

IDENTIFICATION CARDS: Smart cards are being used in applications such as college IDs, driver's licenses, and immigration cards. One advantage of smart cards is that they may contain biometrics, and therefore are extremely difficult to forge.

MULTIPURPOSE CARDS: These cards offer capability of many cards in one i.e. credit card, debit card, cash card, loyalty card. In February 2001, MasterCard International and Korea's Kookmin Card Corp. issued the first multipurpose smart card in the world. It contained credit and debit card features, e-cash (from Mondex), and public transportation fares, all in one card.

VISA CASH: Visa Cash is a contact smart card developed by VISA. Payment for a purchase can be made by inserting the card in a smart card reader. Value can be added to the card via specialized ATMs without entering a PIN or a signature. VISA cash cards work more like debit cards - you must first deposit the money before you can use it. The bank or other lender does not lend you any money.

MONDEX: Mondex is a smart card that can be loaded with value, transfer value to another Mondex card, or transfer value to a merchant during a purchase. At one times, Mondex card can store value in up to five currencies. The security programs contained within the microchip on the Mondex card protect transactions between one Mondex card and another. Value can be added to Mondex cards using Mondex-compatible ATMs. Mondex cards can be utilized across multiple channels, including Internet. All payments using Mondex cards require a Mondex card reader. Using a portable reader, Mondex cards can be used to conduct transactions over the Internet. Mondex cards contain logs to track past transactions, pending transactions, and transactions, which raise exceptions. The Smart Card Alliance (www.smartcardalliance.org) is the organization that promotes the widespread acceptance of multiple-application smart cards. Its members include companies in banking, financial services, computer technology, healthcare, telecommunications, and a number of government agencies.

VISA BUXX: Visa Buxx is a prepaid card that enables its users to load the card online or over the phone. This card is generally used by teenagers and only a limited value can be loaded to the card. Parents can monitor the spending of the card. Users are not responsible for transactions made on a lost or stolen card.

Examples of smart card uses most common in the United States include:

- The U.S. Federal Government Personal Identity Verification (PIV) card being issued by all Federal agencies for employees and contractors.

- The ePassport being issued by the Department of State.

- Payment cards and devices being issued by American Express, Discover Network, MasterCard and Visa.

- Transit fare payment systems currently operating or being installed in such cities as Washington, DC, Chicago, Boston, Atlanta, San Francisco and Los Angeles.

- The Subscriber Identity Module (SIM) used in mobile phones.

- Pay (satellite) TV security cards in set-top boxes for cable and satellite television subscribers.

SMART CARD ALLIANCE

The Smart Card Alliance is a not-for-profit, multi-industry association working to stimulate the understanding, adoption, use and widespread application of smart card technology. The Alliance invests heavily in education on the appropriate uses of technology for identification, payment and other applications and strongly advocates the use of smart card technology in a way that protects privacy and enhances data security and integrity. Through specific projects such as education programs, market research, advocacy, industry relations and open forums, the Alliance keeps its members connected to industry leaders and innovative thought. The Alliance is the single industry voice for smart card technology, leading industry discussion on the impact and value of smart cards in the U.S. and Latin America.

The Alliance is comprised of over 200 members worldwide, including participants from financial, government, enterprise, transportation, mobile telecommunications, healthcare, and retail industries. A mix of issuers and adopters of smart card technology work in concert with leading industry suppliers of the full range of products and services supporting the implementation of smart card based systems for secure payments, identification, access, and mobile communications.

Smart Card Alliance is involved in many activities.

- **Events**–Smart Card Alliance conferences and web seminars that provide the latest information on the development and implementation of smart card initiatives.

- **Industry and Technology Councils**–Member-driven groups that focus on topics in specific industries or market segments to accelerate smart card adoption and industry growth.

- **Education**–Smart Card Alliance conferences, web seminars, Educational Institute courses, training and certification programs, and web content that provide smart card market, application and technology education and information for both Alliance members and the industry. The Smart Card Alliance also publishes reports and white papers on smart card applications, public policy positions, business and implementation case studies and other deliverables focused on helping the industry to understand and overcome deployment issues.

- **Outreach**–Industry and international outreach and advocacy to stimulate the understanding, adoption, use and widespread application of smart card technology.

Payment Gateways

The payment gateway is the infrastructure and a system of computer processes that allows a merchant to accept credit card and other forms of electronic payment. It is usually a third-party service that process credit card transactions on behalf of the merchant through secure Internet connections. In general, payment gateway has two components: 1) a virtual terminal that merchant can use to securely login and key in credit card numbers and 2) the merchant's website shopping-cart that connect to the payment gateway via an API (Application Programming Interface) to allow for real time processing from the merchant's website. In general, a merchant account provider is a separate company from the payment gateway provider. While some merchant account providers do have their own payment gateways, most merchant account providers use third party payment gateways. Merchants have the option to select payment gateway and merchant account from different providers separately. In general, payment gateways are not an absolute requirement for doing online business. A business may take orders from a website and process them manually but it requires a huge time commitment and effort. Most businesses prefer not to go for manual processing and therefore need both a payment gateway and a merchant account provider.

A merchant must provide business information (e.g. a business plan, details about existing bank accounts, and a business and personal credit history) before the merchant account provider (usually a bank) a merchant account. The type of business also influences the merchant's likelihood of getting a merchant account because these businesses increases financial risk for the merchant account provider (e.g. some businesses have a higher likelihood of customers repudiating payment card charges than others). The merchant account provider assesses the level of risk in the business based on the type of business and the credit information that is provided. To ensure that sufficient funds are available to cover charge-backs, a merchant bank might require a company to maintain funds on deposit in the merchant account.

Selection Process for Payment Gateways

Merchant account or payment gateway providers are service provider. The most important thing to consider, before selecting a merchant account or payment gateway provider, is to gain an understanding of the merchant account or payment gateway provider. The service provider's longevity and history should be weighed with as much importance as many other factors. One should also use the wealth of information available on the web to make an informed decision. Following are some common criteria businesses can use to make payment gateway/merchant account selection.

BUSINESS HISTORY: A merchant gateway provider with a long and established business history is preferred by many businesses.

CREDIT HISTORY: A new business without an established credit history might find it difficult to obtain a merchant account quickly.

FRAUD PROTECTION: That includes the type of fraud protection offered by the service provider.

BRANDED PAYMENT INTERFACE: Payment gateway providers can offer you to brand the payment system interface so that it fits on your website, with your logo, colors and design.

INTEGRATION WITH EXISTING BILLING SYSTEMS: Payment systems should be able to integrate into an existing billing mechanism.

PCI-COMPLIANCE: The payment systems should adhere to PCI (Payment Card Industry Security Standards Council standards.

SETTLEMENT PERIOD: This is the time period between the finalization of a sale and the transfer of funds into merchant's bank account.

CONVERTIBILITY: It refers to payment systems ability to handle global currencies and provide conversion when needed.

INTEGRATION WITH SHOPPING CART: That refers to the ability of payment system to integrate seamlessly with your existing shopping cart. Payment gateways generally offer two ways to integrate their payment gateway services into a merchant's Web site. In the first method, which is less complex, a merchant would send the customer to the payment gateway provider website where a secure order page will capture their transaction information. Upon completion of the transaction the customer is returned to the merchant's Web site. This method is usually preferred by small businesses that do not have the technical capabilities to do a more advanced integration. The second method, which is more complex but more powerful, utilizes the payment gateway API to process a transaction without the customer leaving the merchant's Web site. The customer is unaware of how the transaction is processed and the checkout process is seamless to them. This method is usually chosen by larger businesses that wish to present their Web site in a professional manner.

COST STRUCTURE: Make sure to understand how much the payment system costs to you and clarify any confusion about various fees charged.

SWITCHING COST: That refers to the ease of switching from one payment system to another in case we become unsatisfied with the first payment system we selected. Merchants should remember that switching payment gateway providers can result in significant information loss. To counter information loss, merchants must download as many of the reports and details of their transactions with the previous provider as possible before finalizing a changeover.

CUSTOMER SUPPORT: The customer support capabilities of merchant gateway provider should be enough to avoid any after-sale problems.

Electronic Payments (electronic-payments.co.uk) have a toolkit available. You can use this toolkit to calculate the estimated costs of using various payment gateway providers.

Rate and Fee Structures and their Determination

Rates and fees charged by a merchant account provider can vary depending on whether the merchant considers the transaction a standard transaction or not. Definition of standard transaction is up to the merchant but in general it is the transaction where there is little doubt as to the authenticity of the purchaser.

Rates applied to the transactions can be of three types: Qualified, Mid-qualified, and Non-qualified.

QUALIFIED RATE: This is typically the rate you are quoted when you sign up for a merchant account. This rate only applies to swiped regular retail cards.

MID-QUALIFIED RATE: This is typically the rate charged by merchant account provider for credit card transactions that does not qualify for qualified rate. This may happen for several reasons (e.g. when a credit card is keyed into a credit card terminal instead of being swiped or a special kind of credit card is used like a rewards card or business card). A qualified rate is generally lower than a mid-qualified rate.

NON-QUALIFIED RATE: Non-qualified rate is the highest rate charged by merchant account provider for credit card transactions that does not qualify for qualified or mid-qualified rate. This may happen for several reasons such as when proper identity and address verification procedures are not being performed, transaction settlement times not being adhered to, information normally provided in a standard transaction is not provided, or a special kind of credit card is used like a business card and all required fields are not entered.

In general, there are three basic components of payment gateway fees: One-time fees, Recurring monthly fees, and transaction fees. There can be some other fees involved. Following are some of the common fees charged by payment gateways and their brief description.

PCI COMPLIANCE FEE: Payment gateways need to be PCI-compliant. If a business decides not to become PCI-complaint then it will need to pay a monthly or yearly PCI fee. This fee can range from $20 - $30 per month to hundreds of dollars a year.

DISCOUNT RATE: The discount rate charged by payment gateways comprises of a number of dues, fees, assessments, network charges and mark-ups.

TRANSACTION FEE: The fee charged by the payment gateway for each credit card transaction processed for the account. The credit card transaction types that the per-transaction fee is charged for include authorizations, captures, refunds, declines or other related transactions, completed or submitted within the payment gateway. This fee is charged in access to the discount rate. For example, a discount rate of 1.85% plus 30 cents per transaction.

BATCH CLOSURE FEE: The fee assessed per batch of settled credit card transactions

AUTHORIZATION FEE: This is the amount charged to a merchant account each time communication happens between the software or point of sale terminal and the payment authorization network.

INTERCHANGE FEE: It is a fee that a merchant's acquiring bank pays to a customer's card issuing bank when merchants accept cards using card networks such as Visa and MasterCard for purchases.

GATEWAY SETUP FEE: One-time fee charged to establish a payment gateway account.

MONTHLY GATEWAY FEE: It is the monthly fee charged for a payment gateway account.

FRAUD DETECTION FEE: This is the fee charged by payment gateway to detect fraudulent transactions. In most cases, the service is optional and merchants are charged only if they opt for fraud detection service.

Value-Added Services

Payment gateways also provide value-added services e.g. fraud detection and customer information management. Fraud detection services use multiple filters (e.g., address mismatches IP address etc.) and tools to look for indicators of fraud in transactions.

Using Customer Information Management (CIM) service of payment gateway, merchants can store customer's sensitive payment information on payment gateway servers. This could simplify the payments process for returning customers and recurring transactions. You can also issue transactions manually from within the Merchant Interface, or integrate your Web site or other application using an Application Programming Interface (API). Additionally, the merchants can also view and search for customer transaction history.

Payment Gateways: Available Alternatives

Authorize.Net, Cybersource, and PaymentPlus are major players in the payment-systems marketplace. Vendors such as PayPal and Google have also made significant progress. Customers using Google Checkout enter their billing information once and can then make subsequent purchases online with a single click of mosue.

Authorize.Net (authorize.net) is a payment gateway service provider owned by Cybersource. Authorize.Net sells its services to merchants both directly and indirectly through re-sellers. Authorize.Net resellers are typically merchant account providers. If using a reseller pricing for services of their payment gateway will vary. Authorize.Net currently does not offer merchant accounts as a service. To use the payment gateway offered by Authorize.Net a merchant will need to establish a merchant account through a separate company. Authorize.Net provides payment-processing facility to business processing less than $3 million online.

CyberSource (cybersource.com) is a provider of electronic payment services. CyberSource services can be used for electronic payment processing for Web, call centre, and POS environments.

Payment Plus, Inc. (paymentplusinc.com), founded in 1999, is a state-of-the-art electronic payment processing provider, based in USA.

How Payment Gateways Works

A variety of tasks is performed by payment gateway when a customer orders a product, using his credit card, from an online store. In general, two processes take place: Transaction Authorization and Transaction Settlement

---
TRANSACTION AUTHORIZATION

1.   Order is placed by the user via secure Web site connection, retail store, MOTO (Mail Order/Telephone Order) center or wireless device.

2.   The information to be sent between the customer's PC and merchant's web server is encrypted by customer web browser.

3.   Merchant website forwards, via SSL encrypted connection, the transaction details to the payment server hosted by the payment gateway.

4.   The payment server forwards the transaction information to the payment processor used by the merchant's acquiring bank (Merchant's acquiring bank is the merchant account provider).

5.   Payment processor of the merchant bank submits the transaction to the credit card network. A credit card network is a system of financial entities that facilitates the processing, clearing, and settlement of credit card transactions. This network routes the transaction to the bank that issued the customer's credit card. Two available networks include VISA and Master Card. In case an American Express or Discover Card was used, the processor acts as the issuing bank and directly provides a response (approval or disapproval of transaction) to the payment gateway.

6.   The credit card issuing bank receives the authorization request and sends a response back to the payment processor (via the same process as the request for authorization) with a response code. The response code includes a response (approval or disapproval of transaction) and the reason why the transaction failed (such as insufficient funds, or bank link not available)

7.   The payment processor stores the transaction result and forwards the response to the payment gateway.

8.   The payment gateway receives the response, and forwards it to merchant website. This step completes the transaction authorization process.

Figure 8-10 shows the transaction authorization process.

FIGURE 8-10: TRANSACTION AUTHORIZATION CYCLE

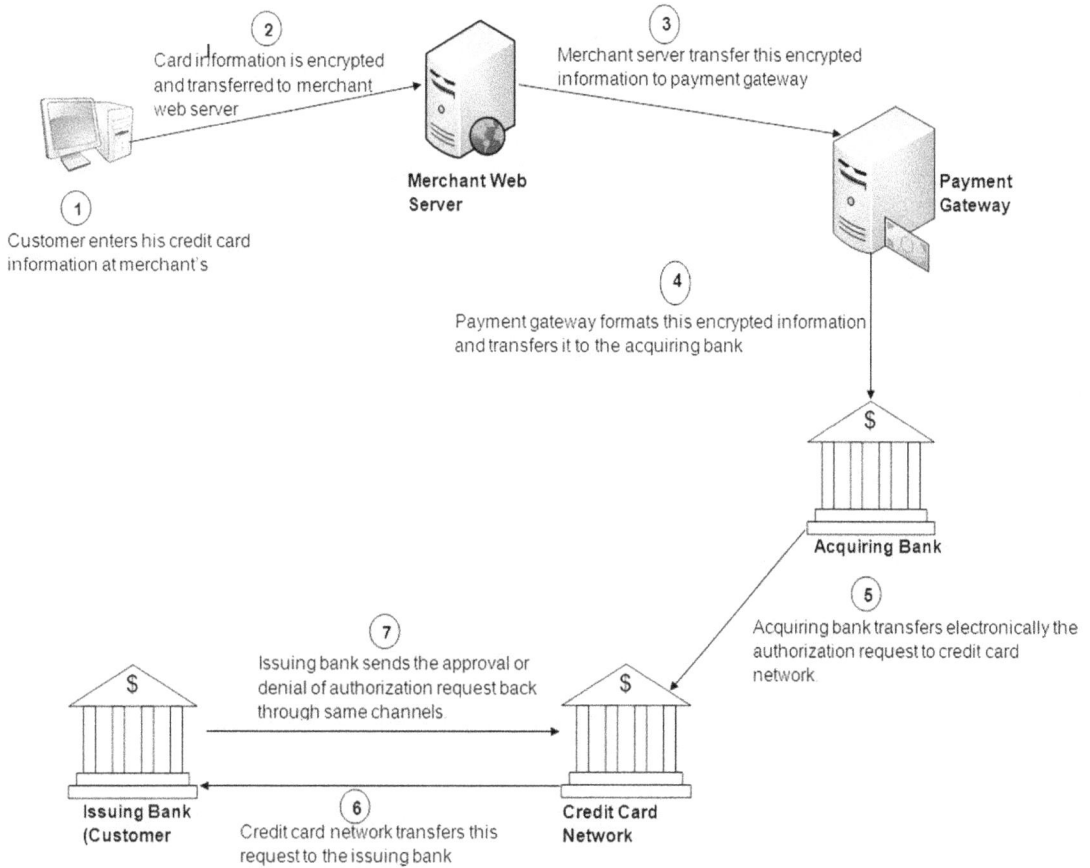

Merchant Web Server

② Card information is encrypted and transferred to merchant web server

③ Merchant server transfer this encrypted information to payment gateway

Payment Gateway

① Customer enters his credit card information at merchant's

④ Payment gateway formats this encrypted information and transfers it to the acquiring bank

Acquiring Bank

⑤ Acquiring bank transfers electronically the authorization request to credit card network

⑦ Issuing bank sends the approval or denial of authorization request back through same channels.

Issuing Bank (Customer

⑥ Credit card network transfers this request to the issuing bank

Credit Card Network

TRANSACTION SETTLEMENT

9. Merchant website interprets the response and if it is valid, it relays it back to the cardholder and the merchant.

10. The merchant submits a batch containing all the approved authorizations to its acquiring bank for settlement.

11. The Customer's Credit Card Issuing Bank sends the appropriate funds for the transaction to the Credit Card Network, which passes the funds to the Merchant's acquiring Bank.

12. The total amount of the approved funds will be deposited by acquiring bank into merchant's nominated bank account. This account could be with the acquiring bank itself, if the merchant does banking through the acquiring bank, or it could be with a different bank. This step completes the transaction settlement process.

The process illustrated in Figure 8-11 offers a "big picture" view of Visa card payment settlement events that can take place. The process may vary slightly depending on your technology requirements and the service providers you use.

FIGURE 8-11: ONLINE TRANSACTION SETTLEMENT

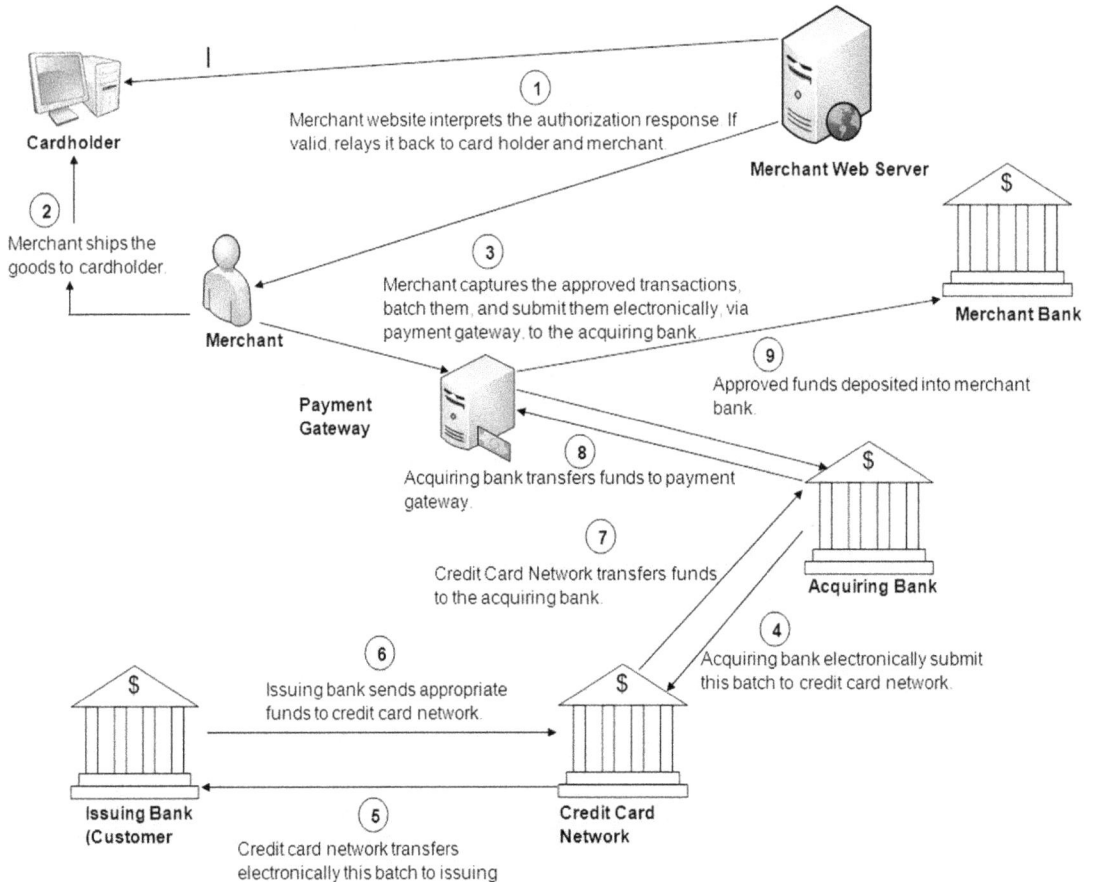

Cardholder

(1) Merchant website interprets the authorization response. If valid, relays it back to card holder and merchant

Merchant Web Server

(2) Merchant ships the goods to cardholder.

Merchant

(3) Merchant captures the approved transactions, batch them, and submit them electronically, via payment gateway, to the acquiring bank

Merchant Bank

Payment Gateway

(9) Approved funds deposited into merchant bank.

(8) Acquiring bank transfers funds to payment gateway.

(7) Credit Card Network transfers funds to the acquiring bank.

Acquiring Bank

(4) Acquiring bank electronically submit this batch to credit card network.

(6) Issuing bank sends appropriate funds to credit card network.

Issuing Bank (Customer

Credit Card Network

(5) Credit card network transfers electronically this batch to issuing

The entire transaction authorization process typically takes 2–3 seconds. The entire process from authorization to settlement to funding can typically take up to 3 days. Many payment gateways also provide tools to automatically screen orders for fraud (e.g. delivery address verification) and calculate tax in real time prior to the authorization request being sent to the processor.

Getting a Merchant Account

Merchant account is an account provided by a bank through which the merchant can process payment card transactions. Typically, a new merchant must supply a business plan, details about existing bank accounts, and a business and personal credit history. The merchant bank wants to be sure that the merchant has a good prospect of staying in business and wants to minimize its risk. An online merchant that appears disorganized is less attractive to a merchant bank than a well-organized online merchant is. The type of business also influences the bank's likelihood of granting the account. In some industries, merchant banks will be reluctant to offer a merchant account because of the type of business; some businesses have a higher likelihood of customers repudiating payment card charges than others. For example, a business that sells a guaranteed weight loss scheme will find many merchant banks unwilling to provide an account. This is because it is a business in which many customers might want their money back. Merchant banks must estimate what percentage of sales is likely to be contested by cardholders. When a cardholder successfully contests a charge, the merchant bank must retrieve the money it placed in the merchant account in a process called a chargeback. For example, a new or risky business that plans to make $100,000 in sales each month might be required to keep $50,000 or more on deposit in its merchant account.

Selecting a Merchant Account Provider

By coupling your retail or online store with a merchant account, you can begin accepting credit cards. Although you are free to continue relying exclusively on cash or checks, adding more payment options will allow you to capture many more types of sales. The merchant services industry is notorious for hidden fees, unethical agents, and expensive contracts. As a result, finding a quality provider can take hours of research. With so many other things on your plate, you might be tempted to simply go with the company that quotes the lowest rates. However, not all merchant account providers are created equally. Below are some essential considerations for a business selecting a merchant account for his business (BluePay.com, 2014; Parker, 2014).

CUSTOMER SUPPORT

Customer service is changing at a rapid pace. Traditional call centers and personal service reps are being replaced by options like FAQ sections, user forums, live chat, email tickets, and social media channels. Most providers do offer some level of technical support. Before you sign up with a provider, you should ask yourself what kinds of customer service are most important to you and what kind of availability you expect from your payment processor. At a minimum, it is best to find companies with phone support during business hours. You never know what emergencies might arise during the business day, and it can be a major frustration to lose sales because you cannot contact a real human being. Beyond basic phone support, you will have to determine which methods you find most effective. It is best if the merchant account provider provides 24/7 support, since issues can crop up at any time. Some merchant account providers offer round-the-clock email support. Many a times, customer wants to either speak to a real person via phone or live chat. Once you have found a company that offers the support you require, you should also be sure to search for user reviews of that company. You should know the satisfaction level of other businesses with the service they receive, the way company responds to public complaints, and any positive testimonials of the company. Settle on a provider that is available when you need it, that takes its reputation seriously, and that will work to retain your business, then you are well on your way to securing peace of mind as a merchant.

BUSINESS HISTORY

You want a provider with a proven record of accomplishment. While new companies may make some attractive offers to you it is not advisable to go with a company that has little to none business history.

ABILITY TO PROCESS ONLINE AND OFFLINE TRANSACTIONS

As you expand your business, you do not want to have to switch merchant account providers because they do not offer the full range of credit card processing services. Stick with a provider who can eventually take your retail business online.

PCI-COMPLIANT PAYMENT SYSTEMS

PCI compliance ensures that the payment systems follow security standards set forth by the credit card industry.

FRAUD PROTECTION SERVICES

New credit card liability rules going into effect in 2015 (BluePay.com, 2014). You want to reduce your exposure to credit card fraud as much as possible. Ask each merchant account provider how it protects sensitive financial data.

AVERAGE PROCESSING TIME FOR FUNDS

Predictable cash flow is essential to the success of your business. You want to know (in advance) how long it takes for funds clearing. It is important so that you do not end up with any surprises.

CONTRACT LENGTH

Many merchant account providers use contracts with predetermined lengths. If you are just starting out, shorter contract lengths are better. Once you feel comfortable, you can explore longer policies with that provider. Most credit card processing contracts last for years and can include harsh fees for merchants who cancel the service early. Month-to-month contracts are increasingly common, and it is getting harder for processors to conceal expensive penalties in their terms and conditions. You should make it clear from the start that you expect a month-to-month agreement with no termination fees, and then you should double-check the actual language of your contract to ensure that you will be able to cancel without consequences if you are unhappy with the service. This is the easiest way to protect yourself from a bad merchant account contract.

CANCELLATION OR EARLY TERMINATION FEE

Most merchant account agreements often have complicated cancellation procedures that must be correctly followed in order to prevent the contract from renewing for additional terms. Some merchant account providers penalize clients for canceling contracts prematurely. Although such fees are standard, you should not have to pay more than a few hundred dollars.

OTHER FEES

Cancellation fees are standard. Make sure that you understand any other charges that may apply.

COMPATIBILITY PAYMENT PROCESSING WITH YOUR SYSTEM

Many merchant account providers use proprietary payment systems that clash with pre-existing online shopping carts. Before signing the contract, make sure that the provider's processing technology is compatible with your system. Or better still, stick with providers that use universally accepted payment systems (like Authorize.net).

REFERENCES

You should ask to speak with other satisfied clients. These clients should preferably those whose business needs are similar to your own.

YOUR SPECIFIC BUSINESS NEEDS

First, ask yourself what you want out of your merchant account provider. Assess your mobile processing and online payments needs to narrow your search from the outset. Many businesses simply find the lowest price and work backward from there. The better option is to first compile a list of providers who specialize in your business needs before you turn your attention to contractual considerations. Do not expect that a provider will operate outside of its comfort zone.

LONG-TERM EQUIPMENT LEASES

Consumer electronics like iPads and iPhones are becoming increasingly viable options for payment processing, but most businesses still require traditional hardware like credit card terminals, PIN pads, printers, or check readers to accept digital payments. If you are in the market for specialized credit card processing hardware, be sure to watch out for predatory equipment leases. In the vast majority of cases, it is cheaper to buy your credit card processing equipment than it is to lease it. While merchant account providers will do their best to convince you that an equipment lease will save you money, but these offers usually require a lot of creative math and more than a little misinformation. Most credit card terminals can be purchased for $200-$300 (at the very most) on eBay or Amazon, while most equipment leases average $40-$50 per month. It is clear that you will dramatically overpay for that equipment over the life of a 48-month lease. Long-term equipment leases are a major source of income for sales agents, so they are often included as part of your merchant account paperwork in the hope that you will not realize what you are committing to. They are also legally binding contracts, so they could leave you on the hook for the full repayment term regardless of when you decide to cancel. To avoid this, you should be sure to select a provider that does not advertise long-term equipment leases, and you should double-check all documentation during the application process to ensure that you are not being duped into a lease you do not wish to sign.

PRICING

It is quite difficult to compare merchant account rates across providers. This is because of a common pricing tactic called tiered pricing. Let us say that one merchant account provider advertises, "rates as low as 1.15%" while another offers "rates starting at 1.25%." Looking at these two options, it is not simple to say which one is cheap. For one thing, these rates only apply to a certain class, or "tier," of credit cards, known as the "qualified" tier. Certain card types like signature debit and non-rewards credit cards will swipe at this rate, but most other cards will process at higher rates known as "mid-qualified" and "non-qualified" rates. Merchant account providers rarely quote these other rate tiers because they are much higher than the qualified rates, with many "non-qualified" rates exceeding 3% per transaction. Most merchant account provider does not tell you which card type fall under which tiers. Processors are free to move card types between tiers as they see fit, and they are under no legal obligation to tell you exactly how they have structured their tiers. This leaves you in the dark as to how much the processors are marking up each transaction and makes it impossible for you to compare your current provider's rates to those offered by a competitor. The solution to this problem is interchange-plus pricing. Interchange-plus pricing is a different kind of rate structure that removes the confusion surrounding tiers. Under an interchange-plus pricing model, providers pass along card costs at predetermined "interchange" rates (a non-negotiable per-transaction rate set by Visa/MasterCard) and then add their own per-transaction markup. For example, assumes one provider quote you a rate of "Interchange plus 0.18% per transaction" while another provider quotes you a rate of "Interchange plus 0.20% per transaction." In this case, it is easy to tell which processor offers cheaper rates, since the processor's markup is clearly listed as the "plus" portion of the quote.

While selecting a merchant account provider, remember that the best overall fit will depend on a combination of factors. You may not find a provider for your business type that meets all of these criteria. Some providers might offer excellent customer service but try to pressure you into equipment leases. Others might offer month-to-month agreements but charge higher-than-expected rates. The key is to stick to your priorities and avoid common hazards like misleading rate quotes or multiyear contracts. If you can find a handful of providers that meet your specific needs and narrow your list down to two or three providers based on the considerations discussed above, then you can negotiate with each remaining provider to secure the best rates and eliminate fees. With some patience and a little bit of research, you should be able to land yourself an excellent deal.

Merchant Account Marketing

Merchant accounts are marketed to merchants by two basic methods: either directly by the processor or sponsoring bank or by an authorized agent for the bank and additionally directly registered with both Visa and MasterCard as an ISO/MSP (Independent Selling Organization / Member Service Provider). Marketing details are by card issuers like Visa and MasterCard, and are enforced by various rules and fines. A few of the largest processors also partner with warehouse clubs to promote merchant accounts to their business members.

MARKETING BY BANKS: A bank that has a merchant processing relationship with Visa and MasterCard, also known as a member bank, can issue merchant accounts directly to merchants. To reduce risk, some banks limit approval to merchants in its geographical area, those with a physical retail storefront, or those that have been in business for 2 years or more.

MARKETING BY INDEPENDENT SALES ORGANIZATION (ISO)/MSPS: To market merchant accounts, an ISO/MSP must be sponsored by a member bank. This sponsorship requires that the bank verify the financial stability and suitability of the company that will be marketing on its behalf. The ISO/MSP must also pay a fee to be registered with Visa and MasterCard and must comply with regulations in how they may market merchant accounts and the use of copyrights of Visa and MasterCard. One way to verify if an ISO/MSP is in compliance is to check a website or any other marketing material for a disclosure "company is a registered ISO/MSP of bank, town, state. FDIC insured". This disclosure is required by both Visa and MasterCard and will cause a fine of up to $25,000 if it is not clearly visible.

Payment Aggregators

Payment Aggregators are service providers that allows merchants to accept credit card and bank transfers without having to setup a merchant account with a bank or card association. Payment aggregators typically hold consumer credit card information to allow for faster purchases (e.g. Google Checkout) or hold money in an account to allow for future purchases (e.g. PayPal). Aggregator makes payment to the merchant. Some examples of payment aggregators include Google Checkout (checkout.google.com), PayPal (paypal.com), and Amazon Payments (payments.amazon.com). Payment aggregator firms differ in their payment aggregator approaches, costs and services delivered to merchants. Payment aggregators are easy to implement, provide easy access to cross border markets, and access to large customer segments, however, they may or may not guarantee against fraudulent charge backs. Payment aggregators are suitable for small to large merchants and for those merchants who may have difficulty establishing a merchant account with a bank.

Payment aggregation can cost significantly higher than direct credit card payments. Before implementing or buying the payment aggregator functionality a business needs to make sure that they understand the costs for the services. Estimated costs vary from aggregator to aggregator but generally include monthly fees, merchant account fee, and interchange fees. Amazon charges higher fees than both PayPal and Google Checkout payment services but effectively acts as a marketplace with a tremendous customer base. Utilizing Amazon, merchants can save marketing and website development costs.

Alternative Payment Systems

Credit card or electronic check payments are very popular in the Internet. However, many times these methods cannot be used to make payments because of several reasons e.g. card issuer restrictions on businesses with the high risk of charge backs or the businesses accepts payments such as checks or money orders through the mail. Therefore, new alternative payment methods are needed that allow the merchants to develop their business where regular credit card payments are not possible. Many systems have been implemented that allow the merchants to accept payments from the buyers without credit cards. Some of these methods include:

Cash on-delivery (COD)

In this system, payment for a good is made at the time of delivery. Upon delivery, if the payment is not made by the buyer the good is returned to the seller. Depending on the terms of shipping contract, the payment can be made through money order, cash, or a certified check. A shipping company is generally used as a third party in the transaction. This way both the seller and the buyer of the product can minimize the risk of fraud or default. Buyer makes the payment to the shipping company, which then sends the payment back to the seller. Good is returned to seller in case of non-payment by buyer. COD is a popular payment method for e-commerce business dealing in consumer electronic (like mobile phones, TV, cameras, laptop, computer etc.) and other businesses dealing in garments, commodities items, cosmetics and medicines. Shipping charges may be applied on cash on delivery depending on the nature of product and location of customers. www.nowtees.com is an online garments business that provides the facility of cash on delivery.

Payment by Telephone using the Payment Code

This system is suitable for sellers who prefer to receive payments from their clients by telephone calls via 900-line. Typical users of the system include Internet sites that provide charged services, TV broadcasts, which accept SMS messages from their viewers, radio stations that accept SMS messages from their listeners, organizers of SMS lotteries, and organizers of social campaigns. One example of this system is webtopay (webtopay.com). When a customer accesses a merchant site, who is a member of webtopay, the customer is asked to either send a message of certain content to a short GSM number or call to a specific 900-line number. When customers call the 900-number, they are given a payment code. Having received an SMS, Webtopay system shall inform merchant website about the received message and transfer it to the script in your website. If the customer calls by 900-number, then the customer shall enter the code given into a form on merchant website; the form verifies the code on Webtopay system. Remuneration to the merchant for a payment shall be included into the seller's account on Webtopay website.

Internet Technologies and Banking Industry

Electronic Checks

Physical check processing by banks requires transporting tons of paper checks around the country. In addition to the transportation costs, another disadvantage of using paper checks is the delay that occurs between the time that a person writes a check and the time that check clears the person's bank. Electronic Check (e-Check) Processing is an electronic payment process that directly debits consumers bank accounts for payment for goods or services without the need for a paper check. E-checks are less expensive than credit cards for merchants, and they are much faster than paper-based traditional checking. E-Check processing is not new to the financial industry. It is a safe and reliable system that uses technology that has been developed and tested to process check information securely. Some major digital checking systems are paybycheck (paybycheck.com), and eCheck.net (authorize.net/solutions/merchantsolutions/merchantservices/echeck).

A check you write may be processed as a check or a merchant may use your check as a source of information to create an electronic fund transfer. In the first case, your rights are governed by check laws and regulations. Consumer rights and laws that govern electronic fund transfers are different as compared with payments with through checks. For example, a customer has the right to ask his financial institution to initiate an investigation if an error occurs in electronic check conversion.

ELECTRONIC CHECK PROCESSING

Providing a standard payment form on merchant website, is one of the most commonly used method for electronic check processing. Following are the steps involved.

Step 1: A customer initiates an electronic check transaction, with authorization to charge their bank account, via Web form or mail order/telephone order. Figure 8-12 shows a web form presented to the customer in which he can enter his check details.

FIGURE 8-12: ELECTRONIC CHECK FORM

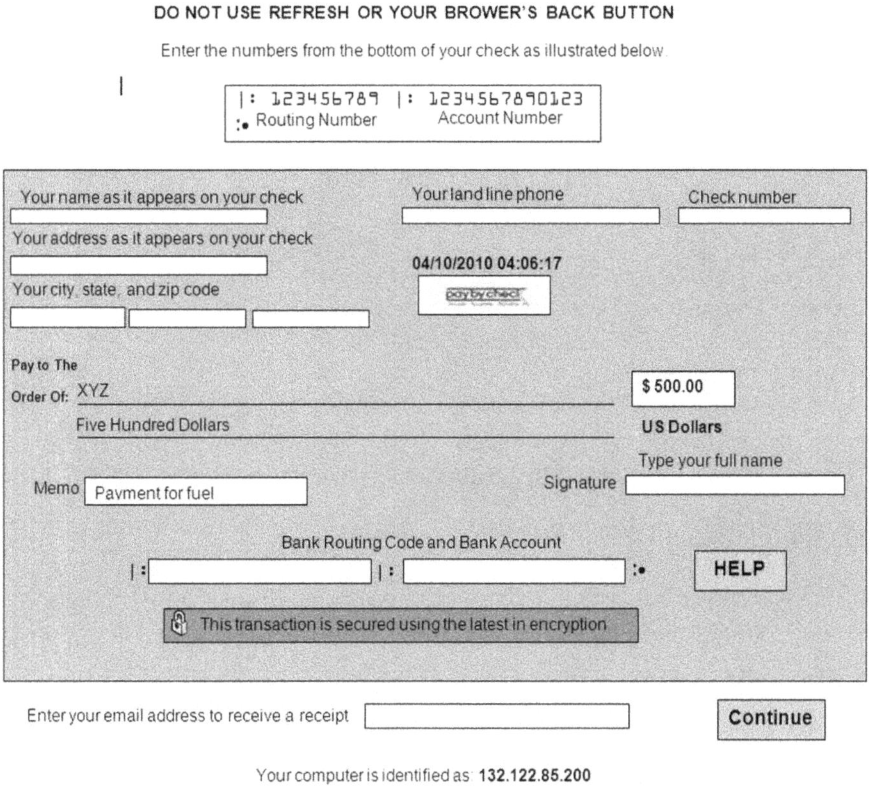

DO NOT USE REFRESH OR YOUR BROWER'S BACK BUTTON

Enter the numbers from the bottom of your check as illustrated below

|: 123456789 |: 1234567890123
:● Routing Number   Account Number

Your name as it appears on your check   Your land line phone   Check number

Your address as it appears on your check
                                               04/10/2010 04:06:17
Your city, state, and zip code   paybycheck

Pay to The
Order Of: XYZ $ 500.00

Five Hundred Dollars   US Dollars

                                                               Type your full name
Memo Payment for fuel   Signature

                    Bank Routing Code and Bank Account
        |: |:   :●   HELP

              🔒 This transaction is secured using the latest in encryption

Enter your email address to receive a receipt [   ]   Continue

Your computer is identified as **132.122.85.200**

Step 2: Once the transaction is submitted, order and payment information is securely transmitted via the Internet to the Payment Gateway. Payment gateway validates the data to either accepts or reject the transaction.

Step 3: A confirmation page is presented to the customer where he can review the entered information and read the required legal information. Customer can either Agree to the transaction and move to the next step or Not Agree which will take him back to the check form to start over. Figure 8-13 shows the confirmation page.

FIGURE 8-13: E-CHECK CONFORMATION PAGE

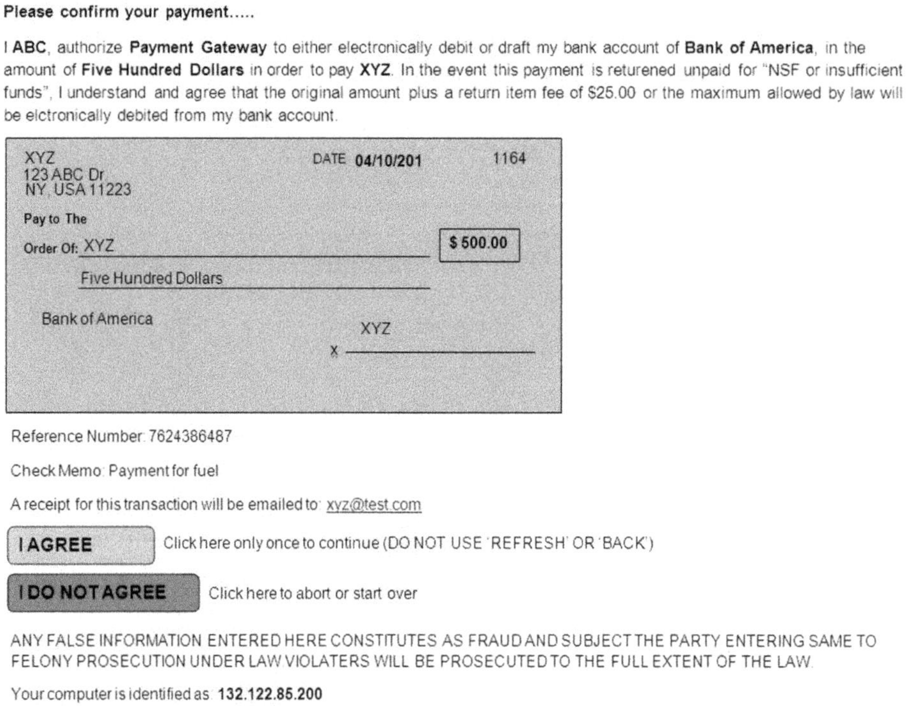

Please confirm your payment.....

I **ABC**, authorize **Payment Gateway** to either electronically debit or draft my bank account of **Bank of America**, in the amount of **Five Hundred Dollars** in order to pay **XYZ**. In the event this payment is retuerned unpaid for "NSF or insufficient funds", I understand and agree that the original amount plus a return item fee of $25.00 or the maximum allowed by law will be elctronically debited from my bank account.

```
XYZ                              DATE 04/10/201        1164
123 ABC Dr
NY, USA 11223

Pay to The
Order Of: XYZ_____    [ $ 500.00 ]

          Five Hundred Dollars_____

     Bank of America
                              XYZ
                          x _____
```

Reference Number: 7624386487

Check Memo: Payment for fuel

A receipt for this transaction will be emailed to: xyz@test.com

I AGREE   Click here only once to continue (DO NOT USE 'REFRESH' OR 'BACK')

I DO NOT AGREE   Click here to abort or start over

ANY FALSE INFORMATION ENTERED HERE CONSTITUTES AS FRAUD AND SUBJECT THE PARTY ENTERING SAME TO FELONY PROSECUTION UNDER LAW VIOLATERS WILL BE PROSECUTED TO THE FULL EXTENT OF THE LAW.

Your computer is identified as: **132.122.85.200**

After the information has been processed by payment gateway, customers are redirected to one of the following places: An Approved page specified by merchant, a Denied page specified by merchant, a referral page specified by merchant, or the main E-check processor website (if merchant did not specify anything).

An e-mail receipt for the transaction or denial message is immediately sent to the customer. The receipt will appear to come from merchant and contains merchant name, contact information, amount, date, check number, and a unique reference number.

Payment gateway also sends an immediate notification of the transaction to the merchant. This notice contains information such as customer's name, transaction amount, a unique reference number, and the transaction status (approved declined). This notice will appear to have been sent the customer so if merchant wishes he can set up an auto responder to send customer additional information

Step 4: Payment Gateway formats the transaction information and sends it as an ACH transaction to its bank with the rest of the electronic check transactions received that day.

Step 5: The bank of payment gateway passes the transaction information to the ACH Network for settlement. The ACH Network looks at the transaction information to find out the customer bank account information.

Step 6: The ACH Network instructs the customer's bank to charge or refund the customer's account.

Step 7: The customer's bank passes funds from the customer's account to the ACH Network. The customer's bank also notifies the ACH Network of any returns (in the event that funds for a transaction could not be collected from the customer's bank account) or chargebacks (in the event that a customer disputes a purchase). In the event of a returned transaction, payment gateway will post the return to the merchant.

Step 8: The ACH Network relays the funds for the transaction to the payment gateway's bank.

Step 9: The payment gateway's bank passes any returns to payment gateway.

Step 10: After the holding period, payment gateway initiates a separate ACH transaction to deposit e-Check funds to the merchant's bank account.

The figure 8-14 shows processing of electronic check transaction.

FIGURE 8-14: E-CHECK PROCESSING

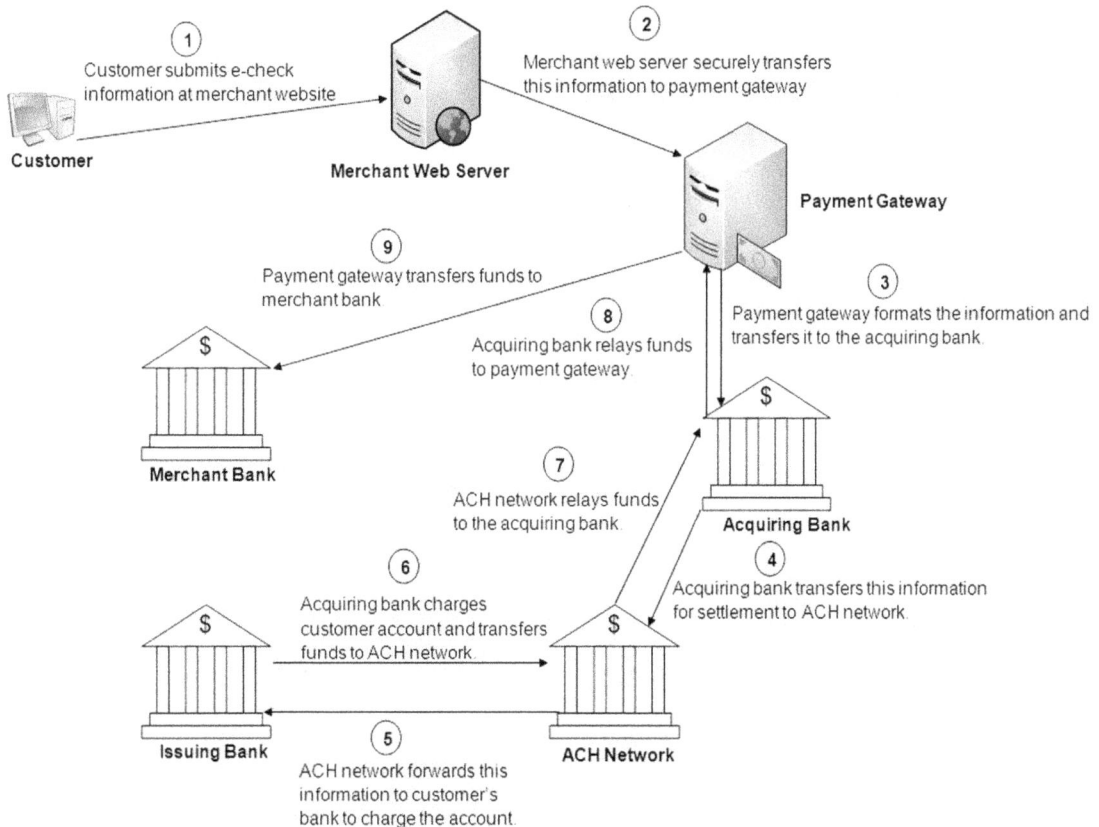

E-banking (or Internet Banking)

E-banking includes familiar and relatively mature electronically based products in developing markets, such as telephone banking, credit cards, ATMs, and direct deposit. It also includes electronic bill payments and products mostly in the developing stage, including stored-value cards (e.g., smart cards/smart money) and Internet-based stored value products. E-banking in developing countries is in the early stages of development. Most banking in developing countries is still done the conventional way. However, there is an increasing growth of online banking, indicating a promising future for online banking (Mohapatra, 2012).

In the years to come, electronic banking is expected to gain more popularity among the general users. E-banking is growing and banks are offering many services via e-banking e.g. personal e-banking (Transfer/ withdrawal of funds, Bill Payment, Account Enquiry, Time Deposit, Foreign Currency, Currency Switching , e-Statement/e-Advice), mobile trading, e-IPO services, and e-market news. With e-IPO services, you can easily subscribe new public offer shares, bonds, and certificates of deposits or view the required information. Mobile trading allows you to trade securities and foreign exchange using your mobile. E-market news provides current market news to your e-mail inbox.

The most common e-banking services include banking inquiry functions, bill payments, credit card payments, fund transfers, share investing, insurance, travel, electronic shopping, and other basic banking services. There exist many barriers in the development of e-banking. Security is one issue that scares many people away from taking their banking online.

GUIDELINES FOR SAFE INTERNET BANKING

As use of the Internet continues to expand, more banks and thrifts are using the Web to offer products and services or otherwise enhance communications with consumers. The Internet offers the potential for safe, convenient new ways to shop for financial services and conduct banking business, any day, any time. However, safe banking online involves making good choices - decisions that will help you avoid costly surprises or even frauds. Following are some tips help an individual maintain his/her security and privacy while conducting online banking (United States Federal Deposit Insurance Corporation, 2015).

CONFIRM THAT AN ONLINE BANK IS LEGITIMATE AND THAT YOUR DEPOSITS ARE INSURED

Whether you are selecting a traditional bank or an online bank that has no physical offices, it's wise to make sure that it is legitimate and that your deposits are federally insured. Here are tips specifically designed for consumers considering banking over the Internet.

READ KEY INFORMATION ABOUT THE BANK POSTED ON ITS WEB SITE

Most bank Web sites have an "About Us" section or something similar that describes the institution. You may find a brief history of the bank, the official name and address of the bank's headquarters, and information about its insurance coverage from the FDIC.

PROTECT YOURSELF FROM FRAUDULENT WEB SITES.

For example, watch out for copycat Web sites that deliberately use a name or Web address very similar to, but not the same as, that of a real financial institution. The intent is to lure you into clicking onto their Web site and giving your personal information, such as your account number and password. Always check to see that you have typed the correct Web site address for your bank before conducting a transaction. Users in USA should also verify the bank's Insurance status. To verify a bank's insurance status, look for the familiar FDIC logo or the words "Member FDIC" or "FDIC Insured" on the Web site.

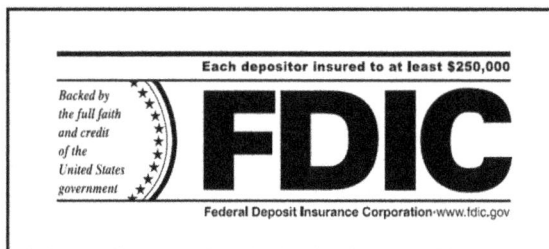

In addition, you should check the FDIC's online database of FDIC-insured institutions. You can search for an institution by going to "Bank Find" (http://research.fdic.gov/bankfind/). Search by name, city, state or zip code of the bank, and click the "Find" button. A positive match will display the official name of the bank, the date it became insured, its insurance certificate number, the main office location for the bank (and branches), its primary government regulator, and other links to detailed information about the bank. If your bank does not appear on this list, contact the FDIC. Some bank Web sites provide links directly to the FDIC's Web site to assist you in identifying or verifying the FDIC insurance protection of their deposits. Remember that not all US banks operating on the Internet are insured by the FDIC. Many banks that are not FDIC-insured are chartered overseas. If you choose to use a bank chartered overseas, it is important for you to know that the FDIC may not insure your deposits. Check with your bank or the FDIC if you are not certain. For insurance purposes, be aware that a bank may use different names for its online and traditional services; this does not mean you are dealing with separate banks.

Audit of E-commerce Transactions

Some believe that the audit trails in e-commerce are generated at the end of the total process, with customer and product information. This is incorrect. Audit trails are used in a variety of systems and equipment. For example, every customer or partner who enters the organization's network with a user ID and password is logged, and these type of transactions must be recorded for later control. Transactions that are sent/received using the external and/or internal networks have integrity and confidentiality categorized as "high." A log must be kept and special attention should be given to these kind of messages because they usually are the core of the business.

An organization should log the customer's transaction from its initiation through collection of the receipt and delivery of the product. Additionally, the organization should keep the security administrator's log because he has the option to assign processing functions, assessed as highly confidential, integrity or availability to the employees. Furthermore, these tasks should be logged beyond the compensating controls implemented (e.g., dual control).

Without a good audit trail, the organization may have difficulty dealing with customer inquiries, questions about the delivery of service, audit investigations, etc., particularly for older transactions. It would be disappointing to a customer if, following the delivery of the purchased goods or service, the organization could not answer a question or complaint because its system would not provide enough useful information (ISACA, 2002). Many logs are need to be reviewed in an IT audit of e-commerce. Therefore, the auditors should first interview appropriate management and staff to gain an understanding of business, organization, roles, policies, laws and management reporting, and to define audit scope. Auditors should then identify information requirements relevant for the business process. The use of computer assisted audit techniques (CAATs) to assess the safeguarding, integrity, effectiveness and efficiency objectives of audit trails also is recommended. The use of CAATs allows for the complete analysis of audit trails, focusing testing on subsets that appear with errors or irregularities and presenting them to managers and/or clients.

The following are some sound practices to help ensure that a clear audit trail exists for e-commerce transactions:

- Sufficient logs should be maintained for all e-commerce transactions to help establish a clear audit trail and assist in dispute resolution.
- E-commerce systems should be designed and installed to capture and maintain forensic evidence in a manner that maintains control over the evidence and prevents tampering with and the collection of false evidence.
- In instances where processing systems and related audit trails are the responsibility of a third-party service provider:
  - The E-commerce firm should ensure that it has access to relevant audit trails maintained by the service provider.
  - Audit trails maintained by the service provider must meet the bank's standards.

Review Questions

1. List some common payment methods used for traditional and B2C E-commerce.

2. What are some advantages of disadvantages of using cash as a method of payment?

3. List some basic problems with electronic payments.

4. List major stakeholders in the electronic payments.

5. List basic methods of electronic payments.

6. What is electronic cash?

7. What are advantages and disadvantages of electronic cash?

8. List some popular electronic cash systems.

9. What is blind signature?

10. What is secret splitting?

11. How SET works?

12. What is a phishing attack? What are various phishing attacks countermeasures?

13. List various B2B electronic payment solutions.

14. What is E-billing?

15. List EFT and its major systems.

16. List major payment cards and their advantages.

17. What is the difference between smart card and stored-value card?

18. What is an electronic wallet? What are its advantages and disadvantages?

19. What is payment gateway?

20. How a merchant can obtain a merchant account?

21. List some alternative payment systems.

22. What is electronic check? How electronic check is processed?

Bibliography

Adams, R. (2014, November 14). Electronic payment systems' biggest problem? Retrieved from http://www.allianttechnology.com/electronic-payment-systems-biggest-problem/

Adyen.com. (2015, April 30). Over 27% of global online transactions are now on mobile devices - Adyen. Retrieved July 6, 2015, from https://www.adyen.com/home/about-adyen/press-releases/mobile-payments-index-april-2015

americanexpress.com. (2010, November 5). American Express Launches SafeKey(SM) to Help Protect Merchants and Cardmembers from Online Fraud. Retrieved July 6, 2015, from http://about.americanexpress.com/news/pr/2010/safekey.aspx

Amtower, M. (2015, January 10). The Federal Purchase Card: Still a Viable Procurement Method for B2B. Retrieved from http://multichannelmerchant.com/marketing/b2b/federal-purchase-card-still-viable-procurement-method-b2b-10012015/

Annamalai, D. (2015, March 10). Battle For Mobile Payments: Guide to Digital Wallets. Retrieved July 6, 2015, from http://thefinancialbrand.com/50720/mobile-payments-digital-wallet-analysis/

bitcoinist.net. (2015, April 10). Micro-payments new avenue for Bitcoin. Retrieved July 6, 2015, from http://bitcoinist.net/bitcoin-future-micropayments/

BluePay.com. (2014a, October 1). Small Businesses at Risk of Missing EMV Credit Card Deadline | BluePay. Retrieved July 24, 2015, from http://www.bluepay.com/blog/small-businesses-risk-missing-emv-credit-card-deadline

BluePay.com. (2014b, November 19). 12 Questions to Ask When Choosing a Merchant Account Provider | BluePay. Retrieved July 24, 2015, from http://www.bluepay.com/blog/12-questions-ask-when-choosing-merchant-account-provider

Boldyreva, A. (2003). Efficient threshold signature, multisignature and blind signature schemes based on the gap-diffie-hellman-group signature scheme. *Public Key Cryptography – PKC. 2567*, pp. 31-46. Springer Berlin/Heidelberg.

Burger, A. K. (2008, January 5). Taking Internet Finance to the Next Level, Part 1 | E-Commerce | E-Commerce Times. Retrieved July 24, 2015, from http://www.ecommercetimes.com/story/60981.html

chips.org. (2009). *What is CHIPS?* Retrieved September 13, 2009, from What is CHIPS?: http://www.chips.org/about/000650f.php

Christine E. Drake, Jonathan J. Oliver, and Eugene J. Koontz, "Anatomy of a Phishing Email", 2004, Retrieved July 15, 2009, from http://www.ceas.cc/papers-2004/114.pdf)

Cipparone, M. (2009). *Digicash Convertibility - a look into the future*. Retrieved September 2009, from www.arraydev.com: http://www.arraydev.com/commerce/JIBC/9601-5.htm

CKGSB. (2014, May 28). The Rise of E-Finance in China – Implications of Money Market Funds, Mobile Payments and Other Trends on China's Transformation | CKGSB. Retrieved July 24, 2015, from http://english.ckgsb.edu.cn/news_content/rise-e-finance-china-%E2%80%93-implications-money-market-funds-mobile-payments-and-other-trends#.VbIlP7WN1sl

Clark, B. (2015, April 3). 8 Things You Probably Didn't Know About Digital Wallets. Retrieved July 6, 2015, from http://www.makeuseof.com/tag/8-things-you-probably-didnt-know-about-digital-wallets/

Crook, J. (2015, April 17). Coin, The One Credit Card To Rule Them All, Is Finally Shipping. Retrieved from http://social.techcrunch.com/2015/04/17/coin-the-one-credit-card-to-rule-them-all-is-finally-shipping/

Delegosoftware.com. (2015, March 3). From Paper to Plastic: The Road to B2B Electronic Payments. Retrieved from http://www.delegosoftware.com/blog/from-paper-to-plastic-the-road-to-b2b-electronic-payments/

emergingequity.org. (2014, December 27). Moscow's Response To Economic Warfare: Central Bank Of Russia Launches SWIFT Alternative For Domestic Payments. Retrieved from http://emergingequity.org/2014/12/27/moscows-response-to-economic-warfare-central-bank-of-russia-launches-swift-alternative-for-domestic-payments/

Fallon, N. (2014, January 27). Why Your Business Should Consider Online B2B Payments. Retrieved July 6, 2015, from http://www.businessnewsdaily.com/5826-online-b2b-payments.html

Fan, C.-I., & Liang, Y.-K. (2008). Anonymous Fair Transaction Protocols Based on Electronic Cash. *International Journal of Electronic Commerce , 13* (1), 131.

federalreserve.gov. (2015). FRB: Automated Clearinghouse Services. Retrieved July 6, 2015, from http://www.federalreserve.gov/paymentsystems/fedach_about.htm

Fraudpractice.com. (2009). *Payment Aggregator Services*. Retrieved July 15, 2009, from Fraudpractice.com: http://www.fraudpractice.com/alt-paymentaggre.html

fundscrip.com. (2015). Canada's Most Successful Gift Card Fundraiser - Home - FundScrip. Retrieved July 5, 2015, from http://www.fundscrip.com/

Gatch, L. (2012). Local money in the United States during the Great Depression. Essays in Economic & Business History, 26(1).

giftcardswapping.com. (2015). Gift Card Swapping | Sell Gift Cards, Gift Card Exchange, Buy Gift Cards | Buy, Sell, Swap. Retrieved July 5, 2015, from http://www.giftcardswapping.com/

Gutzman, A. (2001, June). *An Overview of B2B Payment Systems*. Retrieved September 14, 2009, from E-commerce-guide.com: http://www.ecommerce-guide.com/news/news/article.php/790991

Hamburger, E. (2014, October 7). Plastc wants to replace your entire wallet with a single card. Retrieved July 24, 2015, from http://www.theverge.com/2014/10/7/6926669/plastc-wants-to-replace-your-entire-wallet-with-a-single-card

IBM. (2015, June 4). IBM Using Secure Sockets Layer (SSL) Protocol - United States [CT742]. Retrieved July 6, 2015, from http://www-01.ibm.com/support/docview.wss?uid=swg21680710

ISACA. (2002). Audit Trails in an E-commerce Environment. Retrieved July 24, 2015, from http://www.isaca.org/JOURNAL/ARCHIVES/2002/VOLUME-5/Pages/Audit-Trails-in-an-E-commerce-Environment.aspx

ITU. (2006). *Cybersecurity Awareness Survey*. Retrieved September 15, 2009, from Internet Telecommunication Union: www.itu.int/newsroom/wtd/2006/survey/charts/index.asp

LBI.com. (2010). *LBI.com News*. Retrieved July 2010, from www.LBI.com: http://www.lbi.com/en/News/4856/

Lee, J. (2014, November 14). 3 Online Fraud Prevention Tips You Need To Know In 2014. Retrieved July 6, 2015, from http://www.makeuseof.com/tag/3-online-fraud-prevention-tips-need-know-2014/

Lee, T. B. (2013, November 9). The $11 million in bitcoins the Winklevoss brothers bought is now worth $32 million. The Washington Post. Retrieved from https://www.washingtonpost.com/news/the-switch/wp/2013/11/09/the-11-million-in-bitcoins-the-winklevoss-brothers-bought-is-now-worth-32-million/

Liu, W.-Y., Luo, Y.-A., & Si, Y.-L. (2007). A Security Multi-Bank E-cash Protocol based on Smart Card. *Proceedings of the Sixth International Conference on Machine Learning and Cybernetics*. Hong Kong.

Mohapatra, S. (2012). E-Commerce Strategy: Text and Cases. Springer Science & Business Media.

Murdoch, S. J., & Anderson, R. (2010). Verified by Visa and MasterCard SecureCode: Or, How Not to Design Authentication. In R. Sion (Ed.), Financial Cryptography and Data Security (pp. 336–342). Springer Berlin Heidelberg. Retrieved from http://link.springer.com/chapter/10.1007/978-3-642-14577-3_27

OECD. (2006). *Online Payment Systems for E-Commerce*. Retrieved June 12, 2010, from www.oecd.org: www.oecd.org/dataoecd/37/19/36736056.pdf

OFT. (2007). *Internet Shopping, an OFT market study*. Retrieved July 12, 2009, from oft.gov.uk: www.oft.gov.uk/shared_oft/reports/consumer_protection/oft921.pdf

Orf, D. (2015, March 3). Stratos Card Is Another Smart Payment Card That Wants To Rule Them All. Retrieved July 24, 2015, from http://gizmodo.com/stratos-is-another-credit-card-that-wants-rule-them-all-1689042166

Parker, P. (2014, July 15). How to Choose the Right Merchant Account Provider for Your Business • Expert & User Reviews. Retrieved July 24, 2015, from https://www.cardpaymentoptions.com/credit-card-processing/choose-right-merchant-account-provider/

Pymnts.com. (2015, February 27). ING Boosts B2B E-Payment Services with Basware | PYMNTS.com. Retrieved from http://www.pymnts.com/news/b2b-payments/2015/ing-boosts-b2b-e-payment-services-with-basware/#.VZquCFKN1sk

Rampton, J. (2015, April 14). How Online Fraud is a Growing Trend. Retrieved July 6, 2015, from http://www.forbes.com/sites/johnrampton/2015/04/14/how-online-fraud-is-a-growing-trend/

Rocheleau, J. (2015). 6 Safety Steps To Making Secure Mobile Transactions. Retrieved July 6, 2015, from http://www.hongkiat.com/blog/secure-mobile-payment-tips/

sagepay.co.uk. (2015). 3D Secure explained - Sage Pay. Retrieved July 6, 2015, from http://www.sagepay.co.uk/support/12/36/3d-secure-explained

Sandoval, G. (1999). *E-tailers scramble to win teen buyers*. Retrieved July 12, 2009, from news.cnet.com: http://news.cnet.com/E-tailers-scramble-to-win-teen-buyers/2100-1017_3-233731.html

Schneider, G. (2010). *Electronic Commerce, 9th Ed.* Course Technology.

shopwithscrip.com. (2015). Support. Retrieved July 5, 2015, from https://www.shopwithscrip.com/Support/about-us/our-company

Smart Card Alliance. (2015). About the Alliance : Overview » Smart Card Alliance. Retrieved July 24, 2015, from http://www.smartcardalliance.org/alliance/

Stanford, U. (2009). *Digital Wallets Project*. Retrieved July 14, 2010, from Stanford University: http://infolab.stanford.edu/~daswani/wallets/

tass.ru. (2014, September 10). Russia, China in talks to make alternative to SWIFT — deputy PM. Retrieved July 6, 2015, from http://tass.ru/en/economy/748916

Tewari, H., O'Mahony, D., & Peirce, M. (1998). *Reusable Off-Line Electronic Cash Using Secret Splitting*. Retrieved July 13, 2009, from Technical Report TCD-CS-1998-27, Trinity College Dublin Computer Science Department, Dublin: www.scss.tcd.ie/publications/tech-reports/reports.98/TCD-CS-1998-27.pdf

Turban, E., Lee, J., & Chung, M. (2006). *Electronic Commerce: A Managerial Perspective*. Prentice Hall.

United States Federal Deposit Insurance Corporation. (2015). FDIC: Safe Internet Banking. Retrieved July 24, 2015, from https://www.fdic.gov/bank/individual/online/safe.html

UNITED STATES SECURITIES AND EXCHANGE COMMISSION. (2015). Form 10-K. Retrieved July 24, 2015, from http://www.sec.gov/Archives/edgar/data/1036044/000119312512139749/d269588d10k.htm

VISA.com. (2015). Preventing Fraud - Visa Security Sense. Retrieved July 6, 2015, from http://www.visasecuritysense.com/en_US/preventing-fraud.jsp

Vohra, A. (2015, January 11). E-wallets: Money on the move. Retrieved from http://www.financialexpress.com/article/personal-finance/e-wallets-money-on-the-move/28571/

Wang, Q., & Li, C.-y. (2007). The Model of Anonymous Fair e-Cash Transactions Protocol with Off-line TTP. Proceedings of the Second International Conference on Innovative Computing, Information and Control, (p. 416).

Wisegeek.com. (2015). What are the Different Types of Smart Card Technology? Retrieved July 24, 2015, from http://www.wisegeek.com/what-are-the-different-types-of-smart-card-technology.htm

Wong, I. (2014, February 12). PBOC set to watch over e-finance platforms - The Standard. Retrieved July 24, 2015, from http://www.thestandard.com.hk/news_detail.asp?we_cat=2&art_id=142434&sid=41550824&con_type=1&d_str=20140212&fc=2

9

E-CRM, E-SRM AND E-SCM

Learning Objectives

After reading this chapter, reader should be able to:

- Understand the need of a CRM.

- Define e-CRM and its components.

- Describe the benefits and limitations of e-CRM.

- Discuss social and cloud CRM

- Explain the working of e-CRM.

- Understand implementation of e-CRM.

- Identify various e-CRM solutions.

- Discuss how to choose a particular e-CRM solution.

- Define e-SCM.

- Describe key processes and infrastructure components of e-SCM.

- Discuss key success factors for e-SCM.

- Describe functions and benefits of e-SCM.

- Identify categories of available e-SCM solutions.

- Discuss how to choose a particular e-SCM solution.

Customer Relationship Management (CRM)

In recent years, with increased global competition, many organizations have realized the need to become more customer-facing and therefore customer relationship management has become an essential for many organizational strategies (Bull, 2003). Managers realize that enhanced customer relations brings the benefit of profitable and sustainable revenue growth because of shifting focus from product to customer (Ling & Yen, 2001) and radical changes in industrial marketing environment (Deeter-Schmelz & Kennedy, 2002).

A customer relationship management (CRM) system is a combination of people, processes, and technology that seeks to provide understanding of a company's customers and to support a business strategy to build long-term, profitable relationships with customers [(Chen & Popovich, 2003);(Ling & Yen, 2001)]. CRM is grounded on high quality customer-related data and enabled by information technology (Buttle, 2003). Customer data can be captured from different parts of the company and stored in a centralized database. Customer data can be captured from a variety of sources e.g. from company's business operations software (e.g. sales automation, customer service center operations, and marketing campaigns), from company's Web site and any other touch points which the company has with its existing and potential customers. This data is analyzed and distributed to various touch points in the organizations. Touch points can include company's sales force, call centers, web sites, point-of-sale, direct marketing channels, and any other part of company that interact with the customer.

CRM software uses customer data to conduct analytical activities (e.g. gathering business intelligence, planning marketing strategies, customer behavior modeling, and customizing the products and services) to meet the needs of specific customers or categories of customers. Using CRM, companies can retain existing customers, sell more to existing customers, and find and win new customers. The global revenue of the CRM industry is predicted to grow strongly from 2005 [(He, 2004); (Blery, 2006), with a focus on both technology and organizational efforts to improve customer experiences. In recent years, CRM has become widely accepted as an important management discipline that enables a business to understand better the stated, and especially the implied, requirements of its customers. CRM was ranked by Bain and Company's global survey as one of the top ten tools used by managers (Rigby & Bilodeau, The Bain 2005 management tool survey, 2005).

Electronic Customer Relationship Management (e-CRM)

Repeat customers are many times more profitable than new customers are. To reap the benefits of repeat sales, online businesses need to retain their customers but more than 50 percent of repeat customers seldom complete a third purchase. One reason for this is the inability of on-line vendors to manage customers' changing expectations (Gupta & Kim, 2007). Delivering on-line service is therefore a key component of customer satisfaction and customer retention. The nature of service delivery is changing significantly. Face-to-face communication of organizations with customers is increasingly being replaced by information and communication technology (ICT) systems and applications and ICT-mediated channels. This trend has created both new opportunities and challenges for businesses. Many firms are now implementing a range of customer-centric e-CRM systems (Fjermestad & Nicholas C. Romano, 2008).

(Lee-Kelley, Gilbert, & Mannicom, 2003) defines e-CRM as "the marketing activities, tools and techniques delivered over the Internet (using technologies such as web sites, email, data capture, warehousing, and mining) with a specific aim to locate, build and improve long-term customer relationships to enhance their individual potential". The core aim of e-CRM is that companies do not necessarily sell more, but sell smarter (Tan, Yen, & Fang, 2002). In this regard, e-CRM can be of great assistance to organizations in delivering value by sorting their most profitable customers from the less profitable customers (Ross, 2005). By integrating distributed customer data from different data sources, e-CRM can provide a single, cohesive, and sound picture of the customers and their relationships with organization [for details see (Plessis & Boon, 2004); (Bradshaw & Brash, 2001); (Padmanabhan & Tuzhilin, On the Use of Optimization for Data Mining: Theoretical Interactions and e-CRM Opportunities, 2003)]. E-CRM tools can also be used to collect customer details and background information and later share this information within the organization [(Chen & Ching, 2007); (Jayachandran, Sharma, Kaufman, & Raman, 2005)]. With information and communication technology, the integration of this information is seamless and much easier (Padmanabhan, Zheng, & Kimbrough, An empirical analysis of the value of complete information for e-CRM models, 2006). Using advanced functions of e-CRM, analytical analysis of customers can also be performed. This analysis can be used to identify customers' behavioral models.

Components of e-CRM

According to Gartner, the three main components of e-CRM are an E-commerce sell-side platform, communication infrastructure, and applications.

E-commerce sell-side platform is used to interact, via web, with customers. This platform should include functions like order management, catalog and content management, and secure transfer of data on the web between business and customers.

The communication infrastructure for e-CRM is used to facilitate and analyze interactions with the customers. The communication infrastructure generally contains Multilanguage support, e-mail, messaging and workflow applications, and web measurement tools such as those used to analyze web log files.

E-CRM applications include content management applications, customer support applications (e.g. automated response applications), and sales/marketing automation tools, and marketing campaign applications.

Benefits of e-CRM

Benefits of an e-CRM system include:

Service Level Improvements: e-CRM uses an integrated database that provides consistent and improved customer responses.

Revenue Growth: Revenue increases because e-CRM decreases costs by focusing on retaining customers and selling additional products.

Increased Productivity: With e-CRM, sales and service procedures becomes consistent and create efficient work processes. With e-CRM, sales cycle is shorten and key sales-performance metrics increases (such as revenue per sales representative, average order size and revenue per customer).E-CRM also increases service agent productivity and customer retention

Increased Customer Satisfaction: Automatic customer tracking and detection, provided by e-CRM, ensures inquiries are met and issues are managed. This improves the customer's overall experience in dealing with the organization.

One-Time Data Entry: The core data need only be entered once. All departments share the same data although each can have additional information specific to their processes.

Automation: E-CRM software helps automate campaigns including telemarketing, telesales, direct mail, lead tracking and response, opportunity management, quotes and order configuration.

Limitations of e-CRM

The major limitation of CRM is that it requires integration with a company's other information systems e.g. accounting, inventory, etc. This integration may not be an easy task especially for large enterprises. Providing justification of expense of a CRM project is not an easy task. It is difficult to support certain mobile employees and there is no solid evidence proving that CRM works. Although it is true that CRM may not attract customers that are less than desirable for a company, investments in CRM do not guarantee the attraction of desirable customers, either.

Working of e-CRM

In today's world, customers interact with an organization via multiple communication channels and many organizations have multiple lines of business that interact with the same customers. Using e-CRM systems, customers can do business with the organization the way they want. For example, customers may use telephone for responding to direct marketing. For purchase decision, their contact can be field sales or a telephone-ordering center. Post-sales contact may be via a support help-desk. E-CRM make customers feel that they are dealing with a single, unified organization that recognizes them at every stage of interaction. The e-CRM system does this by creating a central e-CRM database of customer records. All customer touch points across the company and every member of the organization share this central e-CRM database. Through this system, organizations become more knowledgeable about their customers, products and performance results using real time information across their business. Integration between the e-CRM database and the back-office systems e.g. accounts, manufacturing, distribution, ERP etc. may be real time or batch interface depending upon need. Figure 9-1 illustrates an e-CRM system.

FIGURE 9-1: AN E-CRM SYSTEM

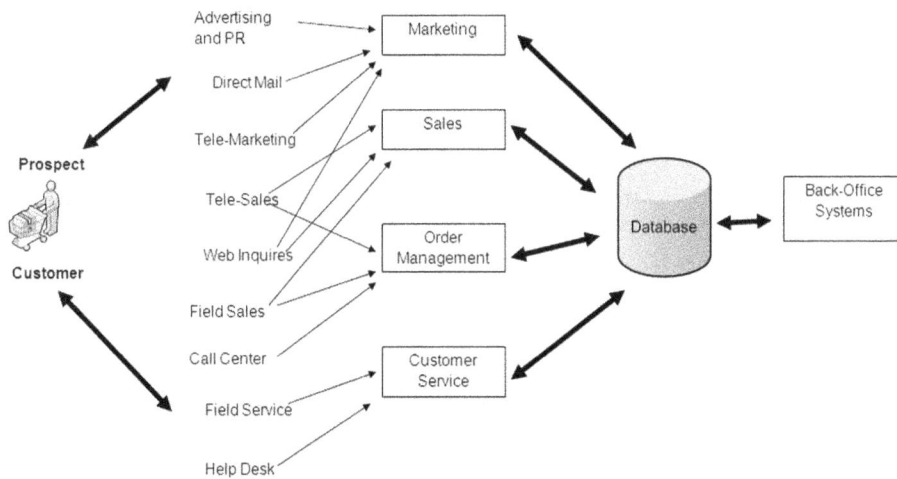

The value and role of E-CRM is that the marketers can deliver cheaper, more flexible and faster CRM to customers through online tools. For example, companies can handle CRM through the social media, blog, text messages (SMS) or emails. E-CRM is that helps marketer to build longer lasting customer relationships. The Figure 9-2 shows E-CRM cycle.

Figure 9-2: e-CRM Cycle

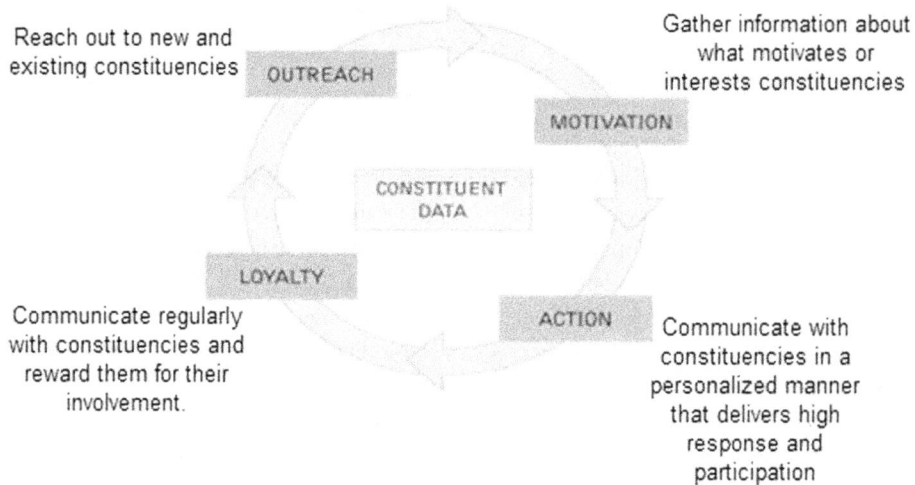

Reach out to new and existing constituencies — OUTREACH

Gather information about what motivates or interests constituencies — MOTIVATION

CONSTITUENT DATA

LOYALTY — Communicate regularly with constituencies and reward them for their involvement.

ACTION — Communicate with constituencies in a personalized manner that delivers high response and participation

Technology (i.e. software, hardware and services) needs to be deployed by an organization before e-CRM can be practiced. These deployments are called CRM system implementations. Effective implementation of CRM is neither easy nor cheap. The historical failure rates for these e-CRM system implementations were high. A global survey of senior executives worldwide in April 2003, conducted by the Economist Intelligence Unit for AT&T, shows only 29% of respondents are satisfied with the performance of their CRM implementations (AT&T, 2003). Moreover, a global survey released in April 2004 by IBM Business Consulting Services revealed that only 15 percent of global companies believed they are fully succeeding with their CRM initiatives, and another 20 to 30 percent are having only partial success (IBM, 2004). Over the years, the situation is changing. The success rate of CRM implementations reached to 53 % in 2009 (Krigsman, 2009). (Ang & Buttle, 2006) reported that companies that did employ CRM software were generally satisfied with their return on investment from the software and the level of satisfaction with software performance varied directly with its reported impact on business profitability.

(Bull, 2003) reported that companies fail in CRM implementation because companies generally underestimate the complexities of CRM, lack clear business objectives and tend to invest inadequately in CRM software. According to (Zablah, Bellenger, & Johnston, 2004), the first step towards successful CRM is to specify a relationship management strategy. This is turn requires firms to prioritize relationships with customers and allocate resources to relationship building based on customers value to the firm. People resistance to using the CRM software may lead to implementation failures [(Crosby, 2002); (Kavanagh, 2003)]. A culture that is conducive for the successful implementation of a CRM system is necessary [(Mack, Mayo, & Khare, 2005); (O'Malley & Mitussis, 2002); (Rigby, Reichheld, & Schefter, Avoid the Four Perils of CRM, 2002); (Wilson, Daniel, & Malcolm, 2002)]. A suitable corporate infrastructure (which includes networks, storage, and data backup, computing platforms etc.) and effective integration of this infrastructure with e-CRM is the key for successful e-CRM implementation. E-CRM package selection should consider not only the specific corporate CRM needs but also the ease of integration with legacy enterprise applications such as the ERP system.

E-CRM Solutions

Once a CRM plan has been developed e-CRM deployment can begin. There are many forms of CRM solution, ranging from hosted CRM and customized CRM solutions to off-the-shelf CRM products. For small business, e-CRM packages can range from $200 for a single user. Cost of customized e-CRM solution depends upon the scope of the e-CRM offering. For a medium-sized business, e-CRM packages can range from approximately $3,000 for multi-user license.

Hosted CRM solutions: This type of CRM solution is managed and housed on a third party's server. This eliminates the need for purchasing and maintaining additional hardware or servers, allowing your company to focus on its core business practices. All upgrades, maintenance and other aspects of managing the software component are handled by the ASP. Many hosted solutions are somewhat flexible, but cannot be customized beyond basic parameters, such as choosing from specific reports and, sometimes, choosing various templates for the solution. Sherweb.com and esalestrack.com are examples of hosted CRM solution providers.

Customized e-CRM Solutions: Customized CRM solutions are built specifically for your business. Because they are completely customized, you can design the program to accommodate any information, reports and other data specifications you wish. These types of programs are usually more expensive than hosted solutions. Customized CRM solutions often require enterprise software, service and support from external providers. A wide range of vendors and consultants provide CRM solutions. These vary from simple applications to the implementation of comprehensive software, hardware and customer relationship ideologies. Some key providers are SalesLogix (www.saleslogix.com), Siebel (www.siebel.com), and SalesForce (www.salesforce.com).

Off-the-shelf e-CRM Solutions: Relatively simple CRM software applications can also be purchased. These are limited in how much they can be customized and require some in-house IT support, but they can be an affordable option. Sugar CRM (www.sugarcrm.com) is an off-the shelf CRM software.

Essential Features of E-CRM Software

Following are some of the key features of E-CRM software your company should look for when purchasing a solution (Business-Software.com, 2012).

Sales Force Automation: CRM systems make it easier for sales professionals to identify potential opportunities and win new deals; effectively manage their contacts, accounts, pipelines, and related activities; generate accurate and timely forecasts; and manage quota performance across teams and territories. Most CRM systems also offer mobile capabilities to provide on-the-go sales staff with anytime, anywhere access to up-to-the-minute customer data via laptops, or Palm, Blackberry, iPhone, iPad, Windows Pocket PC, and other handheld devices.

Marketing Automation: CRM systems provide marketing professional insight to implement more successful lead generation strategies and effectively track and manage prospect databases. Marketing professional can organize and launch initiatives across multiple communication channels and conduct in-depth analysis of campaign results to increase response rates and boost conversions while reducing cost-per-lead and accelerating return on marketing investment.

Customer Service/Help Desk Automation: CRM systems can empower customer service, help desk, and call center teams rapidly respond to customer issues and inquiries and to accurately track and manage problems from the time they are reported until they are resolved. Customer services department become profit centers because now they get access to tools required for active participation in up-sell and cross-sell programs. CRM systems can also enhance field support operations such as scheduling, dispatching, and other related processes.

Reporting and Analysis: Business intelligence (BI) is a key component of most CRM systems today. Using BI, companies can maximize the value of their customer data by using it to make better, more informed decisions. Reports, performance management dashboards, and analytics provide users at all levels with complete visibility into customer-related activities within and across multiple departments. This insight into customer needs, preferences, and behaviors can help business to identify emerging business trends and opportunities and uncover problems or areas in need of improvement.

Collaboration: Effective customer acquisition and retention requires synchronized, well-executed efforts across multiple customer-facing departments. CRM systems provide features that enable sales, marketing, service, and support staff to share information, coordinate activities, and eliminate overlapping or redundant functions. As a result, they can work together to deliver a superior experience and build strong, long-lasting customer relationships.

Feature Comparison of Popular E-CRM Software

	Saas or On-premise solution	Mobile accessible	Analytics	Data import	Email marketing	Email integration	Data cap per user or unlimited	Free trial version
Salesforce Sales Cloud (www.salesforce.com)	√	√	√	√	√	√	√	√
NetSuite CRM+ (www.netsuite.com)	√	√	√	√	√	√	√	
C2CRM (www.c2crm.com)	√	√	√	√	√	√	√	√
Infor CRM (crm.infor.com)	√	√	√	√	√	√	√	√
Sage CRM (na.sage.com)	√	√		√	√	√	√	√

Cloud CRM

Cloud CRM stands for "cloud customer relationship management". It essentially refers to any cloud-based technology that streamlines and harnesses a company's customer data for improved customer service and overall revenue. Cloud CRM provides many benefits to the organizations. Because of the nature of cloud-based systems, a sales professional can login from just about anywhere with Internet and be able to access their account and client/customer information. This can be helpful when making last minute sales calls, closing a sale from a remote location, and finding contact information on the go. Most Cloud CRM providers have a robust enough mobile application that can be downloaded onto an Adroid or iOS device, and the most important information is readily available once the app is activated. Because cloud-based systems are flexible with providing capacity, cloud-based CRMs can allow a company to scale up or down depending on their needs. The way many companies use CRM systems is by keeping it updated and serving as a database of all client/customer information. Other applications or software that provide added services (email clients, etc.) are linked directly to the CRM to allow seamless information exchange between the technologies. With a cloud-based CRM system, these integrations are usually much easier to initiate and maintain. Cloud-based CRM systems allow users to have their own logins and therefore, multiple accounts can be logged in at the same time and working on edits to contacts or other information. This also prevents information from being stored and isolated on individual devices, as it provides an easy-to-understand option for sharing. Cloud-based CRMs pay close attention to their backup policies and usually have a recovery process if data was disturbed or breached.

Social CRM

Social CRM is a strategy that is often supported by various tools and technologies. The strategy is based around customer engagement and interactions, with transactions being a byproduct. Social CRM is a back-end process and system for managing customer relationships and data in an efficient and process-centric way. Social CRM is one component of developing a social or collaborative business, both internally and externally.

Traditional CRM was very much based around data and information that brands could collect on their customers, all of which would go into a CRM system that then allowed the company to better target various customers. In social CRM, that has completely changed. The customer is actually the focal point of how an organization operates. Instead of marketing or pushing messages to customers, brands now talk to and collaborate with customers to solve business problems, empower customers to shape their own experiences and build customer relationships, which will hopefully turn into customer advocates. Social CRM is an evolution of traditional CRM that is significantly different from traditional CRM.

TABLE 9-1: EVOLUTION OF CRM TO SCRM

CRM	Social CRM
Assigned departments	Everyone
Company defined processes	Customer defined processes
Business sets working hours	Customer sets working hours
Defined channels	Customer-driven dynamic channels
Transaction	Interaction
Inside-out	Outside-in

According to a report, 93% of Americans want brands to have a presence on social media sites and 60% of Americans regularly interact with companies on a social media site. Three most influential factors for consumers when deciding which company to do business with are: personal experience (98%), company's reputation or brand (92%), and recommendations from friends and family (88%). 41% of customers believe that companies should use social media tools to solicit feedback on products and services. 43% of consumers say that companies should use social networks to solve the customers' problems (Lardinois, 2008). Only 7% of organizations understand the CRM value of social media (Cespedes, 2014). According to a research by Cone Communication (Cone Communications, 2010), before deciding whether to purchase recommended products or services, more than four out of five consumers (81%) will go online to verify those recommendations, specifically through researching product/service information (61%), reading user reviews (55%) or searching ratings websites (43%). According to American Express (americanexpress.com, 2010), Americans will spend 9% more with companies that provide excellent service. The statistics clearly show consumers are changing and evolving with the growth of social media. The challenge for organizations now is adapting and evolving to meet the needs and demands of these new social customers. However, many organizations still do not understand the CRM value of social media. According to Gartner (Gartner, 2010), social applications offer a great opportunity for CRM practitioners to improve customer experience and influence the customer, particularly in an economic downturn when companies are trying to keep customers and increase wallet share. However, by 2010, more than half of companies that have established an online community will fail to manage it as an agent of change, ultimately eroding customer value.

Mobile CRM

Mobile CRM, or Mobile Customer Relationship Management, is a CRM tool designed for mobile devices including smartphones and tablets. By connecting through mobile CRM, you allow your sales team's access to customer data through a CRM mobile app or through a web-based browser with cloud CRM. By connecting through mobile CRM, you allow your sales team's access to customer data through a CRM mobile app or through a web-based browser with cloud CRM. A key benefit of using mobile CRM is to allow your sales force to access real time data while out in the fields meeting prospects and customers. Research by Forrester Research in 2008 (Forrester Research, 2008) found that 50% of Enterprises and more than 40% of SMBs were piloting, rolling out or currently using smart phone applications for sales force automation. Forbes in 20102 reported that there were 110 CRM applications in the Apple App Store and 47 in the Android App Store (Columbus, 2012). The Gartner Group predicts an exceptional growth rate of 500% by 2014 for mobile CRM. For comparison purposes, mobile usage increase between December 2010 and November 2012 was 225% (Columbus, 2012). Before you begin to implement your mobile CRM, it is important that you define goals on what you want to achieve by having a mobile CRM and what your work force needs in order to take advantage of real time data access. It is important that you get internal buy-in from the sales force. Start out by providing only basic access that the teams use on a day to day basis.

Integration of CRM, ERP, SCM, and other application systems

Integration among CRM, Enterprise Resource Planning (ERP), and Supply Chain Management (SCM) systems is crucial to build a complete customer support system [(Chen & Shang, 2005); (Huang, Yen, Chou, & Xu, 2003); (Tan, Yen, & Fang, 2002); (Kalakota & Robinson, 2000)]. Integration of CRM and ERP systems allows the sales department to be more aware of customer, and the production department is more aware of customers reactions. The information provided by CRM can also be used by ERP system to prioritize its processes and optimize services to preferred customers (Huang, Yen, Chou, & Xu, 2003) Effective SCM can facilitate CRM by providing customers good quality, low-price products through speedy distribution channels and successful CRM can provide information for SCM system demand forecasting and delivery designing (Tan, Yen, & Fang, 2002).

Suppliers may also take advantage of the integration to better schedule the delivery of raw materials and to prioritize material flow to enhance service to profitable customers (Huang, Yen, Chou, & Xu, 2003).

Choosing the Best CRM Application for Your Business

Choosing the best solution for the business depends on budget, the people who will be using the solution, business goals, and IT resources.

Depending on the solution you choose, implementation will include either the cost of software or hosted application-licensing fees. If you choose customized or off-the-shelf applications, you will also need to consider the cost of the hardware needed to run the system. There may also be fees for upgrading software or for customizing interfaces or other aspects of the program, as well as the cost of in-house or contract IT personnel to manage the system. Deployment of the system can take anywhere from two to nine months, depending on the size of the company and the complexity of the system.

Business goals are important. For example, if business goal is to increase customer retention or create more customized promotions based on customer buying habits, the software should have functions that help identify opportunities, execute tactics and examine results in those areas.

It is essential that a business determine the skill level of the employees that will be using the e-CRM system. Knowing this will help you forecast how intensive the implementation will be. Adequate training is essential for efficient adoption and high-quality data. Most e-CRM packages come with vendor-provided training, but additional training may be necessary to ensure that employees fully understand and are comfortable using the system.

The more complex your application, the more you will need to rely on in-house or contract IT services. Hosted CRM solutions largely eliminate this need, while customized software may require intensive IT support.

All these factors can significantly impact the company's bottom line, make CRM system a profit centre and a key driver in running the business better.

Metrics in Customer Service and CRM

Metrics are a set of standards that can be used to determine how much service to provide. Metrics can be either quantitative or qualitative standards. Here are some Web-related metrics a company can use to determine the appropriate level of customer support:

Response Time: This is the time taken by the systems to respond to a user's query.

Site Availability: This is the time when your website is not available for the visitors. This should be as close to zero as possible.

Download Time: This time should not be more than 10 to 21 seconds.

Timeliness: Company websites must contain up-to-date information. Companies should also set a time interval (say, every month) at which information must be revised. Not following either can result in loss of potential sales.

Security and Privacy: Web sites must provide a privacy statement and an explanation of security measures.

On-time Order Fulfillment: Order fulfillment must be fast and on time.

Return Policy: Companies should have a return policy. Having a return policy increases customer trust and loyalty. The ease by which customers can make returns is important to customer satisfaction.

Navigability: A Web site must be easy to navigate. To gauge navigability, companies might measure the number of customers who initiate the order but never completed it.

E-SCM

According to the Global Supply Chain Forum, SCM is defined as "The integration of key business processes from end user through original suppliers that provides products, services, and information that add value for customers and other stakeholders (Chen & Kong, 2007).

E-supply chain refers to the supply chain that is managed electronically, usually with Web technologies. A more precise definition by (Norris, Hurley, Hartley, Dunleavy, & Balls, 2000) is that electronic supply-chain management (e-SCM) is the collaborative use of technology to enhance business-to-business processes and improve speed, agility, real-time control, and customer satisfaction. The E-Supply Chain links trading partners through various information technologies (including the internet and/or electronic data interchange) to allow them to buy, sell and move products, services and cash. E-SCM is about technology change as well changes in organizational culture, management policies, performance metrics, business processes, and organizational structures across the supply chain.

There are many examples of innovative e-supply chains. Dell Computer and Cisco has streamlined their supply chain and business processes while providing their customers with customized product options.

E-SCM: Key Processes and Infrastructure Components

According to (Norris, Hurley, Hartley, Dunleavy, & Balls, 2000), the e-supply chain consists of the following processes:

Supply Chain Replenishment: Real-time replenishment information is shared electronically. Replenishment information can be used by companies for a variety of purposes e.g. to reduce inventories, eliminate stocking points, and facilitate make-to-order and assemble-to-order manufacturing strategies across the enterprise.

E-procurement: E-procurement refers to using Web-based functions (online catalogs, contracts, purchase orders, and shipping notices) to support the key procurement processes (e.g. requisition of good or service, sourcing, contracting, ordering of goods or service, and payment for goods and service purchased). E-procurement can improve supply chain in many ways e.g. by eliminating redesign of components in product development (by using online catalogs), providing quick decision making (through visibility of available parts and their attributes), expediting order processing (through online purchase orders), and streamlining deliveries (through advanced- shipping notifications and acknowledgments).

Collaborative Planning: Collaborative planning refers to activities by buyers and sellers to develop a single shared forecast of demand and a plan of supply to support this demand. This plan is updated regularly based on information shared over the Internet.

Collaborative Product Development: Collaborative product development involves inter-organizational use of product design and development techniques and sharing of various design and development resources. Collaborative product development can improve product launch success and reduce time to market.

E-logistics: E-logistics is the use of Web-based technologies to support the warehouse and transportation processes. E-logistics can optimize routing in distribution and provide inventory-tracking information.

The key processes of E-Supply Chain use a variety of infrastructure and enabling tools. The following are the major infrastructure elements and tools of e-supply chains.

Extranets: Extranets provide support for inter-organizational communication and collaboration.

Intranets: Intranets provide support for internal communication and collaboration in an organization.

Corporate Portals: These provide a single point access for external and internal collaboration, communication, and information search. Companies may build separate portals for outsiders and insiders.

Workflow Systems and Tools: These systems manage the flow of information in organizations by using a set of software programs that provides automation of almost any information-processing task. Some leading vendors of workflow applications are IBM FileNet, Tibco Staffware, and IBM Lotus.

Groupware and Collaboration Tools: Collaboration tools facilitate collaboration within an organization and among organizations. Collectively these tools are called groupware e.g. electronic teleconferencing, group decision support systems etc.

Critical Success Factors for E-supply Chains

The success of an e-supply chain depends on two major factors.

Critical Mass and Support of Participants: Partner collaboration is necessary and it should be a top priority of the company. To do so all companies involved in the supply chain must clearly define their value in the supply chain and their attractiveness as business partners. This way each participant will have sufficient incentives to participate and make the investments needed to help the effort succeed.

Supporting Technology and Processes: Supporting technology (hardware and software) and processes must be in place to implement e-SCM. Information must be managed properly with strict policies, discipline, and daily monitoring. Information about inventories at various segments of the chain, demand for products, delivery times, and any other relevant information must be visible to all members of the supply chain at any given time.

Benefits of E-supply Chains

E-supply chain provides many benefits. Following are some of the benefits.

Price Transparency and Competition: Buyers can take advantage of competition and dynamic pricing to get goods and services at lowest possible price.

Lower Costs of Conducting Business: With E-supply chains, procurement function is automated and burden of purchase is transferred to the users. Integration of automated procurement function (e-procurement) with legacy systems of the company lowers transaction costs and reduces overheads.

Enhanced Buyer-Supplier Relationships: E-SCM provides buyers opportunities to forge strong relationships with suppliers through collaboration in product design and development process.

Enhanced Supplier Capabilities: Suppliers can enhance their capabilities through collaboration with their partners and opportunities to take part in active and large online marketplaces. They can strengthen their forecasting capabilities, improve inventory management, extend market reach, gauge the demand for their products and services, and determine the price market is willing to pay for their products and services.

Functions of E-SCM Software

E-SCM software can perform many functions e.g. capturing data about various supply chain processes and produce information in the form of reports. E-SCM software can also be used for inventory tracking, demand forecasting, and processing order and returns.

E-SCM software may also include radio frequency identification (RFID) technology. RFID tags can be attached to large products or expensive assets or in cases where some other type of inventory identification is impractical. RFID devices can dynamically store and transfer data without necessarily requiring proximity to tagged supply.

RFID devices are being used in a variety of areas including environmental monitoring and management, insurance, and the pharmaceutical industry. Unlike barcodes, RFID tags can be read at any angle and from a distance, providing more flexibility to manage physical space at facilities.

SCM software price can range from thousands to millions of dollars.

Categories of e-SCM Solutions

There are a number of specific e-SCM solutions. Following are the most common classifications.

BASIC SUPPLY CHAIN SYSTEMS

These systems are internal back-office systems that are either custom-developed or packaged software systems. These systems help companies perform various supply chain-related functions (e.g. forecasting, planning and scheduling, order processing, inventory management, logistics, transportation, warehouse management, and fulfillment). Many Software vendors such as i2 Technologies (now part of JD software), SynQuest, McHugh, Logility and Optum provide these systems.

ERP/EDI SYSTEMS

ERP systems are integrated back-end systems that provide access to data generated across the enterprise. ERP systems can be used by the companies to share data with suppliers and partners using similar systems. Many ERP systems have specific supply chain functionality to provide broader exchange of information and collaboration. EDI systems are generally custom interfaces used to exchange data between two parties' dissimilar systems. The type of data exchanged is generally product and inventory information. Software vendors offering ERP/e-SCM solutions include SAP, Baan, Oracle and Microsoft. Many consulting firms e.g. Accenture, PriceWaterhouseCoopers and KPMG provide services to implement ERP packaged solutions and to design custom solutions.

CUSTOM INTERFACE SYSTEMS

These systems are used by companies to share information and strengthen relationships with their trading partners, vendors and customers. These interfaces may rely on extranets or virtual private networks (VPNs). These systems may allow one-to-many interactions with third parties and data exchange. Generally, these solutions are implemented with outside consulting assistance.

E-PROCUREMENT SYSTEMS

These are standalone corporate procurement systems. Typically implemented over an intranet, these systems are used to distribute the purchasing function to end users. These systems contain information about supplies and electronic catalogs from pre-approved vendors. These systems also provide automatic authorizations and approvals to purchasing activities performed by end users. Purchase data can be fed into back-end systems (e.g. accounting systems) which help to reduce overhead and potential data entry errors. These systems do not support dynamic selling or bidding and therefore dynamic pricing is not possible. These systems are costly to implement and generally used by large companies. Many software vendors offer e-procurement systems e.g. Commerce One, i2 Technologies, Trilogy, and Ariba. ASPs offering e-procurement solutions include PurchasePro, Clarus and USInternetworking.

E-MARKETPLACE INTEGRATED SYSTEMS

E-marketplaces support many-to-many relationships between buyers and sellers. Most marketplaces provide dynamic pricing that allow suppliers and buyers to conduct auctions, offers and counteroffers, and other types of price negotiations online. ASPs offering e-marketplace integrated systems include PurchasePro and VerticalNet. Software vendors providing e-marketplace integrated systems include CommerceOne, Ariba, i2 Technologies and JDA. Many consulting firms e.g. Keane, Intelligent Information Systems and Sapient provide services to integrate back-office systems of e-marketplace participants.

COLLABORATION SYSTEMS

These systems are used by companies to increase collaboration and to optimize supply chain operations, particularly in the product design, planning and forecasting stages. One example of such systems is Ford sharing its product design information with its suppliers. Another example is Kmart CPRF (collaborative planning, forecasting, and replenishment) which connects Kmart with suppliers. Many software vendors provide collaboration systems e.g. webPLAN(webplan.com), i2 Technologies (i2.com), Logility (logility.com), and JDA(jda.com).

Preparing for e-SCM

Implementing an e-SCM solution can be simple or complex. More difficult to implement e-SCM solutions are the ones that provide most benefits but require significant investment depending on the state of a company's existing infrastructure and systems. Massive integration with multiple supplier systems, new software systems, hardware upgrades, and consulting assistance are required many a times. A business should carefully analyze the following issues before deciding to implement an e-SCM solution.

Leverage: Leverage is the relative advantage one company has over others in a supply chain. In general, company's size, relative to others in the supply chain, determines the degree of leverage. A company should be aware of its leverage in the supply chain. A company with high degree of leverage is in a better position to dictate its terms.

Systems Integration: Communicating and sharing data between a large numbers of supply chain partners is complex and require a fair amount of integration between supplier systems. Supplier systems may run on different platforms, use different technologies, and store and manipulate data in different ways. Careful planning is required is required to agree on a set of standards to enable accurate and timely data sharing. For data exchange in e-SCM, XML is emerging as the de facto standard. Many software vendors, such as Content DSI (contentdsi.com) and Hostbridge Technology (teubner.com) offer XML translation tools companies can use for translation of internal company data into XML. Several consulting firms, such as Intelligent Information Systems, also specialize in data translation. In addition to XML, there are a number of middleware products to assist companies in communicating with their supply chain participants. Companies offering these middleware products include Oracle, Progress Software, BEA Systems, Neon Systems, Unify and Vitria.

System Selection: Depending on a company's need, it can select a custom-built, packaged or ASP e-SCM solution. Custom-built solution is an expensive option but it is most suited for unique requirements. Packaged solution can be readily implemented and allow multiple companies to share the same functionality and reduce the cost of maintenance. These solutions are appropriate when there are no unique company requirements. E-SCM Solutions provided by ASPs are generally the fastest and least capital-intensive. These solutions are suitable for medium and small-sized companies that require cost-effective e-SCM functionality.

Collaboration and Coordination: Increased complexity of e-SCM solution requires increased collaboration and coordination among the participants. A less affluent but uncooperative participant can cause serious problems in e-SCM implementation. An appropriate management structure in each participating company in e-SCM implementation is necessary to oversee the implementation of e-SCM solution and resolve any conflicts.

Security and Privacy: Companies need carefully devise their strategy to guard their customer, sales and forecasting data while implementing e-SCM solution. Companies also need to make sure that the e-SCM solution is capable to prevent unauthorized access to the data produced and exchanged between e-SCM applications.

Trends in E-SCM

As modern Supply Chain Management (SCM) becomes strategic, adaptive and demand driven, software vendors are developing systems to support the new SCM paradigm. The new supply chain systems will go beyond operating efficiencies to a totally integrated system that improves management, planning and execution at all levels, giving the organization key competitive advantages. Supply chain software vendors are building and integrating the following five revamped best practices into their SCM solutions (Heistand, 2014; Chainanalytics.com, 2015).

Omnichannel

The omnichannel universe is leading to expansion in the physical world, as more stores are built that can serve as fulfillment centers for outbound shipments, as well as customer pick-ups, inventory management and returns. Buy online, pick-up in store has proven to be a breakout success and the interplay between physical and virtual shopping channels strengthens both. The trend of online pure plays expanding to brick and mortar can be expected to continue.

Differentiation

Companies are now focusing on what they do well and perfecting it. For example, Macy's is finding success by leveraging its vast network of stores and quality private label brands while Home Depot is one of the few shippers that truly understands how to facilitate heavy and difficult-to-ship merchandise fulfillment. This competition among retailers results in the acceleration of customer expectations, serving them more efficiently, all while differentiating the experience along the way to deliver what they really want.

Supply Chain and Corporate Strategy

Logistics must be in place before new services can be launched, but they must also be easily scalable to adjust for unplanned-for demand signals. The process should proceed like this: launch, evaluate demand signal, then optimize. The supply chain has to be in place on day one, but also flexible enough to be optimized for what the demand signals indicate.

RFID

Macy's began implementing RFID technology three years ago to maintain accuracy in its inventory. The 95-98 percent confidence level the company has achieved on its inventory levels allows it to fulfill from over 800 stores. RFID is expected to become ubiquitous within five years dramatically cutting down the time for inventory management.

Drop Shipping

Forty percent supply chain headaches are caused by the 20 percent of orders that are drop shipped. Drop shipping allows carrying more SKUs than company would ever be able to hold in inventory. Large retailers with virtually limitless inventory capacity often include drop shipping on items where volume does not warrant internal fulfillment, but demand is strong enough to keep the item available to customers. Smart fulfillment management requires analyzing which SKUs make the most sense to keep in inventory vs. which are wiser to drop ship, and constantly re-evaluating orders to optimize the ratio.

Expedited Shipping

Customers want fast shipping. At the same time, they do not want to pay for it. In the battle between those scenarios, free shipping beats fast shipping. Today, very few companies report a significant number of their orders being sent with shipping that carries extra costs. Home Depot has offered same-day onsite deliveries for professional contractors for years, but very few contractors are willing to pay for 2- or 4-hour delivery windows, instead opting for flexible all-day windows or for doing the pick-up themselves. Locating supply closer to customers makes even more sense in a world where consumers want it tomorrow but with free shipping.

Globalization

The challenges of globalization include mitigating factors such as delays, longer lead times, sub-standard quality, lesser customer service, and, in some cases, higher total cost. The benefits of globalization include higher revenues and lower labor costs. To minimize the challenges and enhance the benefits of globalization, new SCM systems will need real time updates (using GPS, RFID and other types of integrated tracking and identification systems) and interconnected supply chain partners providing real time visibility to product movement and 3) collaboration on planning, execution strategies and mitigating risks.

Risk Management

According to a recent IBM study, the lack of standardized processes, insufficient data and inadequate technologies thwart effective risk management (Ho, 2009). Supply chain software vendors are enhancing management features to better understand risks, the impact of risks and how to mitigate risks. High impact risk management planning is done with software simulation models. To manage common low impact events, supply chain sub-systems will provide real-time updates on events and facilitate the deployment of contingency plans. For example – when an overseas container is lost during shipping, the transportation system will get earlier notification; the demand management team can recalculate the sales forecast, deploy demand-shifting strategies and re-prioritize fulfillment.

Collaborative Relationships

Distribution and supply chain systems are trending toward new ways to facilitate relationships and enhance collaboration. The new supply chain paradigm calls for collaboration that crosses hierarchical, departmental and organizational boundaries. Enterprise software is facilitating departmental, trading partner and 3PL stakeholders to interactively collaborate on operational tasks and planning activities. For example, Sales & Operational Planning software will coordinate and synthesize input from all internal and external stakeholders. In addition, collaborative SCM relationships will be enhanced by tightly integrating supply chain management systems with Customer Relationship Management software and Supplier Relationship Management applications.

Essential Features of Supply Chain Management Software

An optimized supply chain can increase customer satisfaction while a poorly functioning supply chain will ruin relationships with suppliers and customers, and will cause headaches in every aspect of business. Choosing a SCM software solution is not as simple and straightforward as the benefits of utilizing one. Choosing SCM software can be a daunting task, especially for business owners who are not sure what to look for when buying a new solution. Following are some of the key features of SCM software your company should look for when purchasing a solution (Picone, 2014).

Inventory Management: Built-in inventory management ensures that your stock levels are optimized. A good E-SCM software help optimize inventory levels to ensure no overstocking or understocking. Real-time tools let you know where your inventory is at any given moment, allowing you to operate precisely.

Order and Billing Management: Your chosen E-SCM solution should provide a flexible, built-in order management system that is able to create orders and bill clients from one centralized location, all from within the solution. Flexible order management capabilities are best as they ensure that the solution can work with your unique order needs and are highly configurable.

Logistics and Transportation Tools: Transportation and logistics tools ensure that materials are shipped efficiently and cost-effectively and provide the necessary resources to manage fuel costs and various ever-changing government regulations. These tools help reduce overall operational costs and remove any kinks from the supply chain Dispatch management, appointment scheduling and yard management may all be included as part of your built-in logistics and transportation tools.

Supplier Collaboration: Supplier collaboration functions keep suppliers, customers and other key players in the loop and empower partners to ensure a smoother supply chain. Suppliers, employees and customers can use these tools to collaborate from disparate locations in real time, ensuring that every key player is involved in problems or issues as they occur. A self-service portal provides suppliers access to relevant information and ability to submit requests, all directly through E-SCM solution.

Warehouse Management: Warehouse management tools allow your company to optimize warehouse stock and increase warehouse accuracy and efficiency. These capabilities allow companies to process orders from multiple sources, ensuring that all needs are fulfilled in a timely, cost-effective manner. Effective warehouse management tools provides a smooth and efficient overall supply chain process and your company can oversee receiving, put away, replenishment and cross docking.

End-to-End Visibility of Your Supply Chain in Real Time: A deep real-time visibility into the end-to-end functions and processes is necessary to manage your supply chain. Your employees need immediate, real-time access to various aspects of the supply chain to ensure that there are no hiccups, bottlenecks, missing goods or unhappy customers. Real-time capabilities allow your business to react to changes in the supply chain as they arise, as well as access up-to-date analytics that can inform future decisions. This feature is incredibly useful for businesses of any size, though larger businesses must have real-time capabilities to stay afloat.

Feature Comparison of E-SCM Software

Following is a feature comparison of some popular E-SCM software.

	SaaS solution	Inventory management	Logistics & transportation	Warehouse management	On-premise solution	Order & billing management	Supplier collaboration	Real-time supply chain visibility
SAP Supply Chain Solutions (www.sap.com)	√	√	√	√	√	√	√	√
Epicor Supply Chain Management (www.epicor.com)	√	√	√	√	√	√	√	√
Viewlocity Control Tower Platform (www.viewlocity.com)	√	√	√	√		√	√	√
NetSuite ERP (www.netsuite.com)	√	√	√	√		√	√	√
Infor Lawson Supply Chain Management (www.infor.com)	√	√			√	√		√

Electronic Supplier Relationship Management (E-SRM)

It involves electronically managing a close and efficient relationship between a firm and its various suppliers and other components of the supply chain. Electronic supplier relationship management (e-SRM) is important in order to maintain strong, long lasting and beneficial relationship between e-commerce firms and their suppliers. One important function of e-SRM is to predict suppliers who tend to churn such that early "treatment" can be given. In the e-commerce systems that involve suppliers as the websites users, predicting suppliers' churn tendency can be based on analyzing their frequencies in accessing the e-commerce websites. e-SRM is a relatively new concept. According to proponents, the use of E-SRM software can lead to lower production costs and a higher quality, but lower priced end-product.

SRM products are available from a number of vendors, including I2 Technologies (now part of JDA software), Manugistics, PeopleSoft, and SAP. SAP Supplier Relationship Management 7.0 is the E-SRM software provided by SAP. SAP Supplier Relationship Management (SAP SRM) provides you with innovative methods to coordinate your business processes with your key suppliers and make them more effective. With SAP SRM you can examine and forecast purchasing behavior, shorten procurement cycles, and work with your partners in real time. This allows you to develop long-term relationships with all those suppliers that have proven themselves reliable partners. The efficient processes in SAP SRM enable you to cut down your procurement expenses and to work more intensively with more suppliers than ever before. The SAP SRM documentation offers a comprehensive description of the functional scope of SAP SRM, and highlights the relationship between the application and the underlying technologies. PeopleSoft supplier relationship management is a family of applications in Oracle's PeopleSoft product line. PeopleSoft supplier relationship management tightly integrates with other PeopleSoft applications, including PeopleSoft financial management and PeopleSoft supply chain management to deliver a complete procure-to-pay solution. The modular nature of the PeopleSoft supplier-relationship management family enables customers to deploy applications based on their specific needs. The Manguistic SRM Suite offers collaborative design; spend analysis and optimization, strategic sourcing and contract management, procurement execution, and collaborative supply planning modules. The average sale price on a Manugistics suite license currently hovers around US$850,000. JDA's SRM solutions enable companies to bridge the procurement and product design functions for lean supply management; ncrease speed, efficiency and reliability of purchasing activities; gain visibility into cross-functional part and component availability; and reduce cost of goods sold and improve inventory consumption. Companies can also ensure that guidelines for preferred parts, material reuse and suppliers are incorporated into the product design process.

Review Questions

1. What is e-CRM? Why it is important for a company?

2. List components of an e-CRM.

3. List some benefits and limitations of e-CRM.

4. How an e-CRM works?

5. Why e-CRM implementation fails?

6. What is the difference between cloud and social CRM?

7. List types of e-CRM solutions.

8. What is e-SCM?

9. List key processes and infrastructure components in e-SCM.

10. What are benefits of E-supply chains?

11. List various categories of e-SCM solutions.

Bibliography

americanexpress.com. (2010, July 7). Americans Will Spend 9% More With Companies That Provide Excellent Service. Retrieved July 23, 2015, from http://about.americanexpress.com/news/pr/2010/barometer.aspx

Ang, L., & Buttle, F. (2006). CRM software applications and business performance. *Journal of Database Marketing & Customer Strategy Management* , 4–16.

Ang, L., & Buttle, F. (2006). Managing For A Successful Customer Acquisition: An Exploration. *Journal of Marketing Management* , *22*, 295-317.

AT&T. (2003). *More than numbers: CRM in the networked organisation.* Retrieved January 26, 2009, from corp.att.com: http://www.corp.att.com/emea/docs/crm_whitepaper.pdf

Blery, E. (2006). Customer relationship management:A case study of a Greek bank. *Journal of Financial Services Marketing* , *11*, 116-124.

Bradshaw, D., & Brash, C. (2001). Managing customer relationships in the e-business world: how to personalise computer relationships for increased profitability. *International Journal of Retail and Distribution Management* , *29* (11/12), 520-529.

Bull, C. (2003). Strategic issues in customer relationship management (CRM) implementation. *Business Process Management Journal*, *9* (5), 592-606.

Burger, A. K. (2007). *RFID and Wireless Sensor Networks in the Supply Chain*. Retrieved September 25, 2009, from ecommercetimes.com: http://www.ecommercetimes.com/story/55492.html

Buttle, F. (2003). *Customer relationship management: concepts and tools*. Butterworth-Heinemann.

Cespedes, F. V. (2014, August 15). Sales Still Matters More than Social Media. Retrieved July 23, 2015, from https://hbr.org/2014/08/sales-still-matters-more-than-social-media

Chainanalytics.com. (2015). E-Commerce Supply Chain Advice From Top Retailers. Retrieved July 23, 2015, from http://www.chainalytics.com/e-commerce-supply-chain-advice-from-top-retailers/

Chapelfieldsystems.com. (2003). *What is eCRM?* Retrieved March 24, 2009, from Chapelfieldsystems.com: http://www.chapelfieldsystems.com/downloads/Downloads_files/WhatiseCRM.doc

Chatham, B., Temkin, B., Gardiner, K., & Nakashima, T. (2002). *CRM's Future: Humble Growth Through 2007*. Retrieved July 14, 2009, from Forrester Research: http://www.forrester.com/ER/Research/Report/Summary/0,1338,14653,00.html

Chen, H., & Kong, L. (2007). An Analysis on the Implementation of Electronic Supply Chain in International Trade. *Integration and Innovation Orient to E-Society*, *1*, 115-125.

Chen, I. J., & Popovich, K. (2003). Understanding CRM: people, process and technology. *Business Process Management Journal*, *9* (5), 672-688.

Chen, J., & Ching, R. K. (2007). The effects of Information and Communication Technology on Customer Relationship Management and customer lock-in. *International Journal of Electronic Business*, *5* (5), 478-498.

Chen, M., & Shang, S. (2005). Implementing an Enterprise-wide CRM Infrastructure. *Proceedings of the Fourth Workshop on e-Business*. Las Vegas, Nevada, USA.

Christopher, M. (1998). *Logistics and Supply Chain Management: Strategies for reducing cost and improving service*. London: Financial Times Pitman Publishing.

Columbus, L. (2012, April 12). Roundup of CRM Forecasts and Market Estimates, 2012 - Forbes. Retrieved July 23, 2015, from http://www.forbes.com/sites/louiscolumbus/2012/12/04/roundup-of-crm-forecasts-and-market-estimates-2012/

Cone Communications. (2010, July 16). New Cone Research: Consumers Need to Verify Product Recommendations. Retrieved July 23, 2015, from http://www.conecomm.com/consumers-need-to-verify-product-recommendations

Crosby, L. (2002). Guru's view: exploding some myths about CRM. *Managing Service Quality*, *12* (5), 271-277.

Deeter-Schmelz, D., & Kennedy, N. (2002). An exploratory study of the Internet as an Industrial communication tool – Examining buyer's perceptions. *Industrial Marketing Management , 31*, 145-154.

Fjermestad, J., & Nicholas C. Romano, J. (2008). Introduction to the Special Section: Consumer-Focused Processes in E-Commerce. *International Journal of Electronic Commerce , 12* (3), 7.

Forrester Research. (2008, May 9). Forrester Research : Research : Mobilizing Sales To Handheld Devices. Retrieved July 23, 2015, from https://www.forrester.com/Mobilizing+Sales+To+Handheld+Devices/fulltext/-/E-RES45140?objectid=RES45140

Gartner. (2010, February 19). Gartner Says Companies Need to Pursue Four Steps to Harness Social Computing in CRM. Retrieved July 23, 2015, from http://www.gartner.com/newsroom/id/889712

GovernmentofVictoria. (2009). *eCustomer Relationship Management.* Retrieved January 15, 2010, from mmv.vic.gov.au: http://www.mmv.vic.gov.au/Assets/219/1/ElectronicCustomerRelationshipManagement.pdf

Gupta, S., & Kim, H.-W. (2007). The Moderating Effect of Transaction Experience on the Decision Calculus in On-Line Repurchase. *", International Journal of Electronic Commerce , 12* (1), 127.

He, X. (. (2004). The ERP challenge in China: a resource-based perspective. *Information Systems Journal , 14* (2), 153-167.

Heistand, S. (2014). 5 Trends in Supply Chain Management Software. Retrieved July 23, 2015, from http://www.erpsearch.com/supply-chain-software-trends.php

Ho, V. (2009, March 16). Visibility lacking in supply chain | ZDNet. Retrieved July 23, 2015, from http://www.zdnet.com/article/visibility-lacking-in-supply-chain/

Huang, A., Yen, D. C., Chou, D. C., & Xu, Y. (2003). Corporate applications integration: Challenges, opportunities, and implementation strategies. *Journal of Business and Management , 9* (2), 137-145.

IBM. (2004). *Doing CRM right: What it takes to be successful with CRM.* Retrieved July 12, 2009, from ibm.com: http://www-935.ibm.com/services/au/bcs/pdf/ge510-3604-01f_ibm_crm_study_asiapac.pdf

Jayachandran, S., Sharma, S., Kaufman, P., & Raman, P. (2005). The Role of Relational Information Process and Technological Use in Customer Relationship Management. *Journal of Marketing , 69* (10), 177-192.

Kalakota, R., & Robinson, M. (2000). *E-business: Roadmap for success.* Massachusetts: Addison-Wesley.

Kavanagh, S. (2003, August). Planning for CRM success. *Government Finance Review* , 39-45.

Krigsman, M. (2009, August 3). *CRM failure rates: 2001-2009.* Retrieved June 13, 2010, from zdnet.com: http://www.zdnet.com/blog/projectfailures/crm-failure-rates-2001-2009/4967

Lardinois, F. (2008, September 26). Study: 93 Percent of Americans Want Companies to Have Presence on Social Media Sites. Retrieved July 23, 2015, from http://readwrite.com/2008/09/26/study_social_media_presence

Lee-Kelley, L., Gilbert, D., & Mannicom, R. (2003). How e-CRM can enhance customer loyalty. *Marketing Intelligence and Planning*, *21* (4/5), 239.

Ling, R., & Yen, D. C. (2001). Customer relationship management: An analysis framework and implementation strategies. *The Journal of Computer Information Systems*, *41* (3), 82-97.

Mack, O., Mayo, M. C., & Khare, A. (2005). A strategic approach for successful CRM: A European perspective. .*Problems and Perspectives in Management*, *2*, 98-106.

Microsoft. (2009). *Customer Loyalty*. Retrieved January 15, 2010, from microsoft.com: http://download.microsoft.com/download/8/1/e/81e5ca54-452d-4c83-a86d-85f026e2b178/Entrepreneur_eBook_Customer_Loyalty.pdf

Nesbitt, S. (2014, July 8). Features of top five open source customer relationship management systems | Opensource.com. Retrieved July 23, 2015, from http://opensource.com/business/14/7/top-5-open-source-crm-tools

Norris, G., Hurley, J. R., Hartley, K. M., Dunleavy, J. R., & Balls, J. D. (2000). *E-Business and ERP: Transforming the Enterprise*. John Wiley.

Norton, D. (2001). *What you need to know about e-CRM*. Retrieved May 15, 2009, from Techrepublic.com: http://articles.techrepublic.com.com/5100-10878_11-1033131.html

O'Malley, L., & Mitussis, D. (2002). Relationships and technology: strategic implications. *Journal of Strategic Marketing*, *10* (3), 225-238.

Padmanabhan, B., & Tuzhilin, A. (2003). On the Use of Optimization for Data Mining: Theoretical Interactions and eCRM Opportunities. *Management Science*, *49* (10), 1327-1343.

Padmanabhan, B., Zheng, Z., & Kimbrough, S. O. (2006). An empirical analysis of the value of complete information for eCRM models. *MIS Quarterly*, *30* (2), 247-267.

Picone, K. (2014, September 3). What are the Essential Features of Supply Chain Management Software? Retrieved July 23, 2015, from http://www.business-software.com/blog/essential-features-of-supply-chain-management-software/

Plakoyiannaki, E., & Tzokas, N. (2002). Customer Relationship Management (CRM): A Capability Portfolio Perspective. *Journal of Database Marketing*, *9* (3), 228-237.

Plessis, M., & Boon, A. J. (2004). Knowledge management in eBusiness and customer relationship management: South African case study findings. *International Journal of Information Management*, *24* (1), 73-86.

Rigby, D. K., & Bilodeau, B. (2005). The Bain 2005 management tool survey. *Strategy and Leadership , 33* (4), 4-12.

Rigby, D. K., Reichheld, F. F., & Schefter, P. (2002). Avoid the Four Perils of CRM. *," Harvard Business Review , 80* (2), pp. 101–109.

Ross, D. (2005). E-CRM From a Supply Chain Management Perspective. *Information Systems Management , 22* (1), 37-44.

Shaw, R., & Davies, J. (2001, October 11). *Customer Relationship Management (CRM): Overview.* Retrieved July 11, 2009, from Gartner.com: http://ecrmguru.googlepages.com/Gartner-CRM-CustomerRelationshipMana.pdf

Tan, X., Yen, D., & Fang, X. (2002). Internet Integrated Customer Relationship Management - A Key Success Factor for Companies in the E-Commerce Arena. *Journal of Computer Information Systems , 42* (3), 77.

Wilson, H., Daniel, E., & Malcolm, M. (2002). Factors for success in customer relationship management (CRM) systems. *Journal of Marketing Management , 18* (1-2), 193-219.

Zablah, A. R., Bellenger, D. N., & Johnston, W. J. (2004). An evaluation of divergent perspectives on customer relationship management: Towards a common understanding of an emerging phenomenon. . Industrial Marketing Management , 33, 475-489.

10

E-COMMERCE SECURITY

After reading this chapter, reader should be able to:

- Understand the scope of E-commerce security problems.

- Describe the important types of computer security.

- Describe classifications of information assets.

- Understand the basic E-commerce security issues.

- Discuss the governance of E-commerce security in an organization.

- Describe the key elements of E-commerce security governance structure in an organization.

- Identify the key security threats, vulnerabilities, and risks in the E-commerce environment.

- Understand active content and its types.

- Identify various solutions to counter security threats to E-commerce environment.

- Identify and assess major technologies and methods for securing E-commerce communications.

- Describe the major technologies for protection of E-commerce networks.

- Describe how technology helps protect the security of messages sent over the Internet.

- Identify the tools used to establish secure Internet communications channels, and protect networks, servers, and clients.

- Understand phishing and techniques to protect from it.

- Understand the difference between computer and online crime.

- Understand how Risk in E-commerce is assessed and managed.

Overview

When it comes to the Internet users may have many fears in their mind ranging from someone gaining unauthorized access to their emails to some form of identity theft. In today's world, the stakes are much higher and consequences of these threats are far more serious than in the past. For web shoppers the biggest concern is the exposure of their credit card information to millions of people as the information travels across the Internet. There still exists a significant portion of shoppers that have concerns about the security of their credit card information in E-commerce transactions (Ecommercetimes.com, 2010). People are concerned about personal information they provide to companies over the Internet and there are concerns about willingness and ability of these companies for keeping the confidentiality of this personal customer information. Businesses adoption of new business models, rapidly evolving and changing technology, difficulties in identifying and measuring risks and business impacts, potential for widespread and immediate visibility of any problems with E-commerce systems, compliance with legal and regulatory requirements are some of the factors that give rise to the importance of E-commerce security. Failure of securing your E-commerce business can have a significant impact upon business. Some potential business implications of a security incident may include direct financial loss as a consequence of fraud or litigation, subsequent loss as a result of unwelcome publicity, criminal charges if the business is found to be in breach of the data protection or computer misuse laws or other regulation on e-commerce, and loss of market share if customer confidence is affected by unavailability of website because of a Denial of Service (DoS) attack. The business image, together with the brands under which the business trade, are valuable assets. The damage from such data breaches can be severe. If consumers lose faith in a company's ability to keep their personal data safe, the company can ultimately lose customers. Most certainly they stand to lose money, and in some cases, intellectual property. In its most recent analysis, the Ponemon Institute found that in 2013 each lost data record cost companies an average of $145 per record, with companies in Germany losing the most per record for each data breach ($201), followed by the United States ($195), and companies in India the least at $51.2 Based on the global average cost per record, that means a major retailer with millions of leaked credit cards could face more than $100 million in direct costs, including fines. A university that leaked 40,000 records could suffer over $5.4 million in losses (IBM, 2015). Data breaches are currently among the most common and most costly security problems for organizations of all sizes. The 2014 Cyber Security Intelligence Index by IBM shows that companies are attacked around 16,856 times a year, and data breaches are one of the preeminent causes for these attacks (Zaharia, 2015).

It is important to understand that while adoption of E-commerce by business provides various advantages at the same time the adoption of E-commerce creates new ways through which both image and brands of business can be attacked.

Computer Security

Computer security refers to information security as applied to computers and networks. The objectives of computer security are to protect information and property from theft, corruption, or natural disaster while keeping them remains accessible and productive to the intended users (Feruza & Kim, 2007).

There are two types of computer security i.e. Physical security and Logical security.

Physical security consists of any measure to prevent or deter access to a facility, resource, or information stored on physical media. Physical security can include tangible protection devices such as alarms and guards.

Logical security is protecting the assets using non-physical means e.g. software safeguards of an organization's systems, user Identification and password access, authentication, access rights and authority levels.

Classification of Information Assets

Information whether personal or organizational can be treated as an asset. We can divide these information assets into four categories. Each category has an associated risk impact.

Unrestricted Information

This is public information that is available to the public, all employees etc. Examples of such information include job postings and ordinary staff meeting agendas and minutes. This information is created in normal course of business. Loss or change in any such information has little or no impact and has no legal effect. Unavailability of such information cause minimal inconvenience.

Protected Information

This information is sensitive in nature and needs to be protected outside the organization. Access to this information is authorized and on a need-to-know basis for business related purposes. An example of protected information would include information that an accountant learned in his or her capacity as an accountant with a company. Other examples include documents containing personal information and business information applications. There is a low degree of risk if such information is corrupted or modified. Unavailability of protected information can cause disruption in business. If lost, it can cause low level of financial loss to the business.

Confidential Information

This information is sensitive within the organization and is available only to specific function, group or role. Examples of confidential information include a social security number or credit card number. Other examples include industrial trade secrets, and third-party business information submitted in confidence. Loss of this type of information could cause serious loss of privacy, competitive advantage, loss of confidence in organization, and loss of relationships and reputation.

Restricted Information

This information is highly sensitive and is available only to specific, named individuals or specific positions. Examples of restricted information may include e-mails of an employee, investment proposals etc. Loss of restricted information can cause extreme damage (e.g. substantial financial loss) to the integrity or image of the organization.

Basic Security Issues

E-Commerce security involves more than just preventing and responding to cyber-attacks and intrusion. Each party involved in E-commerce transaction has different perspective and concerns about E-commerce security.

Consumer's perspective:

- How can one be sure that the Web server is owned and operated by a legitimate company?

- How does one know that the Web page and form do not contain some malicious or dangerous code or content?

- How does one know that Web server will not distribute the information I provide to some other party?

Merchant's perspective:

- How does know the user will not attempt to break into the Web server or alter the pages and content at the site?

- How does one know that the user will not try to disrupt the server so that it isn't available to others?

Both Merchant and Consumer's perspective:

- How do we know that the network connection is free from eavesdropping by a third party?

- How do we know that the information sent back and forth between the server and the user's browser hasn't been altered?

Some Important Security Terminologies

Following are some important terminologies related to computer security (IBM, 2015)

Access or credentials abuse: Activity detected that violates the known use policy of that network or falls outside of what is considered typical usage.

Breach or compromise: An incident that has successfully defeated security measures and accomplished its designated task.

Denial of service (DoS): Attempts to flood a server or network with such a large amount of traffic or malicious traffic that it renders the device unable to perform its designed functions.

Droppers: Malicious software designed to install other malicious software on a target.

Event: An event is an observable occurrence in a system or network.

Inadvertent actor: Any attack or suspicious activity sourcing from an IP address inside a customer network that is allegedly being executed without the knowledge of the user.

Key loggers: Software designed to record the keystrokes typed on a keyboard. This malicious software is primarily used to steal passwords.

Malicious code: A term used to describe software created for malicious use. It is usually designed to disrupt systems, gain unauthorized access, or gather information about the system or user being attacked. Third party software, Trojan software, key loggers, and droppers can fall into this category.

Outsiders: Any attacks that sourced from an IP address external to a customer's network.

Phishing: A term used to describe when a user is tricked into browsing a malicious URL designed to pose as a website they trust, thus tricking them into providing information that can then be used to compromise their system, accounts, and/or steal their identity.

Security device: Any device or software designed specifically to detect and/or protect a host or network from malicious activity. Such network-based devices are often referred to as intrusion detection and/or prevention systems (IDS, IPS or IDPS), while the host-based versions are often referred to as host-based intrusion detection and/or prevention systems (HIDS or HIPS).

Spear phishing: Phishing attempts with specific targets. These targets are usually chosen strategically in order to gain access to very specific devices or victims.

SQL injection: An attack used that attempts to pass SQL commands through a website in order to elicit a desired response. One that the website is not designed to provide.

Suspicious activity: These are lower priority attacks (or suspicious traffic) that could not be classified into one single type of category. These are usually detected over time by analyzing extended periods of data.

Sustained probe/scan: Reconnaissance activity usually designed to gather information about the targeted systems such as operating systems, open ports, and running services.

Trojan software: Malicious software hidden inside another software package that appears safe.

Unauthorized access: This usually denotes suspicious activity on a system or failed attempts to access a system by a user or users who does not have access.

Wiper: Malicious software designed to erase data and destroy the capability to restore it.

Zero-Day: An unknown vulnerability in an application or a computer operating system.

Security Events, Attacks, and Incidents

A security event is an event on a system or network detected by a security device or application. A security attack is a security event that has been identified by correlation and analytics tools as malicious activity that is attempting to collect, disrupt, deny, degrade or destroy information system resources or the information itself. Security events such as SQL Injection, URL tampering, denial of service, and spear phishing fall into this category.cA security incident is an attack or security event that has been reviewed by security analysts and deemed worthy of deeper investigation. In 2014, there were 81,342,747 security events, 12,017 security attacks, and 109 security incidents (IBM, 2015).

Security Threats, Vulnerability, and Countermeasures

A threat represents the type of action likely to be harmful. Vulnerability represents the level of exposure to threats in a particular context. A countermeasure is an action implemented to prevent the threat. The threat comes from those who would like to acquire the information or limit business opportunities by interfering with normal business processes. The object of security is to protect valuable or sensitive organizational information while making it readily available. Attackers trying to harm a system or disrupt normal business operations exploit vulnerabilities by using various techniques, methods, and tools. System administrators need to understand the various aspects of security to develop measures and policies to protect assets and limit their vulnerabilities.

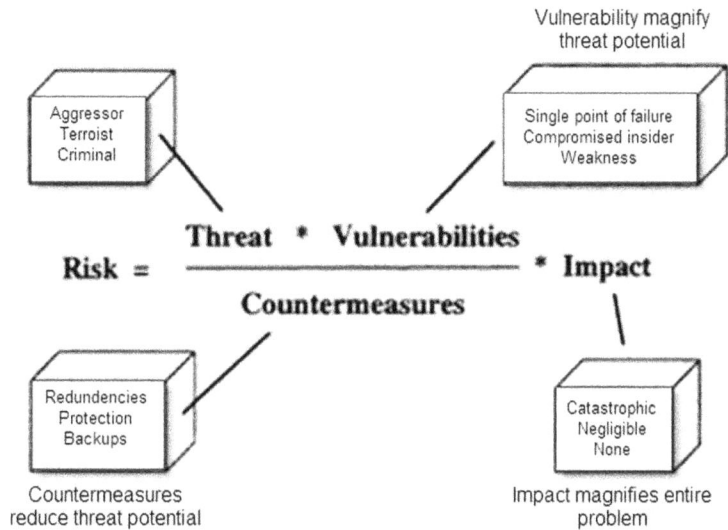

Attackers generally have motives or goals e.g. to disrupt normal business operations or steal information. To achieve these motives or goals, they use various methods, tools, and techniques to exploit vulnerabilities in a computer system or security policy and controls.

Security threats can be broken down in different categories as shows in Figure 10-3.

FIGURE 10-3: CATEGORIES OF SECURITY THREATS

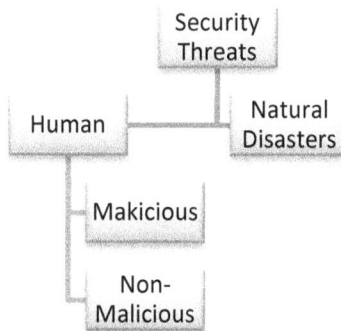

Natural Disasters

Natural disasters, such as earthquakes, hurricanes, floods, lightning, and fire can cause severe damage to computer systems. Information can be lost, downtime or loss of productivity can occur, and damage to hardware can disrupt other essential services. Few safeguards can be implemented against natural disasters. The best approach is to have disaster recovery plans and contingency plans in place. Other threats such as riots, wars, and terrorist attacks could be included here. Although they are human-caused threats, they can still be classified as disastrous.

Human Threats

Human threats are threats that are caused by human activity. Human threats can be classified into two sub categories: Malicious and Non-malicious. Malicious threats consist of inside attacks by disgruntled or malicious employees and outside attacks by non-employees just looking to harm and disrupt an organization. The most dangerous attackers are usually insiders. In 2014, 55 percent of all attacks were carried out by either malicious insiders or inadvertent actors (IBM, 2015). They know many of the codes and security measures that are already in place and have legitimate access to the system. Insiders are likely to have specific goals and objectives. These goals could be to disrupt services and the continuity of business operations and steal useful information that can be sold to competitors. They most likely to know what actions might cause the most damage. The insider attack can affect all components of computer security. By browsing through a system, confidential information could be revealed. Insider attacks can affect availability by overloading the system's processing or storage capacity, or by causing the system to crash.

Security Incidents

There could be a variety of security incidents in which attackers try to achieve their goals. According to IBM (2015), following were the top rated security incidents in 2014. Unauthorized access topped this list.

37 % → Unauthorized Access

20 % → Malicious Code

20 % → Sustained probe/scan

11 % → Suspicious Activity

8 % → Access or credential abuse

4 % → Denial of Service

With reference to incident rates across industries, the finance industry was at the top of the list in 2014 followed by the information and communication. Manufacturing industry stood at third place. Meanwhile, the energy and utilities category narrowly edged out the health and social service category for fifth place. Of this group, only manufacturing experienced fewer incidents in 2014 than in the previous year (IBM, 2015).

35.33% → Finance & Insurance

19.08 % → Information & Communication

17.79 % → Manufacturing

9.37 % → Retail & Wholesale

5.09 % → Energy & Utilities

The largest number of attacks both originated (50 percent) and took place (59 percent) in the United States in 2014. Next in line was China, where 16% of all attacks originated, and Japan, which was the target of 24% of the year's attacks.

Types of Security Attacks

One classification of security attacks is based on whether attack is active or passive. A passive attack is a network attack in which a system is monitored and sometimes scanned for open ports and vulnerabilities. The purpose is solely to gain information about the target and no data is changed on the target. An active attack is a network exploit in which a hacker attempts to make changes to data on the target or data en route to the target. The Figure 10-5 shows the classification of active and passive attacks.

FIGURE 10-5: ACTIVE/PASSIVE ATTACKS CLASSIFICATION

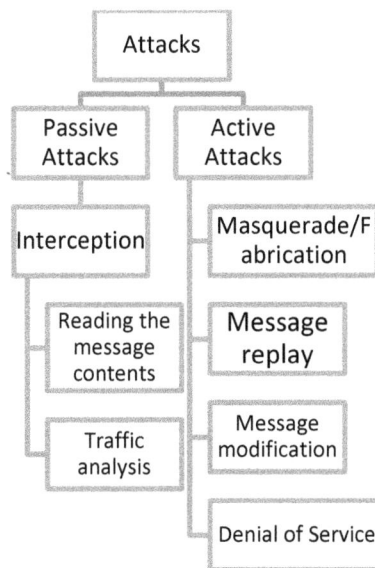

The following Figures describe how an active and passive attack takes place.

FIGURE 10-6: PASSIVE ATTACKS

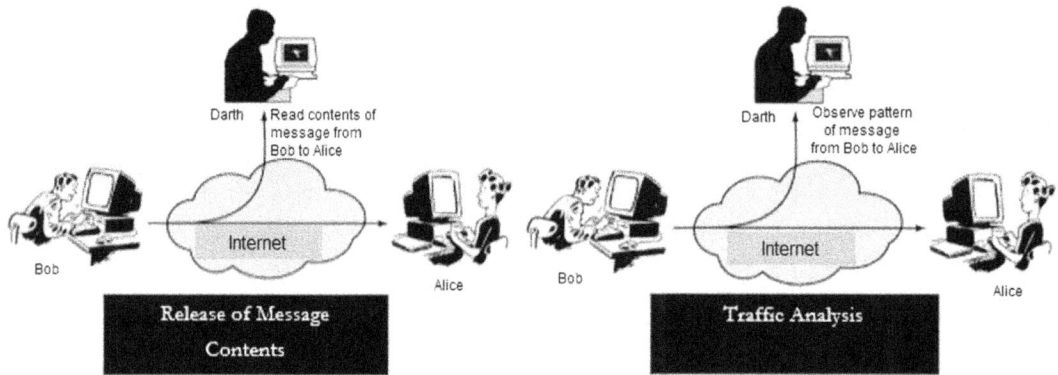

Release of Message Contents

Traffic Analysis

FIGURE 10-7: ACTIVE ATTACKS

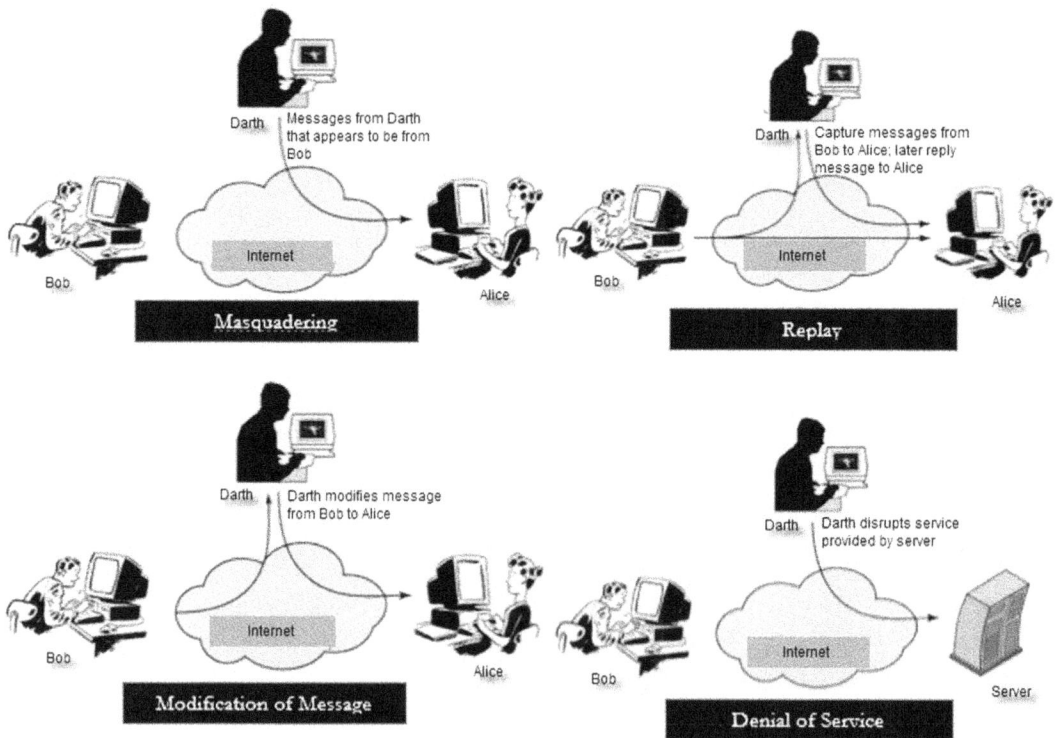

Masquadering

Replay

Modification of Message

Denial of Service

Active attacks present the opposite characteristics of passive attacks. Whereas passive attacks are difficult to detect, measures are available to prevent their success. On the other hand, it is quite difficult to prevent active attacks absolutely, because of the wide variety of potential physical, software, and network vulnerabilities. Instead, the goal is to detect active attacks and to recover from any disruption or delays caused by them. If the detection has a deterrent effect, it may also contribute to prevention.

A masquerade is a type of attack where the attacker pretends to be an authorized user of a system in order to gain access to it or to gain greater privileges than they are authorized for. Spoofing is a general term that is used to describe a variety of ways in which hardware and software can be fooled. Eavesdropping is the unauthorized real-time interception of a private communication, such as a phone call, instant message, and videoconference or fax transmission. A man-in-the-middle attack is an attack where the attacker secretly relays and possibly alters the communication between two parties who believe they are directly communicating with each other. A repudiation attack happens when an application or system does not adopt controls to properly track and log users' actions, thus permitting malicious manipulation or forging the identification of new actions. This attack can be used to change the authoring information of actions executed by a malicious user in order to log wrong data to log files. Media scavenging is the term used for attempting to recover or obtain information from various types of media. A replay attack (also known as playback attack) is a form of network attack in which a valid data transmission is maliciously or fraudulently repeated or delayed.

Governance of E-commerce Security in an Organization

Due to historical difficulty in measuring costs associated due to information security breaches, assessment of business value of IT security is a challenge. On average, each security breach can cause a loss of $1.65 billion in market value of a firm (Cavusoglu, Mishra, & Raghunathan, 2004). Governance of E-commerce Security may be addressed through existing IT and Security Governance structures or through a new framework. In either case, it will require new or enhanced policies and processes to be established to address the new security challenges associated with E-commerce.

Key elements of an E-commerce Security Governance Structure

Key elements of an E-commerce security governance structure include security policies, organizational responsibilities, risk management processes, standards and compliance processes. Each of these items should be reviewed to ensure that adequate guidance and processes are in place to provide the clear management direction necessary to manage the complex risks associated with E-commerce.

SECURITY POLICY

A security policy is a living written document, which is reviewed and updated at regular intervals. It primarily addresses physical security, network security, access authorizations, virus protection, and disaster recovery. The security policy describes which assets are to be protected and why, who is responsible for that protection, and acceptable/unacceptable behaviors. Security policy also emphasizes the importance of E-commerce security and set out or references the specific policies, principles, standards and compliance requirements for achieving this. For example, a company that stores its customers' credit card numbers might decide that those numbers are an asset that must be protected from unauthorized access. Then, the organization must determine the level of access to the system for various people in the organization. Next, the organization determines what resources are available to protect the assets identified. Using all this information, the organization develops a written security policy and commits resources required to implement the security policy. Absolute security is difficult to achieve but deployment of a comprehensive security policy can help avoid most intentional security breaches and reduce their impact.

In order to ensure the minimum level of acceptable security for most E-commerce operations, a comprehensive security policy should fulfill some basic requirements. Following are these requirements.

Secrecy: It refers to preventing unauthorized access to information assets (e.g. credit card information and contents of email messages).

Integrity: It refers to ensuring that ensuring that a communication received has not been altered or tampered with.

Availability: It refers to ensuring access to a resource.

Non-repudiation: It refers to ensuring that none of the parties involved can deny an operation at a later date.

Authentication: It refers to ensuring each party that their partners are truly who they think they are

Organizational Responsibilities: There should be clear and specific accountability for all aspects of E-commerce Security.

CHAPTERS IN SECURITY POLICY

Here are some of the chapters to include in a company's information security policy:

- **Acceptable Use Policy:** An Acceptable Use Policy (AUP), acceptable usage policy or fair use policy, is a set of rules applied by the owner or manager of your company's network that restrict the ways in which the network or system may be used.

- **Internet Access Policy:** This policy applies to all Internet users (individuals working for the company, including permanent full-time and part-time employees, contract workers, temporary agency workers, business partners, and vendors) who access the Internet through the computing or networking resources.

- **Email and Communications Policy:** This policy regulates the way email and other communication channels specific to the company are used.

- **Network Security Policy:** A network security policy, or NSP, is a generic document that outlines rules for computer network access, determines how policies are enforced and lays out some of the basic architecture of the company's security environment.

- **Remote Access Policy:** The remote access policy is a document which outlines and defines acceptable methods of remotely connecting to the internal network. It is essential in large organization where networks are geographically dispersed and extend into insecure network locations such as public networks or unmanaged home networks.

- **Bring your own device (BYOD) Policy:** A BYOD policy, or bring-your-own-device policy, is a set of rules governing a corporate IT department's level of support for employee-owned PCs, smartphones and tablets.

- **Encryption Policy:** The purpose of an encryption policy is to provide guidance that limits the use of encryption to those algorithms that have received substantial public review and have been proven to work effectively.

- **Privacy Policy:** A privacy policy is a statement or a legal document (in privacy law) that discloses some or all of the ways a party gathers, uses, discloses, and manages a customer or client's data. It fulfills a legal requirement to protect a customer or client's privacy.

RISK MANAGEMENT

Risks should be identified and addressed at the earliest stage in an E-commerce project; risks should be reviewed periodically to take account of changes to information systems, infrastructure, and risk levels or to business requirements and priorities.

STANDARDS

Standards provide detailed guidance required to enable the consistent implementation of an effective control structure. Standards can be either flexible in interpretation or rigid. No single best fit-for-all approach exists. The optimum approach for an organization depends on the culture and overall governance structure. The International Standard ISO/IEC 17799–1:2000 "Code of practice for Information Security Management" is a useful reference standard in drawing up E-commerce Security standards.

COMPLIANCE REQUIREMENTS

The design, operation, use and management of E-commerce systems may be subject to a range of statutory, regulatory and contractual security requirements. Compliance requirements to consider include intellectual property rights, software licensing, data protection, and privacy etc.

Threats to E-commerce Systems

Threats to the E-commerce systems can be divided into four groups.

1. Threats to E-commerce Service Side

2. Threats to E-commerce Transactions

3. Threats to Client Side

4. Legal and Regulatory Threats

Threats to E-commerce Service-side

Service-Side includes an organization's infrastructure for supporting an E-commerce service. It typically has two main elements: Front End Systems and Back-s End Systems. Front-End generally consists of one or more web servers connected to the Internet. Back-End systems are needed both to supply information to the front-end systems (such as product information and stock holding) and to extract information from them (such as orders for transferring to logistics systems and payments to be cleared through third parties). Figure 10-9 illustrates the front-end and back-end systems.

FIGURE 10-9: FRONT AND BACK-END SYSTEMS

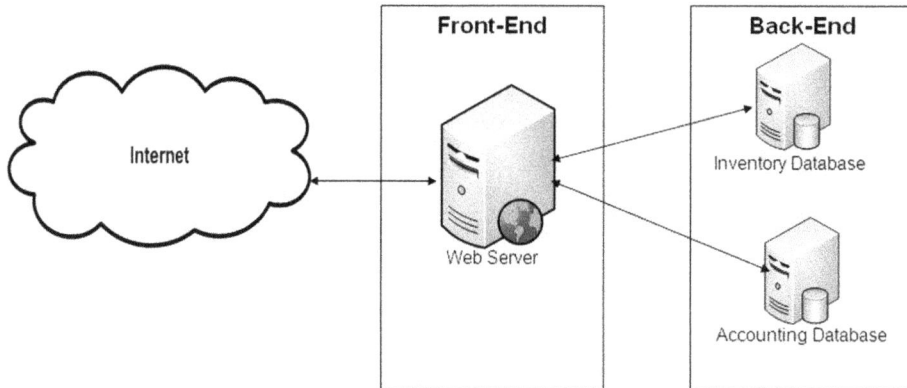

Threats to Front-end systems

Major threats to front-end systems and the resulting business impacts are illustrated below.

Threat	Possible Causes	Possible Effects	Possible Business Impacts
Web server or Internet connection overload	Unforeseen demand for access	Degradation of performance or loss of service	Loss of revenue as customers move to alternative vendors
Web Server failure	Hardware or software failure, operational mistakes	Highly visible unavailability of service	Loss of reputation and revenue.
Inaccurate price or product data on web pages	Operational mistakes, intrusion	Confusion among customers	Loss of profits if products sold at wrong prices
Web pages alteration to include obscene or defamatory content	Intrusion	Customers offended by the content	Loss of customers, loss of reputation
Diversion of potential customers to wrong websites	Cybersquatting	Customer diversion, confusion among existing customers	Loss of potential revenue to competitors
Inappropriate disclosure of customer information stored on the website	Intrusion, mis-configuration of security control,	Customers' data revealed to unauthorized parties	Loss of customer confidence, Loss or reputation, Loss of credibility

Threats to Back-end Systems

Connections made from front-end systems to back-end systems can expose critical business systems to some major threats. Following are the main threats and their impacts.

Threat	Possible Effects	Business Impact
Unforeseen volume of transactions from Web Servers	Systems unavailability and degradation of performance	Disruption in critical business functions
Failure of links between front-end and back-end systems	Inaccurate or outdated information on E-commerce website, unrealistic commitments made to customers	Possible loss of customers
Unauthorized access to internal systems	Corruption of critical information	Serious disruptions in business
Incorrect or unexpected data exchange between front-end and back-end systems	Undesirable behavior of back-end systems	Failure of critical business systems due to inability of handling unexpected data

Threats to E-commerce Transactions

E-commerce transactions between buyers and sellers can include requests for information, quotation of prices, placement of orders and payment, and after sales services. Public perception of insecurity can be a true barrier to E-commerce and organizations must take care to address this. The main threats to E-commerce transactions are listed below.

Threat	Business Impact
Interception of sensitive payment information (e.g. credit card numbers)	Loss of customer confidence
Interception of password or other systems access information	Disruption in business, loss of confidence in organization, unauthorized release of sensitive information.
Customer denial of a purchase agreement at a specified price	Loss of costs incurred in fulfilling the order
Modification or forging of transaction before delivery	Dispatch of goods to wrong person
Delays in transaction due to network congestion	Customer get frustrated and goes away

Threats to Client-Side

Client-Side is the computer operated by the customer or trading partner of an E-commerce business. In most cases, the client environment is outside the direct control of the E-commerce businesses. Some of the main threats resulting from this lack of control of the client-side are shown below.

Threat	Business Impact
Unauthorized access to passwords of other systems access information stored on a client PC	Unauthorized release of sensitive information
Different web browsers that have different features and interact with E-commerce servers in different ways.	Loss of customers that are unable to use these features to make effective use of browsers
Users or organizations blocking cookies or active content technologies (e.g. Java, JavaScript, and ActiveX technologies)	Loss of revenue if users are unable to access services and features provided by active content technologies

---

DENIAL-OF-SERVICE (DOS)

A Denial-of-Service (DoS) attack is an attack meant to shut down a machine or network, making it inaccessible to its intended users. DoS attacks accomplish this by flooding the target with traffic, or sending it information that triggers a crash. In both instances, the DoS attack deprives legitimate users (i.e. employees, members, or account holders) the service or resource they expected.

A zombie (also known as a bot) is a computer that a remote attacker has accessed and set up to forward transmissions (including spam and viruses) to other computers on the Internet. The purpose is usually either financial gain or malice. Attackers typically exploit multiple computers to create a botnet, also known as a zombie army.

Victims of DoS attacks often target the web servers of high-profile organizations such as banking, commerce, and media companies, or government and trade organizations. Though DoS attacks do not typically result in the theft or loss of significant information or other assets, they can cost the victim a great deal of time and money to handle.

There are two general methods of DoS attacks: flooding services or crashing services. Flood attacks occur when the system receives too much traffic for the server to buffer, causing them to slow down and eventually stop. Popular flood attacks include:

- **Buffer overflow attacks:** It is the most common DoS attack. The concept is to send more traffic to a network address than the programmers have built the system to handle. It includes the attacks listed below, in addition to others that are designed to exploit bugs specific to certain applications or networks

- **ICMP flood:** This attack leverages misconfigured network devices by sending spoofed packets that ping every computer on the targeted network, instead of just one specific machine. The network is then triggered to amplify the traffic. This attack is also known as the smurf attack or ping of death.

- **SYN flood:** This attack sends a request to connect to a server, but never completes the handshake. Continues until all open ports are saturated with requests and none are available for legitimate users to connect to.

Other DoS attacks simply exploit vulnerabilities that cause the target system or service to crash. In these attacks, input is sent that takes advantage of bugs in the target that subsequently crash or severely destabilize the system, so that it can't be accessed or used.

An additional type of DoS attack is the Distributed Denial of Service (DDoS) attack. A DDoS attack occurs when multiple systems orchestrate a synchronized DoS attack to a single target. The essential difference is that instead of being attacked from one location, the target is attacked from many locations at once. The distribution of hosts that defines a DDoS provide the attacker multiple advantages. Hacker can leverage the greater volume of machine to execute a seriously disruptive attack. The location of the attack is difficult to detect due to the random distribution of attacking systems (often worldwide). It is more difficult to shut down multiple machines than one. The true attacking party is very difficult to identify, as they disguised behind many (mostly compromised) systems. These compromised systems are called zombies. Figure 10-10 illustrates how a DDoS can take place.

FIGURE 10-10: DDoS ATTACK

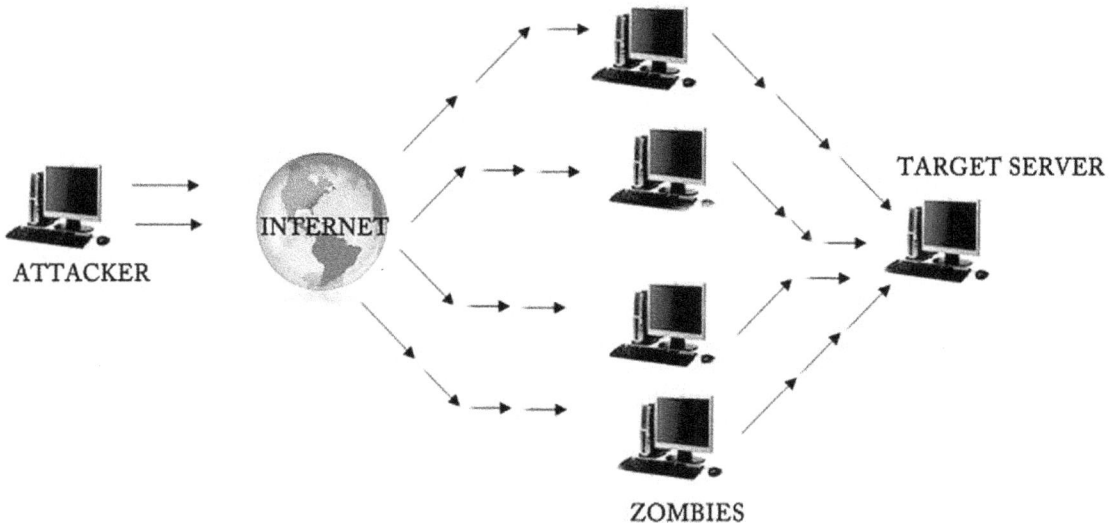

Modern security technologies have developed mechanisms to defend most forms of DoS attacks, but due to the unique characteristics of DDoS, it is still regarded as an elevated threat and is of higher concern to organizations that fear being targeted by such an attack

ACTIVE CONTENT

Active Content is transparently embedded code in a web page that can cause some action to occur e.g. download and play audio. Active content is used in E-commerce to place items into a shopping cart and compute a total invoice amount, including sales tax and shipping/handling costs. Active content can also be used to extend the functionality of Hyper Text Markup Language (HTML) and move some data processing from server to client computer. Active content is a security hazard for client computers because it contains programs that run on the client computer and can damage the client computer. Active content can take various forms e.g. Cookies, Java applets, JavaScript, VBScript, ActiveX controls, graphics, Web browser plug-ins, and e-mail attachments. VBScript and JavaScript are scripting languages. These languages provide scripts, or commands, that can be executed.

A **Java Applet** is a program that is written in the Java programming language and can be included in an HTML web page. Applets typically run within the Web browser and can be automatically downloaded and run. Java applets can pose security threat for client computers because when client views a page that includes java applets the applet's code is transferred to client computer and executed by the browser's Java Virtual Machine (JVM). JavaScript, just like other active content, can be used for attacks by executing code that destroys the client's hard disk, discloses the e-mail stored in client mailboxes, or sends sensitive information to the attacker's Web server. Code written in JavaScript is capable of retrieving the URLs of pages visited and information entered into the web forms by the users. For example, a JavaScript can copy credit numbers when a user enters the credit card information to reserve an airplane ticket. Security can be implemented in these active content technologies so that users can elect not to receive such active content, which may reduce the market that the application can reach. Unlike Java applets, a JavaScript program cannot commence execution on its own and requires intervention e.g. clicking of a button to execute the script. Ultimately, the responsibility rests with web developers to make sure their websites are clean and secure. As an end user, however, you should always keep your browsers up to date and regularly scan for malware.

An **ActiveX Control** is an object containing programs and properties that can be placed on web pages to perform particular tasks. ActiveX controls are mostly constructed using C++ and Visual Basic programming languages however, they can be created in any programming language that recognizes Microsoft's Component Object Model. Unlike Java or JavaScript code, ActiveX controls run only on computers with Windows operating systems. Similar to Java Applets, when a client downloads a Web page containing an ActiveX control, the ActiveX control is executed on the client computer. ActiveX controls can pose security hazard for client computers because once downloaded, they execute like any other program on a client computer and have full access to all system resources, including operating system code. A malicious ActiveX control could reformat a user's hard disk, rename or delete files, send e-mails to all the people listed in the user's address book, simply shut down the computer, and cause secrecy, integrity, or necessity violations. Once executed, actions of ActiveX controls cannot be halted. Most Web browsers can be configured to warn the user when downloading an ActiveX control. Some graphics file formats, by design, contain code or instructions on how to render a graphic. Any web page containing such graphic files can be a security threat for client computer depending on the code inside the graphic.

Browser Plug-ins are programs used by web browsers to handle content that browser cannot handle themselves e.g. playing audio clips, displaying movies, or animating graphics. Apple's QuickTime, for example, is a plug-in that downloads and plays movies stored in a special format. Other popular plug-ins includes Adobe Flash Player and Shockwave Player. Plug-ins can also pose security threats to a client computer. Users download these plug-in programs and install them so their browsers can display content that cannot be included in HTML tags. Many plug-ins execute commands buried within the media being manipulated. That means that hackers could embed destructive commands within a seemingly innocuous video or audio clip. Theses commands hidden within the object that the plug-in is interpreting could damage a client computer by erasing some (or all) of its files. There exist some best practices to protect against malicious browser plug-ins. Users should use a good, trusted Anti-Virus. Scan regularly and keep real-time protection enabled. They should install software only if you really need to. If you only use it from time to time, look for an online service that accomplishes the same task. Users should install browser extensions only from trusted sources (official browser repositories) and trusted developers. Popularity is not a reliable sign because it is easy to fake, and developers are constantly approached by adware companies and criminals who pay to bundle their code into the extension. When working with your site, or any site that involves the exchange of sensitive information, use "Incognito" or "Private Browsing" mode with all extensions disabled.

Cookies are small text files that web servers place on client computers to identify returning visitors and to maintain continuing open sessions with clients. An open session is necessary to do a number of things that are important in online business activity. For example, shopping cart and payment processing software both need an open session to work properly. Cookies can be categorized by time duration and by source. Duration cookies can be either session cookies (i.e. cookies that exist until the Web client ends the connection) or persistent cookies (i.e. cookies that remain on the client computer indefinitely). E-commerce sites use both kinds of cookies. For example, a session cookie can be used to keep information about a particular shopping visit and a persistent cookie can be used to store login information. Source cookies can be either First-Party Cookies (i.e. cookies placed by the website server) or Third-Party Cookies (i.e. cookies placed by a different Web site i.e. other than the site being visited). A cookie itself cannot harm the computer, as it does not and cannot hold code (therefore, the cookie cannot perform an action itself). However, the cookie can support malicious actions to be taken on the respective system. Even more, being a plain text file, cookies are vulnerable, meaning that they can be "harvested" by other applications. Many types of attacks can be performed against a web application. Three that specifically target authentication between the browser and the server are man-in-the-middle (MITM), cross-site request forgery (CSRF), and cross-site scripting (XSS). Plain cookie authentication is vulnerable to all three. Cross-Site Request Forgery (CSRF) is a type of attack that occurs when a malicious Web site, email, blog, instant message, or program causes a user's Web browser to perform an unwanted action on a trusted site for which the user is currently authenticated. Cross-site scripting (XSS) is a type of computer security vulnerability typically found in Web applications. XSS enables attackers to inject client-side script into Web pages viewed by other users. A cross-site scripting vulnerability may be used by attackers to bypass access controls such as the same-origin policy. A man-in-the-middle attack is an attack where the attacker secretly relays and possibly alters the communication between two parties who believe they are directly communicating with each other.

Users can disable cookies completely to protect themselves from revealing private information or being tracked by cookies however, this approach can also block useful cookies. Many sites do not offer their full resources unless the client computers accept cookies. Most Web browsers have settings that allow the user to refuse only third-party cookies or to review each cookie before it is accepted. Some browsers, such as Mozilla Firefox and Microsoft Internet Explorer, provide comprehensive cookie management functions. One other solution is to use one of the many third-party programs, called cookie blockers that prevent cookie storage selectively (e.g. by blocking cookies from the Web servers that load advertising banners into Web pages. Some of these programs, such as WebWasher, plug into a browser and allow users to block cookies from the Web servers that load advertising banners into Web pages or blocking cookies based on IP addresses).

WEB BUGS

Web bugs are small bits of code embedded in Websites that add functionality and share information. They are used by everything from Google Analytics, to ad networks, to popular blogging platforms, to social networks, to affiliate shopping programs. When a Web bug is on a site, it communicates back to its parent site to carry out its functionality. Therefore, when you open Website XYZ, you are also sending information to XYZ analytics provider, XYZ ad network and any other sites XYZ website may be partnering with. In addition, your data can be tracked across every site that uses the bug, even if you block cookies and do everything else possible to protect your privacy. The other scary thing is that Web bugs are rarely mentioned in Website privacy policies, and they seem to exist as a loophole in most site policies. A common statement in most site policies is that they will not share data with third parties. However, in the case of a Web bug, the site is not actually sharing data; your browser is sending the data directly to the third party outside of any action by the main site. Even if you block cookies and constantly flush your cache, the XYZ website could, through its Web bugs, identify your IP address, operating system and browser, and track your movements and activities across every site that uses its Web bugs. Outside of not using the Web, constantly changing your IP address or using an anonymous network such as Tor, there is not much you can do. Users need to become aware and vigilant about Web bugs, and make sure that sites explain how they are using this data (Rapoza, 2009).

ANIMATED GIFS

An animated GIF is a graphic image that moves in a loop. This loop is created by combining a batch of consecutive images. Animated GIFs are commonly used by websites like Imgur, Twitter Twitter, Tumblr, Reddit and BuzzFeed. Animated GIFs are also used in news articles, marketing campaigns and art galleries nowadays. Facebook has decided to begin supporting animated GIF images. Facebook will let you post animated GIFs to the News Feed by pasting a link to one in the status update box (Chowdhry, 2015). Java, Flash, and other tools can be used to achieve the same effects as an animated GIF. However, animated GIFs are generally easier to create than comparable images with Java or Flash and usually smaller in size and thus faster to display, that is why they can easily be used at web page for many purpose.

ADWARE

Generically, adware is any software application in which advertising banners are displayed while the program is running. The authors of these applications include additional code that delivers the ads, which can be viewed through pop-up windows or through a bar that appears on a computer screen. The justification for adware is that it helps recover programming development cost and helps to hold down the cost for the user. Adware has been criticized because it usually includes code that tracks a user's personal information and passes it on to third parties, without the user's authorization or knowledge. This practice has been dubbed spyware and has prompted an outcry from computer security and privacy advocates, including the Electronic Privacy Information Center.

Other than displaying advertisements and collecting data, Adware doesn't generally make its presence known. Usually, there will be no signs of the program in your computer's system tray – and no indication in your program

menu that files have been installed on your machine. There are two main ways in which Adware can get onto your   computer.
Adware can be included within some freeware or shareware programs – as a legitimate way of generating advertising revenues that help to fund the development and distribution of the freeware or shareware program. A visit to an infected website can result in unauthorized installation of Adware on your machine.

Often, Adware programs do not have any uninstall procedures and they can use technologies that are similar to those used by viruses to penetrate your computer and run unnoticed. However, because there may be legitimate reasons why Adware is present on your computer, antivirus solutions may not be able to determine whether a specific Adware program poses a threat to you. Kaspersky Lab's products give you the option on whether to detect Adware – and how to react to it: Many freeware and shareware programs stop displaying adverts, as soon as you've registered or purchased the program. However, some programs use built-in, third-party Adware utilities and, in some cases, these utilities can remain installed on your computer after you have registered or purchased the program. For some programs, if you remove the Adware component that may cause the program to malfunction (Kaspersky, 2015).

RISK WARE

Riskware is the name given to legitimate programs that can cause damage if they are exploited by malicious users – in order to delete, block, modify, or copy data, and disrupt the performance of computers or networks. Riskware can include many types of programs (such as dialer programs, file downloaders, and software for monitoring computer activity) that may be commonly used for legitimate purposes. These programs are not designed to be malicious – but they do have functions that can be used for malicious purposes. With so many legitimate programs that malicious users can employ for illicit purposes, it can be difficult for users to decide which programs represent a risk. For example, remote administration programs are often used by systems administrators and helpdesks for diagnosing and resolving problems that arise on a user's computer. However, if such a program has been installed on your computer by a malicious user – without your knowledge – that user will have remote access to your computer. With full control over your machine, the malicious user will be able to use your computer in virtually any way they wish.

PORNWARE

Pornware is the name given to a class of programs that display pornographic material on a device. In addition to the programs that some users may deliberately install on their computers and mobile devices (e.g. to search for and display pornographic material), Pornware also includes programs that have been maliciously installed, without the user having any knowledge of their presence. Often, the purpose of unrequested Pornware is to advertise fee-based pornographic websites and services. Malware developers can exploit unpatched vulnerabilities – within commonly used applications or the operating system – in order to install Pornware on a user's computer, tablet, or smartphone. In addition, Trojans – such as Trojan-Downloader and Trojan-Dropper – can be used to infect a device with Pornware. Examples of Pornware include porn-dialers and porn-downloaders. Because Pornware may have been deliberately downloaded by a user, antivirus solutions may not be able to determine whether a specific Pornware program poses a threat to the user's device (Kaspersky, 2015).

SOCIAL ENGINEERING

Social engineering is a non-technical method of intrusion that hackers use that relies heavily on human interaction and often involves tricking people into breaking normal security procedures. It is one of the greatest threats that organizations today encounter. A social engineer is a person who performs social engineering attacks. Social engineering includes extensive research information (legal and illicit) about an enterprise, which is gathered and used to exploit people. Successful social engineering results in partial or complete circumvention of an enterprise's security systems. The best firewall is useless if the person behind it gives away either the access codes or the information it is installed to protect. Social engineering principally involves the manipulation of people rather than technology to breach security.

FIGURE 10-11: SOCIAL ENGINEERING

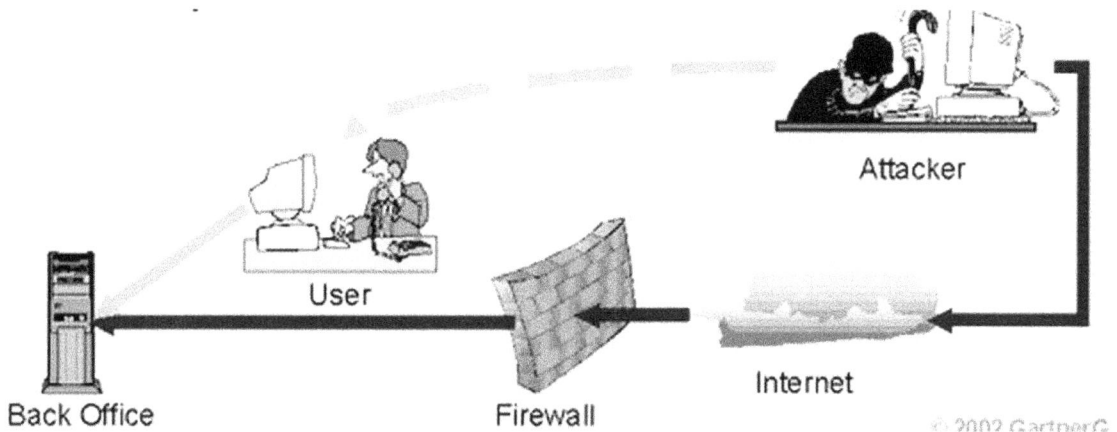

Some of the common social engineering techniques include dumpster diving, shoulder surfing, tailgating/piggybacking, phishing, pretexting, intimidation, and bribery. Dumpster diving involves getting data about a user in order to impersonate that user and gain access to his or her user profiles or other restricted areas of the Internet or a local network. Dumpster diving can mean looking through physical trash for such information, or searching discarded digital data. Shoulder surfing refers to using direct observation techniques, such as looking over someone's shoulder, to get information. It is commonly used to obtain passwords, PINs, security codes, and similar data. Tailgating is actually a form of social engineering, whereby someone who is not authorized to enter a particular area does so by following closely behind someone who is authorized. Phishing is a fraudulent attempt, usually made through email, to steal your personal information. Pretexting is a form of social engineering in which an individual lies to obtain privileged data. A pretext is a false motive. Intimidation means threat of punishment or ridicule for following correct procedures. Bribery is a method of last resort for a social engineer. Bribing a target employee is a mechanism to gain information quickly but it also exposes the social engineer's motives immediately.

Legal and Regulatory Threats

Rapid changes are occurring in the legislative and regulatory regime in response to the development of E-commerce. Organizations need to monitor these developments to adapt their E-commerce strategies appropriately. Legal and regulatory requirement can pose threats to the security aspects of E-commerce. These threats and their impacts are illustrated below.

Issue	Business Impact
Privacy: Many jurisdictions are now implementing laws to protect personal information and prohibit export of personal information to countries that do not have comparable legislation in place. One example of such legislation is 1995 EU directive on data protection. Such legislation has serious implications for E-commerce especially cross-border E-commerce that involves transfer of personal information.	Additional costs involved in complying with data protection legislation, possible reduction in scope of E-commerce
Electronic Documents: Not all countries recognize electronic documents equivalent to paper documents. In many circumstances, paper documents are necessary to provide legal validity. Similarly, digital signatures (that prove the authenticity of electronic transactions) have varying legal acceptability in various jurisdictions.	Lack of legal resources in case of a dispute
Cryptography: Many E-commerce security solutions rely of cryptographic products. However, these products are subject to restrictions on export, import, or use in certain jurisdictions.	The inability of a business to implement desired protection level that can cause the business some unacceptable exposures.

Solutions for E-commerce Security

There exist many security solutions that can be used to address the threats discussed earlier. Many of these threats may hold organizations back from participating in E-commerce. Careful implementation of security solutions can enable businesses to take advantage of the benefits of E-commerce while minimizing the security risks.

Solutions for Service Side Security

To protect the front-end and back-end systems, various methods can be adopted including developing robust applications and support systems, establishing a protected network environment, introducing security as an essential management practices, using web servers with robust software and hardware platforms, and using scalable hardware (in terms of disk space, memory and processing capacity). Software used (e.g. operating systems, web server applications) needs to be well understood and supported by the organization. The network environment needs to be robust enough to avoid service interruptions and protect both web servers and internal systems from intentional attacks. It can be done by estimating likely network traffic volumes as accurately as possible and by limiting the way network connections can be made to the web server and to internal systems. Strong management practices are necessary to maintain secure E-commerce infrastructure. Organization's commitment to strong and robust security practices can also help to build the confidence of customers.

Securing Web Server

Powering over 90% of the World Wide Web, Apache, IIS and nginx are considered the three most important web servers. They are considered easy to get up and running, have an active development team behind them and react quickly to security issues.

SECURING IIS

Move the Inetpub folder to a different drive

The Inetpub folder is the default location for your web content, IIS logs and so on. By default IIS 7 and upwards install the Inetpub folder in the system drive. It is good practice to move the Inetpub folder to a different partition so that the web content is separate from the operating system. This folder can be moved after IIS installation is completed. Thomas Deml, IIS Lead Program Manager provided this batch file to help with the move.

Install the appropriate IIS modules

IIS includes more than 30 modules. You should only install the ones, which are needed by your web applications. Disable any modules that are not required, to minimize the capacity of potential attacks. Periodically review the modules that are installed and enabled and remove any that are no longer required. You can use IIS Manager to list all the modules that are enabled.

Enable Dynamic IP Restrictions

The Dynamic IP Restrictions module helps blocks access to IP addresses that exceed a specified number of requests and thus helps prevent Denial of Service (DoS) attacks. This module will inspect the IP address of each request sent to the web server and will filter these requests in order temporarily deny IP addresses that follow a particular attack pattern. The Dynamic IP Restrictions module can be configured to block IP addresses after a number of concurrent requests or by blocking IP addresses that perform a number of requests over a period. Depending on your IIS version, you will need to enable either the 'IP Security' feature or the "IP and Domain Restrictions".

Enable and Configure Request Filtering Rules

It is also a good idea to restrict the types of HTTP requests that are processed by IIS. Setting up exclusions and rules can prevent potentially harmful requests from passing through to the server, since IIS can block these requests based on the request filtering rules defined. For example, a rule can be set to filter traffic for SQL Injection attempts. Whilst SQL Injection vulnerabilities should be fixed at source, filtering for SQL Injection attacks is a useful mitigation. This can be set from the Rules tab found in the Request Filtering page in IIS Manager.

Enable logging

Configuring IIS logging will cause IIS to log various information from HTTP requests received by the server. This will come in handy and can give a better understanding of issues that might have occurred on your website when things go wrong. It is the place where you will start the troubleshooting process in such situations. The server's logs can also be continuously or periodically monitored in order to review the server's performance and provide optimizations if needed. This can be automated using various server-monitoring tools. Make sure to keep a backup of the logs. Microsoft also provides Log Parser, which is a tool that can be used to query and retrieve specific data from IIS logs. Additionally, log consolidation tools prove useful for consolidating and archiving data from logs in a more meaningful way.

Use the Security Configuration Wizard (SCW) and the Security Compliance Manager (SCM)

Both of these Microsoft tools can be used to test your IIS security. The Security Configuration Wizard (SCW) runs different checks and provides advice and recommendations on how to boost your server's security. The Security Compliance Manager (SCM) tool performs security tests on your server and compares server configurations to predefined templates as per industry best practices and security guide recommendations.

Updates

Finally, ensure that you keep up to date with the latest updates and security patches. It is interesting how often this basic security requirement is missed. The majority of hacks affecting the web server occur on unpatched servers. This just demonstrates how important it is to always keep your IIS web server up to date.

SECURING NGINX SERVER

Disable any unwanted nginx modules

Nginx modules are automatically included during installation of nginx and no run-time selection of modules is currently supported, therefore disabling certain modules would require re-compilation of nginx. It is recommended to disable any modules, which are not required, as this will minimize the risk of any potential attacks by limiting the operations allowed by the web server. To do this, you would need to disable these modules with the configure option during installation.

Disable nginx server_tokens

By default, the server_tokens directive in nginx displays the nginx version number in all automatically generated error pages. This could lead to unnecessary information disclosure where an unauthorized user would be able to gain knowledge about the version of nginx that is being used. The server_tokens directive should be disabled from the nginx configuration file by setting – server_tokens off.

Control Buffer Overflow Attacks

Buffer overflow attacks are made possible by writing data to a buffer and exceeding that buffers' boundary and overwriting memory fragments of a process. To prevent this in nginx we can set buffer size limitations for all clients. This can be done through the Nginx configuration file.

Disable any unwanted HTTP methods

It is suggested to disable any HTTP methods which are not going to be utilized and which are not to be implemented on the web server.

Make use of ModSecurity

ModSecurity is an open-source module that works as a web application firewall. Different functionalities include filtering, server identity masking, and null byte attack prevention. Real-time traffic monitoring is also allowed through this module.

Set up and configure nginx access and error logs

Nginx access and error logs are enabled by default and are located at logs/error.log for error logs and at logs/access.log for access logs. The error_log directive in the nginx configuration file will allow you to set the directory where the error logs will be saved as well as specify which logs will be recorded according to their severity level.

Monitor nginx access and error logs

Continuous monitoring and management of the nginx log files will give a better understanding of requests made to your web server and list any errors that were encountered. This will help to expose any attempted attacks made against the server as well as identify any optimizations that need to be carried out to improve the server's performance. Log management tools, such as logrotate, can be used to rotate and compress old logs in order to free up disk space. In addition, the ngx_http_stub_status_module module provides access to basic status information, and nginx Plus, the commercial version of nginx, provides real-time activity monitoring of traffic, load and other performance metrics.

Updates

As with any other server software, it is recommended that you always update your Nginx server to the latest stable version. These often contain fixes for vulnerabilities identified in previous version.

SECURING APACHE SERVER

Ensure that Apache server-info is disabled

If the <Location /server-info> directive in the httpd.conf configuration file is enabled it would display information about the Apache configuration when the /server-info page is accessed from http://www.example.com/server-info. This could potentially include sensitive information about server settings such as the server version, system paths, database names, library information and so on. Apache server-info lists the server version, which also includes the OpenSSL version. From this information, an attacker could deduce that this server is making use of a version of OpenSSL, which is vulnerable to the Heartbleed Bug and thus could now exploit this vulnerability. This can be disabled by either commenting out the entire mod_info module from the httpd.conf Apache configuration file.

Ensure that Apache server-status is disabled

When enabled, the <Location /server-status> directive lists information about the server's performance, such as server uptime, server load, current HTTP requests, and client IP addresses. An attacker may make use of this information to create an attack against the web server.

Disable the ServerSignature directive

The ServerSignature directive endows server-generated documents with a footer, which includes information about your Apache configuration such as the version of Apache and the OS server name. In order to restrict Apache from displaying this sensitive information the ServerSginature directive in your Apache configuration would need to be disabled.

Set the ServerTokens directive to Prod

The ServerTokens directive controls what information about the server is sent back in the Server response header field. The ServerTokens directive should be set to Prod in order to instruct Apache to return only 'Apache' in the server response headers.

Disable Directory Listing

Directory listing displays a list of the directory contents, which would include all the files from that website. If this is enabled, an attacker can simply discover and view any file. This could potentially lead to the attacker decompiling and reverse engineering an application in order to obtain the application's source code. The attacker can then analyze the source code for possible security flaws or to obtain more information about an application, such as database connection strings, passwords to other systems etc. Directory listing can be disabled by setting the Options directive in the Apache httpd.conf file.

Enable only the modules that are required

A default installation of Apache may include a number of pre-installed and enabled modules, which you might not need. The Apache module documentation lists and explains all the modules available within Apache. Research the modules that you have enabled, and ensure that these are required for the functionality of the website. Unnecessary modules should be disabled.

Make use of an appropriate user and group

By default Apache will run under the daemon user and group, however it is best practice to run Apache in a non-privileged account. Furthermore, if two processes, such as Apache and MySQL for example, are running under the same user and group, issues in one process might lead to exploits in the other process. To change Apache's user and group the User and Group directives in the Apache httpd.conf configuration file need to be changed.

Restrict unwanted services

You may want to disable certain services, such as CGI execution and symbolic links, if these are not needed. You can disable these services with the Options directive from the httpd.conf configuration file and may disable these services for a particular directory only.

Make use of ModSecurity

The mod_security is an open-source module that works as a web application firewall. Different functionalities include filtering, server identity masking, and null byte attack prevention. Real-time traffic monitoring is also allowed through this module.

Updates

You should always keep up to date with the latest versions of Apache, as new updates will contain new fixes and patches that will address past security issues and introduce new security measures. The best way to keep up to date about new versions of Apache is to subscribe to the Apache Server Announcements mailing list.

Enable logging

Apache logging provides detailed information about client requests made on your web server, hence enabling such logging will prove useful when investigating the cause of particular issues. In order to enable logging the mod_log_config module needs to be included from the Apache httpd.conf file.

Securing Database Server

Following are some general best practices to secure your database server.

Separate the Database and Web Servers

Keep the database server separate from the web server. When installing most web software, the database is created for you. To make things easy, this database is created on the same server where the application itself is being installed, the web server. Unfortunately, this makes access to the data all too easy for an attacker to access. If they are able to crack the administrator account for the web server, the data is readily available to them.

Instead, a database should reside on a separate database server located behind a firewall, not in the DMZ with the web server. While this makes for a more complicated setup, the security benefits are well worth the effort.

Encrypt Stored Files

Encrypt stored files. The stored files of a web application often contain information about the databases the software needs to connect to. This information, if stored in plain text like many default installations do, provide the keys an attacker needs to access sensitive data.

Encrypt Your Backups Too

Encrypt back-up files. Not all data theft happens because of an outside attack. Sometimes, it's the people we trust most that are the attackers.

Use a Web Application Firewall (WAF)

Employ web application firewalls. In addition to protecting a site against cross-site scripting vulnerabilities and web site vandalism, a good application firewall can thwart SQL injection attacks as well. By preventing the injection of SQL queries by an attacker, the firewall can help keep sensitive information stored in the database safe.

Keep Patches Current

Keep patches current. This is one area where administrators often come up short. Web sites that are rich with third party applications, widgets, components and various other plug-ins and add-ons can easily find themselves a target to an exploit that should have been patched.

Minimize Use of Third Party Apps

Keep third-party applications to a minimum. We all want our web site to be filled with interactive widgets and sidebars filled with cool content, but any app that pulls from the database is a potential threat. Many of these applications are created by hobbyists or programmers who discontinue support for them.

Do not Use a Shared Server

Avoid using a shared web server if your database holds sensitive information. While it may be easier, and cheaper, to host your site with a hosting provider you are essentially placing the security of your information in the hands of someone else.

Enable Security Controls

Enable security controls on your database. While most databases nowadays will enable security controls by default, it never hurts for you to go through and make sure you check the security controls to see if this was done.

Keep in mind that securing your database means you have to shift your focus from web developer to database administrator. In small businesses, this may mean added responsibilities and additional buy in from management. However, getting everyone on the same page when it comes to security can make a difference between preventing an attack and responding to an attack.

Demilitarized Zone (DMZ)

A DMZ (demilitarized zone) is a physical or logical sub-network that separates an internal local area network (LAN) from other untrusted networks, usually the Internet. External-facing servers, resources and services are located in the DMZ so they are accessible from the Internet but the rest of the internal LAN remains unreachable. This provides an additional layer of security to the LAN as it restricts the ability of hackers to directly access internal servers and data via the Internet. Figure 10-12 shows how a DMZ works. FW represents Firewall.

FIGURE 10-12: WORKING OF A DMZ

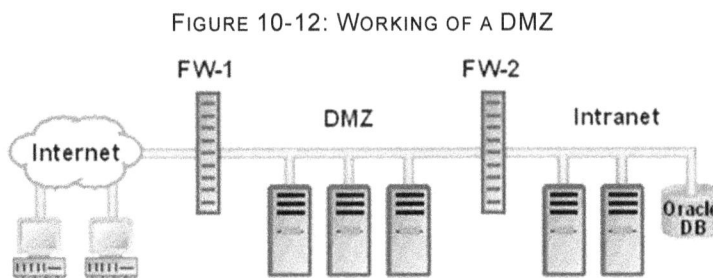

Firewalls

Firewall is a basic tool in network security and its purpose is to protect a local area network (LAN) from intruders outside the network. Each LAN can be connected to the Internet through a gateway, which usually includes a firewall. With increasing reliance of businesses on access to the Internet, an increasing number of security threats are originating outside the firewall. A firewall acts as a safety barrier for data flowing into and out of the LAN. Firewalls can prohibit all data flow not expressly allowed, or can allow all data flow that is not expressly prohibited. The choice between these two models is up to the network security administrator and should be based on the need for security versus the need for functionality.

FIGURE 10-13: BASIC FIREWALL OPERATIONS

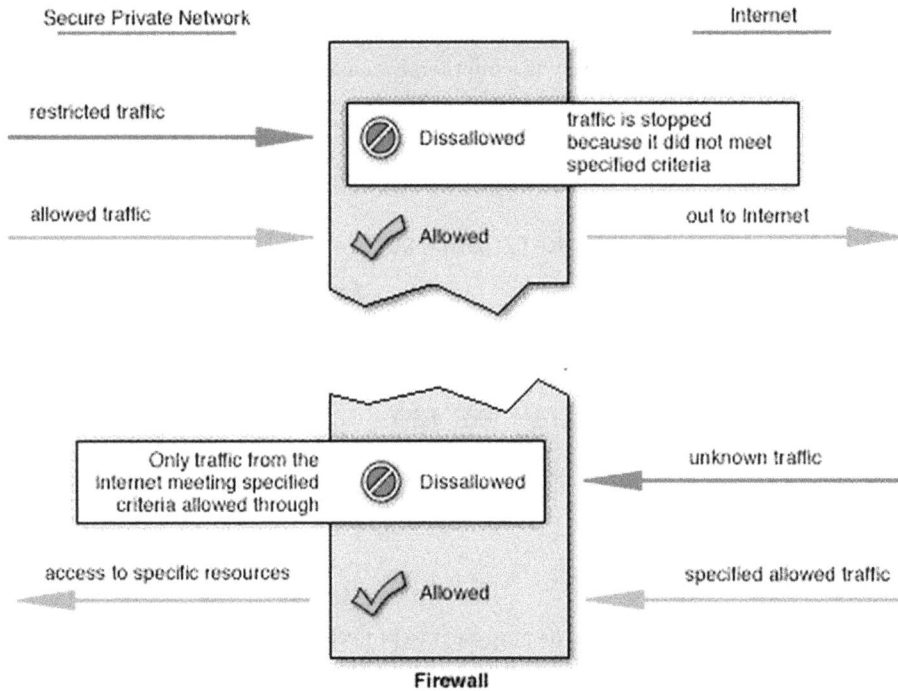

Two main types of firewalls are packet-filtering firewalls and application-level gateways. A packet-filtering firewall (e.g. Cisco Systems PIX firewall) examines all data sent from outside the LAN and automatically rejects any data packets that have local network addresses. For example, a hacker from outside the network may obtain the address of a computer inside the network and tries to slip a harmful data packet through the firewall. A packet-filtering firewall will reject the data packet, since it has an internal address, but originated from outside the network. A problem with packet-filtering firewalls is that they consider only the source of data packets; they do not examine the actual data. As a result, malicious viruses can be installed on an authorized user's computer, giving the hacker access to the network without the authorized user's knowledge. A stateful firewall (any firewall that performs stateful packet inspection (SPI) or stateful inspection) is a firewall that keeps track of the state of network connections (such as TCP streams, UDP communication) traveling across it. The firewall is programmed to distinguish legitimate packets for different types of connections. Only packets matching a known active connection will be allowed by the firewall; others will be rejected. The application-level gateway (e.g. Microsoft ISA Server) screens the actual data. If the message is deemed safe, then the message is sent through to the intended receiver. Using a firewall is probably the single most effective and easiest way to add security to a small network. There are numerous firewall software products available.

There also exist some other types of firewalls. A circuit gateway firewall works at the session layer of the OSI model. This firewall monitors TCP handshaking between packets to determine whether a requested session is legitimate. Information passed to a remote computer through a circuit level gateway appears to have originated from the firewall. Circuit-level firewalls conceal the network itself from the external world. Circuit level gateways are relatively inexpensive and have the advantage of hiding information about the private network they protect.

On the other hand, they do not filter individual packets. A hybrid firewall is a combination of two of the above-mentioned firewalls. The first commercial firewall, the DEC Seal, was a hybrid developed using an application gateway and a filtering packet firewall. This type of firewall is generally implemented by adding packet filtering to an application gateway to quickly enable a new service access to and from the private LAN. Personal firewalls are usually software implementations of an application gateway firewall. Exceptions to this are products such as a router like the Linksys router that contains a packet filtering firewall within it.

Proxy Servers

A proxy server is a computer situated at the access point between a local network and the Internet, or between two different parts of a network. This means that traffic entering and leaving the network must go through the proxy server. Furthermore, the proxy server might handle traffic using only certain communication protocols, such as Web traffic (HTTP) or direct FTP. Home wired or wireless routers often act, or can act, like primitive proxy servers. Figure shows working of a proxy server.

FIGURE 10-14: WORKING OF PROXY SERVER

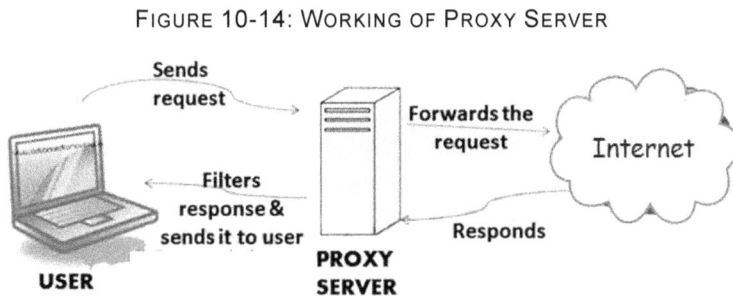

Source: advancecomputing.in

A firewall and a proxy server are both components of network security. To some extent, they are similar in that they limit or block connections to and from your network, but they accomplish this in different ways. Firewalls can block ports and programs that try to gain unauthorized access to your computer, while proxy servers basically hide your internal network from the Internet. It works as a firewall in the sense that it blocks your network from being exposed to the Internet by redirecting Web requests when necessary.

Kerberos

Some sites deploy firewalls to solve their network security problems. It is important to note that while most of the really damaging threats come from inside the network, firewalls do not provide protection to a network from internal security threats. Firewalls also cannot restrict how the network users use the Internet.

Kerberos is a solution to these network security problems. Kerberos provides authentication and strong encryption so that information systems can be secured across enterprise. Kerberos is a freely available, open-source network authentication protocol developed at MIT. MIT also provides Kerberos code to anyone who wishes to use it. Kerberos is also available as commercial product from many different vendors e.g. CyberSafe TrustBroker. The Kerberos uses strong cryptography for identification of client and servers. Client and servers can use Kerberos to prove their identity and to encrypt all of their communications to assure privacy and data integrity.

Biometrics

Biometric systems use unique personal information, such as fingerprints, eyeball iris scans or face scans, to identify a user. This system eliminates the need for passwords, which are much easier to steal. In addition, managing passwords for accessing a variety of resources (e.g. your online course website, online banking, your office network etc.) is a big hassle. In recent years, the cost of biometric devices has dropped significantly. Instead of using passwords, keyboard mounted fingerprint scanning devices are being used to log into systems or access secure information over a network. Each user's iris scan, face scan or fingerprint is stored in a secure database. Each time a user logs in; his or her scan is compared with the database. Upon a successful match user is allowed to login. Two companies that specialize in biometric devices are iridian technologies (www.iriscan.com) and PreciseBiometrics (www. precisebiometrics.com).

Two-factor authentication uses two means to authenticate the user, such as biometrics or a smart card used in combination with a password. Using two methods of authentication is more secure than just using passwords alone. There is a gradual shift towards use of smart cards and Biometrics instead of using passwords. The BioAPI Consortium is a group of over 120 companies and organizations with goal of promoting the growth of the biometrics market. The consortium develops the BioAPI Specification, which is intended to provide a high-level generic biometric authentication model suited for any form of biometric technology. Privacy is one of the major concerns associated with the use of biometric devices. Implementing fingerprint scanners means that organizations will be keeping databases with each employee's fingerprint.

Intrusion Detection Systems (IDS)

It is estimated that 70 to 90 percent of attacks on corporate networks are perpetrated internally (ExpressComputer, 2003). Intrusion Detection System (IDS) is a special type of software that can monitor activity across a network or on a host computer, watch for suspicious activity, and take automated action based on what it sees. IDSs can be either host-based or network-based. In both cases, IDS look for specific patterns that usually indicate malicious or suspicious intent (attack signatures). Each approach has its strengths and weaknesses and each is complementary to the other. A truly effective intrusion detection system employs both host-based and network-based approach.

Network-based IDS use network packets as the source of data. These IDS and typically utilizes a network adapter running in promiscuous mode to monitor and analyze all traffic in real-time and checks for attack signatures in network traffic. Network-based IDS usually consists of a monitor and software agents. The monitor is a software package that resides on a server and scans the network. Software agents reside on various host computers and feed information back to the monitor. Snort (snort.org) is an open-source network-based IDS.

A host-based IDS resides on the host system that is being monitored. Host-based IDS involves not only looks at the network traffic in and out of the host system, but also checks the integrity of host system's files and watch for suspicious processes. Price of commercial host-based IDS can range from $100 to thousands of dollars.

OSSEC is an Open Source Host-based Intrusion Detection System that runs on most operating systems, including Linux and Windows.

FIGURE 10-15: WORKING OF IDS

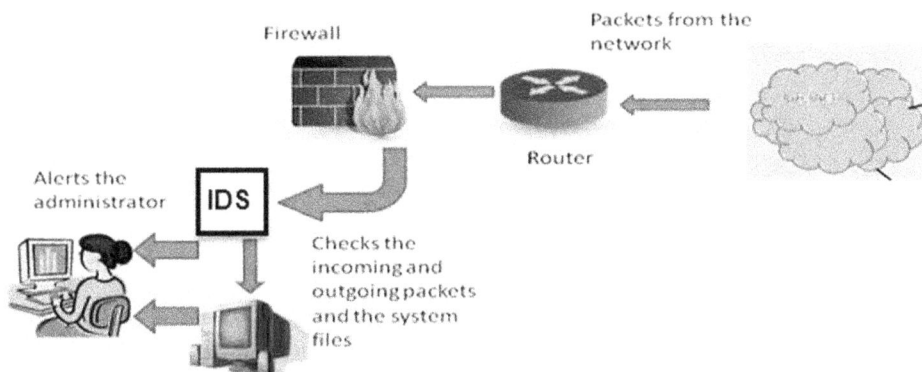

Intrusion Prevention System (IPS)

An Intrusion Prevention System (IPS) is a network security/threat prevention technology that examines network traffic flows to detect and prevent vulnerability exploits. The IPS often sits directly behind the firewall and is provides a complementary layer of analysis. Intrusion Detection System (IDS) is a passive system that scans traffic and reports on threats. On the other hand, the IPS is placed inline (in the direct communication path between source and destination), actively analyzing and taking automated actions on all traffic flows that enter the network. Specifically, these actions include sending an alarm to the administrator (as would be seen in an IDS); dropping the malicious packets; blocking traffic from the source address; and resetting the connection. As an inline security component, the IPS must work efficiently to avoid degrading network performance. It must also work fast because exploits can happen in near real-time. The IPS must also detect and respond accurately, to eliminate threats and false positives (legitimate packets misread as threats).

FIGURE 10-16: WORKING OF IPS

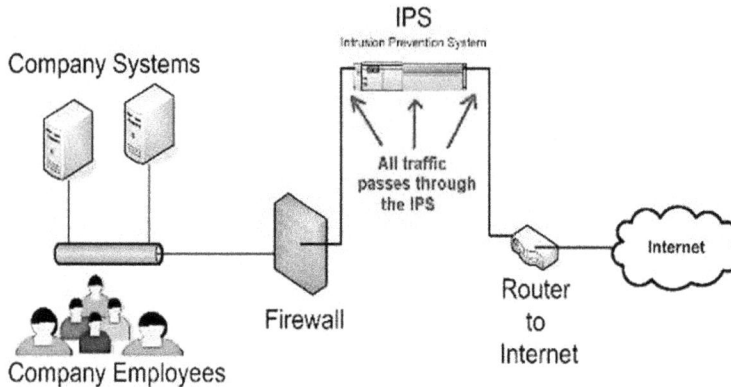

The IPS has a number of detection methods for finding exploits, but signature-based detection and statistical anomaly-based detection are the two dominant mechanisms. Signature-based detection is based on a dictionary of uniquely identifiable patterns (or signatures) in the code of each exploit. As an exploit is discovered, its signature is recorded and stored in a continuously growing dictionary of signatures. Statistical anomaly detection takes samples of network traffic at random and compares them to a pre-calculated baseline performance level. When the sample of network traffic activity is outside the parameters of baseline performance, the IPS takes action to handle the situation.

Solutions for E-commerce Transaction Security

E-commerce transactions take place over Internet, which is an open, un-trusted network largely outside the control of the parties involved in E-commerce transaction. The principle technique to respond to the threats to E-commerce transactions is the use of cryptography.

Cryptography

Cryptography enables you to hide the content of electronic transactions, detect any changes to electronic transactions, and confirm the source of electronic transactions. These capabilities are achieved through a combination of encryption and digital signatures. Documented use of cryptography in writing dates back to circa 1900 B.C. when an Egyptian describe used non-standard hieroglyphs in an inscription. New forms of cryptography came soon after the widespread development of computer communications. In data and telecommunications, cryptography is necessary when communicating over any untrusted medium, which includes just about *any* network, particularly the Internet.

Within the context of any application-to-application communication, there are some specific security requirements, including:

- *Authentication:* The process of proving one's identity. (The primary forms of host-to-host authentication on the Internet today are name-based or address-based, both of which are notoriously weak.)

- *Privacy/confidentiality:* Ensuring that no one can read the message except the intended receiver.

- *Integrity:* Assuring the receiver that the received message has not been altered in any way from the original.

- *Non-repudiation:* A mechanism to prove that the sender really sent this message.

Cryptography, then, not only protects data from theft or alteration, but can also be used for user authentication. There are, in general, three types of cryptographic schemes typically used to accomplish these goals: secret key (or symmetric) cryptography, public-key (or asymmetric) cryptography, and hash functions, each of which is described below. In all cases, the initial unencrypted data is referred to as *plain text*. It is encrypted into *cipher text*, which will in turn (usually) be decrypted into usable plaintext.

Types of Cryptographic Algorithms

There are several ways of classifying cryptographic algorithms. For purposes of this paper, they will be categorized based on the number of keys that are employed for encryption and decryption, and further defined by their application and use. The three types of algorithms that will be discussed are :

- **Secret Key Cryptography** (SKC): Uses a single key for both encryption and decryption (private or symmetric encryption).

- **Public Key Cryptography** (PKC): Uses one key for encryption and another for decryption (asymmetric encryption).

- **Hash Functions:** Uses a mathematical transformation to irreversibly "encrypt" information. (Kessler, 2011)

PRIVATE KEY CRYPTOGRAPHY

In private key encryption, both the merchant and consumer share a private key that is used to encrypt and decrypt data. Private key systems are simpler and faster. The main drawback is the distribution and management of the keys. Imagine having thousands of customers who require their own key. You would need to devise a method that ensures each person receives a key and that the key is managed appropriately. Hence, private key systems are best for small networks where the parties know each other and can trust each other with the keys.

FIGURE 10-17: PRIVATE KEY CRYPTOGRAPHY

HASH FUNCTION CRYPTOGRAPHY

This technique does not involve any key. Rather it uses a fixed length hash value that is computed on the basis of the plain text message. Hash functions are used to check the integrity of the message to ensure that the message has not be altered, compromised or affected by virus. A hash function is any function that can be used to map digital data of arbitrary size to digital data of fixed size. The values returned by a hash function are called hash values, hash codes, hash sums, or simply hashes.

FIGURE 10-18: HASH FUNCTION CRYPTOGRAPHY

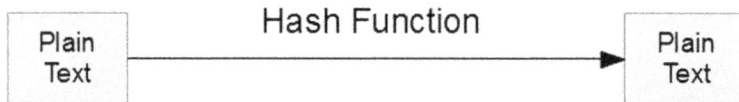

PUBLIC KEY CRYPTOGRAPHY:

Public key encryption uses two keys - a public key that encrypts the message and a private key that decrypts the message. Both the consumer and merchant would have their own pair. The public key is stored in a key repository with a certification authority (trusted third party) and is publicly available, while the private key is retained by the user. For instance, a customer uses his or her credit card to make an online purchase. The merchant's public key is used to encrypt the customer's credit card information. When the merchant receives the encrypted data it is decrypted with the merchant's private key.

FIGURE 10-19: PUBLIC KEY CRYPTOGRAPHY

In Public Key Encryption method, there exist two algorithms, one called the encryption algorithm (E), and the other called decryption algorithm (D). The important characteristics of these two algorithms are:

1. If the decryption algorithm is applied on the encrypted message, the plaintext is obtained.

2. It is difficult to determine the decryption algorithm from the encryption algorithm.

3. The encryption algorithm is difficult to be broken.

The encryption key is made public and the decryption key is kept as a secret. Using this technique, what is being done between the sender and the receiver is as follows:

1. The sender and the receiver make the encryption algorithm public.

2. The sender encrypts the message using the public key and transmits the ciphertext so obtained to the receiver.

3. The receiver, upon receiving the ciphertext, applies the secret decryption key and obtains the message.

Figure 10-20 shows this process.

FIGURE 10-20: PUBLIC KEY CRYPTOGRAPHY

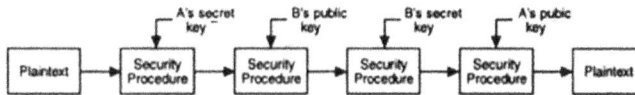

Data Encryption Standard (DES) is one of the most commonly used encryption algorithms. The time taken for breaking a DES encryption depends upon the length of the keys. Table 1-1 shows the theoretical time needed to break the code.

TABLE 10-1: TIME TAKEN TO BREAK DES ENCRYPTION

Key length	Time Taken
40-bit DES	approx. 0.4 second max
56-bit DES	7 hours max
64-bit DES	74 hours and 40 minutes
128-bit DES	157,129.203,952,300.000 years

Apart from DES, there is yet another encryption algorithm called RC-4. Table 10-2 gives the time taken for breaking this code.

TABLE 10-2: TIME TAKEN TO BREAK RC-4 ENCRYPTION

Key length	Time taken
40-bit RC-4	15 days max
56-bit RC4	2.691.-11 years max
64-bit RG4	689,021.57 years max
128-bit RC-4	12,710,204,652,610,000,000,000,000 years max

Other well-known encryption mechanisms are Differential Workfactor Cryptography from Lotus, a combination of DES and Diffie-Hellman key exchange mechanism by Sun Microsystems, Phil Zimmerman's Pretty Good Privacy (PGP), International Cryptography Framework from Hewlett Packard, and RSA Data Security's RSA.

The main advantages of a public key system are that it supports digital certificates and digital signatures, and it provides all security elements required for an E-commerce transaction. The main disadvantages are that it uses more computer resources than private key cryptography, which means it is slower, and it is more costly to implement. They are significantly slower than private-key systems. This extra time can add up quickly as individuals and organizations conduct commerce on the Internet. Public key systems do not replace private-key systems, but serve as a complement to them. Public key systems are used to transmit private keys to Internet participants so that additional, more efficient communication can occur in a secure Internet session.

Encryption using Symmetric and Asymmetric Algorithms

Encryption is the coding of information by using a mathematically based program and a secret key to produce a string of characters that is unintelligible to all others except sender and receiver of the message. Original message or information can be recovered through a corresponding decryption process. Encryption plays a crucial role in maintaining the confidentiality of electronic transactions. In cryptography, plain text or clear text refers to any message that is not encrypted. Data that has been encrypted is called Cipher text. Cipher text cannot be read unless converted back to plain text (decryption) using a decryption key. Figure 10-21 illustrates the encryption process.

FIGURE 10-21: ENCRYPTION PROCESS

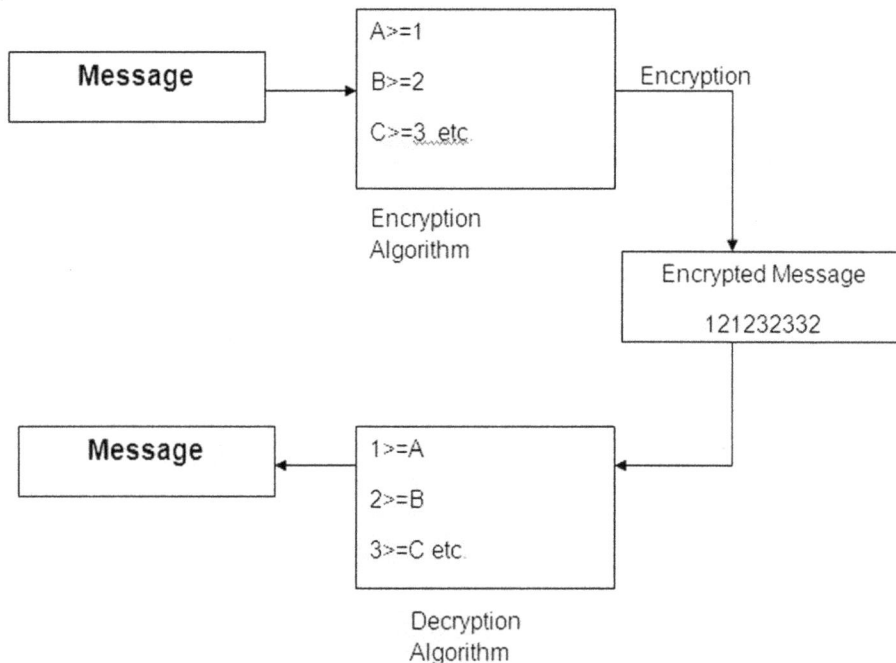

Public-key cryptography is another cryptographic approach which uses asymmetric key algorithms, where the key used to encrypt a message is not the same as the key used to decrypt it. Each user has a pair of cryptographic keys (i.e. a public key and a private key). The private key is kept secret, while the public key may be widely distributed. Messages are encrypted with the recipient's public key and can *only* be decrypted with the corresponding private key. The keys are related mathematically, but the private key cannot be derived from the public key. Since every user has a unique private key therefore public key cryptography is useful in digitally signing a document.

Cryptography is the branch of science that studies encryption. To an unauthorized reader the encrypted message appears as a string of random text characters, numbers, and punctuation.

In cryptography, a key is a sequence of bits used by encryption/decryption algorithms. For example, the following represents a hypothetical 40-bit key:

00001010 01101001 10011110 00011100 01010101

An encryption algorithm uses the bits of this key to alter the original message and creates an encrypted message. Similarly, a decryption algorithm takes an encrypted message and restores it to its original form using one or more keys. Some cryptographic algorithms use a single key for both encryption and decryption. Other algorithms use one key for encryption and a second, different key for decryption. In this case the encryption key can remain public, but decryption requires a private key. In cryptographic algorithm, the key length is the number of bits in the key. Mathematically speaking, 2^n possible values exist for an n-bit key. In order to calculate how long it takes to try all the keys and crack the encryption system we use the following equation. Quite generally, we expect the time needed to guess a key to be proportional to $2^{number\ of\ bits}$, i.e., $2^{key\ length}$. If the key attempt rate is N per second, then the time T needed to crack the key is

$$T = \frac{2^{key\ length}}{2N}\ \text{seconds}$$

Imagine a 56 bit encryption system. If we have a computer that can guess 2^{47} keys per second, then we need about 5 minutes to crack the key. Now consider a 128 bit encryption system. Suppose we have a computer five times faster than the previous one. Using the same formula, we need about 74 billion years to crack it. Point to note is that if computers get faster, we need longer key length to ensure adequate security. However, we cannot just use a very long key just to prevent cracking because long keys are not efficient and the encryption and decryption times can be very long. Thus, we need to set the key long enough to prevent possible cracking, but short enough to allow fast encryption and decryption.

Cryptanalysis

Cryptanalysis refers to the study of cryptosystems to find their weaknesses that could permit the breaking of cipher (i.e. retrieval of the plaintext from the ciphertext without knowing the key or the algorithm). In the most common form of cryptanalytic attacks, the encryption algorithm is analyzed to find relations between bits of the encryption key and bits of the ciphertext. The goal is to determine the key from the ciphertext. Proper key management and expiration dates on keys help prevent cryptanalytic attacks.

Cyrptovirology

Cryptovirology is a field that studies how to use cryptography to design powerful malicious_software to attack the security of a system or security of a user's information. i.e. the art of turning the very methods designed to protect your data into a means of subverting it. The field also encompasses covert attacks in which the attacker secretly steals private information such as private keys. An example of the latter type of attack is asymmetric backdoors. An asymmetric backdoor is a backdoor (e.g., in a cryptosystem) that can be used only by the attacker. Cryptovirology possess a threat on security because it can be used to extort money by file hacking. In addition to this, it attacks and steals private information such as private keys. Private keys are stolen with the use of asymmetric backdoors. The attacker, and no one else, uses an asymmetric backdoor. This asymmetric backdoor differs from traditional symmetric backdoors because anyone that finds it can use it. Kleptography, a subfield of cryptovirology, deals mainly with the study of asymmetric back doors in key generation algorithms, digital signature algorithms, key exchanges, and so on. Attackers use cryptoviruses in deniable password snatching used with cryptocounters, private information retrieval, and to secure communication between different instances of a distributed cryptovirus.

Security using SSL/TLS

Current Internet technology has a built-in mechanism, known as Secure Sockets Layer (SSL) that can be used to encrypt messages sent between web browsers and web servers. SSL is widely used to protect sensitive information during transmission across the Internet. Besides HTTP, SSL can be used to secure many different types of communication between computers.

This mechanism is widely used by on-line merchants to ensure that credit card numbers and other sensitive information sent by customers are protected during transmission across the Internet. In addition to HTTP, SSL can secure many different types of communication between computers. For example, SSL can be used for secure downloading and uploading of sensitive electronic data through FTP. Such FTP protocol is called FTP over TLS/SSL or FTPS (Secure FTP). Using this protocol, password and information transfer is encrypted. FTPS runs over TCP port 21 or 990. Using Telnet, SSL can be used to securely login to remote computers. HTTPS is the protocol that implements SSL. URL preceded by protocol name HTTPS indicate that the communication with remote server will take place through a secure connection. Encryption can also be used to create a virtual private network (VPN). This is effectively an encrypted channel across the Internet between two organizations or two parts of the same organization. VPNs are becoming increasingly used for business-to-business E-commerce.

FIGURE 10-22: WORKING OF FTP OVER TLS/SSL

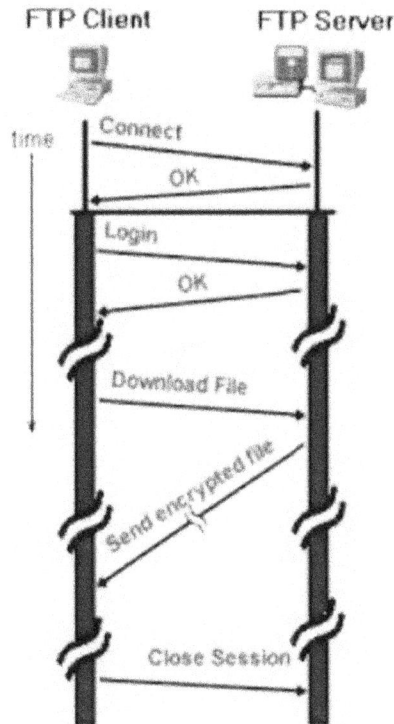

Digital Signatures and Digital Certificates

With rapidly changing business environments and the shift from the traditional face-to-face business models, mechanisms need to be developed to maintain and flourish trusted relationships with customers. Merchants are typically not willing to ship goods or perform services until a payment has been accepted for them. In addition, authentication can allow for a measure of non-repudiation so the customer cannot deny the transaction occurred. Similarly, consumers need assurance that they are purchasing from a legitimate enterprise, rather than a hacker's site whose sole purpose is to collect credit card numbers.

Digital signatures and digital certificates play a major role in securing E-commerce. They can prevent forgery by proving the authenticity of an electronic transaction, prevent imitation on-line by confirming the identity of an individual, and prevent repudiation by providing proof of transmission and receipt of transactions.

A digital signature is an electronic signature that can be used with either an encrypted or plain text message. Digital signature provides two basic capabilities. They allow the source of an electronic message to be confirmed and also permit any changes to the message to be detected. Using digital signatures, receiver of the message can be assured of the sender's identity and integrity of the message. Current Internet technology has built-in digital signature capability through SSL. This built-in capability can help users to prove their identity in a much more reliable way than user-name/password mechanisms, and confirm the identity of the web server with which they are communicating.

WORKING OF DIGITAL SIGNATURES

Suppose a person A wants to send a message to person B with his digital signature. The plain text message will run through a hash function and so given a value called the message digest. This is how the process will work.

1.   "A" writes the message.

2.   The software creates a digest from the message.

3.   The software encrypts the digest with A's private key.

4.   The message together with the digest is sent to "B".

When "B" receives the message, his software will do the following to verify the signature of "A":

1.   The software creates hash of the message received.

2.   The software uses A's public key to decrypt the message hash or summary.

3.   The decrypted hash is compared with the generated one.

4.   If the hashes match, the received message is valid and sent from 'A'.

Note that the hash generation must be a one-way process. It cannot be reversed and we cannot generate the same hash with a different message. Figure 10-23 illustrates working of digital signatures.

FIGURE 10-23: WORKING OF DIGITAL SIGNATURES

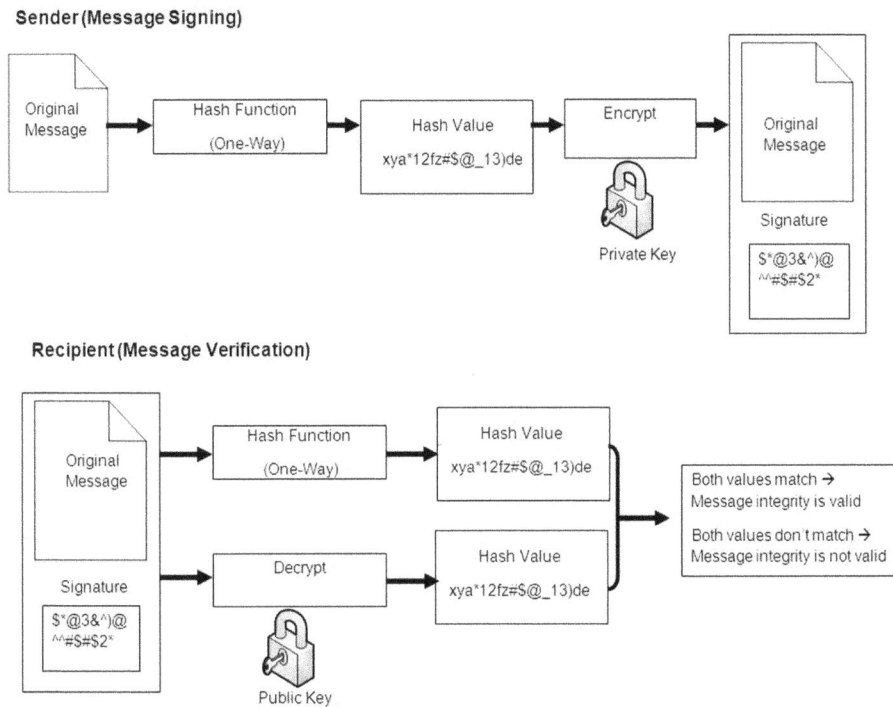

Increasingly, encrypted digital signatures are used in e-commerce and in regulatory filings as digital signatures are more secure than a simple generic electronic signature. The concept itself is not new, with common law jurisdictions having recognized telegraph signatures as far back as the mid-19th century and faxed signatures since the 1980s.

In many countries, including the United States, the European Union and Australia, electronic signatures (when recognised under the law of each jurisdiction) have the same legal consequences as the more traditional forms of executing of documents.

A digital certificate is essentially a bit of information provided by an independent source known as a Certificate Authority (CA). Digital certificate tells the customer that the E-commerce website they are dealing with a valid, secure site and not an impersonation. The digital signature of the certificate issuing authority is attached with the digital certificate so that the genuineness of the certificate can be established.. A digital certificate also includes the name of the subject (the company or individual being certified), the subject's public key, a serial number, an expiration date, the signature of the trusted certification authority and any other relevant information.

A Certificate Authority confirms that each computer is in fact who they say they are and then provides the public keys of each computer to the other. A CA is a financial institution or other trusted third party, such as VeriSign. The CA takes responsibility for authentication, so it must carefully check information before issuing a digital certificate. Digital certificates are publicly available and are held by the certification authority in certificate repositories. VeriSign, Inc. is a leading certificate authority. For more information about VeriSign, visit www.verisign.com.Thawte (thawte.com) is another leading digital certificate vendor.

Public Key Infrastructure (PKI)

Public Key Infrastructure (PKI) is a complex set of hardware, software, people, policies, and procedures used to create, manage, distribute, use, store, and revoke digital certificates in practice. PKI infrastructure provides the following functions:

1. Allows standards to be established so that digital certificates can be created which are valid across different business units, organizations and countries

2. Creates digital certificates and associated cryptographic

3. Store and manage digital certificates securely

4. Renews expired digital certificates

5. Revoke digital certificates e.g. if they have been used fraudulently

The procedure works like this. The merchant server sends a message to the CA, which replies with a digital certificate. The server then presents the certificate to the consumer. The consumer presents this certificate to CA. CA verifies the validity of the certificate and informs the consumer. The consumer is responsible for going through the step of verifying a certificate with a Certificate Authority to figure out if the certificate presented to him is valid. Ultimately, the job of validating the certificate mostly lies with the web browser used by consumers. Figure 10-24 shows the working of PKI.

FIGURE 10-24: WORKING OF PKI

PKI is becoming an important technology for E-commerce because of the role of PKI in strengthening trust in E-commerce. The products and services needed to build or use a PKI are developing rapidly. Identifying best

practice among successful organizations is a good way to start with PKI implementation. Periodically changing key pairs is helpful in maintaining a secure system in case your private key is compromised without your knowledge. The longer you use a given key pair, the more vulnerable the keys are to attack. As a result, digital certificates are created with an expiration date, to force users to switch key pairs. If your private key is compromised before its expiration date, you can cancel your digital certificate and get a new key pair and digital certificate. Canceled and revoked certificates are placed on a certificate revocation list (CRL). CRLs are stored with the certification authority that issued the certificates. Figure 10-25 shows an example of a digital certificate.

FIGURE 10-25: A DIGITAL CERTIFICATE

Source http://www.mysecurecyberspace.com/encyclopedia/index/digital-certificates.html

Steganography

Steganography is the practice of hiding information within other information. Steganography allows you to take a piece of information, such as a message or image, and hide it within another image, message or even an audio clip. Steganography takes advantage of insignificant space in digital files, in images or on removable disks.

Consider a simple example: If you have a message that you want to send secretly, you can hide the information within another message, so that no one but the intended receiver can read it.

One very popular application of steganography is digital watermarks. Using digital watermarking technologies, users can embed a digital code into electronic files (audio, video, still images and printed documents) that is invisible during normal use but readable by computers and software. Digital watermark is usually a company logo, copyright notification or other mark or message that indicates the owner of the document. The major purpose of digital watermarks is to protect intellectual property, which is in digital form (e.g. an audio file or a photograph). The hidden watermark can be shown in a court of law to prove that the watermarked item was stolen. Digital watermarking can significantly impact E-commerce. Music industry, for example, can significantly benefit from digital watermark. Using digital watermarks, music publishers can make indistinguishable changes to a part of a song at a frequency that is not audible to humans, to show that the song was, in fact, copied. Blue Spike's Giovanni® Digital Watermarking Suite uses cryptographic keys to generate and embed steganographic digital watermarks into digital music and images. The watermarks are placed randomly and undetectable by persons not aware of the watermarks embedding scheme, and thus the watermarks cannot be identified and removed. Digital watermarking capabilities are built into some image-editing software applications, such as Adobe PhotoShop (www.adobe.com). Two companies that offer digital watermarking solutions include Digimarc (www.digimarc.com/) and MSI Copy Control (www.msicopycontrol.com). The Digital Watermarking Alliance (DWA) (www.digitalwatermarkingalliance.org) is an international alliance of industry leading organizations that deliver valuable digital watermarking solutions to a broad range of customers and markets around the world.

Solutions for Client-Side

E-commerce businesses lack direct control over the client's computer. It can make it difficult for E-commerce businesses to implement security measures. Thus, compensating mechanisms may be needed in the E-commerce architecture. However, there are some measures that can be adopted, particularly in situations where a degree of control exists over the client environment or where there is user cooperation.

End-user Agreements and Education

Even after successful implementation of technical mechanisms on client-side, security will always be dependent on the correct behavior of users, for example in protecting passwords, PINs and smartcards from misuse. E-commerce businesses should aim to establish end-user agreements through on-screen terms and conditions. Comprehensive and practical advice should also be provided to raise user awareness of security issues and education

Hardware Devices

Standard web browsers include support handling the cryptographic keys and certificates needed to employ digital signatures. However, weak PC security can reduce the reliance on these mechanisms. One potential solution to this issue is to use smartcard. A smart card can provide an ideal solution for both storage (e.g. storage of cryptographic keys) and computer processing on a relatively tamper-resistant hardware platform. Smartcards are already being used successfully in E-commerce applications. The widespread use of smart cards requires well-developed standards and compatible smart card reader devices. Progress is being made on both fronts. The Smart

Card Alliance, is a non-profit association, working to stimulate widespread use of smart cards. Through initiatives by Microsoft, leading PC manufacturers and the Smart Card Alliance, it is hoped that the smartcard may become one of the key technologies for conducting secure E-commerce.

Customized Client Software

Another possible option is the use of customized client software. For example, the standard web browser on client PC can be replaced with a version customized for a specific application. Additional security software can be introduced that will work in conjunction with a conventional browser.

Anti-Virus Software

A virus is software that is capable to attach itself to another program and cause damage when the host program is launched. A worm is a type of virus, which is capable of replicating itself quickly on the computers that it infects. It can spread very quickly through Internet. In order to replicate itself, a virus must be permitted to execute code and write to memory. For this reason, many viruses attach themselves to executable files that may be part of legitimate programs (see code injection). If a user attempts to launch an infected program, the virus' code may be executed simultaneously. Viruses can be divided into two types based on their behavior when they are executed. Nonresident viruses immediately search for other hosts that can be infected, infect those targets, and finally transfer control to the application program they infected. Resident viruses do not search for hosts when they are started. Instead, a resident virus loads itself into memory on execution and transfers control to the host program. The virus stays active in the background and infects new hosts when those files are accessed by other programs or the operating system itself.

Anti-virus software can employ a variety of techniques to detect viruses and malware. Signature based detection is the most common method that antivirus software uses to identify malware. This method is somewhat limited by the fact that it can only identify a limited amount of emerging threats, e.g. generic, or extremely broad, signatures. When antivirus software scans a file for viruses, it checks the contents of a file against a dictionary of virus signatures. A virus signature is the viral code. Finding a virus signature in a file is the same as saying you found the virus itself. If a virus signature is found in a file, the antivirus software can take action to remove the virus. Heuristics-based detection aims at generically detecting new malware by statically examining files for suspicious characteristics without an exact signature match. For instance, an antivirus tool might look for the presence of rare instructions or junk code in the examined file. The tool might also emulate running the file to see what it would do if executed, attempting to do this without noticeably slowing down the system. A single suspicious attribute might not be enough to flag the file as malicious. However, several such characteristics might exceed the expected risk threshold, leading the tool to classify the file as malware. The biggest downside of heuristics is it can inadvertently flag legitimate files as malicious. Behavioral detection observes how the program executes, rather than merely emulating its execution. This approach attempts to identify malware by looking for suspicious behaviors, such as unpacking of malware code, modifying the hosts file or observing keystrokes. Noticing such actions allows an antivirus tool to detect the presence of previously unseen malware on the protected system. As with heuristics, each of these actions by itself might not be sufficient to classify the program as malware. However, taken together, they could be indicative of a malicious program. The use of behavioral techniques brings antivirus tools closer to the category of host intrusion prevention systems (HIPS), which have traditionally existed as a separate product category. Cloud-based detection identifies malware by collecting data from protected computers while analyzing it on the provider's infrastructure, instead of performing the analysis locally. This is usually done by capturing the relevant details about the file and the context of its execution on the endpoint, and providing them to the cloud engine for processing. The local antivirus agent only needs to perform minimal processing. Moreover, the vendor's cloud engine can derive patterns related to malware characteristics and behavior by correlating data from multiple systems. In contrast, other antivirus components base decisions mostly on locally observed attributes and behaviors. A cloud-based engine allows individual users of the antivirus tool to benefit from the experiences of other members of the community (Zeltser, 2011). An emerging technique to deal with malware in general is whitelisting. Rather than looking for only known bad software, this technique prevents execution of all computer code except that which has been previously identified as trustworthy by the system administrator. By following this "default deny" approach, the limitations inherent in keeping virus signatures up to date are avoided. Additionally, computer applications that are unwanted by the system administrator are prevented from executing since they are not on the whitelist. Since modern enterprise organizations have large quantities of trusted applications, the limitations of adopting this technique rests with the system administrators' ability to properly inventory and maintain the whitelist of trusted applications. Viable implementations of this technique include tools for automating the inventory and whitelist maintenance processes. The Figure 10-26 shows how Anti-Virus software have evolved over time.

FIGURE 10-26: EVOLUTION OF ANTI-VIRUS SOFTWARE

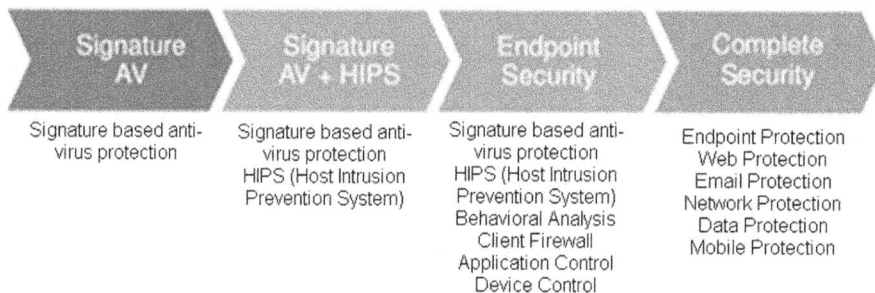

Signature AV	Signature AV + HIPS	Endpoint Security	Complete Security
Signature based anti-virus protection	Signature based anti-virus protection HIPS (Host Intrusion Prevention System)	Signature based anti-virus protection HIPS (Host Intrusion Prevention System) Behavioral Analysis Client Firewall Application Control Device Control	Endpoint Protection Web Protection Email Protection Network Protection Data Protection Mobile Protection

In recent years, the increased severity of the viruses has increased the time and money required for recovery. Anti-virus software can provide protection against viruses and worms. It can detect viruses and worms, prevent access to infected files and quarantine any infected files. A whole industry has grown up around computer viruses. Companies such as Network Associates (owner of McAfee products) and Symantec (Owner of Norton products) exist for the sole purpose of fighting viruses. The antivirus industry is extensive and profitable. Today, it has expanded beyond viruses and now also follows and catalogs worms, macro viruses and macro worms, and Trojan horses.

A macro virus or macro worm is usually executed when the application object (e.g., spreadsheet, word processing document, e-mail message) containing the macro is opened or a particular procedure is executed (e.g., a file is saved). Melissa and ILOVEYOU were both examples of macro worms that were propagated through Microsoft Outlook e-mail.

Anti-Spyware Software

Spyware is a type of software that can be installed by some websites when you visit them. A spyware can be used by its perpetrator to secretly transfer information (e.g. computer usage data) from the user computer. A spyware can also slow down or crash the system. Anti-spyware programs can combat spyware in two ways. They can provide real-time protection in a manner similar to that of anti-virus protection: they scan all incoming network data for spyware and block any threats it detects. Anti-spyware software programs can be used solely for detection and removal of spyware software that has already been installed into the computer. This kind of anti-spyware can often be set to scan on a regular schedule. Such programs inspect the contents of the Windows registry, operating system files, and installed programs, and remove files and entries, which match a list of known spyware. Real-time protection from spyware works identically to real-time anti-virus protection: the software scans disk files at download time, and blocks the activity of components known to represent spyware. In some cases, it may also intercept attempts to install start-up items or to modify browser settings. Earlier versions of anti-spyware programs focused chiefly on detection and removal. Javacool Software's SpywareBlaster, one of the first to offer real-time protection, blocked the installation of ActiveX-based spyware. Like most anti-virus software, many anti-spyware/adware tools require a frequently updated database of threats. As new spyware programs are released, anti-spyware developers discover and evaluate them, adding to the list of known spyware, which allow the software to detect and remove new spyware. As a result, anti-spyware software is of limited usefulness without regular updates. Updates may be installed automatically or manually.

Malware

Malware (or malicious software) is software designed to gain access to a computer system without the owner's knowledge. The term malware is general in nature and can include a variety of unfriendly or interfering, software or code. There are various types of malware including spyware, keyloggers, true viruses, worms, or any type of malicious code that infiltrates a computer. Generally, software is considered malware based on the intent of the creator rather than its actual features.

A Trojan horse (sometimes also called Trojan) is non-self-replicating malware that appears to be legitimate but actually contains another program or block of undesired malicious, destructive code, disguised and hidden in a block of desirable code. Viruses can also be attached with Trojans. A Trojan horse appears to have a useful function but contains a hidden function that presents a security risk (Norton and Stockman 2000). There are many types of Trojan horse programs. One example is a program that makes it possible for someone else to access and control a person's computer over the Internet e.g. Girlfriend Trojan. A rootkit is a collection of tools (programs) that enable administrator-level access to a computer or computer network. Typically, a cracker installs a rootkit on a computer after first obtaining user-level access, by either exploiting a known vulnerability or cracking a password. Once the rootkit is installed, it allows the attacker to mask intrusion and gain root or privileged access to the computer and, possibly, other machines on the network. Crimeware is any computer program or set of programs designed expressly to facilitate illegal activity online. Many spyware programs, browser hijackers, and keyloggers can be considered crimeware, although only those used illicitly. A keylogger is a hardware device or software program that records the real time activity of a computer user including the keyboard keys they press. A dropper is a type of malware developed to launch viruses by "dropping" (installing) them. Dropper viruses may go undetected because they are hidden, difficult to pinpoint and relatively uncommon. Backdoor is a feature of a program, which is used for gaining access to a computer system in such a way that a person who is running it did not intend. It can be installed for accessing a variety of services like online service or an entire computer system.

Software Updates and Patches

Other than using anti-virus software, the threat of virus infection can be minimized by installing software patches provided by the supplier of your operating system to close security loopholes that could be exploited by viruses. Organizations can also use a firewall to prevent unauthorized access to your network, avoid the download of unauthorized programs and documents from the internet and ensure that staff adheres to this policy. Organization should also consider subscribing to a service or supplier who will provide virus alerts. Some services are paid while other are provided free of cost by provided by suppliers of anti-virus software to their customers.

Anti-Spyware software scans your systems and detects known spyware programs. A detected spyware can either be removed or quarantined. Similar to anti-virus software, spyware software needs to be updated on a regular basis.

Solutions for Legal and Regulatory Issues

E-commerce businesses can address the legal and regulatory issues by either passive monitoring of legal and regulatory developments affecting E-commerce or by active participation in influencing legislation related to E-commerce.

It is important for every E-commerce business to closely monitor the evolving legislation and regulations affecting E-commerce. In such rapidly changing environment, any new laws or regulations will certainly influence the mechanisms and practices that need to be adopted to conduct secure E-commerce. Capability to predict developments and react quickly to exploit them will provide businesses with a competitive advantage.

By participating in the development of E-commerce legislation, businesses can minimize the issues that arise due to inappropriate legislation. For example, inappropriate legislation could significantly increase the costs or risks of carrying out E-commerce. Most legislators in many countries do not possess good understanding of the issues at stake and are currently receptive to the opinions of business. Businesses can use external bodies or involve directly lobbying for workable and beneficial legislation.

Training can include raising awareness of the need to protect information, training for skills necessary to operate secure E-commerce systems, educating for specific security measures or best practice methodologies. Training is a critical component of a successful E-commerce security framework.

The increasing complexity of the technologies widely used to secure E-commerce systems giving rise to the need for skilled IT security personnel. Although traditional sources of trained security personnel are available (e.g. programmers, networking professionals, and external consultants) the challenge remains obtaining the right balance of skills and competencies through external recruitment. An ideal candidate is one with an information or technical security background and able to understand the unique business drivers of an E-commerce business. Effective training of the in-house staff that has core competencies is one way to develop staff with the desired skill-set.

E-commerce system's management, design, operation, and use subject to a wide range of statutory, regulatory and contractual security requirements. Legislation governing E-commerce is still in its infancy and is anticipated to change quickly in years to come. Compliance requirement may include protecting Intellectual Property Rights (IPR), Safeguarding of records, Data Protection and Privacy, Prevention of misuse of IT facilities, and Regulation of cryptographic controls. Such regulatory environment poses a new set of risks that are difficult to identify and assess, and dynamic in nature. It is unlikely that business managers and development staff will be aware of the existence and consequences of the full range of risks. Therefore, E-commerce businesses need an effective compliance structure to ensure that management and staff is aware of the nature and range of legal, regulatory and contractual requirements associated with the deployment of E-commerce systems. E-commerce Businesses should also ensure that adequate controls are in place to assess and manage the resultant risks.

COMPUTER FORENSCIS

Computer forensics is a science concerned with the gathering and analyzing of digital evidence. This evidence can take many forms and its analysis has many uses. Digital evidence is found within the computer itself, within the network and within various types of storage devices (for example, USB drives, dash drives, CD-ROMs, DVDs). The analysis of these types of digital evidence is used in court cases--both criminal and domestic--and within companies to monitor use of resources. In spite of the varied applications of computer forensics, the actual techniques employed are nearly identical. Evidence from computer forensics investigations is usually subjected to the same guidelines and practices of other digital evidence. It has been used in a number of high profile cases and is becoming widely accepted as reliable within US and European court systems. Computer forensics is important because it can save your organization money. Angry employees are more likely than independent hackers and competitors to commit a crime against a firm. The individuals who launch these attacks may have a specific type of attack in mind. In recent years, malicious-minded individuals have assaulted numerous e-commerce web sites with denial-of-service attacks and have imposed several other malevolent acts on corporations and governments including, but not limited to, viruses, wiretapping and financial fraud. These attacks can cause financial hardships to companies, especially their e-commerce activities. Cybercrime potentially costs millions, if not billions, of US dollars in unrealized profits and exposes organizations to significant risk. For example, if a disgruntled employee finds an exploit in the company's financial securities and begins stealing the company's money, this could potentially hinder the viability of the company. If someone is suspected of embezzlement, a computer forensics specialist could be hired to analyze and gather evidence against the suspected individual (ISACA, 2015). Increasingly, laws are being passed that require organizations to safeguard the privacy of personal data. It is becoming necessary to prove that your organization is complying with computer security best practices. If there is an incident that affects critical data, for instance, the organization that has added a computer forensics capability to its arsenal will be able to show that it followed a sound security policy and potentially avoid lawsuits or regulatory audits.

Risk Assessment and Risk Management in E-commerce

Barriers to entering e-commerce are comparatively low, but new opportunities can be accompanied by new risks. Today's threats to e-commerce systems include:

1. Physical threats to E-commerce infrastructure (for example, fire or flood)

2. Data threats (Threats of data corruption and destruction)

3. Hoaxes (e.g. warnings about non-existent viruses circulated by email)

4. Technical failure (e.g. software bugs)

5. Infrastructure failures (e.g. server crashes)

6. Credit card and payment fraud.

7. Malicious attacks from inside or outside your business.

8. Hacker threats

9. Risk to corporate information and intellectual property from internal staff and trading partners.

10. Hacker exploitation of errors in software application design, technical implementation or systems operation.

11. Website defacement

12. Denial-of-service attacks

Unless swift action is taken, any problems with your e-commerce site will be immediately obvious to the world. E-commerce customers typically have little loyalty, so if your website is unavailable they will simply move on to one of your competitors. In addition, technical failure can also have a significant impact on your key trading partners.

Risk assessment means listing all of the risks a business might face and assigning varying degrees of importance to them. Risk management means prioritizing these risks and formulating policies and practices to balance and mitigate them. Every business can benefit from a risk assessment of their e-commerce systems, although smaller businesses may not need to implement some of the more sophisticated techniques.

A risk assessment can be carried out to provide an organization with a clear understanding of the risks facing its e-commerce system and associated business processes, and the potential impact if a security incident arises. A key part of a risk assessment is defining the business' information access requirements. This will cover the rules of access for different groups of users. For example, different rules may apply for employees, consultants, managed service providers, suppliers, customers, auditors and government agencies. Any analysis should also take account of how electronic transactions are verified. How do you know that an order has actually come from a known customer? Where contracts are exchanged electronically, who can sign them and how can it be proved which is the signed version?

Risk assessment involves determining:

· The likelihood of a risk occurring

· Its possible impact

Risk assessment can be either qualitative or quantitative. Qualitative risk assessment involves identifying the threats, where your systems are vulnerable, and the controls you can put in place to counter or minimize the threats. Once these have been identified, you should be able to assess whether the risk is high, medium or low. Quantitative assessment assumes a value can be placed on any losses suffered as a result of a security violation. Probability can be used to measure the likelihood of such incidents occurring.

Quantifying the Risks

All risks can be quantified against their probability and potential impact on a high - medium - or low-risk basis. To quantify risks, brainstorm all possible risks with relevant internal and external experts and agree on a probability rating (high, medium or low) for each risk. Then you should get an agreement on an impact rating for each risk (high, medium or low). The most important risks are indicated as 5, with the least important risks rated as 1.Once you have quantified all possible risks, you can assess how much time and money you should spend implementing appropriate security controls. For example, it is not worth implementing controls for events that are unlikely to occur and which would have little impact. On the other hand, you should concentrate resources on developing security controls for events that are likely to occur and would have a big impact on your business.

It is not possible to eliminate all the potential risks to your business. This may be because there is no practical way of removing the threat posed by some risks or eliminating some risks is simply not financially worthwhile. Therefore the risk management framework you design should reflect where the greatest potential risks lie, set out practical measures to reduce risks to their lowest possible level, and reflect the costs and benefits of taking action to reduce or eliminate risks. Figure shows a framework for classifying and managing risks. This framework has two dimensions: likelihood of occurrence, and severity of the potential consequences. These two dimensions form four quadrants, which in turn suggest how we might attempt to mitigate those risks.

FIGURE 10-28: RISK MANAGEMENT FRAMEWORK

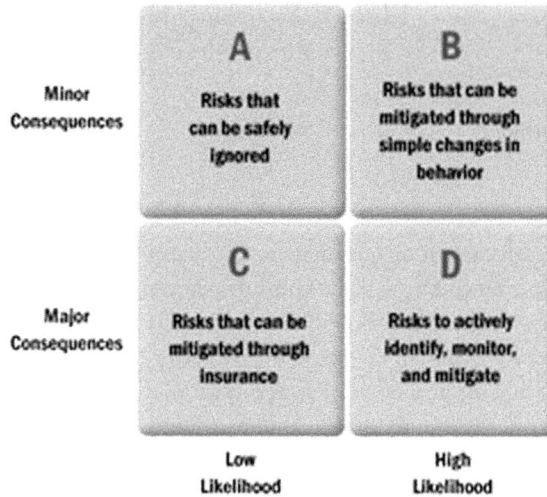

	A Risks that can be safely ignored	**B** Risks that can be mitigated through simple changes in behavior
Minor Consequences		
Major Consequences	**C** Risks that can be mitigated through insurance	**D** Risks to actively identify, monitor, and mitigate
	Low Likelihood	High Likelihood

Quadrant A: Risks that can be ignored

Cost effectiveness is an important consideration in deciding how we face up to risks. In quadrant A, we have risks with relatively minor consequences and a relatively low likelihood of occurring. These risks are not worth spending time and effort. An example of a low-likelihood, minor-consequence risk might be the possibility of a client computer failure. Assuming the IT department performs regular maintenance of all client computers in the organization and only tasks allowed are performed using the client computers, such an incident can occur rarely and not you should not worry about it.

Quadrant B: Risks that can be mitigated

These risks can arise from activities that can often go wrong but whose impacts are easy enough to minimize through straightforward changes in behavior. One example is the case where you did not follow the IT department guideline of installing a good anti-virus software. Your system is got infected and responding too slow causing loss of time and productivity. This risk can simply be avoided by changing your habit of not following the guidelines.

Quadrant C: Risks that should be transferred

These risks can be transferred in two ways. The first is through insurance. This can be problematic in e-commerce as it is often difficult to quantify the business loss following a security incident. It is even more difficult if the impact was due to a security violation within a trading partner's business. The second option is to contract aspects of your e-commerce function out to a third party. This could involve another business hosting your systems or running them on your behalf. The attraction is that many third-party hosting services operate in a more secure technical environment. However, while contractual arrangements can describe the service agreements and any penalties that may be inflicted, the primary impact of any incident will always be on your business. It could also potentially cost you more money. One good example is the 9/11 terrorist attacks that destroyed twin towers. As a result, the crucial data of many companies was lost resulting in heavy irreparable damages.

Quadrant D: Risks that should be prevented

These are the risks with both a relatively high likelihood of occurrence and major consequences. These risks can sink startups and Fortune 500 companies alike. The survival of your E-commerce venture depends on your ability to identify and prevent the company killers from happening. Uncertainty plagues businesses in countless ways, but we can group most company killers into the following categories:

Market Risks: Refer to whether or not there is sufficient demand for what you have to offer at the price you set.

Competitive Risks: Refers to what your competitors might do to try to beat you.

Technology & Operational Risks: These risks broadly cover everything having to do with execution of your E-commerce business.

Financial Risks: Risks of not having sufficient funds to run your business.

People Risks: These are the risks that can arise if you fail to recruit, motivate, and retain the right people for your E-commerce business.

Legal & Regulatory Risks: The risks of unfavorable changes in laws and regulations.

Systemic Risks: Systemic risks are those that threaten the viability of entire markets, not just a single firm within a market.

These categories are neither exhaustive nor mutually exclusive. Some risks span several categories.

Current Computer Security Landscape

Exploiting Vulnerabilities

The most notable trend during 2014 has been the increasing dominance of vulnerability-leveraging malware. This means that unpatched operating systems and applications continue to contribute to these detection statistics, though in far lower numbers than in previous years. One particular form of such malware is exploit kits. Exploit kits are toolkits planted on compromised websites that exploit vulnerabilities on a site visitor's device in order to silently drop malware onto his or her machine. The Angler and Astrum exploit kits surged in our detection statistics in 2014. Reports of detections of the AnglerEK kit have skyrocketed making it a clearly visible prevalent threat. Malware targeting vulnerabilities in the Java platform (collectively identified as Majava) are still effective enough to appear in our top threats list. We can infer from this that people continue to use unpatched versions of the popular development platform. Variants in the Wormlink family, exploit a vulnerability in the Windows operating system. Vulnerability-targeting malware appear most active in North America and Europe. While generalizations are tricky at best, the other regions appear to be more affected by 'older' threats that are no longer effective against newer or more up-to-date operating systems or programs. As such, it is at least plausible that the prevalence of vulnerability-targeting malware in North America and Europe is loosely correlated to a difference in the behavior and machine setup of users in the various regions.

Social Media and Worms

The introduction of Kilim, a family of malicious browser extensions that target Facebook users, in the top threats list also highlights the continuing misuse of social networking sites as a medium for spreading malware. While threats targeting and/or spreading on social media networks are hardly new, this is perhaps the first year in which a threat family targeting a single social media network has gained such widespread prevalence. Kilim's presence in South America, the Middle East and Oceania is more of a testament to Facebook's global reach than anything else, but it nonetheless speaks to the severity of the threat. While nowhere near the same level of prevalence, the Rimecud worm family also uses social media networks to spread its infections across continents.

Ransomware

The rising presence of the Browlock family in the threat statistics reflects the growth of ransomware as a whole. This particular form of malware has been a noticeable problem for users in the past couple of years, and of all the threats seen, may be the most problematic. Though the details may differ between families, this current crop of ransomware typically encrypts files held for ransom, making them effectively impossible to recover without the decryption key held by the attacker(s). In addition to older threats such as Cryptolocker and CryptoWall, new families such as CTB-Locker and SynoLocker are emerging as noteworthy PC-targeted menaces. The emergence of the SynoLocker family, which infects network attached storage (NAS) devices, is also a clear indication that malware developers are expanding their products' targeting capabilities.

On Android, the Koler and Slocker ransom-trojan families have also been busy increasing their count of variants, making them the largest ransomware families on that platform. The extreme difficulty in decrypting affected files without a decryption key, and the various thorny issues involved in paying a ransom (especially if a business is affected), makes ransomware a particularly difficult threat to resolve.

The recommended remediation for recovering from a ransomware infection is to report the incident to the appropriate legal authorities and to restore the affected files from a clean, recent backup onto a cleaned system. There are earlier variants of ransomware that simply hid the ransomed files, left copies of the original files with the Volume Shadow Copy service or left copies of the private encryption keys locally or in memory. It is certainly worth the effort of researching the details of the variant you encountered to see if there are options for you, but for the majority of instances, these options are no longer the case as the threat writers have updated their methods using the funds from earlier rounds of extortion. The current ransomware employ an RSA-2048 bit encryption key. Brute-forcing the key is simply not possible currently. You should install, configure and maintain an endpoint security solution for your business. With the endpoint being the final line of defense from any threat, a multi-faceted security solution should be employed. This solution should have protections for not just file based threats (traditional AV), but should also include download protection, browser protection, heuristic technologies, firewall and a community sourced file reputation scoring system. One of the primary vectors of these threats is "Spear Phishing" attempts, where an unsolicited e-mail will come from an unknown sender with an attachment that is then executed. Educating your users as to proper handling of unknown or suspicious files is crucial. Inbound emails should be scanned for known threats and should block any attachment types that could pose a threat. Exploit kits hosted on compromised websites are commonly used to spread malware. Regular patching of vulnerable software is necessary to help prevent infection. IDS or IPS systems can detect and prevent the communication attempts that the malware uses to create the public and private encryption keys required to encrypt the data. The current ransomware threats are capable of browsing and encrypting data on any mapped drives that the end user has access to. Restricting the user permissions for the share or the underlying file system of a mapped drive will provide limits to what the threat has the ability to encrypt. The fastest way to regain access to your critical files is to have a backup of your data. Backups of data should take place not only for files housed on a server, but also for files that reside localy on a workstation. If a dedicated peice of backup software is not an option, simply copying your important files to some sort of removable media and then removing that media from the system will provide a safeguard for your data being impacted by these types of threats (Sherman, 2015).

Modern System-Crippling Threats and Their Countermeasures

SHELLSHOCK

A more than 20-year-old vulnerability in the GNU Bash shell (widely used on Linux, Solaris and Mac OS system sparked the mobilization of attacks known as ShellShock beginning in late September 2014. The vulnerability could allow an attacker to gain control over a targeted computer if exploited successfully. This vulnerability potentially affects most versions of the Linux and UNIX operating systems, in addition to Mac OS X (which is based around UNIX). Security experts regard regards this vulnerability as critical, since Bash is widely used in Linux and UNIX operating systems running on Internet-connected computers, such as Web servers. Although specific conditions need to be in place for the bug to be exploited, successful exploitation could enable remote code execution. This could not only allow an attacker to steal data from a compromised computer, but enable the attacker to gain control over the computer and potentially provide them with access to other computers on the affected network (Symantec, 2014). The most likely route of attack is through Web servers utilizing CGI (Common Gateway Interface), the widely-used system for generating dynamic Web content. An attacker can potentially use CGI to send a malformed environment variable to a vulnerable Web server. Because the server uses Bash to interpret the variable, it will also run any malicious command tacked-on to it. The Figure 10-29 explains working of ShellShock.

FIGURE10-29: WORKING OF SHELLSHOCK

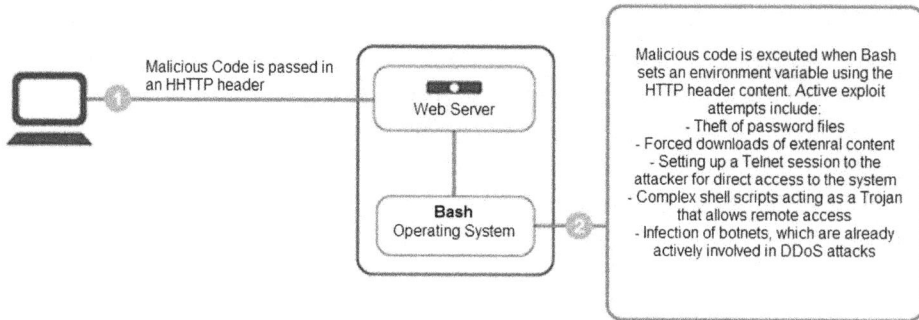

Businesses, in particular website owners, are most at risk from this bug and should be aware that its exploitation may allow access to their data and provide attackers with a foothold on their network. Accordingly, it is of critical importance to apply any available patches immediately. If a patch is unavailable for a specific distribution of Linux or Unix, it is recommended that users switch to an alternative shell until one becomes available. Consumers are advised to apply patches to routers and any other web-enabled devices as and when they become available from vendors. Users of Apple's Mac OS X should be aware that the operating system currently ships with a vulnerable version of Bash. Mac users should apply any patches for OS X when they become available (Symantec, 2014).

HEARTBLEED

The Heartbleed vulnerability is a security bug in OpenSSL, a popular open source protocol used extensively on the Internet. It allows attackers to access and read the memory of systems thought to be protected. Vulnerable versions of OpenSSL allow the compromise of secret keys, user names, passwords and even actual content. It is believed that this vulnerability has been in existence for at least two years and has quite possibly been exploited for just as long. This vulnerability is very significant. Due to the popularity of OpenSSL, approximately 66% of the Internet or two-thirds of web servers could be using this software. The Figure 10-30 shows how Heartbleed works.

FIGURE 10-30: WORKING OF HEARTBLEED

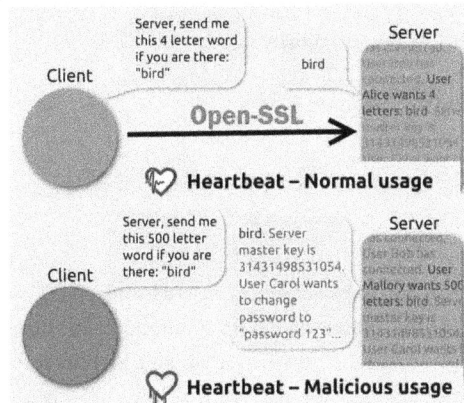

Courtesy: mrunal.org

Businesses using OpenSSL 1.0.1 through 1.0.1f should update to the latest fixed version of the software (1.0.1g), or recompile OpenSSL without the heartbeat extension. Businesses should also replace the certificate on their web server after moving to a fixed version of OpenSSL. Finally, businesses should also consider resetting end-user passwords that may have been visible in a compromised server memory. Consumers should be aware their data could have been seen by a third party if they used a vulnerable service provider. They should monitor any notices from the vendors you use. Once a vulnerable vendor has communicated to customers that they should change their passwords, users should do so. They should also avoid potential phishing emails from attackers asking you to update your password – to avoid going to an impersonated website, stick with the official site domain.

UNICORN

In this type of attack, the attacker changes key data structures used by the program's logic, forcing the control flow into existing parts of the program that would be otherwise unreachable. Discovered in November 2014, this is a complex and rare vulnerability. Attackers can use it in "drive-by attacks" to run programs remotely and take over a user's machine. Similar to ShellShock, it's yet another serious vulnerability going unnoticed for an extremely long time despite all the efforts of the security community. The vulnerability allows a remote attacker to execute arbitrary code via a crafted web site. Microsoft released a patch to fix this vulnerability. However, the patch only applies to Windows Vista and higher, as support for Windows XP ended in April. Therefore, if any computer you are running a 13-year-old operating system, you will have to grapple with a critical bug that is even older.

Organizations Promoting E-commerce Security

There are many computer security organizations, operating in different countries of the world. The following are government, public and private computer security organizations, working for providing advice, creating awareness and helping users to avoid any type of computer threats.

1.   CERT

2. Get safe online

3. European network and information security agency (ENISA)

4. Information system security association (ISSA)

CERT (Community Emergency Response Team)

The CERT Program is part of the Software Engineering Institute (SEI), a federally funded research and development center at Carnegie Mellon University in Pittsburgh, Pennsylvania. Following the Morris worm incident, which brought 10 percent of internet systems to a halt in November 1988, the Defense Advanced Research Projects Agency (DARPA) charged the SEI with setting up a center to coordinate communication among experts during security emergencies and to help prevent future incidents. CERT continue to respond to major security incidents and analyze product vulnerabilities, our role has expanded over the years. Along with the rapid increase in the size of the internet and its use for critical functions, there have been progressive changes in intruder techniques, increased amounts of damage, increased difficulty of detecting an attack, and increased difficulty of catching the attackers. To better manage these changes, the CERT/CC is now part of the larger CERT Program, which develops and promotes the use of appropriate technology and systems management practices to resist attacks on networked systems, to limit damage, and to ensure continuity of critical services.

Get Safe Online

Get Safe Online is a United Kingdom-based campaign and national initiative to teach citizens about basic computer security and internet privacy. It was launched in October 2005 with the backing of several government departments, and the support and sponsorship of many companies. The campaign's centre point is its website where people can go to get information about internet safety, how to protect themselves online, and practical advice. The site offers a 10-minute guide for beginners, as well as more technical information and guidance for businesses as well as home users and Links are also given to the official website for users wanting to find out more.

ENISA (European Union Agency for Network and Information Security)

ENISA is as a **body of expertise**, set up by the EU to carry out very specific technical, scientific tasks in the field of Information Security, working as a "European Community Agency". ENISA's mission is essential to achieve a high and effective level of Network and Information Security within the European Union. Together with the EU-institutions and the Member States, ENISA seeks to develop a culture of Network and Information Security for the benefit of citizens, consumers, and business and public sector organizations in the European Union.

ENISA is helping the European Commission, the Member States and the business community to address, respond and especially to prevent Network and Information Security problems. The Agency also assists the

European Commission in the technical preparatory work for updating and developing Community legislation in the field of Network and Information Security.

ISSA (Information Systems Security Association)

The Information Systems Security Association (ISSA) is a not-for-profit, international organization of information security professionals and practitioners. It provides educational forums, publications, and peer interaction opportunities that enhance the knowledge, skill, and professional growth of its members. The primary goal of the ISSA is to promote management practices that will ensure the confidentiality, integrity, and availability of information resources. The ISSA facilitates interaction and education to create a more successful environment for global information systems security and for the professionals involved. Members include practitioners at all levels of the security field in a broad range of industries such as communications, education, healthcare, manufacturing, financial, and government.

Review Questions

1. Define computer security and its types.

2. List various classifications of information assets.

3. List some basic security issues.

4. List some key elements of E-commerce security governance structure.

5. List classes of threats to E-commerce systems.

6. What is active content?

7. Differentiate between Java applet and ActiveX control.

8. What is a cookie? What are its types?

9. List some solutions for server-side security of E-commerce systems.

10. What is the role of encryption in an E-commerce transaction?

11. How a digital certificate and digital signature works?

12. What is Steganography?

13. List some solutions to protect client-side of an E-commerce system.

14. Describe how risk in E-commerce is assessed and managed?

Bibliography

Bluehawksecurity.com. (2009). *FAQ*. Retrieved August 26, 2009, from Bluehawksecurity.com: http://www.bluehawksecurity.com/FAQ004.html

Cavusoglu, H., Mishra, B., & Raghunathan, S. (2004). The Effect of Internet Security Breach Announcements on Market Value: Capital Market Reactions for Breached Firms and Internet Security Developers. *International Journal of ELectronic Commerce* , 70-104.

Chowdhry, A. (2015, June 1). Facebook Starts Supporting Animated GIFs. Retrieved July 18, 2015, from http://www.forbes.com/sites/amitchowdhry/2015/06/01/facebook-starts-supporting-animated-gifs/

Cybersafe.com. (2010). *About Kerberos*. Retrieved January 26, 2010, from Cybersafe.com: http://www.cybersafe.com/menu_prodsolserv/products_solutions_and_services_technology.htm

Deitel, H. M., & Deitel, P. J. (2001). *e-Business & e-Commerce for Managers*. Prentice Hall.

Dell. (2014, February). Protecting the organization against the unknown. Retrieved from http://software.dell.com/documents/protecting-the-organization-against-the-unknown-whitepaper-27396.pdf

Ecommercetimes.com. (2010). *What Security Fears Cost E-Commerce*. Retrieved June 15, 2010, from Ecommercetimes.com: http://www.ecommercetimes.com/rsstory/69667.html

ExpressComputer. (2003). *Have you recognised the importance of information security?* Retrieved June 26, 2009, from ExpressComputeronline.com: http://www.expresscomputeronline.com/20030929/security09.shtml

Feruza, S., & Kim, T.-h. (2007). IT Security Review: Privacy, Protection, Access Control,Assurance and System Security. *International Journal of Multimedia and Ubiquitous Engineering , 2* (2).

GovernmentofAlberta. (2005). *Information Security Classification*. Retrieved June 26, 2009, from www.im.gov.ab.ca: www.im.gov.ab.ca/publications/pdf/InfoSecurityClassification.pdf

HowStuffWorks.com. (2008). *What is a digital signature*. Retrieved July 02, 2010, from www.HowStuffWorks.com: http://computer.howstuffworks.com/digital-signature.htm

HP. (2015). Cyber Security Vulnerability and Risk Management – HP Cyber Risk Report 2015 | HP® Official Site. Retrieved July 18, 2015, from http://www8.hp.com/us/en/software-solutions/cyber-risk-report-security-vulnerability/

IBM. (2015, May 29). IBM 2015 Cyber Security Intelligence Index. Retrieved July 18, 2015, from http://www-01.ibm.com/common/ssi/cgi-bin/ssialias?subtype=WH&infotype=SA&htmlfid=SEW03073USEN&attachment=SEW03073USEN.PDF

Infosecwriters.com. (2006). *Stagenography FAQ*. Retrieved July 16, 2009, from www.Infosecwriters.com: http://infosecwriters.com/text_resources/pdf/Steganography_AMangarae.pdf

ISACA. (2015). Computer Forensics: An Overview. Retrieved July 19, 2015, from http://www.isaca.org/Knowledge-Center/ITAF-IS-Assurance-Audit-/IT-Audit-Basics/Pages/Computer-Forensics-An-Overview.aspx

Kaspersky. (2015). Kaspersky Personal & Family Security Software. Retrieved July 18, 2015, from http://usa.kaspersky.com/

MIT. (2010). *Kerberos: The Network Authentication Protocol*. Retrieved July 27, 2010, from MIT: http://web.mit.edu/Kerberos

Mitchell, B. (2009). *Encryption*. Retrieved August 12, 2010, from about.com: http://compnetworking.about.com/od/networksecurityprivacy/l/aa011303a.htm

Nasir, M. A. (2009). *Legal Issues Involved in E-Commerce*. Retrieved June 26, 2009, from ACM: http://www.acm.org/ubiquity/views/v4i49_nasir.html

Onlinegrowthprogram.com. (2009). *Why and how your business should be e-commerce enabled.* Retrieved June 26, 2009, from onlinegrowthprogram.com.au: http://onlinegrowthprogram.com.au/AnnouncementRetrieve.aspx?ID=32135

Petra, U. (2009). *Ecommerce Security.* Retrieved July 26, 2009, from www.uop.edu.jo: www.uop.edu.jo/download/research/members/e-commerce_security_en.pdf

PWC. (2014). PwC's 18th Annual Global CEO Survey. Retrieved July 18, 2015, from http://www.pwc.com/gx/en/ceo-survey/2015/index.jhtml

Rapoza, J. (2009, June 4). Web Bugs Can Make Your Privacy Sick. Retrieved July 18, 2015, from http://www.eweek.com/c/a/Web-Services-Web-20-and-SOA/Web-Bugs-Can-Make-Your-Privacy-Sick-170960

Schneider, G. (2010). *Electronic Commerce, 9th Ed.* Course Technology.

Sherman, M. (2015, February 18). Ransomware Do's and Dont's: Protecting Critical Data. Retrieved July 19, 2015, from http://www.symantec.com/connect/blogs/ransomware-dos-and-donts-protecting-critical-data

Sun.com. (2001). *Public Key Infrastructure Overview.* Retrieved Januray 26, 2009, from sun.com: www.sun.com/blueprints/0801/publickey.pdf

Symantec. (2014, September 25). ShellShock: All you need to know about the Bash Bug vulnerability. Retrieved July 18, 2015, from http://www.symantec.com/connect/blogs/shellshock-all-you-need-know-about-bash-bug-vulnerability

Tse, M. (2009). *e-Commerce Infrastructure and Processes.* Retrieved January 26, 2010, from www.eie.polyu.edu.hk: http://www.eie.polyu.edu.hk/~cktse/eie417-2001/lecture9-10.html

WillametteUniversity. (2009). About Digital Watermarking. Retrieved July 27, 2009, from www.willamette.edu: www.willamette.edu/wits/idc/mmcamp/watermarking.htm

Zaharia, A. (2015, April 1). Corporate Security Checklist – a CEO's Guide to Cyber Security. Retrieved July 18, 2015, from https://heimdalsecurity.com/blog/corporate-security-checklist-a-ceos-guide-to-cyber-security/

Zeltser, L. (2011, October). How antivirus software works: Virus detection techniques. Retrieved July 19, 2015, from http://searchsecurity.techtarget.com/tip/How-antivirus-software-works-Virus-detection-techniques

11

MOBILE COMMERCE

Learning Objectives

After reading this chapter, reader should be able to:

- Understand definition of mobile commerce.

- Understand mobile commerce terminology.

- Discuss the specific characteristics and value-added attributes of mobile commerce.

- Describe building blocks of mobile commerce (hardware, software, mobile networks).

- Discuss pros and cons of mobile commerce.

- Describe various applications of mobile commerce.

- Describe collaborative commerce.

- Describe pervasive computing

- Describe ubiquitous commerce

- Discuss various trends and challenges of M-commerce

- Describe various methods of mobile payments

- Understand mobile payments

Introduction

M-commerce or mobile commerce refers to the buying and selling of goods and services through wireless handheld devices e.g. mobile phones. M-commerce transactions can be conducted in full or in part in a wireless environment (Wang & Wang, 2005). M-commerce is a major new frontier in E-commerce that provides many opportunities. Emergence of M-commerce is creating new markets and new organizational forms. Businesses are developing new strategies to take advantage of m-commerce. With rapid developments in Wireless Application Protocol (WAP) and wireless content delivery systems, M-commerce is becoming increasingly popular. M-commerce is affecting many industries e.g. financial services, telecommunications, service, and information. In financial services, we see increasing use of mobile banking, where customers use their handheld devices to access their accounts, and brokerage services (e.g. stock trading using PDA). In telecommunications, mobile devices can be used from change of service request to bill payment and account reviews. Retail consumers are now able to place their orders from their mobile devices. Information services e.g. delivery of financial news, sports figures and traffic updates, can be provided to a mobile device.

M-Commerce Terminology

We will begin our discussion of m-commerce by first defining some common terms related to m-commerce.

Global positioning system (GPS): GPS is a satellite-based tracking system that can be used to determine the location of a GPS device. GPS can also be used to deliver a coupon or a message to a mobile phone, display local ads to a person visiting a website from a specific location; deliver detailed product information when someone is standing in front of the product, offer incentives for location-based activities such as visiting a store multiple times, find nearby things such as stores or ATM, provide event, meet-up and social opportunities based on a physical location, and share location-based information with others in a social network.

Global System for Mobile Communications (GSM): GSM is a mobile digital data communication technology used to provide wireless LAN and Internet access to mobile users. GSM is widely used technology in Europe and other parts of world.

Enhanced Data GSM Environment (EDGE): EDGE is a faster version of GSM that is built on existing standards of GSM. EDGE is designed to deliver data at rates up to 384 Kbps and can be used to provide multimedia and other broadband applications to mobile phone and computer users.

High Speed Circuit Switched Data (HSCSD): A mobile data communication technology used to provide wireless LAN and Internet access to mobile users of GSM networks. Combined with compression and filtering technology, this high-speed and high performance technology offers higher data transfer speed. HSCSD is designed for existing GSM networks and is very easy to implement and rollout.

Universal Mobile Telecommunications Services (UMTS): UMTS is a wireless internet access technology. UMTS is a major new 3G mobile technology, which integrates packet and circuit data transmission.

General Packet Radio Services (GPRS): GPRS, an enhancement to existing GSM and TDMA networks, is a mobile data communication technology used to provide high-speed wireless LAN and Internet access to mobile users. GPRS is a packet-linked technology enables always on mobility.

1G: The first generation of wireless technology that used radio signals which were analog-based.

2G: The second generation of wireless technology that used radio signals which were digital. 2G technology mainly accommodates text.

2.5G: It is an interim wireless technology between the 2G and 3G and is based on GPRS and EDGE. This technology can accommodate limited graphics.

3G: It is the third generation of digital wireless technology. It supports rich media such as video clip. 3g technology is a mix of packet and circuit switching.

4G: The fourth-generation wireless technology that is expected to provide users with higher data transmission rates that could support high-quality streaming video. 4G technology will use packet switching.

Wi-Fi: Wi-Fi is a wireless technology, based on IEEE 802.11 standard, which is used to connect devices (such as mobile phones, computers and many such electronic devices) over a wireless network. The places or spots, which offer Wi-Fi access to the Internet, are called Wi-Fi hotspots. Range of Wi-Fi signals are about 120 feet indoors and 300 feet outdoors. When distance increase between the user and the signal, the connection speed decreases.

WiMAX (Worldwide Interoperability for Microwave Access): WiMAX is a digital wireless broadband access technology with performance similar to 802.11/Wi-Fi networks but provides better quality of service and coverage.

Personal Digital Assistant (PDA): PDA refers to any small mobile hand-held device that provides computing and information storage and retrieval capabilities for personal or business use.

Short Message Service (SMS): It refers to the transmission of short text messages to and from a mobile phone, fax machine and/or IP address. These messages contain no images or graphics. SMS is known as the e-mail of m-commerce.

Enhanced Messaging Service (EMS): EMS is an extension of SMS that supports simple animation, tiny pictures, and short melodies. EMS works on all GSM networks. User will be able to see only the text of the message if the message is sent to a non-EMS phone.

Multimedia Messaging Service (MMS): This refers to the transmission of messages that allows users to exchange multimedia communications (images, audio, and video clips), in addition to text, between capable mobile phones and other devices. A common current application of MMS is picture messaging. MMS can also be used to send graphic presentations of stock quotes, sports news, and weather reports.

Wireless Application Protocol (WAP): WAP is a technology that allows users to access information instantly via handheld wireless devices such as mobile phones, pagers, two-way radios, smartphones and communicators.

Mobile Commerce Service Provider (MCPS): MCPS is a company or organization that specializes in all aspects of m-commerce and can provide any combination of consulting, software and computer systems, and other support for m-commerce. Digby (http://www.phunware.com/platform/digby/) provides retailers with a mobile commerce solution, offering branded applications for iPhone, BlackBerry, Windows Mobile, Android and the mobile web. Digby is now part of Phunware Location Marketing. Businesses can use services of MCPS to distribute and sell products and services over both the internet and Mobile internet and manage their online businesses. MCPs provide service in areas such as mobile device databases, billing systems, text messaging services, hardware/software design, mobile payments, brand recognition, distribution control, web site development and hosting, web site performance monitoring, fulfillment management, online marketing, and order processing and delivery.

Why M-Commerce

Currently, mobile phone subscribers can use their devices to perform a variety of activities including purchase and download of content (e.g. movies, music, ring-tones, or games), play online games, gamble online, access information available on a mobile screen (e.g. weather forecasts, news, mobile TV etc.), obtain location-based information tailored (e.g. address of a nearby café), access online banking and financial services, make financial transactions, and make payments for mobile activities through either credit cards or mobile phone bills. The development of the 3G mobile services, complete with audio and higher quality graphics, has expanded consumers' interest in mobile devices and opened up the potential of new commercial applications.

M-commerce can help both businesses and consumers in a variety of ways. One example is using mobile wireless networks to provide continued and extended access to information for mobile users. Another example is mobile bill payment facility, enabled by mobile networks, for mobile users. Wireless devices can have profound impacts for people with disabilities. For example, consider a hearing-impaired person in shopping mall. Using Bluetooth technology, the shopping mall can broadcast the sale announcement to the person's Bluetooth-enabled PDA device. The PDA can then convert the audio message to text and display the text message for the hearing-impaired person. There can be many other examples.

In the past 2 years, the percentage of U.S e-commerce sales via mobile devices has risen from 2% to 8%. Percentage of total U.S. e-commerce dollars spent via tablet or smartphone was 8% in first quarter of 2012. There were more than 150 million smartphone users in USA in 2014. In March 2012, more than 40 million U.S. consumers visited Amazon.com via tablet or smartphone. Unique visitors to mobile web properties of U.S retailers were in this order: Amazon (41.2 million), eBay (27.2 million), Netflix (13.4 million), Wal-Mart (12.7 million), Target (6.9 million). The percentage of mobile visits to online retailers' websites by device in 2011 was Smartphone (6%), PC (90%), and Tablet (4%). Tablet owners are more important to online retailers because they are more affluent. In USA, 38% of Tablet owners in USA had household income of $100k+. The average order value by device was Smartphone ($80), PC ($102), and Tablet ($124). Customer conversion rate with respect to device was Smartphone (0.6%), PC (2.5%), and Tablet (2.4%). Online retailers had different objectives for tablets. 87% were looking to drive revenue/sales to websites, 65% were looking to improve customer satisfaction, 65% were looking to develop brand loyalty, 61% wanted to provide a source of product and price information, and 57% wanted to improve customer acquisition (eMarketer.com, 2014).

Latest Developments in M-commerce

Many latest developments have sparked growth and ongoing evolution of M-commerce.

Rise of iPad: Over 92% of non-desktop online sales originated from an iPad or other iOS enabled device (Bosomworth, 2015).

Mobile-Optimized Sites: 36% of consumers plan to make future purchases from their mobile devices (Bosomworth, 2015).

Mobile Apps: 43% of current mobile shoppers have downloaded a retail app (Phimister, 2015).

Text-based Mobile Coupons: Over half of consumers are willing to share their location in exchange for more relevant content/information, including mobile deals (Accenture.com, 2015).

QR Codes: The number of consumers who added QR code-scanning capabilities to their smartphones during first quarter of 2011 increased to 938%, compared to new users added in first quarter of 2010. However, only 15% of smart device users know how to scan QR codes properly, which is an enormous drawback, considering a much bigger chunk of the world's population uses these devices and could already be taking advantage of the information provided by the codes (Visualead.com, 2014).

Personalization: 2/3 of smartphone users say they will share personal information with marketers in order to reap the benefits of personalized services (Cmocouncil, 2015).

Geolocation & Location-based Services: According to the Pew Institute, "74% of adult smartphone owners ages 18 and older say they use their phone to get directions or other information based on their current location." Use of a geolocation feature is more than a nice-to-have option on a smartphone or tablet — its use has become a routine part of a consumer's daily life. With U.S. smartphone penetration at 74% and rising, this represents a significant market share of the U.S. buying population (Zickuhr, 2013; Sahagian, 2013.

Augmented Reality: Augmented reality-focused apps, such as the Magic Leap or the Microsoft Hololens, are set to boost consumption. Majority of augmented reality apps are to be used on smartphones and tablets. The use of mobile augmented reality apps will rake in revenues and will increase from just 184 million in 2014 to 1.3 billion by 2019 (Jennelyn Magdirila, 2015).

Siri & Product Content Management: The iPhone 4S personal assistant Siri is expected to revolutionize the way consumers use smartphones to interact with retailers. Best Buy has been early adopter, making its products easily searchable via Siri.

Price Check: Launched for Android-powered devices in November 2011, Amazon's Price Check enables users to scan a barcode, snap a picture, or even just say the name of any item they find in a retail store to quickly and easily compare its cost to similar products available for sale on Amazon.com

Specific Characteristics of M-Commerce

M-commerce has two major characteristics that differentiate it from other forms of e-commerce. These are mobility and broad reach.

Mobility: Every M-commerce user carries a cell phone or other mobile device everywhere he goes. Users can establish a real-time connection with commercial and other systems from wherever.

Broad Reach: With m-commerce, users carrying an open mobile device can be reached at any time instantly.

According to (Turban, Lee, & Chung, 2006), these two broad characteristics of M-commerce create many value-added attributes.

Value-Added Attributed of M-commerce

Ubiquity: Ubiquity means be everywhere at the same time. A mobile device can provide the user with location independent real-time information and communication facility.

Convenience: With mobile devices, user can conveniently access real-time information and communicate.

Instant Connectivity: Mobile devices provide users with instant connectivity to Internet and other networks.

Personalization: Personalization refers to providing users such content that matches with their individual needs. The goal of personalization is increased user satisfaction.

Products and Services Localization: It refers to providing users such products and services that match with their individual need e.g. providing location of a coffee shop near to a customer. Such products and services are known as location-based commerce (L-commerce).

Location-based Commerce: Location-based Commerce (L-Commerce) refers to the localization of products and services through mobile commerce and context aware computing technologies. Location-based commerce is an evolving marketing and commercial technology that is a mix of mobile commerce and context aware computing. L-commerce focuses over five key service areas: **location (**determining the basic position of a person or a thing), **navigation (**plotting a route from one location to another), **tracking (**monitoring the movement of a person or a thing), **mapping (**creating maps of specific geographical locations), and **timing (**determining the precise time at a specific location). Recently, major players and small start-up companies have launched services that engage user communities in updating their existing content and creating new dynamic real-time content. Many of these initiatives are embedded within a mobile social networking context in which real-time experiences are shared.

Building Blocks of M-Commerce

The major building blocks of mobile commerce include hardware, software, and mobile networks.

M-Commerce Hardware

Several hardware devices are used in M-commerce. The major ones include the following:

Cellular (mobile) Phones: A cellular phone is a hand-held mobile radiotelephone. Cell phones are improving with time, adding more features, larger screens, keyboards, and provide support for small Java applications (such as games), and more. Most cell phone today is Internet-enabled. Mobile phones use a protocol known as Wireless Application Protocol (WAP) to access network services. Mobile phones have limited processing capability. WAP provides filtering of information so that it can be displayed on a mobile phone.

Attachable Keyboard: It is a keyboard that can be attached to the cell phone for executing transactions. Attachable keyboards can also be used with other wireless devices, such as PDAs. LG VX9600 Versa is a mobile phone with an attachable keyboard.

PDA: PDA uses a variety of special purpose operating systems (e.g. Windows Mobile). PDAs have significant processing capability as compared to cellular phones. Today Internet-enabled PDAs are available and their capabilities are increasing. These PDAs can also be connected to Internet using special software and a wireless modem.

Two-way Pagers: These pagers can be used to send and receive text messages (such as stock market orders). These pagers are suitable for conducting limited M-commerce activities on the Internet. Some examples of these pagers include Motorola Talkabout T900 Two-Way Pager.

Screenphones: A screenphone (e.g. HTC Touch 3G handset) is an Internet-enabled phone equipped with a color screen and a keyboard. These phones are mainly used for e-mail.

E-mail Handheld: These devices (e.g. BlackBerry) provide integrated Internet access. There is no need to dial into an Internet provider for access. These devices include a keypad and a variety of services for data communication so users can receive and send messages from anywhere.

Smartphone: A smartphone is a handheld device that integrates mobile phone capabilities with the more common features of a handheld computer or PDA. Smartphones can store user information, provide e-mail facility and other applications. The iPhone is an Internet and multimedia-enabled smartphones by Apple Inc.

Position-Determining Equipment (PDE): This equipment identifies the location of the mobile device (either through GPS or by locating the nearest base station) and then sends the position information to the mobile positioning center.

Mobile Positioning Center (MPC): The Mobile Positioning Center is a server that manages the location information sent from the Position-Determining Equipment.

GPS Locator: It is used to determine the location of the person carrying the mobile computing device. GPS locator is essential for providing location-based applications. The GPS locator can be attached or inserted into a mobile device.

Geographical Information System (GIS): The location provided by GPS is expressed in terms of latitude and longitude. For this information to be useful for businesses and consumers, it has converted into a certain place or address. A GIS does this by inserting the latitude and longitude onto an electronic map.

In addition to the hardware described above, M-commerce also requires the following hardware. Most of this hardware is not seen by the M-commerce users however, they are essential for connectivity purposes.

- A wireless modem or a wireless LAN adapter

- A Web server with wireless support, a WAP gateway, a communications server, and/or a mobile communications server switch (MCSS). Handheld devices use this web server to communicate with the Internet or intranet infrastructure.

- An application or database server providing E-commerce functionality.

M-Commerce Software

Software development for wireless devices is challenging because widely accepted standards for wireless application development are still developing. Wireless software applications need to be customized for each type of device with which the application may communicate. The major software products required for M-commerce include the following:

Microbrowsers: A microbrowser, also called a mobile browser or mini-browser, is a type of browser used on small screens e.g. mobile phones. Mobile browsers are designed to be small and efficient to accommodate the low memory capacity and low-bandwidth of wireless handheld devices. They can access the Web via the wireless Internet. Microbrowsers have limited bandwidth and linked memory requirements. Some examples of microbrowser are jB5 Mobile Browser, Firefox mobile, and Internet Explorer Mobile.

Mobile-Client Operating System: This is the operating system software that resides in the mobile device e.g. Windows Mobile, Android (from Google), and iOS (from Apple). Android is software for mobile devices that includes an operating system, middleware and key applications. The Android SDK provides the tools and APIs (Application Programming Interface) necessary to begin developing applications on the Android platform using the Java programming language. Android has a large community of developers writing application (or apps) that extend the functionality of the devices. There are currently over 200,000 apps available for Android. Android Market is the online app store run by Google; though apps can be downloaded from third party sites. iOS (previously iPhone OS) is a mobile operating system developed and distributed by Apple Inc. Originally unveiled in 2007 for the iPhone, it has been extended to support other Apple devices such as the iPod Touch, iPad, iPad Mini and second-generation Apple TV. Unlike Microsoft's Windows Phone and Google's Android, Apple does not license iOS for installation on non-Apple hardware. As of October 2013, Apple's App Store contained more than 1 million iOS applications, 500,000 of which were optimized for iPad. Windows Mobile is a compact mobile operating system developed by Microsoft, and designed for use in smartphones and mobile devices.

Bluetooth: It is a short-range, low-power radio technology that simplifies communications among Internet devices and between devices and the Internet. It also simplifies data synchronization between Internet devices and other computers. Products with Bluetooth technology are required to qualify and pass interoperability testing.

Wireless Middleware: Wireless middleware is software that shields applications from the underlying wireless network. Use of middleware makes it easier to develop new wireless applications and to port existing applications to the wireless environment. CORBA and JMS (Java Messaging Services) are examples of wireless middleware.

Wireless Application Protocol (WAP): WAP is a set of communication protocols that is designed to enable different kinds of wireless devices to talk to a server installed on a mobile network so that users can access the Internet. WAP was designed especially for small screens and limited bandwidth mobile devices. WAP has become the de facto standard for delivering Web content to mobile devices.

In WAP architecture, there are two servers, a WAP gateway and a Web server. A WAP gateway receives a request from a mobile device via a communications tower. It then translates it into a form understandable by a Web server, and passes it on through the wired Internet to the appropriate Web server. When web server sends the response back to the WAP gateway, the WAP gateway translates it into a form that is understandable by the mobile device. The translated communication is then sent to the communication tower and back to the mobile device.

Wireless Markup Language (WML): WML is the scripting language used to create content to be delivered to wireless handheld devices. It is based on XML and it is fast because it removes unnecessary content, such as animation, from web pages. WML works with WAP to deliver content. WML tags specify how the web page should be formatted on a wireless device.

Voice XML: Voice XML is an extension of XML designed to accommodate voice.

Web clipping: Web clipping refers to the extraction of relevant pieces of static information from a Web site and format it in form that could be displayed on a Web-enabled mobile device. Web clipping eliminates excess content and graphics that can make browsing the website on a wireless device cumbersome. For example, you can clip the headlines from online news portal, clip sports scores or clip stock quotes for specific companies. Web clipping conserve mobile device resources by extracting the static data once and storing that data on the mobile device. The mobile device then makes a wireless connection to a web server in order to retrieve any dynamic content. Palm™, the leading manufacturer of PDAs, has designed Web-clipping applications for many of the most popular Web sites. PalmWeb clipping uses a proxy server to respond to queries for Web pages. A proxy server lies between the client (such as a Web browser) and the regular Web server. The main reason web clipping was created was to allow mobile palm users access to information on the internet virtually wherever and whenever they wanted. The main advantage of using web clipping is efficiency. It is all about getting the need information as fast as possible and with the least amount of effort. It allows users to specifically locate what they want on the internet with just the use of their wireless enabled PDA. Web clipping literally clips the important information from the rest of the "noise" on the internet. Whether a stock investor wants up to the minute stock quotes or a caffeine addict looking for the nearest Starbucks, all the information is available instantaneously and accurately. Web Clipping is perfect to accommodate the limited bandwidth of the wireless PDA's. Instead of browsing a rich content web page, which takes quite a long time, web clipping retrieves exactly what the user wants in a fraction of that time. Some of the main problems with web clipping come from the lack of bandwidth of PDA devices. The limited bandwidth does not allow very much data to be passed. Therefore, these clippings are scaled down from normal web pages. In addition, bandwidth is very expensive. This leads to a restriction for data that can be sent in both the query and the response. In addition, users may not want to pay extra when they exceed their bandwidth quota so they may reduce their use of web clipping. Advertising is also not realistic since users would not want to use their limited bandwidth for advertisements, although this could also be seen as an advantage. In addition to low bandwidth, there is high latency. Therefore, the requests take a long time to process (about 10 seconds).

Wireless Video Technology: This technology allows streaming video and rich multimedia on mobile devices. PacketVideo (www.packetvideo.com) specializes in wireless video technology for mobile devices and its technology is used worldwide.

Mobile Networks

Most mobile networks are operated by global communications and cellular phone companies. A mobile phone converts voice, text, or multimedia into radio signals. Mobile phone base stations transmit and receive these transmissions and connect the mobile phones to other phones and other networks. The various base station controllers are connected to mobile switching centers that connect the mobile network with the public wired phone network. Mobile phone networks are divided into thousands of overlapping, individual geographic areas (called cells) each with a base station. The size of a cell depends on coverage area and number of calls made in coverage area. When a mobile phone moves from one area to another, transmissions from mobile phone are automatically routed from a base station in one cell to a base station in another, thus providing the best signal and available capacity.

One of the major problems facing the mobile communication system providers is how to service extremely large numbers of users given limited communication bandwidth. This is done through multiplexing protocols. There are three main multiplexing protocols

Frequency Division Multiple Access (FDMA): FDMA is the division of the frequency band allocated for wireless cellular telephone communication into 30 channels, each of which can carry a voice conversation or, with digital service, carry digital data. FDMA is a basic technology in the analog Advanced Mobile Phone Service (AMPS), which is the most widely installed cellular phone system in North America. FDMA assign each channel to only one user at a time. FDMA is used by 1G system.

Time Division Multiple Access (TDMA): TDMA is a technology used in digital cellular telephone communication (2G systems). This technology assigns different users different time slots on a given communications channel (e.g., every one-eighth time slot). The division of cellular channel is done to increase the amount of data that can be carried.

Code Division Multiple Access (CDMA): CDMA, a form of multiplexing, allows many signals to occupy a single transmission channel, thereby optimizing the use of available bandwidth. The technology is used in ultra-high-frequency (UHF) cellular telephone systems in the 800-MHz and 1.9-GHz bands. CDMA is used with most 3G systems.

Pros and Cons of M-commerce

For Users: M-commerce provides anytime, anywhere access to commerce services. Users can access Internet using smartphone or other Internet-enabled wireless device avoiding the need to buy a PC. Even though the cost of a PC can be as low as $300 (or even less), it is still a major expense for the vast majority of people in the world. Furthermore, one needs to learn how to operate a PC, service it, and replace it every few years to keep it up-to-date. According to (CnetNews, 2010), prices of Mobile and broadband were dropping worldwide. From 2008-2009, the prices of information and communication technologies dropped by 42%. Over the years, the price of wireless devices is declining, and the per-minute pricing of mobile services was expected to decline by 50 to 80 percent before 2005. At the same time, functionalities are increasing.

One disadvantage of mobile devices is small size of mobile device screens and their relatively limited multimedia capabilities. In addition, the costs/tariff on mobile data transmission over public mobile networks is relatively high. Input technologies are developing and markets may be less open for M-commerce than E-commerce. The standardization of security methodologies for wireless-enabled websites and Wi-Fi hotspots is still developing. This could hamper the M-commerce development because of low customer trust on wireless devices security to make mobile payments. Users of mobile devices may experience multipath interference, weather and terrain problems. Reception of mobile devices may be poor in tunnels and certain buildings. GPS does not work well in cities with tall buildings. With increase in bandwidth, the power consumption of mobile devices increases which reduces the battery life.

For Network Operators: For network operators, the biggest advantage is huge number of predicted wireless users: 650 Million 3G wireless users by 2010 (Venkataram, 2003). Since 80 % of worldwide wireless users are still on GSM technology, this means a widespread deployment and acceptance of 3G technologies. That could lead to make mobility a commodity (widespread use, low-cost, and little differentiation in service). According to (CBSNews, 2010), the number of mobile phone subscriptions worldwide reached 4.6 billion in 2009 and was expected to increase to five billion in 2010. The number of mobile devices with Internet access has increased sharply over last many years. Number of mobile devices accessing the Internet was expected to surpass one billion by 2013 (IDC, 2009). However, network providers must open up their network services and information to external service providers if they are to provide a wide range of commerce services. That can make network operators concerned about losing control of their networks and the value that they generate.

For Service Providers: For service providers, getting information about user location means that they can use location-aware advertisements. Since M-commerce provides convergence of information standards and communications networks, service providers can create completely new services. However, this could also means closer mobile networks as network operators seek to encourage and/or limit network users to choose preferred service providers. Use of mobile phones is becoming a social phenomenon, especially among teens. It is expected that teen-aged group, once begin to make and spend reasonable amounts of money, will be a major driver of M-commerce growth. To properly conduct M-commerce, it is necessary to have sufficient bandwidth for transmitting text, voice, video, and multimedia. The 4G technology is expected to provide that, at a data rate of up to 2 mbps. One major concern is that wireless websites based on WAP are able to provide rudimentary applications compared with standard websites. Overall, there were fewer than 50,000 WAP sites worldwide in 2003 compared with millions of standard websites. Percentage of WAP phones shipped has steadily increased over the years. Most modern models of cellular phones are WAP capable. As a result, the number of WAP sites is increasing. Therefore, there is need for and development of services specifically tailored to mobile users employing WAP phones (Chittaro & Cin, 2002). National culture also influences the perception of mobile Internet services by the users who have adopted these services. Therefore promoting positive perceptions by mobile service adopters is necessary to stimulate continuing use, loyalty, and the broadening of the initial relationship (Lee, Choi, Kim, & Hong, 2007).

Service providers can also take advantage of new-and-flexible-service delivery methods supported by new pricing schemes. Offering users an opportunity to pay according to their usage (e.g. low price for regular usage at lower speeds and high price for fast download of large files at high speed) can be attractive choice for users. A service, attractive to users, would also increase the profit of the service provider. With the proliferation of large new domains of infrastructure use (e.g. mobile telecommunications and video delivery) these new pricing scheme has great potential (Bandyopadhyay & Cheng, 2006).

Applications of M-Commerce

Mobile Financial Applications

These services include banking, brokerage, and payments for mobile users. Mobile banking allows bank customers to access account information and makes transactions (e.g. purchasing stocks and remitting money). For instance, Citibank has a diversified mobile banking service that consumers can use to access account balances, pay bills, and transfer funds using SMS. Many banks in Japan allow all banking transactions to be done via cell phone. A number of companies are now providing their customers with the option of paying their bills directly from a cell phone. For example, HDFC Bank of India (hdfcbank.com) and Citibank, Malaysia provides customers the service to pay their utility bills through SMS. Similarly, a number of mobile service and network providers such as Airtel of India allow their customers to pay their bills by phone.

Mobile Marketing and Advertising

Mobile Marketing Association defines mobile marketing as "a set of practices that enables organizations to communicate and engage with their audience in an interactive and relevant manner through any mobile device or network" (MMA, 2009).Mobile marketing can include marketing on or with a mobile device, and marketing in a moving fashion e.g. moving billboards. The importance of mobile marketing can be realized by increasing trend of mobile phone devices. A research indicates that millions of mobile phones are being used in world. The smart phones are widely and which use application which can show mobile marketing ads. A very large number of people use mobile phones for socializing through Twitter and Facebook by using internet on mobile phones. These social websites can also be used for advertising.

What differentiates mobile marketing from most other forms of marketing communication is the fact that mobile marketing is often initiated by consumer and future communication requires the express consent of the consumer. With technological advancement, companies are increasingly using mobile devices for new forms of advertising. Many consumers carry mobile devices wherever they are. Advertisers are now able to send them not only personalized ads but also provide some of the services not possible using other platforms (i.e. PCs) (MMA, 2008). These services include SMS, MMS, banners, advertising within applications (such as games), and downloads (such as wallpapers, themes, and ringtones). According to a 2006 study, approximately 43% of US mobile video viewers (more than 19 million consumers) were willing to watch advertisements on their handsets in return for free access to mobile TV and video services (MAG Mobile Advertising Task Force, 2007, p. 17).

Mobile marketing provides many advantages to businesses. Message can be received instantly when advertisers sent it to target customers, provided the mobile phone is turned on. It is less expensive as compare to when you create advertising content for other advertising types. Mobile devices are convenient to use and mobile marketer can directly send their ads to those whom they want to send and can get feedback through SMS. Mobile advertiser can analyze the target customer behavior immediately and based on customer behavior they can improve their product and service. Mobile marketing has an advantage of multiple effect for example the content in mobile phones can be shared easily so when mobile marketer send an ad of information as a marketing campaign then it is big possibility that these people might share it to their friends and family. Most of the people have their mobile phone with them. Therefore, it is a big advantage that you can cover lot of area or you can share your information to the distant areas of country. Mobile marketers also have advantage to identify the location or can target niche market by using geo location, GPS and Bluetooth technology. People use Facebook and Twitter on their mobile phones so it is advantage for marketer that they can place mobile formatted ads.

There exist some disadvantages of mobile marketing as well. There are many types of mobile phones and have different screen which may destroy your marketing campaign because the ad might not display to all these mobile phone which you are targeting. Mobile marketers or advertiser must take care of privacy of their target market. People might not feel comfortable to share their information online so mobile marketer must take permission. Mobile phones have no mouse and navigation is difficult even in touch screen mobiles. Mobile marketers must create content for their campaign too short and to the point.

Location-based Services (LBS)

Location-based services (LBS) are used to send custom advertising and other information to mobile phone subscribers based on their current location. The location information is obtained through a GPS chip built into the mobile phone. For non-GPS enabled phones, the location information can be obtained using radiolocation based on the signal-strength of the closest cell-phone towers. Companies are now offering location-based services (such as local maps, local offers, local weather, and people tracking/monitoring) to mobile users. LBS can be beneficial for both consumers and businesses. Consumers can use this service to pinpoint their location in an emergency, and businesses can use them to provide services that meet customers' needs.

SMS Marketing

Over the years, SMS has become a legitimate advertising channel in many parts of the world. , The IAB and the Mobile Marketing Association have established guidelines for SMS marketing. SMS marketing is especially expanding rapidly in Europe and Asia as a new channel to reach the consumer. SMS marketing is rapidly growing as one of the best marketing communication tool to reach masses using bulk-messaging services. A SMS can be sent to a number of recipients at the same time in a simple, cheap and cost effective way. Use of SMS marketing is increasing with ongoing toughened economic conditions. SMS messages, on average can be read within few minutes and provider a very effective way for business to keep their customers well informed and up-to-date about various promotional activities.

In E-mail marketing, up to 50% or more of your customers may not receive your messages due to spam filters on their end or unreliability of e-mail relaying through the internet. SMS messages almost reach to every customer since you are changed for every SMS message sent. There are rare cases in which SMS message is not reached otherwise it is safe to say that majority of your recipients will receive it. The receiver of SMS message not necessarily knows that what the message is and who has sent it. However, he must open it to find out what the message is. This reason usually guarantees that SMS message is always read by the receiver. Nowadays, users carry mobile phones wherever they go. This also makes SMS messages convertible. If marketing message constructed well, with an incentive to respond quickly via return SMS, they are more likely to respond to your offer quickly than via e-mail marketing.

MMS Marketing

Multimedia Messaging Service, or MMS, is a standard way to send messages that include multimedia content to and from mobile phones. It is most popular use is to send photographs from camera-equipped handsets, although it is also popular as a method of delivering news and entertainment content including videos, pictures, text pages and ringtones. The standard is developed by the Open Mobile Alliance (OMA), although during development it was part of the 3GPP and WAP groups.

MMS marketing can contain slideshows of images, text, audio and video. Most new phones are produced with a color screen and able to send and receive standard MMS message. MMS marketing allows companies to interact with their customers lively and grab their attention. One example of MMS marketing is House of Blues. At ongoing campaigns at House of Blues venues, the brand allows the consumer to send their mobile photos to the LED board in real-time as well as blog their images online (Tasner, 2010).

A good example of MMS mobile originated Motorola's ongoing campaigns at House of Blues venues where the brand allows the consumer to send their mobile photos to the LED board in real-time as well as blog their images online. Mobile MMS videos increase user interest and boost direct response rates offering marketers and advertisers a viable mobile channel to deliver richer media content. Since the release of the iPhone and other smart phones with high resolution displays, mobile marketing and advertising firms, such as Crisp Wireless, have developed media-rich advertising. However, SMS text messaging still dominates.

Mobile Marketing via Games

Mobile game marketing can involve interactive real-time 3D games, massive multi-player games, and social networking games. With predicted 650 Million 3G wireless users by 2010, the potential audience for mobile games is substantially large. Because of the market potential, Nokia is producing not only the phone/console, but also the games that will be delivered on memory cards. Companies can deliver messages within mobile games to reach their target audiences. Promotional messages are now being delivered within mobile games and companies are sponsoring entire games.

Mobile Marketing via Applications

It refers to marketing or advertising through mobile application. Mobile applications mean software on mobile, which performs different functions. This software may be pre-installed or users can install this software themselves. When this software is developed, the advertiser can also put their message.

Mobile Marketing on Web

This refers to advertising on web pages specifically meant for access by mobile devices. Google and Yahoo are some major mobile content providers selling their web space for advertising placement.

Bluetooth Marketing

In Bluetooth marketing, the consumer is asked to receive a message when a consumer turns on his Bluetooth mobile device. In most form, the customer is provided with some form of promotion code if he accepts to receive the message. Consumer can use this promotion code to get some benefit.

Mobile Viral Marketing

Mobile viral marketing relies primarily on consumers to transmit adverting to their contacts who are also mobile phone users.

Mobile Targeted Advertising

With the knowledge of current location of mobile users (using GPS) and their preferences or surfing habits, marketers can send user-specific advertising messages via SMS messages or short paging messages. These advertisements can also be location sensitive e.g. to inform a user about restaurants close to where he is. When more wireless bandwidth becomes available, rich-media advertising involving audio, pictures, and video clips will be generated for individual users with specific needs, interests, and inclinations.

Messages sent to a mobile phone are more likely to be seen than email sent to a PC, which can get caught in the spam filter. A cell phone is a very personal device that people take with them wherever they go, making it easy for marketers to develop a relationship with customers through this medium. Carriers have customer data and location information potentially available for targeting. Personalization, immediacy, and interactivity of mobile ads encourage response by consumers on the go. Mobile marketing can help build a customer's database. Once customers opt in to receive an ad, you can use information for loyalty marketing and customer's retention.

Mobile Auction

This includes services for mobile customers to buy/sell certain items.

Mobile Booking and Ticketing

These services allow customers to book, tickets for travel, hotel and events. A variety of techniques (e.g. by using a simple application or accessing a WAP-enabled portal of ticket provider) can be used to deliver tickets on the mobile phones. Users can then show these tickets at the venue. Same techniques can be used to book or cancel the tickets. Vouchers and coupons can also be distributed through mobile phones. A voucher or coupon is sent to a mobile phone in the form of a virtual token. Customer can present his mobile phone to show token at any point of sale.

Mobile CRM

Mobile CRM can extend a company's reach by providing access to customers as well as to employees and partners alike on a 24/7 basis. The sales force automation and field service functions of mobile CRM can provide salespersons and field service personnel mobile access to required information (e.g. recent billing history for a particular customer or current availability of various parts in order to fix a piece of machinery) from anywhere anytime.

Salesforce.com products (Sales Cloud 2 and Service Cloud 2) provide CRM solution with mobile capability. A combination of CRM and voice portal can be connected to company's legacy systems to provide enhanced customer service or to improve access to data for employees. For examples, customers on remote locations can use a vendor's voice portal to check on the status of deliveries to a job site. Service technicians could he provided with diagnostic information, enabling them to diagnose more difficult problems. Salespeople could check on inventory status during a meeting to help close a sale.

Mobile Storefronts

A mobile storefront, such as MartMobi, helps businesses quickly create mobile sites and apps, a valuable service to small businesses that have limited physical reach on their own or resources to independently spend on coding, but could potentially sell to the world if showcased online. A mobile storefront can automatically generates mobile and tablet websites, or mobile storefronts, as well as apps for iPhones, iPads, Android smartphones and tablets, saving development costs for online businesses, which could use its technology to build websites and apps in a way that required no coding skills.

Mobile Brokerage

Stock market services offered via mobile devices have also become more popular.

Mobile Vouchers

Mobile ticketing technology can also be used for the distribution of vouchers, coupons, and loyalty cards.

Mobile ATM

Mobile ATMs have been specially engineered to connect to mobile money platforms and provide bank grade ATM quality.

Mobile Ticketing

Tickets can be sent to mobile phones using a variety of technologies. Users are then able to use their tickets immediately.

Mobile Purchase

Catalog merchants can accept orders from customers electronically via the customer's mobile device.

Mobile Content

Mobile content refers to any type of media, which is viewed or used on mobile phones and can include text or multimedia content hosted on standard websites or websites developed for mobile devices. Common type of mobile content include ringtones, games, graphics, any audio file that is played on a mobile phone, video, mobishows (a program produced, directed, edited and encoded for the mobile phone), streaming radio, streaming TV, and live video. A mobile game is a game, which can be played on PDA, Smartphone, mobile phone, handheld computer or portable media player. This excludes the games, which are played on handheld video game systems for example, Nintendo Ds or Play station Portable. Mobile images can be used as a wallpaper or screensavers. It can also be used to display when a particular person calls. Users can download the free content from the sites like adg.ms. Mobile music is an audio file, which can be played on a mobile phone. It is usually formatted as an MP3 or an AAC (Advanced Audio Coding) file. There are different forms of video in mobile such as RTSP, MPEG-4, 3GPP and Flash Lite. Mobile streaming radio is an application, which streams on-demand live radio stations or audio channels to the mobile phone. Mobile video also appears in the type of streaming TV over the network of mobile. The network must be of 2.5 or 3G network, it stimulates a television station in a way that users cannot choose their desired channel but had to watch whatever channel is streaming at that time. Mobile broadcast TV broadcast the matter like a traditional television station to different spectrum. This allows handling data usage and calls due to "one-to-many" nature. The quality of video is far better than "one-to-one" system.

From cell phone, live videos can be streamed & transferred. It can also be shared to friends through emails and social networking sites. It also stream live videos via network of cell or Wi-Fi.

Recent mobile phone are usually developed with WML,HDML, WML Script and XHTML basic for WAP and JAVA based on the CLDC/MIDP platform. HDML (Handheld Device Markup Language) is a language, which permits the text section of web pages to be displayed on personal digits assistant (PDA) and cellular telephones through the access of wireless. HDML is an open language, which offers free royalty. The difference between the WML and HDML is that HDML is not based on XML whereas WML is XML-based. Secondly, WML permits its own JavaScript version known as WML Script whereas HDML does not permit scripting.

With increasing capabilities of modern mobile phones, the use of mobile content has grown accordingly. Users can use these phones to make calendar appointments, send and receive text messages, listen to music, watch videos, shoot videos, redeem coupons for purchases, view Microsoft Word documents etc. With higher bandwidths available, music vendors are offering instant delivery of songs from their music libraries for online purchase e.g. Amazon MP3 (AmazonMP3.com) and apple's iTunes Store (itunestore.com). Location-based services can even be integrated to target subscribers with location-sensitive streaming content (e.g. trailers of movies being shown at nearby theaters). Currently, the most popular mobile content being purchased and delivered includes ring-tones, wallpapers, and games for mobile phones. The convergence of mobile phones, mp3 players and video players into a single device is increasing the purchase and delivery of full-length music tracks and video. Download bandwidths, if increased to 4G levels, will make it possible for users on the go to buy a movie on a mobile device in a couple of seconds.

According to a report, revenues of UA mobile content will go up to $1.54 billion, stimulated by the increasing consumer demand for mobile games, video and music. According to research companies, there will be a further growth because of increasing number of smart phone users and tablet devices present in the market. According to e-marketer's forecast of 2014, revenues will reach up to $3.53 million. $849 million of revenues has been generated from 64 million mobile gamers in country. Mobile TV and video services will generate revenues of $719 million in current year and $1.3 billion in 2014. Currently, the main reason of driving revenues to publisher is subscription or paid content. From 2014, growth in ad-supported revenues is expected to grow at more than double the rate of paid mobile content.

Mobile Entertainment

Mobile entertainment comprises a range of activities associated with mobile electronics. The explosive growth of new entertainment forms such as blogs, personal web pages, mobile communities and user-generated content has seen the consumption of traditional media decline rapidly in recent years. The mobile phone is playing an increasingly important role in the delivery and billing of entertainment content. Sophisticated mobile phones are now common among subscribers and the scope and demand for delivery of compelling consumer services (such as TV, Music, Video, ringtones, games and social networking) is increasing daily. Along with basic web surfing, mobile users are also finding entertainment via their handhelds, specifically in the form of mobile video. Mobile Video will gain more and more momentum, as mobile users get lots of useful content including information, films and other entertainment related content on their mobiles. Many mobile network operators worldwide have rolled out commercial mobile TV services already, and consumers are becoming more discerning in their demand. Mobile TV will generate additional revenue stream for service providers and content providers. Mobile music is another major part of mobile entertainment. Mobile music will remain the largest single sector of the mobile entertainment industry over the next five years. Statistics form Statista.com shows that revenues from music reached $6.9 billion in 2014 bolstered by the increasing availability of full-track download and streamed services, the former in both paid-for and rental formats. IPhone and its App Store have reinvigorated the mobile games industry at every level, including developers and publishers, but also handset makers and operators who are shaking up their own services and making games a priority.

Mobile Purchases

Mobile purchase allows customers to shop online at any time in any location. Customers search products, compare prices, order, and view the status of their order using their mobile devices. Customers can receive from retailers a products list directly on their mobile devices or they can visit mobile version of a retailers E-commerce site. Additionally, retailers can use location-based services to track customers and provide them discounted offers at local stores near their location. Many vendors allow customers to shop from mobile devices. For example, customers with Internet-ready cell phones can shop at certain sites such as mobile.yahoo.com or amazon.com. Mobile customers can also participate in online auctions. For example, eBay offers eBay mobile service (pages.ebay.com/mobile). Using this service, eBay account holders can access their accounts, browse, search, bid, and rebid on items from any Internet-enabled phone or PDA.

Mobile Payments

A mobile phone can be used for payment of various good and services e.g. Music, videos, ringtones, online game subscription etc. Some primary methods for mobile payments include premium SMS-based payments, direct mobile billing, mobile web payments, and Contactless Near-Field Communication (NFC) (Eagle, 2010). According to (Gartner, 2010) the number of mobile payment users worldwide will exceed 108.6 million in 2010. According to Juniper, mobile payments will increase 275% to reach $670 billion by 2015. In 2017, nearly half of all mobile payments will fall into the mobile proximity category reaching $41 billion. Global mobile payments will reach $1.3 trillion in 2017.Worldwide physical goods sales from mobile/nomadic devices will account for 30% of e-retail by 2017. More than 2/3 of mobile transactions use alternative payment methods like PayPal instead of credit or debit cards.

Premium SMS: In premium SMS method, the consumer sends a SMS message containing a payment request to a short code. The service provider applies a premium charge to consumer phone bill and informs the merchant of the payment success. Upon receiving the information, merchant can then release the goods.

Direct Mobile Billing: In direct mobile billing, the consumer uses his mobile account to pay for goods purchased at an E-commerce site. In this process, the consumer's mobile account is charged for the purchase. It is also called in-app billing or in-app purchases. Examples include PaymentOne Boku, and mopay. After two-factor authentication involving a PIN and One-Time-Password, the consumer's mobile account is charged for the purchase. Direct mobile billing is a true alternative payment method that does not require the use of credit/debit cards or pre-registration at an online payment solution such as PayPal. This method bypasses banks and credit card companies altogether. This type of mobile payment method, which is extremely prevalent and popular in Asia, provides many benefits including security (Two factor authentication), convenience (no pre-registration and no new mobile software is required), ease of use, and speed.

Mobile Web Payment: In mobile web payment, the consumer uses web pages displayed or additional applications downloaded and installed on the mobile phone to make a payment.

Contactless Near-Field Communication: This method is mostly used for purchases in physical stores or transportation services. A customer using a special mobile phone equipped with a smartcard waves his phone near a reader device. The payment is then deducted from prepaid account or charged to mobile or bank account owned by the customer. By 2015, a full 50% of smartphones will be NFC-enabled. About 4% of UK population uses contactless payment cards and an additional 22% are interested in using one. Within five years, 78% of POS terminal will be NFC-enabled in Europe (NFC Bootcamp, 2015).

Other Methods of Mobile Payments

There are some other methods of mobile payments.

Mobile at the Point-of-Sale (The Mobile Wallet): It is paying for a good/service at a store with a mobile device using NFC and Tap-and-Go method. Examples include Google Wallet, MasterCard, VISA, and ISIS.

Mobile as the Point-of-Sale: In this, merchants use a mobile device to process credit card payment. This type of payment is not same as using mobile wallet. Examples include Verifone and Square.

Mobile Payment Platform: This is a method in which consumers can send payments to the merchants or to themselves. This payment can be done at point-of-sale, online, or using a text message. Examples include PayPal, Serve, and ApplyPay.

Closed-Loop Mobile Payments: In this method, companies develop their own wallet of mobile payment platform that customers can use to make mobile payments. A good example is Starbucks. The Starbucks mobile payment apps let you conveniently pay for purchases, earn stars, and redeem rewards with My Starbucks Rewards, find stores etc. The app has helped boost My Starbucks Rewards, its loyalty program, because it makes it easier to track purchases to get free drinks and other perks. In the first quarter, Starbucks added 900,000 new rewards members, for 9 million, up 23% from a year ago (Hof, 2015).

FIGURE 11-1: WORKING OF MOBILE PAYMENT SYSTEM

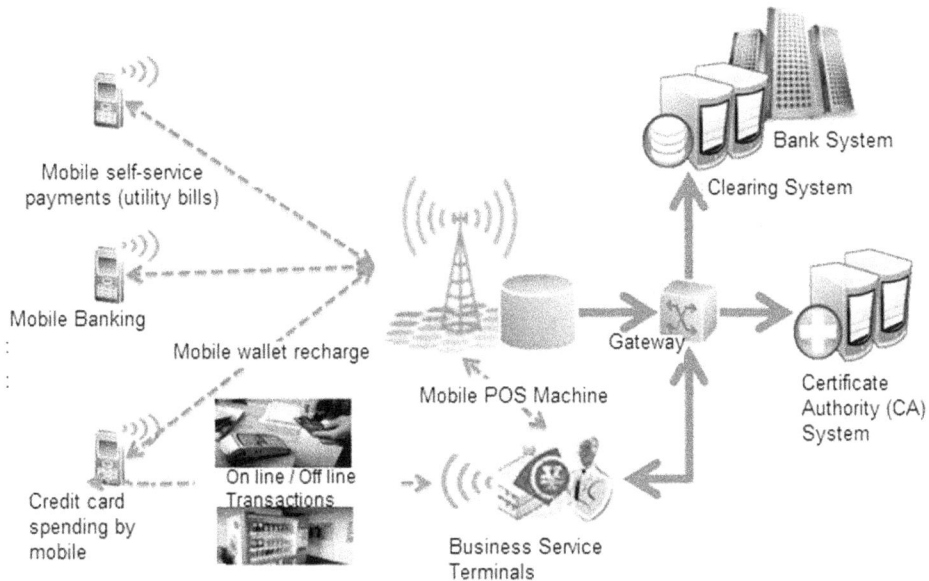

Mobile payments face some specific challenges. Merchants must design NFC into services, buy and upgrade technologies. Disparate merchants and technology solutions will initially slow customer adoption of NFC. Personal identification is managed by governments and will be slow to see smart phone integration.

Wireless Wallets

A wireless wallet is software that can store an online shopper's credit card numbers and other personal information. Users can then use wireless wallet to automatically provide this information every time they make an online purchase. Mobile Wallet (mobilewallet2u.com) is a mobile enabled payment service provider that can be used by users to make payments conveniently to any designated affiliates or Mobile Wallet's appointed network of merchants. Major players in mobile wallets are traditional credit card companies (Visa), Tech giants (Google) and Underdog startups (Square). To use a mobile wallet, user registers and input his phone number. The provider then sends him a SMS with a PIN. User enters the received PIN, authenticating the number. User inputs his credit card information (if not already registered user of mobile wallet) and validates payments.

Wireless Telemedicine

Wireless telemedicine is currently being used for two purposes: for storage and forwarding of digital images from one location to another and for real-times consultation between patient and physician at remote locations (through videoconferencing). New wireless and mobile technologies, especially the next generation, are opening up a number of new and novel application opportunities.

Wireless Application Portal

A site that offers variety of services (content, messaging etc.) to wireless device (e.g. smart phone) users. These portals allow multiple users to access large amount of information. Wireless portals are gaining popularity with increasing popularity of wireless PDAs and Web-enabled cell phones, wireless portals can have variety of users. Companies can use them to give online access to on-the-road sales staff, enabling them to obtain real-time inventory status, and track customer orders.

M-Commerce Security Issues

Today, most of the cell phones have their applications, operating systems, and other functionality burned right into the hardware. Since these phones do not store any application, it is not possible to propagate a virus, worm, or other rogue program from one phone to another. For a malicious program, it is difficult to permanently alter the operation of a cell phone. However, with increasing capabilities and functionality of cell phones and convergence of cell phones and PDS functionality, the threat of attack from malicious code will certainly increase. (Raina & Harsh, 2001) state that mobile devices and mobile transactions produce some unique security challenges. These challenges include the following.

Physical Security Issues: Due to their small size, mobile devices are very attractive target for theft. A stolen mobile device can deprive owner of valuable data and digital credentials that can be used to compromise an m-commerce network.

Transactional Security Issues: Eventually all M-commerce transactions go through some public network and therefore end-to-end security (from the consumer to the M-commerce server) must be maintained. The situation becomes complex because you not only have to take care of security issues specific to wireless world but also the security issues facing the wired world.

Post-Transaction Issues: The overall M-commerce system must be able to provide some method of proving that a particular transaction has occurred. This proof is necessary for problem resolution after a transaction has occurred. This proof can be in the form of a digital receipt or some other type of proof.

M-Commerce Security Techniques

M-commerce transactions eventually go through the wired Internet. Therefore, many of the processes, procedures, and technologies used to secure E-commerce transactions also can he applied in M-commerce. E-commerce security is discussed in detail in Chapter on E-commerce security. Here we will discuss only some of the techniques that apply directly to mobile devices and networks.

SIM-Based Authentication

GSM, 2.5G, 3G technologies all include Subscriber Identification Module (SIM). A SIM is an extractable smart card that holds an authentication key and other vital information about the subscriber. The authentication key also is stored in a database of GSM network. SIM is used on mobile handsets to identify the user, to provide customer location information, and transaction processing. SIM also makes it possible for a handset to work with multiple phone numbers. SIM cards protect against unauthorized use of a particular subscriber's account but they do not prevent the use of a stolen cell phone. A stolen cell phone can be used by replacing the existing SIM. By using a cell phone's International Mobile Equipment Identity number it possible to track such a stolen phone (Evers, 2001).

Upon turning on the phone, user is asked to enter a PIN code. This protects the cell phone against illegal use if it happens to be stolen or lost. By providing the PIN, the cell phone and the network engage in a challenge-response process of authentication. If the process is successful, cell phone is authenticated and communication can take place. Although SIM cards protect against unauthorized use of a particular subscriber's account, they do not prevent the use of a stolen cell phone. If a thief steals a phone, the thief can simply replace the existing SIM card with another one and sell it on the open market. A cell phone's International Mobile Equipment Identity number can be used to track a stolen phone (Evers, 2001).

Wireless Transport Layer Security (WTSL)

You can secure the transmissions between the WAP gateway and the Web server through the wired Internet security protocols (e.g., PKI, SSL/ TSL, the Secure Socket Layer and Transport Layer Security). TSL enables encrypted communication. However, these protocols cannot be used in communication between WAP gateway and mobile device. For this communication, WAP uses WTSL protocol (Wireless Transport Layer Security). WTSL enables encrypted communications between a mobile device and the WAP gateway. In addition, WTLS supports the key elements of PKI i.e. public and private encryption keys, digital certificates, digital signatures, and the like.

Wireless Identity Module (WIM)

A WIM is a smart card device that can hold the security keys and digital certificates used by the gateway and the Web server to encrypt/decrypt communications. WIM is much like a SIM and can be used in combination with WTLS.WIM can be issued by a bank or other financial institution to handle M-commerce payments and transactions.

Voice Technologies for M-Commerce

Voice and data can be combined to develop useful applications. Voice applications can enable hand- and eyes-free operations to increase the productivity, safety, and effectiveness of mobile computer users. These technologies can assist disabled persons to perform various tasks. Two popular voice applications are interactive voice response and voice portals.

Interactive Voice Response (IVR)

IVR systems enable users to interact with a computerized system to request and receive information and to enter and change data. The communication is conducted through regular telephone lines or through 1G cell phones. Using this system, for example, patients can schedule doctor appointments and users can request a pick-up from UPS. Examples of companies providing IVR solutions include 2ergo.com and angel.com.

Voice Portals

A voice portal is a Web site with an audio interface accessed through a standard or a cell telephone. Customers connect to the participating site using a certain phone number and then make their requests verbally. The system finds the required information, translates into computer-generated voice reply, and replies to the user. For example, AOL provides many voice services including AOL By Phone (http://voice.aol.com/). This service allows consumers to access their emails through phone. iPing.com (iping.com) is a reminder and notification service that allows users to enter information via the Web and receive reminder calls. Eckoh (eckoh.com) is the UK's leading developer of speech recognition solutions for customer contact centers.

M-Commerce: Trends, Challenges, and Future

Today, mobile commerce is playing a growing role in many countries but its development varies from country to country. With increased access to broadband and advances in mobile device technology, the development may increase even more rapidly. Mobile devices have provided growing opportunities for both consumers and businesses to participate in E-commerce. Mobile technology is changing the way that consumers research, purchase and pay for goods and services. Smartphones will replace feature phones as the price point drops around the world. According to a forecast, the adoption rate of smartphones will effectively double to 6 billion by 2020 and number of smartphone connections will grow three-fold over the next six years (Molt, 2015). A consumer behavioral shift is equally as important to note. The next billion consumers will find out about your product first on mobile phone. These consumers will use their mobile phone to find out where to buy your product and will use the mobile phone to buy a product. It will be critical to know how to effectively reach consumers on mobile phones. Multiple factors are contributing to consumer belief that mobile payments will be adopted by a majority of mobile phone users. Consumers expect value-added services such as enhanced shopping experiences, real-time proximity offers, improved loyalty and other benefits. Mobile payments can save time by enabling a payment process within a mobile-only environment. While 2G is still the dominant technology in emerging markets, it is quickly being replaced by 3G service. The two main factors contributing to the shift to 3G are a rapid consumer adoption of smartphones and the decrease in smartphone prices. The expansion to 3G will help facilitate the growth of commerce in emerging markets. The mobile payment landscape remains fragmented with tech giant firms (such as Apple, Google and Amazon) and carriers battling/partnering with established retailers, financial institutions and other providers, all vying for broader consumer and merchant acceptance and adoption. Customers expect seamless and easy interactions with brands via mobile devices. A company's ability to optimize the customer experience across all channels will drive its m-commerce success. The user interaction will need to be designed as a mobile-first experience. The shopping and product experience needs to provide value to drive to purchase. The purchase experience needs to be as quick and simple as possible. A one-click experience is optimal. Customers want the same content, delivered to them on whatever device they want, so they do not have to waste time searching. Companies need to tailor the interface and content to easily allow customers to find what they are looking for. New media planning and marketing tools will be needed to plan, measure and deliver the right message, at the right time to influence purchase of your products. The mobile phone is the most personal and responsive path to the consumers. The investment in new media planning and marketing tools will need to be integrated in a new type of CRM tool (Hodges, 2014). Businesses should consider segment testing their target consumers with different marketing approaches to see when and how those consumers convert. Many major retail companies, such as Apple and Nordstrom, have introduced dedicated mobile point-of-sale (POS) devices in their stores to make transactions simpler and more convenient for employees and customers. Mobile POS is a valuable tool for small businesses, because it increases operational efficiencies. Mobile POS also provides small businesses with benefits that help improve their bottom line, such as accurate business reporting, improved projections and the ability to accept nearly any type of customer payment. Despite a lot of hype, mobile wallets do not seem to be catching on as quickly as projected. One possible reason is unclear value proposition. Mobile wallets need to have a concrete value proposition before small merchants will adopt them, and widespread availability needs to exist before consumers will start using the tools. With mobile design becoming more intuitive and responsive, it is expected that both the number of customers who make mobile purchase and response rate to mobile promotional campaign would increase. For companies considering creating a mobile app, be sure to look for one that makes in-app conversions easy. While expanding their mobile offerings, businesses should not forget their core business. Businesses should recognize and participate in some of the mobile trends that are changing their businesses, like online ordering, social media marketing and review sites. It is important to realize that mobile commerce is both enormous and irrelevant at the same time. A business still sells the same thing it always has, but almost all of its customers are going to find them online or via mobile first. So while mobile does matter, it should only serve to enhance, not distract from, core product (Fallon, 2014). However, use of mobile devices has given rise to some challenges as well. Some specific challenges include consumers' difficulty to access full information about the products and transaction due to limited size of mobile device screen and limited storage capacity; protection of children from certain forms of advertising, unauthorized access to customer personal data, lack of clear dispute resolution mechanism for complex contracts, and issue of privacy and security. Ensuring that children benefit from mobile devices' opportunities while receiving effective protection against aggressive, inappropriate and abusive mobile marketing practices and offers represents a key challenge for all stakeholders.

Collaborative Commerce (C-Commerce)

In collaborative commerce, a community of businesses in an industry supply chain collaborates with each other. The collaboration is done for two reasons: to improve the business processes of each firm within the community and to create a customer-centric environment with a large selection of good quality products and services readily available to the customer (Shuman, Twombly, & Rottenberg, 2001). By exposing its internal systems to participating businesses, each participating business helps the other businesses to become better at what they. Knowledge and information sharing are key aspects in a collaborative commerce. Other typical collaboration areas include product design, resource planning, forecasting, and replenishment, supply and demand visibility, and, marketing coordination.

C-Commerce is a realistic new business option which is already taken into account when planning new business models by the most progressive and innovative companies. The largest problem is the lack of standards. Another very important aspect that needs to be taken into consideration is the massive changes that need to occur within the organization. The human factor and the necessary change management methods are to be planned carefully. An example of a working C-Commerce solution is the company www.transora.com.br. Consisting of 49 companies (e.g. Coca-Cola, Kraft Foods, Sara Lee Corporation, Proctor and Gamble, and Unilever). Transora is a global trading community of Manufacturers, Retailers and GS1 Member Organizations with the objective to develop technology solutions to lower supply chain costs.

Pervasive Computing

Pervasive computing refers to an environment where almost any everyday use device (clothing, tools, appliances, cars etc.) can be imbedded with chips to provide an unobtrusive and always on connectivity to an infinite network of other devices. Pervasive computing combines current network technologies with wireless computing, voice recognition, Internet capability and artificial intelligence.

Ubiquitous Commerce

Ubiquitous commerce, also referred to as "u-commerce" or "über-commerce," is the combination of electronic, wireless/mobile, television, voice, and silent commerce. The core of the u-commerce vision is overcoming spatial and temporal boundaries. U-commerce has some specific characteristics.

Ubiquity: It means that computers will be everywhere, and every device will be connected to the Internet (Watson et al., 2002). It represents the ability to be connect at any time and in any place as well as the integration of human-computer interaction into most devices and processes, e.g. household objects.

Universality: The devices used in U-commerce are universal devices. A universal device will make it possible to stay connected at any place and any time (Junglas & Watson, 2006),

Uniqueness: U-commerce will add uniqueness of information. Uniqueness means that the information provided to the users will be easily customized to their current context and particular needs in specific time and place (Junglas & Watson, 2006). Uniqueness stands for the unique identification of each customer or user regarding his identity, current context, needs and location resulting in an individual service.

Unison: In a u-commerce environment, it is possible to integrate various communication systems such that there is a single interface or connection point to them (Junglas & Watson, 2003b; Junglas & Watson, 2006).

Junglas and Watson (2003a) view u-commerce as a conceptual extension of e-commerce and m-commerce. Besides E-commerce and m-commerce, other components of u-commerce are voice, television, and silent commerce. Voice commerce make use of computerized voice technologies: speech recognition, voice identification, and text-to-speech that businesses can use to reduce call-center operating costs and improve customer service. Television commerce is mainly used as an end-consumer channel. Silent commerce refers to the business opportunities created by making everyday objects intelligent and interactive. For example, radio frequency identification (RFID) chips allow the tagging, tracking, and monitoring of objects along an organization's supply chain.

Several characteristics of u-commerce will drive its growth. Ubiquitous continuous presence, the ability to capture context through sensors, and the ability to communicate with service providers make u-commerce attractive to businesses (Gershman & Fano, 2005). There are few global phenomena that will accelerate the growth of u-commerce (Schapp & Cornelius, 2001). These include pervasiveness of technology and growth of wireless. The explosive growth of nanotechnology, and the continuing capital investments in the technology at the enterprise level, increase the pervasiveness of the technology and expand the platform on which to leverage innovation and new applications. Wireless is one of the fastest-growing distributed bases: wireless networks have expanded around the globe; mobile phone usage and new applications have also exploded. Wireless commerce is, therefore, a critical component of u-commerce.

Review Questions

1. What is mobile commerce?

2. What is the importance of m-commerce?

3. List some specific characteristics of m-commerce?

4. List building blocks of m-commerce.

5. List various m-commerce software.

6. What is web clipping?

7. Differentiate among FDMA, CDMA, and TDMA.

8. What are some pros and cons of m-commerce?

9. Describe some applications of M-commerce.

10. What is collaborative commerce?

11. List some notable trends in M-commerce.

12. List some mobile wallets and their characteristics.

13. What do you understand by the term wireless wallet?

Bibliography

Accenture.com. (2015, March 9). U.S. Consumers Want More Personalized Retail Experience and Control Over Personal Information, Accenture Survey Shows | Accenture Newsroom. Retrieved August 15, 2015, from https://newsroom.accenture.com/industries/retail/us-consumers-want-more-personalized-retail-experience-and-control-over-personal-information-accenture-survey-shows.htm

Bandyopadhyay, S., & Cheng, H. K. (2006). Liquid Pricing for Digital Infrastructure Services. *International Journal of Electronic Commerce* , *10* (4), 47.

Bosomworth, D. (2015, July 22). Mobile marketing statistics 2015. Retrieved August 15, 2015, from http://www.smartinsights.com/mobile-marketing/mobile-marketing-analytics/mobile-marketing-statistics/

CBSNews. (2010, February). *Number of Cell Phones Worldwide Hits 4.6B*. Retrieved June 14, 2010, from www.cbsnews.com: http://www.cbsnews.com/stories/2010/02/15/business/main6209772.shtml

Chittaro, L., & Cin, P. D. (2002). Evaluating Interface Design Choices on WAP Phones: Navigation and Selection. *Journal of Personal and Ubiquitous Computing* , *6* (4), 237-244.

Cmocouncil. (2015, April). Mobile Marketing Facts & Statistics. Retrieved August 15, 2015, from https://www.cmocouncil.org/facts-stats-categories.php?view=all&category=mobile-marketing

CnetNews. (2010). *Mobile, broadband prices dropping worldwide.* Retrieved July 23, 2010, from news.cnet.com: http://news.cnet.com/8301-1023_3-10458877-93.html

Eagle, L. (2010). *Transitioning to Mobile Banking What Every Financial Institution Should Know.* Retrieved July 23, 2010, from Banker's Academy Briefings: http://www.bankersacademy.com/pdf/Transitioning-to-Mobile-Banking.pdf

eMarketer.com. (2010, August 31). Mobile Content Revenues to Top $1.5 Billion - eMarketer. Retrieved August 12, 2015, from http://www.emarketer.com/Article/Mobile-Content-Revenues-Top-15-Billion/1007903

eMarketer.com. (2014, June 11). Worldwide Smartphone Usage to Grow 25% in 2014 - eMarketer. Retrieved August 12, 2015, from http://www.emarketer.com/Article/Worldwide-Smartphone-Usage-Grow-25-2014/1010920

Evers, J. (2001). *Dutch Police Fight Cell Phone Theft with SMS Bombs.* Retrieved July 14, 2009, from www.accessmylibrary.com: http://www.accessmylibrary.com/article-1G1-73281158/dutch-police-fight-cell.html

Fallon, N. (2014, July 3). 5 Trends That Are Shaping Mobile Commerce. Retrieved July 20, 2015, from http://www.businessnewsdaily.com/6720-mobile-commerce-trends.html

Gartner. (2010). *Gartner Press Release.* Retrieved August 12, 2010, from www.gartner.com: http://www.gartner.com/it/page.jsp?id=1388914

Hodges, D. (2014, December 1). Changing landscape of mobile commerce and what it means for brands - Mobile Commerce Daily - Columns. Retrieved from http://www.mobilecommercedaily.com/changing-landscape-of-mobile-commerce-and-what-it-means-for-brands

Hof, R. (2015, January 22). Once Again, Starbucks Shows Google And Apple How To Do Mobile Payment. Retrieved July 21, 2015, from http://www.forbes.com/sites/roberthof/2015/01/22/once-again-starbucks-shows-google-and-apple-how-to-do-mobile-payment/

IDC. (2009, December 09). *Press Release.* Retrieved March 24, 2010, from idc.com: http://www.idc.com/getdoc.jsp?containerId=prUS22110509

Jennelyn Magdirila, P. (2015, April 2). Juniper Research: Augmented reality set to increase by 2019 | ITProPortal.com. Retrieved August 15, 2015, from http://www.itproportal.com/2015/04/02/juniper-research-augmented-reality-set-increase-2019/

Lee, I., Choi, B., Kim, J., & Hong, S. (2007). Culture-Technology Fit: Effects of Cultural Characteristics on the Post-Adoption Beliefs of Mobile Internet Users. *International Journal of Electronic Commerce , 11* (4), 11-51.

MMA. (2009, November 17). *MMA Updates Definition of Mobile Marketing*. Retrieved March 26, 2010, from mmaglobal.com: http://mmaglobal.com/news/mma-updates-definition-mobile-marketing

Molt, S. (2015, June 3). 70 Percent of Population Will Have Smartphones by 2020 | News & Opinion | PCMag.com. Retrieved July 20, 2015, from http://www.pcmag.com/article2/0,2817,2485277,00.asp

NFC Bootcamp. (2015, April). April 2015 News - NFC Bootcamp. Retrieved August 12, 2015, from https://www.nfcbootcamp.com/industry/april-2015-news/

Phimister, H. (2015, February 18). 10 Reasons Retailers Need A Mobile App | NN4M – Mobile Commerce App Provider. Retrieved August 15, 2015, from http://www.nn4m.co.uk/10-reasons-retailers-need-a-mobile-app/

Raina, K., & Harsh, A. (2001). *Mcommerce Security*. New York: Osborne.

Sahagian, J. (2013, October 16). Study: U.S. Smartphone Penetration Is at 74 Percent. Retrieved August 15, 2015, from http://www.cheatsheet.com/technology/study-u-s-smartphone-penetration-is-at-74-percent.html/

Shuman, J., Twombly, J., & Rottenberg, D. (2001). *Collaborative Communities: Partnering for Profit in the Networked Economy*. Dearborn Financial Publishing.

Subhash. (2010, February 10). ROCK THE WORLD: Next Generation Mobile Communication Services. Retrieved from http://subhash-rockband.blogspot.com/2010/02/blog-post.html

Tasner, M. (2010). *Marketing in the Moment:The Practical Guide to Using Web 3.0 Marketing to Reach Your Customers First*. FT Press.

Turban, E., Lee, J., & Chung, M. (2006). *Electronic Commerce: A Managerial Perspective*. Prentice Hall.

Venkataram, P. (2003). *Keynote Address :"Mobile Commerce – Vision and Challenges*. Retrieved August 15, 2009, from ITPC, Nepal: http://pet.ece.iisc.ernet.in/sunil/mcommerce-keynote.ppt

Visualead.com. (2014, January). Are QR Codes Dead- Are QR Codes Still Relevant Today? | Visual QR Code Generator Blog | Visualead. Retrieved August 15, 2015, from http://www.visualead.com/blog/qr-codes-dead-qr-codes-still-relevant-today/

Wang, S., & Wang, H. (2005). A location-based business service model for mobile commerce. International Journal of Mobile Commerce , 3 (4), 339-349.

Zickuhr, K. (2013, September 8). Location-Based Services. Retrieved from http://www.pewinternet.org/2013/09/12/location-based-services/

12

E-COMMERCE ENVIRONMENT: MAJOR ISSUES

Chapter Outline

LEGAL FRAMEWORK FOR E-COMMERCE

ISSUES OF E-COMMERCE

ACCOUNTABILITY IN E-COMMERCE

DISPUTE RESOLUTION AND REDRESS

ROLE OF GOVERNMENTS

INTERNATIONAL AND REGIONAL CO-OPERATION INITIATIVES

COMPUTER AND ONLINE CRIME

HACKER AND CRACKER

GREEN E-COMMERCE AND IT

Learning Objectives

After reading this chapter, reader should be able to:

- Understand why E-commerce raises various issues.

- Recognize the main financial, legal, ethical, social, market access, cultural, and language Issues raised by E-commerce.

- Understand basic issues related to customs, taxation, and electronic payment systems.

- Understand the various forms of intellectual property and the challenges involved in protecting it.

- Understand basic issues related to privacy.

- Describe the different approaches used to protect online privacy.

- Understand basic issues related to customer data security.

- Understand the difference between hacker and cracker.

- Describe the different approaches used to protect customer data online.

- Describe civil and common law.

- Describe the law enforcement challenges raised by E-commerce.

- Explain defamation on Internet, its challenges, and how to deal with it.

- Describe types of fraud on the Internet and how to protect against them.

- Describe E-commerce related ethical issues.

- Describe E-commerce related social issues.

- Describe E-commerce related market access issues.

- Describe E-commerce related cultural and language issues.

- Discuss the role of governments in the growth and development of E-commerce.

- Discuss various international and regional initiatives to promote global E-commerce.

- Describe Green E-commerce and IT.

Introduction

Today, Internet is affecting almost every aspect of daily life. Internet is empowering citizens, democratizing societies, and changing the classic business and economic paradigms. New business models is being developed as more and more businesses and consumers continuing to participate in E-commerce.

The unique nature of the Internet as a medium, widespread competition, and increased consumer choices are the defining features of the new digital marketplace. E-commerce has no geographical boundaries and a business that uses the Web immediately becomes an international business subject to many more laws. The increased efficiency and speed of communication provided by web enables much more interactive and complex relationships of customers with online businesses and within themselves. Violation of laws or breach of ethical standards by online businesses therefore can result in rapid and intense reactions from their customers and stakeholders. With expanding use of Internet, E-commerce businesses and users continue to face extensive regulations on the Internet and E-commerce.

Legal Framework for E-commerce

New models of commercial interaction are developing and business and consumers are using innovative ways to participate in transactions and pay for goods and services over the internet displacing the need to handle physical cash. However, the emergence of electronic commerce has brought with it a number of legal and socio-economic issues. The bigger issue is that the internet lacks the clear and fixed geographic lines of transit that traditionally characterize the physical trade in goods and services. The laws should not only be applicable to E-commerce but should also accommodate legal development in E-commerce. The developments in E-commerce have provided us with new types of contracts and goods such as virtual goods, digital contracts, online transactions etc. As a result, certain considerations lost their relevance such as the medium of the transaction or the geographic location of the parties. It also raised unsettling implications for tax, conflict of law, etc. While the basic idea of contracts applies to both offline and online transactions, current laws of commerce need to be changed to accommodate E-commerce (Graeys & Perchstone, 2014).

E-commerce requires customer confidence and trust on the E-commerce transactions. There is need for a guaranteed level of privacy/confidentiality with respect to information. In an electronic transaction, the original of a data message is almost indistinguishable from a copy and bears no handwritten signature. This increases the likelihood of fraud and evidential issue as to the admissibility of such electronic information in a court. A great challenge to admissibility of electronic evidence under the old Evidence Act related to the definition of the word 'document' under the Act. It is still uncertain whether electronic documents such as PDF copies, emails, emails etc. can be regarded as equivalent to the paper documents. In some countries, such as UK, the Evidence Acts have been amended to include any device by means of which information is recorded, stored or retrievable including computer output. Such adaptations can facilitate performing all the functions attributed to commerce in the conventional paper-based commerce can be validly performed in E-commerce. The E-commerce legal frameworks should be adapted to address both the commercial aspect of the transaction and its corollary technological issues.

Two central issues in e-commerce contracts are documentation and signature. There are a number of specific statutory requirements that certain contracts be evidenced in writing, and which require a signature. The courts have also held in a number of cases that an unsigned document is a worthless document. However, electronic commerce presents some peculiarities in this regards. A first issue is whether the use of emails may suffice as contracts in writing within the various legislations on the subject; and secondly, whether an electronic mark will constitute a valid signature for executing a contract. Electronic signatures or marks in emails are generally considered sufficient to satisfy the traditional requirements of writing and execution but the situation varies in many parts of the world. For establishing proof of electronic signature, the use of passwords, identification, usernames etc. may suffice. Hence, one may safely posit that electronic signatures for the purpose of execution are admissible in evidence provided it is certified and incorporated in an electronic communication in the course of an e-transaction. There is a need to assure consumers of the authenticity, reliability and legality of electronic transactions. Public confidence in electronic transactions must be boosted and those entering into such transactions must be assured that the law will not discriminate against the sanctity of their agreements merely because it is in electronic form. This mandates an enabling legal environment on principal and ancillary issues surrounding electronic commerce or transactions. The national Government in countries should also go beyond creating laws, to activate utilization of electronic delivery platforms in its commercial interactions and for the delivery of government services.

Issues of E-commerce

Doing E-commerce or E-business cannot be a competitive alternative to traditional commerce or maximize the benefits of E-commerce/E-business unless a number of technical as well as enabling issues are considered. Increasing size of E-commerce is forcing governments to adopt a non-regulatory, market-oriented approach to E-commerce that will support a transparent and predictable legal environment to support global E-commerce. However, there are several major issues of E-commerce where agreements are needed to achieve this goal.

Financial Issues

CUSTOMS AND TAXATION

Taxation in E-Commerce is a controversial and extremely important issue. This is because it is related to global E-Commerce as well as to fairness in competition between E-Commerce and conventional offline businesses. The issue becomes more important due to the current and forecasted large transaction volumes of E-commerce. E-Commerce transactions are multiplying at an exponential rate, and Forrester research predicts that by 2013, the US online retails sales will reach $229 billion. Therefore, there exists lots of potential sales tax and cities, states, and countries all would be interested to get a piece of the pie.

Critics against the introduction of tariffs on goods and services delivered over the Internet argue that Internet is a truly global medium and free trade on Internet will benefit economies and citizens of all nations. In addition, the Internet makes possible the trade of both hard and soft goods and services. While tariff can be introduced for hard goods ordered over the Internet but ultimately delivered via surface or air transport same cannot be done for soft goods and services because the structure of the Internet makes it difficult to do so.

However, nations looking for new sources of revenue may introduce tariffs on global E-commerce. Worldwide efforts are being made by governments and other organizations (e.g. worldwide web consortium) to ensure that taxation of E-commerce is consistent with the principles of international taxation, avoid inconsistent local tax laws and double taxation, simple to administer, easy to understand. International business has suggested that tax neutrality should be the guiding principle in government policy towards electronic commerce. Since 1998, WTO Ministers have agreed not impose customs duties on electronic transactions. In principle, if electronic sales are not taxed it could give this type of transaction a considerable advantage over taxed forms of commerce, raising issues of equity and fairness. Traditional domestic commerce within national markets and cross-border electronic trade involving physical goods that pass through customs will continue to face customs duties and value-added taxes but products delivered electronically will not. In general, the geographic fluidity of electronic commerce creates jurisdictional and other issues that are difficult to address. A similar pattern may emerge internationally if electronic commerce is taxed in some countries but not in others. The same applies to corporate taxation. Firms operating on the Internet could relocate to tax havens and shield themselves from income tax. As electronic commerce increases in volume and popularity, there will be major implications for the ability of governments to raise revenues through traditional mechanisms like sales and corporate taxes, value-added taxes, and tariffs. It is highly likely that as governments obtain less revenue as tariffs decrease as a result of trade liberalization, they will be tempted to shift to sales taxes and various consumption taxes to make up for losses at the border.

IMPORTANT E-COMMERCE TAXATION ISSUES FOR POLICY MAKER

Following are some of the important E-commerce taxation issues for policy makers to consider

Source of Income: As the physical location of an activity become less important, it becomes more difficult to determine where an activity is carried out and hence the source of income.

User identification: In general, proof of identity requirements for internet use is very weak. The pieces of an internet address (or domain name) only indicate who is responsible for maintaining that name. It has no relationship with the computer or user corresponding to that address or even where the machine is located.

No Intermediary institutions: Traditionally taxing statues have imposed reporting and withholding on financial institutions that are easy to identity. In contrast, one of the greatest commercial advantages of EC is that is often eliminates the need for intermediary institutions. The potential loss of these intermediary functions poses a problem for the tax administration.

Double taxation: EC also gives to rise to new issues concerning the characterization of payments under the double tax treaties. Moreover, though EC does not give rise to any fundamentally new issues relating to transfer pricing, there may be some difficulties in applying traditional transaction methods, establishing comparability, deciding the tax treatment of integrated businesses and complying with documentation and information reporting requirements. Unless these issues are addressed, an erosion of the tax base may result, especially for developing and under developed countries.

The taxation of e-commerce is a complex issue, which cuts across city, country and state borders within the national borders and tax types. The growth of e-commerce raises complex issue associated with the taxation of multi-jurisdictional transactions and the sourcing of sales of, or income from, services or intangible property transactions. E-commerce is borderless, with transactions slowing seamlessly across the globe. The taxation of cross-border transactions is one of the most challenging areas of taxation. Remote vendors selling directly to customers in other jurisdictions without wholesale distributors or retail outlets, referred to as disintermediation, may encounter the necessity to contend with tax requirements in numerous additional jurisdictions for the first time. The continued growth in borderless commerce will lead to a corresponding increase in both business taxes compliance efforts in multiple jurisdictions and jurisdictional disputes over which jurisdiction can impose taxes. E-commerce has contributed to the ease with which companies can enter into joint ventures, partnership, out sourcing agreements, and other arrangements, transforming them into virtual organizations. The emergence of virtual organizations will put pressure on taxing authorities to develop new rules for apportioning the income of these more mobile and dynamic businesses.

ELECTRONIC PAYMENT SYSTEMS

Electronic payment systems have many associated issues e.g. security during the transaction and storage, double spending, counterfeiting, and issue of liability. The business and technological environment, in which the electronic payment systems are being developed is changing rapidly. Therefore, developing a policy that is both timely and appropriate is difficult and governments need to develop flexible and non-rigid regulations. Government intervention is needed in long run to ensure the security and reliability of electronic payment systems and to protect consumers.

Legal Issues

Laws are written and enacted by government. All citizens within jurisdiction of laws must act according to the laws. Any activity against the laws can be held liable for punishment by the legal system. Using the Internet in general and E-Commerce in particular raises many legal issues. A main characteristic of e-commerce is its independence of geographical boundaries. Whereas this unarguably has many economic advantages, it makes it difficult to determine the jurisdiction. With this respect, e-commerce poses two difficulties: The first is the choice of the forum; the second is the choice of the law. Once the forum is chosen, the judge has to determine which law should be applied. This is a major issue when a commercial dispute arises between two contracting parties of a cross-border electronic contract. Has the merchant created a virtual storefront in the buyer's jurisdiction to make a sale, or has the purchaser virtually traveled to the seller's jurisdiction to make a purchase? In cross-border disputes, the eventual inability for national jurisdictions to enforce foreign judgments may be an additional complication.

ELECTRONIC CONTRACTS

A legally binding contract requires a few basic elements: offer, acceptance, and consideration. In an online transaction, these requirements are difficult to establish because there is no human involvement and the contracting is performed electronically. In general, contracts are valid and enforceable even if they are not in writing or signed by parties involved. A signature is any symbol executed or adopted for authenticating writing. The US Electronic Signatures in Global and National Commerce Act gives contracts signed online the same legal status as a contract signed with pen on paper. Similar laws have been enacted in several European and Asian countries.

People are asked to "sign" an electronic contract in various ways such as typing in their name or initials, scanning and copy-pasting their signature, clicking a box that says "I agree" (or something similar) (a "clickwrap" agreement), using a mouse to scrawl an approximation of a signature directly on the screen, and using e-signature software. Companies can improve their chances of having their online contracts enforced if they follow these guidelines (Leff, 2014):

• Make the Terms of Use easy to find and easy to read.

- Use a click-wrap agreement rather than a browserwrap. In a click-wrap agreement, a user must agree to terms and conditions prior to using the product or service. In a browser-wrap agreement, it is not necessary.

- Allow users to print the terms or receive an email with the terms. The Uniform Electronic Transactions Act (UETA) of USA provides that the terms of an electronic agreement will not be enforceable against a person who was unable to print or store the agreement.

- Keep a record of agreements. The "I accept" button should not just open the door to the next step. A company needs to save the date, time, user information, and form version in order to establish who agreed, when, and to what, in case a dispute later arises.

- Avoid unilateral updates, as these are likely to be found unenforceable. When you update your form, give users notice (usually via email) and the chance to opt out or opt in.

- Use identity verification to establish that the persons agreeing are those who they say they are.

INTELLECTUAL PROPERTY PROTECTION

World Intellectual Property Organization (WIPO) defines intellectual property as "creations of the mind: inventions, literary and artistic works, and symbols, names, images, and designs used in commerce". Intellectual property protection is the major concern of those who own intellectual property. Sellers must be assured of the protection of their intellectual property and buyers must be assured of receiving the authentic products. Without intellectual property protection rights, the movie, music, software, publishing, pharmaceutical, and biotech industries would collapse (Claburn, 2001. Intellectual property laws deals with protecting the rights of those who create original works. These laws protect these rights to encourage new technologies and inventions while promoting economic growth. There are four main types of intellectual property in E-commerce: copyrights, trademarks, domain names, and patents. Table 12-1 lists these intellectual properties (IPR) and what these IPR protect.

TABLE 12-1: TYPES OF INTELLECTUAL PROPERTY RIGHTS

Intellectual Property	What is Protected
Utility Patents	any new and useful process, machine, manufacture, or composition of matter, or any new and useful improvement (chemical, mechanical, or electrical)
Design Patents	a new, original, and ornamental design for an article of manufacture (the appearance of a functional product)
Plant Patents	any distinct and new variety of plant invented or discovered and asexually reproduced
Trade Secrets	a formula, pattern, process, or device that a company keeps secret to give it an advantage over the competition
Trademarks	a word, name, symbol, or device that is used in trade to indicate the source of goods and to distinguish them from the goods of others
Copyright	original works of creative expression fixed in any tangible medium

COPYRIGHTS

A copyright is an exclusive grant from the government that confers on its owner an essentially exclusive right to: (1) reproduce a work, in whole or in part, and (2) distribute, perform, or display it to the public in any form or manner, including the Internet. In general, the owner has an exclusive right to export the copyrighted work to another country (Delgado-Martinez, 2003).

Copyrights usually exist in literary works (e.g., books and computer software), musical works (e.g., compositions), dramatic works (e.g., plays), artistic works (e.g., drawings and paintings), sound recordings, films, broadcasts, and cable programs. On the Internet, copyrights can also be used to protect images, photos, logos, text, HTML, JavaScript, and other materials.

A court order may be sought by a copyright owner to prevent or stop any violation of copyright and to claim damages. Certain kinds of copyright violations also incur criminal liabilities e.g. selling or dealing in infringing works and manufacturing and selling technology for defeating copyright protection systems.

Many treaties e.g. Berne Convention for the Protection of Literary and Artistic Works establish international norms for the protection of copyrights. The WIPO Copyright Treaty and the WIPO Performances and Phonograms Treaty include provisions relating to technological protection, copyright management information, and the right of communication to the public.

Software tools are available to produce digital content that cannot be copied e.g. CrypKey(crypkey.com). Cryptography and tracking of copyright violations are two approaches used to design effective electronic copyright management systems [see (Piva, Barni, Bartolini, & De Rosa, 2001)]. Another technique is using digital watermarks. A digital watermark is a unique identifier embedded in the digital content. Digital watermark can be used to identify pirated works but they cannot prevent illegal reproduction of the digital content. Third-party services are available to prevent illegal distribution of digital content. For example, Digimarc (digimarc.com) provides such services. If a pirated copy is placed on the Internet, digimarc software tools can locate the illegal copies and notify the rightful owner.

A copyright does not last forever and is good for a fixed number of years after the death of the author or creator (e.g., 50 years in the United Kingdom, and 70 years in the United States). After that time, the copyright of the work reverts to the public domain.

The Digital Millennium Copyright Act (DMCA) is a controversial United States digital rights management (DRM) law that was enacted to create an updated version of copyright laws to deal with the special challenges of regulating digital material. Broadly, the aim of DMCA is to protect the rights of both copyright owners and consumers. The law complies with the World Intellectual Property Organization (WIPO) Copyright Treaty and the WIPO Performances and Phonograms Treaty, both of which were ratified by over 50 countries around the world in 1996. DMCA would outlaw many entirely ethical and even necessary activities. For example, security-related tasks that involve circumventing security systems, encryption research, or reverse engineering software would be illegal. Revisions were made to DMCA to allow specified exceptions, such as encryption and security research. Industry, consumer, and civil rights groups continue to appraise the law, and many states are considering their own versions. In April 2003, a group called the Broadband and Internet Security Task Force produced an update to the law, sometimes referred to as "Super DMCA." This later version adds important concepts, such as "the intent to defraud," to the stipulations of the original law.

PATENTS

E-commerce development and growth will both depend upon and stimulate innovation in many fields of technology, including computer software, computer hardware, and telecommunications. Patent is a document that grants the holder exclusive rights on an invention for a fixed number of years (e.g., 17 years in the United States and 20 years in the United Kingdom). Patents are designed to protect tangible technological inventions, especially in traditional industrial areas. They are not designed to protect artistic or literary creativity. Examples of E-Commerce patents are IBM's patent # 5870717 (a system for ordering from electronic catalogs) and patent # 5926798 (a system for using intelligent agents to perform online commerce).

TRADEMARKS

A trademark is a symbol used by businesses to identify their goods and services. A trademark can be composed of words, designs, letters, numbers, shapes, a combination of colors, or other. Trademark owner must register the trademark in a country to be eligible for legal protection. To be eligible for registration, a trademark must be distinctive, original, and not deceptive. Once registered, a trademark lasts forever, as long as a periodic registration fee is paid. The owner of a registered trademark has exclusive rights to use the trademark on goods and services for which the trademark is registered and take legal action to prevent anyone else from using the trademark without consent on goods and services (identical or similar) for which the trademark is registered. Trademark rights are generally restricted to a certain jurisdiction and conflicts may arise in case where different companies operating in different countries own similar trade marks for similar goods or services. The rules for determining trademark infringements also vary from country to country.

Genericide refers to the phenomenon where a trademark can no longer be protected under trademark laws because it has become a generic term that describes all related goods or services. Generic marks cannot be protected under trademark law because they describe a whole group of goods or services in which one product in the group cannot be distinguished from other e.g. you cannot trademark CARS to be used with the sale of a type of car because CARS describes a whole group of cars. Some famous examples of once-protected but now generic marks include ASPIRIN, ESCALATOR and ZIPPER. Genericide presents an interesting problem to companies especially the ones with strong recognized brands. The popularity of the brand can make the trademark of company become generic. For example, Xerox Corporation spent a substantial amount of time and money to educate the public about using the term XEROX as an identifying mark rather than as a general term (for example, "Copy this paper on a Xerox copier" rather than "make a Xerox copy of this paper).

Aspirin used to be a trademarked item owned by Bayer before the trademark was revoked in 1919. Escalator was owned by Otis and the trademark was revoked in 1950. When that trademark is revoked, it is essentially a death sentence for the original product made by the original company. That is because revoking the trademark makes it fair game for any competing company to use the same name in marketing and branding. Brand genericide has been a problem for some companies since it became part of the USA Lanham Trademark Act of 1946, but it's become an even greater cause for concern in today's digital world, as the time it takes for brand names to be revoked has decreased dramatically. Many of the victims of brand genericide had previously held trademarks for at least two decades, if their registration year was before 1980 and before the advent of many electronic messaging systems. Today, popular brand names invented in the modern age of messaging (e.g. hacky sack in 1985, wine cooler in 1988, and pilates in 2001) lost trademark rights in under 10 years. In Australia, Apple lost an appeal to use its "App Store" trademark (Campbell, 2014). There are two ways businesses can prevent brand genericide. First, they should market in a way that prevents product names from becoming ubiquitous. Second, they can find a way to measure rates and exposure to brand genericide for court battles. The first strategy was used by Xerox when the company was driven to tell customers to copy items using a Xerox copier instead of using Xerox as a verb. The second incentive is already being explored by business analytics professionals where they scrape data from the Web to gather statistics about brand name linguistic usage. For example, they could scrape millions of messages on Twitter to find statistics about how brand names are used in speech and in what context. With this data, he could theoretically prove how close a brand is to genericide, and that information could bolster a case for, or against, a company trademark in a court of law (Bentley.edu, 2015). Another common practice amongst trademark owners is to follow their trademark with the word brand to help define the word as a trademark. Johnson & Johnson changed the lyrics of their Band-Aid television commercial jingle from, "I am stuck on Band-Aids, 'cause Band-Aid's stuck on me" to "I am stuck on Band-Aid brand, 'cause Band-Aid's stuck on me." Google has gone to lengths to prevent this process, discouraging publications from using the term 'googling' in reference to web-searches. In 2006, both the Oxford English Dictionary and the Merriam Webster Collegiate Dictionary] struck a balance between acknowledging widespread use of the verb coinage and preserving the particular search engine's association with the coinage, defining *google* (all lower case, with -le ending) as a verb meaning "use the Google search engine to obtain information on the Internet" (Revolvy.com, 2015).

There are other examples of Genericide. Table 12-2 lists some of these terms and what we actually mean.

TABLE 12-2: GENERICIDE TERMS

Term Used	What We Actually Meant
Band-Aid	adhesive bandage
Chap Stick	Lip balm
PowerPoint	electronic presentation
Google	looking up something in a search engine
iPod	any type of personal portable media player
Photoshop	photo manipulation
Polaroid	instant photograph or instant camera
Tylenol	over-the-counter pain reliever

DOMAIN NAMES

Domain name is a variation of a trademark. A domain name functions as a source identifier on the Internet. Generally, source identifiers are not protected intellectual properties (i.e., a trademark) per se. Conflicts arise when third parties register domain names that are the same as, or similar to, registered trademarks. As the internet is getting popularity, more and more companies are coming online and therefore acquisition of unique domain name is getting even more difficult. Therefore, the war for acquiring good and desirable domain names has given rise to never ending clashes. These disputes are mainly between trademark owners and domain name owners. As there is no agreement within the internet community that would allow organizations that register domain names to pre-screen the filing of potentially problematic names, domain name disputes are arising. Domain name disputes arise largely from the practice of cybersquatting, which involves the pre-emptive registration of trademarks by third parties as domain names. Cyber squatters exploit the first-come, first-served nature of the domain-name registration system to register names of trademarks, famous people or businesses with which they have no connection. Since registration of domain names is relatively simple and inexpensive – less than US$100 in most cases – cyber squatters often register hundreds of such names as domain names

As internet is global, its nature is bound less whereas the nature of trademark laws is territorial. Due to this, internet is unable to apply trademark laws of which one reason is its variability in different countries. This has given rise to various disputes between trademark owners and domain name owners. Both, the trademark owner and the domain name owners have their own perspective regarding the ownership of domain names. Trademark owners propose that internet acts as just another medium of promotion than radio, television and billboards etc. and therefore domain names should be covered by trademark protection whereas domain name owners on the other hand highlights the principle of "first come, first serve basis".

Cybersquatting refers to registering or using a domain name with wrong intentions to profit from the goodwill of a trademark that belongs to someone else. Cybersquatting can be of two main types. In first type, it involves making or registering a domain name similar to a well-established trademark with the hope of selling it to the trademark owner at a profit. In the second type, it involves making or registering a domain name similar to a well-established trademark with the intent of drawing the traffic to the site. This practice is also called "cyber piracy". Network Solutions, Inc. (NSI), a subsidiary of VeriSign, has created a domain-name registration system that it shares with several other competing companies. The registration system is handled by ICANN. Internet domain names are obtained from one of several registries. These registries make not attempt to assess the eligibility of the ownership of domain name. As a result, many individuals and companies have obtained domain names that they think someone else will want, either now or in the future. Generally, cybersquatters reserve domain names consisting of common English words and mistyped spellings of popular web sites.

There is couple of ways for a company to check if they are victims of cybersquatting. First, using a search engine, type in a name, which has been trademarked, preceded by "www" and followed by ".com", ".net" or ".org". Generally, first check to see if the domain name takes you to a website. If it does not take you to a functioning website, but instead takes you to a site stating "this domain name for sale", or "under construction" or "can't find server," the likelihood increases that you are dealing with a cyber-squatter. Second, if the functioning web site you encountered comprised primarily of advertisements for products or services related to your trademark, you may also have a case of cybersquatting. For example, if your company is well known for providing audio-visual services and the web site you encounter is packed with ads for other company's audio-visual services, the likelihood is very strong that the site is operated by a cybersquatter who is trading off your company's popularity to sell Google ads to your competitors.

Misuse of a domain name could significantly violate, dilute, and weaken valuable trademark rights. The use of domain names as source identifiers is increasing and courts have begun to attribute intellectual property rights to them. Legal actions in various jurisdictions can be initiated for domain-name dispute resolution. Legal action is a more compelling protection because it cannot only win you right to use a certain domain name but also get monetary damages. However, legal action can be a long process. You can sue to get your domain name -- and possibly some money damages -- under a 1999 federal law known as the Anti-Cybersquatting Consumer Protection Act.

To avoid legal actions, ICANN has created a Uniform Domain-Name Dispute-Resolution Policy (UDRP). According to this policy, in order for a registrar to cancel, suspend, or transfer a domain name, the dispute over the domain name must first be resolved by arbitration, court action, or agreement. Disputes that arise from abusive registrations of domain names (for example, cybersquatting) may be addressed by expedited administrative proceedings. These proceedings can be initiated by holder of trademark rights by filing a complaint with an approved dispute-resolution service provider.

To invoke the policy, a trademark owner should either file a complaint in a court of proper jurisdiction against the domain-name holder or submit a complaint to an approved dispute-resolution service provider. Three arbitration organizations have been given the authority to make determinations regarding domain name disputes. They are Disputes.org Consortium, the National Arbitration Forum, and the WIPO. If one of these organizations makes a determination regarding a domain name dispute, NSI/ICANN will respect the decision and will transfer the registrations of the disputed domain name in accordance with the arbitration determining. The procedure is faster compared with dispute resolution through legal actions.

Note that this UDNDRP results in an arbitration of the dispute, not litigation. An action can be brought by any person who complains that a domain name is identical or confusingly similar to a trademark or service mark in which the complaint has rights, the domain name owner has no rights or legitimate interests in the domain name, and the domain name has been registered and is being used in bad faith. The UDNDRP does not require a lawyer and generally considered to be faster and less expensive that suing under the ACPA.

PROTECTING YOUR E-COMMERCE SITES

Many parts of your website may be protected by different types of intellectual property (IP) rights. For example:

• **E-commerce systems, search engines or other technical Internet tools** may be protected by patents or utility models;

• **Software**, including the text-based HTML code used in websites, can be protected by copyright and/or patents, depending on the national law;

• Your **website design** is likely to be protected by copyright;

• Creative website **content**, such as written material, photographs, graphics, music and videos, may be protected by copyright;

• **Databases** can be protected by copyright or by *sui generis* database laws. A sui generis database right is considered to be a property right, comparable to but distinct from copyright, that exists to recognize the investment that is made in compiling a database, even when this does not involve the 'creative' aspect that is reflected by copyright (Legislation.gov.uk, 2015).

• **Business names, logos, product names, domain names and other signs** posted on your website may be protected as trademarks;

• **Computer-generated graphic symbols, screen displays, graphic user interfaces (GUIs)** and even Web pages may be protected by industrial design law;

• **Hidden aspects of your website** (such as confidential graphics, source code, object code, algorithms, programs or other technical descriptions, data flow charts, logic flow charts, user manuals, data structures, and database contents) can be protected by trade secret law, as long as they are not disclosed to the public and you have taken reasonable steps to keep them secret.

PRIVACY

One of Internet's major functions is facilitating the collection, re-use, and instantaneous transmission of information. Therefore, careful management of Internet is required to ensure personal privacy. E-commerce will flourish only if the privacy rights of individuals are balanced with the benefits associated with the free flow of information.

Recognizing the importance of the issue, various jurisdictions are taking steps to safeguard consumer privacy. Countries have enacted laws, implemented industry self-regulation, or introduced administrative solutions designed to protect consumer privacy. Technology and industry self-regulation is expected to offer solutions to many privacy concerns on the Internet. If not, the governments will face increasing pressure to play a more direct role in protecting consumer privacy online. Some global principles have been set to govern the collection, processing, storage, and re-use of personal data on the Internet. For example, organizations that collect consumer data are being bound to provide disclose to consumers of their policy of data collection and its intended use (i.e. why information is being collected, what the information will be used for, what steps will be taken to protect that information, the consequences of providing or holding back information, and any rights of remedy that consumers may have). Such disclosure will be helpful for customers in judging the level of privacy protection and whether they would like to participate or not.

A consumer's on-line social awareness[14] can mitigate his/her privacy concerns on the Internet (Dinev & Hart, 2005). However, higher level of social presence has varying effects depending on the product being sold on-line. Web sites selling products for which consumers seek fun and entertaining shopping experiences (e.g. apparel) benefit from high level of social presence. Web sites selling products for which consumers primarily seek detailed product information (e.g. headphones) does not appear to benefit from higher levels of social presence (Hassanein & Head, 2005).

Privacy Rights and Obligations

Every customer has its privacy right on his/her personal information and so as the businesses. Securing your confidential or personal data in online business is quite difficult. Many businesses and customers suffered losses

---

[14] On-line social awareness is described as a social process of communication and transactions between one person and other persons over the networked environment.

just because they ignored online privacy issues. Privacy obligations are very hard to be classified in such a way that it satisfies all the environments. Different types of privacy obligations are been identified for financial institutions, health care, enterprises and e-commerce. Privacy obligations can be very abstract. For example, the Gramm-Leach- Bliley Act of 1999 says, "Every financial institution has an affirmative and continuing obligation to respect customer privacy and protect the security and confidentiality of customer information". Privacy obligations may require that personal data cannot be used after certain period of time like 30 years or not more than one or two days if the customer's consent is not included.

Privacy Laws

Privacy laws and regulations may restrict the use of personally identifiable information of the customers. These laws may vary country to country. USA Patriot Act was enacted to make law enforcement investigatory tools efficient, investigate online activities and to restrict terrorist activities both within the USA and around the world. This act led the law enforcement to search various methods and tools of communication like telephone, emails, personal records etc. The Electronic Communications Privacy Act of 1986 is the law to update the existing law that prevented interception of audio signals transmissions so that any type of electronic transmissions would be given same type of protections. This law makes it unlawful under certain conditions like one clause permits ISP to view private email if the sender is suspected that he/she might cause damage to the other users. Children's Online Privacy Protection Act (COPPA)was enacted to protect children ages 13 and under from sharing personal information like email address, residential addresses, hobbies or any other personally identifiable information without the consent of the parents. California Online Privacy Protection Act of 2003 requires all the owners of online businesses to post their online privacy policy. This law is for the protection of citizens of California specifically. Disclose all types of personally identifiable information collected and the purpose of that data to be collected. This law also states that businesses should provide a description of process in order to delete the information given.

Online Privacy Policy

Nowadays consumers and government regulators are keeping check on online business's websites that how well they are protecting consumer privacy. The first thing they look for is online privacy policy, which should be written clearly. Online privacy alliance encourages all websites to post such policies. A good privacy policy informs customers clearly that why such information is collected, where it will be used and for what purpose. A good online privacy policy will state clearly the practices of a web business.

SECURITY

Unless Internet users feel confident about security of their communication and data from unauthorized access or modification, it is unlikely they will use the Internet on a routine basis for E-commerce. No single technology can ensure security and reliability of Internet. Effective and consistent use of a range of technologies (encryption, authentication, password controls, firewalls, etc.) globally supported by trustworthy key and security management infrastructures and market-driven standards are required to achieve this goal.

Another issue is the development of trusted certification services. These services depend on cryptographic keys and support digital signatures that allow users to know whom they are communicating with on the Internet. Efforts are being made for the development of a voluntary, market-driven key management infrastructure.

Encryption techniques protect the confidentiality of stored data and electronic communications by making them unreadable without a decryption key. However, encrypted data could be lost forever if the decryption key is lost. Depending upon the value of the information, the loss could be quite significant. If gets into wrong hands, encryption can also be used to reduce law enforcement capabilities to read communications among criminals. Governments are working on legislations that would facilitate development of voluntary key management infrastructures and would govern the release of recovery information to law enforcement officials pursuant to lawful authority. United States government is working within the OECD to develop international guidelines for encryption policies.

CIVIL LAW

Civil law defines and enforces the duties or obligations of persons to one another. By principle, civil law provides a written set of laws that is accessible to all citizens. These laws are applied to all citizens and judges must follow them.

COMMON LAW

Common law is the judge-made law. These are laws, which exist and apply to a group based on historical legal precedents developed over hundreds of years. Since it is written by judges, it is also referred to as "unwritten" law.

BORDERS AND JURISDICTION ON THE INTERNET

Territorial borders in the physical are a very clear way to mark the end of one culture and applicable laws and the beginning of another. When a person crosses a territorial border, one set of rules is replaced by a different set of rules. Effective laws require their enforcement by respective governments. Jurisdiction refers to government's ability to exert control over a person or corporation. In order to enforce laws regarding conduct of online businesses, governments must first establish jurisdiction over that conduct. People or businesses that wish to enforce their rights based on either contract[15] or tort[16] law must file their claims in courts with jurisdiction to hear their cases. A court will have sufficient jurisdiction in a matter if the court has both subject matter jurisdiction and personal jurisdiction.

---

[15] A contract is a promise or set of promises between two or more legal entities - persons or corporations - that provides for an exchange of value (goods, services, or money) between or among them.

[16] A tort is an intentional or negligent action taken by a legal entity that causes harm to another legal entity.

SUBJECT MATTER JURISDICTION

Subject-matter jurisdiction is a court's authority to hear and decide the particular dispute before it. For examples, if all parties to a contract are located in one state, the state court will have jurisdiction over all disputes that arise from terms of that contract.

PERSONAL JURISDICTION

Personal jurisdiction refers to the power of a court to hear and determine a lawsuit involving a defendant by virtue of the defendant's having some contact with the place where the court is located. However, without any marked contact with other states, no state can exercise personal jurisdiction and authority over persons outside its territory. Determining personal jurisdiction in case of a person is not difficult. Difficulties arise when courts have had to decide whether corporations are subject to personal jurisdiction because corporations have a legal existence and a legal identity but no tangible existence. In United States, courts established that a corporation is always subject to the jurisdiction of the courts in the state where it was incorporated. US States also require corporations to file written consents to personal jurisdiction before they can conduct business within the state. Other US states require that either the corporation designate an agent to accept legal process in the state or that the state attorney general be authorized to accept process for all out-of-state corporations doing business within the state.

ENFORCEMENT CHALLENGES

With powers constrained by jurisdictional boundaries, law enforcement agencies face a big challenge when dealing with laws enforcement in E-commerce. Some governments, e.g. UK, cannot currently act against infringements affecting non-EU consumers even if committed in the UK, because of legal restrictions. Consumer protection authorities in some countries also face difficulty in determining the location of an online seller. Such evidence gathering may be time consuming, costly and would require personnel trained in using the technology. Countries need to engage in reciprocal enforcement agreements with foreign enforcement agencies to effectively combat rogue traders in e-commerce.

To enhance law enforcement co-operation at domestic, regional and international levels, many initiatives have been taken. Through US Safe Web Act of 2006, United States FTC obtained significant new powers to engage in information sharing and investigative assistance with its foreign counterparts.

Some enforcement tools have been developed in recent years to enhance enforcement co-operation at domestic, regional and international levels. In the United States, the US FTC has obtained significant new powers to engage in information sharing and investigative assistance with its foreign counterparts pursuant to the US Safe Web Act of 2006. It has also confirmed its authority to sue in the United States when foreign consumers are harmed, and can recover monetary restitution on behalf of foreign as well as domestic consumers. EU established an IT system called Consumer Protection Cooperation System (CPCS) that permits communication between authorities. The Competition Bureau Canada launched Project FairWeb, a dedicated Internet surveillance and enforcement program aimed at combating online deceptive advertising.

DEFAMATION ON INTERNET

A defamatory statement is a statement that is false and that hurts the reputation of another person or company. Such statements can be in the form of postings on consumer review websites. If a statement hurts the reputation of a product or service rather than a person, it is called product disparagement. Product disparagement may arise even in a true and honest comparison of products. Commercial Web sites should avoid making negative, evaluative statements about other persons or products because many a times the difference between justifiable criticism and defamation can be hard to determine.

Defamatory statements concerns business owners because they can hurt their business and they want them to be removed. However, businesses need to keep many things in mind before attempting to remove such statements from the websites where they appear. First, publication of customer reviews, whether negative or positive, is the nature of their business. Removing negative reviews could arguably hurt their credibility and, in many cases, jeopardize their survival. These websites may or may not have incentive to cooperate with the complaining business under the law. For example, the United States Communications Decency Act states that a website will generally not be liable for its posters' speech. Second, businesses need to know their rights under law before seeking damages against a customer who has posted a negative review. The available rights and remedies are dependent on the specific information conveyed in the posting and the unique context facing the victim business and the customer who posted a negative review. Third, just because a review statement is negative does not necessarily make it defamatory statement. A complex analysis of all statements and underlying facts is required determine whether the reviews posted by the customer qualify as defamatory statement. To qualify as a defamatory statement, the statement must not only be damaging, but it also must be false. Defamation laws are analyzed in accordance with the right of free speech and the right to protect from defamation. Defamation law protects those who are hurt by lies. A business or person has legal protection from being defamed but customers also have the right to speak about his opinions and experiences. A bad experience, based on recitation of true history, reported by a customer will not likely qualify as a defamatory statement. Similarly, opinions are protected as well. If a customer review simply indicates that customer does not like the business owner or did not like the business's products or services than this is arguably an opinion and therefore cannot qualify as defamatory statement. However, if a posting is made by someone who has never been a customer of a business or by someone who falsifies facts about the business or its products and services, the victim business may claim for defamation.

E-COMMERCE FRAUD

Faceless transaction, especially cross-border, in E-commerce increases the probability of fraud. Buyer and seller protection is therefore critical to the success E-commerce. Fraud in E-commerce can be of various types e.g. online auction fraud which accounts for 87 percent of all incidents of online crime (OECD, 2009 Conference on Empowering E-Consumers: Strengthening Consumer Protection in the Internet Economy, Washington, 8-10 Dec. 2009, Background report, 2009), sale of bogus investments, ghost business opportunities, and selling poor-quality products and services.

Buyers can protect themselves against E-Commerce fraud in a variety of ways.

- By going directly to the site, rather than through a link.

- By shopping for reliable brand names at online stores.

- By checking business credibility with a local chamber of commerce, public organizations (e.g. Better Business Bureau in USA, National Fraud Information Center (fraud.org), or a third party assurance service e.g. TRUSTe (truste.org).

- By checking the seller's website site organization, policies about consumer data collection and its use, testimonials, remedies is available in case of a dispute, and security measures available.

- By checking any money-back guarantees, warranties, and service agreements offered by the online store before making a purchase.

- By comparing prices of online stores with a traditional store. A large difference is generally suspicious.

The United States FTC (ftc.gov) enforces consumer protection laws in the United States. United States and European Union are working to develop joint consumer protection policies.

Fraud can be committed against sellers. For example, customers can deny placing an order or give false payment information in payment for products and services provided. Sellers can protect themselves against E-Commerce fraud in a variety of ways.

- By using intelligent software to identify suspicious customers. One technique used to identify such customers involves comparing credit card billing and requested shipping addresses.

- By identifying warning signals for possibly fraudulent transactions.

- By using third-party escrow and trust companies

ADVANCE FEE FRAUD

Advance Fee Fraud is a type of Internet fraud in which criminals impostors convince a victim that the victim has won a prize or been selected for a business deal with easy money to be gained. The victim is told that the only condition to obtain the money or gain is that the victim should pay a small amount of money in advance. When the victim pays the fee, the imposters will be encouraged to re-contact the victim to request the payment of additional fees for the same transaction. Criminal impostors use a variety of creative but still credible explanations for payment of additional fees e.g. the advance payment of bank fees for money transfers, payment of courier services to send the check, legal fees to have an attorney prepare documents needed for the money transfer, or the advance payment of taxes on the prize. To make their claim more convincing, the criminals can also establish fake courier companies, law firms, and provide them with domain names, phone numbers, and email addresses. These criminals also counterfeit trusted brands to convince the victim that the prize is plausible and real. Enforcing laws against such fraud is complicated. First, because criminals commit this fraud across borders and question of jurisdiction becomes complicated. Second, although on aggregate basis such frauds involve large monetary amounts, the individual cases may not be large enough to initiate an international investigation and gather evidence to identify the fraudsters.

GUIDELINES FOR BUSINESSES TO MINIMIZE E-COMMERCE FRAUD

Following are some techniques for merchants to protect from fraud.

Held orders: Create a "held orders" department where orders can be reviewed manually. Set certain guidelines for what orders will be hld. For example, you can review orders over $250, which might be raised to $500 or higher.

In-house Database: Create an in-house database of all fraudulent orders by address. Take the time to run all orders through this database.

Shared Database: Establish a network with other e-merchants in the same business as yours. Share fraudulent order information with them.

Carefully Review Orders: Whenever you receive an order, take some extra time and review the order carefully. Make sure the consumer filled out all the information correctly, and that all of the customer's information matches. If the order is a fraudulent order, in most cases you can catch anything that does not seem right by just carefully reviewing the entire order. Check orders booked using an anonymous proxy server, used a mailbox or ship forward service, used a free email address., was used to place the order- Anonymous proxy servers allow Internet users to hide their actual IP address.

Contact, Shipping & Credit Card Information: The customer's contact information should match up with the information they used for the shipping address and the credit card. If this information does not match up, then you need to find out why they want the products shipped to another address or use a credit card with different contact information.

Address Verification Service (AVS): Use AVS and the card company as your first levels of fraud screening. It is provided by most merchant processors, you can run the AVS service on all your transactions to ensure that the information they gave you matches with the information on the file with the card-issuing bank. This step will not help with international orders. However, for domestic orders, AVS can help you determine, among other things, whether the address you have been given matches the credit card company records for the billing address on the account.

ONLINE IDENTITY THEFT

ID theft occurs when a party acquires, transfers, possesses, or uses personal information of a natural or legal person (individual or entity) in an unauthorized manner, with the intent to commit, or in connection with, fraud or other crimes. Identity theft (or ID theft) is a very old problem. Traditionally, ID theft has been committed by accessing information acquired from public records, stealing personal belongings, improper use of databases, credit cards, and checking and saving accounts and misusing that information. The most common type of identity theft is credit card fraud. Following are the types of personal information that identity thieves most want to obtain.

- ° Social Security or Social Insurance Numbers

- ° Driver's License Number

- ° Credit Card Numbers

- ° Credit Card Verification Numbers (3-4 digit numbers printed on back of credit cards)

- ° Password/PINs

- ° Date of Birth

- ° ATM Card Numbers

- ° Telephone Numbers

- ° Home Address

Some common techniques used for off-line ID theft include payment card theft, shoulder surfing, skimming, and business record theft. With the development of Internet and E-commerce, online ID theft has become a significant problem. While in most cases, the scope of online ID theft appears to be limited but its implications are significant and growing risk of such ID theft can undermine consumer confidence in using the Internet for E-commerce. Educating customers about ways to protect their identities can contribute significantly to lowering the risk of, or preventing, online ID theft. Recognizing the implications of the problem for E-commerce growth, governments have acted to fight against such fraud (both online and offline) at the domestic and international levels. In 2008, OECD established Policy Guidance on Online Identity Theft. United States government has established a website, which maintains a website providing information on ways to protect personal information and avoid Internet fraud and ID theft (http://onguardonline.gov/index.html). The 2007 report of the US President's Task Force on Identity Theft provides a strategic plan for addressing the challenges presented by ID theft (FTC, 2008).

In recent years, the problem has grown significantly and covered many more areas such as medical identity theft, child identity theft, and corporate identity theft. According to the Medical Identity Fraud Alliance (MIFA), MIT is defined as the fraudulent theft of an individual's protected health information (PHI) and personally identifiable information (PII) (such as a name or Social Security number) to obtain medical goods and services or for financial benefit. Additionally, the MIFA states that synthetic identities have been used to commit MIT in which the PHI of several individuals may be mixed to create separate identities (Alfonso, 2015). In Child Identity Theft, identity thieves use stolen Social Security number of child to apply for government benefits, open bank and credit card accounts, apply for a loan or utility service, or rent a place to live. Business identity theft is the newest threat to small businesses all across America. In the case of a business, a criminal will seize a company's identity and use it to acquire credit in the company's name. According to Norton, a leading computer security company, 9 millions identities are stolen in United States each year (Kovacs, 2014). Following are some techiques that can be used to prevent identity theft.

- Transact financial business online only with secure websites with URLs that begin with "https:" or that are authenticated by companies like VeriSign or the Norton Secured Seal.
- Install personal firewall, antivirus, antispyware, and antispam software.

- Avoid clicking on links or opening files from unknown senders.

- Use unique secure passwords for each site you visit.

- Be careful about what personal information you divulge via social networks. Be sure to check your privacy settings on your social accounts to be sure that information is only visible to trusted friends and family.

- Watch out for "shoulder surfers" that are looking over your shoulder while you are on your computer or phone in a public place.

- Never send personal information such as Social Security numbers or credit card numbers via email, instant messages and across social networks. Not even in a private message.

- Keep your web browsers up to date to avoid vulnerabilities that will allow hackers to access your personal information.

- Don't store any sensitive information about yourself or your bank accounts on your computer.

- When disposing of old technology, be sure to completely wipe all information from the device. Reset the device to factory settings(in case of mobile phone or tablet) or erase the hard drive by installing a clean version of the operating system on the hard drive.

FRAMING

Framing is technique used on World Wide Web in which a composite webpage can be created that consists of specifically identified areas. These specifically identified areas are called frames. The frames may contain either highlighted URL addresses of other Web pages or other pages within the same Web site. When users click on URL address in the frame the users browser software is directed to the URL containing the content and does not import the content into the frame. Users will not feel that they had accessed a new URL page directly, but instead the content will appear within the frame. Each frame functions independently, therefore the content that is displayed into one frame will fill up only that frame, and it will not overwrite or affect the contents of the other frames on the framing web page. Using frames, a user can simultaneously view different web pages within a single web page without losing the user's connection to the framing page site. Furthermore, the URL address in the address bar of user's browser continues to display only the address of the framing page. It is also important to understand that the content appearing in a frame is only temporarily available.

Both frames and framing technology offers unique opportunities to web site owners. Web page owner can capitalize on the design layout of their web site by keeping advertising and certain web site material fixed within a particular frame. It is frequently done by placing advertisements in one section of the screen and a including a navigation that provides contents of a page in scrollable frame. This way web site can keep their advertising in sight of the user while displaying the content from another URL.

Though framing is easy to do, it has its associated legal risks. These legal issues could include copyright and trademark infringement, unfair competition, commercial misappropriation, breach of contract, fraud, defamation, right of privacy and right of publicity.

One good example is the case of Washington Post Vs. TotalNews. A group of prominent news organizations (including The Wall Street Journal, The Washington Post, and CNN) filed a lawsuit against the TotalNews (a web-based news gathering site). TotalNews was using frame technology to display the news organization's information on the TotalNews Web site and was surrounding the frames with its own advertising. The news organizations alleged that the TotalNews Web site by its use of framing and hyperlinks to the news organization own Web sites changed the manner in which one read their Web sites because the TotalNews frame contained paid advertising of TotalNews advertisers instead of the advertising that was incorporated on the new organization's web sites. The TotalNews lawsuit was settled and the terms of the settlement provided, among other things, that TotalNews would stop framing the news organization's Web sites, and that TotalNews would only link to the news organization's sites with permission.

Many issues raised in this case are still unsettled and as a result, legal guidelines related to framing have yet not been established. Websites willing to used framing technology should evaluate its use. They should make sure to obtain necessary permission before framing the content of another URL and avoid framing the URL within their Web sites advertising.

DEEP LINKING

On World Wide Web, deep linking is referred to making a hyperlink that points to a specific page or image on another website, instead of that website's main or home page. For examples the link: http://www.cnn.com/breaking_news is an example of a deep link. The URL contains all the information needed to point to a particular item, in this case the breaking news section of the CNN instead of CNN home page.//www.cnn.com/. Deep linking can be objectionable to many sites because it can bypasses advertising on their main pages. In the past, deep linking has resulted into legal actions e.g. the 1997 case of Ticketmaster vs. Microsoft. In this case, Microsoft Corporation created a deep-link to Ticketmaster's site from its Sidewalk service. Ultimately, Microsoft and Ticketmaster agreed on a licensing agreement to settle the case. Another case, similar to the case of Ticketmaster vs. Microsoft, was Ticketmaster versus Tickets.com. In this case, the court ruled that such linking was legal as long as it was clear to whom the linked pages belonged. The court also ruled that URLs themselves could not be copyrighted.

Ethical Issues

Ethics is supported by common agreement in a society and deals with what is considered right and wrong. Though many ethical guidelines exist, yet what is unethical activity is not necessarily illegal activity (except when it overlaps with activities that are also illegal). Something unethical in one culture may be perfectly ethical in another culture. Many Western countries, for example, are much concerned about individuals and their rights to privacy while in Asian countries more emphasis is placed on the benefits to society rather than on the rights of individuals. Some countries, such as Sweden and Canada, have very strict privacy laws; others have none. In E-commerce the definition of right and wrong is not always clear [see (Hamelink, 2001)]. The conflict in rules and regulations can pose serious challenges for companies operating globally especially with E-commerce. There are many E-Commerce and Internet-related ethical issues. We discussed some issues earlier (e.g. channel conflict and trust). Two other important ethical issues are non-work-related use of computing facilities and code of ethics.

NON-WORK-RELATED USE OF COMPUTING FACILITIES

In some companies, non-work related use of Internet is extremely out of proportion with its work-related use [see (Anandarajan, 2002)]. For businesses, the main concern is time wasted by employees while using corporate computing facilities (e.g. Internet and email) for non-work related activities. Non-work related use of computing facilities can be limited by developing formal computing resources usage policies and making them known to employees [see (Siau, Nah, & Teng, 2002)]. Without a formal policy, it is much more difficult to enforce desired behavior and deal with violators. Companies may send monthly newsletters on corporate intranet about the usage policy. Companies may also send notice to employees that their usage of computing facilities may be monitored.

CODE OF ETHICS

Code of ethics is the formal rules and expected behavior and action. Code of ethics should specify the acceptable usage policy of computing resources (e.g. use of E-mail and Internet etc.), offensive content and graphics as well as proprietary information, whether employees can setup their web pages on the corporate intranet. Most importantly, code of ethics should provide company employees a rational for the policies outlined in code of ethics.

Social Issues

Growth and development of E-commerce has given rise to many social issues. Following are some of the important social issues related to E-commerce.

PROTECTING CHILDREN ONLINE

A growing number of children today are increasingly using Internet, computers, and mobile devices to engage in mobile-commerce, for entertainment, playing games, information and research, and social networking [see (Livingstone & Haddon, 2009); (OFCOM, Children's web access, 2009)]. Such tend has raised new opportunities and challenges for parents and businesses alike. The most serious concerns are that the Internet and m-commerce may expose children to in appropriate commercial content, allow children to purchase unsuitable products (e.g. drugs, weaponry, alcohol, adult literature) and facilitate overspending through the purchase of goods or services, or by generating large network usage charges. Many countries, including United States, have well established rules (self- regulatory and otherwise) for protecting children from certain harmful advertising practices. In most legal system of OECD countries, minors cannot enter into commercial transactions but children often purchase goods or services online in the absence of parent consent or knowledge (OECD, OECD Policy Guidance for Addressing Emerging Consumer Protection and Empowerment Issues in Mobile Commerce, 2008).

Research indicates that children are particularly vulnerable to advertising. Another serious concern is aggressive advertising that target children and obtain personal information from children. Children's growing participation in social networking websites has raised privacy concerns. A profile on a social network site has now become an essential part of children online lives despite nominal age restrictions on these sites that restrict pre-teens from using such sites. Among UK children with access to Internet, more than a quarter of 8 to 11-year-old children claimed to have a profile on a social networking website (OFCOM, 2008).

Another concern is about inadequate age identification systems, which make it difficult to determine whether buyers meet the age criterion. Currently, many online retailers simply ask consumers to confirm their age by ticking a box and employ no other measures to verify buyers' age. The situation is of particular concern in the case of m-commerce because credit cards may not be required make a purchase.

Protecting children online has become a key priority worldwide. Many initiatives at domestic, regional or international levels have been carried out to ensure coordinated efforts. Most OECD countries do not have laws specific to tackle privacy issues involving children. United States passed the Children's Online Privacy Protection Act (COPPA), in 1998. In 2008, the ITU launched the Child Online Protection (COP) project. The goal of the project was to educate children and other stakeholders about the many risks children are facing online. In 2009, the ITU and Japan's Ministry of Internal Affairs and Communications (MIC) organized a Strategic Dialogue on Safer Internet Environment for Children. The dialogue was intended to provide a platform to exchange good practices on key policy and strategy among policy makers, regulators, industry representatives, and academia.

Children have a need for privacy online as much as offline. They need, for example, to be able to socialize online with peers without constant supervision by parents and other guardians. There might be a "privacy protection paradox" if children are subject to "friendly" surveillance by adults as a way to protect them from offline and online risks, including privacy risks. For example, certain parental control technologies can provide detailed reports on online activities. Schools or libraries also increasingly monitor children's online behavior as part of a cyber-safety strategy.

THE DIGITAL DIVIDE

Digital divide is defined as a gap between those who have and those who do not have the ability to use the technology (Adams, 2001;Venkat, 2002). Despite the factors and trends that contribute to future E-Commerce growth, digital divide is increasing. The gap exists both within and between countries. Governments are trying to reduce digital divide within countries by encouraging training and supporting education and infrastructure. However, digital divide among countries may be widening rather than narrowing. Many government and international organizations are trying to close the digital divide.

EDUCATION

E-commerce has had a major impact on education and learning by imparting virtual or e-learning. Virtual universities are helping to reduce the digital divide. Homebound individuals can get degrees from good institutions, and many vocational professions can be learned from home. Companies are using the Internet to retrain employees much more easily.

PUBLIC SAFETY AND CRIMINAL JUSTICE

Various E-commerce tools can help increase consumer safety at home and in public. These include collaborative commerce (for collaboration among national and international. law enforcement units); E-procurement (of unique equipment to fight crime); E-government (for coordination, information sharing, and expediting legal work and cases); and E-training of law enforcement officers.

HEALTH ASPECTS

Generally speaking, online shopping is considered safer and healthier though it may take years before the truth of this claim is established. E-Commerce technologies such as collaborative commerce can help improve health care. For example, using the Internet, the approval process of new drugs has been shortened, saving lives and reducing suffering. Intelligent systems facilitate medical diagnoses. Health-care advice can be provided from a distance.

Market Access Issues

TELECOMMUNICATION INFRASTRUCTURE AND INFORMATION TECHNOLOGY

Global E-commerce depends upon a modern, seamless, global telecommunications network and upon the computers and information systems that connect to it. However, telecommunications policies in many countries, trade barriers on imported information technology, and different levels of telecommunication infrastructure development are hindering the development of advanced digital networks and availability of computers and information systems needed to participate in E-commerce.

In most cases, data networks of most online service providers are constructed with leased lines that must be obtained from national telephone companies. These national telephone companies are often monopolies or governmental entities. In many cases, service providers are dependent on these national telephone companies for interconnections. In absence of effective competition, these national telephone companies may overprice their services offerings and may refuse to provide service at all.

Some jurisdictions consider real time services provided over the Internet as similar to traditionally regulated voice telephony and broadcasting. As a result, these services are subject to the same regulatory restrictions that apply to those traditional services. In some jurisdictions, providers of these services must obtain a license so that both the carriage and control of the content can be exercises by the government. Efforts are being made to ensure global competition in the provision of basic telecommunication services and address the many underlying issues affecting online service providers. Regulatory authorities in many countries are undertaking initiatives to stimulate bandwidth expansion, especially to residential and small/home office customers.

Bilateral exchanges with individual foreign governments, regional forums (such as APEC and CITEL), and multilateral forums (such as the OECD and ITU), and various other forums (e.g. international alliances of private businesses, the International Organization of Standardization (ISO), the International Electrotechnical Commission (IEC)) are important for international discussions on telecommunication-related Internet issues and removing trade barriers that inhibit the export of information technology.

PARALLEL IMPORT

Parallel import refers to import of a non-counterfeit product from another country without the permission of the product owner. Parallel imports are also called grey products. Parallel imports occur because of two reasons. First, a company may produce different versions of the same product to be sold in different markets. Second, companies may set different prices for the same product in different markets. Parallel importers can buy the product at a cheap rate in one country import it to another country and sell it at a higher price.

Brand owners want to be able to prevent parallel import for various reasons:

- Brand owners lose control over the way in which their goods are sold and risk having their brand image undermined if the goods are sold in inferior markets or in markets with poor after-sales care.

- Brand owners lose exclusivity that will result from designer goods being sold in supermarkets will eventually undermine the brand and make it worthless.

- Due to the loss of value of designer brands, brand owners will be discouraged from investing in developing new products, which will lead to a loss of choice for consumers.

- Damage to brands and consequent loss of investment by brand owners, together with brand owners being unwilling to meet the higher costs of manufacture will damage country's economy. It is especially true for pharmaceutical industry.

CONTENT

Internet facilitates free flow of information across international borders through World Wide Web pages, news and other information services, virtual shopping malls, and entertainment features, such as audio and video products. In contrast to traditional broadcast media, Internet provides its users greater opportunities to protect themselves and their children inappropriate content. With the availability of effective filtering technology, there will be no need to impose the traditional content regulations on the Internet.

Companies wishing to do business over the Internet, and to provide access to the Internet (including online service providers with foreign affiliates or joint ventures) are concerned about liability based on the different policies of every country through which their information may travel. Each country that has adopted or planning to adopt laws restricting access to Internet content has different concern that arise from the cultural, social, and political differences. These different laws can impede E-commerce in the global environment.

Through web advertising, companies are able to offer more affordable products and services to a wider, global customer base. Some countries stringently restrict the language, amount, frequency, duration, and type of teleshopping and advertising spots used by advertisers.

Many organizations including U.S Department of State, OECD, the G-7 Information Society and Development Conference, the Latin American Telecommunications Summits, and the Summit of the Americas, and APEC Telecommunications Ministerials are participating in the process to find pragmatic solutions to issues related to content control.

TECHNICAL STANDARDS

Standards are critical to the long-term commercial success of the Internet because standards allow products and services from different vendors to work together. Standards encourage competition and reduce uncertainty in the marketplace.

To ensure the growth of global E-commerce standards will be needed to assure reliability, interoperability, ease of use and scalability in areas such as electronic payments, security, security services infrastructure (e.g., public key certificate authorities), electronic copyright management systems, video and data-conferencing, and high-speed network technologies (e.g., Asynchronous Transfer Mode).

Internet is dominated by voluntary standards and Internet's consensus- based process of standards development and acceptance are stimulating rapid growth of these standards. These standards flourish because they are developed through a non-bureaucratic development system managed by technical practitioners working through various organizations. A small number of countries allow private sector standards development. Most of these standards rely on government- mandated solutions and therefore fall behind the technological cutting edge.

Language Issues

Since 2000, the language representation on the Internet has evolved from then monolingual with one English language into now multilingual with more than one thousand languages [(Crystal D. , 2001); (GlobalReach, 2004)].

The United States Census Bureau estimates that the world population exceeded 7 billion on March 12, 2012. Most of them are from non-English speaking language zones. The Figure shows a detailed breakdown of top ten language groups in global population online.

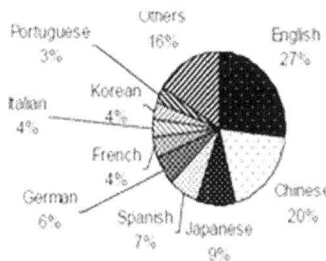

Globalization of ecommerce has made a phenomenal growth of content in languages other than English and increasing online subscriptions from new users in developing countries such as China and India where telecommunication infrastructure is rapidly being improved.

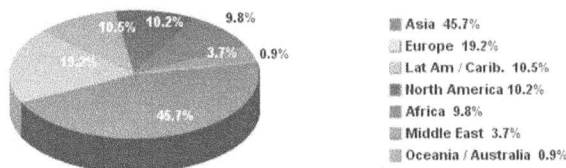

Internet Users in the World
Distribution by World Regions - 2014 Q2

- Asia 45.7%
- Europe 19.2%
- Lat Am / Carib. 10.5%
- North America 10.2%
- Africa 9.8%
- Middle East 3.7%
- Oceania / Australia 0.9%

Source: Internet World Stats - www.internetworldstats.com/stats.htm
Basis: 3,035,749,340 Internet users on June 30, 2014

The language barriers may refer to the historical dominance of the English language but also refer to the current increasing multilingualism on the Internet [(Crystal D. , 2001); (He, Multilingual Web Sites in Global Electronic

Commerce, 2008)]. These language barriers can seriously hinder the ability of multilingual users to effectively browse, navigate, access, filter, process and retrieve multilingual information on the Internet [(Berner, 2003); (Cheon, 2005); (MacLeod, 2000); (Perrault & Gregory, 2000); (Seilheimer, 2004); (Spethman, 2003)]. Language barriers would create obstacles to effective communication on a website, such as a) differences in display, alphabets, grammar, and syntactical rules b) regional variants of the same languages and c) different meanings of idioms and expression when translated into other languages [(MSDNLibrary, 2007); (Goswami, 2003)]. Language may represent a double barrier: first, the number of native speakers determines the number of web-hosts, and hence the amount of information and the inter-connectedness of information sources. Second, to access information on a particular website the languages offered are an even more important factor than network effects: non-native speakers and links from websites in other languages are always underrepresented. In traditional media there seems to be a preference for ads in the mother tongue, however this preference may not be the case with new media (such as Internet). If a business has already developed a Web site, it is important to translate the site into the target language of a particular foreign market. Cultural translation goes beyond mere linguistic translation, since it involves designing a Web site that is sensitive to the cultural differences, between the originated country and the target country. Language translation and localization for an online business is very important because once a business goes online, its website is not necessarily restricted to a particular locality but most online businesses are international once they make an appearance on the web. It is important for an online business utilize language translation and localization services to ensure the whole audience understands their business and the goods and services the business is selling.

The number of multilingual websites is increasing. Many multinational companies, e.g., Cisco, IBM, Intel and Microsoft, have developed their websites in different languages to accommodate various cultures and business practices of their users in the world. From 1999 to 2000, the number of Fortune 100 companies with multilingual websites increased from 33 to 57 (AberdeenGroup, 2001). According to (thebigword.com, 2006), 58% of the Fortune 500 companies currently have multilingual websites and 70% of the largest 20 Fortune 500 companies have localized content on the websites.

Many governments have multilingual websites to meet the needs of their users. For example, the website of Government of Canada is available in both English and French. The website of Government of China is available in both Chinese and English.

Cultural adaptability is essential to do business in a multiple cultures. Cultural adaptability is essential for businesses that operate in multiple cultures. Researchers have found that the likelihood of customer purchase of a product or service is high for those websites that are in their own language. Localization of web content i.e. translation that considers multiple elements of the local environment, is very important (Schneider, 2010).

Cultural Issues

An important element of business trust is anticipation about the expected behavior, under specific circumstances, of other party in the transaction. A culture is a combination of language and customs of people in a certain place. Culture varies across national boundaries and across regions within nations. Cultural background of consumer is very important in web interface design. When a company decides to go global through the Internet, it should identify the potential problems concerning cultural differences across borders. Some specific elements should be taken into account while thinking of potential cultural barriers. These factors may include language, cultural values, and infrastructure. These factors are broad in nature and further sub classification is possible. Here infrastructure refers to an environment consisting of e payment system, logistic system, laws, taxation, etc. Culture can be reflected in many ways. Some forms of representations include symbols, rituals, behaviors, values and communication values (Hofstede, 1997). Resources are currently available for web designers who wish to maximize the usability of a web site including usability guidelines [(Keeker, 1997); (Hubert, Accessibility and usability guidelines for mobile devices in home health monitoring, 2006)].

Collections of human factor references, web sites intended as a gateway to human factors resources and companies offering web-focused human factors consultancy (Davies, 2000). Also Human Computer Interaction (HCI) theories, methods, techniques and tools may be applied to the study of computer–meditated communication (CMC) in general, and web sites in particular [(Forysthe, Grose, & Ratner, 1998); (Kellog & Nielson, 1997); (Shneiderman, 1997)].

Accountability in E-commerce

To strengthen consumer confidence and trust in E-commerce developing accountability measures in the E-commerce marketplace are necessary. Governments and businesses are working together and have taken a range of actions (e.g. consumer laws, policies, and practices limiting fraudulent, misleading and unfair commercial conduct) to achieve this goal. Businesses are also required to provide disclosures about their businesses. These disclosures aim to provide accurate, clear and easily accessible information about businesses. In this regard, domain-name registration information can also serve as a useful compliment. Law enforcement agencies and business typically use Whois services (a public directory of domain name information) to get domain registration information. ICANN's Registrar Accreditation Agreement specifies that information about businesses must be accurate and publicly available online.

Dispute Resolution and Redress

Effective dispute resolution and redress mechanisms are key to building consumer trust in E-commerce. Lack of such mechanisms can hinder consumers from taking full advantage of E-commerce especially cross-border transactions. An E-commerce report showed that some 71% of surveyed consumers believed that it would be harder to resolve problems involving cross-border purchases, than purchases made within home country (EuropeanComission, 2009). OECD issued a Recommendation on Consumer Dispute Resolution and Redress In 2007 (OECD, Recommendation on Consumer Dispute Resolution and Redress, 2007). The recommendation contained principles for resolving B2C disputes (at domestic and cross-border levels), both offline and on-line.

A mediation mechanism is one in which neutral parties can help the parties reach a settlement and where no party to a dispute enjoys any home-court advantage (Giardina & Zampetti, 1997). Many countries have taken actions on this front. In France, the Forum des Droits de l'Internet developed an online mediation system. In Mexico, an online platform, Concilianet, has been developed to process consumer complaints and settle the dispute. In 2003, the International Chamber of Commerce developed guidance for resolving B2C and C2C disputes on-line (ICC, 2003).

Role of Governments

Governments, by their actions, can have a profound effect on the growth of E-commerce. Governments can facilitate as well as inhibit E-commerce. Knowing when governments should act and when not act will be crucial to the development of E-commerce. Following certain principles, governments can play a major role in growth of E-commerce.

The first principle is that governments should allow private sector to take lead in guiding the growth and development of E-commerce. Innovation, expanded services, broader participation, and lower prices are key factors for global E-commerce development. These are possible only in market-driven E-commerce and not in government-regulated E-commerce. The private sector should also be part of the E-commerce policy making especially in the matters where government action or intergovernmental agreements are required e.g. taxation.

Second principle is that government intervention, when required, should be minimal and its goal should be to ensure competition, protect intellectual property and privacy, prevent fraud, foster transparency, support commercial transactions, and facilitate dispute resolution.

Third principle is that governments should strive to create a legal framework for supporting E-commerce transactions, which is governed by consistent principles across state, national, and international borders. This framework should be able to lead to predictable results regardless of the jurisdiction in which a particular buyer or seller resides.

International and Regional Co-operation Initiatives to Promote E-commerce

Non-harmonized consumer protection efforts across the globe are a major barrier to a global E-commerce development. A number of agreements (Free Trade Agreements, Joint Statements, Memorandums of Understanding, partnerships, information sharing protocols, and specific alliances) have been concluded among countries to enhance E-commerce policy and enforcement co-operation. Some examples include joint Consumer Policy Consultative meeting between Korea and China, Joint Statements on Electronic Commerce, which was concluded by Canada with Australia, Costa Rica, and the EU, and the Seoul-Melbourne Multilateral Anti-spam Agreement which Australia signed with a number of countries in the Asia-Pacific region on co-operation in countering spam.

Regional Initiatives

The European Union Consumer Protection Cooperation Network (EUCPCN): The EU CPC Regulation aims to exchange information and prevent any cross-border breaches to consumer protection laws in European Union countries.

The Australasian Consumer Fraud Task Force (ACFTF): This task force is responsible for consumer protection regarding frauds and scams.

The Trans Atlantic Consumer Dialogue (TACD): The TACD is a forum of US and EU consumer organizations, which develops joint consumer policy recommendations to the US government and the European Union.

The Asia-Pacific Economic Cooperation (APEC) forum: The E-commerce Steering Group of APEC promotes the development and use of predictable e-commerce regulatory and policy environments in the Asia-pacific region.

International Initiatives

The International Consumer Protection and Enforcement Network (ICPEN): Enforcement authorities use ICPEN to exchange enforcement information. Its website (www.econsumer.gov) allows consumers to register cross-border e-commerce complaints and offers tips for safe online shopping.

The London Action Plan (LAP): London Action Plan (www.londonactionplan.com) is a global network of governments and private sector representatives to fight spam.

The Messaging Anti-Abuse Working Group (MAAWG): With member companies from Asia, Europe, North America and South America, the MAAWG (www.maawg.org) is working on many initiatives to address ongoing and emerging spam-related issues.

Digital Phish Net (DPN): Digital PhishNet (www.digitalphishnet.org) is a collaborative enforcement operation to combat phishing. DPN assist law-enforcement agencies by providing real-time critical data needed to fight phishing.

ISO COPOLCO: ISO COPOLCO is the Committee on Consumer Policy of the International Organization for Standardization (ISO) that includes members from some 105 national standards bodies from countries around the world. Members use COPOLCO as a forum to exchange information and experience on standards.

The United Nations Commission on International Trade Law (UNCITRAL): UNCITRAL (www.uncitral.org) is a body of the United Nations whose objective is to promote harmonization and unification of the law of international trade.

The United Nations Conference on Trade (UNCTAD): UNCTAD (unctad.org) is an intergovernmental body dealing with trade, investment, and development issues. UNCTAD provides its Member countries with training and capacity building on the legal aspects of E-commerce.

The World Trade Organization (WTO): WTO is an international organization. In 1998, WTO issued the Declaration of Global E-Commerce that proclaimed a need for the establishment of a work program and a moratorium on new Internet restrictions.

Computer and Online Crime

A computer crime defined as any illegal activity that involves accessing computer data without permission. This unauthorized access does not necessarily have to result in loss or modification of data. Computer crime is often attributed to hackers and crackers, but increasingly organized crime groups have become active.

Online crime refers to the crimes in the physical world that are committed on Internet. These crimes may include ID theft, gambling etc. A new variant of online crimes is commandeering one computer to launch attacks on other computers, are new.

Law enforcement agencies face many difficulties combating online crimes. The first obstacle they face is the issue of jurisdiction. For example, assume that a person X residing in country A commits a crime using Internet against a person Y residing in country B. It is unclear which part of the crime could establish sufficient violation of laws of country B so that law enforcement agencies in country A could contact their counterparts in country B to take action against person X. It is possible that the actions considered criminal under country B laws might not be considered criminals in country A. Another issue is the difficulty of applying old laws written before the advent of Internet on the online crimes.

Hacker and Cracker

A hacker is a computer enthusiast, i.e. a person that possesses an incredible curiosity about how computers work. While hackers, on many occasions, can be involved in questionable activities, e.g. breaking into computer systems but their intention is not to cause any harm by doing this. An ethical hacker possesses the skills, mindset, and tools of a hacker but is also trustworthy. Ethical hackers perform the hacks as security tests for their systems. Ethical hackers (or good guys) protect computers against illicit entry. Ethical hacking is also known as penetration-testing or white-hat hacking.

A cracker can be defined as malicious hacker who involves into questionable activities e.g. breaking into computer systems or cracking copyrighted software with an intention to cause harm or steal.

Green E-commerce and IT

Green computing or green IT, refers to environmentally sustainable IT. San Murugesan, (2008) defines the field of green computing as "the study and practice of designing, manufacturing, using, and disposing of computers, servers, and associated subsystems—such as monitors, printers, storage devices, and networking and communications systems—efficiently and effectively with minimal or no impact on the environment". The broad goals of green computing are to reduce hazardous materials use, promote recyclability of out of use products and production waste, and maximization of energy efficiency of a product.

Environmental concerns are affecting E-commerce and many consumers are viewing online shopping as a "green" alternative to shopping at traditional bricks-and-mortar stores. Green shopping sites have emerged that help consumers find environmentally friendly products. Increasingly manufacturers are displaying information about product ingredients and publicizing environmentally friendly products and practices on-line. In 2007, Amazon.com launched an E-commerce page dedicated to green products and stepped up posts on Green Life (Amazon's green blog). In the same year, Office Depot launched an e-commerce site for green office products (See http://www.officedepot.com/a/browse/greeneroffice/N=5+11332/).

IMRG (imrg.org) identified six reasons why it believes E-commerce is green (IMRG, 2006). E-commerce is green because it results in less vehicle miles, lower inventory requirements, less printed materials, less packaging, less waste, and dematerialization (i.e. digitization of products).

Review Questions

1. List some major issues of E-commerce.

2. List some strategies to protect copyrights online.

3. What is Genericide?

4. What is Cybersquatting?

5. List some steps a consumer can take to protect its privacy online.

6. Differentiate between personal-matter jurisdiction and subject-matter jurisdiction.

7. How companies can deal with defamation on Internet?

8. How buyers and consumers can protect themselves from fraud in E-commerce?

9. What legal issues can arise by use of deep linking and framing on a website?

10. List some ethical issues in E-commerce.

11. List some social issues in E-commerce.

12. How parallel important can be disadvantageous for E-commerce?

13. List some language issues in E-commerce.

14. List some cultural issues in E-commerce.

15. How governments can help promoting the global E-commerce?

16. Differentiate between green E-commerce and green IT.

Bibliography

AberdeenGroup. (2001, May 18). *Multilingual sites imperative for large firms*. Retrieved April 12, 2009, from www.nua.com: http://www.nua.com/surveys/index.cgi?f=VS&art_id=905356776&rel=true

Adams, A. (2001). Introduction: beyond numbers and demographics: 'Experience-near' explorations of the digital divide. *Computers and Society* , 5-8.

Alfonso, S. D'. (2015, February 3). The Growing Problem of Medical Identity Theft. Retrieved from http://securityintelligence.com/the-growing-problem-of-medical-identity-theft/

Anandarajan, M. (2002). Internet Abuse in the Workplace. *Communications of the ACM, 45*.

Bentley.edu. (2015, May 19). The popular brands that had their trademarks revoked by law | Masters in Marketing Analytics News & Insights | Business Degree | Bentley University. Retrieved July 23, 2015, from http://www.bentley.edu/graduate/ms-programs/marketing-analytics/news-insights/popular-brands-had-their-trademarks-revoked

Berner, S. (2003). *Lost In Translation": Cross-Lingual Communication, and Virtual Academic Communities.* Retrieved January 15, 2009, from general.rau.ac.za: http://general.rau.ac.za/infosci/www2003/Papers/Berner,%20S%20Lost%20in%20Translation.pdf

Campbell, M. C. (2014, December 2). Apple loses appeal for "App Store" trademark in Australia. Retrieved July 23, 2015, from http://appleinsider.com/articles/14/12/02/apple-loses-appeal-for-australian-app-store-trademark

Chakrabarti, R. (2002). *The Asian Manager's Handbook of E-commerce.* McGrawhill.

Cheon, K. (2005). Multilingualism and the Domain Name System. *Reforming Internet Governance: Perspectives from the Working Group on Internet Governance*, (pp. 67-72).

Chin, W. (2001). Electronic Business and Education: Recent Advances in Internet Infrastructures. *Springer* .

Claburn, T. (2001). Pursuing Pirates. *Ziff Davis Smart Business , 14* (9), 38.

Crystal, D. (2001). *Language and the Internet.* Cambridge: Cambridge University Press.

Crystal, D. (2001). *Weaving a Web of linguistic diversity.* Retrieved January 12, 2009, from http://www.guardian.co.uk/GWeekly/Story/0,3939,427939,00.html

D'Alfonso, S. (2015, February 3). The Growing Problem of Medical Identity Theft. Retrieved from http://securityintelligence.com/the-growing-problem-of-medical-identity-theft/

Davies, A. (2000). A framework for understanding human factors in web-based electronic commerce. *International Journals of Human-Computer Studies* , 131-163.

Day. (1998). Editorial: Shared Values and Shared Interfaces: The Role of Culture in the Globalisation of Human-Computer Systems. *Special Issue: The Role of Culture in the Globalisation of Human-Computer Systems / Interacting with Computers* , 269-274.

DefamationLawBlog. (2009). *Internet Defamation Law Blog.* Retrieved April 12, 2010, from www.internetdefamationlawblog.com: www.internetdefamationlawblog.com

Delgado-Martinez. (2003). *What is Copyright Protection.* Retrieved April 23, 2009, from /www.whatiscopyright.org: http://www.whatiscopyright.org

Dinev, T., & Hart, P. (2005). Internet Privacy Concerns and Social Awareness as Determinants of Intention to Transact. *International Journal of Electronic Commerce* , 7.

Dubois, C. (1997). Multilingual Information Systems: Some Criteria for the Choice of Specific Techniques. *Journal of Information Science* , 5-12.

Editorial. (2003). Globalisation of Human-Computer Systems. *Interacting with Computers* , 269-274.

Ess, D. L., & Sudweeks, F. (1998). *Proceedings International Cultural Attitudes Towards Technology and Communication-CaTac'98*. Retrieved April 23, 2009, from University of Sydney, Australia.

EuropeanComission. (2009, March 5). *European Commission's report on cross-border e-commerce in the EU, Brussels.* Retrieved April 23, 2010, from ec.europa.eu: http://ec.europa.eu/enterprise/newsroom/cf/document.cfm?action=display&doc_id=2277&userservice_id=1&request.id=0

Forysthe, C., Grose, E., & Ratner, J. (1998). *Human Factors and Web Development.* NJ: Lawrence Erlbaum Associates, Publishers.

FTC. (2008, October 21). *President's Identity Theft Task Force Issues Report on Steps Taken to Implement Strategic Plan.* Retrieved October 25, 2010, from ftc.gov: http://www.ftc.gov/opa/2008/10/idtaskforce.shtm

Giardina, A., & Zampetti, A. B. (1997). Settling Competition-Related Disputes: The Arbitration Alternative in the WTO Framework. *Journal of World Trade* .

GlobalReach. (2004). *Evolution of Online Populations.* Retrieved April 12, 2009, from global-reach.biz: http://global-reach.biz/globstats/evol.html

Goswami, A. (2003). *Perspectives on E-globalization.* Retrieved May 12, 2009, from Digital Web Magazine: http://www.digital-web.com/articles/perspectives_on_e_globalization

Graeys, & Perchstone. (2014, February 19). Towards A Legal Framework For The Development Of E-Commerce In Nigeria: Issues & Prospects - Corporate/Commercial Law - Nigeria. Retrieved July 23, 2015, from http://www.mondaq.com/Nigeria/x/294344/Contract+Law/Towards+A+Legal+Framework+For+The+Development+Of+ECommerce+In

Hamelink, C. J. (2001). *Ethics of Cyberspace.* California: Sage Publishers.

Hassanein, K., & Head, M. (2005). The Impact of Infusing Social Presence in the Web Interface: An Investigation Across Product Types. *International Journal of Electronic Commerce* , 31.

He, S. (2008). Multilingual Issues in Global E-commerce Web Sites. In C. Wankel, *The Handbook of 21st Century Management* (pp. 391-400). Sage Publications.

He, S. (2008). Multilingual Web Sites in Global Electronic Commerce. *Electronic Commerce: Concepts, Methodologies, Tools and Applications* , 1187-1194.

Hofstede, G. (1997). *Cultures and Organisations: Software of the mind.* UK: McGraw-Hill International Ltd.

Hubert, R. (2006). Accessibility and usability guidelines for mobile devices in home health monitoring. *ACM SIGACCESS Accessibility and Computing* , 26-29.

Hubert, R. (2006). *Usability field study of older adults using multi-modal home health monitoring devices.* Retrieved August 25, 2009, from ETD Collection for Pace University: http://digitalcommons.pace.edu/dissertations/AAI3225086

ICC. (2003, November). *Resolving Disputes Online; Best practices for Online Dispute Resolution (ODR) in B2C and C2C Transactions.* Retrieved June 12, 2010, from www.iccwbo.org: www.iccwbo.org/uploadedFiles/ICC/policy/e-business/pages/ResolvingDisputesOnline.pdf

IMRG. (2006). *Valuing Home Delivery; A Cost – Benefit Analysis.* Retrieved March 12, 2010, from www.imrg.org: http://www.imrg.org/80257418006E81C9/%28httpInfoFiles%29/011D343B14922454802574780039D02C/$file/Valuing%20Home%20Delivery%202008%20Report.pdf

ITU. (2009). *ICT Legislations - FAQ.* Retrieved May 10, 2010, from International Telecommunication Union, Asia Pacific Centre of Excellence - Telecommunications: http://itu-coe.ofta.gov.hk/vtm/ict/faq/q6.htm

Keeker, K. (1997). *Improving web site usability and appeal.* Retrieved July 12, 2009, from Microsoft MSDN: http://msdn.microsoft.com/en-us/library/cc889361%28office.11%29.aspx

Kellog, W. A., & Nielson, J. (1997). Argumented Conceptual Analysis of the Web. *CHI '97 Extended Abstracts on Human Factors in Computing Systems: Looking To the Future*, (p. 228). Atlanta, Georgia.

Kovacs, N. (2014, September 30). How To Avoid Identity Theft Online | Norton Community. Retrieved July 23, 2015, from https://community.norton.com/en/blogs/norton-protection-blog/how-avoid-identity-theft-online

Leff, B. (2014, August 26). Electronic Signatures and Enforcing Online Contracts. Retrieved from http://shaked-law.com/electronic-signatures-enforcing-online-contracts/

LegalDictionary. (2009). *Personal Jurisdiction.* Retrieved August 12, 2010, from /legal-dictionary.thefreedictionary.com: http://legal-dictionary.thefreedictionary.com/Personal+Jurisdiction

Legislation.gov.uk. (2015). The Copyright and Rights in Databases Regulations 1997. Retrieved July 23, 2015, from http://www.legislation.gov.uk/uksi/1997/3032/contents/made

Livingstone, S., & Haddon, L. (2009). *EU Kids Online: Final report; LSE, London: EU Kids Online; EC Safer Internet Plus Programme Deliverable D6*. Retrieved August 12, 2010, from www.lse.ac.uk: www.lse.ac.uk/collections/EUKidsOnline/Reports/EUKidsOnlineFinalReport.pdf

MacLeod, M. (2000). Language Barriers. *Supply Management* , 37-38.

MSDNLibrary. (2007). *Language Issues*. Retrieved January 26, 2009, from http://msdn2.microsoft.com: http://msdn2.microsoft.com/en-us/library/aa292134(VS.71).aspx

Murugesan, S. (2008, January-February). Harnessing Green IT: Principles and Practices. *IEEE Professional* , 24-33.

OECD. (2007). *Recommendation on Consumer Dispute Resolution and Redress*. Retrieved June 26, 2010, from www.oecd.org: www.oecd.org/dataoecd/43/50/38960101.pdf

OECD. (2008). *OECD Policy Guidance for Addressing Emerging Consumer Protection and Empowerment Issues in Mobile Commerce*. Retrieved February 12, 2010, from www.oecd.org: www.oecd.org/dataoecd/50/15/40879177.pdf

OECD. (2009, December). *2009 Conference on Empowering E-Consumers: Strengthening Consumer Protection in the Internet Economy, Washington, 8-10 Dec. 2009, Background report*. Retrieved June 12, 2010, from oecd.org: http://www.oecd.org/dataoecd/44/13/44047583.pdf

OFCOM. (2008). *Social Networking: A quantitative and qualitative research report into attitudes, behaviours and use*. Retrieved July 15, 2010, from www.ofcom.org: www.ofcom.org.uk/advice/media_literacy/medlitpub/medlitpubrss/socialnetworking/report.pdf

OFCOM. (2009). *Children's web access*. Retrieved September 9, 2010, from www.ofcom.org: www.ofcom.org.uk/consumer/2009/10/more-children-have-broadband-in-the-bedroom

Perrault, A. H., & Gregory, V. L. (2000). Think Global, Act Local: The Challenges of Taking the Website Global. *International Journal of Special Libraries* , 227-237.

Piva, A., Barni, M., Bartolini, F., & De Rosa, A. (2001). A new decoder for the optimum recovery of nonadditive watermarks. *IEEE Transactions, Image Processing* , 755-766.

Ratner, J. (2002). *Human Factors and Web Development*. CRC Press.

Revolvy.com. (2015, March 4). Generic trademark - Expand Your Mind. Retrieved July 23, 2015, from http://www.revolvy.com/main/index.php?s=Generic%20trademark

Rogerson, K., & Lynch, S. (2005). Rebuilding Borders around the Internet: Global Attempts at Internet Taxation. *annual meeting of the International Studies Association*. Hilton Hawaiian Village, Honolulu, Hawaii.

Schneider, G. (2010). *Electronic Commerce, 9th Ed.* Course Technology.

Seilheimer, S. (2004). Productive Development of World Wide Web Sites Intended for International Use. *International Journal of Information Management* , 363-373.

Shneiderman, B. (1997). Designing information-abundant web sites: issues and recommendations. *International Journal of Society* , 5-29.

Spethman, M. (2003). Web Site Globalization. *World Trade* , 56-57.

thebigword.com. (2006). *Fortune 500 - Multilingual Websites.* Retrieved March 2009, from thebigword.com: http://www.thebigword.com/Fortune500MultilingualWebsites.aspx

Venkat, K. (2002). Speakout: delving into the digital divide. *IEEE Spectrum* , *39* (2), 14-16.

w3.org. (1997). *A framework for global electronic commerce by White House.* Retrieved April 12, 2009, from www.w3.org: http://www.w3.org/TR/NOTE-framework-970706

YahooSmallBusiness. (2009). *Cybersquatting: What It Is and What Can Be Done About It.* Retrieved May 11, 2010, from smallbusiness.yahoo.com: http://smallbusiness.yahoo.com/r-article-a-40986-m-1-sc-12-cybersquatting_what_it_is_and_what_can_be_done_about_it-i

Yunker, J. (2003). *Beyond Borders: Web Globalization Strategies.* New Riders Press.

13

BUILDING E-COMMERCE SITES

Introduction

Building an E-commerce site is a complex task, similar to building an entirely new information system, which requires investment of major physical and human resources. For many firms the cost-effective solution is to outsource in part or full the development effort to specialized firms. Companies must first layout a methodology for approaching the problem, consider the major issues in building a site, and analyze the available options carefully.

Method for Building E-commerce Site

According to (Laudon & Traver, 2009), System Development Life Cycle (SDLC) approach can be applied to building an E-commerce Web site. SDLC is a methodology for understanding the business objectives of any system so that an appropriate solution can be designed. There are five major steps in the System Development Life Cycle for an E-commerce site. These steps are systems analysis, systems design, building the system, testing the system, and implementation.

Systems Analysis

In this step, the business objectives for the E-commerce site are identified. These business objectives are necessary capabilities for the site. Business objectives are then translated into lists of the types of information systems and the elements of information that will be needed to achieve them.

System Design

In this step, the main components in the E-commerce system and their relationships to one another are identified. There are two types of design i.e. logical web site design and physical web site design. A logical design for a Web site describes the following:

- The flow of information at the site including the processing functions that must be performed and the databases that will provide information.

- Description of the security and emergency backup procedures

- Description of the controls that will be used in the system

A data flow diagram is created is created in the logical design process. A physical design, on the other hand, translates the logical design into the physical components that will be needed such as the servers, software, and size of the telecommunications link, backup servers, and security system.

Building the System

Physical components needed are purchased.

System Testing

Three types of testing are performed i.e. unit testing, systems testing, and acceptance testing. In unit testing, the program modules are tested one at a time. Once unit testing is completed, system testing is performed in which the site is tested as a whole. System testing examines every conceivable path a customer might try to make sure that customers can find what they want easily and quickly and complete a purchase without a problem. In the end acceptance testing is performed. Acceptance testing requires the firm's key personnel and managers to use the system to verify that the business objectives as originally conceived are being met.

Implementation

In this step continuing maintenance is performed. Continuing maintenance is needed over the life of the site to keep it functional. It includes correcting mistakes and continuing to improve, update, and modify links and other site features. E-commerce site maintenance is more expensive as these sites are in a continuous process of change and improvement. Every time a change is made code must be debugged, hyperlinks must be continually tested and repaired, reports, data files and links to backend databases must be maintained and updated as necessary, and the general administrative tasks of the site, such as updating the products and prices, must be attended to.

Six Web Pages of Good E-commerce Sites

Every good E-commerce site should include the following six pages.

ABOUT US: This page provides information about the firm, its mission, vision, objectives, values, biographies of top management etc.

CONTACT US: This page describes how a customer can contact the company, name/designation and contact information of relevant persons etc.

SITE MAP: A site map is a hierarchal map of the web pages on corporate website. A search engine uses the site map to rank your site. This page is important from SEO point of view.

WHY BUY FROM US: This page provide evidence to convince users buy products/services from E-commerce site.

LEGAL INFORMATION: This page includes information such as Terms of use, privacy policy etc.

TESTIMONIALS: This page provides customers comments about the E-commerce site. Comments can be negative or positive.

Components of a Well-functioning E-commerce Site

A well-organized web site should be easy for a customer to use and easy for the owner to maintain. It takes a lot to build a well-organized web site. Below is a list of the most important elements that a good ecommerce web site should have (cs.wellseley.edu, 2015).

A WELL-ORGANIZED COLLECTION OF PRODUCTS AND/OR SERVICES

Smaller web sites can just list all their products on one or several web pages. Larger sites provide indices of products and search engines so that customers can find what they need. There should be a way for customers to get all necessary information about products, compare several products, get an advice on related products that they might want to get, etc. Making changes (such as add or remove products based on every day's availability, change in prices, add product cross-references, etc.) should be easy.

E-commerce site should provide short and detailed product descriptions with a clear and high-quality product picture. In case of many products, an industry standard classification system can be used to categorize products. It is also important to include sales terms and after sales support information.

CONVENIENT PRODUCT SELECTION BY CUSTOMERS

This feature is usually implemented as a shopping cart. The customer should be able to select and delete products while browsing the web site.

CONVENIENT ORDER FORMS

The form should be flexible enough to specify a different address for the product delivery, a gift message, etc. It should have as few required fields as possible. A returning customer should be provided default information so that he does not need to type it every time. There should not be any typos in the forms.

CONVENIENT WAYS OF PAYMENT

E-commerce site should provide as many payment methods as possible. Every customer has different preference of payment methods. Having different payment methods can facilitate customers' and increase sales. PayPal has become a popular payment service for small payments. There should be options of paying by a credit card, by a check, electronic check, and by a credit card over the phone if the customer is not comfortable sending his/her credit card number. The options may include some electronic payment systems. The site should provide business some quick way to verify the credit card information or in some other way to check that, the payment is valid.

SECURE COMMUNICATION SYSTEM

A secure system not only protects transmission of credit card number but also guarantee privacy of the customer (including details of the purchase). A web site might have a user registration system with a password, in which case all transactions by the user should be private. It is also important to prevent unauthorized access to the web site.

SHOPPING CART WITH TAX CALCULATION

Shopping carts on E-commerce site should be capable to provide relevant tax calculations. This capability is important to avoid any tax related issues with the government.

REPOSITORY OF CUSTOMER INFORMATION

An E-commerce website should provide ways to store information about customers. This is convenient for customers so that they do not need to reenter their information every time they access the site. It also allows to "customize" the web site for someone's interests. This can be done via customer registration or by means of cookies. This customer information can provide many benefits to the business. The business can customize ads based on the customer's profile or send an e-mail advertising a new product. However, the greatest benefit is that the business can monitor customer's behavior (e.g. which pages have the customer visited and which purchases (if any) he/she has made afterwards).

All E-commerce sites should support SSL to encrypt information that needs to remain secure. This is especially true for credit card and payment information, but also any customer information like address, phone number, email, etc. Customers have an expectation that their personal information will remain secure when they make a purchase online, so ensuring that SSL is implemented is not just a good idea, but also something, that is essential for ensuring that your customers trust that their information will remain secure. In addition, security is required to meet PCI compliance for any business, which accepts credit card payments. This also will reassure customers about the commitment the online business has about keeping the customers' personal and sensitive information secure. Do not store credit card numbers in your site's database. While it might seem like a good idea to keep card information on file to make purchases easier for customers, keeping this information stored on your servers is a huge security risk, and if your database is compromised, you will be liable for that loss.

REPOSITORY OF ORDER INFORMATION

This repository keeps information about orders. This allows customers to track their orders, and for business to get all kinds of financial and statistical information. It is also important to keep order information in case of later disputes.

CUSTOMER SUPPORT AND FEEDBACK

There should be online documentation for all products ever sold on the web site, various FAQs, and, ideally, a way of customers to post their opinion about the product. Easy access to this information may make a difference between a frequently visited web site and a lonely online looser. Note that this aspect of E-commerce website cannot be completely mechanized. You will need a human being has to answer e-mail, judge the relevance of customer's comments, organize comments by topics, and so on. However, there is a lot that can be computer-aided in this process, for instance sorting incoming messages by their title and/or return address to forward them to an appropriate customer support person.

OPTION OF SITE SEARCH

Statistics show that 30% of visitors to E-commerce sites use search to find the products they are looking for (Jacobson, 2015), so it is important to make sure the search functionality is available and easy to use. In addition, it is a good idea to utilize features like autocomplete to help users find popular products or items related to their searches. Faceted search is another important way to help users find products. This functionality allows them to narrow their search in a variety of ways, including by department, size, price range, manufacturer, etc. Providing this functionality give users more power to find what they need, letting them limit their searches to exactly what they are looking for. In order to make sure searches return the best results, make sure product information is fully defined and well organized, which will allow for faceted search and better search results in general.

SUPPORT FOR GUEST CHECKOUTS

Companies with E-commerce sites will often want to require users to create an account in order to make a purchase, since this allows for follow-up communication that encourages future sales, as well as tracking customers' demographic information to analyze sales. However, it is important to remember that not everybody wants to go through the process of creating an account in order to buy a product. Repeat customers will want to register and get the benefits of having an account, such as saving their information for future purchases and receiving notifications about upcoming sales, but it is still a good idea to provide an option for people who just want to make a one-time order. Since you will still want to try to encourage users to create an account, you might try to design the purchase process to allow users to complete a guest checkout, and upon completion, allow them to create an account using the information that they just entered.

GUARANTEES

Money back guarantees are becoming increasingly common on online retailer sites. If your site offers money-back, guarantees make sure these guarantees provide a validity period especially for intangible products (e.g. e-books).

POLICIES

Some of the policies that an E-commerce site should include are privacy policy and accessibility rules. The privacy policy is a public statement that describes how the customers' information gathered at the site will be used. Many Web sites can help you build a privacy policy. Users (i.e., e-business owners) can visit the site and submit their use of personal information into a questionnaire. The Web site then generates a graph in HTML that e-business owners can plug into their code, making it easier for users to educate themselves on the site's policies. Accessibility rules describes how users with a disability can effectively use the E-commerce site.

DISCLAIMERS

An online business provides these disclaimers to inform site users about the limits of products and services offered by the online business. For example, small spelling or editing errors, such as an inaccurate statistic or product price could result in product misrepresentation, angry consumers and lost business. Dated information and inaccurate links can also lead to problems for Web-site owners. Disclaimers should be easy to locate and information should be presented in a clear and concise manner.

WAYS FOR SITE PERFORMANCE OPTIMIZATION

If your site is slow, you are likely to lose customers. Statistics show that 40% of users will abandon a website that takes more than 3 seconds to load (Kickassmetrics.com, 2011). This is especially true for mobile users, who are often multi-tasking as they access websites and are more likely to move on to something else if a site is too slow. In order to keep from losing customers due to slow load times, you will want to make sure your site is optimized to run as quickly as possible. Here are a few ways to help your site run more smoothly. You can combine a site's JavaScript or CSS resource files into single files. That will speed up their interaction with the site, because users will only have to download one JavaScript file or style sheet rather than five or ten. You can compress images, which will allow them to provide the best visuals at the smallest possible size, reducing download times. You can use caching to reduce the time spent sending data between the web server and the database server.

LOAD BALANCING MECHANISM

Load balancing involves adding extra servers to distribute the workload and traffic among them. This server management technique can help when you have only one server handling all the HTTP requests for your website. As traffic grows, this single server may no longer be able to deal with the traffic influx and as a result, the site can load slowly or worse -- it crashes or times out.

Keeping these aspects of an E-commerce site in mind during development will help you ensure that you are providing the best experience for your customers.

Compliance with Web standards

Web standards are the formal, non-proprietary standards and other technical specifications that define and describe aspects of the World Wide Web. Web standards were developed by the World Wide Web Consortium (W3C). These standards are best practices for web coding. Of most relevance to businesses looking to get online are the standards for XHTML and CSS coding. In recent years, the term has been more frequently associated with the trend of endorsing a set of standardized best practices for building web sites, and a philosophy of web design and development that includes those methods. Web standards include many interdependent standards and specifications, some of which govern aspects of the Internet, not just the World Wide Web. Even when not web-focused, such standards directly or indirectly affect the development and administration of web sites and web services. Considerations include the interoperability, accessibility and usability of web pages and web sites.

Web standards, in the broader sense, consist of the following:

- Recommendations published by the World Wide Web Consortium (W3C)
- Internet standard (STD) documents published by the Internet Engineering Task Force (IETF)
- Request for Comments (RFC) documents published by the Internet Engineering Task Force
- Standards published by the International Organization for Standardization (ISO)
- Standards published by Ecma International (formerly ECMA)
- The Unicode Standard and various Unicode Technical Reports (UTRs) published by the Unicode Consortium
- Name and number registries maintained by the Internet Assigned Numbers Authority (IANA)

Web standards are not fixed sets of rules, but are a constantly evolving set of finalized technical specifications of web technologies. Web standards are developed by standards organizations. These standards organizations are groups of interested and often competing parties chartered with the task of standardization. It is crucial to distinguish those specifications that are under development from the ones that already reached the final development status (in case of W3C specifications, the highest maturity level). The web standards project (www.webstandards.org) is a coalition dedicated to make standards, which ensure simple, affordable access to web technologies for all.

When a web site or web page is described as complying with web standards, it usually means that the site or page has valid HTML, CSS and JavaScript. The HTML should also meet accessibility and semantic guidelines. Full standard compliance also covers proper settings for character encoding, valid RSS or valid Atom news feed, valid RDF, valid metadata, valid XML, valid object embedding, valid script embedding, browser- and resolution-independent codes, and proper server settings.

Advantages of Web-standards Compliant Website

There are many benefits for developing a website in accordance with the web standards. Some key advantages of a standards compliant website are discussed below (The Web Standards Project, 2015).

A FASTER WEBSITE

Use of web standards reduces the amount of code used to create each page. Your website files are smaller to download making speed of browsing for your customers much faster. A secondary advantage of this is reduced hosting bandwidth costs.

IMPROVED SEARCH ENGINE RANKINGS

Complying with web standards can give your web pages greater visibility in web searches. The structural information present in compliant documents makes it easy for search engines to access and evaluate the information in those documents, and they are indexed more accurately. As search engines do not understand layout code, a website that does not comply with standards can negatively impact your search engine rankings. In a standards compliant website your important website content is given much more weighting as the search engines are presented with pages to spider which use only a small amount of page markup that they understand - the rest of the page being your content.

WEBSITE SEARCH

Because use of web standards makes it easier for server-side as well as client-side software to understand the structure of your document, adding a search engine to your own site becomes easier and gives better results.

COMPATIBILITY

Standards are written so that old browsers will still understand the basic structure of your documents. Even if they cannot understand the newest and coolest additions to the standards, they will be able to display the content of your site.

EASIER TO UPDATE

With a web standards compliant website, it is easier to make changes to your website. As the design is kept separate from the content alterations to the look and feel of the website is a lot quicker and easier, usually involving just updating 1 CSS file instead of having to alter code on every page!

MORE ACCESSIBLE

Web standards are a great move in making your website accessible for all your users. For example, users with visual impairments often rely on screen readers which have problems reading the old 'tables based' websites. As a website owner, you have a legal obligation to make your website accessible. Proper use of the web standards goes a long way towards meeting these requirements. As the variety of web access methods increases, adjusting or duplicating websites to satisfy all needs will become increasingly difficult. Following standards is a major step towards solving this problem. Making your sites standards-compliant will help ensure not only that traditional browsers, old and new, will all be able to present sites properly, but also that they will work with unusual browsers and media.

CODE VALIDATION

Compliant code gives you the opportunity of validating your page with a validation service. Validators process your documents and present you with a list of errors. This makes finding and correcting errors a lot easier, and can save you a lot of time.

CONVERSION AND MIGRATION

Compliant documents can easily be converted to other formats, such as databases or Word documents. This allows more versatile use of the information within documents on the web and simplified migration to new systems (hardware as well as software) including devices such as TVs and PDAs.

STABILITY

Most web standards are generally designed with forward- and backward-compatibility in mind. It is done so that data using old versions of the standards will continue to work in new browsers, and data using new versions of the standards will "gracefully degrade" to produce an acceptable result in older browsers. Because a website may go through several teams of designers during its lifetime, it is important that those people are able to comprehend the code and to edit it easily. Web standards offer a set of rules that every Web developer can follow, understand, and become familiar with. When one developer designs a site to the standards, another will be able to pick up where the former left off.

Developing Accessible Websites

Web accessibility refers to the inclusive practice of removing barriers that prevent interaction with, or access to websites, by people with disabilities. When sites are correctly designed, developed and edited, all users have equal access to information and functionality (American Foundation for the Blind, 2015). You need to make sure that your site is prepared to catch all potential shopper traffic to help maximize your sales. Though a good marketing strategy is imperative, a brilliant pull strategy will not mean much if your website is not optimized to both accept and guide visitors through your checkout process. There are many ways to make your E-commerce site more accessible (Hallock, 2013; Berkeley.edu, 2015).

Guidelines to Develop Accessible Websites

SWITCH TO CSS

With new advances in standards compatibility, it is really time to adopt this best practice and start using CSS instead. This step makes page content easier to understand when it is being read aloud by a screen reader, too.

WRITE SEMANTICALLY-VALID CODE

You should use some validation services (the W3C Validator is a popular one) when you are building out a site. The code that does not validate should be the exception rather than the rule. This also means understanding the proper usage of tags for your version. HTML5 has a number of new tags that allow for a more fine-grained native control of the document object model, so if you're not current, that would be a good place to start with your efforts. One caveat: be sure to consider browser support, as your demographic may not typically have browsers that support certain new features of HTML5.

USE ARIA ROLES AND LANDMARKS

ARIA (Accessible Rich Internet Applications) is an easy and powerful technical specification for ensuring your site structure is accessible. By assigning ARIA roles and landmarks to web elements, you enhance the ability of screen reader users to navigate and interact with your content. ARIA roles and landmarks can be easily added to your HTML, in the same way that you add classes to HTML in order to load attributes from CSS. Examples of common ARIA usage include:

- **Roles of "navigation" and "menu"**: Used to identify the site's primary navigation and individual page menus.
- **Landmarks of "banner" and "main content"**: Used to identify the header and main content sections of a webpage.
- **Alerts of "live" and "atomic":** Used to help screen reader users with dynamic page changes, such as stock tickers and search filters.
- **Forms:** ARIA allows the addition of descriptive text to a form field ("described by" or "labeled by") and the identification of buttons and required controls (more information under tip #6).

USE PROPER HEADINGS FOR STRUCTURE OF YOUR CONTENT

Screen reader users can use heading structure to navigate content. By using headings (<h1>, <h2>, etc.) correctly and strategically, the content of your website will be well organized and easily interpreted by screen readers. Be sure to adhere to the correct order of headings, and separate presentation from structure by using CSS (Cascading Style Sheets). Do not pick a header just because it looks good visually (which can confuse screen reader users); instead, create a new CSS class to style your text. There are some examples of proper usage of headings. First, you should use <h1> for the primary title of the page. Avoid using an <h1> for anything other than the title of the website and the title of individual pages. Second, use headings to indicate and organize your content structure. Third, do not skip heading levels (e.g., go from an <h1> to an <h3>), as screen reader users will wonder if content is missing.

USE COLORS WISELY

The very first thing your website visitors are confronted with is the background color of your website, the color of your header, the color of your text and the headlines, etc. The most common form of color deficiency, red-green color deficiency, affects approximately 8% of the population (National Eye Institute, 2015). Using only colors such as these (especially to indicate required fields in a form) will prevent these individuals from understanding your message. Other groups of people with disabilities, particularly users with learning disabilities, benefit greatly from color when used to distinguish and organize your content. To satisfy both groups, use color, but also be sure to use other visual indicators, such as an asterisk or question mark. Be sure to also distinguish blocks of content from one another using visual separation (such as whitespace or borders). There are several tools you can use to evaluate color contrast, which will assist you in making your page as visually usable as possible to individuals with low vision or varying levels of color blindness. For a list of these tools, see (Berkeley Web Access, 2015).

Colors sometimes have a greater degree of emotional impact on the viewer's senses. The following list provides common colors and the types of first-impression, psychological emotion they may invoke upon the website viewer (Price Wharton, 2015).

- ° **RED** invokes feeling of love, passion, danger, warning, excitement, food, impulse, action, adventure.
- ° **BLUE** can invoke feelings of trustworthiness, success, seriousness, calmness, power, professionalism.
- ° **GREEN** can invoke feelings of money, nature, animals, health, healing, life, harmony.
- ° **ORANGE** can invoke feelings of comfort, creativity, celebration, fun, youth and affordability.
- ° **PURPLE** can invoke a feeling of royalty, justice, ambiguity, uncertainty, luxury, fantasy, dreams.
- ° **WHITE** can invoke feelings of innocence, purity, cleanliness, simplicity.
- ° **YELLOW** can invoke feelings of curiosity, playfulness, cheerfulness, amusement.
- ° **PINK** can invoke feelings of softness, wetness, innocence, youthfulness, tenderness.
- ° **BROWN** can invoke feelings of earth, nature, tribal, primitive, simplicity.
- ° **GREY** can invoke feelings of neutrality, indifference, reserved.
- ° **BLACK** can invoke feelings of seriousness, darkness, mystery, secrecy.

Colors help sell your website products or ideas. Colors can distract, even turn off the website viewer. Therefore, choose your colors wisely. Your chosen colors should appeal to the widest possible demographics of website viewers.

CAPTION YOUR MULTIMEDIA

Alt text should be provided for images, so that screen reader users can understand the message conveyed by the use of images on the page. This is especially important for informative images (such as infographics). When creating the alt text, the text should contain the message you wish to convey through that image, and if the image includes text, that text should also be included in the alt. The exception to this rule is when an image is used purely for decoration; in this case, the alt text can be left empty so that the screen reader user is not distracted from the more important content on the page.

Similarly, deaf users will miss audio cues, and video can be problematic for those with both challenges. Make sure to include some plaintext description (a caption, a transcript, or a simple explanation of content) with media items. This one is an SEO benefit in addition to an accessibility one. Be sure to use of the appropriate HTML tags in your descriptive content.

DESIGN ACCESSIBLE FORMS

When form fields are not labeled appropriately, the screen reader user does not have the same cues available as the sighted user. It may be impossible to tell what type of content should be entered into a form field. Each field in your form should have a well-positioned, descriptive label. For example, if the field is for a person's name, it should be labeled appropriately as either "Full Name" or have two separate fields labeled as "First Name" and "Last Name." Use the <label> tag or an ARIA property to associate the label text with the form field. As you are going through a form field, a person should be able to tab through the form and fill out all the fields before getting to the "Submit" button, or they may not even realize that additional fields exist. Essentially, the tab order should follow the visual order. If you have fields that are related or similar, consider grouping them together using fieldsets. For example, fields like "Full Name" and "Date of Birth" could be grouped together as "Personal Information." This type of form organization can help a screen reader user keep track of progress, and can provide the context that might be lost while filling out the form.

If certain form fields are required, the field should be labeled accordingly, and configured to alert the screen reader user. Commonly, required fields are noted as such with an asterisk, which will not be spoken by some screen readers. Asterisks (or similar visual indications) should still be used for sighted users, people with learning disabilities or people who speak English as a second language. Indicating that a field is required to a screen reader can be accomplished by adding ARIA required="true" and ARIA required="false" for optional fields. After submitting the form, user will need to be alerted to submission confirmation and any submission errors. We recommend including any error counts in the page title (after the user has submitted), so the user will immediately be informed that there are errors on the page. If a user submits a form with errors, the user should be brought to a submission page that indicates what the errors are, and provides an easy way to navigate to those errors. Finally, the use of CAPTCHA is inaccessible and should not be used to validate submissions. WebAIM (WebAIM, 2015) provides a helpful summary of accessible alternatives to CAPTCHA to keep forms free of spam submissions.

KEEP REPETITIVE CONTENT TOWARD THE BOTTOM OF THE MARKUP

Instead of placing your menus and such first in the document, place them below the main content and position them using CSS. This makes it easier for people making use of screen readers to find the content they are looking for in the document.

MAKE DYNAMIC CONTENT ACCESSIBLE.

When content updates dynamically (i.e. without a page refresh), screen readers may not be aware. This includes screen overlays, lightboxes, in-page updates, popups, and modal dialogs. Keyboard-only users may be trapped in page overlays. Magnification software users might be zoomed in on the wrong section of the page. These functions can easily be made accessible. Options include ARIA roles and alerts, as well as front-end development frameworks that specifically support accessibility. Ensure that video players do not auto-play (non-consensual sound), and that the players can be used with a keyboard. Additionally, all videos must have options for closed captioning and transcripts for the hearing-impaired. If your site contains a slideshow, make sure that each photo has alt text and can be navigated via the keyboard. If you are using any unique widgets (such as a calendar picker or drag-and-drops), be sure to test for accessibility.

USE TABLES FOR TABULAR DATA ONLY

Using tables for page layout adds additional verbosity to screen reader users. Whenever a screen reader encounters a table, the user is informed that there is a table with "x" number of columns and rows, which distracts from the content. In addition, the content may be read in an order that does not match the visual order of the page. Do not create the layout of a website using a table; instead, use CSS for presentation. When a data table is necessary (i.e. you have a set of data that is best interpreted in a table format, such as a bank statement), use headers for rows and columns, which helps explain the relationships of cells. Complex tables may have several cells within the table that have a unique relationship to each other, and these should be identified by using the "scope" attribute in HTML. Table captions (HTML5) can be used to give additional information to users about how best to read and understand the table relationships.

DON'T USE GENERIC HYPERLINKS

A major usability issue that also translates to an accessibility issue is the idea of the generic hyperlink. Users should always be able to tell you where they are going by reading the thing they are about to click on. When including links in your content, use text that properly describes where the link will go. Using "click here" is not considered descriptive, and is ineffective for a screen reader user. Just as if sighted users scan the page for linked text, visually impaired users can use their screen readers to scan for links. As a result, screen reader users often do not read the link within the context of the rest of the page. Using descriptive text properly explains the context of links to the screen reader user. The unique content of the link should be presented first, as screen reader users will often navigate the links list by searching via the first letter. For example, if you are pointing visitors to a page called "About Us": Try not to say: "Click here to read about our company." Instead, say: "To learn more about our company, read About Us."

ENSURE ALL CONTENT CAN BE ACCESSED WITH THE KEYBOARD ALONE

Users with mobility disabilities, including repetitive stress injuries, may not be able to use a mouse or trackpad. These people are able to access content using a keyboard by pressing the "tab" or "arrow" keys, or using alternative input devices such as single-switch input or mouth stick. As a result, the tab order should match the visual order, so keyboard-only users are able to logically navigate through site content. Long pages with lots of content should be broken up with anchor links (jump lists), allowing keyboard-only users to skip to relevant portions of the page without having to negotiate through other content. "Skip to main content" should be provided at the top of each page, so keyboard-only users will not have to tab through the page navigation in order to get the main content. For pages with local menus and multiple levels and sub-items, the menus should be configured so that all menu items can be accessed with the keyboard. Do not use elements that only activate when a user hovers over items with a mouse, as keyboard-only or screen readers users will not be able to activate them.

CHOOSE A CONTENT MANAGEMENT SYSTEM THAT SUPPORTS ACCESSIBILITY

There are many content management systems available to help you build your website. Common examples include Drupal and Wordpress, but there are many other options available. Once you have chosen a CMS that suits your needs, make sure to choose an accessible theme/template. Consult the theme's documentation for notes on accessibility and tips for creating accessible content and layouts for that theme. Be sure to follow the same guidelines when selecting modules, plugins, or widgets. For elements like editing toolbars and video players, make sure that they support creating accessible content. For example, editing toolbars should include options for headings and accessible tables, and video players should include closed captioning. The CMS administration options (such as creating a blog post or posting a comment) should be accessible as well.

Examples of Good Accessible Website Designs

CouchSurfing.org

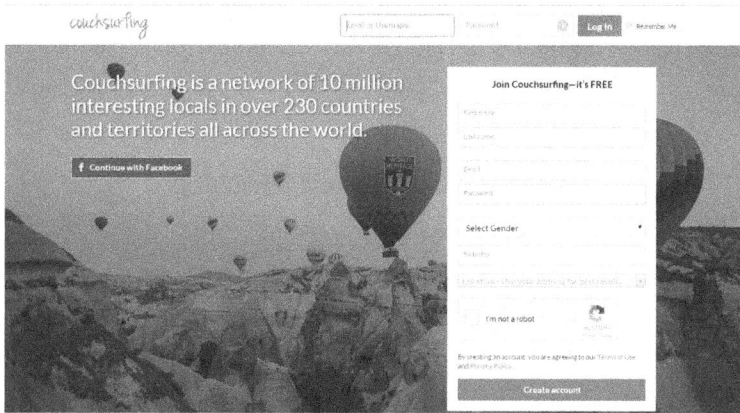

The website provides excellent heading structure and a clean, simple interface. Labels are placed on all form fields. There is a language picker at the bottom of the page. When users select a new language, the page is reloaded with all text translated. The lang attribute on the <html> element also changes to reflect the new language, so screen readers will pronounce the text correctly. This is a nice feature, and makes the site accessible to millions of people who would otherwise be excluded due to language.

U.S. Government Sites

US government sites are required to be accessible via Section 508 of the Rehabilitation Act. Some good examples of U.S. government accessible sites are the Center for Disease Control and Social Security Administration sites.

The accessibility features of these sites are very similar. Both sites include a full set of ARIA landmark roles, including (in various places) banner, main, navigation, search, complementary, and contentinfo. Both provide good heading structure. Both sites include hidden same-page links at the top of the page for skipping to particular content. The CDC site includes five of these links, roughly corresponding with each of the landmark regions. These links become visible when keyboard users tab into the page. Non-mousers with eyesight can benefit from skip links like the ones on these government sites. Both sites support multiple languages (English and Spanish)

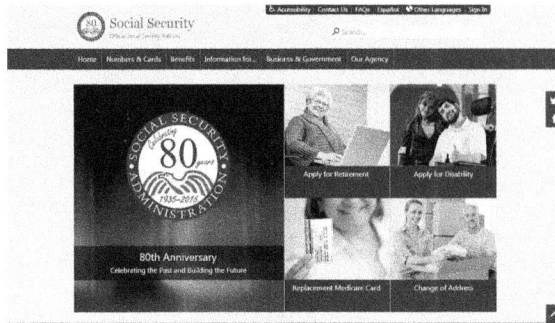

USA.gov

This site, the U.S. Government's official web portal, is fully controllable by keyboard. Users can tab into it, visibly see which control has focus, and execute that control by pressing enter.

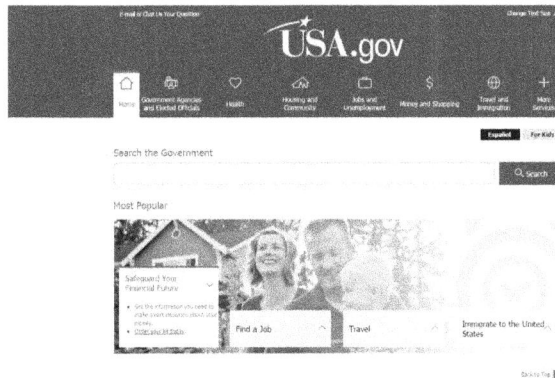

Canada.ca

The Canadian government site is highly accessible. All content appears in both English and French. The home page is a simple page, beautiful in its perfect symmetry between English and French.

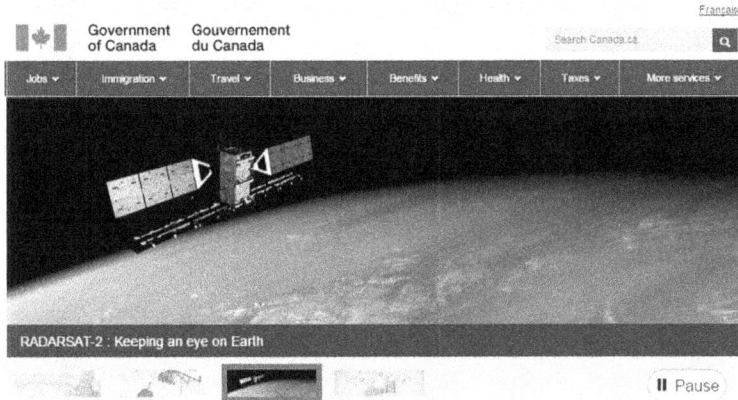

Once a user selects their preferred language, the content inside the site features several exemplary features. There is extensive use of ARIA, not just for landmarks but for communicating roles and properties of other content as well. This is particularly important for making dynamic content accessible to screen reader users. There is a slideshow that has comparable that automatically advances. There is a pause button available. A keyboard-accessible dropdown menu is here that can be navigated not just with the tab key, but also with a full keyboard model including tab, arrow keys, and escape.

Service-Public.fr

The French government site has perfect heading structure, a full set of "Aller au" ("Skip to") links. All images have alternate text and all form fields have labels. The websites is available in different languages.

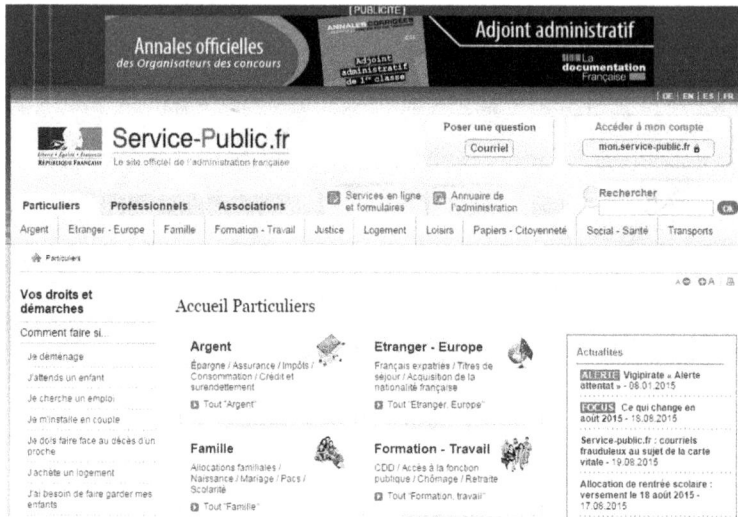

Mozilla.jp

Mozilla Japan has many accessibility features such as a clean, simple interface, good heading structure, and a keyboard-accessible dropdown menu that includes support for arrow keys in addition to tab.

Developing E-commerce Site for Mobile Users

The number of people browsing the web on mobile devices (tablets and smartphones) is increasing at an incredibly fast rate (Musil, 2013). Unfortunately, many E-commerce websites and platforms were built with only the desktop browsing experience in mind, and if you are one of them, you can say good-bye to a fair deal of new sales. One way to gain an edge on your competition and be accessible by consumers on whatever device they decide to use is to make sure your current website is mobile friendly or create a separate mobile site. If your mobile presence is optimized well, you may pick up many more sales for simply being convenient to visitors and returning customers.

First, Check Your Current E-commerce Site on Mobile Devices. To make sure your current E-commerce store is already optimized for mobile web browsing, check your site on Apple iPhone and Apple iPad, various Android Smartphones, and couple of different android tablets. Android makes up approximately half of the smartphone market. However, the mix of different phones and operating system versions makes Android a very inconsistent browsing experience so check thoroughly. In checking your site try conducting a multiple product purchase, conduct a multiple product purchase and test a coupon code, test your contact page to wee if it sends out a message and check to see if you get it, and see how your product images zoom and pop out (if you have pop-out functionality installed). Repeat all of these tests on different mobile browsers (Chrome, Safari and any default mobile browser that may come preinstalled on the device you are testing) (Shopify.com, 2015).

If, for whatever reason, your current E-commerce site does not work well with the mobile web, there are a couple routes you can take. The first step is to check out how your existing site works on a mobile device. There are emulators and testers you can use that simulate mobile devices on a PC, but it's much better to get yourself a modern cellphone with a browser or a tablet computer and see how your website works for real. Following are the options if you want to design your website as mobile-ready.

Guidelines for Developing Mobile-ready Websites

REDESIGN YOUR EXISTING SITE IN SUCH A WAY THAT IT WORKS ON BOTH DESKTOP AND MOBILE

This is called responsive design. For example, the text is deliberately quite large so that when you view it on a mobile device, it remains readable. The pages are designed to work at more or less any width and wherever possible. Responsive design is simply web design that expands and contracts nicely for different devices, whether you view the websites on a desktop computer, smartphone or tablet. A good web developer can usually tweak your website code to make this happen. In some cases, your site might be coded in such a way that responsive design is not possible. You are also going to want to take a closer look at how your buttons, forms and other interactive elements work with your mobile users. In most cases if you optimize these elements for mobile users, it will work fine with desktops users as well. This would also be a great time to simplify your checkout process to make it easier for both mobile and desktop users.

USE COMMON WEB PAGES BUT FORMAT THEM WITH A DIFFERENT STYLE SHEET FOR DESKTOP AND MOBILE DEVICES

You can use JavaScript or PHP to test for a mobile device and then serve a desktop or mobile style sheet that reformats your pages accordingly. If you want, you can set a cookie to remember which version of the site to serve as users hop from page to page. It is a good idea to give mobile users the option to view your full site, if they prefer.

DETECT WHETHER USERS ARE ON A DESKTOP PC OR A HANDHELD DEVICE AND THEN REDIRECT TO A COMPLETELY SEPARATE MOBILE VERSION OF YOUR SITE

This means you can optimize your site for mobile users without compromising usability for people on the desktop, but the drawback is having to maintain a second, duplicate version of your site. Search engines such as

Google dislike duplicate content, so use techniques like canonical URLs to help them understand where to find the definitive version of each page.

DEVELOP A FREE APP FOR MOBILE DEVICES THAT EFFECTIVELY BYPASSES YOUR WEBSITE ALTOGETHER

There is an upfront expense in having apps developed, but they generally give a much better user experience that mobile websites. Not only that, but they prevent users getting distracted, clicking other links, going off to competitor sites

REDIRECT YOUR MOBILE VISITORS TO A SUBDOMAIN

If you cannot build a "one-size fits all" responsive website, you can build another website structure that resides in a separate subdomain. For example if your website is hi-topheaven.com, you can make a mobile friendly website that resides in m.hi-topheaven.com. By inserting a simple redirect statement in the <head> of your home page, you can redirect anyone on a mobile device to your mobile websites. The goal would not be to build a completely new website where you would have to reinsert all your copy and images. Instead, you would build a new mobile friendly website structure that can "call" the same images and copy that you keep current on your main website. That way you will not be responsible for managing two different websites every time you need to update your inventory.

Depending on what E-commerce platform you are using, check Google to see what other "etailers" are doing for their mobile solutions. Plenty of other business owners have already gone down this path and have most likely vetted good solutions for your store. The important thing here being that it's no secret that consumers are increasingly turning to anything but their desktop to browse the web, which means if they stumble upon your site, it better satisfy their need to browse through touch and make a purchase on their tablet or mobile device, otherwise, you'll be missing out on a lot of sales.

Developing Responsive Websites

The use of mobile devices to access websites is continually growing, and E-commerce sales are a large portion of this traffic. In fact, statistics show that by 2017, over one fourth of E-commerce sales will be made via mobile (Bennette, 2014). This means that it is incredibly important to make sure your E-commerce website is optimized for mobile, providing the best experience for users no matter what device they are using to access your site. Implementing Responsive Design to make a website accessible and usable on every device is important for the success of an E-commerce site.

An E-commerce business probably needs both a responsive site and mobile apps. Web sites are discoverable, but apps are more engaging for frequent users. However, you may not need the extra features of a native app, so start by optimizing for the mobile Web, then monitor usage rates. If you have a group of customers who regularly engage with your Web site multiple times a week, they might benefit from the seamless experience of a mobile app, which can remember them and where they left off. You can use that data to estimate future use and do a straightforward cost/benefit analysis to figure out how much investing in app development would really be worth to your company. If you do both, make sure that your apps and Web site work well together. In addition, it is essential that every feature in the app have an equivalent on the Web site and vice versa. You do not want users to think the app is in any way inferior to the Web site. So be sure to use custom URIs (Uniform Resource Identifiers), so links to your site from search results, texts, or email messages are redirected to the corresponding information in your app.

Whether you should go for a native mobile app or responsive site design depends on the strategies and objectives of the business. In most cases, a native mobile app is the way to go because it offers a more powerful, cleaner, faster experience. In addition, it provides push-notification options and delivers personalized content directly to the user. However a well-designed, well-executed Web site is often a good way to go. Installing a mobile app takes time and usually requires registration. It's better to have a responsive E-commerce site that allows fast one-to two-touch conversions. Further, browser support is quite good, so building a separate solution for each mobile platform would make little sense. With a single Web application, you can cover an entire market and deliver instant updates, because there is no delay for app-store approvals or customer updates.

Responsive Web design takes time and money that some companies do not have. Take a closer look and determine whether a possible increase in business from a Web site that supports mobile users would pay for the design and implementation work. If time and budget are extremely constrained, start by implementing a responsive Web site, making sure that you take advantage of WURFL (Wireless Universal Resource FiLe) technologies in automatically resizing your site's pages and changing the layout of columns as necessary. Even though this would be only a minimal and less-than-ideal solution for an E-commerce site, it would at least make it apparent that your company cares about mobile users. In addition, if you implement a mobile site that offers limited functionality, be sure to add a Go to Full Site link on every page of your Web site, so power users who are familiar with your desktop site can do anything they might want to do. It would be best to create a complete mobile Web site or native mobile apps if your company has the time and money. In any case, ensure that mobile users feel they can accomplish everything they need to do on their mobile device, just as they could on your desktop Web site. If your design solution at least achieves that, your customers will not walk away frustrated, taking their money with them (Six, 2015).

Best Practices of Responsive Website Development

Responsive web designing is an entirely different designing version than traditional web designing. Pages that include data tables pose a special challenge to the responsive web designer. Data tables are extremely wide by default, and when someone zooms out to see the whole table, it becomes too small to read. When one tries to zoom in to make it readable, he or she is supposed to scroll both horizontally and vertically to look through it. There are several ways to avoid this problem. Reformatting the data table as a pie or mini-graph is an approved solution. The mini-graph fixes even in narrow screens. Images in responsive web designs are called context-aware. This particular technique serves the purpose of responsive designing in true sense as the images serve at different resolutions, ranging from larger screens to smaller ones. The scaled images appear to change fluidly with the help of updated developer tools and coding languages, allowing designs to look sharp in every context. Responsive web designing is remarkably different from traditional designing in terms of technical and creative issues, and a careful use of this can do wonders while designing.

Make sure that the site is built so that the browsing experience is evenly the same for all users across the board. This means that your site's appearance and visual structure should change without creating content and function losses for users of any specific device or screen size. A visitor accessing your pages from their desktop should be getting the same sort of browsing experience as a visitor coming in through their smart phone or their tablet. This means flexible everything and requires that you ensure your entire image, content and grids are fully fluid and will reconfigure accordingly on a wide assortment of screen sizes. Avoid completely things such as overly complex divs, useless absolute positioning, and fancy Javascript or Flash elements that will just complicate site adjustment overall. Resolutions can be defined in an assortment of breakpoints, but there are several major sizes that you need to focus on more than any others are. Use a program such as GZIP to compress your page resources for easier transmission across networks. You will have lowered the number of bytes sent per page or element and made your content easier to browse and access from devices with varying or low bandwidth. Furthermore, you can speed things up even further by removing any unnecessary white space and line breaks. Doing this will reduce file sizes overall and keep things flowing more smoothly. Some content and content elements are never used in a mobile context. If you have these elements at play in your website or potential site layout, then get rid of them immediately for any mobile setting. You can simply get rid of such elements permanently from all versions of your site (Jukic, 2013).

Examples of Good Responsive Website Designs

8 Faces (http://8faces.com/)

Typography magazine 8 Faces' E-commerce website features big images and even bigger typography. This E-commerce website is fully responsive and allows readers to buy the magazine, prints and back issue PDFs.

Currys (www.currys.co.uk)

Currys is a UK-based retailer. Their site provides a good experience on a range of devices while dealing with a massive amount of content and products of all shapes and sizes, and includes an easy-to-use buying process – all within a single codebase.

Kershaw knives (http://www.kershaw.kaiusaltd.com/)

Kershaw has been designing and manufacturing high-quality and affordable knives since 1974. Their website is a good responsive design with a clean aesthetic to showcase the products. The product range benefit from the well-executed filtering system on the catalogue page. The list of filters is displayed across multiple columns on large viewports making the best use of the space. The filters translate to smaller displays, folding down into a horizontal, scrollable list with off-canvas list items that can be scrolled into view.

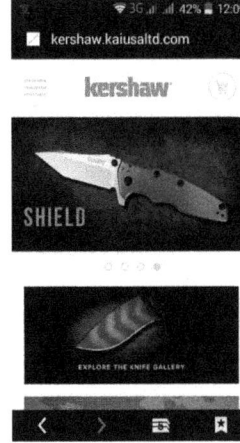

SALT (http://saltsurf.com/)

SALT is a website that sells surfboards. The full canvas photography on the home page is beautiful. The site uses an inspiring aesthetic of considered typography, a purposeful grid and seductive product shots.

A Book Apart (http://abookapart.com/)

A Book Apart's E-commerce website makes a visual feature of its book covers. Whilst many E-commerce websites rely on large beautiful photography to sell their products A Book Apart instead use the block color covers to vividly display their series of "brief books for people who make websites". This works especially well when displaying their entire collection of books, available as a bundle.

Hiut Denim (http://hiutdenim.co.uk/)

This website uses a brilliantly responsive grid-based layout. This site published more in-depth feature articles and for visitors to gain a greater insight into the inspirational fashion brand and the people behind the products.

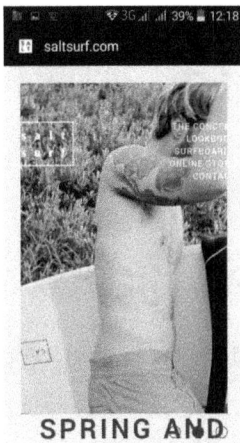

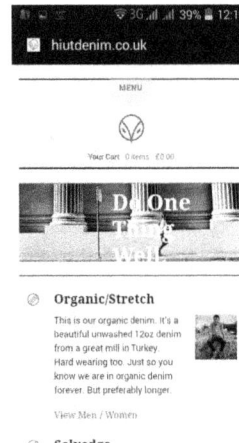

Starbucks (http://www.starbucks.com/)

This site is elegant, light, and fresh. This E-commerce site represents the brand well. In terms of features, this site has most things you would expect a modern coffeehouse to have: nice big videos on the homepage, along with news items listed vertically below. For core navigation, the mobile view condenses the header nicely, consolidating items into a single, now-almost-universal responsive menu icon, which shows and hides vertically when tapped. The richness of the menu items and their descriptions are maintained across the board, which is better than most responsive menu degradations.

Kiwibank (http://www.kiwibank.co.nz/)

Kiwibank is a New Zealand-based bank. Its site design rests heavily on good copy, beautifully typeset in Meta Serif. Navigation on smaller screens is accessed via an 'off-canvas' menu that slides in from the right, similar to apps such as Facebook and Path.

Indochino (http://www.indochino.com/)

Indochino make high quality custom menswear and its responsive store features beautiful product photography combined with a minimal design aesthetic. The whole site responds extremely well to various screen sizes. The way in which the product pages are handled is particularly worth noting.

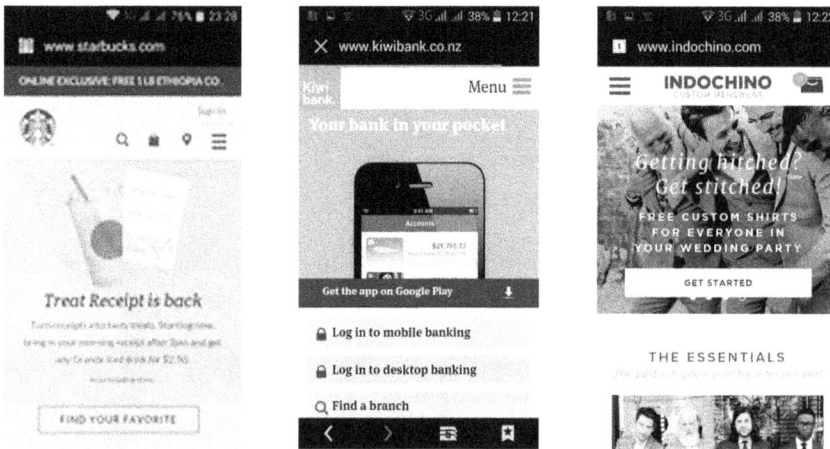

Skinny Ties (http://skinnyties.com/)

Skinny Ties is a niche online retailer that sells nothing but retro-style skinny ties and have been doing so since 1997. The layout of the site is carefully structured and complemented by the use of great product photography and concise and helpful information on the individual product pages. The site navigation also works seamlessly across a variety of screen sizes and allows users to quickly search for products via a variety of options, such as material and color.

Fit For A Frame (http://fitforaframe.com/)

Fit For A Frame is an Online print shop that sells gorgeous prints created by talented designers and illustrators. Their online store features simple navigation and an equally simple layout, both of which condense beautifully for smaller screens. There are some interesting tricks on show here too, such as the way the prints on the brickwall background on the homepage switch to a vertical layout when viewed on a smaller screen.

United Pixel Workers (http://unitedpixelworkers.com/)

United Pixel Workers sells T-shirts and accessories created by the web design community and given the tech savvy nature of their customers, it is not surprising to see that their website is responsive. The site utilizes a simple grid layout and large typography that adjusts exceptionally well when viewed on smaller screens. The way in which the cart and menu buttons are arranged on narrower devices is particularly well thought out too.

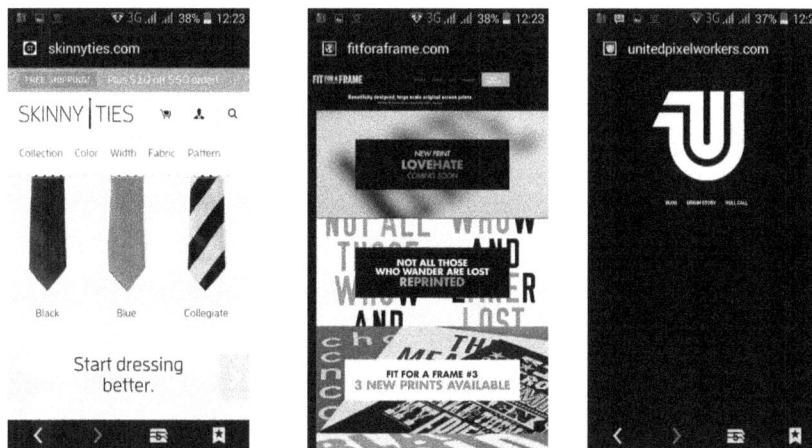

Tattly (http://tattly.com/)

Tattly sell 'Designy Temporary Tattoos' created by leading designers and illustrators. Its online store utilizes a simple grid structure throughout, which is complemented by a monochromatic color palette, which prevents the user interface from detracting from the products on show. The flexibility of the underlying grid means the layout resizes beautifully when the site is viewed on smaller screens.

Nixon (http://www.nixon.com)

Nixon is one of few mainstream watch brands to have a responsive E-commerce store. The design is exceptionally well done. It works so well due to the simple and minimalistic layout and like many other sites in this showcase, a grid structure is employed throughout. Nixon's great product and lifestyle images play a big part in pulling the design together. When viewed on a smaller screen the flyout navigation and checkout process works particularly well too.

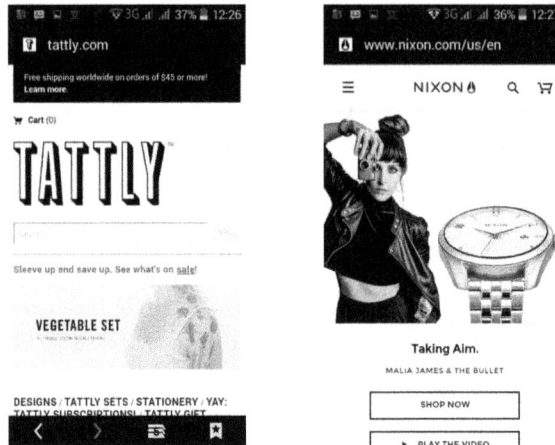

Adaptive Web Design

Companies often design mobile sites for the lowest common denominators, such as screen size, device, browser and OS. Adaptive Web Design is another term for progressive enhancement of a web site. Adaptive Web Design also encompasses a range of other strategies, which can be combined with responsive web design. Adaptive sites are created in several different versions beforehand. The server detects factors like device and OS, and then sends the correct version of the site. Adaptive design is a lot more complex and expensive, but it is the only way to reach the broadest mobile audience. That is because responsive design uses CSS media queries, a relatively recent technology that is not supported by older mobile phones (Charlton, 2014). Adaptive web design or content adaptation is a website structure in which different website code is tailored for different buckets of users. This can be done using a single URL (Dynamic Serving), or with mobile-specific URLs, such as m-dot or dot-mobi. In both cases, server-side device detection is utilized allowing the web server to verify what should and what shouldn't be sent to the requesting device.

The adaptive approach helps with catering to the needs of different groups of users coming up with device-optimized website features, menus, images, texts, etc. You can adjust the website speed and page weight to suit users on varying connectivity and data plans (Deviceatlas, 2015).

Adaptive and responsive web designs have some key differences. With responsive design, the layout decision is made on a user's browser, so the same file is sent to all consumers but significant parts of the file may be hidden from the user. With adaptive design, layout decisions are made on the web server, not the client or browser. The server detects factors like device and OS, and then sends the correct version of the site on the fly, making it quicker for the consumer.

Adaptive websites are much better for load time performance and overall user experience. This is because adaptive delivery works by only transferring those assets necessary for the specific device and optimising images and multimedia content on the fly to suit display resolution and size. The experience on an adaptive site can be finely tuned to the device, so that it is intuitive, super quick and takes advantage of things like location, voice and HTML5. Developers do not have to go back to the drawing board and re-code the existing website from scratch.

This is important because many websites are complex, with a lot of legacy code built up over time, and scratching all the effort that has gone into it is generally not an option.

Adaptive design requires a large team of developers and the budget to handle the complexity that comes with choosing to develop and support an adaptive site. This makes it more expensive, so it is not the solution for everyone. Creating too many separate designs takes a lot of work and can defeat the purpose of trying to use one set of content on one URL.

Examples of Good Adaptive Design

Following are some examples of good responsive site designs. For comparative purpose, each design is shown in both traditional and mobile version.

Lufthansa (www.lufthansa.com)

Lufthansa's use of adaptive design shows how experiences can be tailored according to likely user behavior. Their desktop site is more geared towards flight search and holiday planning, with prominence give to the search box and destinations available. However, the adaptive mobile site is more geared towards customers who have already booked, with flight status, check-in and my bookings some of the most prominent options.

Avenue 32 (http://www.avenue32.com/)

Avenue 32 is an online luxury-shopping destination. Their website is a good example of a retailer employing adaptive Web design for its smartphone and tablet Web experience. The brand, which features emerging as well as established designers, created a mobile and tablet experience that was seamless, intuitive and visually engaging. With a desktop site filled with content-rich pages including product images, designer details, curated looks and more, it was essential that the brand create a multichannel Web offering that was visually and functionally consistent with this desktop experience.

USA Today (http://www.usatoday.com/)

Offering an adaptive Web design, USA Today chose an adaptive approach. The design, allows the brand to detect the specific device by taking operating system and screen size into account to serve up a tailored experience. Thus, providing a richer news experience than responsive may have allowed.

Amazon (http://www.amazon.com/)

Amazon is embracing adaptive design in order to deliver their sites on mobile. Amazon's adaptive site provides mobile users with the opportunity to use "Amazon.com Full Site" on their mobile devices, which some users prefer and responsive does not offer.

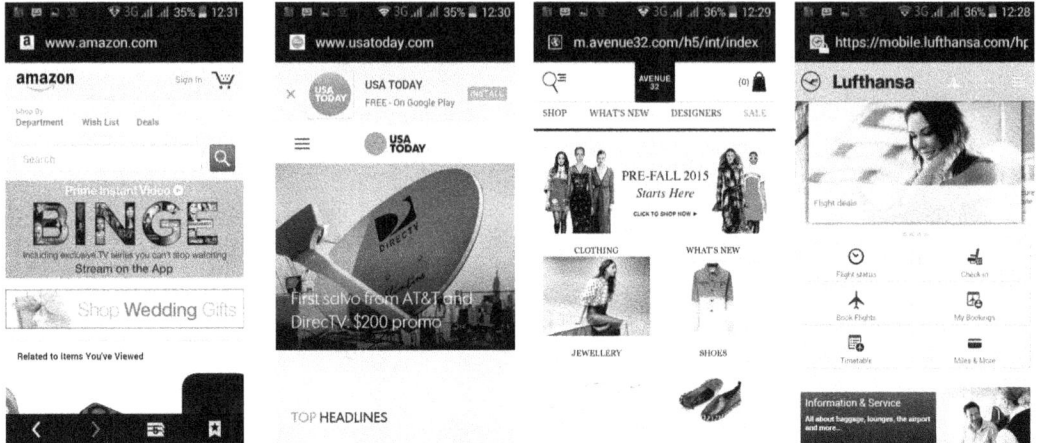

Apple (www.apple.com)

Apple's website is a good example of adaptive design that change according to device type and its functionalities.

About.com

About.com opted an adaptive approach and does not deliver anything that is unnecessary to the user.

Maplin

Maplin.co.uk, a UK-based electronics retailer, has chosen an adaptive approach. Mobile devices are detected and served a different version of the Maplin website optimized for browsing the products and making purchases on the go. The mobile website is lighter and faster than desktop version of the site.

Home Depot

Home Depot also addresses its mobile visitors am adaptive m-dot website optimized for their needs. The website uses location sharing to detect a customer's nearest store. It also allows users to access real-time store inventory and details on the aisle location of products.

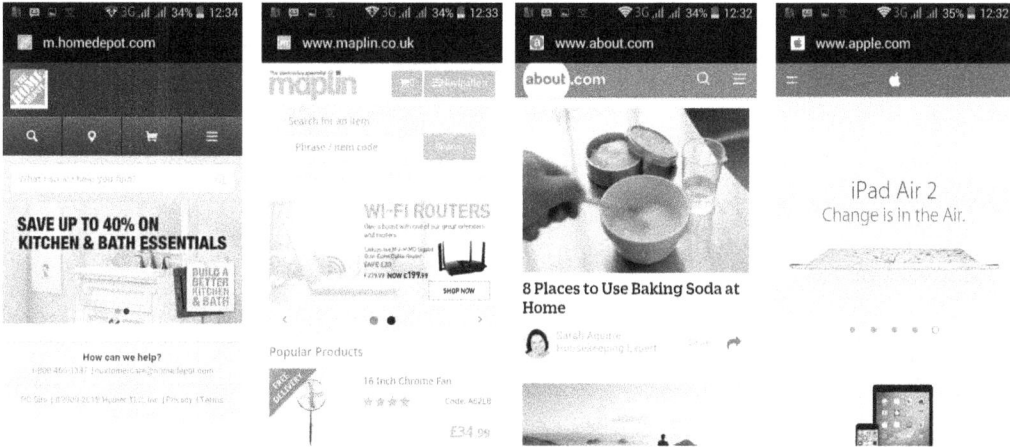

Corcoran

Corcoran, a New York real estate firm, utilizes location features to provide potential home-buyers and renters with a handy list of amenities near the searched property.

InterContinental Hotels Group (IHG)

IHG mobile website encourages on-the-go local bookings through location services, serving up nearby hotels to customers based on GPS data from their phones. The mobile site is action-oriented, with quick-links to find a hotel nearby, review your reservations, and view current offers.

Opentable.com

The Open Table mobile site highlights location and contact details on mobile, while the desktop site emphasizes reviews and browsing.

Turkish Airlines

Turkish Airlines offers its mobile visitors the choice of three different experiences. Users choose between desktop website, mobile website, and mobile app. Mobile website highlights features sought by travelers (online check-in, travel details), as well as all mobile users looking to book a flight.

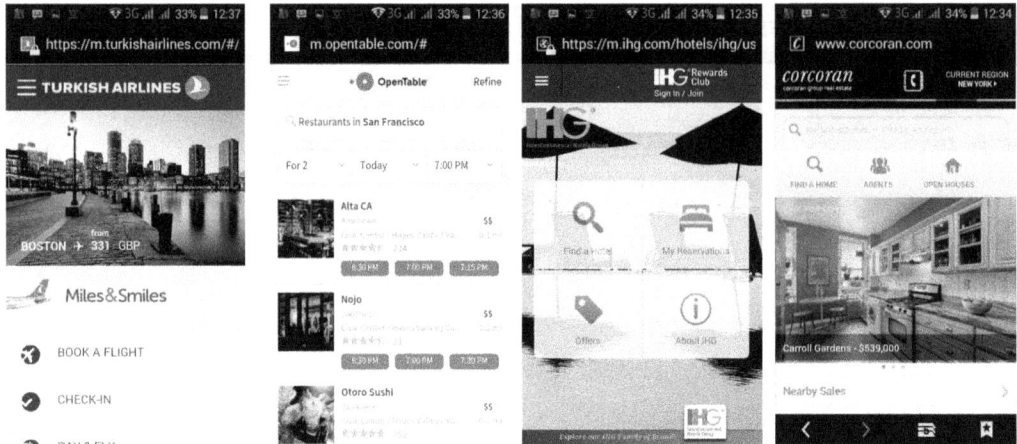

IKEA

IKEA's mobile website utilizes an m-dot smartphone-tailored version with streamlined menus and product browsing, also highlighting the shopping list option, which on the desktop site is much less conspicuous.

AccuWeather

AccuWeather serves mobile visitors an m-dot version of its website. It comes with a revamped menu, more bandwidth-sensitive graphics, and a GPS-based 'use current location' option for checking the weather forecast.

HostelWorld.com

Hostelworld.com is a travel-focused adaptive website addressing mobile visitors with an optimized, adaptive experience. Hostelworld's mobile website features a mobile-optimized search engine including a GPS-powered 'Current location' feature.

Adidas

Adidas provides its mobile visitors with an m-dot website that mimics the desktop website making a seamless experience. The site provide lighter images and simplified shopping process.

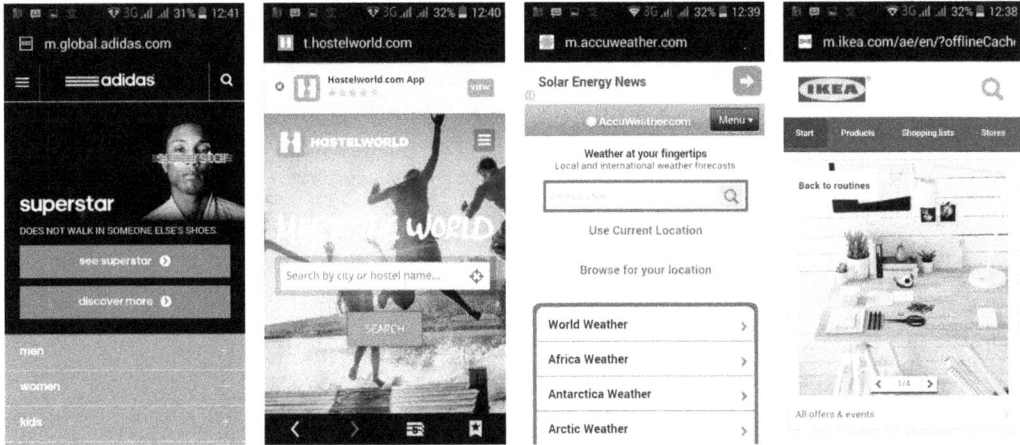

E-commerce Site Design Trends

Smartphones and tablets have changed the approach toward design and user experience. Before the proliferation of mobile devices with advanced web-browsing capability, web designers had only one primary challenge to deal with – keeping the same look and feel of their websites in various desktop computer browsers. Therefore, Responsive Design still be a great trend in next time.

In 2015, responsive, mobile-friendly E-commerce websites will stop being a competitive advantage and start to become a necessity for doing business online. This change will also impact how online retail and business-to-business sites are designed. Competitive advantage is something that a company is faster, better, or more efficient than its competitors. For the past few years, having a mobile optimized website was something of a competitive advantage for online sellers, but in 2015 mobile-ready design is no longer a way to get ahead of the competition. It is a requirement. Mobile design, then, is becoming so prominent and important that it is likely to influence E-commerce design in general. What follows are some web design trends that could impact how E-commerce sites are built or redesigned in 2015 and later (Roggio, 2014; Rocheleau, 2015; Mitsakis, 2015).

HIDDEN MENUS

Mobile site design has required that some common and necessary site elements be hidden until they needed. As an example, global navigation or even product navigation is vital for an E-commerce site, but it can also be in the way on a small screen when a shopper is doing something other than navigating. Thus, hidden menus (sometimes called hamburger menus because of the icon most often used to represent them) have risen in popularity. For example, one can find hidden menus on Starbucks site and on the responsive site for Suitsupply, a men's apparel company.

There also E-commerce sites, designed for larger screens of desktops and laptops that use hidden menus. Trendy design firms are already starting to do this. As an example, London-based branding agency Pollen displays a hidden menu on its home page even when the site is viewed at 1,920-by-1,080 pixels.

RESPONSIVE DESIGN FOR LARGE SCREENS

While responsive design for mobile devices has nearly become a requirement for success in E-commerce, there are some compelling reasons to design for large screens too. For example, many smart televisions or game consoles, like the Xbox One, transform the television into an Internet-enabled device. Other sites, particularly B2B sites, may also have visitors using relatively large screens. Take example of Firebox, an online gift shop. The Firebox website's responsive design is not just good for mobile devices; it also fills up larger screens.

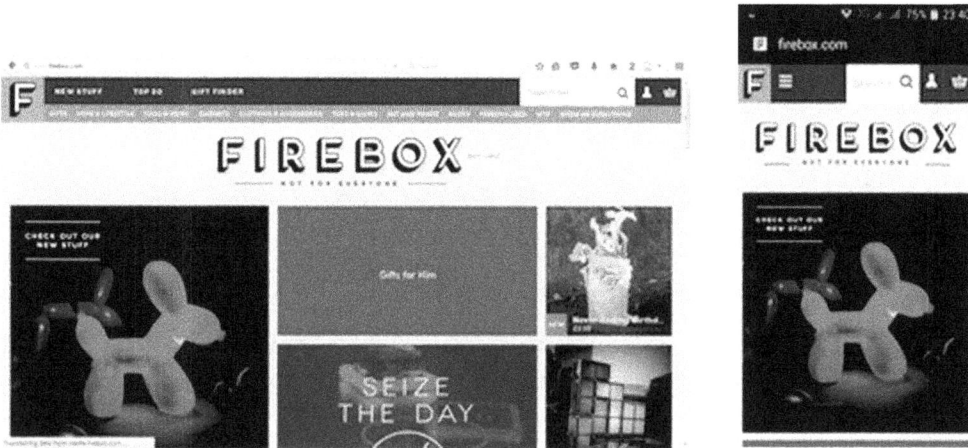

Flexible and Large Typography

It is believed by many in the web design industry that 2015 could be a notable year for typography, particularly flexible and large typography that works well in the context of a responsive design. These designs will also need to be managed within the context of responsive design so that the text flows well whether it is displayed on a relatively small mobile screen or on a large desktop monitor. Large typography is not only essential for readability on mobile screens, but also for not inadvertently losing online sales to those who are aged. Pelican Books is an example of a site that is using large, flexible typography.

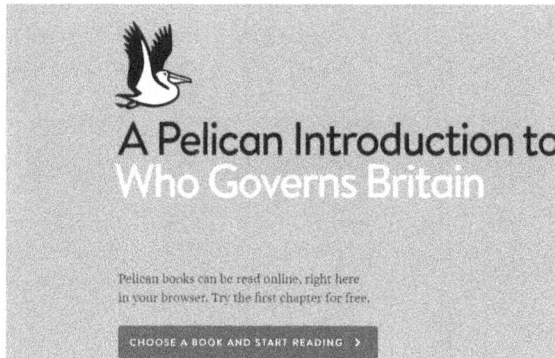

Large Photography

Site speed is extremely important, and many website designers have, in recent years, done away with large images, in part, to make sites load faster on mobile devices. However, with improved responsive design techniques and adaptive images, it is possible to serve relatively fast-loading but large images. These large images can help E-commerce businesses sell more. First, home pages will include very large images. The FALVE site (a manufacturer of handcraft clothing and accessories), uses large, attractive images on its home page.

Many websites are also using large photographs on their product detail pages with some reaching dimensions of more than 1,000 pixels wide. Greats, an online shoe store, use large images on its product detail pages, and one can even rotate those images 360 degrees.

Rich Media

Realizing the benefits of rich media, including videos that help to sell products and content marketing videos that help to attract customers, E-commerce sites are using video on product detail pages, in blogs, on YouTube, and even as background images for home pages. Wiggle uses video to help demonstrate products.

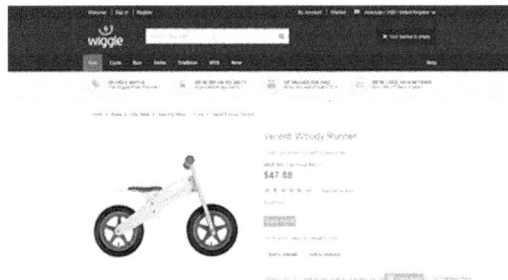

The Gridbooks (http://gridbooks.com/) is a brand of notebooks designed with grid-formatted pages for sketching wireframe mockups of websites and mobile apps. Everything that a new visitor needs to know is located above-the-fold in a sliding carousel widget.

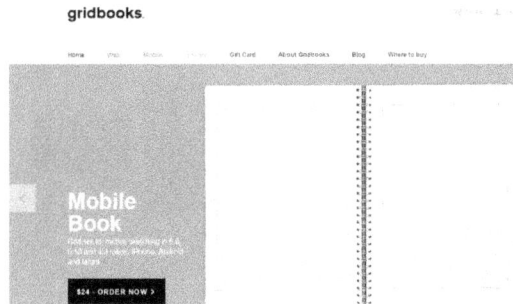

Tinkering Monkey (http://www.tinkeringmonkey.com/) uses a very similar content strategy. Their homepage includes a list of featured products in a big full screen rotating slider. The homepage features a video clip.

Material Design

Material design is a "visual language" that Google developed to combine classic design principles and innovative technologies. It is a way of unifying user experience across platforms so that there are similar looks, conventions, and interactions regardless of what sort of Internet enabled device someone is using. An example for a well-implemented material design is the website of PA Design.

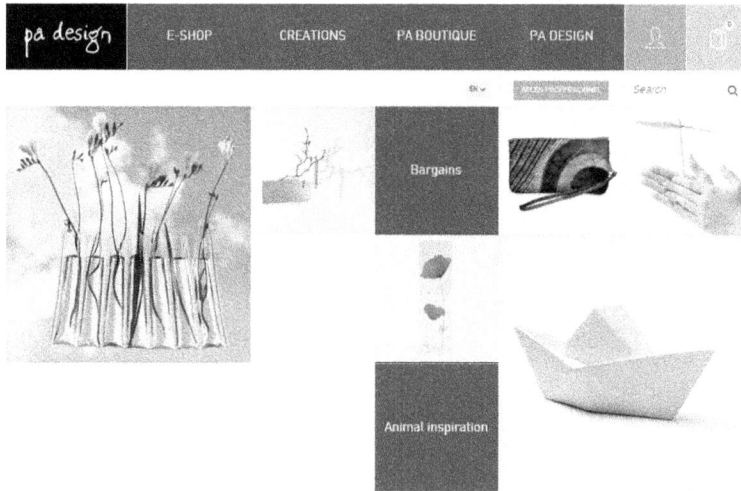

Large Category Lists

It is a browsing method without full reliance on the navigation menu. You split each item into categories and subcategories, which are then listed on a single page. Take example of Zazzle.com. From the Zazzle homepage, visitors might get curious and wander right onto the shop page (http://www.zazzle.com/shop). This interface is a perfect example of organized content structure and how you might list important categories. It helps tremendously to include a small photo of each major category – visuals plus good content are two decisive factors.

The Adobe KnowHow course site (https://www.adobeknowhow.com/) uses a very similar interface catering instead to their different products. Each course is listed with a matching icon, price, and related information. Although the interface takes up a majority of the homepage, everything still feels balanced.

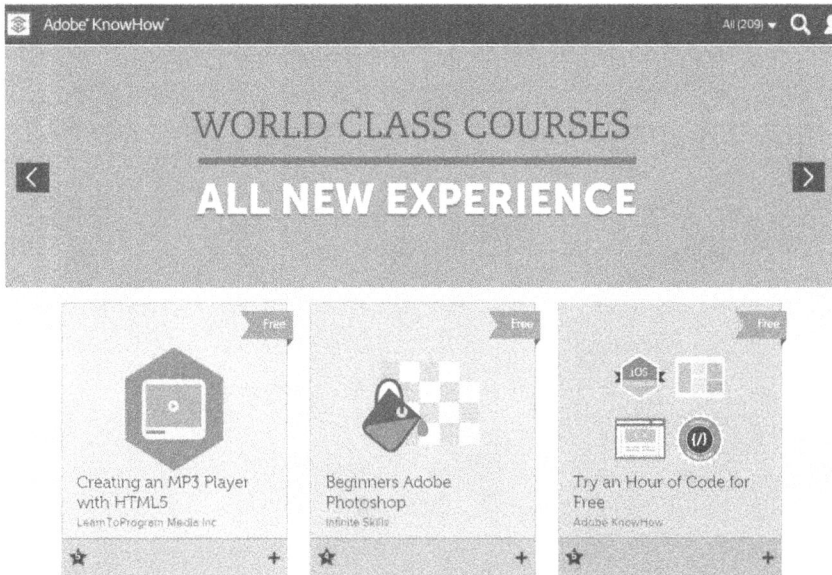

Expanding Navigation

When you are designing a website jam-packed with information, you'll need to use some finesse with navigation. Visitors obviously want to find whatever they are looking for in a quick yet concise manner. When it comes to large header navigation links, dropdown menus are quite the popular trend. As an example, the Caliroots homepage (http://caliroots.com/) features a top navigation using both icons and text labels. Hovering any of these links will animate a dropdown menu with a metric ton of content.

The navigation for Skinny Ties (http://skinnyties.com/) also has a dropdown listing for each primary link. Their layout is fully responsive and still manages to support the dropdown menus regardless of browser resolution.

Dynamic Product Information

Individual product pages are oftentimes the most intricate and messy piece to any e-commerce website. There are so many elements to consider like the item price, quantity, product photos, reviews, colors or sizes for clothing, and plenty of similar qualities. Making a dynamic product page is the simplest way to fit all this information together and make it easier on the visitor. Dynamic content might include automatically updating the price for non-singular quantities or displaying multiple photos from unique vantage points to really sell each product. Take example of the product page for JOCO Cups (https://jococups.com). The design is both functional and aesthetically brilliant. Textures and buttons all match the website design perfectly. Yet the interface is blatantly simple. Scrolling down a bit on the page you will notice dynamic tabs for switching between product details, warranty details, and delivery estimates. This tabbed user interface could also include reviews or related accessories to purchase. At the very bottom is a diagram of the product displaying a few of the prominent features. This graphic offers useful information in a creative and fun way so that it catches your attention.

Filterable Search Criteria

Filterable search items allow users to limit search results based on item size, quantity, manufacturer, any number of related properties. In addition, these filters can work both physical and digital goods. Any e-commerce website selling a handful of different products would greatly benefit from this user interface. Take example of Amazon.com. When you search an item, such as laptop, you will notice sorting options (on the left hand side) for type, price etc. Thus, a user might limit the products down to only laptops under $500. This feature could be very specific and very useful. The user interface is dynamically animated so all the sorting happens without any page refresh.

Another example is Jenier World of Teas (http://www.jenierteas.com). Instead of using a dropdown menu, each filter is shown in full listed in the left-side column. Users merely check a box and the product display automatically updates accordingly.

Card Design

The card design is probably one of the biggest upcoming trends and offers tremendous advantages in terms of design. User-friendliness is excellent as the visitor gets to see all relevant information at a glance. These cards contain short texts, an image/or several share buttons, as well as the most relevant information about the product. It helps potential customers to decide at a glance if the product is right for them. A great example is The Verge, which has an informative and well-implemented card design.

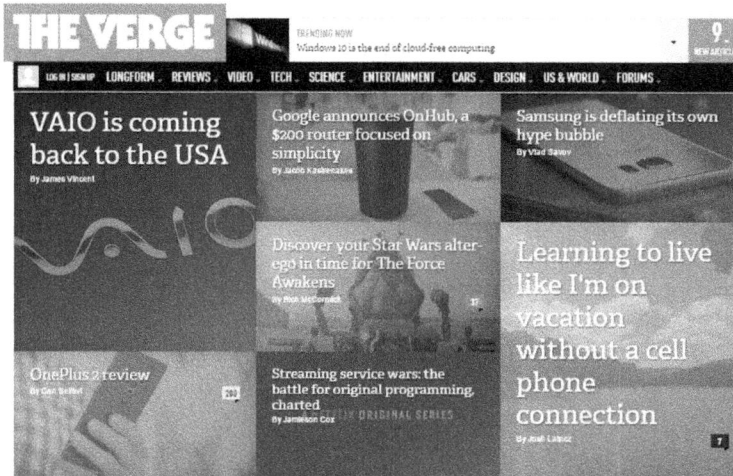

Parallax Scrolling and Background Effects

Parallax scrolling and background effects can be impressive if applied correctly. However, they easily distract from what is important and should be, for this reason, implemented thoroughly and with caution. Parallax effects are only recommended if they highlight the actual content – that is your product. Some good examples are Ditto (https://www.ditto.com/), IWC Schaffhausen, L'unità, and The Royal British Legion.

Minimalist Buttons (Ghost Buttons)

Minimalist ghost buttons work perfectly as call to action buttons on sliders or full page hero backgrounds without distracting too much from the foreground and its message. Minimalist buttons fit perfectly in modern, fresh designs. Some good examples of this design are:

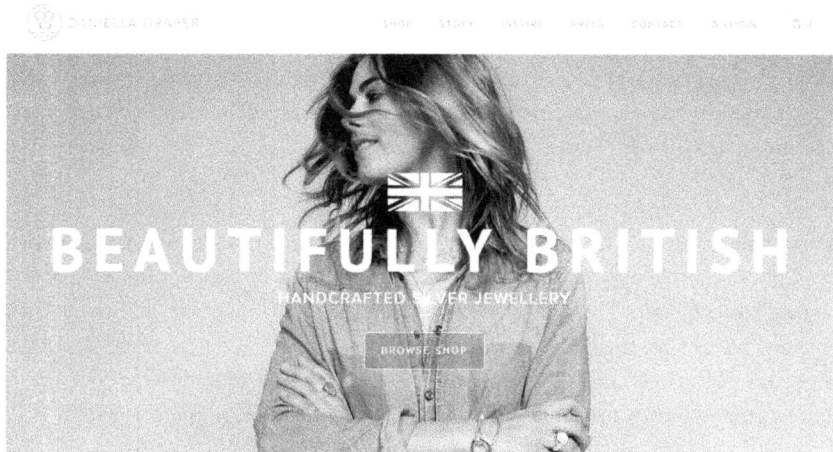

Developing Your E-commerce Website: The Steps

In this section, we will discuss how an entrepreneur can create his own do-it-yourself (DIY) e-commerce web site. Before you start creating your site, you must create your E-commerce business plan. After you have your business plan in place and are ready to focus on your new online venture, it's time to start creating your e-commerce web site.

Choosing Your Web Design Tools

The first step is selecting the tools you want to use to design and code your site. You can start from scratch and code everything by hand, but this requires solid knowledge of HTML, PHP, MySQL, and other programming languages. To shorten the learning curve, you might want to invest in web design software like Adobe Dreamweaver or Microsoft Expression Web. To cut costs, there are also several free open source options to choose from, including KompoZer and BlueGriffon. While they may not offer all of the sophisticated features that paid software does, these open source programs are still powerful and easy to use.

There are some key elements you should look for in any software. An easy-to-use interface is important, especially if you are less familiar with HTML and would prefer to work with a WYSIWYG (What You See Is What You Get) application. WYSIWYG HTML editors allow you to design your site by clicking and dragging elements and inserting text and links like you would in a word processing program. While you design your site, the WYSIWYG editor will generate the HTML code for you. Many of these editors will allow you to toggle between the visual layout and the site code, giving you flexibility to tweak the HTML to see how it affects the page layout, and vice versa.

If your web site is small and simple, HTML/CSS is fine. On the other hand, if the content or structure of your e-commerce site is going to change frequently, you may want to consider an open source content management system (CMS) such as Joomla and Drupal. A CMS is software that allows users to add new web pages or update content on existing pages quick and easily. With a CMS in place, you will not need to worry about coding new content in HTML because the system will do it for you.

Make sure that the applications you choose support all of the functionalities that you want in your web site. For example, if you plan to use a MySQL database for your online product catalog, be sure that you can integrate it into your site using the program you decide to go with. Aside from product catalogs, many e-commerce sites offer newsletter registration forms, customer support widgets, and other elements that make a site more interactive. After you have a list of features you want on your site, do a little research to learn more about the technologies behind them.

You should at least familiarize yourself with HTML, PHP, ASP, SQL, CSS, and AJAX technologies. There are a wide range of resources available at your local library, bookstore, and of course online. The World Wide Web Consortium is a great place to start. Even if you decide to create a DIY web site, you can use a wide variety of free HTML templates. These templates can be tweaked to meet your needs. These templates are available at sites such as 4templates.com or freewebtemplates.com. Before you start these templates for your site, be sure that the designer really has given permission for others to use and change the look of the template.

Selecting Your Web Host for E-commerce

There are hundreds of different web hosting providers to choose from. There are few key issues you need to pay attention to. The first is price. The cheapest host you can find may not always be the best. Providers that have lower prices sometimes skimp in other areas, like customer service or technical support. You should also be on the lookout for providers that may offer low prices up front, but then run up charges with hidden fees.

In addition to price, you should also consider the type of features offered by a hosting provider. Since PHP and MySQL currently play an integral role in creating dynamic, interactive e-commerce web sites, picking a host that offers compatibility with these languages should be at the top of your list. If you are interested in having email addresses that match your domain name access to an email server is critical. This is an important factor when considering how you will communicate with your customers and any employees you may have. Host-based security services, like firewalls and virus detection, are also must-haves.

It is also a good idea to check how much uptime a hosting provider guarantees. A high uptime percentage helps to ensure that your web site will be available when a visitor types your web address into a browser or finds your site via a search engine. Choosing a host that offers frequent backups is also a safe bet just in case your site goes down and you need to recover your files. In the end, find a hosting provider that offers solid customer and technical support. A good hosting provider will be able to answer any questions you may have during your site launch and should be able to help you quickly address any day-to-day problems that may arise.

Picking a Payment Solution

There are two kinds of payment systems that you should consider for your site: a payment processor and a payment gateway. Payment processors, like PayPal Website Payments Standard and Google Checkout, will send a customer to a checkout page that is hosted by the processing company. After customers submit their credit card information, they will be sent back to your web site. In contrast, payment gateways integrate directly with your shopping cart and the transaction is essentially invisible to your customer.

There are pros and cons for payment processors and payment gateways that you should weigh. For some web site owners, payment processors can be easier to use because they do not require any backend integration with web sites. When you use a payment processor, you also do not have to worry about securing financial transactions because the processing web site will take care of it for you. However, a payment processor will also take your customers away from your site during checkout, a process that may cause confusion and may give them second thoughts about completing their transaction. Payment gateways require more technical expertise, and you will have to obtain an SSL certificate to protect credit card transactions on your site (learn more about SSL security below). Even with these additional considerations, a payment gateway will keep customers at your web site during the transaction, making the user experience smoother and more professional.

Before you decide on a payment processor or gateway, you should also research the transaction, set up, and service fees that each provider charges. Average charges vary widely across the industry, with some processors charging a cash fee plus a percentage of each transaction (for example, PayPal charges 30 cents plus 2.9 percent of every transaction for merchants who receive $3,000 or less per month in payments). Setup fees for gateways can cost anywhere from $99 to $299, and fees typically range from 10 to 50 cents per transaction. Merchants with high sales volumes can often find providers that offer a fixed monthly rate for transactions. It is often best to start with your own bank to see what merchant solutions for credit card processing they recommend.

While set up costs and fees and important, there are a few more features that you should think about before making a decision. Check to see whether the payment services provider you are interested in offers automatic tax calculations. This will make it much easier to figure out how much sales tax you should be collecting and will cut down on accounting headaches during tax season. Also, be sure to find out if the processor or gateway you want to use can automatically calculate shipping charges. With this kind of service, you will know exactly how much to charge your customers so shipping costs won't eat into your bottom line.

Shopping Cart: Build vs. Buy

When choosing a shopping cart to go with your payment processing solution, make sure it's supported by your hosting provider and look to see if the cart offers the payment and shipping options you need. There are a number of open source carts with active communities like osCommerce and Zen Cart, as well as GeoTrust Partner solutions like 3dcart.com. You can also create your own shopping cart using web page authoring programs like Dreamweaver or programming it from scratch using PHP, MySQL, and other web programs and languages.

SSL Security

SSL security is critical to your web site. With an SSL certificate, you will be able to prevent cybercriminals from intercepting financial data as it is transmitted over the internet. You will also be able to show your customers that your business is legitimate and can be trusted. After an SSL certificate is installed on a site, it uses an extensive system of security checks to establish a domain and server as trustworthy. The vast majority of businesses obtain SSL certificates from third-party providers called Certificate Authorities (CAs). In addition to providing the SSL certificate, a CA will also authenticate a business to help ensure that the company represented by the web site actually exists. SSL is now considered a standard internet security technology, but that does not mean that all SSL certificates and providers are the same. Just like picking your design tools and your web host, there are several important factors you should think about before you select a CA and an SSL certificate.

Find a credible SSL provider that has a strong reputation for online security. Without legitimate SSL security, you run the risk of losing business because customers will not recognize that your site is safe. Even worse, it can take just one fraudulent incident to damage your business reputation, even if it was not your fault. Customers may still blame you and tell their social circle (including their network on social media) that your site cannot be trusted. Along with the reputation of your SSL provider, you also need to research what type of SSL certificate will be best for your site. There are three types of SSL certificate: Extended Validation, organization validation, and domain validation.

Domain validated SSL is a basic certificate that provides encryption and only verifies that the person applying for the certificate has the right to use a specific domain name. These certificates are ideal for business owners who want to get a certificate as quickly as possible, however, they represent the lowest level of SSL security. Organization validated SSL, on the other hand, also confirms that a company is a confirmed legal entity and establishes a physical location for that organization, giving customers more assurance that the business is legitimate. When people click an on organization validated certificate, it will display your company name, and trust marks provided by CAs may also display your business name. GeoTrust offers a domain validated SSL product called QuickSSL Premium as well as an organization validated certificate called True BusinessID SSL.

For sites that handle financial transactions, Extended Validation (EV) is your best bet. In addition to the encryption and authentication that organization validation provides, EV SSL also offer a clear sign that your web site is safe: the address bar in high-security browsers will turn green and show the name of your company as well as the name of your SSL provider. EV SSL requires a more thorough authentication process, but many businesses experience increased conversions and higher sales thanks to the green address bar and recent research shows that top performing sites are more likely to use EV SSL.

Usually, certificates range from 40-bit to 256-bit encryption. 256-bit encryption is the strongest. It would take a hacker a trillion years to break into a session protected by a 128-bit SSL certificate, and even longer to hack into a session secured with 256-bit encryption. When you're comparing CAs, also check to see if they're using 2048-bit roots that support up to 256-bit encryption. This means that the CA follows the latest recommendations developed by the Certificate Authority Browser Forum to help ensure that SSL certificates are as strong as they can be. The Certificate Authority Browser Forum is an industry oversight organization.

Here is a quick recap that will help you put everything into place:

1. Choose your web design tools: Determine what tools you are going to use to design and your site.

2. Select your web host: Find a hosting provider that can accommodate all of the tools you want to use and offer the features you need.

3. Pick a payment solution: Weigh your options to see if a payment processor or gateway is right for you.

4. Choose or create a shopping cart: Your customers are going to need a way to select items and buy them, so find a shopping cart solution that meets your needs.

5. Obtain credible SSL security: Work with a provider like GeoTrust to get credible SSL protection for your site.

With the rise of identity theft and malware, item five is particularly important. If customers do not believe that a site is secure, they will not choose to spend their hard-earned money there. By using a reliable SSL certificate provider, you will be able to arm your site against online threats and send your customers a clear message that your company can be trusted. Not all SSL certificates are equal, however. Be sure to invest in a solution from a credible provider that will help you build confidence in your site, communicate your commitment to online safety, and protect your business. Taking the time to find the right SSL certificate for your site now will pay off in the future, helping to ensure that your business is a success.

Popular E-commerce Platforms: Features Comparison

Here a comparison of features of some popular E-commerce platforms is provided. This comparison is helpful for businesses looking to select an E-commerce platform for the first time or for replatforming.

Security and Protection

	Bigcommerce	Magento	Mozu	Netsuite
SSL	Yes	Can be purchased from third parties	Yes	Yes
PCI DDS Compliant	Yes	Yes	Yes	Yes
Third party SSL installation support	Free	No	Yes (on fee payment)	Free

Analytics and Reporting

	Bigcommerce	Magento	Mozu	Netsuite
Out-of-the box solution	Included	Included	Included	Included
Reports Include	Store overview, real-time revenue reports, purchase funnel report, abandoned cart report, sales, customer, order and product reporting, marketing and web analytics, merchandising analytics	varies based on integration partner chosen	Order history, customized attributes, product references, customer lifetime value, sales, customer and product reporting, marketing and web analytics	Marketing and web analytics, sales, customer and product reporting, order history, abandoned cart report

Performance

	Bigcommerce	Magento	Mozu	Netsuite
Uptime	99.99%	Varies	99.98%	99.98%
Service Level Agreement (SLA) offered?	Yes at 99.5%	No	No	Yes at 99.5%

Customization

	Bigcommerce	Magento	Mozu	Netsuite
Integration & Customization capabilities	Unlimited	Unlimited	Unlimited	Unlimited
Built-in integration	200+	0	60+	14+
Popular integration	Square, Endicia, HubSpot, Intuit, Avalara	Custom	Avalara, BazzzrVoice, Bronto, ShipWorks	UPS, Other Netsuite platforms (NetCRM, NetERP)

Service and Support

	Bigcommerce	Magento	Mozu	Netsuite
Dedicated Account Management	Yes	Yes	No	Yes
24/7 Support	Yes	No	Yes	Yes

Total Cost of Ownership and Return on Investment (ROI)

On-premise Solution

Hosting	-	$ 2,500
License & Setup	-	$ 32,000
Development/Design	$150 per hour	$ 20, 000
Quality Assurance	$ 100 per hour	$ 2,000
Maintenance/Support	$ 150 per hour	$ 5,000
Project Management		$ 4,000
Total Cost		**$ 66,000**

SaaS Solution

Hosting	-	$ 1,500
License & Setup	-	$ 0
Development/Design	$ 150 per hour	$ 7,000
Quality Assurance	$ 100 per hour	$ 500
Maintenance/Support	$ 150 per hour	$ 0
Project Management	-	$ 500
Total Cost		**$ 9,500**

Review Questions

1. Explain the method of developing an E-commerce site.
2. What is responsive site design? List some examples of good responsive site design.
3. What is accessible site design? List some examples of good accessible site design.
4. What is adaptive site design? List some examples of good adaptive site design.
5. List the steps involved in developing an E-commerce site for a business.

Bibliography

American Foundation for the Blind. (2015). Creating Accessible Websites - American Foundation for the Blind. Retrieved July 14, 2015, from http://www.afb.org/info/programs-and-services/technology-evaluation/creating-accessible-websites/123

Bennette, S. (2014, June 10). Social, Mobile, Search – Amazing E-Commerce Stats, Facts & Figures [INFOGRAPHIC] | SocialTimes. Retrieved July 14, 2015, from http://www.adweek.com/socialtimes/ecommerce-stats/499256?red=at

Berkeley Web Access. (2015). Self Assessment: Tools for Finding Issues | Web Access. Retrieved July 14, 2015, from http://webaccess.berkeley.edu/evaluating/self-assessment/tools

Berkeley.edu. (2015). Top 10 Tips for Making Your Website Accessible | Web Access. Retrieved July 14, 2015, from http://webaccess.berkeley.edu/resources/tips/web-accessibility

Charlton, G. (2014, May 14). Adaptive web design: pros and cons. Retrieved July 17, 2015, from https://econsultancy.com/blog/64833-adaptive-web-design-pros-and-cons/?utm_campaign=bloglikes&utm_medium=socialnetwork&utm_source=facebook

Creativebloq.com. (2013, September 18). 18 brilliantly responsive ecommerce sites | Web design | Creative Bloq. Retrieved July 15, 2015, from http://www.creativebloq.com/web-design/responsive-ecommerce-websites-12121456

cs.wellseley.edu. (2015). CS349 Introduction to E-commerce: lecture 1. Retrieved July 14, 2015, from http://cs.wellesley.edu/~ecom/lecture/overview.html

Hallock, B. (2013, November 15). 5 Ways to Make Your E-Commerce Site More Accessible | 2Checkout Blog. Retrieved July 14, 2015, from https://www.2checkout.com/blog/article/5-ways-to-make-your-e-commerce-site-more-accessible

Jacobson, B. (n.d.). Site Search Functionality for eCommerce – UX Best Practices | Aidalicious | Curated Ecommerce Since 2012 Powered by AbandonAid. Retrieved from http://www.abandonaid.com/blog/site-search-functionality-for-ecommerce-ux-best-practices/

Jukic, S. (2013, November 13). 7 Best Practices of Responsive Web Design. Retrieved from http://webdesignledger.com/tips/7-best-practices-of-responsive-web-design

Kickassmetrics.com. (2011, April). How Loading Time Affects Your Bottom Line. Retrieved July 14, 2015, from https://blog.kissmetrics.com/loading-time/

Mitsakis, L. (2015, February 11). 6 eCommerce Design Trends 2015 for Your Online Shop. Retrieved from http://www.noupe.com/design/six-ecommerce-design-trends-for-2015-to-level-up-your-online-shop-88102.html

Musil. (2013, January 9). Shopping via mobile devices increased 81 percent in 2012 - CNET. Retrieved July 14, 2015, from http://www.cnet.com/news/shopping-via-mobile-devices-increased-81-percent-in-2012/

National Eye Institute. (2015). Facts About Color Blindness. Retrieved August 1, 2015, from https://nei.nih.gov/health/color_blindness/facts_about

Oneclickhere. (2014, December). 10 Ecommerce Design Trends of 2015 | Oneclickhere.co.za. Retrieved July 17, 2015, from http://oneclickhere.co.za/10-ecommerce-design-trends/

Price Wharton, C. (2015, February 1). Psychological Impact Of Website Colors. Retrieved August 1, 2015, from http://www.webdesign.org/web-design-basics/color-theory/psychological-impact-of-website-colors.3459.html

Rocheleau, J. (2015, June 14). Modern Design Ideas & Trends for eCommerce Websites. Retrieved from http://marketblog.envato.com/trends/modern-ui-trends-ecommerce-websites/

Roggio, A. (2014, December 23). 6 Ecommerce Design Trends for 2015. Retrieved from http://www.practicalecommerce.com/articles/77729-6-Ecommerce-Design-Trends-for-2015

Shopify.com. (2015). How to Make a Mobile Ecommerce Website. Retrieved July 14, 2015, from https://www.shopify.com/guides/make-your-first-ecommerce-sale/mobile-design

Six, J. M. (2015, February 16). Responsive Web Design for eCommerce Web Sites :: UXmatters. Retrieved July 15, 2015, from http://www.uxmatters.com/mt/archives/2015/02/responsive-web-design-for-ecommerce-web-sites.php

The Web Standards Project. (2015). Frequently Asked Questions (FAQ) - The Web Standards Project. Retrieved from http://www.webstandards.org/learn/faq/#p3

WebAIM. (2015). WebAIM: Spam-free accessible forms. Retrieved July 14, 2015, from http://webaim.org/blog/spam_free_accessible_forms/

Surprisingly, this book does not have an index. There are strong reasons for that. First, the table of contents is pretty detailed. As such, a reader would not need an index that much. Second, most importantly, including an index would have cost 20-25 pages of valuable content on E-commerce. Academicians purchasing this book would be eligible to receive high quality PowerPoint prsenentations of all chapters that would include high-resolution images used in this book.

www.ingramcontent.com/pod-product-compliance
Lightning Source LLC
Chambersburg PA
CBHW080335220326
41598CB00030B/4513